Response of Field Crops to Abiotic Stress

Response of Field Crops to Abiotic Stress: Current Status and Future Prospects is a collection of useful scientific resources for students, researchers, and academicians on diverse aspects of abiotic stress responses in field crops. The book provides its readers with a vivid understanding of abiotic stress responses in field crops by covering diverse aspects. It offers exhaustive explanations of the impact and responses of field crops to abiotic stresses.

This book offers comprehensive coverage of:

- Climate change impact on field crops.
- Arsenic and aluminium stress responses in field crops.
- Drought, high temperature, and flooding stress responses in field crops.
- Salinity and osmotic stress responses in field crops.
- Heavy metal stress responses in field crops.
- UV stress responses.
- Elemental biofortification.
- Reactive oxygen species (ROS) metabolism.
- Nutraceutical and human health.
- Computational modelling approaches for abiotic stresses in plants.

Shuvasish Choudhury is a Ph.D. graduate from Assam University, Silchar, and Assistant Professor of Plant Biology in the Department of Life Science and Bioinformatics, Assam University, Silchar, India. He is extensively working on understanding the biochemical, molecular, and metabolomic basis of abiotic stress responses and tolerance in field crops. His research interests include arsenic and acid soil stress responses in rice, biochar for arsenic and heavy metal mitigation in agro-ecosystem, and drought stress responses in field crops. His current work focuses on abiotic stress mitigation through nutrient management strategies, seed priming technology, and environmentally sustainable approaches.

Debojyoti Moulick is a Ph.D. graduate from University of Kalyani, India, and a Research Associate in the Plant Stress Biology and Metabolomics Laboratory (CIL), Department of Life Science and Bioinformatics, Assam University, Silchar, India, under the Research Associateship (RA) program of Department of Biotechnology (DBT), Government of India. His research interest includes understanding agro-ecotoxicology and proposing suitable mitigation of options. He has extensively worked on arsenic stress mitigation in rice using seed priming technology and related mitigation approaches.

Response of Field Crops to Abiotic Stress

Current Status and Future Prospects

Edited by
Shuvasish Choudhury and Debojyoti Moulick

CRC Press
Taylor & Francis Group
Boca Raton London New York

CRC Press is an imprint of the
Taylor & Francis Group, an **informa** business

Designed cover image: © Shutterstock

First edition published 2023
by CRC Press
6000 Broken Sound Parkway NW, Suite 300, Boca Raton, FL 33487-2742

and by CRC Press
4 Park Square, Milton Park, Abingdon, Oxon, OX14 4RN

CRC Press is an imprint of Taylor & Francis Group, LLC

Library of Congress Cataloging-in-Publication Data
Names: Choudhury, Shuvasish, editor. | Moulick, Debojyoti, editor.
Title: Response of field crops to abiotic stress: current
status and future prospects / edited by Shuvasish Choudhury and
Debojyoti Moulick.
Description: First edition. | Boca Raton, FL : CRC Press, 2023. | Includes
bibliographical references and index.
Identifiers: LCCN 2022029321 (print) | LCCN 2022029322 (ebook) | ISBN
9781032191966 (hardback) | ISBN 9781032191973 (paperback) | ISBN
9781003258063 (ebook)
Subjects: LCSH: Crops–Effect of stress on. | Field crops–Effect of stress on.
Classification: LCC SB112.5 .R48 2023 (print) | LCC SB112.5 (ebook) | DDC
631.4/52–dc23/eng/20220822
LC record available at https://lccn.loc.gov/2022029321
LC ebook record available at https://lccn.loc.gov/2022029322

ISBN: 9781032191966 (hbk)
ISBN: 9781032191973 (pbk)
ISBN: 9781003258063 (ebk)

DOI: 10.1201/9781003258063

Typeset in Times
by codeMantra

Contents

Preface

Plants are sessile organisms, which makes them susceptible to environmental stresses and constraints. For the past few decades, the world is witnessing the devastating impact of climate change. Climate-change-driven alteration in global precipitation patterns has led to the emergence of recurrent drought and uncontrolled flooding in many parts of the world. Besides this, environmental pollution has largely contributed to instability in the diverse ecosystems. All these factors are the controlling forces for the decline and loss of agricultural productivity and crop yield globally. Such a significant change in agricultural productivity emerged as a threat to global food security and human nutrition. Thus, it is imperative to dissect strategies by which responses of crops to diverse environmental changes and abiotic stresses could be comprehended.

Crops have inbuilt machinery to cope up with environmental stresses throughout their entire lifecycle. Among these, abiotic factors of the agro-environment often termed as abiotic stress are the most crucial for the existence and maintenance of steady yield to feed the increasing world population. Alone abiotic stress is responsible for approximately yield loss by 50%. In order to ensure food security for the world population in a sustainable fashion, better understanding of the concept, and interdisciplinary and transdisciplinary approaches are required to conceptualize suitable mitigation options for crops. It has been assessed that crops can display ~30% of their actual genetic potentiality when exposed to certain abiotic stress condition. So, it is essential to know the stress-induced responses in order to facilitate researchers (physiologists, breeders, biotechnologists) to propose suitable stress-tolerant and stress-resilient crop varieties. A comprehensive approach to understanding the biochemical and molecular aspects of plants is crucial to device the knowledge of tolerance mechanisms under abiotic stresses condition. This book will provide an in-depth scenario about UV stress, climate change as well as elevated CO_2, O_3, and salinity on productivity and yield of major field crops, submergence or sudden floods, consequences of aluminium, arsenic and chromium contamination in soil-water-plant system, and drought and elevated temperature as a major challenge to achieve sustainability in crop production. In addition, to the above-mentioned aspects related to abiotic stress "computational biology" and application of nanotechnology (seed priming) in recent studies offer an opening at the molecular level to interpret abiotic stresses. Besides these, signalling aspects have also been incorporated in this book. Thus, a cell-to-field scenario of crop plants with respect to abiotic stress has been amalgamated.

This edited volume is a cordial effort to summarize in the form of chapters about the current development of plant biology, agronomic as well as development in molecular approaches. The aim of this book is to familiarize a wide range of readers (students, scholars, and senior researchers) with the in -depth understanding and to devise suitable mitigation plants to combat abiotic stress in a wide range of crops.

Dr. Shuvasish Choudhury
Silchar, Assam, India

Dr. Debojyoti Moulick
Kolkata, West Bengal, India

Editors

Dr. Shuvasish Choudhury is an Assistant Professor of Life Science and Bioinformatics, Assam University, Silchar, India, since 2009. He obtained his M.Sc., M.Phil. and Ph.D. in Life Science (Botany) from Assam University, Silchar and did his Post Doctorate from Oklahoma State University, USA. He has published more than 65 publications in the form of research and review articles in reputed international journals as well as in the form of book chapters on abiotic stress, phytochemistry and computational biology. He has also edited three books addressing important aspects such as Biodiversity Conservation, Medicinal Plants, Sustainable Agriculture, Abiotic Stress and Climate Change. He is also enlisted as a reviewer for various journals published by reputed publishing houses such as Springer Nature, Elsevier, American Chemical Society, Nature Publishing Group, BioMed Central etc. Dr. Choudhury was awarded the Young Botanist Award in 2010 by the Indian Botanical Society and thereafter with Fast Track Young Scientist and Overseas Associateship for Post-Doctoral Research by SERB and DBT, Government of India. He has undertaken research projects funded by agencies such as the University Grants Commission, DBT, SERB and Council of Scientific and Industrial Research (CSIR) of Government of India. Dr. Choudhury's publications have received more than 2200 citations. His research interest includes understanding the molecular physiology and metabolomic responses of field crops to abiotic stresses. His current research involves metabolomic and transcriptomic aspects to comprehend arsenic mitigation and drought responses in rice and legumes through seed priming technology and cattle manure biochar application. Besides this, Dr. Choudhury's research also includes the identification of bioactive markers compounds in higher plants and computational biology.

Dr. Debojyoti Moulick worked as a Post-Doctoral Fellow at Assam University, India, for three years. In 2018, he bagged the prestigious DBT-RA award from DBT, Government of India. Earlier he also enjoyed a Senior Research Fellowship from Ministry of Environment and Forest (2013) for supporting his Ph.D. research. Besides these, in 2011, he also awarded with ICAR-SRF (from the Indian Council for Agricultural Research, Government of India). Dr. Moulick exposed himself to a wide range of subjects. He earned his Ph.D. and M.Phil in Environmental Science, M.Sc in Biotechnology and B.Sc in Microbiology. He carried out his Ph.D. work under the supervision of Prof. S.C. Santra, a well-recognized Environmental Biotechnologist and an expert on arsenic toxicity throughout the world. Before joining to M.Phil courses, he worked on gas-hydrates and geochemical exploration in Geochemistry Division of KDMIPE of ONGC in Dehradun, India, under the supervision of Dr. A.K. Jain. Dr. Moulick has gathered extensive experience in biomonitoring, seed priming technology and phytotoxicity, grain quality analysis etc. in soil-water-crop agro-ecosystem. From 2016 onwards, he has authored 22 papers and 13 book chapters in highly reputed journals like Journal of Hazardous Material, Ecotoxicology and Environmental Safety, Environmental Science and Pollution Research, Plant Physiology and Biochemistry, Biocatalysis and Agriculture Biotechnology, Annals Plant and Soil Research, Frontiers in Environmental Science, Agronomy, Molecules as well as in various books published by Springer Nature, Astral Publications etc. Besides these, from 2017, he also reviewed 160 papers for publishing houses as a Certified Reviewer such as ACS, Elsevier, Springer, Nature Publication Group, and Taylor and Francis journals. Recently, he edited a volume for CRC Press and worked as a Guest Editor for Frontiers also. Apart from these, Dr. Moulick also has the credit to be the Guest Editor for Frontiers in Sustainable Food System Journal (Frontiers Media SA). In 2018, Dr. Moulick's research work has been recognized by the Department of Science and Technology (Government of India) in Vigyan Prasar (India Science Wire) and several media houses (both print and e-media) as a breakthrough in proposing an alternative, cheap and farmer-friendly mitigation option for minimizing arsenic-induced phytotoxicity and transmission through rice plant and grain.

Contributors

Khasru Alam
Cytogenetics & Plant Biotechnology Unit
Department of Sericulture (Centre for Applied Biology)
Raiganj University
Raiganj, India

Ananya Ali
Department of Zoology
Serampore Girls' College
Serampore, India

Meenakshi Attri
Division of Agronomy
Sher-e-Kashmir University of Technology and Management
Jammu, India

Mirza Jaynul Baig
ICAR-National Rice Research Institute
Cuttack, India

Madhusmita Barik
ICAR-National Rice Research Institute
Cuttack, India

Jijnasa Barik
Department of Biodiversity and Conservation of Natural Resources
Central University of Odisha
Koraput, India

Dipankar Barman
Division of Plant Physiology
ICAR-Indian Agricultural Research Institute
New Delhi, India

Martin Leonardo Battaglia
The Nature Conservancy
Arlington, Virginia

Prafulla K. Behera
Department of Biodiversity and Conservation of Natural Resources
Central University of Odisha
Koraput, India

Urjashi Bhattacharya
Department of Agronomy
Bidhan Chandra Krishi Viswavidyalaya
Mohanpur, India

Malay K. Bhowmick
International Rice Research Institute
South Asia Regional Centre
Varanasi, India

Debasis Chakrabarty
Biotechnology and Molecular Biology Division
CSIR-National Botanical Research Institute
Lucknow, India
and
Academy of Scientific and Innovative Research (AcSIR)
Ghaziabad, India

Bhaben Chowardhara
Department of Botany
Arunachal University of Studies
Namsai, India

Tanmay Chowdhury
Cytogenetics & Plant Biotechnology Unit
Department of Sericulture (Centre for Applied Biology)
Raiganj University
Raiganj, India

Umakanta Chowra
Department of Botany
Guwahati College
Guwahati, India

Debapriya Rajlakshmi Das
Department of Botany
Taki Government College
Taki, India

Diyadyuti Das
School of Environmental Studies
Jadavpur University
Kolkata, India

Kuntal Das
International Rice Research Institute
South Asia Regional Centre
Varanasi, India

Goutam Kumar Dash
Centurion University of Technology and Management
Paralakhemundi, Odisha, India

Archita Dey
School of Environmental Studies
Jadavpur University
Kolkata, India

Neelam Gautam
Biotechnology and Molecular Biology Division
CSIR-National Botanical Research Institute
Lucknow, India
and
Academy of Scientific and Innovative Research (AcSIR)
Ghaziabad, India

Dibakar Ghosh
ICAR-Indian Institute of Water Management
Bhubaneswar, India

Parthadeb Ghosh
Cytogenetics and Plant Biotechnology Unit
Department of Botany
University of Kalyani
Nadia, India

Swati Hazra
School of Agricultural Sciences
Sharda University
Greater Noida, India

Akbar Hossain
Department of Agronomy
Bangladesh Wheat and Maize Research Institute
Dinajpur, Bangladesh

Sk. Mosharaf Hossain
International Rice Research Institute
Bhubaneswar, Odisha, India

Yuvraj Indoliya
Biotechnology and Molecular Biology Division
CSIR-National Botanical Research Institute
Lucknow, India
and
Academy of Scientific and Innovative Research (AcSIR)
Ghaziabad, India

Sagar Sudam Jadhav
National Institute of Plant Genome Research
New Delhi, India

Vartika Jain
Plant Stress Biology Laboratory
Institute of Environmental and Sustainable Development
Banaras Hindu University
Varanasi, India

Jagdish Jena
Siksha O Anusandhan University
Odisha, India

Jince Mary Mundonayil Joy
Department of Agronomy and Agroforestry
Centurion University of Technology and Management
Odisha, India

Rajesh Shriram Kalasare
Department of Agronomy and Agroforestry
Centurion University of Technology and Management
Odisha, India

Renu Kumari
National Institute of Plant Genome Research
New Delhi, India

Vinita Lamba
National Institute of Plant Genome Research
New Delhi, India

Subrata Mahanta
Department of Chemistry
National Institute of Technology (NIT) Jamshedpur
Jharkhand, India

Himanshu Mahawar
ICAR-Directorate of Weed Research
Jabalpur, India

Chirag Maheshwari
Division of Biochemistry
ICAR-IARI
New Delhi, India

Sanjeet Kumar Mahtha
National Institute of Plant Genome Research
New Delhi, India

Ibandalin Mawlong
Division of Biochemistry
ICAR-Directorate of Rapeseed-Mustard Research
Bharatpur, India

Arnab Majumdar
Department of Earth Science
Indian Institute of Science Education and Research
Kolkata, India

Jayjit Majumdar
Department of Ecological Studies
University of Kalyani
Nadia, India

Brajesh Kumar Mishra
ICAR-DRMR Guwahati
Guwahati, India

Sagar Maitra
Department of Agronomy and Agroforestry
Centurion University of Technology and Management
Paralakhemundi, India

Brahmarshi Mondal
School of Environmental Studies
Jadavpur University
Kolkata, India

Triptesh Mondal
Department of Agronomy and Agroforestry
Centurion University of Technology and Management
Odisha, India

Deepanjan Mridha
School of Environmental Studies
Jadavpur University
Kolkata, India

Arkabanee Mukherjee
Indian Institute of Tropical Meteorology
Pune, India

Swati Nayak
International Rice Research Institute
South Asia Regional Centre
Varanasi, India

Kritika Sinha Nigam
Indian Academy Degree College (Autonomous)
Bangalore, India

Jnana Bharati Palai
Department of Agronomy and Agroforestry
Centurion University of Technology and Management
Paralakhemundi, India

Chetan Kumar Panda
Department of Agricultural Biotechnology
Odisha University of Agriculture and Technology
Bhubaneswar, India

Debabrata Panda
Department of Biodiversity and Conservation of Natural Resources
Central University of Odisha
Koraput, India

Vijai Pandurangam
Department of Plant Physiology
Institute of Agricultural Sciences
Banaras Hindu University
Varanasi, India

Selukash Parida
ICAR-National Rice Research Institute
Cuttack, India

Binaya Kumar Pattnaik
Symbiosis Institute of Geoinformatics (SIG)
Symbiosis International Deemed University
Pune, India

Anindita Paul
Molecular Biophysics Unit
Indian Institute of Science
Bangalore, India

Birendra K. Padhan
Division of Plant Physiology
ICAR-Indian Agricultural Research Institute
New Delhi, India

Subhashisa Praharaj
Department of Agronomy and Agroforestry
Centurion University of Technology and Management
Paralakhemundi, India

Biswajit Pramanick
Department of Agronomy
Dr. Rajendra Prasad Central Agricultural University
Samastipur, Bihar, India

Ravi Kiran Purama
National Institute of Plant Genome Research
New Delhi, India

Rajiv Rakshit
Department of Soil Science and Agricultural Chemistry
Bihar Agricultural University
Sabour, Bihar, India

Dibakar Roy
ICAR-Directorate of Weed Research
Jabalpur, India

Priyabrata Roy
Department of Molecular Biology and Biotechnology
University of Kalyani
Nadia, India
and
Centre of Excellence in Rice Value Addition
International Rice Research Institute
South Asia Region Centre
Varanasi, India

Shyamashree Roy
Department of Agronomy
Regional Research Station (Old Alluvial Zone)
Uttar Banga Krishi Viswavidyalaya
Dinajpur, West Bengal, India

Tarit Roychowdhury
School of Environmental Studies
Jadavpur University
Kolkata, India

Lalichetti Sagar
Department of Agronomy and Agroforestry
Centurion University of Technology and Management
Paralakhemundi, India

Bedabrata Saha
Plant Pathology and Weed Research Department
Newe Ya'ar Research Centre
Agricultural Research Organization
Ramat Yishay, Israel

Soumen Saha
Cytogenetics & Plant Biotechnology Unit
Department of Sericulture (Centre for Applied Biology)
Raiganj University
Raiganj, India

Tanima Saha
Centre of Excellence in Rice Value Addition
International Rice Research Institute
South Asia Region Centre
Varanasi, India

Soumya Kumar Sahoo
Siksha O Anusandhan University Odisha, India

Upasana Sahoo
Department of Agronomy and Agroforestry
Centurion University of Technology and Management
Paralakhemundi, India

Chandan Sahu
School of Biotechnology
Gangadhar Meher University
Sambalpur, India

Masina Sairam
Department of Agronomy and Agroforestry
Centurion University of Technology and Management
Odisha, India

Monolina Sarkar
Department of Botany
Sammilani Mahavidyalaya
Kolkata, India

Lekshmy Sathee
Division of Plant Physiology
ICAR-Indian Agricultural Research Institute
New Delhi, India

Tanmoy Shankar
Department of Agronomy and Agroforestry
Centurion University of Technology and Management
Paralakhemundi, India

Jaya Shree
Department of Chemistry
National Institute of Technology (NIT) Jamshedpur
Jharkhand, India

Anuj Kumar Singh
Discipline of Plant Physiology
Department of Genetics and Plant Breeding
C.P. College of Agriculture
S.D. Agricultural University
Gujarat, India

Sultan Singh
Department of Agronomy
Sri Karan Narendra Agriculture University
Jobner, India

Sudhanshu Singh
International Rice Research Institute
South Asia Regional Centre
Varanasi, India

Uma Maheshwar Singh
International Rice Research Institute
South Asia Regional Centre
Varanasi, India

Vikas Kumar Singh
International Rice Research Institute
South Asia Hub, Hyderabad, Telangana, India

Sapana Sinha
Department of Chemistry
National Institute of Technology (NIT) Jamshedpur
Jharkhand, India

Dasari Sreekanth
ICAR-Directorate of Weed Research
Jabalpur, India

J.P. Srivastava
Department of Plant Physiology
Institute of Agricultural Sciences
Banaras Hindu University
Varanasi, India

Sudhakar Srivastava
Plant Stress Biology Laboratory
Institute of Environmental and Sustainable Development
Banaras Hindu University
Varanasi, India

Padmini Swain
ICAR-National Rice Research Institute
Cuttack, India

Madhu Tiwari
Biotechnology and Molecular Biology Division
CSIR-National Botanical Research Institute
Lucknow, India

Aruna Tyagi
Division of Biochemistry
ICAR-IARI
New Delhi, India

Munish Kumar Upadhyay
Institute of Environment and Sustainable Development
Banaras Hindu University
Varanasi, India

Pawar Deepak Vishwanath
ICAR-Directorate of Weed Research
Jabalpur, India

Archana Watts
Division of Plant Physiology
ICAR-Indian Agricultural Research Institute
New Delhi, India

Gitanjali Yadav
National Institute of Plant Genome Research
New Delhi, India

Marko Zupanič
Faculty of Agriculture and Life Sciences
University of Maribor
Pivola, Slovenia

1

Abiotic Stress Management through Elemental Biofortification in Field Crops

Sudhakar Srivastava and Vartika Jain
Banaras Hindu University

Kritika Sinha Nigam
Indian Academy Degree College

CONTENTS

1.1 Introduction

Stress is any factor, biotic or abiotic, that exerts pressure on individual plant or community causing it to deviate from normal system functioning and alters its growth and development (Rhodes and Nadolska-Orczyk 2001). The plant life is dependent on its interactions with environment, including individual components of soil, water and air. Since plants are static organisms, when components of the environment like temperature, elemental concentrations and water availability are beyond the optimum range, they suffer from stress as they cannot move/migrate to another place of optimum conditions. The well-known abiotic stresses include salinity, drought, extreme temperatures and metal(loid) concentrations. Abiotic stresses interfere with various physiological processes, affect biochemical machinery and alter molecular functions to alter the growth and metabolic activities of plants leading to reduced biomass accumulation and yields. Today's increasing demand for crop products has led to the requirement of management of crop growth by the available resources effectively (Ahmed 2020).

Nevertheless, owing to continuous exposure to one or other type of abiotic stress during its life cycle, plants have devised several defense strategies to effectively cope with the stresses and grow and reproduce. These strategies can operate temporarily for a specific duration causing short-term modifications, such as to tackle extremes of temperature during summer and winter, respectively. On the other hand, to tackle lifelong stresses, like in salinity-affected area or to low water availability in a desert, plants have adapted themselves with development of new or modified organs and physiological mechanisms (Tester and Bacic 2005; Pandey et al. 2017). Root growth or root architecture is a common feature that shows changes in response to a lot of abiotic stresses. In the case of drought conditions, root grows profusely to increase the surface area and root hairs for water uptake (Pavia et al. 2019). For metalloid stress like arsenic (As), the growth of roots may be reduced or specifically altered with induced growth of lateral roots or root hairs so as to reduce As uptake (Srivastava et al. 2021). The production of reactive oxygen species (ROS) as a byproduct of oxygen-dependent metabolism is a normal process. In fact, there are several pro-oxidant enzymes, which produce different ROS in their reaction, and these enzymes have a crucial role in normal plant growth (Grob et al. 2013). However, in the presence of most of the abiotic stresses, the production and quenching of ROS are disturbed and levels of ROS increase rapidly. ROS are highly reactive and can interact with proteins, lipids and DNA to start chain reaction of radicals and disturb structural features of cell (Mittler 2017). To fine-tune the ROS levels, plants are equipped with several enzymes and molecules having antioxidant functions. These include superoxide dismutase (SOD), enzymes of Asada-Halliwell pathway [ascorbate peroxidase (APX), dehydroascorbate reductase, monodehydroascorbate reductase, glutathione reductase (GR)], catalase (CAT) and peroxidases among enzymatic antioxidants and ascorbic acid (ASA), glutathione (GSH), carotenoids, proline, phenolics etc. among molecular antioxidants (Alscher et al. 2002; Shigeoka et al. 2002; Kováčik et al. 2011; Srivastava et al. 2016a; Awasthi et al. 2018). Another important feature of stresses is the reduction in photosynthetic efficiency due to altered gaseous and water vapor exchange that may be caused by stomatal closure or due to effects on

DOI: 10.1201/9781003258063-1

chlorophyll concentrations, chloroplast ultrastructure and activity of photosynthetic enzymes. While stresses like salinity, drought and high temperature may predominantly operate through stomatal closure to reduce photosynthesis, metal(loid) stresses can affect photosynthetic reactions and decrease chlorophyll concentrations (Srivastava et al. 2013a; Kumar et al. 2020). To mitigate the loss of photosynthetic efficiency, plants reduce their growth and may lead to an early maturity of flower and fruit. To fine-tune the stress tolerance pathways, a number of regulatory mechanisms are constitutively expressed or induced in a sequential and intricate manner (Roy et al. 2014; Pradhan et al. 2019).

The stress tolerance of plants can be induced and/or supported by the optimum nutrition of different elements and, therefore, introduction of nutrients to the plants has been considered a viable option to cope with different types of abiotic stresses (Hasanuzzaman et al. 2018). Optimal elemental nutrition is the key to the growth and development of plants under normal conditions and also under stressed conditions. A total of 17 elements are considered to be essential (carbon (C), hydrogen (H), oxygen (O), nitrogen (N), phosphorus (P), potassium (K), calcium (Ca), magnesium (Mg), sulfur (S), iron (Fe), manganese (Mn), zinc (Zn), copper (Cu), boron (B), molybdenum (Mo), nickel (Ni) and chlorine (Cl)). These are further categorized into micro- (0.1–100 mg/kg) and macronutrients (>1–150 g/kg). There are some other elements like Al, Si and Se, which are not considered essential but perform crucial biological functions in plants. Such elements are known as beneficial or functional elements. The optimum supply of mineral elements or additional supply of a particular element in a few abiotic stresses helps in maintaining metabolic processes, energy reactions and source-to-sink relationship allowing it to have good growth and productivity (Hasanuzzaman et al. 2018). Micronutrients enhance stress tolerance due to their involvement in electron movement through electron transport chains, maintenance of cellular osmotic potential and plasma membrane integrity (Brown et al. 2002). Thus, micronutrients reduce ROS increase and membrane damage. Their role in antioxidant enzyme (SOD, CAT) functioning also helps in ROS quenching (Apel and Hirt 2004). Micronutrients also play a role in lignification of the cell wal,l which is an adaptive mechanism of stress tolerance (Stiles 2013). Agronomic approach of applying fertilizers to maintain nutrient balance in plants is a sustainable approach for enhancing stress tolerance, and quality and quantity of crops (Almendros et al. 2015). The approach also manages the nutrient deficiency in humans, which is a burgeoning issue worldwide, and promises to enhance agricultural yields for feeding the growing population (Bilski et al. 2012). Agronomic management of crops through biofortification for sustainable growth in stressed environments avoids the cost and complications of developing new varieties through breeding or genetic engineering (Gonzali et al. 2017). Further, agronomic nutrient biofortification can be optimized for different areas of varying climatic conditions and having different combination of abiotic stresses. Hence, while breeding is successful for a particular zone and environment, agronomic biofortification is amenable to change as per the conditions in space and in time. This chapter discusses a few important abiotic stresses and their management via elemental supplementation. This chapter also envisages the wide applicability of these approaches in near future.

1.2 Heavy Metal Stress

Heavy metals are one of the common contaminants of soil and water resources across the world. These are the elements with metallic properties and have a density of >5 g/cm^3. However, there are variations in this definition and a more precise classification is based on their reactivity with different chemical groups and groups metals into A, B and intermediate classes (Nieboer and Richardson 1980). Heavy metals comprise both essential and non-essential (toxic) metals. The metals and metalloids of concern include cadmium (Cd), arsenic (As), lead (Pb), mercury (Hg) etc. whose contamination is a problem in one or other areas of the world (Shukla et al. 2020; Afonne and Ifediba 2020). The major sources of metal(loid)s in agricultural fields include the use of agricultural chemicals (fertilizers, pesticides, fungicides, weedicides etc.) and irrigation with wastewater or contaminated groundwater (Antoniadis et al. 2019; Afonne and Ifediba 2020; Shukla et al. 2020). Agricultural soil also gets contaminated due to soil erosion and its subsequent deposition, infiltration of leachates from nearby landfill sites, sewage sludge and fly-ash deposition (Dwivedi et al. 2007; Rai et al. 2019). Industrial activities, automobile emissions, as well as natural geological processes are also the sources of metal(loid) contamination. Toxic metal(loid)s may alter the ecosystem with changes in microbial communities involved in the metal(loid) metabolism. The accumulation of these metal(loid)s to levels higher than the maximum permissible limits in crop produce or grains poses threat to human health. There are several ailments and diseases associated with metal(loids) contamination including various cancers (Srivastava 2020). The excess concentration of metal(loid)s may be removed by various physicochemical methods and with the use of plants (Shukla and Srivastava 2018). However, the affected areas are many times huge, such as As-contaminated Gangetic plains of India, and therefore, their removal from the soil and water becomes difficult due to huge financial requirements, years to decades of time needed and unpredictable impacts of the treatment procedures on agricultural ecosystem).

The toxic metal(loid)s have no known function in plants and are therefore toxic even at very low concentrations. Plants not only take up metal(loid)s from soil through roots via a number of transporters, which are actually meant for the uptake of essential metals, but also allow entry of toxic metals too due to their inability to differentiate between the two, for example, Cd and Zn, and As and silica (Si) (Rai et al. 2019; Upadhyay et al. 2019a; Shukla et al. 2020). The total metal uptake by plans depends on its bioavailability and plant root architecture. Plants can actively change their root architecture with features like root hairs, lateral roots, crown roots, etc. to reduce the uptake of toxic metals (Srivastava et al. 2019, 2021). The expression of transporters mediating the uptake and transport of metals can also change, which has been explained in a number of transcriptomic studies (Awasthi et al. 2017; Kumari et al. 2018; Chauhan et al. 2020; Srivastava et al. 2021). The bioavailability of metals is dependent on soil pH and redox conditions, presence of other interacting elements, adsorptive surfaces (soil particle type), etc. (Majumdar and Bose 2018; Upadhyay et al. 2019a). Inside the plants, metal(loid)s induce the production of ROS, impact photosynthesis, transpiration, water status, several biochemical and molecular pathways to reduce biomass accumulation and impede growth of plants (Srivastava et al. 2013b; Andresen and Kupper, 2013; Awasthi et al. 2017). The toxicity of metal(loid)s is also dependent on their species like in the case of Cr, As, Se and Hg (Chauhan et al. 2019; Upadhyay et al. 2019b). Apart from the antioxidant defense machinery as discussed in Section 1.1, metal(loid)s' detoxification is carried out by their complexation with ligands like GSH, phytochelatins (PCs), metallothioneins, histidine, nicotianamine, etc. (Mendoza-Cózatl et al. 2011; Benatti et al. 2014; Kumar et al. 2015). The changes in expression of a number of genes and proteins have also been observed in response to various metals (Chakrabarty et al. 2009; Yu et al. 2012; Srivastava et al. 2013c, 2015; Chauhan et al. 2020).

The presence of metal(loid)s in agricultural ecosystem should be managed not only to achieve high-yielding crops but also to reduce their concentration, considering their subsequent consumption by humans. Since metal(loid)s enter into plants and compete with other elements to affect various plant processes, it is worthwhile to examine the influence of additional supply of essential/beneficial elements to plant responses toward toxic metals. In the case of As accumulation in rice, it is known that As enters through Si transporters. Since rice is a Si hyperaccumulator, Si is a beneficial element for rice plants. Hence, the external supply of Si has been found to impart metal stress tolerance and reduce metal accumulation in rice plants (Chen et al. 2021). Selenium is a beneficial element for plants and is involved in the activity of some enzymes. It acts as an antioxidant and thus protects the plants against oxidative damage. The supply of Se has been found to play a protective role and improve the growth of plants subjected to different metal(loid) stresses including As and Cd (Chauhan et al. 2020; Hussain et al. 2020). Similarly, the application of Zn, which is a constituent of several enzymes in plants, has been found to enhance tolerance in plants to different metal stresses (Hussain et al. 2018). Table 1.1 summarizes some recent research findings demonstrating the beneficial effects of elemental supplementation on metal(loid) tolerance of crop plants.

1.3 Salinity Stress

Soil salinization is an extremely detrimental abiotic stress leading to inhibitory effects on crop production. About 20% of agricultural land comes under the category of salt-affected area which is continuously increasing with time resulting in intensive agriculture and climatic changes (Daliakopoulos et al. 2016). The electrical conductivity (EC) of <0.7 dS/m is acceptable due to no significant effect on crops while EC over 3.0 dS/m causes severe effects on growth and crop yield. Salt stress or salinity affects plant growth causing annual yield losses in the range of 4%–6.5% across the world. Salinity is characterized by high concentration of salts, especially sodium chloride (Shi-Ying et al. 2018). The high salt level also reduces soil porosity and water conductance. Further, rhizospheric microbiome profiles of crop plants have also been found to alter in response to salt stress. Excessive levels of sodium ions make water in the soil physiologically unavailable to plants due to large differences in osmotic potentials (Srivastava et al. 2016a). Salt accumulation with the plants affects a number of morphological (root architecture, number of stomata, leaf growth), physiological (water uptake and movement, transpiration, gaseous exchange, photosynthesis), biochemical (ROS production, enzymatic activities) and molecular (salt overly sensitive response, expression of transcription factors, hormonal and signaling network responses) activities in plants (Srivastava et al. 2016b). Salt stress causes a decline in pigment levels and induces the inhibition of photosystem II (PSII) (Kalaji et al. 2011). Overall, salt stress reduces the growth of plants, which in turn respond at various levels to minimize the damages (Kumar and Verma 2019; Shi-Ying et al. 2018).

One of the major effects of salt stress is water uptake and upward movement. This also affects the uptake and homeostasis of nutrient elements (Zn, Fe, Cu and Mn). To mitigate the impacts of salt stress in plants, the use of plant-growth-promoting bacteria has been successfully employed (Naveed et al. 2020; Kumar et al. 2020). The application of some hormones and chemicals like GSH, melatonin and L-glutamate was also found to improve salt stress tolerance of plants (Zhang et al. 2021; Fardus et al. 2021; Abd El-Hameid and Sadak, 2020; Hasanuzzaman et al. 2021; Miao et al. 2020). Another approach is to supply elements like Se, Zn and Si to enhance stress tolerance. The elemental supplementation was found to work via increased growth, antioxidant potential and reduced salt accumulation in plants (Dhiman et al. 2021). Table 1.2 summarizes recent studies in this direction.

1.4 Drought Stress

The ever-growing global population demands a sustained increase in crop yields. However, abiotic stresses and their combinations have become more prevalent making it difficult to achieve high crop yields (Chakraborty et al. 2014). Global warming, being an ongoing phenomenon, has added to the drought condition around the world and has impacted the aim to meet the increasing crop demands. Further, rainfall patterns have also been grossly disturbed in the past few decades, triggering drought conditions in low rainfall areas (Lobell et al. 2011). Drought conditions reduce growth of plants, flowering time and reproduction and yield. A study examining long-term crop yields from 1980 to 2017 has estimated that wheat, tomato and maize yields have been reduced to an extent of 45% due to drought (Fahad et al. 2017). Drought conditions reduce water availability for the plants that may occur as short episodes during the life cycle of crop plants. During the high temperatures during hot summer months, water can be lost quickly through transpiration and plants can experience water stress (Anjum et al. 2011). High temperature during summer may also induce rapid water loss from soil, which depends on water-holding capacity of soil.

Water stress caused due to drought induces stomatal closure to avoid water loss and this, in turn, affects photosynthesis resulting in reduced gaseous exchange. Further, reduced water

TABLE 1.1

Effect of Supplementation of Essential or Beneficial Elements on Metal(loid) Stress Tolerance and Accumulation in Crop Plants

S. No.	Amendment	Stress	Crop Plant	Result/Effect	References
1.	Foliar application of 1 and 2 mM Si	Cd stress (25 and 50 mg/kg soil)	Basil (*Ocimum basilicum* L.)	Attenuated Cd toxicity and promoted basil growth by reducing Cd accumulation, root-to-shoot Cd transport, MDA content, electrolyte leakage and increasing activity of antioxidant enzymes, concentration of phenolics and flavonoids	Gheshlaghpour et al. (2021)
2.	Media enrichment with Si (5.0 mmol/L)	Cd stress (50 μmol/L) and As(III) stress (25 μmol/L)	Rice (*Oryza sativa* L.)	Si enrichment reduced Cd and As toxicity by decreasing their translocation from roots to shoots and ROS generation in rice seedlings	Chen et al. (2021)
3.	FeNPs (25–50 mg/L) supply	As(III) stress (25–50 μM)	Rice (*O. sativa* L.)	FeNPs strengthened antioxidant defense to reduce oxidative damage, increased As complexation and vacuolar sequestration via GSH and PCs, reduced As uptake while increased that of Fe through changes in expression of transporters	Bidi et al. (2021)
4.	Soil and foliar application of Si (3–6 mg/kg and 2.5–5 mM) and Se (0.5–1 mg/kg and 0.125–0.25 mM)	Cd in soil (1.12 mg/kg)	Wheat (*Triticum aestivum* L.) (Soft and durum wheat)	Both Si and Se treatments reduced Cd in wheat grains (16%–61%) in soil in both wheat but only in soft wheat (20%–36%) in medium; both treatments increased antioxidant system, reduced Cd toxicity; expression of Cd influx transporter was decreased while that of efflux transporter increased	Zhou et al. (2021)
5.	Se NPs, 50–100 μmol/L, foliar spray (four times from tillering to booting)	Cd (3 mg/kg)-Pb (300 mg/kg) spiked soil	Rice (*O. sativa* L.)	Optimum dose of Se NPs was found to be 50 μmol/L that downregulated Cd transporters-related genes in leaves and node, reduced Cd accumulation in grains, improved photosynthesis, reduced oxidative stress and protein carbonylation	Wang et al. (2021)
6.	Sulfur (S); 200 mg/kg	As (5.24 mg/kg); two irrigations (continuous flooding; intermittent drainage)	Rice (*Oryza sativa* L.)	S supply reduced As level in soil and rice; total As and As(III) content in rice grains were decreased both under continuous flooding (62% and 79%) and intermittent drainage (50% and 76%)	Liu et al. (2021)
7.	S-containing materials [chicken manure, elemental S]; 200 mg/kg	As (84.4 mg/kg); continuous flooding and intermittent flooding	Rice (s*O. sativa* L.)	As in rice grains decreased by 65% and 70% in continuous and intermittent flooding, respectively; As speciation was also changed	Wisawapipat et al. (2021)
8.	ZnO NPs (75 mg/L)	Cd (12.5 mg/L) and Pb (150 mg/L)	Cotton (*Gossypium hirsutum* L.)	Growth and biomass of plants increased, photosynthetic pigments and antioxidant enzymes (SOD, CAT, peroxidase, APX) enhanced	Priyanka et al. (2021)
9.	Foliar application of Se and SiNPs (5, 10 and 20 mg/L)	Cd stress (0.84 mg/kg soil) and Pb stress (86.12 mg/kg soil)	Rice (*O. sativa* L.)	Reduced the concentrations of the Cd and Pb contents in brown rice and also enhanced rice growth and grain yield and also improved grain quality (with increased Se reduced phytic acid)	Hussain et al. (2020)
10.	Media enrichment with Se (25 μM Se(IV))	As stress (25 μM As(III))	Rice (*O. sativa* L.)	Selenium enhanced plant growth and diminished structural anomalies via transcriptional and proteomic changes	Chauhan et al. (2020)
11.	Boron (B) supply (46.2 and 462 μM)	Cd treatment (0.5 and 50 μM)	Wheat (*T. aestivum* L.)	High B supply decreased Cd content of wheat plants through down-regulation of genes involved in Cd uptake and transport (CAX, NRAMP and PDR)	Qin et al. (2020)
12.	ZnO NPs (100 mg/kg)	As (6.76 mg/kg) and Cd (1.13 mg/kg)	Rice (*O. sativa* L.)	ZnO NPs reduced As (up to 80%) and Cd (about 26%) accumulation in rice shoots; As speciation was also altered in plant tissues	Ma et al. (2020)
13.	S (2 and 4 mM)	Cd (50 μM)	Pakchoi (*Brassica chinensis* L.)	Excessive S application (4 mM) improved growth of plants due to reduced Cd accumulation that was in turn achieved increase in vacuolar sequestration through high GSH and PC levels; the activity of ATP sulfurylase and PC synthase was also increased	Li et al. (2020)

(*Continued*)

TABLE 1.1 (*Continued*)

Effect of Supplementation of Essential or Beneficial Elements on Metal(loid) Stress Tolerance and Accumulation in Crop Plants

S. No.	Amendment	Stress	Crop Plant	Result/Effect	References
14.	Se (2.25 kg/ha)	Cd (0.16 mg/kg)	Rice (*O. sativa* L.)	The total Cd level in rice grains was decreased to lower than safe limits prescribed	Xue et al. (2020)
15.	Foliar spray of Se and Zn (10–40 mg/L) at tillering elongating and, heading stages	Cd (0.3 mg/kg)	Wheat (*T. aestivum* L.)	Both sprays lead to increase in biomass via elevated photosynthesis and reduced oxidative stress via increased antioxidant enzyme activity and decreased Cd in wheat plants and grains; levels of Se and Zn were also enhanced in plants	Wu et al. (2020)
16.	Soil (20.6–41.2 mg/kg) and foliar (0.1%–0.2%) Zn application	Cd (1.12 mg/kg)	Wheat (*T. aestivum* L.); high and low-Cd varieties	Zn application reduced grain Cd in high Cd variety by 25%–52% and improved grain yield; Cd toxicity was alleviated, expression of Cd transporters and antioxidant enzymes activities were altered	Zhou et al. (2020)
17.	Priming of seeds in ZnO NPs (25–100 mg /L) or Fe NPs (5–20 mg/ L)	Cd stress (0.93 mg/kg soil)	Wheat (*T. aestivum* L.)	Cd levels in wheat decreased at higher concentrations of ZnO and Fe NPs; while Zn and Fe concentrations in roots, shoots and grains were enhanced	Rizwan et al. (2019)
18.	Si (2 mM)	Cd (10–100 mg/kg)	*Pennisetum glaucum* and *Pennisetum glaucum x P. purpureum*	Si application decreased Cd uptake in plants through Cd immobilization in soil via increased pH	Dong et al. (2019)
19.	Si (9 μM)	Cd (40 μM)	*Solanum tuberosum* L.	Si mitigated Cd toxicity and enhanced plants growth	Shahid et al. (2019)
20.	SiO_2 NPs (4 mg/kg)	Al (12 and 25 mg/kg)	Maize (*Zea mays* L.)	Al toxicity resulted in reduced growth, photosynthesis and increased production of ROS and cell damage. Si NPs ameliorated Al toxicity; reduced photorespiratory enzymes and NADPH oxidase and increased antioxidant enzymes and metabolites and organic acids	De Sousa et al. (2019)
21.	Media enrichment with Se (20 μM)	As(III) stress (150 μM)	Rice (*O. sativa* L.)	Se ameliorated reduced As accumulation and oxidative stress parameters, cell death and increased activity of thiol metabolism enzymes and expression of sulfur-related genes; nutrient levels (N, P, K, Ca, Mn, Fe, Zn, Cu, Mg) were augmented in Se-supplemented plants	Pandey and Gupta (2018)
22.	Se (1 and 5 μM)	Cd (5 and 10 μM)	Wheat (*T. aestivum* L.)	Cd stress significantly reduced growth, root volume and surface area, Se concentration, and photosynthetic parameters due to an increase in Cd accumulation. Se treatments restored the growth and increased root volume and surface area by decreasing the shoot Cd concentration; Se1 and Se5 treatments performed variably against two Cd stresses	Qin et al. (2018)
23.	S (2.64 and 5.28 mM)	Cd (50 μM)	Rice (*O. sativa* L.)	S supply increased plant growth parameters via decreasing Cd concentrations in rice plants and grains; iron (Fe) plaque formation and GSH and PC synthesis helped in increased Cd chelation and vacuolar sequestration	Cao et al. (2018)
24.	Foliar and soil application of ZnO NPs (25–100 mg/L)	Cd stress (7.38 mg/kg soil)	Wheat (*T. aestivum* L.)	ZnO NPs reduced Cd concentration in wheat while improving the Zn concentration, growth, photosynthesis and grain yield	Hussain et al. (2018)
25.	Foliar application of Se (3 μM Na_2SeO_3)	Cd stress (50 μM $CdCl_2$)	Cucumber (*Cucumis sativus* L.)	Attenuated Cd-induced growth inhibition, reduced Cd concentration, while ameliorated SPAD value and photosynthetic performance	Sun et al. (2016)

TABLE 1.2

Effect of Supplementation of Essential or Beneficial Elements on Salinity Stress Tolerance and Accumulation in Crop Plants

S. No.	Amendment	Stress	Crop Plant	Result/Effect	References
1.	Foliar application of L-ASA (200 ppm ASA) and Zn (0.03% $ZnSO_4$ $7H_2O$)	Salinity stress (200 mM NaCl)	Barley (*Hordeum vulgare* L.)	Ameliorated the salt stress tolerance and yield attributes like vegetative growth, photosynthetic pigments, synchronized ion uptake, the synthesis of enzymatic and non-enzymatic antioxidants and the harvest index	Noreen et al. (2021)
2.	Media enrichment with Si solution (1.5 mM)	Salinity stress (50 mM NaCl)	Rice (*O. sativa* L.)	Alleviated salt stress-induced K deficiency by stimulating K uptake and root-to-shoot translocation	Yan et al. (2021)
3.	Application of FeO nanoparticles (NPs) (25–100 mg/kg)	Saline soil (EC: 8.07 dS/m) containing Cd (4.75 mg/kg)	Wheat (*T. aestivum* L.)	Improved growth, nutrient concentration (N, P, K), while decreased Na^+ and Cl^- accumulation and Cd uptake, reduced oxidative stress by improving antioxidant defense	Manzoor et al. (2021)
4.	Si (10 mg/kg)	Osmotic stress (PEG; 15%)	Finger millet (*Eleusine coracana* (L.) Gaertn)	Growth of seedlings increased, MDA content decreased, Si accumulation increased that upregulated lipid biosynthesis while downregulated TCA catabolism; responses were mediated by JA	Mundala et al. (2021)
5.	Si (0.3 mM)	Salinity (50 mM)	Cucumber (*C. sativus* L.)	Si supplementation ameliorated the NaCl toxicity, reduced oxidative stress and increased biomass; decreased proline content by changes in proline biosynthetic enzymes through interaction with phytohormones (ABA, IAA, cytokinins and SA)	Zhu et al. (2020)
6.	Si (1 mM)	Salinity (100 mM)+Cd (200 µM)	Date palm (*Phoenix dactylifera* L.)	Si application enhanced growth of plants and plant growth, as well as physiology by reducing Cd uptake on one hand and increasing macronutrient uptake (P, K, Ca, Mg) on the other, also reduced oxidative stress; Si effects were coordinated through changes in hormone levels (SA, JA, ABA)	Khan et al. (2020)
7.	Foliar Application of Se (5 and 10 µM)	Salinity stress (100 mM NaCl)	Wheat (*T. aestivum* L.)	Ameliorated wheat growth, synthesis of photosynthetic pigments, proline, and sugars while decreasing the H_2O_2 contents, Na^+ uptake, and Na^+/K^+ ratio	Elkelish et al. (2019)
8.	Si (2 mM)	Salinity stress (25–100 mM)	Rice (*O. sativa* L.); two varieties	Salinity stress either increased or decreased organic acid production and enzyme activities of TCA cycle in the contrasting varieties; Si application induced changes in organic acids and enzyme activities and imparted tolerance to salinity; Na^+ and Cl^- level decreased and K^+ level increased	Das et al. (2019)
9.	Soil application of zeolite (4 and 8 ton/ha) and foliar application of Se (0.5 and 1 kg/ha) and Si (200 and 400 kg/ha)	Salinity stress (electrical conductivity soil 8 dS/m)	Onion (*Allium cepa* L.)	Ameliorated bulb yield, dry matter, soluble solids, water and chlorophyll and protein contents; give rise to bulb Se and K enrichment; while reducing Na^+ ion concentration	Bybordi et al. (2018)
10.	Foliar Application of Se (1, 5 and 25 µM Na_2SeO_3)	Salinity stress (100 mM NaCl)	Maize (*Z. mays* L.)	Ameliorated growth and biomass, gas exchange attributes and the shape of thylakoids via attenuation of damage in the ultrastructure of chloroplasts and also upregulates ZmMPK5, ZmMPK7, ZmCPK11, and ZmNHX1 genes transcript levels in roots	Jiang et al. (2017)
11.	Si application (2 mM)	Salinity stress (60 mM)	Maize (*Z. mays* L.)	K^+ accumulation was increased while that of Na^+ was decreased; water use efficiency was improved and osmotic and oxidative stress was alleviated, quantum yield was also enhanced	Khan et al. (2017)

(Continued)

TABLE 1.2 (*Continued*)

Effect of Supplementation of Essential or Beneficial Elements on Salinity Stress Tolerance and Accumulation in Crop Plants

S. No.	Amendment	Stress	Crop Plant	Result/Effect	References
12.	K (50 ppm) and Zn (20 ppm)	Salt stress (500 ppm)	Wheat (*T. aestivum* L.)	Salt stress induced increase in MDA and reduced antioxidant enzyme activities were reversed by K^+ Zn; increase in root and shoot length and biomass, photosynthetic pigments, and osmolytes was also observed	Jan et al. (2017)
13.	Si ions and nano-Si	Salinity stress (100–800 mM)	Rice (*O. sativa* L.)	Si ions acted through increase in JA and decrease in ABA to increase Si uptake, antioxidant defense, osmolytes, macronutrients (Ca, K) and reduce oxidative stress and Na^+ accumulation	Abdel-Haliem et al. (2017)

uptake by roots also leads to a decrease in nutrient acquisition and disturbs ionic balance and cellular osmotic potential (Praba et al. 2009; Takahashi et al. 2020). The effects of drought are therefore similar to salinity, however, in the case of salinity plants also suffer from high salt accumulation. The severity of drought on the growth of plants depends not only on the extremity of drought/water stressed conditions and its duration but also on the developmental stage of crop plants. For example, drought stress at early life would reduce the growth of plants, which may recover later on; however, drought stress during the reproductive phase might impact the proper grain development and crop yield (Kaya et al. 2006; Estrada-Campuzano et al. 2008). Decreased germination percentage, germination vigor, early seedling growth and biomass accumulation have been reported in crop plants due to less availability of water. The size and number of cells have been found to reduce as a consequence of reduced water level and decreased turgor pressure. Improper development of flag leaf, changes in assimilates distribution and alteration in the activity of enzymes like starch synthase resulted in reduced yields (Anjum et al. 2011; Craven et al. 2016). Nutrient homeostasis is significantly affected under drought stress due to reduced water uptake and upward movement.

Plants, native to desert habitat, have adaptations that allow them grow normally in the conditions of water stress. For example, thorns like leaves and succulent stems are morphological adaptations of desert plants, while the photosynthesis process takes place via crassulacean acid metabolism (CAM). Other plants cope with drought stress with altered physiological processes, responses at biochemical and molecular levels and anatomical and morphological modifications (Yang et al. 2009). The fluidity of plasma membranes and cell membranes rigidity are changed in drought stress. The production of osmolytes like proline and glycinebetaine has been found to increase. The increased ROS production is managed by antioxidant defense mechanisms (Qi et al. 2018; Takahashi et al. 2020). The C4 crop plants like maize and sorghum can comparatively manage water stress well. A number of chemical effectors like salicylic acid (SA), 24-epibrassinolide, jasmonic acid (JA) etc. have been found to augment plants' defense mechanisms and improve growth under drought stress. In addition, external elemental (Se, Zn, Si) supplies have also been found to improve drought stress tolerance of plants (Table 1.3). These elemental and chemical effectors improve antioxidant defense mechanisms, protect photosynthetic apparatus and pigment content, enhance germination and augment nutrient content of plants. Table 1.3 summarizes recent findings of drought stress mitigation of crop plants in presence of different amendments.

1.5 Heat (High Temperature) Stress

Global warming is a serious issue threatening the quality of human, animal and plant life. The inherent reason for global warming is the continuous increase in the emission of greenhouse gases include mainly carbon dioxide (CO_2) and methane (Friedlingstein et al. 2010). The effects of an increase in greenhouse gases in environment have been aggravated by rapid urbanization, agricultural expansion and reduced forest cover; human activities being the driver of these changes. Global warming is marked by the increase in mean temperature of earth which has resulted in severe losses in crop yields in past decades.

Heat/high-temperature stress causes damage to plants by reducing/inactivating their enzyme activities, decreasing their protein synthesis, inducing the production of ROS and damaging cellular membranes (Hasanuzzaman et al. 2013; Asthir 2015). The folding of proteins is disturbed by heat stress forming the protein aggregates leading to dysfunction of proteins. Heat stress leads to stomatal closure and reduces water loss through transpiration in turn leading to reduced photosynthesis (Hassan et al. 2020; Hu et al. 2020). To manage heat stress, plants increase the production of heat shock proteins and chaperones under the regulation of heat shock transcription factor to restore protein synthesis and its proper folding. The induction of antioxidant defense mechanisms and osmoprotectants has also been recorded in heat-stressed plants (Hasanuzzaman et al. 2013; Haq et al. 2019; Hassan et al. 2020; Hu et al. 2020). Plants have also been found to respond at molecular level with changes in gene expression of phytohormones, osmolytes, kinases, antioxidants, etc. The increase in uptake of beneficial elements like Si and Se has also been noticed (Hasanuzzaman et al. 2013; Waraich et al. 2012; Barnabás et al. 2008). The researchers have also found that the addition of certain chemical

TABLE 1.3

Effect of Supplementation of Essential or Beneficial Elements on Drought Stress Tolerance and Accumulation in Crop Plants

S. No.	Amendment	Stress	Crop Plant	Result/Effect	References
1.	Priming and soaking of seeds in Zn (10 mM) and Se (0.075 mM) solutions	Drought stress (60% field capacity)	Maize (*Z. mays* L.)	Ameliorated drought tolerance via enhancing germination indices, upregulating enzymatic antioxidants (CAT, APX, GPX, SOD). Also increasing the Se and Zn content due to efficient translocation	Nawaz et al. (2021)
2.	Foliar spray of $ZnSO_4$ (0.1%, and 0.2% $ZnSO_4 \cdot 7H_2O$) at vegetative and reproductive stages	Drought stress (60% field capacity)	Wheat (*T. aestivum* L.)	Ameliorated productivity and grain nutrient content; growth, yield traits, photosynthetic pigments under water deficit stress	Anwar et al. (2021)
3.	FeO NPs and hydrogel NPs (25–100 mg/kg)	Cd in soil (10 mg/kg); for drought, irrigation was given 35% of WHC	Rice (*Oryza saiva* L.)	NPs treatment reduced ROS levels by increased antioxidant enzymes, improved growth by enhancing photosynthesis efficiency and augmented nutrient levels along with reduced Cd accumulation by change in expression of transporters ($OsHMA_2$, $OsHMA_3$ and $OsLCT_1$) under normal and drought stress conditions	Ahmed et al. (2021)
4.	Pretreatment with SO_2 (10 mg/m^3)	Drought stress (water stopped for 10 days, 30% of FC	Wheat (*T. aestivum* L.)	Survival rate and relative water content of seedlings increased, proline accumulation and SOD and peroxidase activities enhanced, H_2O_2 and MDA decreased; effects were achieved via H_2S signaling	Li et al. (2021)
5.	Seed priming with Se (0.075 mM) and Zn (10 mM) for 12 and 24 hours, individually and in combination	Drought stress (60% field capacity)	Maize (*Z. mays* L.)	Se$^+$ Zn primed seeds showed increased germination indices, enzymatic antioxidants (CAT, GPX, SOD, APX) and higher Se and Zn content in maize seedlings	Nawaz et al. (2021)
6.	Foliar exposure of ZnO NPs (25–100 mg/L)	Water stress (35% of WHC)	Wheat (*T. aestivum* L.)	ZnO NPs showed the best results at 100 mg/L with improved growth and lowest Cd and maximum Zn in grains, chlorophyll levels and SOD and peroxidase activities increased	Adrees et al. (2021)
7.	Zn (20–80 kg/ha)	Drought (rainfed until 40 days after sowing, then rainout shelter conditions)	Maize (*Z. mays* L.)	Zn (50 kg/ha) improved plant performance in terms of chlorophyll content, stem sap flow rate, root hydraulic conductance, leaf relative water content, water use efficiency and grain yield	Zhang et al. (2020)
8.	ZnO NPs (1–5 mg/kg)	Drought (40% FC)	Sorghum (*Sorghum bicolor*)	The delay in emergence of flag leaf and grain head was reduced from 6–17 days to 4–5 days, grain yield and grain-N, K, P and Zn levels increased	Dimpka et al. (2019)
9.	Zn (0.4%) seed priming alone and coupled with Zn (0.1%) foliar spray	Severe drought during anthesis (volumetric soil water at 5%) followed by recovery	Wheat (*T. aestivum* L.)	Severe drought inhibited recovery of plants was found to be better in Zn-primed and sprayed plants; however, yield loss could not be recovered	Pavia et al. (2019)
10.	Se (0.5–2 mg/kg)	Soil water (irrigated: water near field capacity and water deficit: water at −50 kPa)	Rice (*Oryza saiva* L.)	Se application (0.5 mg/kg) increased the plant height, photosynthesis, SOD activity and Cu, S and Se accumulation; water deficit caused a decrease in growth and yield was reversed by Se treatment	Andrade et al. (2018)
11.	Si (0.1 g/kg)	Drought (45%–50% and 30%–35% field capacity) and salt (6 g/kg) stress	Chinese herb plant (*Glycyrrhiza uralensis* Fisch. ex DC)	The stress treatments decreased growth of plants, which was alleviated by Si; the positive effects included reduced Na$^+$ content and ROS, MDA levels while increased K$^+$/Na$^+$ ratio, soluble sugar, proline, CAT, APX and GR activities and GSH levels	Zhang et al. (2018)

(Continued)

TABLE 1.3 (*Continued*)

Effect of Supplementation of Essential or Beneficial Elements on Drought Stress Tolerance and Accumulation in Crop Plants

S. No.	Amendment	Stress	Crop Plant	Result/Effect	References
12.	Si (2–6 mM)	Drought (50% field capacity), salinity (200 mM), and Cd (2 mM)	Wheat (*T. aestivum* L.)	Si treatments increased shoot length, leaf area, fresh and dry weight, photosynthetic rate, water use efficiency, soluble sugars, proline, GSH and ASA, SOD, CAT and peroxidase activities and reduced oxidative stress against all stresses; Si also reduced Cd and Na^+ accumulation in Cd and salinity stresses	Alzahrani et al. (2018)
13.	High Zn-containing seeds obtained by spraying, additional Zn (5 mg/kg supplied)	Drought stress (30% WHC); Zn deficient conditions (0 mg/kg)	Durum wheat (*Triticum durum* Desf.)	High Zn-containing seeds showed better germination and seedling growth under zinc deficiency and drought stress	Candan et al. (2018)
14.	Foliar application of Se (150 mg)	Drought stress (300 mL/kg water-holding capacity)	Maize (*Z. mays* L.)	Ameliorated drought-induced tolerance of plants via limiting the decrease in plant biomass and oxidative damage due to water shortage condition; while enhancing proline, K concentrations and nitrogen metabolism in aerial parts	Bocchini et al. (2018)
15.	Foliar application of Zn (14 mg/kg $ZnSO_4{\cdot}7H_2O$)	Drought stress (55% ± 5% relative water content)	Wheat (*T. aestivum* L.)	Attenuated drought stress via ameliorating photosynthetic pigment, active oxygen scavenging substances, and decreasing lipid peroxidation	Ma et al. (2017)

TABLE 1.4

Effect of Supplementation of Essential or Beneficial Elements on High Temperature/Heat Stress Tolerance and Accumulation in Crop Plants

S. No.	Amendment	Stress	Crop Plant	Result/Effect	References
1.	Foliar application of Nano-Se (25 mg/L), Si (200 mg/L) and H_2O_2 (20 mmol/L)	Saline soil (EC 4.49 dS/m), heat stress (ambient temperature ranging from 26°C to 41°C)	Cucumber (*C. sativus* L.)	The foliar applied amendments improved growth and productivity of cucumber subjected to combined saline and heat stress. Nano-Se showed maximum improvement in growth, while Si had the highest effect on \ antioxidant enzymes, fruit yield and yield quality due to high nutrient uptake (N, P, K, Mg, Se and Si)	Shalaby et al. (2021)
2.	Foliar spray of 1% potassium phosphate on seedlings	Heat stress condition (37°C/29°C, 16 hours light/8 hours dark cycle)	Potato (*S. tuberosum* L.)	Ameliorated thermo-tolerance of plants via alleviating oxidative stress and DNA damage and enhanced biosynthesis of osmolytes and defense metabolites	Xi et al. (2020)
3.	N application (1.5, 4.5 and 7.5 g/plant)	High temperature (35°C)	Maize (*Z. mays*)	Medium N (4.5 g/plant) improved grain yield and quality, kernel number, weight, moisture, and starch contents were the maximum at MN	Wang et al. (2020)
4.	Nitrate ($Ca(NO_3)_2$) priming (0.2%)	Heat stress (35°C)	Eggplant (*Solanum melongena* L.)	Seed germination and radicle emission were improved, seedling dry mass was also increased	Salles et al. (2019)
5.	Zn (3 mg/kg)	Heat (35/30°C day/night temperature) and drought (35% WHC) and	Chickpea (*Cicer arietinum* L.)	Zn supply ameliorated toxicity of drought and heat; increased CO_2 assimilation, Fv/Fm, leaf area and leaf water content, proline, antioxidants, and Zn concentration, MDA content was reduced	Ullah et al. (2019)
6.	Foliar spray (10 cm^3) and soil application (50 cm^3) of a solution containing 50 mg/ dm^3 Se	Heat stress 35/22°C; day/night	Lamb Lettuce (*Valerianella locusta* L.)	Ameliorated thermo-tolerance of plants via antioxidant enzymes (CAT; GPX) and alleviated heat stress-induced disorders in the growth and metabolism of plants	Hawrylak-Nowak et al. (2018)
7.	KNO_3 (5 and 10 mM)	Heat stress (40°C)	Wheat (*T. aestivum* L.)	Increase in root and shoot length, fresh and dry weights and leaf area	Mirza et al. (2016)
8.	Foliar application of Se (2 and 4 mg/L)	High stress (38°C ± 2°C) at heading stage	Wheat (*T. aestivum* L.)	High Se concentration (4 mg/L) effectively increased grains per spike and grain yield, enzymatic (CAT, APX) and non-enzymatic antioxidants and reduced oxidative stress	Iqbal et al. (2015)

effectors and elements to the medium (soil/water) or via foliar spray augments plants' potential to tolerate heat stress. These chemical/elements work via stimulation of various tolerance pathways (Tang et al. 2021; Ahmad et al. 2021; Tonhati et al. 2020). However, compared to other stresses, studies on heat stress alleviation via elemental supplementation are few. Table 1.4 depicts the responses of plants to heat/high-temperature stress under the influence of external supply of elements.

REFERENCES

Abd El-Hameid, A.R., and Sadak, M.S. 2020. Impact of glutathione on enhancing sunflower growth and biochemical aspects and yield to alleviate salinity stress. *Biocatalysis and Agricultural Biotechnology* 29:101744.

Abdel-Haliem, M.E.F., Hegazy, H.S., Hassan, N.S., and Naguib, D.M. 2017. Effect of silica ions and nano silica on rice plants under salinity stress. *Ecological Engineering* 99:282–289.

Adrees, M., Khan, Z.S., Hafeez, M., et al. 2021. Foliar exposure of zinc oxide nanoparticles improved the growth of wheat (*Triticum aestivum* L.) and decreased cadmium concentration in grains under simultaneous Cd and water deficient stress. *Ecotoxicology and Environmental Safety* 208:111627.

Afonne, O.J., and Ifediba, E.C. 2020. Heavy metals risks in plant foods: Need to step up precautionary measures. *Current Opinion in Toxicology* 22:1–6.

Ahmad, A., Shahzadi, I., Mubeen, S., et al. 2021. Karrikinolide alleviates BDE-28, heat and Cd stressors in *Brassica alboglabra* by correlating and modulating biochemical attributes, antioxidative machinery and osmoregulators. *Ecotoxicology and Environmental Safety* 213:112047.

Ahmed, M. 2020. *Systems Modeling*. Springer Nature Singapore Pvt. Ltd, Singapore.

Ahmed, T., Noman, M., Manzoor, N., et al. 2021. Nanoparticle-based amelioration of drought stress and cadmium toxicity in rice via triggering the stress responsive genetic mechanisms and nutrient acquisition. *Ecotoxicology and Environmental Safety* 209:111829.

Almendros, P., Obrador, A., Gonzalez, D., and Alvarez, J.M. 2015. Biofortification of zinc in onions (*Allium cepa* L.) and soil Zn status by the application of different organic Zn complexes. *Scientia Horticultare* 186:254–265.

Alscher, R.G., Erturk, N., and Heath, L.S. 2002. Role of superoxide dismutases (SODs) in controlling oxidative stress in plants. *Journal of Experimental Botany* 53:1331–1341.

Alzahrani, Y., Kusvuran, A., Alharby, H.F., Kusvuran, S., and Rady, M.M. 2018. The defensive role of silicon in wheat against stress conditions induced by drought, salinity or cadmium. *Ecotoxicology and Environmental Safety* 154:187–196.

Andrade, F.R., da Silva, G.N., Guimaraes, K.C., et al. 2018. Selenium protects rice plants from water deficit stress. *Ecotoxicology and Environmental Safety* 164:562–570.

Andresen, E., and Kupper, H. 2013. Cadmium toxicity in plants. *Metal Ions in Life Sciences* 11:395–413.

Anjum, S.A., Xie, X.Y., Wang, L.C., Saleem, M.F., Man, C., and Lei, W. 2011. Morphological, physiological and biochemical responses of plants to drought stress. *African Journal of Agricultural Research* 6:2026–2032.

Antoniadis, V., Golia, E.E., Liu, Y., Wang, S., Shaheen, S.M., and Rinklebe, J. 2019. Soil and maize contamination by trace elements and associated health risk assessment in the industrial area of Volos, Greece. *Environment International* 124:79–88.

Anwar, S., Khalilzadeh, R., Khan, S., Bashir, R., Pirzad, A., and Malik, A. 2021. Mitigation of drought stress and yield improvement in wheat by zinc foliar spray relates to enhanced water use efficiency and zinc contents. *International Journal Plant Production* 15:377–389.

Apel. K., Hirt. H. 2004. Reactive oxygen species: metabolism, oxidative stress, and signal transduction. *Annu Rev Plant Biol* 55: 373–399.

Asthir, B. 2015. Protective mechanisms of heat tolerance in crop plants. *Journal of Plant Interactions* 10:202–210.

Awasthi, S., Chauhan, R., Srivastava, S., and Tripathi, R.D. 2017. The journey of arsenic from soil to grain in rice. *Frontiers in Plant Science* 8:1008.

Awasthi, S., Chauhan, R., Dwivedi, S., Srivastava, S., Srivastava, S., and Tripathi, R.D. 2018. A consortium of alga (*Chlorella vulgaris*) and bacterium (*Pseudomonas putida*) for amelioration of arsenic toxicity in rice (*Oryza sativa*): A promising and feasible approach. *Environmental and Experimental Botany* 150:115–126.

Barnabás, B., Jäger, K., and Fehér, A. 2008. The effect of drought and heat stress on reproductive processes in cereals. *Plant Cell & Environment* 31:11–38.

Benatti, R.M., Yookongkaew, N., Meetam, M., et al. 2014. Metallothionein deficiency impacts copper accumulation and redistribution in leaves and seeds of Arabidopsis. *New Phytologist* 202:940–951.

Bidi, H., Fallah, H., Niknejad, Y., and Tari, D.B. 2021. Iron oxide nanoparticles alleviate arsenic phytotoxicity in rice by improving iron uptake, oxidative stress tolerance and diminishing arsenic accumulation. *Plant Physiology and Biochemistry* 163:348–357.

Bilski, J., Jacob, D., Soumaila, F., Kraft, C., and Farnsworth, A. 2012. Agronomic biofortification of cereal crop plants with Fe, Zn, and Se, by the utilization of coal fly ash as plant growth media. *Advances in Bioresearch* 3:130–136.

Bocchini, M., D'Amato, R., Ciancaleoni, S., et al. 2018. Soil selenium (Se) biofortification changes the physiologicaL., biochemical and epigenetic responses to water stress in *Zea mays* L. by inducing a higher drought tolerance. *Frontiers in Plant Science* 9:389.

Brown, P.H., Bellalovi, N., and Wimmer, M.A., et al. 2002. Boron in plant biology. *Plant Biology* 4:205–223.

Bybordi, A., Saadat, S., and Zargaripour, P. 2018. The effect of zeolite, selenium and silicon on qualitative and quantitative traits of onion grown under salinity conditions. *Archives of Agronomy and Soil Science* 64:520–530.

Candan, N., Cakmak, I., and Ozturk, L. 2018. Zinc-biofortified seeds ameliorated seedling growth under zinc deficiency and drought stress in durum wheat. *Journal of Plant Nutrition and Soil Science* 181:388–395.

Cao, Z.Z., Qin, M.L., Lin, X.Y., Zhu, Z.W., and Chen, M.X. 2018. Sulfur supply reduces cadmium uptake and translocation in rice grains (*Oryza sativa* L.) by enhancing iron plaque formation, cadmium chelation and vacuolar sequestration. *Environmental Pollution* 238:76–84.

Chakrabarty, D., Trivedi, P.K., Misra, P., et al. 2009. Comparative transcriptomic analysis of arsenate and arsenite stresses in rice seedlings. *Chemosphere* 74:688–702.

Chakraborty, S., Pattanayak, A., Mandal, S., Das, M., and Roychowdhury, R. 2014. An overview of climate change: Causes, trends and implications. In: *Crop Improvement in the Era of Climate Change*, R. Roychowdhury (ed.), pp. 1–29. IK Int Publ House, New Delhi.

Chauhan, R., Awasthi, S., Srivastava, S., et al. 2019. Understanding selenium metabolism in plants and its role as a beneficial element. *Critical Reviews in Environmental Science and Technology* 49:1937–1958.

Chauhan, R., Awasthi, S., Indoliya, Y., et al. 2020. Transcriptome and proteome analyses reveal selenium mediated amelioration of arsenic toxicity in rice *(Oryza sativa* L.). *Journal of Hazardous Materials* 390:122122.

Chen, H., Liang, X., Gong, X., et al. 2021. Comparative physiological and transcriptomic analyses illuminate common mechanisms by which silicon alleviates cadmium and arsenic toxicity in rice seedlings. *Journal of Environmental Sciences* 109:88–101.

Craven, D., Isbell, F., Manning, P., et al. 2016. Plant diversity effects on grassland productivity are robust to both nutrient enrichment and drought. *Philosophical Transactions of the Royal Society B* 37:20150277.

Daliakopoulos, I.N., Tsanis, I.K., Koutroulis, A., et al. 2016. The threat of soil salinity: A European scale review. *Science of the Total Environment* 573:727–739.

Das, P., Manna, I., Sil, P., Bandyopadhyay, M., and Biswas, A.K. 2019. Exogenous silicon alters organic acid production and enzymatic activity of TCA cycle in two NaCl stressed indica rice cultivars. *Plant Physiology and Biochemistry* 136:76–91.

De Sousa, A., Saleh, A.M., Habeeb, T.H., et al. 2019. Silicon dioxide nanoparticles ameliorate the phytotoxic hazards of aluminum in maize grown on acidic soil. *Science of the Total Environment* 693:133636.

Dhiman, P., Rajora, N., Bhardwaj, S., et al. 2021. Fascinating role of silicon to combat salinity stress in plants: An updated overview. *Plant Physiology and Biochemistry* 162:110–123.

Dimpka, C.O., Singh, U., Bindraban, P.S., Elmer, W.H., Gardea-Torresdey, J.L., and White, J.C. 2019. Zinc oxide nanoparticles alleviate drought-induced alterations in sorghum performance, nutrient acquisition, and grain fortification. *Science of the Total Environment* 688:926–934.

Dong, Q., Fang, J., Huang, F., and Cai K 2019. Silicon amendment reduces soil Cd availability and Cd uptake of two pennisetum species. *International Journal of Environmental Research and Public Health* 16:1624.

Dwivedi, S., Tripathi, RD., Srivastava, S., et al. 2007. Growth performance and biochemical responses of three rice (*Oryza sativa* L.) cultivars grown in fly-ash amended soil. Chemosphere 67:140–151.

Elkelish, A.A., Soliman, M.H., Alhaithloul, H.A., and El-Esawi, M.A. 2019. Selenium protects wheat seedlings against salt stress-mediated oxidative damage by up-regulating antioxidants and osmolytes metabolism. *Plant Physiology and Biochemistry* 137:144–153.

Estrada-Campuzano, G., Miralles, D.J., and Slafer, G.A. 2008. Genotypic variability and response to water stress of pre- and post-anthesis phases in triticale. *European Journal of Agronomy* 28:171–177.

Fahad, S., Bajwa, A.A., Nazir, U., et al. 2017. Crop production under drought and heat stress: Plant responses and management options. *Frontiers in Plant Science* 8:1147.

Fardus, J., Hossain, M., and Fujita, M. 2021. Modulation of the antioxidant defense system by exogenous L-glutamic acid application enhances salt tolerance in lentil (*Lens culinaris* Medik.). *Biomolecules* 11:587.

Friedlingstein, P., Houghton, R.A., Marland, G., et al. 2010. Update on CO_2 emissions. *Nature Geoscience* 3:811.

Gheshlaghpour, J., Asghari, B., Khademian, R., and Sedaghati, B. 2021. Silicon alleviates cadmium stress in basil (*Ocimum basilicum* L.) through alteration of phytochemical and physiological characteristics. *Industrial Crops and Products* 163:113338.

Gonzali, S., Kiferle, C., and Perata, P. 2017. Iodine biofortification of crops: Agronomic biofortificatio N., metabolic engineering and iodine bioavailability. *Current Opinion in Biotechnology* 44:16–26.

Grob, F., Durner, J., and Gaupels, F. 2013. Nitric oxide, antioxidants and prooxidants in plant defence responses. *Frontiers in Plant Science* 4:419.

Haq, S.U., Khan, A., Ali, M., et al. 2019. Heat shock proteins: Dynamic biomolecules to counter plant biotic and abiotic stresses. *International Journal of Molecular Sciences* 20:5321.

Hasanuzzaman, M., Nahar, K., Alam, M., Roychowdhury, R., and Fujita, M. 2013. Physiological, biochemical, and molecular mechanisms of heat stress tolerance in plants. *International Journal of Molecular Sciences* 14:9643–9684.

Hasanuzzaman, M., Nahar, K., and Fujita, M. 2018. *Mechanisms of Arsenic Toxicity and Tolerance in Plants.* Springer Nature, Singapore.

Hasanuzzaman, M., Inafuku, M., Nahar, K., Fujita, M., and Oku, H. 2021. Nitric oxide regulates plant growth, physiology, antioxidant defense, and ion homeostasis to confer salt tolerance in the mangrove species, *Kandelia obovata. Antioxidants* 10:611.

Hassan, M.U., Chattha, M.U., Khan, I., et al. 2020. Heat stress in cultivated plants: Nature, impact, mechanisms, and mitigation strategies: A review. *Plant Biosystems* 155:211–234.

Hawrylak-Nowak, B., Dresler, S., Rubinowska, K., Matraszek-Gawron, R., Woch, W., and Hasanuzzaman, M 2018. Selenium biofortification enhances the growth and alters the physiological response of lamb's lettuce grown under high temperature stress. *Plant Physiology and Biochemistry* 127:446–456.

Hu, S., Ding, Y., and Zhu, C. 2020. Sensitivity and responses of chloroplasts to heat stress in plants. *Frontiers in Plant Science* 11:375.

Hussain, A., Ali, S., Rizwan, M., et al. 2018. Zinc oxide nanoparticles alter the wheat physiological response and reduce the cadmium uptake by plants. *Environmental Pollution* 242:1518–1526.

Hussain, B., Lin, Q., Hamid, Y., et al. 2020. Foliage application of selenium and silicon nanoparticles alleviates Cd and Pb toxicity in rice (*Oryza sativa* L.). *Science of the Total Environment* 712:136497.

Iqbal, M., Hussain, I., Liaqat, H., Ashraf, M.A., Rasheed, R., and Rehman, A.U. 2015. Exogenously applied selenium reduces oxidative stress and induces heat tolerance in spring wheat. *Plant Physiology and Biochemistry* 94:95–103.

Jan, A.U., Hadi, F., Midrarullah, Nawaz, M.A., and Rahman, K. 2017. Potassium and zinc increase tolerance to salt stress in wheat (*Triticum aestivum* L.). *Plant Physiology and Biochemistry* 116:139–149.

Jiang, C., Zu, C., Lu, D., et al. 2017. Effect of exogenous selenium supply on photosynthesis, Na^+ accumulation and antioxidative capacity of maize (*Zea mays* L.) under salinity stress. *Scientific Reports* 7:1–14.

Kalaji, H.M., Bosa, K., Koscielniak, J., and Zuk-Gołaszewska, K. 2011. Effects of salt stress on photosystem II efficiency and CO_2 assimilation of two Syrian barley landraces. *Environmental and Experimental Botany* 73:64–72.

Kaya, M.D., Okçu, G., Atak, M., Cikili, Y., and Kolsarici, Ö. 2006. Seed treatments to overcome salt and drought stress during germination in sunlower (*Helianthus annuus* L.). *European Journal of Agronomy* 24:291–295.

Khan, W.U.D., Maqsood, A.M.A., Farooq, M., Abdullah, Y., Ramzani, P.M.A., and Bilal, H.M. 2017. Silicon nutrition mitigates salinity stress in maize by modulating ion accumulation, photosynthesis, and antioxidants. *Photosynthetica* 56:1047–1057.

Khan, A., Bilal, S., Khan, A.L., et al. 2020. Silicon-mediated alleviation of combined salinity and cadmium stress in date palm (*Phoenix dactylifera* L.) by regulating physio-hormonal alteration. *Ecotoxicology and Environmental Safety* 188:109885.

Kováčik, J., Klejdus, B., Hedbavny, J., and Zon, J. 2011. Significance of phenols in cadmium and nickel uptake. *Journal of Plant Physiology* 168:576–584.

Kumar, A., and Verma, J.P. 2019. The role of microbes to improve crop productivity and soil health. In: *Ecological Wisdom Inspired Restoration Engineering*, V. Achal, and A. Mukherjee (eds.), pp. 249–265. Springer Nature, Singapore.

Kumar, S., Dubey, R.S., Tripathi, R.D., Chakrabarty, D., and Trivedi, P.K. 2015. Omics and biotechnology of arsenic stress and detoxification in plants: Current updates and prospective. *Environment International* 74:221–230.

Kumar, A., Singh, S., Gaurav, A.K., Srivastava, S., and Verma, J.P. 2020. Plant growth-promoting bacteria: Biological tools for the mitigation of salinity stress in plants. *Frontiers in Microbiology* 11:1216.

Kumari, P., Rastogi, A., Shukla, A., Srivastava, S., Yadav, S. 2018. Prospects of genetic engineering utilizing potential genes for regulating arsenic accumulation n plants. *Chemosphere* 211:397–406.

Li, H., Pu, P., Li, X., et al. 2020. Sulfur application reduces cadmium uptake in edible parts of pakchoi (*Brassica chinensis* L.) by cadmium chelation and vacuolar sequestration. *Ecotoxicology and Environmental Safety* 194:110402.

Li, L.H., Yi, H.L., Liu, X.P., and Qi, H.X. 2021. Sulfur dioxide enhance drought tolerance of wheat seedlings through H_2S signaling. *Ecotoxicology and Environmental Safety* 207:111248.

Liu, Z., Zhuang, Z., Yu, Y., Wang, Q., Wan, Y.N., and Li, H.F. 2021. Arsenic transfer and accumulation in the soil-rice system with sulfur application and different water managements. *Chemosphere* 269:128772.

Lobell, D.B., Schlenker, W., and Costa-Roberts, J. 2011. Climate trends and global crop production since 1980. *Science* 333:1204531.

Ma, D., Sun, D., Wang, C., et al. 2017. Physiological responses and yield of wheat plants in zinc-mediated alleviation of drought stress. *Frontiers in Plant Science* 8:860.

Ma, X., Sharifan, H., Dou, F., and Sun, W. 2020. Simultaneous reduction of arsenic (As) and cadmium (Cd) accumulation in rice by zinc oxide nanoparticles. *Chemical Engineering Journal* 384:123802.

Majumdar, A., and Bose, S. 2018. A glimpse on uptake kinetics and molecular responses of arsenic tolerance in rice plants. In: *Mechanisms of Arsenic Toxicity and Tolerance in Plants,* M. Hasanuzzaman, K. Nahar, M. Fujita (eds.), pp. 299–315. Springer Nature, Singapore.

Manzoor, N., Ahmed, T., Noman, M., et al. 2021. Iron oxide nanoparticles ameliorated the cadmium and salinity stresses in wheat plants, facilitating photosynthetic pigments and restricting cadmium uptake. *Science of the Total Environment* 769:145221.

Mendoza-Cózatl, D.G., Jobe, T.O., Hauser, F., and Schroeder, J.I. 2011. Long-distance transport, vacuolar sequestration, tolerance, and transcriptional responses induced by cadmium and arsenic. *Current Opinion in Plant Biology* 14:554–562.

Miao, Y., Luo, X., Gao, X., Wang, W., Li, B., and Hou, L. 2020. Exogenous salicylic acid alleviates salt stress by ameliorated leaf photosynthesis and root system architecture in cucumber seedlings. *Scientia Horticulturae* 272:109577.

Mirza, S.R., Ilyas, N., Batool, N., et al. 2016. Seed priming: Technology to improve wheat growth under abiotic stress condition. *International Journal of Biosciences* 9:268–275.

Mittler, R. 2017. ROS are good. *Trends in Plant Science* 22:11–19.

Mundala, P.S., Barvkar, V.T., Umdale, S.D., Kumar, S.A., Nikam, T.D., and Ahire, M.L. 2021. An insight into the role of silicon on retaliation to osmotic stress in finger millet (*Eleusine coracana* (L.) Gaertn). *Journal of Hazardous Materials* 403:124078.

Naveed, M., Ramzan, N., Mustafa, A., et al. 2020. Alleviation of salinity induced oxidative stress in *Chenopodium quinoa* by Fe biofortification and biochar—endophyte interaction. *Agronomy* 10:168.

Nawaz, F., Zulfiqar, B., Ahmad, K.S., et al. 2021. Pretreatment with selenium and zinc modulates physiological indices and antioxidant machinery to improve drought tolerance in maize (*Zea mays* L.). *South African Journal of Botany* 138:209–216.

Nieboer, E., and Richardson, D.H.S. 1980. The replacement of the nondescript term 'heavy metals' by a biologically and chemically significant classification of metal ions. *Environmental Pollution Series B* 1:3–26.

Noreen, S., Sultan, M., Akhter, M.S., et al. 2021. Foliar Fertigation of ascorbic acid and zinc improves growt H., antioxidant enzyme activity and harvest index in barley (*Hordeum vulgare* L.) grown under salt stress. *Plant Physiology and Biochemistry* 158:244–254.

Pandey, C., and Gupta, M. 2018. Selenium amelioration of arsenic toxicity in rice shows genotypic variation: A transcriptomic and biochemical analysis. *Journal of Plant Physiology* 231:168–181.

Pandey, P., Irulappan, V., Bagavathiannan, M.V., and Senthil-Kumar, M. 2017. Impact of combined abiotic and biotic stresses on plant growth and avenues for crop improvement by exploiting physio- morphological traits. *Frontiers in Plant Science* 8:537.

Pavia, I., Roque, J., Rocha, L., et al. 2019. Zinc priming and foliar application enhances photoprotection mechanisms in drought-stressed wheat plants during anthesis. *Plant Physiology and Biochemistry* 140:27–42.

Praba, M.L., Cairns, J.E., Babu, R.C., and Laitte, H.R. 2009. Identiication of physiological traits underlying cultivar differences in drought tolerance in rice and wheat. *Journal of Agronomy and Crop Science* 195:30–46.

Pradhan, S., Babar, M.A., Robbins, K., et al. 2019. Understanding the genetic basis of spike fertility to improve grain number, harvest index, and grain yield in wheat under high temperature stress environments. *Frontiers in Plant Science* 10:1481.

Priyanka, N., Geetha, N., Manish, T., Sahi, S.V., and Venkatachalam, P. 2021. Zinc oxide nanocatalyst mediates cadmium and lead toxicity tolerance mechanism by differential regulation of photosynthetic machinery and antioxidant enzymes level in cotton seedlings. *Toxicology Reports* 8:295–302.

Qi, J., Song, C.P., Wang, B., et al. 2018. Reactive oxygen species signaling and stomatal movement in plant responses to drought stress and pathogen attack. *Journal of Integrative Plant Biology* 60:805–826.

Qin, X., Nie, Z., Liu, H., Zhao, P., Qin, S., and Shi, Z 2018. Influence of selenium on root morphology and photosynthetic characteristics of winter wheat under cadmium stress. *Environmental and Experimental Botany* 150:232–239.

Qin, S., Liu, H., Rengel, Z., et al. 2020. Boron inhibits cadmium uptake in wheat (*Triticum aestivum)* by regulating gene expression. *Plant Science* 297:110522.

Rai, P.K., Lee, S.S., Zhang, M., Tsang, Y.F., and Kim, K.H. 2019. Heavy metals in food crops: Health riskS., fatE., mechanisms and management. *Environment International* 125:365–385.

Rhodes, D., and Nadolska-Orczyk, A. 2001 Plant stress physiology. eLS. https://doi.org/10.1038/npg.els.0001297.

Rizwan, M., Ali, S., Ali, B., et al. 2019. Zinc and iron oxide nanoparticles ameliorated the plant growth and reduced the oxidative stress and cadmium concentration in wheat. *Chemosphere* 214:269–277.

Roy, S., Negrao, S., and Tester, M. 2014. Salt resistant crop plants. *Current Opinion in Biotechnology* 26:115–124.

Salles, J.S., de Lima, A.H., Binotti, F.F.D.S., et al. 2019. Calcium nitrate priming increases the germination rate of eggplant seeds. *Journal of Agricultural Science* 11:181–186.

Shahid, M.A., Balal, R.M., Khan, N., et al. 2019. Selenium impedes cadmium and arsenic toxicity in potato by modulating carbohydrate and nitrogen metabolism. *Ecotoxicology and Environmental Safety* 180:588–599.

Shalaby, T.A., Abd-Alkarim, E., El-Abidy, F., et al. 2021. Nano-seleniu M., silicon and H_2O_2 boost growth and productivity of cucumber under combined salinity and heat stress. *Ecotoxicology and Environmental Safety* 212:111962.

Shigeoka, S., Ishikawa, T., Tamoi, M., et al. 2002. Regulation and function of ascorbate peroxidase isoenzymes. *Journal of Experimental Botany* 53:1305–1319.

Shi-Ying, Z., Cong, F., Yong-xia, W., Yun-sheng, X., Wei, X., and Xiao-Long, C. 2018. Salt-tolerant and plant growth-promoting bacteria isolated from high-yield paddy soil. *Canadian Journal of Microbiology* 64:968–978.

Shukla, A., and Srivastava, S. 2018. A review of phytoremediation prospects for arsenic contaminated water and soil. In: *Phytomanagement of Polluted Sites: Market Opportunities in Sustainable Phytoremediation,* V.C. Pandey and K. Bauddh (eds.), pp. 243–254. Elsevier, Netherlands.

Shukla, A., Awasthi, S., Chauhan, R., and Srivastava, S. 2020. The status of arsenic contamination in India. In: *Arsenic in Drinking Water and Food*, S. Srivastava (ed.), pp. 1–12. Springer Nature, Singapore.

Srivastava, S. 2020. *Arsenic in Drinking Water and Food.* Springer Nature, Singapore.

Srivastava, S., Srivastava, A.K., Singh, B., Suprasanna, P., and D'Souza, S.F. 2013a. The effect of arsenic on pigment composition and photosynthesis in *Hydrilla verticillata* (L.f.) Royle. *Biologia Plantarum* 57:385–389.

Srivastava, S., Srivastava, A.K., Suprasanna, P., and D'Souza, S.F. 2013b. Quantitative real-time expression profiling of aquaporin-isoforms and growth response of *Brassica juncea* under arsenite stress. *Molecular Biology Reports* 40:2879–2886.

Srivastava, S., Srivastava, A.K., Suprasanna, P., and D'Souza, S.F. 2013c. Identification and profiling of arsenic stress-induced microRNAs in *Brassica juncea. Journal of Experimental Botany* 64:303–315.

Srivastava, S., Srivastava, A.K., Sablok, G., Deshpande, T., and Suprasanna, P. 2015. Transcriptomics profiling of Indian mustard (*Brassica juncea*) under arsenate stress identifies key candidate genes and regulatory pathways. *Frontiers in Plant Science* 6:646.

Srivastava, S., Akkarakaran, J.J., Sounderajan, S., Shrivastava, M., and Suprasanna, P. 2016a. Arsenic toxicity in rice (*Oryza sativa* L.) is influenced by sulfur supply: Impact on the expression of transporters and thiol metabolism. *Geoderma* 270:33–42.

Srivastava, A.K., Pasala, R., Minhas, P.S., and Suprasanna, P. 2016b. Plant bioregulators for sustainable agriculture: Integrating redox signaling as a possible unifying mechanism. *Advances in Agronomy* 137:237–278.

Srivastava, S., Pathare, V.S., Sounderajan, S., and Suprasanna, P. 2019. Nitrogen supply influences arsenic accumulation and stress responses of rice (*Oryza sativa* L.) seedlings. *Journal of Hazardous Materials* 367:599–606.

Srivastava, A.K., Pandey, M., Ghate, T., et al. 2021. Chemical intervention for enhancing growth and reducing grain arsenic accumulation in rice. *Environmental Pollution* 276:116719.

Stiles, W. 2013. *Trace Elements in Plants.* Cambridge University Press, New York.

Sun, H., Dai, H., Wang, X., and Wang, G. 2016. Physiological and proteomic analysis of selenium-mediated tolerance to Cd stress in cucumber (*Cucumis sativus* L.). *Ecotoxicology and Environmental Safety* 133:114–126.

Takahashi, F., Kuromori, T., Urano, K., Yamaguchi-Shinozaki, K., and Shinozaki, K. 2020. Drought stress responses and resistance in plants: From cellular responses to long-distance intracellular communication. *Frontiers in Plant Science* 11:556972.

Tang, C., Xie, J., Lv, J., et al. 2021. Alleviating damage of photosystem and oxidative stress from chilling stress with exogenous zeaxanthin in pepper (*Capsicum annuum* L.) seedlings. *Plant Physiology and Biochemistry* 162:395–409.

Tester, M., and Bacic, A. 2005. Abiotic stress tolerance in grasses from model plants to crop plants. *Plant Physiology* 137:791–793.

Tonhati, R., Mello, S.C., Momesso, P., and Pedroso, R.M. 2020. L-proline alleviates heat stress of tomato plants grown under protected environment. *Scientia Horticultare* 268:109370.

Ullah, A., Romdhane, L., Rehman, A., and Farooq, M. 2019. Adequate zinc nutrition improves the tolerance against drought and heat stresses in chickpea. *Plant Physiology and Biochemistry* 143:11–18.

Upadhyay, M.K., Majumdar, A., Barla, A., Bose, S., and Srivastava, S. 2019a. An assessment of arsenic hazard in groundwater-soil-rice system in two villages of Nadia district, West Bengal, India. *Environmental Geochemistry and Health* 41:2381–2395.

Upadhyay, M.K., Shukla, A., Yadav, P., and Srivastava, S. 2019b. A review of arsenic in crops, vegetables, animals and food products. *Food Chemistry* 276:608–618.

Wang, J., Fu, P., Lu, W., and Lu, D. 2020. Application of moderate nitrogen levels alleviates yield loss and grain quality deterioration caused by post-silking heat stress in fresh waxy maize. *The Crop Journal* 8:1081–1092.

Wang C., Cheng T., Liu H., et al. 2021. Nano-selenium controlled cadmium accumulation and improved photosynthesis in indica rice cultivated in lead and cadmium combined paddy soils. *Journal of Environmental Sciences* 103:336–346.

Waraich, E.A., Ahmad, R., Halim, A., and Aziz, T. 2012. Alleviation of temperature stress by nutrient management in crop plants: A review. *Journal of Soil Science and Plant Nutrition* 12:221–244.

Wisawapipat, W., Chooaiem, N., Aramrak, S., et al. 2021. Sulfur amendments to soil decrease inorganic arsenic accumulation in rice grain under flooded and nonflooded conditions: Insights from temporal dynamics of porewater chemistry and solid-phase arsenic solubility. *Science of the Total Environment* 779:146352.

Wu, C., Dun, Y., Zhang, Z., Li, M., and Yu, G. 2020. Foliar application of selenium and zinc to alleviate wheat (*Triticum aestivum* L.) cadmium toxicity and uptake from cadmium-contaminated soil. *Ecotoxicology and Environmental Safety* 190:110091.

Xi, Y., Han, X., Zhang, Z., et al. 2020. Exogenous phosphite application alleviates the adverse effects of heat stress and improves thermotolerance of potato (*Solanum tuberosum* L.) seedlings. *Ecotoxicology and Environmental Safety* 190:110048.

Xue, T., Liao, X., Wang, L., et al. 2020. Effects of adding selenium on different remediation measures of paddy fields with slight-moderate cadmium contamination. *Environmental Geochemistry and Health* 42:377–388.

Yan, G., Fan, X., Zheng, W., et al. 2021. Silicon alleviates salt stress-induced potassium deficiency by promoting potassium uptake and translocation in rice (*Oryza sativa* L.). *Journal of Plant Physiology* 258:153379.

Yang, F., Xu, X., Xiao, X., and Li, C. 2009. Responses to drought stress in two poplar species originating from different altitudes. *Biologia Plantarum* 53:511.

Yu, L.J., Luo, Y.F., Liao, B., et al. 2012. Comparative transcriptome analysis of transporters, phytohormone and lipid metabolism pathways in response to arsenic stress in rice (*Oryza sativa*). *New Phytologist* 195:97–112.

Zhang, W., Yu, X., Li, M., Lang, D., Zhang, X., and Xie, Z. 2018. Silicon promotes growth and root yield of *Glycyrrhiza uralensis* under salt and drought stresses through enhancing osmotic adjustment and regulating antioxidant metabolism. *Crop Protection* 107:1–11.

Zhang, L., Yan, M., Li, H., et al. 2020. Effects of zinc fertilizer on maize yield and water-use efficiency under different soil water conditions. *Field Crops Research* 248:107718.

Zhang, T., Wang, J., Sun, Y., Zhang, L., and Zheng, S. 2021. Versatile roles of melatonin in growth and stress tolerance in plants. *Journal of Plant Growth Regulation*. doi: 10.1007/s00344-021-10317-2.

Zhou, J., Zhang, C., Du, B., et al. 2020. Effects of zinc application on cadmium (Cd) accumulation and plant growth through modulation of the antioxidant system and translocation of Cd in low- and high-Cd wheat cultivars. *Environmental Pollution* 265:115045.

Zhou, J., Zhang, C., Du, B., et al. 2021. Soil and foliar applications of silicon and selenium effects on cadmium accumulation and plant growth by modulation of antioxidant system and Cd translocation: Comparison of soft vs. durum wheat varieties. *Journal of Hazardous Materials* 402:123546.

Zhu, Y., Jiang, X., Jiang, J., et al. 2020. Silicon confers cucumber resistance to salinity stress through regulation of proline and cytokinins. *Plant Physiology and Biochemistry* 156:209–220.

2

Availing Engineered and Biosynthesized Metal-Based Nanoparticles to Alleviate Abiotic Stress

Monolina Sarkar
Sammilani Mahavidyalaya

Anindita Paul
Indian Institute of Science

Debapriya Rajlakshmi Das
Taki Government College

CONTENTS

Abbreviations

Nanoparticles, NPs; Reactive oxygen species, ROS; Water use efficiency, WUE; Catalase, CAT; Ascorbate peroxidase, APX; Glutathione peroxidase, GPX; Peroxidase, POX or POD; Superoxide dismutase, SOD; ribulose bisphosphate carboxylase, RuBisCO; glutathione, GSH; malondialdehyde, MDA; poly(acrylic acid)-coated cerium oxide nanoparticles or nanoceria, PNC; Indole-3-acetic acid, IAA; Gibberellin, GA; Jasmonic acid, JA; Salicylic acid, SA; Abscisic acid, ABA.

2.1 Introduction

The insatiable global demand for safe, nutritionally enriched, and sufficient food crops is constantly challenged by the ever-increasing global population, anthropogenic activities, climatic variability, and political-economic systems. The population growth has led to excessive industrialization, increasing demand for housing, and adoption of unsustainable agricultural practices that eventually lead to deforestation, over-farming, air and water pollution, and global climate change (Molotoks, Smith, and Dawson 2021). The agricultural and livestock management systems in use, such as farming the same crop on the same land, excessive pesticide or fertilizer application, and the use of modern tools and technologies, negatively affect the environment and biodiversity. Depreciating soil and water quality, diminished nutrient availability, water scarcity, heavy metal contamination, greenhouse gas emissions, change in soil microbiome, and pest-pathogen threats impact global food production cycles which are dependent on a limited number of crop species (Chakraborty and Newton 2011; Pareek, Dhankher, and Foyer 2020). Elevated global temperatures, increased CO_2 and O_3 levels, unpredictable rainfall patterns, and climatic extremes such as heatwaves, drought, chilling periods and floods directly reflect climate change (Pachauri and Reisinger 2008). High salinity, water deficit or excess, heavy metals, drastic temperature changes, and UV radiation, collectively called abiotic stresses, are hostile for agriculture and the ecosystem. The global agricultural sector is affected

DOI: 10.1201/9781003258063-2

by abiotic stresses, which are direct or indirect consequences of natural or human interventions (He, He, and Ding 2018). For example, salinity stress is interlinked with drought conditions that can worsen with high temperatures. Therefore, the focus of modern agricultural research should be to design and develop environment-friendly, inexpensive, and user-friendly strategies to enhance multistress tolerance in plants, such that they can grow in limited resource availability and diverse climatic conditions.

2.1.1 Can Nanoparticles (NPs) Be the Right Choice?

Nanotechnology can be a sustainable alternative to remediate the damaging consequences of abiotic stresses. Nanotechnology has provided a wide array of nanomaterials, for example, metal and non-metal NPs, nanorods, nanotubes, liposomes, dendrimers, quantum dots (QDs), nanocarriers, and nanosensors, which have been extensively used for mitigation of abiotic stress and plant disease and health management (Rajput et al. 2021; Mahto et al. 2021; Elmer, Ma, and White 2018; Elmer and White 2018). NPs with all three dimensions in the range of 1–100 nm have several physiochemical properties, which are advantageous for facilitating plant absorption and NP-plant interactions. The small size, shape diversity, size dispersity, large-surface area-to-volume ratio, net surface charge (zeta potential, ζ), and distinct magnetic, optical, and electrical properties of NPs ensure improved solubility and reactivity compared to their bulk counterparts, enhanced uptake by plant cells, lesser input amounts, low consumption per crop area and extended residual activity (Figure 2.1a) (Mourdikoudis, Pallares, and Thanh 2018; Pérez-de-Luque 2017). This reduction in the load of environmentally hazardous, active chemicals lowers phytotoxicity levels and causes minimal exposure to regions of the plant other than target tissue. The same properties of NPs allow them to be effective carriers for large macromolecules like proteins, nucleic acids, and small molecules, as well as function as nanobiosensors, nano-biozymes, stimulants or nutritional supplements (Figure 2.1b). Several approaches of NP synthesis are available which are rapid, cheap, and can be easily scaled up (Iravani et al. 2014; Singh et al. 2016a). Additionally, hassle-free delivery options of these nanoscale materials, such as seed priming, foliar sprays, and soil application, make NPs a sustainable choice for their intended purpose (Figure 2.1c). The small size and surface charge composition of NPs also enable uptake and translocation across the plant structures and cellular barriers of varying size exclusion limits (Figure 2.1d).

2.1.2 Engineered and Biosynthesized NPs

Due to the unique physicochemical properties of NPs and their extensive application in various fields, improved methodologies are being used for NP synthesis. Physical approaches, such as evaporation-condensation, laser ablation, and chemical methods, for example, chemical reduction by reducing agents (followed by protection and stabilization), microemulsion techniques, electrochemical synthesis, photochemical reduction, UV-initiated photoreduction, and microwave-assisted synthesis, were methods of choice for fabricating NPs (Iravani et al. 2014). Although these physiochemical approaches generated NPs in high concentrations, they are associated with numerous limitations. These synthesis methods are expensive, environmentally hazardous, energy-demanding, cause toxic chemical and heavy metal contamination, are less reproducible, non-sustainable, cannot be easily scaled up, cause problems in separation of the final product from concoction, and can lead to structural deformation and polydispersity in NPs.

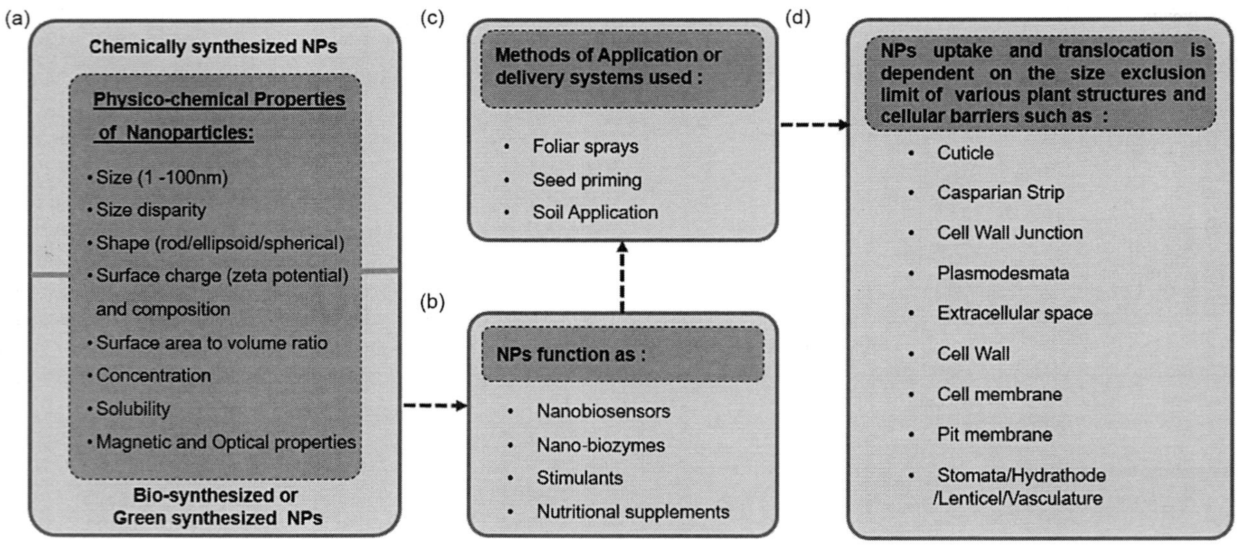

FIGURE 2.1 Schematic representation of factors governing NP use and targeted delivery systems: (a) NPs, chemically synthesized or biosynthesized, possess diverse physicochemical properties, which determine its entry, transport and functioning. (b) The use of NPs in stress detection and alleviation can be manifold – nanobiosensors, nano-biozymes, stimulants, nutritional supplements, nanofertilizers, etc. (c) Delivery systems for NP application differ based on the kind of abiotic stress and the plant organ which is exposed or most affected. (d) Besides surface charge and composition, NP size is a major criterion determining uptake and translocation inside cells. Intracellular entry is controlled by the size exclusion limit of the plant structures and cellular barriers.

'Green' synthesis of NPs provides an alternative choice that is cheap, sustainable, eco-friendly, rapid, reproducible, and scalable and reduces energy demand and generation of toxic by-products (Iravani et al. 2014; Keat et al. 2015). Polysaccharide method, Tollens method, irradiation method, polyoxometalates method, and biological methods are the diverse 'green' synthesis approaches. Several microorganisms (bacteria, fungi, algae, and yeast) and plants or plant tissues have the potential to synthesize NPs of narrow size distribution (less polydispersity) as well as enhanced stability and biocompatibility (Singh et al. 2016a; Paul and Roychoudhury 2021). These organisms have enzymes to reduce metal salts to NPs, detoxify heavy metals and degrade other toxic by-products. Moreover, the plethora of biomolecules and phytochemicals perform the role of natural reducing, stabilizing, and capping agents that prevent NP aggregation, provide additional stability, and improve functional efficacy. The simple methodology, reduced number of processing steps, occasional elimination of downstream processing steps (extracellular biosynthesis), and easy recovery of the final product enable rapid biosynthesis of effective, bioactive, and well-structured NPs.

2.2 NPs Enable Detection and Alleviation of Abiotic Stress

NPs can function as nanosensors and nanofertilizers (Figure 2.1b) (Chhipa and Joshi 2016; Liu and Lal 2015). The small size, customizable surface modifications, and environmentally responsive slow release of nutrients enable metal NPs to enhance the efficiency of nutrient use in plants and lower the transport of these nutrients into runoff or groundwater. In wheat, chitosan NPs induced biosynthesis and transport of indole-3-acetic acid (IAA), reduced the activity of IAA oxidase and augmented the concentration of IAA in shoots and roots (Li et al. 2019). The chitosan NPs also had a positive effect on the germination of seed and seedling growth at a lower concentration as compared to chitosan. They increase nutrient bioavailability and diminish the application of chemical fertilizers. Moreover, the unique surface chemistry, thermal, and electrical attributes, in addition to enhanced sensitivities of NPs, are suitable for monitoring chemical signalling for evaluation of pathogens, nutritional deficiencies, drought, salinity, and thermal stress to improve plant growth and yield (Chhipa and Joshi 2016; Honeychurch 2014). Nanosensors used to perceive abiotic stress in plants include plasmonic nanosensors, carbon-based, fluorescence resonance energy transfer-based, nucleic acid, peptide, antibody, electrochemical, and nanowire-based nanosensors (Figure 2.2a and b) (Zhu et al. 2017; Li et al. 2018b; Kordrostami, Mafakheri, and Al-Khayri 2021). NPs functioning as DNA scaffolds cross the plant cellular impediment and integrate genetically encoded nanosensors (Giraldo et al. 2019). The sensors, generally tagged with fluorescent proteins, are studied by fluorescence imaging. They have high sensitivity toward specific analytes in biological systems. NPs have minimal photobleaching and autofluorescence, thereby permitting the spatiotemporal resolution of

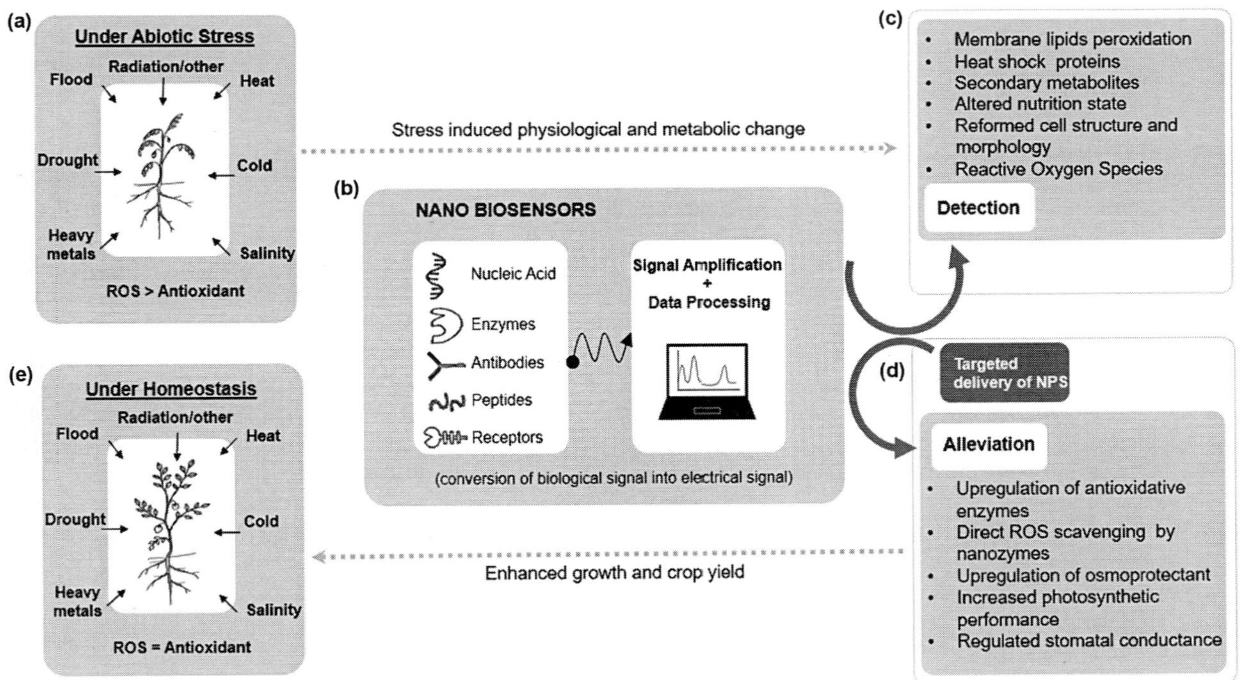

FIGURE 2.2 NPs in detection and alleviation of abiotic stress: (a) Plants under abiotic stress exhibit elevated levels of ROS. (b) Nanobiosensors are responsible for detection of stress-specific metabolites that generate an electronic signal from the biological signal, with the aid of a computational unit. (c) Nanobiosensors detect stress-induced physiological and metabolic changes occurring within the plant. (d) Targeted NP delivery systems alleviate abiotic stress by upregulating antioxidant activities, reprogramming cellular signalling, and subsequently decreasing ROS levels. (e) Elevated antioxidant activity reduces ROS levels, thereby attaining homeostasis.

signalling molecules. The nanobiosensors can be delivered to plants using agroinfiltration, needle-free injection via the leaf lamina, and topical delivery to detect abiotic stress signalling molecules (Kordrostami, Mafakheri, and Al-Khayri 2021). Nanobiosensors enable the real-time monitoring of abiotic stress based on the detection of stress-specific phytohormones as well as signalling molecules including calcium, reactive oxygen species (ROS), NO, glucose, and sucrose (Giraldo et al. 2019) (Figure 2.2c). For instance, single-walled carbon nanotubes are used for near-infrared fluorescence monitoring of NO in *Arabidopsis thaliana* (Giraldo et al. 2014). Probe developed by thioglycolic acid capped QDs and boronic-acid-conjugated QDs enable the *in vivo*, glucose-specific, optical sensing in plant tissues (Li et al. 2018a). The *in vivo* fluorescence response can be identified by confocal microscopy and standoff Raspberry Pi camera. Localized surface plasmon resonance that uses aptamer-functionalized gold NPs (Au NPs) has been reported to detect abscisic acid (ABA) during salt stress in rice (Wang et al. 2017).

Recently, nanozymes have been used as an excellent alternative to biological enzymes and to construct novel biosensors. Catalytic activities of nanozymes can be controlled by pH, hydrogen peroxide (H_2O_2), and glutathione (GSH) concentrations in response to changes in different environmental stimuli. For instance, gas chromatography-mass spectrometry (MS)-based metabolomics showed that foliarly-applied Mn_3O_4 NPs altered the metabolomes in cucumber leaves (Lu et al. 2020). Mn_3O_4 NPs mimic catalase (CAT), glutathione peroxidase (GPX), and superoxide dismutase (SOD) in the leaves of cucumber (*Cucumis sativus*) in response to salt stress. Both precursors (phenylalanine and tyrosine) and downstream products (dihydroxycinnamic acid, resveratrol, chlorogenic acid, benzenetriol, trihydroxybenzene, catechin, quinic acid, and hydroxybenzoic acid) in phenylpropanoid and shikimate pathways were upregulated by Mn_3O_4 NPs, indicating the enhancement of endogenous antioxidants. In addition, Mn_3O_4 NPs significantly enhanced photosynthetic pigment content in leaf, net photosynthesis, and biomass. Under salinity stress, iron oxide NPs induce a substantial enhancement in shoot length and total chlorophyll content (TCC) in *Eucalyptus tereticornis* (Singh et al. 2021). At 300 mM NaCl, 25 ppm iron oxide NPs, significantly enhanced activity of SOD, CAT, and peroxidase (POX), increased concentration of proline and total soluble sugar besides decreasing concentration of malondialdehyde (MDA) in *E. tereticornis*. Iron oxide NPs can upregulate the transcription of salt-responsive genes encoding Na^+/K^+ transporters (HKT1, NHX1, and SOS1) that regulate Na^+/K^+ homeostasis. Maghemite NPs have been reported to act as CAT and POX, improve the chlorophyll concentration and decrease both H_2O_2 and lipid peroxidation following water scarcity in *Brassica napus* (Palmqvist et al. 2017). In addition to having an enzymatic effect, yttrium doping-stabilized γ-Fe_2O_3 NPs were suggested to provide the micronutrient iron. Future studies are being targeted to analyse the possible interaction of iron oxide NPs with proteins, lipids, and other biomolecules, or the possible absorption of native iron ions onto the surface and reduction of toxic Fenton reactions.

2.3 Delivery, Absorption, and Translocation of NPs in Plants

Targeted delivery of the appropriate dose of NPs to plants at the onset of abiotic stress in the suitable physicochemical form is challenging. The NPs can function as nutrients or the platform to deliver the nutrients in a stimuli-responsive manner. A vital challenge for targeted delivery is to design surface coatings that are not affected by adsorbed bio-macromolecules such as natural organic matter or proteins in soils or plant cells. For instance, potassium chloride (KCl) and ammonium thiosulphate ($(NH_4)_2S_2O_3$), in presence of silver sulphide (Ag_2S) NPs and Ag NPs, significantly increased the concentration of Ag in *Lactuca sativa* shoots and roots, while organic matter in soil diminished cerium dioxide (CeO_2) NPs in maize shoots (Zhao et al. 2012; Doolette et al. 2015). As soil application might lead to untargeted broadcasting of NPs, targeted methodologies like foliar application or the use of NP-coated seeds are proposed to be significantly proficient (Lowry, Avellan, and Gilbertson 2019). However, foliar spraying of NPs can be hindered by ultraviolet degradation of NPs and weathering. The surface chemistry of NPs can promote leaf adhesion and improve uptake through plant barriers like cuticle, or translocation through mesophyll cells (Avellan et al. 2019). Leaf surface characteristics like stomatal density or cuticle thickness can vary with the plant's age, growing conditions, and species, thus influencing NP adhesion, uptake, and mobility. For example, drought-resistance response in plants develops tighter cell walls, cuticles, Casparian strips that regulate NP transport. NPs like poly(lactic-co-glycolic) acid NPs can use the clathrin-independent endocytic pathway for entry (Mittal et al. 2020). In seeds, NPs can be directly absorbed by the seed coat through parenchymatic intercellular spaces along with diffusion in the cotyledon. In addition, hydroponic medium exhibits a higher uptake of NPs as compared to sand (Solanki et al. 2015; Mittal et al. 2020). For example, ZnO NPs lead to the highest accumulation of Zn in different forms like Zn-phosphate is observed in the roots and shoots of maize under hydroponic conditions (Lv et al. 2015; Rajput et al. 2020). The observed accumulation may have been brought about by the better solubility of the ionic form of Zn in the rhizosphere and subsequent amplification of absorption and translocation in plants.

The uptake and translocation of NPs through cellular barriers are also regulated by the physicochemical properties of NPs like size, shape, surface charge, and potential biotransformation (Mittal et al. 2020) (Figure 2.1a and d). For instance, NPs with diameters less than 5 nm efficiently cross the undamaged cell wall. Those with a size range of 4–100 nm have been reported to pass through the cuticular barrier. NPs can create enlarged pores in the cell wall and enter the cell. NPs with a size range of 10–50 nm transport via the symplastic route while those between 50 and 200 nm take the apoplastic route. NPs are transferred through the sieve tubes along with the sugar flow and accumulate in the potential sap sinks like roots, stems, young leaves, fruits, and grains. Consequently, the transport is regulated by forces that limit phloem loading as well as transport. The NPs that are transported apoplastically cross through the extracellular spaces till the central vascular cylinder and are translocated up via the xylem. However, before entering the central vascular system, the NPs need to cross the Casparian strip barrier symplastically by binding to the carrier proteins of endodermal cell membranes. The limiting steps of *in planta* transport include the rate at which photosynthate enters the phloem and their flux, an optimum size and charge density for endocytosis, as well as the formation of a corona around the NPs that may regulate their transport. The surface coating of NPs also determines their ability to pass through the cuticle, enter the mesophyll, and transport through the vascular system. For example, positively charged cerium oxide (CeO_2) NPs exhibit better root surface adsorption, while CeO_2 NPs that are negatively charged overcome repulsive forces and exhibit higher root-to-shoot translocation (Spielman-Sun et al. 2019; Liu et al. 2019). Moreover, positively charged CeO_2 NPs form a better association with dicotyledonous plants with a greater root surface area such as lettuce and tomato. Additionally, in monocotyledonous plants, generally, pectins are less abundant as well as loosely bound to the cellulose and have dumbbell-shaped stomata as opposed to kidney-shaped stomata in dicotyledonous plants that result in differential degrees of NP absorption and translocation competences (Mittal et al. 2020).

2.4 Mitigation of Abiotic Stress Using NPs

Heat, cold, salinity, metal toxicity, drought, waterlogging, submergence, excess light, nitrogen excess or deficiency, and UV radiation are abiotic stresses that adversely affect plant anatomy, morphology, physiology, and biochemistry (Figure 2.2a) (He, He, and Ding 2018; Haak et al. 2017). Both vegetative and reproductive development are negatively impacted under stress conditions. In many stressed plants, these effects are manifested through leaf wilting, leaf abscission, decreased leaf area, poor stomatal conductance, diminished transpiration-mediated water loss, and diminished plant health. Seed quality, development, germination, as well as post-germination seedling development are equally impacted. This indicates that fundamental processes like mitosis, cell division, cell expansion, cell growth, and cell elongation are impaired under stress conditions. Along with growth, development, and crop yield, processes like photosynthesis, respiration, metabolism, and homeostasis signalling are affected, which eventually cause plant death. Accumulation of ROS, ROS-induced oxidative and cellular damage, increased production of stress-related solutes, protein denaturation, and lipid peroxidation are the various underlying causes that affect physiological processes (Sachdev et al. 2021). Loss of turgor pressure, electrolyte leakage, reduced nutrient availability, and perturbations in cellular water potential further aggravate the conditions.

Metal, metal oxide, and non-metal-based NPs have been widely studied for their potential in abiotic stress management (Mahto et al. 2021; Rajput et al. 2021). Both engineered and biosynthesized NPSs of varying physiochemical properties have efficiently alleviated the effects of different abiotic stresses (Table 2.1). The outlook of these studies has been either to probe into (i) the potentiality of one type of NPs for management of a single type of stress, (ii) the potentiality of a combination of NPs for alleviation of a single type of stress, and (iii) the potentiality of single type/combination of NPs for promotion of multiple stress tolerance. The following sections discuss the use of NPs for the mitigation of a particular type of abiotic stress on economically important crop species. NPs alleviate the harmful effects of stress by hijacking and enhancing the signalling pathways for stress tolerance and defence pathways. For example, activation of stress-responsive transcription factors (MYB, WRKY, ERF, DREB) regulates stress-related genes, etc. (Figure 2.2d and e). Stress tolerance is increased by the activation of non-enzyme antioxidants and ROS-scavenging enzymes (CAT, APX, GPX, SOD). Compatible osmolytes (sugars, proline, glycine-betaine), phytohormones (ABA, IAA, GA, JA, SA, ethylene), secondary metabolites, ion homeostasis, Ca_2^+ signaling, and melatonin work in harmony to ensure optimum chlorophyll levels and photosynthetic performance in stressful environments.

2.4.1 NPs and Drought Stress

Insufficient supply of physiologically acceptable quality water causes drought stress and affects the growth and yield of crops (Hussain et al. 2018). The frequency and duration of drought periods have increased over decades, affecting especially tropical countries. Drought stress can limit crop production, lead to significant yield losses, and result in food shortage, thereby threatening global food security and sustainability of agriculture. Water shortage can result in reversible and irreversible changes in physiology, during vegetative and reproductive stages of plant development (Hussain et al. 2018; Kapoor et al. 2020). Depending on the duration of drought, certain plants can accumulate metals under water deficit conditions. Additionally, severe drought can adversely affect the water use efficiency (WUE) of plants (Osakabe et al. 2014). Plants, when exposed to drought conditions, exhibit various molecular, biochemical, physiological, and morphological adaptive responses. For example, restricted photosynthesis because of lipid peroxidation, reorganization of chloroplast ultrastructure, changes in levels of chlorophyll, and other photosynthetic pigments result in decreased accumulation of starch granules (Hussain et al. 2018; Osakabe et al. 2014). While plant growth is halted, cell ultrastructure is protected to ensure optimum plant performance under drought stress. Additionally, osmoprotective metabolites are synthesized and accumulated within plant tissues, to maximize water extraction capacity from the surrounding environment (Kapoor et al. 2020). Using nanopreparations to develop the adaptive potential of plants in response to drought stress has been investigated in recent times to delineate the mechanism of their action and sustainability in agricultural environments. Cu/Zn NPs had a significant positive influence on seedling growth of steppe ecotype of winter wheat compared to the forest-steppe under drought conditions (Taran et al. 2017). The physiologically adaptive effects of Cu/Zn NPs encompassed decreased concentration of thiobarbituric acid reactive substances and increased activity of antioxidative enzymes. Differences in the content of photosynthetic pigments and morphometric

TABLE 2.1

Engineered and biosynthesized nanoparticles of varying physiochemical properties used to alleviate abiotic stresses in plants

Type of NPs	Mode of Synthesis	Size	Zeta Potential	Plant Species	Abiotic Stress	Mode of Application	NP Concentration Used	References
Ag NPs	Leaf extract of *M. oleifera*	8–28 nm	Neutral charge	*Triticum aestivum* (var. Pakistan-2013)	Heat	Soil application	25, 50, 75, and 100 mg/L	Iqbal et al. (2019b)
CeO2 NPs (nanoceria)	Chemical method	PNC, 10.3 nm ANC, 12.6 nm	PNC, −16.9 mV ANC, 9.7 mV	*A. thaliana*	Excess light, heat, dark chilling	Leaf infiltration	~50 mg/L	Wu, Tito, and Giraldo (2017)
Ag NPs	Plant extract of *M. oleifera*	7–14 nm	–	*T. aestivum* (var. Pakistan-2013)	Heat	Soil application	25, 50, 75, and 100 mg/L	Iqbal et al. (2019a)
ZnO NPs, Fe3O4 NPs (magnetite)	Wet chemical method (ZnO NPs), hydrothermal method (Fe_3O_4 NPs)	ZnO NPs, 80±5 nm, Fe_3O_4 NPs, 50±5 nm	ZnO NPs, −11.5 mV, Fe_3O_4 NPs, −21.0 mV	*T. aestivum* Both wheat cultivars Sids1 (thermo-tolerant) and Gimmeza7 (thermo-sensitive)	Heat	Sprayed on plants	0, 0.25, 0.50, 0.75, 1.0, and 10 ppm	Hassan et al. (2018)
Ag NPs	Commercially purchased	2, 15, and 50–80 nm	–	*G. max* L. cv. Enrei	Flooding	Supplemented in RO-water	0.2, 2, and 20 ppm	Mustafa et al. (2015)
Ag NPs	Commercially purchased	15 nm	–	*G. max* L. cv. Enrei	Flooding	A combination of Ag NPs, nicotinic acid and KNO_3	5 ppm	Hashimoto et al. (2020)
Ag NPs	–	–	–	*Cr. sativus*	Flooding	Pre-treatment of saffron corms	40, 80, and 120 ppm	Rezvani, Sorooshzadeh, and Farhadi (2012)
Al2O3 NP	Commercially purchased	5 nm, 30–60 nm, 135 nm	–	*G. max* L. cv. Enrei	Flooding	Supplemented in RO-water	50 ppm	Mustafa and Komatsu (2016)
Si NPs	Sol-gel synthesis method	170±10 nm	–	*Z. mays* L. cv. Nootan and *Z. mays* L. hyb. Shaktiman-4	Arsenate toxicity	Added to nutrient media	10 µM	Tripathi et al. (2016)
Fe NPs	Commercially purchased	50–100 nm	–	*T. aestivum* (var. Lasani-2008)	Cadmium toxicity, drought	Soil application	25, 50, and 100 mg/kg	Adrees et al. (2020)
ZnO NPs	Commercially purchased	~50 nm	–	*Lycopersicon esculentum* cv. PKM-1	Cadmium toxicity	Foliar application	50 mg/L	Faizan et al. (2021)
FeO NPs	Extracellular synthesis using culture supernatant of *Bacillus strain* RNT1	15–31 nm	–	*Oryza sativa* cv. Super Basmati Rice	Cadmium toxicity, drought	Soil application	25, 50, and 100 mg/kg of soil	Ahmed et al. (2021)
Fe3O4 NPs	Chemical reduction of colloidal iron hydroxide under hydrothermal conditions to obtain USIRONs	6.8 nm	–	*T. aestivum*	Pb, Zn, Cd, and Cu toxicity	NP suspensions on petri plate	2,000 mg/L	Konate et al. (2017)

(Continued)

TABLE 2.1 ***(Continued)***

Engineered and biosynthesized nanoparticles of varying physiochemical properties used to alleviate abiotic stresses in plants

Type of NPs	Mode of Synthesis	Size	Zeta Potential	Plant Species	Abiotic Stress	Mode of Application	NP Concentration Used	References
Cu NPs, Zn NPs	Using electric impulses of amplitude of 100–2,000 A, copper and zinc granules were dispersed	–	–	*T. aestivum*, ecotype forest-steppe (*Stolichna*) and steppe (*Acveduc*), both varieties grown in winter	Drought	Seed priming	100 parts of water was mixed to 1 part of mother colloid solution	Taran et al. (2017)
Cu NPs	Provided under collaboration	30–40 nm	–	*Z. mays* (L.), Vietnamese maize elite (LVN10)	Drought	Seed priming	69.4 μM (4.444 mg/L)	Van Nguyen et al. (2021)
Fe, Cu, Co, ZnO NPs	Provided under collaboration	20–60 nm	–	*G. max* (L.) Merrill cultivar DT26	Drought	Seed priming	Fe ZnO, and Cu NPs of concentration 50 mg/L was applied along with 0.05 mg/L Co NP	Linh et al. (2020)
Se-, SiO2-, and Se/ SiO2 NPs	Physiochemical method	10–45 nm Se NPs, 20–30 nm SiO_2 NPs, SiO_2: 50–80 nm, Se: 5–20 nm for Se/SiO_2 NPs	–	*Fragaria× ananassa* Duch., Strawberry	Drought	Foliar spray	25 mg/L Se NPs, 125 mg/L SiO_2 NPs, and 50 and 100 mg/L Se/ SiO_2 NPs	Zahedi et al. (2020)
ZnO NPs	Commercially purchased	37.7 ± 15.5 nm	14 mV	*Z. mays* L. cv. Jidan 27	Drought	Soil application	100 mg/L	Sun et al. (2020)
TiO2 NPs	Commercially purchased	7–40 nm	–	*Cicer arietinum* L. genotypes (chill-tolerant, Sel11439 and chill-, ILC533)	Chilling	Foliar spray	5 mg/L	Amini et al. (2017)
TiO2 NPs	Commercially purchased	7–40 nm	–	*C. arietinum* L. genotypes (cold sensitive, ILC533 and tolerant, Sel11439)	Chilling	Foliar spray	0, 2, 5, and 10 ppm	Mohammadi, Maali-Amiri, and Abbasi (2013)
ZnO NPs	Commercially purchased	30 nm	6.697 mV	*O. sativa* ssp. *Japonica* cv. Nipponbare	Chilling	Foliar spray	50 mg/L of TX-10 was used alongside ZnO NPs of concentrations 25, 50, and 100 mg/L	Song et al. (2021)
TiO2 NPs	Commercially purchased	7–40 nm	–	*C. arietinum* L. genotypes (cold tolerant, Sel11439 and cold sensitive, ILC533)	Chilling	Foliar spray	5 mg/L	Hasanpour, Maali-Amir, and Zeinali (2015)

(Continued)

TABLE 2.1 (*Continued*)

Engineered and biosynthesized nanoparticles of varying physiochemical properties used to alleviate abiotic stresses in plants

Type of NPs	Mode of Synthesis	Size	Zeta Potential	Plant Species	Abiotic Stress	Mode of Application	NP Concentration Used	References
TiO2 NPs	Commercially purchased	7–40 nm	–	*C. arietinum* L. genotypes (chill-tolerant, Sel11439 and chill-, ILC533)	Chilling	Foliar spray	5 mg/L	Mohammadi, Maali-Amiri, and Mantri (2014)
CeO2 NPs (nanoceria)	Commercially purchased	23.5±6 nm	−14.7±1.2 mV in nutrient solution; −12.3±1.7 mV in deionized water	*O. sativa* (Y900)	Nitrogen stress	Seedling treatment	100 and 500 mg/L	Wang et al. (2020)
Si NPs	Sol-gel synthesis method	20–95 nm	−0.02 σ(C/m^2)	*T. aestivum* (L.) var. HP 176	UV-B radiation	Supplemented in nutrient media	30 mL of 1 mM stock solution added to each pot	Tripathi et al. (2017)
Ag NPs	Green-synthesized from fungi *Aspergillus terreus* MALEX	15–29 nm	–	*T. aestivum* L. cv. Sods 1	Salinity	Seed priming	2, 5, and 10 mM	Mohamed et al. (2017)
ZnO NPs, Fe2O3 NPs	Commercially purchased	20 nm	–	*Z. mays* L. Bulk and S.C. 704	Salinity	Foliar spray	2 g/L	Fathi, Zahedi, and Torabian (2017)
ZnO NPs	Chemical precipitation method	21.3 nm	–	*Lupinus termis*	Salinity	Seed priming	20, 40, and 60 mg/L	Latef, Alhmad, and Abdelfattah (2017)
Zn NPs	–	–	–	*Gossypium barbadense* L.	Salinity	Exogenous spraying	100 and 200 ppm	Hussein and Abou-Baker (2018)
Zn NPs, B NPs, Si NPs, Zeolite NPs	Chemical method (Zn NP, B NP, Si NP), hydrothermal synthesis (zeolite NP)	Various sizes	–	*Solanum tuberosum* L., cv Diamont	Salinity	Soil application (Zn NP, B NP, Si NP); through irrigation network (zeolite NP)	20, 12, and 15 ppm (Zn NP, B NP, Si NP, respectively); 1.3 L/ha (zeolite NP)	Mahmoud et al. (2020)
Se NPs	Chemical synthesis method and stabilization by chitosan (CS)-PVA	2–20 nm	−29.4 mV	*L. esculentum*	Salinity	Added along with nutrient solution in a directed irrigation system.	1, 5, 10, and 20 mg/L	Morales-Espinoza et al. (2019)
Ag NPs	Green-synthesized from leaf extract of *T. aestivum*	27–33 nm	−20.3 mV	*T. aestivum* L. cv. Pusa Kiran	Salinity	Foliar spray	300 ppm	Wahid et al. (2020)

(Continued)

TABLE 2.1 (*Continued*)
Engineered and biosynthesized nanoparticles of varying physiochemical properties used to alleviate abiotic stresses in plants

Type of NPs	Mode of Synthesis	Size	Zeta Potential	Plant Species	Abiotic Stress	Mode of Application	NP Concentration Used	References
Si NP, Zn NP Se NP, GNRs	Chemical method	($nSiO_2$; 5–15 nm) nZnO < 100 nm (nSe; 100 mesh), graphene nanoribbons (GNRs) (2–15 μm × 40–250 nm)	–	*S. officinarum*	Chilling stress and photoprotection	Foliar application	300 ppm ($nSiO_2$) 50 ppm (nZnO) 15 ppm (nSe) 50 ppm (GNRs)	Elsheery et al. (2020)
Poly(acrylic acid)-coated CeO2 NPs (nanoceria)	Chemical synthesis method	1.8 ± 0.3 nm	−51.7 ± 11.5 mV	*Gossypium hirsutum* L. cv Acala 1517-08	Salinity	Seed priming	500 mg/L	An et al. (2020)
TiO2 NPs	Physiochemical synthesis	20–30 nm	–	*D. moldavica*	Salinity	Added along with nutrient solution	100 mg/L	Gohari et al. (2020)
ZnO NPs	Leaflets extract of *Phoenix dactylifera L. var. Khalas* was used to derive green-synthesized ZnO NP	16–35 nm	–	*Abelmoschus esculentus* L. Moench cv. Hasawi	Salinity	Foliar application	10 mg/L	Alabdallah and Alzahrani (2020)
CNPs	Provided under collaboration	90–110 nm (diameter)	–	*L. sativa*	Salinity	Seed priming	0.3%	Baz et al. (2020)
Hydroxyapatite Ca10(PO4)6(OH)2)	Chemical method	75–125 nm length, 15–5 nm width (needle like)	−1.7 ± 0.1 mV	*Z. mays*	Salinity	Soil application	1 g/kg	Yoon et al. (2020)
Fe3O4 NPs	Biosynthesis	10–18 nm	–	*Setaria italica*	Drought stress	Seed priming	5, 10, 15, 20, 50, 90, and 120 mg/L	Sreelakshmi et al. (2021)
Se NPs, Cu NPs	Chemical method	2–20 nm (Se NPs), 50 nm (Cu NPs)	–	*C. annuum*	Salinity	Plants were drenched with five doses of 10 mL each	10 and 50 mg/L (Se NPs) 100 and 500 mg/L (Cu NP)	González-García et al. (2021)
Cu NPs	Chemical method	50 nm	–	*Solanum lycopersicum*	Salinity	Foliar application	250 mg/L	Pérez-Labrada et al. (2019)
Si NP, Zn NP Se NP, GNRs	Chemical method	($nSiO_2$; 5–15 nm) nZnO < 100 nm (nSe; 100 mesh), graphene nanoribbons (GNRs) (2–15 μm × 40–250 nm)	–	*S. officinarum*	Chilling stress and photoprotection	Foliar application	300 ppm ($nSiO_2$) 50 ppm (nZnO) 15 ppm (nSe) 50 ppm (GNRs)	Elsheery et al. (2020)

indexes of leaves are other adaptive mechanisms adopted by these plants. In a similar study, seeds of Vietnamese maize elite were pre-treated with Cu NPs and cultivated under standard and drought environments (Van Nguyen et al. 2021). Cu NPs promoted maize growth and productivity as indicated by increased shoot height and biomass, reduced wilting and rolling of leaves, increased average seed dry weight per plant in NP-primed plants, compared to water-treated controls. Additionally, Cu NPs augmented the concentration of anthocyanin, chlorophyll, and carotenoid pigments, maintained higher leaf water content, and induced activity of ROS scavenging enzymes, thereby decreasing ROS accumulation. Foxtail millet seeds primed with green-synthesized Fe_3O_4 NP extracted from *Chaetomorpha antennina* (green marine algae) showed enhanced content of chlorophyll, soluble sugars, and proline and exhibited better drought stress adaptation (Sreelakshmi et al. 2021). Spraying strawberry plants under drought stress with SiO_2 NPs, selenium NPs (Se NPs), and Se/SiO_2 NPs enhanced growth, improved drought tolerance, and positively affected fruit quality and yield (Zahedi et al. 2020). Exogenous spraying of NPs resulted in (i) increased photosynthetic capacity by protecting photosynthetic pigments, (ii) increased osmolyte accumulation (carbohydrate, proline) to preserve osmosis, (iii) increased relative water content (RWC), WUE, and membrane stability index (MSI), (iv) activation of antioxidant enzymes (including CAT, APX, GPX, SOD), diminished lipid peroxidation and H_2O_2 levels, and (v) enhanced concentrations of anthocyanin, total phenolics, vitamin C, and 2,2-diphenyl-1-picrylhydrazyl that improve fruit nutritional value and quality.

The application of NPs also triggered drought-associated gene expression. The responses of soybean plants, under drought conditions, upon application of metal-based NPs (Fe NPs, Co NPs, Cu NPs, ZnO NPs) were monitored (Linh et al. 2020). Priming soybean seeds with Fe NPs emerged to be effective compared to others, on the basis of shoot monitoring analysis, root growth, RWC, tolerance to drought, and rate of decrease in biomass. Quantitative reverse-transcriptase polymerase chain reaction (qRT-PCR) analysis of leaves and roots of soybean treated by Fe NPs showed significant elevated expression of drought-tolerant marker genes. The upregulation of the marker gene, *GmERD1* in both leaves and roots in all NP-treated plants indicated it to be the most sensitive candidate marker under drought, post-application of NPs. This study further established that genes participating in both ABA-dependent and independent pathways are upregulated in response to drought. Nano-ZnO soil application activated the ROS scavenging antioxidant enzymes that alleviated damages to mitochondria and chloroplast ultrastructure, caused by drought. This is supported by higher abundances of Fe/Mn SOD, Cu/Zn SOD, APX, and CAT transcripts in plants treated with nano-ZnO. Additionally, relatively higher expression of melatonin biosynthesis genes (e.g. tryptophan decarboxylase, N-acetylserotonin O-methyltransferase, caffeic acid O-methyltransferase, and serotonin N-acetyltransferase) and higher concentrations of metabolites (e.g. tryptophan, tryptamine, serotonin, melatonin) in ZnO NP-treated maize plants indicates upregulation of endogenous melatonin synthesis for improving drought tolerance.

2.4.2 NPs and Heavy Metal Toxicity

Increasing anthropogenic activities and expanding industrialization, coupled with developing agricultural practices have culminated in contamination of agricultural soils with heavy metals like As, Cd, Co, Cr, Cu, Ni, Pb, and Zn. Heavy metal toxicities can arise from prolonged use of fertilizers, release of sewage sludge, industrial and municipal waste, waste from smelting factories and mines, or poor watering practices in cultivable lands (Singh et al. 2016b; Yadav 2010). As a response to heavy metal-rich soils, plants generate ROS directly or indirectly. The indirect consequences are mediated by competition with nutrient cations for absorption, the interaction of heavy metals with antioxidant or ROS scavenging system, rendering functional proteins inactive by altering the structure or displacing cofactors, disruption of electron transport chain, membrane degradation by lipid peroxidation, and perturbed metabolism of essential elements (Yadav 2010). Other effects include a negative impact on the cell cycle and plant cytoskeleton, hormonal changes including auxin redistribution, and alterations in membrane potential. Morphological and physiological effects of heavy metal toxicity include decreased growth and yield, suppressed root and shoot growth, decrease in biomass, reduced photosynthesis, alterations in membrane ultrastructure and permeability, water imbalance, poor nutrient uptake and assimilation, leaf chlorosis, and senescence, leading to plant death (Singh et al. 2016b). Heavy metals interfere with photosynthetic activities by decreasing levels of photosynthetic pigments, photochemical efficiency of photosystems, activity of photosynthetic enzymes such as ribulose bisphosphate carboxylase (RuBisCO), stomatal conductance, and CO_2 assimilation. Photosynthetic parameters like photosynthetic rate, quantum yield, and intracellular CO_2 concentration are therefore negatively impacted.

Majority heavy metals are non-essential, having no physiological significance, and are toxic even at low concentrations. On the contrary, micronutrients like Cu, Zn, Mn, Fe, Ni, and Co, which are also heavy metals, are crucial for basic metabolic processes in angiosperms (Ernst 2006). The use of these micronutrients, especially silicon (Si) as supplements to mitigate heavy metal stresses, has been demonstrated earlier (Bhat et al. 2019; Huang et al. 2018; Kim et al. 2014). Concomitantly, employing micronutrients as nanomaterials to relieve metal stress has been one of the essences of agri-nanotechnological research. Alleviation of arsenate (As^V) toxicity in maize cultivar and hybrid varieties using engineered silicon NPs (Si NPs) was investigated (Tripathi et al. 2016). Si NPs physiochemically synthesized from sodium silicate (Na_2SiO_3) by sol-gel synthesis method were added to nutrient media of control and As^V pre-treated seedlings. Si NP treatment led to a decline in accumulated As concentration and oxidative stress, improvement in the ascorbate-GSH cycle, mitigation of stress-induced lipid peroxidation, and enhancement in levels of photosynthetic pigments and photosynthetic efficiency. Mitigation of toxicity imparted to wheat seedlings by heavy metals has been reported when administered with magnetic nano-Fe_3O_4 (Konate et al. 2017). NP treatment reduced oxidative stress and amplified the activity of antioxidant enzymes, confirmed by the decline in MDA levels. The decrease in toxicity with nano-Fe_3O_4 treatment was confirmed to be connected with differing adsorption capacities of heavy metals. In a separate study, commercially purchased ZnO NPs were used to reduce Cd toxicity in tomato seedlings. ZnO NP treatment improved plant growth and biomass, positively affected photosynthetic attributes (transpiration rate, photosynthetic rate, internal CO_2 concentration, and stomatal conductance), increased activities of enzymes like nitrate reductase and carbonic anhydrase, and decreased MDA and superoxide radical contents (Faizan et al. 2021). Owing to similar morphophysiological effects of drought and Cd toxicity stress, and the effect on metal accumulation by plants linked with inadequate water supply, engineered and biosynthesized iron oxide NPs treated soil mixtures were used to propagate and grow wheat and rice, respectively, under standard and stressed environments (Adrees et al. 2020; Ahmed et al. 2021). In both cases, plant biomass and photosynthetic efficiency improved, while oxidative stress was reduced. Additionally, in NP-treated rice plants, Cd transporter genes (*OsHMA2*, *OsHMA3*, and *OsLCT1*) exhibited diminished expression (Ahmed et al. 2021).

2.4.3 NPs and Heat Stress

Extreme variations in temperature harmfully affect plant health and reproduction, thus imposing a damaging effect on agricultural production and challenging global food security (Bita and Gerats 2013; Hussain et al. 2018). Increasing temperatures coupled with decreased precipitations cause a catastrophic effect on crop production. Elevated CO_2 levels, heightened amounts of other greenhouse gases, extreme fluctuations in temperature, severity during such periods (duration, frequency, and rise in average temperature), higher mean annual temperatures, rise in seasonal temperatures and global warming add to such climatological extremes, resulting in a heat-stressed environment with shorter seasons (Bita and Gerats 2013). Heat affects plant growth, starting from germination to maturity and thus affecting all vegetative and reproductive stages.

Heat stress direly affects plant's physiological, cellular, and molecular functioning (Bita and Gerats 2013; Hasanuzzaman et al. 2013). Physiological responses involve changes in total dry weight of the plant, which get affected due to many factors: scorching or sunburn of stems and leaves, leaf senescence and abscission of plant organs, inhibition of root and shoot growth, changes in plant architecture such as elongated petioles and hypocotyls, fruit discolouration and damage, etc (Wahid et al. 2007). These are coupled to changes at the cellular and molecular level, for example, cytoskeletal reorganization, changes in organelle ultrastructure, membrane remodelling and increased membrane fluidity, elevated production of HSPs, and rise in levels of phytohormones, antioxidants, and osmolytes (Hasanuzzaman et al. 2013; Bita and Gerats 2013). High temperatures bring about perturbations in respiration, photosynthesis, and homeostasis in general. Structural alterations of chloroplast, reduced biosynthesis of photosynthetic pigments, changes in thylakoid membrane fluidity and permeability, changes in mitochondrial and chloroplast electron transport, photochemical changes to carbon flux, impact on the activity of temperature-sensitive enzymes such as RuBisCO, coupled with the injurious effects of ROS curtail both respiratory and photosynthetic performances. Along with ROS production, heat stress induces alterations in primary and secondary metabolite levels and changes in expression patterns of HSPs, chaperones, transcriptional factors (heat shock transcription factors), and activators. Heat stress has a massive impact on plant fertility and sexual reproduction, at various stages. High temperatures affect flowering (e.g. flower bud initiation), fertilization, embryo development, and fruit set. Male sterility is a major consequence in heat stress-challenged plants, with delay in onset of anthesis, tapetum degeneration, impaired development of pollen, pollen sterility, and interruption of sucrose and starch turnover aggravated by diminished nutrient and water transport (Bita and Gerats 2013). Elevated temperatures can halt embryo development and seed development, resulting in reduced germination, loss of vigour, reduced grain number and grain weight, poor yield and grain quality. To combat the adverse effects of heat stress, several techniques including genetic engineering, QTL mapping, marker-assisted breeding, and hybridization, have been adopted that are costly and require technical expertise for unhindered operation.

Nanotechnology offers practical solutions, which are cost-effective and feasible, considering the use of NPs to ameliorate heat stress. *Triticum* at trifoliate developmental stage was subjected to Ag NPs synthesized from *Moringa oleifera* leaf extracts (Iqbal et al. 2019a, b) to examine the influence of green-synthesized Ag NPs in regulating morphological development of *Triticum* under high temperature (Iqbal et al. 2019a). There was an overall increase in plant biomass in presence of Ag NPs under high temperatures. Increase in length of root and shoot, number of root and leaf, as well as leaf area, suggested the heat stress-relieving properties of Ag NPs. In a subsequent analysis, the same group investigated how green-synthesized Ag NPs boosted tolerance to high temperatures in wheat by assessing biochemical and physiological properties (Iqbal et al. 2019b). The application of 50 mg/L Ag NPs at the trifoliate stage negated the effects of heat stress by significantly increasing the RWC, MSI, TCCs, chlorophyll a, and chlorophyll b levels. Additionally, Ag NPs normalized MDA levels and amplified the activity of antioxidant enzymes (SOD, POX, CAT, APX, and GPX). Ag NPs lowered the concentration of superoxide radical, superoxide anion, hydrogen peroxide, and lipid peroxidation, thereby rendering a protective effect on the heat-stressed plants. Ag NPs also enhanced the activity of non-enzymatic antioxidants under the influence of heat stress, as estimated from the levels of total ascorbates, total phenolic contents, and total flavonoid contents. Field experiments with thermo-tolerant and thermo-sensitive cultivars of wheat were conducted to understand the enhancing effect of NPs in the alleviation of heat stress (Hassan et al. 2018). Spraying of ZnO NPs synthesized by wet chemical method and Fe_3O_4 NPs prepared by hydrothermal method had a positive influence on plant height and grain yield. ZnO NP and Fe_3O_4 NP application increased the catalytic activities of antioxidant enzymes like GST, SOD, POX, and CAT and decreased the levels of MDA. As stated previously, the generation of ROS and diminished photosynthetic performance constitute the key outcomes of abiotic stress. In a sophisticated study, it was established that if spherical cerium oxide NPs (also known as nanoceria) could be localized within the chloroplasts of mesophyll cells of leaf, ROS accumulation (like superoxide anion, hydroxyl radicals, and hydrogen peroxide) is significantly reduced (Wu, Tito, and Giraldo 2017). Concomitantly, photosynthetic performance was also improved in heat-stressed, nanoceria-treated plants, as indicated by increase in RuBisCO carboxylation rates, quantum yield of photosystem II, and carbon assimilation rates, as compared to heat-stressed control plants. This study additionally inspected the probable mechanisms of nanoceria uptake into chloroplast by using either negatively charged (PNC, −16.9 mV) and positively charged aminated nanoceria (ANC, 9.7 mV). Two-fold higher PNC colocalization in chloroplasts relative to ANC and temperature-independent nanoceria localization confirmed that nanoceria transport and uptake depend on the electrochemical gradient of plasma membrane potential and occur via endocytosis-independent pathways. Furthermore, the study revealed that the ROS scavenging efficiency and the capacity to protect photosynthetic reactions from oxidative damage by anionic PNC with a low Ce^{3+}/Ce^{4+} ratio (35%) is superior to PNC with high Ce^{3+}/Ce^{4+} ratio (60.8%).

2.4.4 NPs and Chilling/Cold Stress

Similar to high-temperature exposures, in many regions around the world, crop plants are exposed to periods of extreme low temperatures that are unfavourable for plant growth and productivity (Hussain et al. 2018). Chilling stress refers to low, non-freezing temperatures (0°C–15°C). The adverse effect of low ambient temperatures is further accentuated by drought conditions when crops can avail limited water due to decreased root activity and disturbed root hydraulic conductance (turgor stress). Low temperatures have adverse effects on vegetative and reproductive stages of the plant, affecting germination to maturity. However, the intensity of such effects depends on the plants' susceptibility to temperature extremes, genotype, and intra- and inter-specific variations. Low temperatures during cold stress negatively affect the metabolic reactions by thermodynamically decreasing the reaction kinetics. Additionally, oxidative damage due to excessive ROS production and accumulation disturbs membrane integrity, inactivates enzymes, degrades and denatures proteins, chemically modifies cellular macromolecules, and causes ionic imbalance. Chilling stress limits the photosynthetic performance of plants by several mechanisms: damage of chloroplast lipid membrane, aberrant chloroplast development, impaired functioning of photosystems, reduced quantum yield as well as CO_2 assimilation rates (Liu et al. 2018). It also hinders electron transport chain, negatively impacts leaf gas exchange and intercellular conductance of CO_2, besides affecting transpiration rate and metabolite transport. Chilling stress severely affects water uptake and nutrient assimilation. Decrease in root water uptake, root hydraulic conductivity, along with reduced leaf water potential and transpiration rate are common responses to cold stress. Poor root growth (short root length, poor branching, low hydraulic conductance, and bulky root axis) not only affects water uptake but also restricts the uptake of mineral nutrients from soil, causing nutrient deficiency in stressed plants (Hussain et al. 2018). Osmotic imbalance has also been demonstrated for cold stress. Sub-optimal temperatures affect morphological processes like seed germination, seedling vigor as well as growth of root, shoot, and leaf. It also inhibits assimilation of water and nutrients, as well as immunity against pathogens. This culminates in drastic yield losses in crop plants and severely affects yield-related attributes.

The result of chilling stress has been mitigated by applying NPs through the adoption of innovative agricultural practices. Titanium dioxide NPs (TiO_2 NPs) used for tackling cold stress responses in both tolerant and sensitive chickpea genotypes were investigated (Mohammadi, Maali-Amiri, and Abbasi 2013; Mohammadi, Maali-Amiri, and Mantri 2014; Hasanpour, Maali-Amir, and Zeinali 2015). Foliar spraying of NPs (especially at 5 ppm) prevented membrane damage in the cold-stress sensitive chickpea genotype exposed to sub-optimal temperatures, as estimated from the assessment of electrolyte leakage index (ELI) and MDA levels (Mohammadi, Maali-Amiri, and Abbasi 2013). Further investigation of physio-biochemical responses under cold stress conditions revealed MDA content reducing the capacity of TiO_2 NPs accompanied by reduced hydrogen peroxide levels and lowered ELI in the tolerant genotype. TiO_2 NP application on chickpea caused the enhanced activity of antioxidant enzymes, induced lipoxygenase activity, improved stability of photosynthetic pigments, and increased photosynthetic activity and phosphoenolpyruvate carboxylase activity, indicating an orchestration of metabolic processes (Mohammadi, Maali-Amiri, and Mantri 2014; Hasanpour, Maali-Amir, and Zeinali 2015). Identification of differentially expressed transcripts in TiO_2 NP-treated plants using cDNA amplified fragment length polymorphism method, followed by qRT-PCR confirmed differential expression of ten putative genes in response to cold stress (Amini et al. 2017). These included APX6, a receptor-like kinase, RuBisCO large subunit-binding protein subunit alpha (CHN), vacuolar-sorting receptor 6, ATP-dependent chromatin-remodelling complex, nucleosome assembly protein 1, Ethylene-responsive transcription factor (ERF), to name a few. Foliar sprays of Si NPs, Zn NPs Se NPs, and graphene nanorods (GNRs) on chilling sensitive sugarcane plants (*Saccharum officinarum var. Guitang 49*) aid in amelioration of chilling effect accompanied by the betterment of photosynthetic efficiency and accumulation of carotenoids (Elsheery et al. 2020). Recently, a study showed that spraying ZnO NPs on the leaves of rice seedlings grown under hydroponic conditions mitigated chilling stress (Song et al. 2021). NP application boosted morphological parameters like plant height, root length, dry weight, and improved stability of chlorophyll pigments. Decline in levels of H_2O_2, MDA, and proline, coupled with enhanced antioxidant enzyme activities (e.g. SOD, CAT, and POD) helped in easing stress responses. Further, several genes related to antioxidant systems such as *OsCu/ZnSOD1*, *OsCu/ZnSOD2*, *OsCu/ZnSOD3*, *OsPRX11*, *OsPRX65*, *OsPRX89*, *OsCATA*, and *OsCATB* were induced in response to ZnO NP application in cold-stressed plants. Also, chilling-responsive transcription factors, for example, *OsbZIP52*, *OsMYB4*, *OsMYB30*, *OsNAC5*, *OsWRKY76*, and *OsWRKY94*, are expressed in the NP-treated, chilling-stressed rice seedlings.

2.4.5 NPs and Flooding Stress

Flooding stress is frequently encountered by plants growing in flood-prone, irrigated, or heavy rainfall ecosystems or wetland habitats, challenging the global crop production. This excess water stress in the form of either submergence (when water covers the aerial parts of the plant) or waterlogging (when surplus water covers roots and restricts gas diffusion, resulting in hypoxia) can result in huge losses related to agro-economy (Fukao et al. 2019; Pan et al. 2020). The type of stress (waterlogging or submergence) depends on the depth of the groundwater table. Heavy rainfall over prolonged periods, groundwater table depth, or overflowing of a water body can adversely affect soil hydrology. Different strategies are adapted by plants at morphological, physiological, cellular, metabolic, and molecular levels to tolerate this stress. The level and degree of acclimatization depend on species-specific or climate-related variations. Morphological adaptations include formation of adventitious roots, development of aerenchyma (lysigenous or schizogenous) and secondary aerenchyma, elongation of apical meristematic tissue resulting in internode elongation, altered spatial arrangement of roots, formation of radial oxygen loss barrier, and gas films on leaf cuticle (Pan et al. 2020). Photosynthesis and aerobic respiration are considerably hindered, leading to energy and carbohydrate deprivation. This is further accentuated by hypoxia and inhibited root respiration. The net photosynthetic rate and capacity decrease due to multiple factors, including low stomatal conductance, decrease in gas exchange, increased stomatal resistance and closure, decline in photosynthetic pigment levels, downregulated expression of RuBisCO and RuBisCO activase genes, and low RuBisCO activity. Associated morphological adaptations are yellowing of leaves, leaf senescence, and aborted new leaf development. Insufficient oxygen results in increase in intracellular ROS levels, which is tackled by enhanced activity of antioxidant enzyme systems and augmented ROS scavenging by active antioxidants. Additionally, several pathways are orchestrated at the cellular level to manage ATP metabolism, redox status, starch metabolism, transporter function, and elemental toxicity. Abiotic stress events continue even in the post-flood period when plants are suddenly exposed to reoxygenation with subsiding flood waters. Reoxygenation and dehydration can also cause harm to the plants post-submergence (Fukao et al. 2019). Interplay of phytohormones and transcriptional control is one of the adaptive mechanisms in waterlogging stress. Endogenous plant hormones like ethylene, ABA, auxin, gibberellin, SA, JA, brassinosteroid, and melatonin play critical roles in waterlogging tolerance. Transcription factors like ERFs regulate gene expression to aid in adaptive response mechanisms.

A handful of investigations highlight the molecular mechanism underlying NP and plant interaction under flooding-stress conditions. The role of silicon (not in NP form) in mitigating submergence stress in rice has been demonstrated using wild-type and Si-defective mutant varieties (Pan et al. 2021). Rezvani, Sorooshzadeh, and Farhadi (2012) demonstrated the consequences of Ag NPs application on corms of saffron (*Crocus sativus*), which were planted under control or flooding stress conditions. In flood-stressed saffron corms, AgNP treatment improved the number, length, dry weight, and fresh weight of roots. It also enhanced leaf bud number along with leaf dry weight, and fresh weight. Analysis of morphological characters of soybean seedlings exposed to flooding stress revealed that 15 nm Ag NPs at 2 ppm enhance the seedling growth as evaluated from seedling weight (Mustafa et al. 2015). A time-resolved, gel-free, label-free quantitative nanoLC-MS/MS analysis of proteins extracted from root revealed that the expression profiles of 107 root proteins significantly changed. The stress-responsive proteins and those related to signalling and cell metabolism were divided into three clusters based on hierarchical clustering. Subsequently, six flood-responsive proteins that exhibited contrasting trends in protein abundance levels under flooding stress conditions and NP treatment were chosen for qRT-PCR analysis – *glyoxalase II 3* (Glyma15g39370.2), *alcohol dehydrogenase 1* (Glyma06g12780.1), *pyruvate decarboxylase 2* (Glyma13g30490.1), *sucrose synthase 4* (Glyma15g20180.2), *aluminium-induced protein with YGL and LRDR motifs* (Glyma12g35070.1), and *thiamine pyrophosphate-dependent pyruvate decarboxylase family protein* (Glyma01g29190.1). Expression profiles of *alcohol dehydrogenase 1* and *pyruvate decarboxylase 2* genes were heightened during flooding and dampened in response to NPs. Lower transcript level of *glyoxalase II 3* after Ag NP treatment showed that production of glycolysis-related cytotoxic by-products was lowered and thereby decreased contact of soybean seedlings to oxygen deprivation. A similar proteomic study from the same group examined the effects of aluminium oxide (Al_2O_3) NPs of different sizes (5 nm, 30–60 nm, 135 nm) in soybeans under submergence stress (Mustafa and Komatsu 2016). NP treatment (30–60 nm) increased root length (including hypocotyl), suppressed proteins related to glycolytic pathways, regulated the GSH/ascorbate pathway, and enhanced expression of ribosomal proteins. Under flood conditions, Al_2O_3 NPs targeted the mitochondrial proteome and upregulated the expression of voltage-dependent anion channel proteins (NP dimension of 135 nm) and isocitrate dehydrogenase (NP dimension of 5 nm). This confirmed that Al_2O_3 NPs of a number of sizes regulate membrane permeability and tricarboxylic acid cycle proteins to diminish the negative effects of flooding stress. The combinatorial influence of Ag NPs with organic (nicotinic acid) and inorganic (potassium nitrate) chemicals under flooding stress in *Glycine max*, were studied using gel-free, label-free, quantitative nanoLC-MS/MS analysis. Protein-degradation and synthesis-related proteins were differentially expressed as a result of this combined treatment, indicating regulation of protein quality control within the endoplasmic reticulum. This treatment improved the growth of soybean seedlings under submergence stress, as indicated by root length/weight (including hypocotyl) (Hashimoto et al. 2020).

2.4.6 NPs and Salinity Stress

One of the key abiotic stresses restraining crop growth and productivity globally is salt stress. Excessive accumulation of salts in the soil inhibits the normal growth of the plant and eventually leads to death. The suspected causes underlying salt stress are poor quality of water used for irrigation, for instance, brackish water. The detrimental effects of salt stress are further intensified due to changes in climate and other anthropogenic activities like excessive use of chemical fertilizers and salt leaching in irrigated lands. This further increases the area of salinity zones. Salinity or high-salt stress impairs plant growth and development in a multitude of ways, for example, osmotic stress, ion toxicity (such as Na^+ and Cl^-), water deficit, imbalances in nutrient uptake and assimilation, altered metabolic processes including respiration, photosynthesis, nucleic acid, and protein metabolism, and disrupt cellular homeostasis (Gupta and Huang, 2014; Isayenkov and Maathuis, 2019; Valderrama et al., 2007). Moreover, it increases production and build-up of ROS leading to oxidative stress, accumulation of NO and NO_2-Tyr (indicators of nitrosative stress), genotoxicity, changes in membrane ultrastructure, reduced cell division and expansion, etc. (Valderrama et al. 2007; Gupta and Huang 2014; Isayenkov and Maathuis 2019). Analogous to drought, salt stress lowers water potential in the rhizosphere. Plant responses to high-salt stress have been temporally classified into two distinct phases (Isayenkov and Maathuis 2019). The first phase (ranging from minutes to days) causes ion-independent reduction in growth by stomatal closure and inhibition of cell division and expansion. The second phase (ranging from days to weeks) leads to a pick-up in ion levels that are cytotoxic enough to impair metabolic processes and initiate premature senescence and cell death. Plants adopt different morphophysiological mechanisms to cope with soil salinity (Gupta and Huang 2014; He, He, and Ding 2018). Ionic tolerance or maintenance of ionic homeostasis is achieved by regulating ion uptake, transport, and compartmentalization. The excess salt is compartmentalized in vacuoles or isolated in older tissues, which are eventually shed off. Osmotic stress is managed by increased biosynthesis of compatible osmolytes. These osmoprotectants include proline, glycine betaine, sugar, and polyols. For example, proline accumulation inside cells provides osmotic tolerance during salinity stress and also is a potential source of organic nitrogen, important for stress recovery. Glycine betaine increases cellular osmolarity during stress, improves enzyme and protein stability by interacting with hydrophilic and hydrophobic domains, reduces oxidative damage by ROS, and provides protection to the photosynthetic machinery. Increase in intracellular polyamines has also been reported during salt stress. The low-molecular-weight polyamines have been demonstrated to be important for cell proliferation, differentiation, morphogenesis, seed development and germination, embryogenesis, flower and fruit development and senescence. Additionally, the enhanced activity of antioxidant enzymes (CAT, GPX, APX, SOD, and glutathione reductase) and non-enzymatic antioxidants (anthocyanin and ascorbate) is associated with ROS detoxification during salinity stress (He, He, and Ding 2018). The antioxidant properties of NO are also crucial for stress management. Phytohormones, such as SA, ABA, and brassinosteroids, can lessen the disadvantageous effects of salt stress and regulate the expression of several genes.

Mitigation of salt stress in economically important plants by NP application has been well-studied. NP-mediated relief of salinity stress is reflected at morphological, physiological, biochemical, and molecular levels. For example, seed priming or pre-treatment of lettuce (*L. sativa*) seeds with water-soluble carbon NPs (CNPs) positively affected germination of seeds, enhanced lateral root growth as well as chlorophyll content of seedlings (Baz et al. 2020). Seed priming with 2–5 mM of Ag NPs enhanced salt tolerance in wheat plants via increment of fresh and dry weights of shoot, enhanced antioxidant activity, increased osmoprotectant levels, like soluble sugars and proline, and decreased MDA and H_2O_2 content (Mohamed et al. 2017). Similarly, priming lupine seeds with 60 mg/L ZnO NPs was a successful approach to improve salt tolerance (Latef, Alhmad, and Abdelfattah 2017). Cu NPs applied as foliar sprays on tomato plants improved antioxidant content of fruits in plants grown under salt stress (Pérez-Labrada et al. 2019). ZnO and Fe_2O_3 NPs application as foliar sprays improved root growth, sub-stomatal CO_2 concentration, net CO_2 assimilation rate, and leaf Fe/Zn content in salt-stressed maize cultivars (Fathi, Zahedi, and Torabian 2017). Solution of 0.1% hydroxyapatite NPs applied to potted *Zea mays* was also noted to relieve salt stress (Yoon et al. 2020). Foliar spraying of cotton plants with 200 nm nano-zinc augmented whole plant dry weight, boll yield, and mineral status, compared to salt-stressed cotton plants (Hussein and Abou-Baker 2018). In a recent study, spraying green-synthesized ZnO NPs on leaves, rather than ZnO bulk, was more proficient in alleviating the damaging effects of salinity stress in okra, by increasing the level of photosynthetic pigments, enhancing antioxidant activity, lowering proline and cumulative sugar concentrations (Alabdallah and Alzahrani 2020). Besides mitigating the impact of salinity stress, NP application can also enhance characteristics of commercial importance in economically important plants. Nano-zinc improves boll yield in cotton plants, Se NPs improve fruit quality of tomato (including total soluble solid content, acidity, pH, and fruit firmness), while TiO_2 NPs affected essential oil composition (geranial, z-citral, geranyl acetate, and geraniol) of *Dracocephalum moldavica*. In *Capsicum annuum*, the undesired reduction of GSH and flavonoids under salinity stress could be alleviated by drenching the plants in either Se, Si, or Cu NPs (González-García et al. 2021). In a separate approach, the synergistic effect of a combination of NPs (Zn, B, Si, and Zeolite NPs) was investigated in potato plants, where combined nano-micronutrients affected growth parameters, phytohormone profile (ABA, gibberellin A3), antioxidant enzyme content in tubers, osmolytes in leaves, and mineral status of salt-stressed plants (Mahmoud et al. 2020). To recognize the molecular mechanisms behind tolerance to high-salt stress in cotton seedlings which germinated from seeds primed using PNC, high-throughput RNA sequencing of seedling roots was done (An et al. 2020). Transcriptomic profiles revealed that differentially expressed genes associated with PNC seed priming are related to ROS enzymatic pathways, ion homeostasis, Ca^{2+} and Mg^{2+} signalling pathways, and terpene biosynthesis.

2.4.7 NPs and Other Abiotic Stresses

The effect of NPs in alleviating certain other forms of abiotic stress has also been studied. The negative impact of excessive light (2,000 $\mu mol/m^2 s$ of photosynthetically active radiation or PAR) for 1.5 h on *A. thaliana* was alleviated by application of nanoceria (Wu, Tito, and Giraldo 2017). PNC augmented quantum yield and maximum yield (F_v/F_m) of photosystem II and enhanced CO_2 assimilation, quantum efficiency of CO_2 uptake, and RuBisCO carboxylation activity, thereby improving photosynthetic performance under stress conditions. The prospective of nanoceria has also been examined in rice hydroponically grown under nitrogen-deficient and excess conditions (Wang et al. 2020). Cerium oxide NPs improve plant tolerance to nitrogen stress by up- or down-regulating the activities of antioxidant enzymes (glutamine synthetase, glutamine oxoglutarate aminotransferase, glutamate dehydrogenase), important for nitrogen assimilation. Besides reducing oxidative damage, nanoceria regulates the levels of proline and phytohormones (ABA, IAA, GA, zeatin riboside) and improves rice growth. Ultraviolet radiation also affects morphological growth, declines photosynthetic performance, generates ROS, increases lipid peroxidation, and inhibits antioxidant enzymes. Si NPs applied to wheat seedlings exposed to UV-B radiation rendered protection from oxidative stress. Si NPs regulated ROS signalling by enhancing the activities of antioxidants and subsequently preventing ROS-induced damage to photosynthetic apparatus. NO levels maximally increased in (Si NPs+UV-B) treatment, which suggested that NO-mediated upregulation of antioxidant activities conferred protection from UV stress (Tripathi et al. 2017).

2.5 Future Perspectives and Scopes

The global crop production is challenged by the detrimental effects of abiotic stress, making it imperative to develop suitable and sustainable technologies for efficient stress management. NP application via seed priming, foliar sprays, and nutritional supplements emerges to be a promising option and deserves to be further explored. The use of a single type or a combination of metal/non-metal-based NPs to protect plants from multiple abiotic stresses is usually investigated, with some insight into their possible modes of action. These include activation of ROS scavenging pathways, improvement of photosynthetic performance, accumulation of compatible osmoprotectants, and regulation of phytohormone signalling, ion homeostasis signalling, and secondary metabolite pathways. All these responses finally alter the expression profile of genes related to stress responses. But the scope of this research needs to be expanded to address certain issues.

First, it is essential to recognize the mechanism of uptake, distribution, and internalization of nanomaterials inside plant cells and the factors that influence it (Pérez-de-Luque 2017). Nanomaterials are distributed via apoplast and/or symplast pathways, translocated up and down the plant or radially within the plant, and are internalized through several mechanisms, including endocytosis, via carrier proteins, pore formation, and through plasmodesmata. Second, NP properties, including the minimum concentration dose and choice of plant species also influence the final fate and effectiveness of NPs. The size, shape, morphology, chemical and electrical properties, zeta potential, and coating on NP surface alter the uptake and accumulation of NPs inside plants. PNC with a sub-11 nm hydrodynamic diameter, negative ζ-potential, and a low Ce^{3+}/Ce^{4+} ratio performed better in protecting crops from abiotic stress, compared to PNC characterized by higher Ce^{3+}/Ce^{4+} ratio or cationic ANC of analogous dimensions (Wu, Tito, and Giraldo 2017). The physiological and morphological differences in diverse plant species also cause variations in NP uptake, absorption, and accumulation. Also, the modes of delivery of NPs and the target tissues determine NP internalization. Root and leaf exposure of wheat and rapeseed plantlets to TiO_2 NPs resulted in elevated Ti content in rapeseed than wheat during root exposure (Larue et al. 2012). Additionally, organic and inorganic constituents of soil can improve stability and bioavailability of NPs or cause precipitation and hamper uptake (Pérez-de-Luque 2017). Third, the mechanism of NP action inside plant cells needs further investigation in order to obtain a comprehensive overview. Study of the mechanism at cellular and molecular levels is required to identify (i) tissue types, organelles, membranes, and other cellular structures, (ii) the signalling molecules (antioxidants, phytohormones, osmolytes, secondary metabolites, secondary messengers), (iii) the signalling pathways (ion homeostasis, metabolic, phytohormone), (iv) the physiological processes (respiration, photosynthesis, water retention), and (v) the stress-responsive genes and transcription factors that are impacted. Fourth, it needs to be ensured if the unrestrained use of NPs for agriculture can have unintended effects on crop productivity, soil microbiome, agro-ecosystem health, and the environment. Metatranscriptome analysis of maize treated with NPs revealed multiple negative effects on soil microbiota, including, decrease in relative abundance of archael species, compromised nitrogen cycling, selection of phytopathogenic fungal genera, and increase in transcript abundance of phytohormone and aquaporin genes, indicating a trigger of stress responses (Sillen et al. 2020). Also, soil microbiota (bacteria, fungi, actinorhiza) can affect the uptake of NPs, especially if they can establish symbiotic interactions with plants. For example, arbuscular mycorrhizae affect zinc accumulation in maize plants (Wang et al. 2016). Furthermore, it is undetermined how foliar administration of NPs, like those with antimicrobial activity, could influence rhizospheric, phyllospheric, and endophytic microbiome. Fifth, the cost for NP production should be affordable and similar to the cost of other agricultural products that are commonly used. Finally, the application of NPs to protect plants from climatic extremes and other abiotic stress, as a potential technology needs acceptance among the agricultural community. Consideration and inclusion of all these issues within the scope of current NP research will provide a holistic perspective of the feasibility, sustainability, and effectiveness of this technology.

REFERENCES

Adrees, Muhammad, Zahra Saeed Khan, Shafaqat Ali, Muhammad Hafeez, Sofia Khalid, Muhammad Zia Ur Rehman, Afzal Hussain, Khalid Hussain, Shahzad Ali Shahid Chatha, and Muhammad Rizwan. 2020. "Simultaneous mitigation of cadmium and drought stress in wheat by soil application of iron nanoparticles." *Chemosphere* 238:124681.

Ahmed, Temoor, Muhammad Noman, Natasha Manzoor, Muhammad Shahid, Muhammad Abdullah, Liaqat Ali, Gang Wang, Abeer Hashem, Al-Bandari Fahad Al-Arjani, and Abdulaziz A Alqarawi. 2021. "Nanoparticle-based amelioration of drought stress and cadmium toxicity in rice via triggering the stress responsive genetic mechanisms and nutrient acquisition." *Ecotoxicology and Environmental Safety* 209:111829.

Alabdallah, Nadiyah M., and Hassan S. Alzahrani. 2020. "The potential mitigation effect of ZnO nanoparticles on [*Abelmoschus esculentus* L. Moench] metabolism under salt stress conditions." *Saudi Journal of Biological Sciences* 27(11):3132–3137.

Amini, Saeed, Reza Maali-Amiri, Rahmat Mohammadi, and Seyyedeh-Sanam Kazemi-Shahandashti. 2017. "cDNA-AFLP analysis of transcripts induced in chickpea plants by TiO_2 nanoparticles during cold stress." *Plant Physiology and Biochemistry* 111:39–49.

An, Jing, Peiguang Hu, Fangjun Li, Honghong Wu, Yu Shen, Jason C White, Xiaoli Tian, Zhaohu Li, and Juan Pablo Giraldo. 2020. "Emerging investigator series: Molecular mechanisms of plant salinity stress tolerance improvement by seed priming with cerium oxide nanoparticles." *Environmental Science: Nano* 7(8):2214–2228.

Avellan, Astrid, Jie Yun, Yilin Zhang, Eleanor Spielman-Sun, Jason M Unrine, Juergen Thieme, Jieran Li, Enzo Lombi, Garret Bland, and Gregory V. Lowry. 2019. "Nanoparticle size and coating chemistry control foliar uptake pathways, translocation, and leaf-to-rhizosphere transport in wheat." *ACS Nano* 13(5):5291–5305.

Baz, Hanna, Matthew Creech, Jianjun Chen, Haijun Gong, Kent Bradford, and Heqiang Huo. 2020. "Water-soluble carbon nanoparticles improve seed germination and post-germination growth of lettuce under salinity stress." *Agronomy* 10(8):1192.

Bhat, Javaid Akhter, S.M. Shivaraj, Pritam Singh, Devanna B. Navadagi, Durgesh Kumar Tripathi, Prasanta K. Dash, Amolkumar U. Solanke, Humira Sonah, and Rupesh Deshmukh. 2019. "Role of silicon in mitigation of heavy metal stresses in crop plants." *Plants* 8(3):71.

Bita, Craita, and Tom Gerats. 2013. "Plant tolerance to high temperature in a changing environment: Scientific fundamentals and production of heat stress-tolerant crops." *Frontiers in Plant Science* 4:273.

Chakraborty, Sukumar, and Adrian C. Newton. 2011. "Climate change, plant diseases and food security: An overview." *Plant Pathology* 60(1):2–14.

Chhipa, Hemraj, and Piyush Joshi. 2016. "Nanofertilisers, nanopesticides and nanosensors in agriculture." *Nanoscience in Food and Agriculture* 1, 247–282. Springer.

Doolette, Casey L., Michael J. McLaughlin, Jason K. Kirby, and Divina A. Navarro. 2015. "Bioavailability of silver and silver sulfide nanoparticles to lettuce (Lactuca sativa): Effect of agricultural amendments on plant uptake." *Journal of Hazardous Materials* 300:788–795.

Elmer, Wade, and Jason C. White. 2018. "The future of nanotechnology in plant pathology." *Annual Review of Phytopathology* 56:111–133.

Elmer, Wade, Chuanxin Ma, and Jason White. 2018. "Nanoparticles for plant disease management." *Current Opinion in Environmental Science & Health* 6:66–70.

Elsheery, Nabil I., V.S.J. Sunoj, Y. Wen, J.J. Zhu, G. Muralidharan, and K.F. Cao. 2020. "Foliar application of nanoparticles mitigates the chilling effect on photosynthesis and photoprotection in sugarcane." *Plant Physiology and Biochemistry* 149:50–60.

Ernst, Wilfried H.O. 2006. "Evolution of metal tolerance in higher plants." *Forest Snow and Landscape Research* 80(3):251–274.

Faizan, Mohammad, Ahmad Faraz, Anayat Rasool Mir, and Shamsul Hayat. 2021. "Role of zinc oxide nanoparticles in countering negative effects generated by cadmium in Lycopersicon esculentum." *Journal of Plant Growth Regulation* 40(1):101–115.

Fathi, Alireza, Morteza Zahedi, and Shahram Torabian. 2017. "Effect of interaction between salinity and nanoparticles (Fe_2O_3 and ZnO) on physiological parameters of Zea mays L." *Journal of Plant Nutrition* 40(19):2745–2755.

Fukao, Takeshi, Blanca Estela Barrera-Figueroa, Piyada Juntawong, and Julián Mario Peña-Castro. 2019. "Submergence and waterlogging stress in plants: A review highlighting research opportunities and understudied aspects." *Frontiers in Plant Science* 10:340.

Giraldo, Juan Pablo, Markita P. Landry, Sean M. Faltermeier, Thomas P. McNicholas, Nicole M. Iverson, Ardemis A. Boghossian, Nigel F. Reuel, Andrew J. Hilmer, Fatih Sen, and Jacqueline A. Brew. 2014. "Plant nanobionics approach to augment photosynthesis and biochemical sensing." *Nature Materials* 13(4):400–408.

Giraldo, Juan Pablo, Honghong Wu, Gregory Michael Newkirk, and Sebastian Kruss. 2019. "Nanobiotechnology approaches for engineering smart plant sensors." *Nature Nanotechnology* 14(6):541–553.

Gohari, Gholamreza, Asghar Mohammadi, Ali Akbari, Sima Panahirad, Mohammad Reza Dadpour, Vasileios Fotopoulos, and Seisuke Kimura. 2020. "Titanium dioxide nanoparticles (TiO_2 NPs) promote growth and ameliorate salinity stress effects on essential oil profile and biochemical attributes of Dracocephalum moldavica." Scientific reports 10 (1):1–14.

González-García, Yolanda, Claribel Cárdenas-Álvarez, Gregorio Cadenas-Pliego, Adalberto Benavides-Mendoza, Marcelino Cabrera-de-la-Fuente, Alberto Sandoval-Rangel, Jesús Valdés-Reyna, and Antonio Juárez-Maldonado. 2021. "Effect of three nanoparticles (Se, Si and Cu) on the bioactive compounds of bell pepper fruits under saline stress." *Plants* 10(2):217.

Gupta, Bhaskar, and Bingru Huang. 2014. "Mechanism of salinity tolerance in plants: Physiological, biochemical, and molecular characterization." *International Journal of Genomics* 2014: 701596.

Haak, David C., Takeshi Fukao, Ruth Grene, Zhihua Hua, Rumen Ivanov, Giorgio Perrella, and Song Li. 2017. "Multilevel regulation of abiotic stress responses in plants." *Frontiers in Plant Science* 8:1564.

Hasanpour, H., R. Maali-Amir, and H. Zeinali. 2015. "Effect of TiO_2 nanoparticles on metabolic limitations to photosynthesis under cold in chickpea." *Russian Journal of Plant Physiology* 62(6):779–787.

Hasanuzzaman, Mirza, Kamrun Nahar, Md Alam, Rajib Roychowdhury, and Masayuki Fujita. 2013. "Physiological, biochemical, and molecular mechanisms of heat stress tolerance in plants." *International Journal of Molecular Sciences* 14(5):9643–9684.

Hashimoto, Takuya, Ghazala Mustafa, Takumi Nishiuchi, and Setsuko Komatsu. 2020. "Comparative analysis of the effect of inorganic and organic chemicals with silver nanoparticles on soybean under flooding stress." *International Journal of Molecular Sciences* 21(4):1300.

Hassan, Nahla S., Taher A. Salah El Din, Mohamed H. Hendawey, Ibrahim H. Borai, and Asmaa A. Mahdi. 2018. "Magnetite and zinc oxide nanoparticles alleviated heat stress in wheat plants." *Current Nanomaterials* 3(1):32–43.

He, Mei, Cheng-Qiang He, and Nai-Zheng Ding. 2018. "Abiotic stresses: General defenses of land plants and chances for engineering multistress tolerance." *Frontiers in Plant Science* 9:1771.

Honeychurch, Kevin C. 2014. *Nanosensors for Chemical and Biological Applications: Sensing with Nanotubes, Nanowires and Nanoparticles*. Elsevier, Amsterdam, Netherlands.

Huang, Fei, Xiao-Hui Wen, Yi-Xia Cai, and Kun-Zheng Cai. 2018. "Silicon-mediated enhancement of heavy metal tolerance in rice at different growth stages." *International Journal of Environmental Research and Public Health* 15(10):2193.

Hussain, Hafiz A., Saddam Hussain, Abdul Khaliq, Umair Ashraf, Shakeel A. Anjum, Shengnan Men, and Longchang Wang. 2018. "Chilling and drought stresses in crop plants: Implications, cross talk, and potential management opportunities." *Frontiers in Plant Science* 9:393.

Hussein, M.M., and N.H. Abou-Baker. 2018. "The contribution of nano-zinc to alleviate salinity stress on cotton plants." *Royal Society Open Science* 5(8):171809.

Iqbal, Muhammad, Naveed Iqbal Raja, Mubashir Hussain, Muhammad Ejaz, and Farhat Yasmeen. 2019a. "Effect of silver nanoparticles on growth of wheat under heat stress." *Iranian Journal of Science and Technology, Transactions A: Science* 43(2):387–395.

Iqbal, Muhammad, Naveed Iqbal Raja, Zia-Ur-Rehman Mashwani, Feroza Hamid Wattoo, Mubashir Hussain, Muhammad Ejaz, and Hafiza Saira. 2019b. "Assessment of AgNPs exposure on physiological and biochemical changes and antioxidative defence system in wheat (*Triticum aestivum* L) under heat stress." *IET Nanobiotechnology* 13(2):230–236.

Iravani, Siavach, Hassan Korbekandi, Seyed Vahid Mirmohammadi, and Behzad Zolfaghari. 2014. "Synthesis of silver nanoparticles: Chemical, physical and biological methods." *Research in Pharmaceutical Sciences* 9(6):385.

Isayenkov, Stanislav V., and Frans J.M. Maathuis. 2019. "Plant salinity stress: Many unanswered questions remain." *Frontiers in Plant Science* 10:80.

Kapoor, Dhriti, Savita Bhardwaj, Marco Landi, Arti Sharma, Muthusamy Ramakrishnan, and Anket Sharma. 2020. "The impact of drought in plant metabolism: How to exploit tolerance mechanisms to increase crop production." *Applied Sciences* 10(16):5692.

Keat, Cheah Liang, Azila Aziz, Ahmad M. Eid, and Nagib A. Elmarzugi. 2015. "Biosynthesis of nanoparticles and silver nanoparticles." *Bioresources and Bioprocessing* 2(1):1–11.

Kim, Yoon-Ha, Abdul Latif Khan, Duk-Hwan Kim, Seung-Yeol Lee, Kyung-Min Kim, Muhammad Waqas, Hee-Young Jung, Jae-Ho Shin, Jong-Guk Kim, and In-Jung Lee. 2014. "Silicon mitigates heavy metal stress by regulating P-type heavy metal ATPases, Oryza sativa low silicon genes, and endogenous phytohormones." *BMC Plant Biology* 14(1):1–13.

Konate, Alexandre, Xiao He, Zhiyong Zhang, Yuhui Ma, Peng Zhang, Gibson Maswayi Alugongo, and Yukui Rui. 2017. "Magnetic (Fe_3O_4) nanoparticles reduce heavy metals uptake and mitigate their toxicity in wheat seedling." *Sustainability* 9(5):790.

Kordrostami, Mojtaba, Mohammad Mafakheri, and Jameel M. Al-Khayri. 2021. "Contributions of nano biosensors in managing environmental plant stress under climatic changing era." *Nanobiotechnology*, 117–137. Springer.

Larue, Camille, Giulia Veronesi, Anne-Marie Flank, Suzy Surble, Nathalie Herlin-Boime, and Marie Carrière. 2012. "Comparative uptake and impact of TiO_2 nanoparticles in wheat and rapeseed." *Journal of Toxicology and Environmental Health, Part A* 75(13–15):722–734.

Latef, Arafat Abdel Hamed Abdel, Mona Fawzy Abu Alhmad, and Khaled Ebnalwaled Abdelfattah. 2017. "The possible roles of priming with ZnO nanoparticles in mitigation of salinity stress in lupine (Lupinus termis) plants." *Journal of Plant Growth Regulation* 36(1):60–70.

Li, Jinming, Honghong Wu, Israel Santana, Mackenzie Fahlgren, and Juan Pablo Giraldo. 2018a. "Standoff optical glucose sensing in photosynthetic organisms by a quantum dot fluorescent probe." *ACS Applied Materials & Interfaces* 10(34):28279–28289.

Li, Yali, Qianwen Li, Yanan Wang, Joohee Oh, Sila Jin, Yeonju Park, Tieli Zhou, Bing Zhao, Weidong Ruan, and Young Mee Jung. 2018b. "A reagent-assisted method in SERS detection of methyl salicylate." *Spectrochimica Acta Part A: Molecular and Biomolecular Spectroscopy* 195:172–175.

Li, Ruixin, Jinxia He, Hongguo Xie, Wenxia Wang, Santosh Kumar Bose, Yeqing Sun, Jianen Hu, and Heng Yin. 2019. "Effects of chitosan nanoparticles on seed germination and seedling growth of wheat (*Triticum aestivum* L.)." *International Journal of Biological Macromolecules* 126:91–100.

Linh, Tran My, Nguyen Chi Mai, Pham Thi Hoe, Le Quynh Lien, Ninh Khac Ban, Le Thi Thu Hien, Nguyen Hoai Chau, and Nguyen Tuong Van. 2020. "Metal-based nanoparticles enhance drought tolerance in Soybean." *Journal of Nanomaterials* 2020: 4056563.

Liu, Ruiqiang, and Rattan Lal. 2015. "Potentials of engineered nanoparticles as fertilizers for increasing agronomic productions." *Science of the Total Environment* 514:131–139.

Liu, Xiaomin, Yunlin Zhou, Jianwei Xiao, and Fei Bao. 2018. "Effects of chilling on the structure, function and development of chloroplasts." *Frontiers in Plant Science* 9:1715.

Liu, Mengyao, Sheng Feng, Yuhui Ma, Changjian Xie, Xiao He, Yayun Ding, Junzhe Zhang, Wenhe Luo, Lirong Zheng, and Dongliang Chen. 2019. "Influence of surface charge on the phytotoxicity, transformation, and translocation of CeO_2 nanoparticles in cucumber plants." *ACS Applied Materials & Interfaces* 11(18):16905–16913.

Lowry, Gregory V., Astrid Avellan, and Leanne M. Gilbertson. 2019. "Opportunities and challenges for nanotechnology in the agri-tech revolution." *Nature Nanotechnology* 14(6):517–522.

Lu, Li, Min Huang, Yuxiong Huang, Philippe F-X Corvini, Rong Ji, and Lijuan Zhao. 2020. "Mn_3O_4 nanozymes boost endogenous antioxidant metabolites in cucumber (Cucumis sativus) plant and enhance resistance to salinity stress." *Environmental Science: Nano* 7(6):1692–1703.

Lv, Jitao, Shuzhen Zhang, Lei Luo, Jing Zhang, Ke Yang, and Peter Christie. 2015. "Accumulation, speciation and uptake pathway of ZnO nanoparticles in maize." *Environmental Science: Nano* 2(1):68–77.

Mahmoud, Abdel Wahab M., Emad A. Abdeldaym, Suzy M. Abdelaziz, Mohamed BI El-Sawy, and Shady A. Mottaleb. 2020. "Synergetic effects of zinc, boron, silicon, and zeolite nanoparticles on confer tolerance in potato plants subjected to salinity." *Agronomy* 10(1):19.

Mahto, Rakhi, Priyanka Rani, Reshu Bhardwaj, Rajesh Kumar Singh, Saroj Kumar Prasad, and Amitava Rakshit. 2021. "Nanotechnology: A potential approach for abiotic stress management." In *Advances in Nano-Fertilizers and Nano-Pesticides in Agriculture*, 249–264. Elsevier, Amsterdam, Netherlands.

Mittal, Deepti, Gurjeet Kaur, Parul Singh, Karmveer Yadav, and Syed Azmal Ali. 2020. "Nanoparticle-based sustainable agriculture and food science: Recent advances and future outlook." *Frontiers in Nanotechnology* 2:10.

Mohammadi, Rahmat, Reza Maali-Amiri, and Alireza Abbasi. 2013. "Effect of TiO_2 nanoparticles on chickpea response to cold stress." *Biological Trace Element Research* 152(3):403–410.

Mohammadi, R., R. Maali-Amiri, and N.L. Mantri. 2014. "Effect of TiO_2 nanoparticles on oxidative damage and antioxidant defense systems in chickpea seedlings during cold stress." *Russian Journal of Plant Physiology* 61(6):768–775.

Mohamed, Abdel Kareem S.H., Muhammad Farooq Qayyum, Ahmed Mustafa Abdel-Hadi, Rabia Abdur Rehman, Shafaqat Ali, and Muhammad Rizwan. 2017. "Interactive effect of salinity and silver nanoparticles on photosynthetic and biochemical parameters of wheat." *Archives of Agronomy and Soil Science* 63(12):1736–1747.

Molotoks, Amy, Pete Smith, and Terence P. Dawson. 2021. "Impacts of land use, population, and climate change on global food security." *Food and Energy Security* 10(1), e261.

Mourdikoudis, Stefanos, Roger M. Pallares, and Nguyen T.K. Thanh. 2018. "Characterization techniques for nanoparticles: Comparison and complementarity upon studying nanoparticle properties." *Nanoscale* 10(27):12871–12934.

Morales-Espinoza, Mónica Carolina, Gregorio Cadenas-Pliego, Marissa Pérez-Alvarez, Alma Delia Hernández-Fuentes, Marcelino Cabrera de la Fuente, Adalberto Benavides-Mendoza, Jesús Valdés-Reyna, and Antonio Juárez-Maldonado. 2019. "Se nanoparticles induce changes in the growth, antioxidant responses, and fruit quality of tomato developed under NaCl Stress." *Molecules* 24(17):3030.

Mustafa, Ghazala, and Setsuko Komatsu. 2016. "Insights into the response of soybean mitochondrial proteins to various sizes of aluminum oxide nanoparticles under flooding stress." *Journal of Proteome Research* 15(12):4464–4475.

Mustafa, Ghazala, Katsumi Sakata, Zahed Hossain, and Setsuko Komatsu. 2015. "Proteomic study on the effects of silver nanoparticles on soybean under flooding stress." *Journal of Proteomics* 122:100–118.

Osakabe, Yuriko, Keishi Osakabe, Kazuo Shinozaki, and Lam-Son Phan Tran. 2014. "Response of plants to water stress." *Frontiers in Plant Science* 5:86.

Pachauri, R.K., and A. Reisinger. 2008. "Climate change 2007, Synthesis report, Contribution of Working Groups I, II and III to the fourth assessment report." Cambridge University Press, Cambridge.

Palmqvist, N.G. Martin, Gulaim A. Seisenbaeva, Peter Svedlindh, and Vadim G. Kessler. 2017. "Maghemite nanoparticles acts as nanozymes, improving growth and abiotic stress tolerance in Brassica napus." *Nanoscale Research Letters* 12(1):1–9.

Pan, Jiawei, Rahat Sharif, Xuewen Xu, and Xuehao Chen. 2020. "Mechanisms of waterlogging tolerance in plants: Research progress and prospects." *Frontiers in Plant Science* 11, 627331.

Pan, Taowen, Jian Zhang, Lanmengqi He, Abdul Hafeez, Chuanchuan Ning, and Kunzheng Cai. 2021. "Silicon enhances plant resistance of rice against submergence stress." *Plants* 10(4):767.

Pareek, Ashwani, Om Parkash Dhankher, and Christine H. Foyer. 2020. *Mitigating the Impact of Climate Change on Plant Productivity and Ecosystem Sustainability.* Oxford University Press, Oxford, UK.

Paul, Anindita, and Aryadeep Roychoudhury. 2021. "Go green to protect plants: Repurposing the antimicrobial activity of biosynthesized silver nanoparticles to combat phytopathogens." *Nanotechnology for Environmental Engineering* 6(1):1–22.

Pérez-de-Luque, Alejandro. 2017. "Interaction of nanomaterials with plants: What do we need for real applications in agriculture?" *Frontiers in Environmental Science* 5:12.

Pérez-Labrada, Fabián, Elsy Rubisela López-Vargas, Hortensia Ortega-Ortiz, Gregorio Cadenas-Pliego, Adalberto Benavides-Mendoza, and Antonio Juárez-Maldonado. 2019. "Responses of tomato plants under saline stress to foliar application of copper nanoparticles." *Plants* 8(6):151.

Rajput, Vishnu, Tatiana Minkina, Mahmoud Mazarji, Sudhir Shende, Svetlana Sushkova, Saglara Mandzhieva, Marina Burachevskaya, Victor Chaplygin, Ajeet Singh, and Hanuman Jatav. 2020. "Accumulation of nanoparticles in the soil-plant systems and their effects on human health." *Annals of Agricultural Sciences* 65(2):137–143.

Rajput, Vishnu D., Tatiana Minkina, Arpna Kumari, Vipin Kumar Singh, Krishan K. Verma, Saglara Mandzhieva, Svetlana Sushkova, Sudhakar Srivastava, and Chetan Keswani. 2021. "Coping with the challenges of abiotic stress in plants: New dimensions in the field application of nanoparticles." *Plants* 10(6):1221.

Rezvani, N., A. Sorooshzadeh, and N. Farhadi. 2012. "Effect of nano-silver on growth of saffron in flooding stress." *World Academy of Science, Engineering and Technology* 6(1):517–522.

Sachdev, Swati, Shamim Akhtar Ansari, Mohammad Israil Ansari, Masayuki Fujita, and Mirza Hasanuzzaman. 2021. "Abiotic stress and reactive oxygen species: Generation, signaling, and defense mechanisms." *Antioxidants* 10(2):277.

Sillen, Wouter M.A., Sofie Thijs, Gennaro Roberto Abbamondi, Roberto De La Torre Roche, Nele Weyens, Jason C. White, and Jaco Vangronsveld. 2020. "Nanoparticle treatment of maize analyzed through the metatranscriptome: Compromised nitrogen cycling, possible phytopathogen selection, and plant hormesis." *Microbiome* 8(1):1–17.

Singh, Priyanka, Yu-Jin Kim, Dabing Zhang, and Deok-Chun Yang. 2016a. "Biological synthesis of nanoparticles from plants and microorganisms." *Trends in Biotechnology* 34(7):588–599.

Singh, Samiksha, Parul Parihar, Rachana Singh, Vijay P. Singh, and Sheo M. Prasad. 2016b. "Heavy metal tolerance in plants: Role of transcriptomics, proteomics, metabolomics, and ionomics." *Frontiers in Plant Science* 6:1143.

Singh, Davinder, Devendra Sillu, Anil Kumar, and Shekhar Agnihotri. 2021. "Dual nanozyme characteristics of iron oxide nanoparticles alleviate salinity stress and promote the growth of an agroforestry tree, Eucalyptus tereticornis Sm." *Environmental Science: Nano* 8(5):1308–1325.

Solanki, Priyanka, Arpit Bhargava, Hemraj Chhipa, Navin Jain, and Jitendra Panwar. 2015. "Nano-fertilizers and their smart delivery system." In Mahendra Rai, Caue Ribeiro, Luiz Mattoso, Nelson Duran (eds) *Nanotechnologies in Food and Agriculture*, pp. 81–101. Springer, Berlin, Germany.

Song, Yue, Meng Jiang, Huali Zhang, and Ruiqing Li. 2021. "Zinc oxide nanoparticles alleviate chilling stress in rice (Oryza Sativa L.) by regulating antioxidative system and chilling response transcription factors." *Molecules* 26(8):2196.

Spielman-Sun, Eleanor, Astrid Avellan, Garret D Bland, Ryan V Tappero, Alvin S Acerbo, Jason M Unrine, Juan Pablo Giraldo, and Gregory V Lowry. 2019. "Nanoparticle surface charge influences translocation and leaf distribution in vascular plants with contrasting anatomy." *Environmental Science: Nano* 6(8):2508–2519.

Sreelakshmi, B., S. Induja, P.P. Adarsh, H.L. Rahul, S.M. Arya, S. Aswana, R. Haripriya, B.R. Aswathy, P.K. Manoj, and Dalia Vishnudasan. 2021. "Drought stress amelioration in plants using green synthesised iron oxide nanoparticles." *Materials Today: Proceedings* 41:723–727.

Sun, Luying, Fengbin Song, Junhong Guo, Xiancan Zhu, Shengqun Liu, Fulai Liu, and Xiangnan Li. 2020. "Nano-ZnO-induced drought tolerance is associated with melatonin synthesis and metabolism in maize." *International Journal of Molecular Sciences* 21(3):782.

Taran, Nataliya, Volodymyr Storozhenko, Nataliia Svietlova, Ludmila Batsmanova, Viktor Shvartau, and Mariia Kovalenko. 2017. "Effect of zinc and copper nanoparticles on drought resistance of wheat seedlings." *Nanoscale Research Letters* 12(1):1–6.

Tripathi, Durgesh Kumar, Swati Singh, Vijay Pratap Singh, Sheo Mohan Prasad, Devendra Kumar Chauhan, and Nawal Kishore Dubey. 2016. "Silicon nanoparticles more efficiently alleviate arsenate toxicity than silicon in maize cultiver and hybrid differing in arsenate tolerance." *Frontiers in Environmental Science* 4:46.

Tripathi, Durgesh Kumar, Swati Singh, Vijay Pratap Singh, Sheo Mohan Prasad, Nawal Kishore Dubey, and Devendra Kumar Chauhan. 2017. "Silicon nanoparticles more effectively alleviated UV-B stress than silicon in wheat (Triticum aestivum) seedlings." *Plant Physiology and Biochemistry* 110:70–81.

Valderrama, Raquel, Francisco J. Corpas, Alfonso Carreras, Ana Fernández-Ocaña, Mounira Chaki, Francisco Luque, María V. Gómez-Rodríguez, Pilar Colmenero-Varea, A. Luis, and Juan B. Barroso. 2007. "Nitrosative stress in plants." *Febs Letters* 581(3):453–461.

Van Nguyen, Dong, Huong Mai Nguyen, Nga Thanh Le, Kien Huu Nguyen, Hoa Thi Nguyen, Huong Mai Le, Anh Trung Nguyen, Ngan Thi Thu Dinh, Son Anh Hoang, and Chien Van Ha. 2021. "Copper nanoparticle application enhances plant growth and grain yield in maize under drought stress conditions." *Journal of Plant Growth Regulation* 41(1): 364–375.

Wahid, Abdul, Saddia Gelani, M. Ashraf, and Majid R. Foolad. 2007. "Heat tolerance in plants: An overview." *Environmental and Experimental Botany* 61(3):199–223.

Wahid, Iram, Sarika Kumari, Rafiq Ahmad, Sofi J Hussain, Saud Alamri, Manzer H Siddiqui, and M Iqbal R Khan. 2020. "Silver Nanoparticle Regulates Salt Tolerance in Wheat Through Changes in ABA Concentration, Ion Homeostasis, and Defense Systems." Biomolecules 10 (11):1506.

Wang, Fayuan, Xueqin Liu, Zhaoyong Shi, Ruijian Tong, Catharine A. Adams, and Xiaojun Shi. 2016. "Arbuscular mycorrhizae alleviate negative effects of zinc oxide nanoparticle and zinc accumulation in maize plants–a soil microcosm experiment." *Chemosphere* 147:88–97.

Wang, Shun, Wei Li, Keke Chang, Juan Liu, Qingqian Guo, Haifeng Sun, Min Jiang, Hao Zhang, Jing Chen, and Jiandong Hu. 2017. "Localized surface plasmon resonance-based abscisic acid biosensor using aptamer-functionalized gold nanoparticles." *PloS One* 12(9):e0185530.

Wang, Yaoyao, Peng Zhang, Mingshu Li, Zhiling Guo, Sami Ullah, Yukui Rui, and Iseult Lynch. 2020. "Alleviation of nitrogen stress in rice (Oryza sativa) by ceria nanoparticles." *Environmental Science: Nano* 7(10):2930–2940.

Wu, Honghong, Nicholas Tito, and Juan P. Giraldo. 2017. "Anionic cerium oxide nanoparticles protect plant photosynthesis from abiotic stress by scavenging reactive oxygen species." *ACS Nano* 11(11):11283–11297.

Yadav, S.K. 2010. "Heavy metals toxicity in plants: An overview on the role of glutathione and phytochelatins in heavy metal stress tolerance of plants." *South African Journal of Botany* 76(2):167–179.

Yoon, Ho Young, Jeong Gu Lee, Lorenzo Degli Esposti, Michele Iafisco, Pil Joo Kim, Seung Gu Shin, Jong-Rok Jeon, and Alessio Adamiano. 2020. "Synergistic release of crop nutrients and stimulants from hydroxyapatite nanoparticles functionalized with humic substances: Toward a multifunctional nanofertilizer." *ACS Omega* 5(12):6598–6610.

Zahedi, Seyed Morteza, Faezeh Moharrami, Saadat Sarikhani, and Mohsen Padervand. 2020. "Selenium and silica nanostructure-based recovery of strawberry plants subjected to drought stress." *Scientific Reports* 10(1):1–18.

Zhao, Lijuan, Jose R. Peralta-Videa, Armando Varela-Ramirez, Hiram Castillo-Michel, Chunqiang Li, Jianying Zhang, Renato J. Aguilera, Arturo A. Keller, and Jorge L. Gardea-Torresdey. 2012. "Effect of surface coating and organic matter on the uptake of CeO_2 NPs by corn plants grown in soil: Insight into the uptake mechanism." *Journal of Hazardous Materials* 225:131–138.

Zhu, Qingdong, Li Wang, Qianli Dong, Shu Chang, Kexin Wen, Shenghua Jia, Zhilin Chu, Hanmeng Wang, Ping Gao, and Heping Zhao. 2017. "FRET-based glucose imaging identifies glucose signalling in response to biotic and abiotic stresses in rice roots." *Journal of Plant Physiology* 215:65–72.

3

Arsenic Stress Sensitivity, Adaptation, and Mitigation Strategies in Field Crops

Madhu Tiwari
CSIR-National Botanical Research Institute

Neelam Gautam, Yuvraj Indoliya, and Debasis Chakrabarty
CSIR-National Botanical Research Institute
Academy of Scientific and Innovative Research (AcSIR)

CONTENTS

3.1 Introduction

Arsenic (As) poisoning is a global menace, placed as a type I carcinogen by the World Health Organization. Historically, As is used as a pharmaceutical. Inorganic As was used in bronchitis, leukemia, and psoriasis, while organic As is used as antibiotics during protozoal diseases (Chou & Harper, 2007). The physical property of As is intermediate between metal and non-metal; thus As is considered a metalloid. Toxicologically, there are three forms of As – organic, inorganic, and gaseous – of which the inorganic compounds are highly toxic for various life forms. There are four oxidation states of inorganic As, +3, 0, −3, and +5, among them +3 (arsenite) and +5 (arsenate) predominate the environment (WHO, 2001; IARC, 2004). Natural sources of increasing As concentration in agricultural fields are due to the bedrock weathering which can impact not only the near but also distant agricultural fields (Saunders et al., 2005). Leaching of As from As-contaminated bedrock in water bodies (Smedley & Kinniburgh, 2002) as well as its biovolatization (Mestrot et al., 2011) serves as one of the natural resources of As contamination in agricultural fields. In arid regions, the surface evaporation of water bodies resulted in As-enriched water bodies and it is used for irrigation purposes that ultimately result in high As content in crops (Smedley & Kinniburgh, 2002; Lawgali & Meharg, 2011). The dominant anthropogenic source of As contamination is mineral mining, especially iron pyrite (FeS2) (Fendorf & Kocar, 2009). In China, the high level of As content in rice was due to mineral mining (Ma et al., 2016). Coal combustion is another source of As contamination. In the USA, 24 mg/kg As concentration is recorded in coals (Kolker & Nordstrom, 2001). The coal ash generated during coal combustion is an abundant industrial waste (Harkness et al., 2016) and a direct source of As contamination in soil and food products. Also, the use of these coal ash in soil amendments further enriches the As contamination in the soil as well as crops resulting in a loss of soil and crop fertility (Schumann & Sumner, 2004). In the 20th century, the use of As-based pesticides/herbicides in numerous crops such as apple, sweet potato, cotton, and many vegetables (Murphy & Aucott, 1998) boosted the As concentration in

DOI: 10.1201/9781003258063-3

these crops. The As-based wood preservative such as chromate copper arsenate also serves as localized As contamination in soil (Pouschat & Zagury, 2006).

Plants showed a toxicity effect of As when grown under As-contaminated soils. It is important to emphasize that As toxicity not only affects plant's health and yield (Panaullah et al., 2009; Finnegan & Chen, 2012) but the intake of As-contaminated food products ultimately affects the various life forms on earth. Arsenic can be taken up by humans while consuming As-contaminated water/food/food products. When drinking water As concentration is low, the consumption of As-enriched food products is the major factor for As exposure (Carlin et al., 2016). Till now, it has been estimated that As affected ~200 million human population worldwide including the major population of southeast Asian countries. An acute As poisoning is rare but chronic As poisoning is very common and prolonged chronic As exposure affects human health by affecting the cardiovascular, gastrointestinal, nervous, and respiratory systems, and also developed skin lesions (Naujokas et al., 2013). Globally, the average soil As (total) concentration is 5 mg/kg with a large variation in different geographical areas (Koljonen, 1989). Below toxic As concentration, the As uptake by the plant is directly proportional to its soil concentration as studied in many crops such as rice (anaerobic condition) (Adomako et al., 2009; Williams et al., 2007), aerobic horticulture (Norton et al., 2013), and agriculture systems (Williams et al., 2007).

Understanding the As toxicity and its mitigation strategy in crops to produce low As content in grain/food products would be a good initiative for public health.

3.2 Crop Sensitivity under As Stress

Arsenic is a prevalent toxic metalloid that affects the crop's physiological, biochemical, and overall productivity. In addition, physicochemical properties of soil, plant developmental stage, and plant type are the crucial determinants of As toxicity in plants. A systemic study by Yoon et al. (2015) gave a clear picture of As toxicity in different crops grown with two physiochemically different soil. Except for mung bean and cucumber all other eight crops, pea, wheat, sorghum, barley, Chinese cabbage, broccoli, mustard, and kale, showed inhibition in seed germination after treatment with As(III) and As(V). However, in the presence of organic As, dimethylarsinic acid (DMA), the toxicity depends on the physiochemical property of soil. For instance, in soil type A there was no inhibitory effect observed but in soil type B the DMA was found to be toxic. In addition, the growth of all seedlings was affected in presence of either inorganic or organic As species but the trend of As toxicity is different for these two types of As compounds. Yoon et al. (2015) observed the relative toxicity of As compounds on various crops followed the same trend of As toxicity, i.e., As(III) > DMA > AsV. In contrast, As toxicity on wheat was majorly affected by soil type and it was observed that in B-type soil, DMA was more toxic than As(III).

Previous studies indicated that As in low concentration improved the plant growth but at higher concentration becomes detrimental for plants (Azam & Gousul, 2015). It was observed that *Helianthus annuus* seedling showed the best growth under low As supplementation in soil (4 mg/kg); however, increased As concentration affects the overall properties of seedling that ultimately cause the death of the seedlings (Imran et al., 2013). Contrastingly, Swarnakar (2016) measured the mung bean seed germination, seedling length, biomass with different As(V) concentrations (5, 10, and 20 μM) in petridish and reported decreased values in these parameters. Arsenic-induced seedling inhibition has been observed in various crops like rice (Rahman et al., 2007), wheat (Liu et al., 2005b), and *H. annuus* (Imran et al., 2013). Meharg and Hartley-Whitaker (2002) observed the organ level toxicity in rice under As(III) and As(V) toxicity and found that As(V) is more toxic for root development while As(III) is more toxic for seedling germination. Similarly, Bhattacharya et al. (2012) observed that a negative impact of As(III) was more deleterious than As(V) on *Cicer arietinum* under the range of As concentration, i.e., 200, 400, 600, and 800 ppb. Liu et al. (2005a, b) observed the impact of As(V) and As(III) on seed physiology including seed germination in rice, growth and amylase activity. It was found the decline in amylase activity with increasing As concentrations with a more deleterious effect of As(III) rather than As(V). Similarly, Sanal et al. (2014) in barley, i.e., a significant decrease in seed physiology and α-amylase activity. Liu et al. (2005a, b) suggested that abiotic stress indirectly causes dehydration in plants resulting in a decrease in cell growth and division. Also, the reduction in α-amylase activity can be directly correlated with inhibition in seed germination as it is a key enzyme for starch degradation during seed germination.

Arsenic severely affects the crop productivity. Arsenic exposure is directly related to suppressing the leaf area, number, biomass of the plant (Nath et al., 2014). In *C. arietinum* the remarkable decrease of some important amino acids like lysine, methionine, proline, threonine, tryptophan, and valine were observed (Malik et al., 2011). Arsenic affects the photosynthesis of the plant either at photochemical or biochemical step. Arsenic is reported to disorganize the chloroplast membrane (Rafiq et al., 2017) and also affects the chloroplast biosynthesis pathways in maize (Emamverdian et al., 2015). The chloroplast serves as light-harvesting system and chloroplast damage affects the light-harvesting of the photosystem II complex that overall affects photosynthetic rate (Anjum et al., 2011). Arsenic also disturbs the gas exchange, and fluorescence parameters that ultimately result in leaf wilting and overall low photosynthetic rate (Chandrakar et al., 2016; Debona et al., 2017; Musil et al., 2014). Recently, a crop-specific interaction between elevated CO_2 (eCO_2; 620 ppm) and As has been studied in barley (C3) and maize (C4). Barley was found to be more sensitive under ambient CO_2 concentration, but interestingly, at eCO_2 concentration, both species showed differentially mitigated for As-induced damages (AbdElgawad et al., 2021). Being a Pi analog, As(V) is reported to decline the ATP synthesis and affects the normal metabolism of the plant (Tawfik & Viola, 2011; Bhattacharya et al., 2012). The membrane integrity is an important aspect to analyze the tolerance capacity of plant facing As toxicity (Rafiq et al., 2017; Sharma et al., 2012; Gill & Tuteja, 2010). Some studies suggested that the conversion of As(V) to As(III) inside the plant causes a generation of reactive oxygen species (ROS) (Talukdar, 2013). Also, during As(V) to As(III) conversion, the electrons are released that affect the key enzymes of ROS homeostasis and thus lead to high ROS production (Kostecka-Gugała & Latowski, 2018). The root tip mitosis and cytokinesis were affected under disturbed ROS homeostasis in *Triticum turgidum* (Livanos et al., 2012) and it might be a better explanation for reducing root growth during As stress. Carbohydrate metabolism has been deleteriously affected under As stress. For instance, the shoots of As-treated rice showed decreased reduced to non-reduced sugars. Also, α and β amylases which are the key starch degrading enzymes were also inhibited under As toxicity (Jha & Dubey, 2004). Overall, As-mediated toxicity results in compromised germination, root and shoot growth, carbohydrate and sugar metabolism, membrane integrity, etc. which ultimately resulted in low crop productivity (Garg & Kashyap, 2017).

Arsenic affects several biochemical aspects of the plant. Arsenic exposure causes a burst in ROS, but the ROS scavenging system should be very tightly regulated, if not it results in oxidative stress. ROS possesses a free or unpaired electron and is very reactive thus causing cell damage (Sharma et al., 2012). Reportedly, conversion of As(V) to As(III) releases an extra electron and enhanced the production of ROS (Kostecka-Gugała & Latowski, 2018). Excessive ROS damages the membrane, inactivates enzymes, non-specific oxidation of lipid, protein, and carbohydrate (Shahid et al., 2014). Arsenic deleteriously affects the carbohydrate (starch, sugar) metabolism (Chandrakar et al., 2016). Arsenic stress also results in the accumulation of soluble sugar (Gramss, 2012).

3.3 Molecular Mechanism of Adaptation and Mitigation during As Stress in Crop

3.3.1 In Silico Analysis

Plant, under As stress, modulates overall metabolic as well as physiological signal cascade at various developmental stages through regulating gene expression pattern directly or indirectly. Such an array of molecular regulation involves gene expression pattern in a complex way where one intermediate affects the regulation of another signaling intermediates. With the advancement of computational analysis-based high-throughput prediction algorithms, total RNAs may be identified using RNA-Seq analysis. Several transcriptome studies using the Illumina-based RNA-Seq platform have explained dynamic regulation of transcriptional machinery upon As stress. Initially, Chakrabarty et al. (2009), using microarray-based analysis, reported genome-wide expression pattern and its effect under As stress in rice (*Oryza sativa* L., ssp. Indica). They reported differential expression pattern of a group of stress-related genes, cellular transporters, several metallothioneins, TFs as well as other regulatory and stress-related genes with As(V) and As(III) treatment compared to control. Subsequently, As-mediated comparative transcriptome analysis revealed modulation of genes related with phytohormones, transporters, and lipid metabolism pathway in rice (Yu et al., 2012). They performed illumina sequencing to decipher the overall transcriptome pattern as well as microRNA (miRNA) modulations in As(III)-treated rice. The group explained dynamic regulation of genes related to heavy metal transport, and jasmonate biosynthesis pathway and jasmonate signaling under As toxicity. They analyzed the modulation in expression of As(III) transporters and found that 273 (out of 473) transporters were differentially regulated under As(III) toxicity in root and/or shoot. Several genes such as (*OsABCG43, OsABCC9, OsABCB5*: ABC subfamily transporters), (*OsACA1, OsALA4*: P-type ATPase), and (*OsPT19*: phosphate transporter) were differentially regulated under As(III) toxicity. Additionally, they looked into expression pattern of As-responsive phytohormones. They found several jasmonic acid biosynthesis and signaling-pathway-related genes showing differential expression pattern under As stress. For instance, *OsDAD1;2, OsDAD1;3*, and *OsLOX2;1* genes showing differential expression pattern. Another gene, *OsLOX2;3*, also showed differential expression pattern. They further reported the significant upregulation of *OsAOC, OsAOS1*, and *OsAOS2* in root/shoot under As(III) toxicity. They further observed the expression pattern of plant hormones such as auxin, cytokinins (CK), gibberellins (GAs), brassinosteroids (BR), abscisic acid (ABA), and ethylene. They found induction of *OsNCED2*, an ABA biosynthesis pathway gene, in the roots under As(III) toxicity. In addition, expression of GA2ox3 (a GA biosynthesis pathway gene) showed upregulation in root samples treated with As(III). In contrast to it, other GA signaling genes were downregulated under As treatment. Similarly, BR biosynthesis genes were also found to be downregulated under the same stress condition. Nevertheless, two BR deactivation-pathway–related genes (*Os06g39880, Os02g11020*) were showing upregulation under As stress, which indicated reduced BR accumulation in roots under As exposure. A gene related to BR signaling pathway, *OsBRI1*, showed downregulation in root under As toxicity. In addition to it, genes related with biosynthesis of CK (*OsLOG*), auxin (*OsASA2, OsASB1*), and ethylene (*OsACS2*) were also upregulated in As treated root, indicating accumulation of such hormones under As(III) stress. They also looked into expression analysis of the genes related with lipid metabolism. They reported out of 59 lipid biosynthesis Differentially expressed genes (DEGs), 27 and 17 were found to be differentially expressed in roots and shoots respectively, whereas 15 DEGs were common in both roots and shoots under As treatment. Notably, genes related with lipid metabolic pathway, including plastid-specific 3-hydroxyacyl-ACP dehydrase, malonyl-CoA ACP transacylase, as well as enoyl-ACP reductase were differentially expressed under As treatment. Similarly, endoplasmic-reticulum–specific genes, i.e., lysophosphatidate acyltransferase, phosphatidate phosphatase, and 3-hydroxyacyl-CoA dehydrogenase as well as peroxisome-specific COMATOSE (CTS) and multifunctional protein also showed differential expression pattern in As treatment. Moreover, they observed 36 miRNAs that are As(III) responsive and 14 out of 36 miRNAs may possibly be involved in the regulation of gene expression related with signaling, transportation, as well as metabolism. For comparative analysis, Rai et al. (2015) performed transcriptional profiling of rice genotypes with contrasting features. They used six low and high As accumulating rice genotypes, i.e., low-arsenic accumulating rice genotype and high-arsenic accumulating rice genotype, and analyzed differential expression pattern with response to As stress. They reported differential expression pattern of the genes mainly with regulatory genes, transporters, as well as other stress-responsive genes. Similarly, comparative transcriptome profiling in Indian mustard showed the gene network re-patterning under As toxicity (Sapna Thakur et al., 2019). They found upregulation of differentially expressed genes related with stress, metabolism, transporter activity, signaling, and regulation of transcription activity. In contrast to it, several genes related with developmental, photosynthesis, and cell growth/division were downregulated under As treatment. Moreover, they analyzed the pathway using Kyoto encyclopedia of genes and genomes (KEGG) through Automated Annotation server and revealed reprogramming of a large proportion of genes involved in genetic as well as processing of environmental information. They reported maximum KEGG orthology related with carbon metabolism, amino acid biosynthesis, glutathione (GSH) metabolism, and signal transduction of plant hormones. Recently, Pandey et al. (2020) reported temporal expression pattern of Osa-miR156j expression in various stages of plant development, i.e., seedlings formation, tillering, and flower formation as well as different tissues (root, stem, and leaf) under As(III) stress. They also highlighted the possible target/s of Osa-miR156j using bioinformatic tools such as co-expression analysis. Notably, more pronounced downregulation showed in tissues of root at different developmental stages such as seedling, flowering, and tillering stages under arsenite exposure. The most prominent modulation in gene expression was found in the seedlings stage compared to tillering and flowering stage. They further used in silico approach for target prediction and found the presence of regulatory elements that are stress-responsive. This observation confirms the role of Osa-miR156j in rice as a regulator of As-responsive gene expression during As toxicity. Hwang et al. (2017) reported transcriptome profiling and dynamic regulation of As(V)-responsive genes in an As tolerant type 1 (ATT1) mutant of rice. They explained that As toxicity inhibited expression of putative genes involved in photosynthesis and lipid biosynthesis in WT-mutant rice (Hwang et al., 2017). They also identified mutant-specific transcriptional regulation of several genes. Also compared to WT the mutant rice was identified with 50 genes that showed DNA polymorphisms in their upstream regions suggesting the involvement of upstream regions in order to increase As tolerance in the mutants. Dynamic regulation of gene expression and the involvement of miRNA under As stress are key to further investigate a comprehensive signaling approach to develop efficient strategies in order to cope up against As stress as well as its hazardous impact on health and environment.

3.3.2 Study of Various Transporters/Transcription Factors/Genes in Field Crop

3.3.2.1 Arsenate As(V) Uptake Transporters in Crop Plant

In aerobic soil, As(V) predominates the other As species. Structurally As(V) analogous to inorganic phosphate (Pi) can enter plant roots by phosphate (Pi) transporters. Some of Pi transporter families that showed high affinity with Pi have also been characterized as As(V) transporters. In rice, *OsPht1;8* has a high affinity for both As(V) and Pi and found to be involved in Pi as well as As(V) uptake. Wu et al. (2011) demonstrated that overexpression of *OsPht1;8* in rice distinctly increases As(V) uptake (Jia et al., 2011; Wu et al., 2011). In addition, lesser As uptake was found in *OsPHT1;8* mutant compared to WT suggesting the As tolerance of *OsPHT1;8* mutant (Wang et al., 2016). Similarly, the mutant ospht1;1 showed decreased Pi and As(V) uptake suggesting its role in As(V) uptake (Chen et al., 2011; Wu et al., 2011; Sun et al., 2012, Kamiya et al., 2013). Additionally, OsPHR2, a transcription factor, found to be involved in OsPht1;8 expression (Wu et al., 2011). Moreover, *OsPHT1;4* overexpressing lines of rice showed more sensitivity against As toxicity and accumulate more As in rice. A *OsPHT1;4* rice mutant showed approx 44% lesser As accumulation in grains suggesting the role of *OsPHT1;4* as a As transporter.

3.3.2.2 Arsenite Uptake Transporters in Crop Plant

In anaerobic environments, As(III) is the dominant form of As, like submerged soils. According to various studies, As(III) uptake has been done by plant aquaporins, membrane channels involved in water/neutral molecular transport in plant (Li et al., 2014; Mukhopadhyay et al., 2014). Also, various Nodulin 26-like Intrinsic Protein (NIPs) function as transporters for As(III) in plants. As an example, *AtNIP5;1* and *AtNIP6:1* in Arabidopsis, *LjNIP5;1* and *LjNIP6;1* in lotus, and *OsNIP2;1* and *OsNIP3;2* in rice (Bienert et al., 2008). Notably, the As(III) transporters/channels are bidirectional and, therefore, are involved both in assimilation and efflux to the external environment. On the lateral roots of rice, the *OsNIP3;2* is found located on the plasma membrane. Based on the expression pattern analysis, it is revealed that the *OsNIP3;2* channel is suppressed under low As(III) concentrations while it gets upregulated under high As(III) concentrations. In general, rice lateral roots take up As(III) through the *OsNIP3;2* channels under high concentrations of As(III) (Chen et al., 2017). Moreover, researchers found an aquaporin group in rice with higher As(III) permeability, *OsNIP1;1* and *OsNIP3;3*. Neither of the overexpression of two genes changes As(III) uptake, but a marked reduction in root-to-shoot translocation was apparent. The rice plant overexpressing *OsNIP1;1* and *OsNIP3;3* showed low As accumulation. These observations suggest that both genes serve to prevent As(III) efflux from the xylem (Sun et al., 2018). Aquaporins as well as NIP are known to engage plasma membrane intrinsic proteins (PIPs) in As(V) uptake. It has been reported that rice PIPs (*OsPIP2;4*, *OsPIP2;6*, and *OsPIP2;7*) are resistant to *Xenopus laevis* As(III) toxicity when expressed heterologously. Overexpressing PIPs (*OsPIP4;6*, *OsPIP7*, and *OsPIP2;7*) in Arabidopsis reduced As(III) toxicity with improved biomass accumulation (Mosa et al., 2012.).

3.3.2.3 Arsenic Influx and Efflux Transporters in Crop Plant

In rice, As(III) translocation is mediated by silicon (Si) transporters, such as Lsil (silicon inflow transporter). Meanwhile, Lsi2 (silicon efflux transporter) is important for As(III) translocation in plants. The mutant lsi2 rice lines displayed decreased

As(III) transport in the xylem, resulting in decreased As accumulation in grains (Ma et al., 2008). One study discovered that rice nodes may also be involved in storage. The Lsi2 expression profile in Lsi2 mutant rice showed the higher As accumulation in root and nodes but not in grains (Chen et al., 2015). Also, the As-associated transporters genes *OsLsi1, OsLsi2*, and *OsLsi6* are regulated by the MYB transcription factor R2R3 (arsenite-responsive MYB1) (Wang et al., 2017). Thus *OsARM1* is responsible for As(III) uptake and its translocation to the upper parts of the rice. The *OsARM1* knockout lines showed higher expression of *OsLsi1, OsLsi2*, and *OsLsi6* while the expression of these genes is lower in the *OsARM1* overexpressing lines. The overexpression and knockout lines of *OsARM1* showed sensitivity to As(III) and tolerance respectively by regulating genes involved in transport. In addition, one of the ABCC genes for rice, *OsABCC7*, has been discovered to be downregulated under the impact of As(III) (Yu et al., 2012). Further investigation revealed that *OsABBC7* is localized on the root stelar tissues (specifically on plasma membrane) facilitating the upward As translocation in plants (Tang et al., 2019). Furthermore, when plants were supplemented either with As(V) or As(III), the only As species that predominates in external medium is As(III) (Liu et al., 2010; Xu et al., 2007; Zhao et al., 2010). In addition, this observation also confirms that efflux transporters for As(III) are present on root cells. The process occurs when excess as in the cell burden the cell. Again, the mutation in lsi1 results in only 20 percent of As(III) efflux to the external medium. Hence, it can be predicted that Lsi1 has not abundant role in As(III) efflux but it is a bidirectional transporter (Zhao et al., 2010). In addition to that, OsLsi2 is mostly found in exodermis and endodermis of lateral root and root cells respectively. So according to findings, it is observed that, although OsLsi2 is an efflux transporter, the direction of As(III) translocation for root-to-shoot translocation is to the vascular bundle rather than to the external medium.

3.3.2.4 Genes Involved in As Transformation to Its Various Forms

During As detoxification process, firstly, inorganic As transforms into other inorganic forms that further transform into the least toxic organic forms. Shi et al. (2016) reported that in rice, *OsHAC1;1, OsHAC1;2*, and *OsHAC4* possess As(V) reductase activity (i.e., converts As(V) to As(III)). *OsHAC1;1* and *OsHAC1;2* genes showed dual localization in cell, i.e., cytoplasm as well as nucleus, and they are predominantly expressed in rice roots. When these genes are overexpressed in rice plants, they had increased As(V) reduction and increased As(III) efflux to the external medium (Shi et al., 2016). Compared to the WT plants, *hac1;1* and *hac1;2* knockout mutant lines had lower As(V) reduction in the root but higher As accumulation in root as well as shoot. Field studies also suggested that overexpression lines had a 20% decreased amount of As in grain while knockout lines showed more As content in rice grain compared to WT plants (Shi et al., 2016). *OsHAC4*, another As(V) reductase gene in rice, also showed tolerance against As(V) toxicity (Xu et al., 2017). Its overexpression resulted in higher As(V) reduction and thus lesser As gets accumulated in root and shoot. While the mutant lines with decreased As(V) reductase activity, its higher occurrence was found in root, xylem sap, and shoot. Additionally, mutant lines of this gene have a higher As(III)/As(V) ratio than the wild type, indicating that its mutation affects the As speciation in xylem sap as well (Xu et al., 2017). This indicates that reduction of As(V) is one of the effective mechanisms affecting As(III) eventual fate, either back to the external media or the plant part.

Further, glutaredoxins (*Grx*) also showed As stress (Sundaram et al., 2009). Two rice *Grx* genes, *OsGrx_C7* and *OsGrx_C2.1*, were overexpressed in *Arabidopsis thaliana* that showed lesser As accumulation in seeds and also had more As tolerance compared to WT plants (Verma et al., 2016a). In recent studies, a chickpea glutaredoxin, *CaGrx* was identified by Kumar et al. (2020). Glutaredoxin *CaGrx* expressing Arabidopsis lines showed better As tolerance than the wild-type plant under As toxicity by accumulating lesser As content.

Till date, in plants, there is no evidence for As transformation to its organic/volatile form. Meng et al. (2011) generated a high expressing S-adenosyl methyltransferase gene (arsM) transgenic plants and found a 10-fold more methylated form of As species, MMA DMA(V) in these lines. Verma et al. (2016b) overexpressed a fungus (*Westerdykella aurantiaca*) As methyltransferase gene (*WaarsM*) in rice and observed more As tolerance in rice under As stress due to higher As volatilization by *WaarsM*.

3.3.2.5 Study of Arsenic Detoxification Genes in Crop Plant

The presence of As in the plants activates a variety of detoxification genes that confer tolerance in the plants. The differential gene expression/modulation/responses under As toxicity have been reported. Several genes were considered as candidates for detoxification of As, including glutathione reductase (*GR*), glutathione synthetase, phytochelatin synthase (*PCS*), and metallothionine-2 (*MT-2*). Singh et al. (2021) suggested that *GR*, arsenate reductase (*RcAR*), *PCS* 1 (*RcPCS1*), c-type ABC transporter (*RcABCC*), and metallothionein type 2 (*RcMT2*) have potential to confer tolerance against As(V) in WM castor genotype. Singh et al. (2021) analyzed two contrasting genotypes of castor, i.e., As(V) tolerant (WM) and As(V) sensitive (GCH 2), and identified a differential regulation of several genes including *RcMT2, RcPCS1X2, GR*, and *RcABCC*. They found higher expression of *RcPCS1X2, RcMT2* in leaves, and *RcPCS1X1* in root. While *RcABCC* showed differential regulation in roots as well leaves of WM compared to the GCH2 genotype under 200 μM As(V) treatment. They also showed significantly upregulated *GR* activity and increased GSH/GSSH ratio in the WM genotype.

There is evidence that intracellular As(III) forms stable complexes with phytochelatins (*PCs*) and GSH that contain thiols. Das et al. (2017) proved the role of two genes, i.e., *OsPCS1* and *OsPCS2* in As accumulation in rice. By using seed-specific promoters, the *OsPCS1* and *OsPCS2* genes were suppressed with RNA-mediated gene silencing in seeds in rice, resulting in a 35% decline in As content as compared to WT. During increased As concentration, upregulation of an *AtABCC1*, ortholog of rice; *OsABCC1* was found. But it was not observed to be upregulated under low As concentration. Song et al. (2010) suggested that when As toxicity is present, *OsABCC1* sequesters more As in the phloem companion cells, inhibiting As translocation to grain. By examining knockout *OsABCC1* transgenic rice lines, this observation was further confirmed. Compared to WT plants, the knockout lines showed sensitivity during As stress (Song et al., 2010). *OsABCC1* (*has1* rice mutant) and *OsPCS1* (*has2* rice mutant) have been identified by Hayashi et al. (2017) showing an increased As content in grains and decreased As content in node I than wild type. Thus *OsABCC1* and *OsPCS1* inhibit the As translocation to the grain. The declined As content was found in high expressing *OsPCS* lines in rice (Hayashi et al., 2017).

Metallothionein, another metal-binding protein, forms metal complexes and gets sequestered in the vacuoles by the vacuolar transporters. Gautam et al. (2012) reported that rice metallothioneins (*OsMTs*) have a potent role in conferring As tolerance. Pan et al. (2018) observed the differential expression of 16 *Brassica napas MTs* under As stress by RNA-seq and quantitative real-time PCR analysis in plants. Pan et al. (2018) showed the role of *BnaMT3C* under As stress in *Brassica napus*. An additional cysteine-rich protein *OsGST1* confers enhanced tolerance under As toxicity in Arabidopsis. A tau class GST (OsGSTU5) that has previously been identified as a gem for biotic stress tolerance in rice (Tiwari et al., 2020, 2022a) also has a potent role in As detoxification by accumulating more As in rice roots (Tiwari et al., 2022b). Kidwai et al. (2019) suggested that rice class III peroxidase had a key role in As tolerance. Overexpressing lines of class III peroxidize in Arabidopsis showed lower As accumulation because the root apoplastic region contains more lignin formation, which acts as a barrier to As entry.

3.3.3 Other Mitigation Strategies

Increasing As toxicity in crops is a threatening alarm in agricultural productivity and food safety. There is a need for an effective strategy to mitigate As toxicity in the crops.

3.3.3.1 Water and Fertilizer Management

Shortly after planting the rice seedlings, the flooding condition is maintained. It is important to emphasize that flooding condition is not required for seedling germination or growth. Under the flooded condition, the nutrient mobilization is enhanced and it also discourages the growth of nematodes and other insects. The flooded condition (in rice) is suitable for better As uptake. As under reducing conditions, the As(V) is transformed to its more toxic form, i.e., As(III) that can be readily taken up by the plant. Flooding with intermittent irrigation is one of the successful approaches to reduce As dissolution in water and thus can discourage the As uptake by the plant. Also, the intermittent flooding proved to produce more grain yield than the plants with a no-flooding state or constant flooding state (Li et al., 2009; Grassi et al., 2009; Arao T, et al., Hu et al., 2013a, b). Soil amendment is also utilized to make the As less bioavailable for plants. For instance, the addition of Fe (iron)- or silica (Si)-based additives in fertilizers. Fe additives chelate more As and reduce their mobility in soil. In contrast, Si additives increase the Si concentration in soil and act as a competitive inhibitor for As uptake by the plant.

3.3.3.2 Arbuscular Mycorrhizal (AM) Fungi Mediated As Tolerance in Crops

Reportedly, AM fungi increase the nutritional quality of crops. Very recently, the meta-analysis of various crop plants showed enhanced tolerance against As as well as increased productivity. Except *H. annuus* the As concentration was decreased to −12% to −26% in the tissues of maize, rice, and other crops but the highest reduction was noted in *Glycine max* −56% to −40% and *M. sativa* −51% to −59%. The study showed that As-mediated loss in crop yield can be managed by the use of these AM fungi (Neidhardt, 2021). The AM-mediated enhanced productivity is not only dependent on the plant but also AM fungi and the degree of colonization achieved by the AM fungi (Chan et al., 2013; Lekberg & Koide, 2005). Here also the best biomass was achieved by *Diversispora versiformis* and *Funneliformis mosseae* while *Rhizophagus irregularis* showed the highest As reduction in crop plants (Neidhardt, 2021).

3.3.3.3 Nanoparticles

Increasing research on nanoparticles might be useful in the field of agriculture. Titanium (Ti) engineered nanoparticle (TiNP) has been proved to alleviate the detoxification response under As stress in *Vigna radiata*. It was observed that under the presence of TiNPs, the plant showed better growth indices, membrane integrity and had a good antioxidant defense under As toxicity (Katiyar et al., 2020). Similarly, the Iron (III) oxide nanoparticles were also found to be effective against As toxicity in *V. radiata* by lowering the uptake of As. The Fe nanoparticle absorbed As(III) as a result it become less available for plants. Moreover, their nanoparticles do not show inhibition in plant growth and development. These nanoparticles might be utilized effectively under the field of As-contaminated areas (Shabnam et al., 2019).

3.4 Concluding Remark and Future Prospect

Arsenic contamination to the crops severely affects the plant as well as human health. Some crop cultivars are sensitive while others are tolerant. To see As adaptation of tolerant cultivars, the transcriptome analysis has been performed that identified a number of differentially regulated genes suggesting the genetic framework of the plant is an important determinant of As-tolerance/sensitivity. Further, several genes related to As detoxification has been unraveled in crop plants. The As detoxification involved the various influx/efflux As transporters and As chelation. PCs and GSH are related to As chelation inside the crop plants. Overall, plants by utilizing various genes confer tolerance to As but there are other strategies like water management, fungal-based, and nanoparticle-based detoxification. Overall, crop plants utilized various approaches to cope up with As toxicity. However, some other approaches might be utilized to make the crop plant more tolerant against As, i.e., identification of alleles-based detoxification that is related to transporters or detoxification genes that will allow for marker-assisted breeding in crops. Also, a lot of studies have been performed on rice but not on other crops. Thus there is a number of genes are required to be functionally characterized in other crops.

REFERENCES

AbdElgawad, H., Schoenaers, S., & Zinta, G. (2021). Soil arsenic toxicity differentially impacts C3 (barley) and C4 (maize) crops under future climate atmospheric CO_2. *Journal of Hazardous Materials*, *414*, 125331.

Adomako, E. E., Solaiman, A. R. M., Williams, P. N., Deacon, C., Rahman, G. K. M. M., & Meharg, A. A. (2009). Enhanced transfer of arsenic to grain for Bangladesh grown rice compared to US and EU. *Environment International*, *35*(3), 476–479.

Anjum, S. A., Xie, X. Y., Wang, L. C., Saleem, M. F., Man, C., & Lei, W. (2011). Morphological, physiological and biochemical responses of plants to drought stress. *African Journal of Agricultural Research*, *6*(9), 2026–2032.

Azam, S. M., & Gousul, G. (2015). Arsenic removal from contaminated waters by Fe-based (hydr) oxides and its phytoavailability in soil-plant system. Università degli Studi di Napoli Federico II.

Bhattacharya, S., De Sarkar, N., Banerjee, P., Banerjee, S., Mukherjee, S., Chattopadhyay, D., & Mukhopadhyay, A. (2012). Effects of arsenic toxicity on germination, seedling growth and peroxidase activity in Cicer arietinum. *International Journal of Agriculture and Food Science*, 2, 131–137.

Bienert, G. P., Thorsen, M., Schüssler, M. D., Nilsson, H. R., Wagner, A., Tamás, M. J., & Jahn, T. P., (2008). A subgroup of plant aquaporins facilitate the bi-directional diffusion of As $(OH)_3$ and Sb $(OH)_3$ across membranes. *BMC Biology*, *6*(1), 1–15.

Cao, Y., Sun, D., Ai, H., Mei, H., Liu, X., Sun, S., Xu, G., Liu, Y., Chen, Y., & Ma, L. Q. (2017). Knocking out OsPT4 gene decreases arsenate uptake by rice plants and inorganic arsenic accumulation in rice grains. *Environmental Science & Technology*, *51*(21), 12131–12138.

Carlin, D. J., Naujokas, M. F., Bradham, K. D., Cowden, J., Heacock, M., Henry, H. F., & Suk, W. A. (2016). Arsenic and environmental health: State of the science and future research opportunities. *Environmental Health Perspectives*, *124*(7), 890–899.

Chakrabarty, D., Trivedi, P. K., Misra, P., Tiwari, M., Shri, M., Shukla, D., Kumar, S., Rai, A., Pandey, A., Nigam, D., Tripathi, R. D., & Tuli, R. (2009). Comparative transcriptome analysis of arsenate and arsenite stresses in rice seedlings. *Chemosphere*, *74*(5), 688–702.

Chan, W. F., Li, H., Wu, F. Y., Wu, S. C., & Wong, M. H. (2013). Arsenic uptake in upland rice inoculated with a combination or single arbuscular mycorrhizal fungi. *Journal of Hazardous Materials*, *262*, 1116–1122.

Chandrakar, V., Naithani, S. C., & Keshavkant, S. (2016). Arsenic-induced metabolic disturbances and their mitigation mechanisms in crop plants: A review. *Biologia*, *71*(4), 367–377.

Chen, J., Liu, Y., Ni, J., Wang, Y., Bai, Y., Shi, J., Gan, J., Wu, Z. and Wu, P., 2011. OsPHF1 regulates the plasma membrane localization of low-and high-affinity inorganic phosphate transporters and determines inorganic phosphate uptake and translocation in rice. *Plant Physiology*, 157(1), pp. 269–278.

Chen, Y., Moore, K. L., Miller, A. J., McGrath, S. P., Ma, J. F., & Zhao, F. J. (2015). The role of nodes in arsenic storage and distribution in rice. *Journal of Experimental Botany*, *66*(13), 3717–3724.

Chen, Y., Hua, C. Y., Jia, M. R., Fu, J. W., Liu, X., Han, Y. H., Liu, Y., Rathinasabapathi, B., Cao, Y., & Ma, L. Q. (2017). Heterologous expression of Pteris vittata arsenite antiporter PvACR3; 1 reduces arsenic accumulation in plant shoots. *Environmental Science & Technology*, *51*(18), 10387–10395.

Chou, C. H., & Harper, C. (2007). Toxicological profile for arsenic.

Das, N., Bhattacharya, S., Bhattacharyya, S., & Maiti, M. K. (2017). Identification of alternatively spliced transcripts of rice phytochelatin synthase 2 gene OsPCS2 involved in mitigation of cadmium and arsenic stresses. *Plant Molecular Biology*, *94*(1), 167–183.

Debona, D., Rodrigues, F. A., & Datnoff, L. E. (2017). Silicon's role in abiotic and biotic plant stresses. *Annual Review of Phytopathology*, *55*, 85–107.

Emamverdian, A., Ding, Y., Mokhberdoran, F., & Xie, Y. (2015). Heavy metal stress and some mechanisms of plant defense response. *The Scientific World Journal*, *2015*.

Fendorf, S., & Kocar, B. D. (2009). Biogeochemical processes controlling the fate and transport of arsenic: Implications for South and Southeast Asia. *Advances in Agronomy*, *104*, 137–164.

Finnegan, P., & Chen, W. (2012). Arsenic toxicity: The effects on plant metabolism. *Frontiers in Physiology*, *3*, 182.

Garg, N., & Kashyap, L. (2017). Silicon and Rhizophagus irregularis: Potential candidates for ameliorating negative impacts of arsenate and arsenite stress on growth, nutrient acquisition and productivity in Cajanus cajan (L.) Millsp. genotypes. *Environmental Science and Pollution Research*, *24*(22), 18520–18535.

Gautam, N., Verma, P. K., Verma, S., Tripathi, R. D., Trivedi, P. K., Adhikari, B., & Chakrabarty, D. (2012). Genome-wide identification of rice class I metallothionein gene: Tissue expression patterns and induction in response to heavy metal stress. *Functional & Integrative Genomics*, *12*(4), 635–647.

Gill, S. S., & Tuteja, N. (2010). Reactive oxygen species and antioxidant machinery in abiotic stress tolerance in crop plants. *Plant Physiology and Biochemistry*, *48*(12), 909–930.

Gramss, G. (2012). Potential contributions of oxidoreductases from alfalfa plants to soil enzymology and biotechnology: A review. *Journal of Nature Science and Sustainable Technology*, *6*(3), 169.

Grassi, C., Bouman, B. A. M., & Castañeda, A. R. (2009). Aerobic rice: Crop performance and water use efficiency. *Journal of Agriculture and Environment for International Development (JAEID)*, *103*(4), 259–270.

Harkness, J. S., Sulkin, B., & Vengosh, A. (2016). Evidence for coal ash ponds leaking in the southeastern United States. *Environmental Science & Technology, 50*(12), 6583–6592.

Hayashi, S., Kuramata, M., Abe, T., Takagi, H., Ozawa, K., & Ishikawa, S. (2017). Phytochelatin synthase Os PCS 1 plays a crucial role in reducing arsenic levels in rice grains. *The Plant Journal, 91*(5), 840–848.

Hu, P., Li, Z., Yuan, C., Ouyang, Y., Zhou, L., Huang, J., Luo, Y., Christie, P. and Wu, L. (2013a). Effect of water management on cadmium and arsenic accumulation by rice (Oryza sativa L.) with different metal accumulation capacities. *Journal of Soils and Sediments, 13*(5), 916–924.

Hu, P., Huang, J., Ouyang, Y., Wu, L., Song, J., Wang, S., Li, Z., & Christie, P. (2013b). Water management affects arsenic and cadmium accumulation in different rice cultivars. *Environmental Geochemistry and Health, 35*(6), 767–778.

Hwang, S. G., Chapagain, S., Lee, J. W., Han, A. R., & Jang, C. S. (2017). Genome-wide transcriptome profiling of genes associated with arsenate toxicity in an arsenic-tolerant rice mutant. *Plant Physiology and Biochemistry, 120*, 40–51.

IARC Working Group on the Evaluation of Carcinogenic Risks to Humans, World Health Organization, & International Agency for Research on Cancer. (2004). Some drinking-water disinfectants and contaminants, including arsenic (Vol. 84). IARC.

Imran, M. A., Ch, M. N., Khan, R. M., Ali, Z., & Mahmood, T. (2013). Toxicity of arsenic (As) on seed germination of sunflower (*Helianthus annuus* L.). *International Journal of Physical Sciences, 8*(17), 840–847.

Jha, A. B., & Dubey, R. S. (2004). Carbohydrate metabolism in growing rice seedlings under arsenic toxicity. *Journal of Plant Physiology, 161*(7), 867–872.

Jia, H., Ren, H., Gu, M., Zhao, J., Sun, S., Zhang, X., Chen, J., Wu, P., & Xu, G. (2011). The phosphate transporter gene OsPht1; 8 is involved in phosphate homeostasis in rice. *Plant Physiology, 156*(3), 1164–1175.

Kamiya, T., Islam, R., Duan, G., Uraguchi, S., & Fujiwara, T. (2013). Phosphate deficiency signaling pathway is a target of arsenate and phosphate transporter OsPT1 is involved in As accumulation in shoots of rice. *Soil Science and Plant Nutrition, 59*(4), 580–590.

Katiyar, P., Yadu, B., Korram, J., Satnami, M. L., Kumar, M., & Keshavkant, S. (2020). Titanium nanoparticles attenuates arsenic toxicity by up-regulating expressions of defensive genes in *Vigna radiata* L. *Journal of Environmental Sciences, 92*, 18–27.

Kidwai, M., Dhar, Y. V., Gautam, N., Tiwari, M., Ahmad, I. Z., Asif, M. H., & Chakrabarty, D. (2019). Oryza sativa class III peroxidase (OsPRX38) overexpression in Arabidopsis thaliana reduces arsenic accumulation due to apoplastic lignification. *Journal of Hazardous Materials, 362*, 383–393.

Koljonen, T. (1989). 12th international geochemical exploration symposium and the 4th symposium on methods of geochemical prospecting geochemical atlas of Finland: Preliminary aspects. *Journal of Geochemical Exploration, 32*(1), 231–242.

Kolker, A., & Nordstrom, D. K. (2001). Occurrence and micro-distribution of arsenic in pyrite. US Geol Surv website (http://wwwbrr.cr.usgs.gov/Arsenic/).

Kostecka-Gugała, A., & Latowski, D. (2018). Arsenic-induced oxidative stress in plants. In M. Hasanuzzaman, K. Nahar, & M. Fujita (eds), *Mechanisms of Arsenic Toxicity and Tolerance in Plants* (pp. 79–104). Springer, Singapore.

Kumar, A., Dubey, A.K., Kumar, V., Ansari, M.A., Narayan, S., Kumar, S., Pandey, V., Shirke, P.A., Pande, V. and Sanyal, I., 2020. Over-expression of chickpea glutaredoxin (CaGrx) provides tolerance to heavy metals by reducing metal accumulation and improved physiological and antioxidant defence system. Ecotoxicology and Environmental Safety, 192, p.110252.

Lawgali, Y. F., & Meharg, A. A. (2011). Levels of arsenic and other trace elements in Southern Libyan agricultural irrigated soil and non-irrigated soil projects. *Water Quality, Exposure and Health, 3*(2), 79–90.

Lekberg, Y., & Koide, R. T. (2005). Is plant performance limited by abundance of arbuscular mycorrhizal fungi? A meta-analysis of studies published between 1988 and 2003. *New Phytologist, 168*(1), 189–204.

Li, G., Santoni, V., & Maurel, C. (2014). Plant aquaporins: Roles in plant physiology. *Biochimica et Biophysica Acta (BBA)-General Subjects, 1840*(5), 1574–1582.

Li, R. Y., Stroud, J. L., Ma, J. F., McGrath, S. P., & Zhao, F. J. (2009). Mitigation of arsenic accumulation in rice with water management and silicon fertilization. *Environmental Science & Technology, 43*(10), 3778–3783.

Liu, X., Zhang, S., Shan, X., & Zhu, Y. G. (2005a). Toxicity of arsenate and arsenite on germination, seedling growth and amylolytic activity of wheat. *Chemosphere, 61*(2), 293–301.

Liu, Y., Zhu, Y. G., Chen, B. D., Christie, P., & Li, X. L. (2005b). Yield and arsenate uptake of arbuscular mycorrhizal tomato colonized by Glomus mosseae BEG167 in As spiked soil under glasshouse conditions. *Environment International, 31*(6), 867–873.

Liu, W. J., Wood, B. A., Raab, A., McGrath, S. P., Zhao, F. J., & Feldmann, J. (2010). Complexation of arsenite with phytochelatins reduces arsenite efflux and translocation from roots to shoots in Arabidopsis. *Plant Physiology, 152*(4), 2211–2221.

Livanos, P., Galatis, B., Quader, H., & Apostolakos, P. (2012). Disturbance of reactive oxygen species homeostasis induces atypical tubulin polymer formation and affects mitosis in root-tip cells of Triticum turgidum and Arabidopsis thaliana. *Cytoskeleton, 69*(1), 1–21.

Ma, J. F., Yamaji, N., Mitani, N., Xu, X. Y., Su, Y. H., McGrath, S. P., & Zhao, F. J. (2008). Transporters of arsenite in rice and their role in arsenic accumulation in rice grain. *Proceedings of the National Academy of Sciences, 105*(29), 9931–9935.

Ma, L., Wang, L., Jia, Y., & Yang, Z. (2016). Arsenic speciation in locally grown rice grains from Hunan province, China: Spatial distribution and potential health risk. *Science of the Total Environment, 557*, 438–444.

Malik, J. A., Goel, S., Sandhir, R., & Nayyar, H. (2011). Uptake and distribution of arsenic in chickpea: Effects on seed yield and seed composition. *Communications in Soil Science and Plant Analysis, 42*(14), 1728–1738.

Meharg, A. A., & Hartley-Whitaker, J. (2002). Arsenic uptake and metabolism in arsenic resistant and nonresistant plant species. *New Phytologist, 154*(1), 29–43.

Meng, X. Y., Qin, J., Wang, L. H., Duan, G. L., Sun, G. X., Wu, H. L., Chu, C. C., Ling, H. Q., Rosen, B. P., & Zhu, Y. G. (2011). Arsenic biotransformation and volatilization in transgenic rice. *New Phytologist, 191*(1), 49–56.

Mestrot, A., Feldmann, J., Krupp, E. M., Hossain, M. S., Roman-Ross, G., & Meharg, A. A. (2011). Field fluxes and speciation of arsines emanating from soils. *Environmental Science & Technology, 45*(5), 1798–1804.

Mosa, K. A., Kumar, K., Chhikara, S., Mcdermott, J., Liu, Z., Musante, C., White, J. C., & Dhankher, O. P. (2012). Members of rice plasma membrane intrinsic proteins subfamily are involved in arsenite permeability and tolerance in plants. *Transgenic Research, 21*(6), 1265–1277.

Mukhopadhyay, R., Bhattacharjee, H., & Rosen, B. P. (2014). Aquaglyceroporins: Generalized metalloid channels. *Biochimica et Biophysica Acta (BBA)-General Subjects, 1840*(5), 1583–1591.

Murphy, E. A., & Aucott, M. (1998). An assessment of the amounts of arsenical pesticides used historically in a geographical area. *Science of the Total Environment, 218*(2–3), 89–101.

Musil, S., Matoušek, T., Currier, J. M., Stýblo, M., & Dědina, J. (2014). Speciation analysis of arsenic by selective hydride generation-cryotrapping-atomic fluorescence spectrometry with flame-in-gas-shield atomizer: Achieving extremely low detection limits with inexpensive instrumentation. *Analytical Chemistry, 86*(20), 10422–10428.

Nath, S., Panda, P., Mishra, S., Dey, M., Choudhury, S., Sahoo, L., & Panda, S. K. (2014). Arsenic stress in rice: Redox consequences and regulation by iron. *Plant Physiology and Biochemistry, 80*, 203–210.

Naujokas, M. F., Anderson, B., Ahsan, H., Aposhian, H. V., Graziano, J. H., Thompson, C., & Suk, W. A. (2013). The broad scope of health effects from chronic arsenic exposure: Update on a worldwide public health problem. *Environmental Health Perspectives, 121*(3), 295–302.

Neidhardt, H. (2021). Arbuscular mycorrhizal fungi alleviate negative effects of arsenic-induced stress on crop plants: A meta-analysis. *Plants, People, Planet, 3*(5), 523–535.

Norton, G., Deacon, C., Mestrot, A., Feldmann, J., Jenkins, P., Baskaran, C., & Meharg, A. A. (2013). Arsenic speciation and localization in horticultural produce grown in a historically impacted mining region. *Environmental Science & Technology, 47*(12), 6164–6172.

Pan, Y. Zhu., M.Wang., S. Ma., G. Huang. Qiao., C.Wang., R., Xu., X. Liang., Y. Lu., K. Li., J., & Qu., C. (2018). Genome-wide characterization and analysis of metallothionein family genes that function in metal stress tolerance in *Brassica napus* L. *International Journal of Molecular Sciences, 19,* 2181.

Panaullah, G. M., Alam, T., Hossain, M. B., Loeppert, R. H., Lauren, J. G., Meisner, C. Ahmed, Z. U., & Duxbury, J. M. (2009). Arsenic toxicity to rice (*Oryza sativa* L.) in Bangladesh. *Plant and Soil, 317*(1), 31–39.

Pandey, A. K., Gedda, M. R., & Verma, A. K. (2020). Effect of Arsenic Stress on expression pattern of a rice specific miR156j at various developmental stages and their allied co-expression target networks. *Frontiers in Plant Science*, 11, 752.

Pouschat, P., & Zagury, G. J. (2006). In vitro gastrointestinal bioavailability of arsenic in soils collected near CCA-treated utility poles. *Environmental Science & Technology, 40*(13), 4317–4323.

Rafiq, M., Shahid, M., Abbas, G., Shamshad, S., Khalid, S., Niazi, N. K., & Dumat, C. (2017). Comparative effect of calcium and EDTA on arsenic uptake and physiological attributes of Pisum sativum. *International Journal of Phytoremediation, 19*(7), 662–669.

Rahman, M. A., Hasegawa, H., Rahman, M. M., Islam, M. N., Miah, M. M., & Tasmen, A. (2007). Effect of arsenic on photosynthesis, growth and yield of five widely cultivated rice (Oryza sativa L.) varieties in Bangladesh. *Chemosphere, 67*(6), 1072–1079.

Rai, A., Bhardwaj, A., Misra, P., Bag, S. K., Adhikari, B., Tripathi, R. D., Trivedi, P. K. & Chakrabarty, D. (2015). Comparative transcriptional profiling of contrasting rice genotypes shows expression differences during arsenic stress. *The Plant Genome*, 8(2), 1–14.

Sanal, F., Şeren, G., & Güner, U. (2014). Effects of arsenate and arsenite on germination and some physiological attributes of barley *Hordeum vulgare* L. *Bulletin of Environmental Contamination and Toxicology, 92*(4), 483–489.

Saunders, J. A., Lee, M. K., Uddin, A., Mohammad, S., Wilkin, R. T., Fayek, M., & Korte, N. E. (2005). Natural arsenic contamination of Holocene alluvial aquifers by linked tectonic, weathering, and microbial processes. *Geochemistry, Geophysics, Geosystems, 6*(4).

Schumann, A. W., & Sumner, M. E. (2004). Formulation of environmentally sound waste mixtures for land application. *Water, Air, and Soil Pollution, 152*(1), 195–217.

Shabnam, N., Kim, M., & Kim, H. (2019). Iron (III) oxide nanoparticles alleviate arsenic induced stunting in Vigna radiata. *Ecotoxicology and Environmental Safety, 183*, 109496.

Shahid, M., Pourrut, B., Dumat, C., Nadeem, M., Aslam, M., & Pinelli, E. (2014). Heavy-metal-induced reactive oxygen species: Phytotoxicity and physicochemical changes in plants. *Reviews of Environmental Contamination and Toxicology Volume 232*, 1–44.

Sharma, P., Jha, A. B., Dubey, R. S., & Pessarakli, M. (2012). Reactive oxygen species, oxidative damage, and antioxidative defense mechanism in plants under stressful conditions. *Journal of Botany*, 2012, 1–26.

Shi, S., Wang, T., Chen, Z., Tang, Z., Wu, Z., Salt, D. E., Chao, D. Y., & Zhao, F. J. (2016). OsHAC1; 1 and OsHAC1; 2 function as arsenate reductases and regulate arsenic accumulation. *Plant Physiology, 172*(3), 1708–1719.

Singh, R., Misra, A. N., & Sharma, P. (2021) Differential responses of thiol metabolism and genes involved in arsenic detoxification in tolerant and sensitive genotypes of bioenergy crop Ricinus communis. *Protoplasma, 258*(2), 391–401.

Smedley, P. L., & Kinniburgh, D. G. (2002). A review of the source, behaviour and distribution of arsenic in natural waters. *Applied Geochemistry, 17*(5), 517–568.

Song, W. Y., Park, J., Mendoza-Cózatl, D. G., Suter-Grotemeyer, M., Shim, D., Hörtensteiner, S., Geisler, M., Weder, B., Rea, P. A., Rentsch, D., & Schroeder, J. I. (2010). Arsenic tolerance in Arabidopsis is mediated by two ABCC-type phytochelatin transporters. *Proceedings of the National Academy of Sciences, 107*(49), 21187–21192.

Sun, S., Gu, M., Cao, Y., Huang, X., Zhang, X., Ai, P., Zhao, J., Fan, X., & Xu, G. (2012). A constitutive expressed phosphate transporter, OsPht1; 1, modulates phosphate uptake and translocation in phosphate-replete rice. *Plant Physiology, 159*(4), 1571–1581.

Sun, S. K., Chen, Y., Che, J., Konishi, N., Tang, Z., Miller, A. J., Ma, J. F., & Zhao, F. J. (2018). Decreasing arsenic accumulation in rice by overexpressing Os NIP 1; 1 and Os NIP 3; 3 through disrupting arsenite radial transport in roots. *New Phytologist, 219*(2), 641–653.

Sundaram, S., Wu, S., Ma, L. Q., & Rathinasabapathi, B. (2009). Expression of a Pteris vittata glutaredoxin PvGRX5 in transgenic Arabidopsis thaliana increases plant arsenic tolerance and decreases arsenic accumulation in the leaves. *Plant, Cell, & Environment, 32,* 851–858.

Swarnakar, A. (2016). Mitigation of toxic effects of sodium arsenate on germination, seedling growth and amylolytic enzyme of mungbean seedlings with macronutrients, micronutrients and organic acids. *International Journal of Current Microbiology and Applied Sciences 5*(12), 151–160.

Talukdar, D. (2013). Arsenic-induced changes in growth and antioxidant metabolism of fenugreek. *Russian Journal of Plant Physiology, 60*(5), 652–660.

Tang, Z., Chen, Y., Miller, A. J., & Zhao, F. J. (2019). The C-type ATP-binding cassette transporter OsABCC7 is involved in the root-to-shoot translocation of arsenic in rice. *Plant and Cell Physiology, 60*(7), 1525–1535.

Tawfik, D. S., & Viola, R. E. (2011). Arsenate replacing phosphate: Alternative life chemistries and ion promiscuity. *Biochemistry, 50*(7), 1128–1134.

Thakur, S., Choudhary, S., Dubey, P., & Bhardwaj, P. (2019). Comparative transcriptome profiling reveals the reprogramming of gene networks under arsenic stress in Indian mustard. *Genome, 62*(12), 833–847.

Tiwari, M., Srivastava, S., Singh, P. C., Mishra, A. K. and Chakrabarty, D., 2020. Functional characterization of tau class glutathione-S-transferase in rice to provide tolerance against sheath blight disease. *3 Biotech, 10*(3), 1–7.

Tiwari, M., Gautam, N., Indoliya, Y., Kidwai, M., Mishra, A. K., & Chakrabarty, D. (2022a). A tau class GST, OsGSTU5, interacts with VirE2 and modulates the Agrobacterium-mediated transformation in rice. *Plant Cell Reports,* 41(4), pp.873-891.

Tiwari, M., Kidwai, M., Dutta, P., Narayan, S., Gautam, N., Chawda, K., Shirke, P. A., Mishra, A. K., & Chakrabarty, D. (2022b). A tau class glutathione-S-transferase (OsGSTU5) confers tolerance against arsenic toxicity in rice by accumulating more arsenic in root. *Journal of Hazardous Materials, 426,* 128100.

Verma, P. K., Verma, S., Pande, V., Mallick, S., Deo Tripathi, R., Dhankher, O. P., & Chakrabarty, D. (2016a). Overexpression of rice glutaredoxin OsGrx_C7 and OsGrx_C2. 1 reduces intracellular arsenic accumulation and increases tolerance in Arabidopsis thaliana. *Frontiers in Plant Science, 7,* 740.

Verma, S., Verma, P. K., Pande, V., Tripathi, R. D., & Chakrabarty, D. (2016b). Transgenic Arabidopsis thaliana expressing fungal arsenic methyltransferase gene (WaarsM) showed enhanced arsenic tolerance via volatilization. *Environmental and Experimental Botany, 132,* 113–120.

Wang, P., Zhang, W., Mao, C., Xu, G., & Zhao, F. J. (2016). The role of OsPT8 in arsenate uptake and varietal difference in arsenate tolerance in rice. *Journal of Experimental Botany, 67*(21), 6051–6059.

Wang, F. Z., Chen, M. X., Yu, L. J., Xie, L. J., Yuan, L. B., Qi, H., Xiao, M., Guo, W., Chen, Z., Yi, K., & Zhang, J. (2017). OsARM1, an R2R3 MYB transcription factor, is involved in regulation of the response to arsenic stress in rice. *Frontiers in Plant Science, 8,* 1868.

WHO (World Health Organization). (2001). Arsenic and arsenic compounds: Environmental Health Criteria EHC 224.

Williams, P. N., Villada, A., Deacon, C., Raab, A., Figuerola, J., Green, A. J., Feldmann, J. & Meharg, A. A. (2007). Greatly enhanced arsenic shoot assimilation in rice leads to elevated grain levels compared to wheat and barley. *Environmental Science & Technology, 41*(19), 6854–6859.

Wu, Z., Ren, H., McGrath, S. P., Wu, P., & Zhao, F. J. (2011). Investigating the contribution of the phosphate transport pathway to arsenic accumulation in rice. *Plant Physiology, 157*(1), 498–508.

Xu, X. Y., McGrath, S. P., & Zhao, F. J. (2007). Rapid reduction of arsenate in the medium mediated by plant roots. *New Phytologist, 176,* 590–599.

Xu, J., Shi, S., Wang, L., Tang, Z., Lv, T., Zhu, X., Ding, X., Wang, Y., Zhao, F. J., & Wu, Z. (2017). OsHAC4 is critical for arsenate tolerance and regulates arsenic accumulation in rice. *New Phytologist, 215*(3), 1090–1101.

Ye, Y., Li, P., Xu, T., Zeng, L., Cheng, D., Yang, M., Luo, J., & Lian, X. (2017). OsPT4 contributes to arsenate uptake and transport in rice. *Frontiers in Plant Science, 8,* 2197.

Yoon, Y., Lee, W. M., & An, Y. J. (2015). Phytotoxicity of arsenic compounds on crop plant seedlings. *Environmental Science and Pollution Research, 22*(14), 11047–11056.

Yu, L. J., Luo, Y. F., Liao, B., Xie, L. J., Chen, L., Xiao, S., Li, J. T., Hu, S. N. & Shu, W. S. (2012). Comparative transcriptome analysis of transporters, phytohormone and lipid metabolism pathways in response to arsenic stress in rice (Oryza sativa). *New Phytologist, 195*(1), 97–112.

Zhao, F. J., Ago, Y., Mitani, N., Li, R. Y., Su, Y. H., Yamaji, N., McGrath, S. P., & Ma., J. F. (2010). The role of the rice aquaporin Lsi1 in arsenite efflux from roots. *New Phytologist, 186*(2), 392–399.

4

Heavy Metals' Stress Responses in Field Crops

Munish Kumar Upadhyay
Banaras Hindu University

Arnab Majumdar
Indian Institute of Science Education and Research-Kolkata

CONTENTS

4.1 Introduction

Usually, a wide range of elements is found in agricultural soil with moderate to higher concentrations. Among them, some are considered toxic while remaining are beneficial to crops/plants and organisms. These elements are referred to as heavy metals (HMs)/metalloids and essential elements respectively. These toxic elements are also categorized as non-essential to various metabolic functions in plants. HM's (Cd, Cu, Zn, Ni, Co, Cr, Pb and As) concentration is found in soil mostly due to long-term application of sewage sludge, dust from smelter's industries, phosphate fertilizer and over exploitation of groundwater for irrigation in agricultural soil (Anjum et al., 2015). Further, agricultural practices, such as uncontrolled nitrogen fertilization and water management, may also affect the mobility and bioavailability of respective HMs in agricultural soil and collectively impact their accumulation in various crop plants (Upadhyay et al., 2020; Awa and Hadibarata, 2020). This scenario is predominant in developing countries like India and China that have higher population density and increasing industrialization.

The agricultural soil can be considered a major source and sink for HMs. Thus, accumulation of HMs in crop plants always started from soil and can damage the plant subsistence. Upon HMs' exposure, plants get poorly impacted by generating reactive oxygen species (ROS) which may cause modifications in biomolecules, genes' expression at transcriptional base, cell division and differentiation and eventually leads to oxidative stress. Further, it may also interact with the antioxidant defense system (Upadhyay et al., 2022), impair the electron transport chain and imbalance the overall redox homeostasis in crop plants (Srivastava et al., 2014; Srivastav et al., 2016). Crop plants have also developed various possible mechanisms to combat the negative impacts of HMs. Crop plants start production of thiols with low-molecular-weight metal–binding ligands such as cysteine and glutathione (GSH) which are having high affinity to HMs that helps in its vacuolar sequestration (Dixit et al., 2015a). Further, a small and cysteine-rich polypeptide compound, phytochelatins (PCs), which is present in plants and animals, also plays a crucial role in detoxification process of HMs (Dixit et al., 2015b). GSH acts as a substrate for PC synthesis along with PC synthase enzyme. Further, different phytohormones such as auxin, cytokinin, gibberellins, abscisic acid, salicylic acid (SA) and ethylene act as signaling molecules and regulate various metabolic processes which help to crop plants in various tolerance mechanisms against HMs (Thakur et al., 2022). Other organic acids such as citrate and malate also help in HMs' sequestration process in crop plants.

An understanding of soil and crop translocation mechanisms of HMs is essential for deciphering the sustainable

DOI: 10.1201/9781003258063-4

remediation methodologies in realistic field condition. Further, it is also an emerging issue for global concern, food security and its association with human health. Humans are also exposed to toxic metals significantly through different staple food crops such as rice, wheat and maize. Thus, by considering these facts, this chapter is a compendium of various mitigation strategies by enlightening the various physiological/molecular aspects of crop plants.

4.2 Sources and Translocation of HM Pollution from Soil to Crop Plants

HMs in agricultural soil have become a serious concern due to their adverse ecological impacts. Atmospheric deposition, irrigation with naturally contaminated groundwater, pesticides/herbicides and phosphate-based fertilizers are some major reasons of HM concentration in soil system. HM's transport occurs in different ionic forms through different transporters from soil to plant (McLaughlin et al., 2011). The crops' roots play an important role in the uptake and translocation of HMs. This uptake process of HMs is considered an energy-dependent active physicochemical process and regulated at both extracellular and intracellular levels through various physiological and molecular machineries which may also affect the essential nutrients' uptake (Hossain et al., 2012). Plant organizes different methods to influence the HMs uptake from agricultural soil by phyto-siderophores' (PSs') production, organic acids, phenolics, enzymes and acidification of the rhizospheric zone. These processes help in improving the bioavailability of HMs to the crop plants. Further, cell wall components and other extracellular parts also play a significant role in the uptake and translocation of HMs from cytoplasm to the vacuoles. This process leads to sequestration by forming HM complexes with organic acids, metallothionein and PCs which leads to altering the biochemical, molecular and other physiological activities including gene expressions and protein variation. All these processes collectively trigger the defense and tolerance system in plants under HMs' stress condition.

Various characterized transporters in different crop plants help in the uptake of HMs. These transporters can be classified as high-affinity and low-affinity transporters. Yellow strip like 2 (YSL2) transports Fe into the xylem of plant. Further, Zn^{2+} movement occurs from pericycle to xylem through OsHMA2 and AtHMA4 in rice and Arabidopsis, respectively (Yamaji et al., 2013; Hussain et al., 2004). Arabidopsis plasma membrane HM ATPase 4 (HMA4) transporter helps in xylem loading of Cd and Zn. The di-cysteine motifs also help in this process of loading and export in shoot of plant system. Numerous transporter families viz., ZIP (zinc-regulated transporter and iron-regulated transporter-like protein) family, NIP (nodulin 26-like extrinsic proteins), PHT (phosphate transporter), YSL (Yellow Stripe-like) family, COPT (copper cation transporter) family, NRAMPs (natural-resistance–associated macrophage proteins), CDF (cation diffusion facilitator), HMAs and ABC (ATP binding cassette) have been already characterized and proved to participate in the uptake and transport of different HMs. Further, distribution of HMs across the cellular organelles (Guerinot, 2000) is mainly regulated through the process of transporter kinetics.

4.3 Various Mechanisms Involved in HMs' Stress Tolerance in Crop Plants

4.3.1 *Vacuolar Sequestration*

The tolerance process involves the complexation of HMs with GSH and PCs' establishment which eventually leads to vacuolar sequestration. It causes oxidative stress and the ratio of GSH/GSSG gets disturbed which causes ROS generation. The stress tolerance can develop through HMs' homeostasis along with some genes like ZRT, ABCC and IRT-like protein (ZIP) transporters (Babula et al., 2009; Song et al., 2014; Clemens and Ma, 2016). Under normal conditions, bioavailability of mineral nutrients in rhizospheric zone is regulated through root exudation. The antioxidant defense system of plant helps to remove the ROS by altering the antioxidant defense markers' activity and concentration inside the plants. Superoxide dismutase (SOD), ascorbate peroxidase (APX), catalase, peroxidase and GSH/GSSG are considered some important enzymatic and non-enzymatic defense markers in plants (Babula et al., 2009; Yadav et al., 2021; Upadhyay et al., 2022).

4.3.2 *Production of Different Organic Acids*

Organic acids are considered crucial players in homeostasis of cellular metabolism and HMs' stress tolerance (Panchal et al., 2021) in plants. Citrate, oxalate, succinate and malate are some of the important organic acids which form in plant through different metabolic cycles. However, specificity of organic acid production depends on the specific HMs stress condition. For instance, Luo et al. (2014) showed the importance of oxalic acid, threonic acid, phosphoric acid and tetradecanoic acid in *Sedum alfredii* specifically in Cd stress condition. De La Luz Mora et al. (2009) showed the involvement of oxalate and citrate to alleviate the Mn stress in rye grass plant. Additionally, Chen et al. (2015) also performed higher root malate exudation to tackle Mn toxicity. Further, the role of malate and succinate was established in different field-grown crops under different HMs stress condition. The chelating property of citrate, oxalate and malate was also established earlier and organic-acid–mediated stress tolerance mechanism in plant results higher mineral assimilation at rhizospheric soil indices.

4.3.3 *Phyto-Siderophore Biosynthesis*

Phyto-siderophores (PSs) are considered chelating agent which are produced by plants' roots under Fe deficiency. It is a well-known stress tolerance mechanism of plants to squeeze more Fe under Fe-limiting condition. However, apart from Fe, PS also helps to chelate other elements like Zn, Ni, Cu and Cd. The root exudates of plants are involved in bioavailability of Fe in rhizospheric zone. The Fe bioavailability is also found to be reduced under HMs stress condition (Upadhyay et al., 2020). Apart from this, many other primary and secondary metabolites are also secreted which collectively attracts the microorganism which plays a greater role in dissolution and complexation process (Asgari Lajayer et al., 2017). Thus, all these processes in simultaneous manner make Fe bioavailable to plants. Further, nicotianamine synthase and nicotianamine aminotransferase are important enzymes responsible for the synthesis of PS and their activity was also noticed to be increased under Fe-limiting condition (Thakur et al., 2022).

4.4 Strategies to Regulate HM Pollution in Crop Plants

HM pollution in the agricultural soil causes a serious problem of crop loss and environmental deterioration by compromising the food chain and in turn negatively affecting the consumer health issues (Upadhyay et al., 2019). Depending on the metals involved, the degree of toxicity varies in plants. Some of the metals are considered 'broadline metals' (Fe, Mn, Cu, Zn, Ni, Pb, Cd) and Class A metals (Na, K, Mg, Ca, AL) can variably bind to the different chemical ligands and groups found in plant sap and can become toxic if accumulated more than the needed dosage (Inouhe et al., 2012). On the other hand, naturally toxic elements like As, Hg etc. are accumulated in the crops depending on both geogenic and anthropogenic reasons (Majumdar and Bose, 2017). To minimize the impact of such metal pollution in crop plants, several mitigation strategies, as mentioned below, are in practice which need to be implemented on large scale.

4.4.1 *Application of Plant Bio-Regulators*

The mitigation of HMs and enhancement of crop production can be managed together by using some chemical substances that are termed 'plant bio-regulators' (PBRs) (Srivastava et al., 2016a). Low External Input and Sustainable Agriculture system allows such chemical implementation in lower amount during foliar treatments other than the priming process where greater amount of PBRs is used both during the seed soaking as well as foliar treatments (Srivastava et al., 2016). Common PBRs can be divided into two groups – chemical and hormonal. Among these chemical forms, thiourea (TU), silicic acid (Si), potassium nitrate, putrescine, hydrogen peroxide, nitric oxide and hydrogen sulfide are common (Upadhyay et al., 2021; Majumdar et al., 2019; Zheng et al., 2010; Shu et al., 2013; Ishibashi et al., 2011; Gill et al., 2013; Ali et al., 2014), whereas among hormonal PBRs, Salicylic acid (SA), gibberellic acid (GA) and epibrassinolide (BA) are common (Wu et al., 2014; Kong et al., 2014; Siddiqui et al., 2011).

TU is an organo-sulfur compound that is similar to the urea but the oxygen atom is replaced by the sulfur (Majumdar et al., 2020). In recent years, the TU is being widely used for plant abiotic stress tolerance, enhances plant's growth and yields by acting at the cellular level while scavenging the ROS like peroxide and superoxide radicals generated during metal toxicities (Upadhyay et al., 2020; Srivastava et al., 2021). TU also minimizes the ROS effects by associating with calcium and abscisic acid signaling pathways (Srivastava et al., 2010) which improves the plant-water homeostasis and sulfur metabolism (Srivastava et al., 2014) by restricting HM toxicity. Silicon (Si) or soil silica gene rally remains as polymers of silicate minerals or as lattice form, and due to weathering or chemical redox changes or biological activities, it gets converted into water-soluble ortho/monosilicic acid (H_4SiO_4) (Majumdar et al., 2020). Si is one such important element which is required for strengthening plant's internal ultra-structure. Crops can uptake Si to a greater extent via several transporter proteins if there is a limit of soil As (Majumdar and Bose, 2018). High Si accumulation in crop plants can confer the tolerance ability toward radiation, salt stress, drought, high temperature/freezing, HM toxicity or even other nutrient imbalance (Srivastava et al., 2016). Application of Si to reduce the metal toxicity is also associated with other redox signaling processes and hormonal crosstalk (Farooq et al., 2013). Among the essential nutrients, potassium (K) is an abundant element in plants and it plays role in managing multiple aspects of molecular to physiological changes due to any cellular stress occurred. Cellular turgor pressure, cell elongation, osmoregulation, phloem solute transport, leaf to stomatal transportations, ionic balance, enzymatic activation, pH maintenance in the cytosol etc. are some of the crucial platforms where K regulates the cellular homeostasis and stress management (Srivastava et al., 2016). Polyamines (PAs) including diamine putrescine (Put^{2+}), which is a precursor molecule for other PAs, are straight chain low-molecular-weight aliphatic amines. PAs have a polycationic nature at cellular pH range which allows these molecules to interact with nucleic acids, proteins, cell wall and membrane phospholipids to maintain the cellular homeostasis. PAs are also being researched for the specific anti-stress properties resulting in ROS management. Among the common PA synthetic genes, arginine decarboxylase, S-adenosyl methionine decarboxylase, Spd synthase and ornithine decarboxylase were over-expressed and transfected to promote the stress tolerance against HMs and other abiotic stresses (Marco et al., 2012). PAs are successfully tested against salinity, high/low temperature, drought stress and proved to be effective in crop productivity (Roychoudhury et al., 2011; Radhakrishnan and Lee, 2013; Parvin et al., 2014). Hydrogen peroxide (H_2O_2) was earlier demonstrated as a toxic metabolic by-product of aerobic system, but later, it has been established that H_2O_2 can act as a secondary messenger triggering different plant responses under stressed conditions (Bailey-Serres and Mittler, 2006). Other stress conditions like HMs or salinity are better regulated with exogeneous application of H_2O_2 (Hu et al., 2009; Ishibashi et al., 2011). The targeted modules of H_2O_2 are calcium homeostasis, MAP kinases, ion channel proteins, several transcription factors and phosphatases (Dietz, 2014; Schmidt and Schippers, 2015). H_2O_2 also influences crosstalk between auxin, PA, nitric acid and salicyclic acid signaling pathways. Such triggered pathways are resulted due to exogeneous H_2O_2 application and enhanced cell cycle modulation (Ma et al., 2014). Nitric oxide (NO) is a redox active chemical agent which acts as a secondary messenger like H_2O_2 to modulate different physiological and cellular activities in plants (Yu et al., 2014). Primarily, NO works during the activation of heme oxygenase signaling which triggers root formation or myo-inositol phosphate synthase signaling which improves cellular metabolism and plant growth (Srivastava et al., 2016). NO regulates iron metabolism and homeostasis in root cells by modulating the iron reductase enzymes and iron transporter proteins

(Meiser and Bauer, 2012). In some other reports, NO helps in maintenance of K^+ balance under stressed conditions as well as inactivating the NADPH oxidase to minimize the oxidative damage and activating APX-1 enzyme (Yun et al., 2011; Correa-Aragunde et al., 2013). Hydrogen sulfide (H_2S) is a well-known toxic substance although, in plant cells, regulated amount of H_2S can be a signaling molecule and also considered gasotransmitter like NO and CO. Endogenous H_2S can alter the cellular redox status toward both the reduced and oxidized conditions (Calderwood and Kopriva, 2014). External application of H_2S as sodium hydrosulfide (NaHS) helps in decreased calcium influx in plant cells via H_2S-induced plasma membrane–bound Ca-channels (Sun et al., 2013).

Among hormonal PBRs, brassinosteroids (BRs) are known as a steroid category which help in regulation of plant's growth, differentiation, leaf bending and epinasty, root and stem elongation, xylem differentiation, tolerance toward abiotic and biotic stresses conditions (Vriet et al., 2012). Signaling of active BRs can further trigger the NADPH oxidase expression to modulate the H_2O_2 generation and upregulate the NO production (Cui et al., 2011) and both of these secondary agents can induce plant stress tolerance. Other plant hormones like auxin, abscisic acid, PAs and SA functions are also regulated by these BRs (Choudhary et al., 2010). Reports also suggest that a combined application of BR with salicylate or PAs can provide much greater metal or other abiotic stress tolerance in plants (Hayat et al., 2012). SA is a phenolic compound which has been researched as a plant hormone in recent decades. SA can certainly add pathogen resistance value to the plant species although, in recent reports, it is now proved to be a signal transducer for promoting plant tolerance toward variable abiotic stresses (Miura and Tada, 2014). Within the cellular and nuclear activities, SA modulates different transcriptional regulator and also alter other protein functions involved in photosystem II and mitochondrial oxidases to minimize ROS accumulation (Durner and Klessig, 1996). At higher concentration of SA, although, plant growth can get hindered and it can promote the senescence due to the reduction of cellular antioxidant ability. The balanced content of SA can effectively control metallic stress tolerance, regulate flowering and influence the crop yield (Kong et al., 2014). Gibberellic acid (GA) is a well-known plant hormone that promotes growth in flowering plants. GA-stimulated transcript (GAST1) is one of the GA-induced genes which contains cysteine in its product which helps in the reduction of H_2O_2 and ROS-dependent signaling resulting in the tolerance against abiotic stresses (Rubinovich and Weiss, 2010). GA can also enhance the signaling activity and accumulation of SA along with a crosstalk with jasmonic acid pathway. GA triggers the degradation of DELLA protein which restrict plant growth and hence GA enhances the plant elongation. Also, exogeneous GA application has been reported to increase the crop yield (Siddiqui et al., 2011).

4.4.2 *Elemental-Application–Based Management Strategy*

The agro-ecosystem has been found to play a determining role in regulating HMs' concentrations in various crops (Upadhyay et al., 2020). Hence, management of local agricultural fields along with the identification of a suitable cultivar of different crops can also achieve the goal of reducing HMs' concentration. In this direction, several approaches have been already attempted. A reasonable effect of silica (Si) addition on As uptake has been tried as both follow the same transporters in rice. Li et al. (2009) conducted a pot experiment and tested the impact the Si (in the form of silica gel, SiO_2) on As concentration. The As concentration reduction was observed in both straw and grain total As. Further the inorganic As and DMA (di-methylarsinic acid) in rice grains were observed to be reduced and increased respectively. Majumdar et al. (2019) also conducted an experiment in natural As-contaminated fields in West Bengal, India and showed the positive impact of Si on growth and overall physiology of rice plant. It was also proved to be reduced As concentration significantly. Seyfferth and Fendorf (2012) evaluated the influence of two silicate minerals (diatomaceous earth and silica gel) on As concentrations in rice and found conflicting results. While diatomaceous earth Si addition either did not change or increased As in rice grains, Si gel addition decreased As in rice grains owing to adaptable mineralogy. Similarly, Lin et al. (2017) demonstrated that external silica supplementation could reduce the Cd accumulation and improve tolerance in rice plant. The involved mechanism was related with antioxidant enzyme activity and *Lsi1* expression. However, some studies are also conducted on the negative impact of Si also which are contrary to these positive reports. For instance, Lee et al. (2014) studied Si application in field conditions as per the recommended application rate of Si fertilizer and As and Si concentrations were found to be increased in soil solution. It leads to increase the As accumulation rate. Therefore, Si application should be optimized extensively especially in field conditions as this may have different impacts on fields having different conditions.

The phosphate application was also verified by several workers owing to potential of phosphate to replace adsorbed As(V) or As(III) from soil and iron plaque on roots. Iron is an important element in determining As bioavailability in rice fields. In the form of iron oxides/hydroxides, it strongly influences As biogeochemistry. While on the surface of rice roots, iron plaque formation plays a crucial role in As availability for uptake by plants. Li et al. (2022) found that addition of phosphorus (P) fertilizer with biochar alleviated Cd toxicity in maize plants. However, this strategy should be further tested at field scale for the practical approach.

Sulfur (S) helps to detoxify arsenic (As) through complexation of arsenite (AsIII) with thiol-rich peptides. Plants deficient in sulfur have a greater root-to-shoot translocation of As (Zhang et al., 2011). Further, S can influence the iron plaque formation (Hu et al., 2007) and thus affect As levels. The uptake and subsequent distribution of S is regulated in response to its demand and various environmental factors (Talukder et al., 2011). Dixit et al. (2015a) conducted a study showing higher As accumulation pattern in root along with reduced translocation into shoot upon higher S application while antagonistic result was noticed by applying low concentration of S. Thus, study revealed that S plays a crucial role in As chelation and restricts its entry to shoot. Similar phenomenon was also observed in previous studies (Zhang et al., 2011; Fan et al., 2013). Zhang et al. (2016) found that S application decreased As concentration in rice grains by 44% and that too of As(V) and As(III) while no effect was seen in DMA concentrations. It helps to reduce the As accumulation by modulating proteome, amino acids and thiol metabolism in rice leaves which may also be correlated with the changes in expression of a number of genes including transporters (Dixit et al., 2015b; Srivastava et al., 2016). Similarly, S supply found to be significantly reduced the Cd concentration in rice grain through enhances antioxidant defense and photosynthesis system (Liu et al., 2020).

Recently Huang et al. (2021) proved that use of Se and Si collectively alleviates the Cd toxicity in rice plant through enhanced GSH, PC content along with downregulation of *OsHMA2* and upregulation of *OsNrampl* and *OsMHA3* genes.

Thus, these elemental fertilizations may have potential to minimize HMs accumulation and toxicity in different plants during different field or pot experiments.

4.4.3 *Phyto and Eco-Remediation Strategies*

Phytoremediation is one of the most common and efficient approaches to remove soil pollutants, being organic or inorganic or radioactive in nature. With the proper disposal mechanism phytoremediation aspects can be a great virtue (Majumdar et al., 2018). Several macro and micro hydrophyte plants can be considered a potential phytoremediating agent those are either water emergent or submergent or wetland plants. Some of the hyperaccumulator plants are able to take up pollutant content 10–500 times greater than any other non-accumulator plants. Several categorical phyto-applications are known with specific terminology. Phytoextraction is dependent on the absorption of dispersed/dissolved organic or inorganic pollutants through roots and sprigs of the associated plants with an ability to produce high biomass for the storage of the contaminants like *Thlaspi* sp., *Morus* sp. (Suman et al., 2018). The effectiveness of the phytoextraction is assessed based on the ratio of the metal(loid)s accumulated in plant to the soil content. Further, the plants can be incinerated and used for the landfill as a disposal (Awa and Hadibarata, 2020). Precipitating HMs on the root surface or in the rhizospheric zone and restricting the metal mobility in the soil by the associated plants are termed phytostabilization (Basharat et al., 2018). In this process, the elemental redox status is changed by the enzymes secreted by the plant root system making the soil contaminant less toxic (Ali et al., 2013). Plants like *Deschampsia* sp., *Picea* sp. and *Typha* sp. are used for phytostabilization and effective in preventing the seepage of metals to the groundwater (Sarkar et al., 2017). Phytovolatization, is a process in which HMs, transoports inside the plants through xylem and converts into less toxic forms and eventually release into aerosphere. Metals like selenium and mercury can be phytovolatilized (Cristaldi et al., 2017), although heat, light intensity and transpiration rates can influence effectiveness of this process. From surface water bodies, contaminants can be removed effectively by rhizofiltration process, another means of phytoremediation, where roots and submerged rhizomes accumulate the majority of the pollutants. Plants like *Eichornia* sp. and *Helianthus* sp. can remove arsenic, mercury, uranium as much as 90% or more using such filtration (Majumdar et al., 2018; Awa and Hadibarata, 2020). These are some of the primary modes of phytoremediation followed by some other secondary means like rhizodegradation and phytodesalination. The former indicates the removal process of some organic pollutants by a combined action of released plant root exudates and microbial community nearby, whereas the latter shows the removal of excess salts from saline soils by plants (Awa and Hadibarata, 2020). To enhance the phytoremediation ability, some exogenous amendments are also getting popularized including biochars and nanoparticles (NPs) of different origins. Biochar is a pyrolyzed, low-cost product which contains mainly carbon and a few other chemical groups with a high adsorption ability due to the porous structure (Gholami et al., 2020; Hussain et al., 2022). Due to the association of COOH, OH-, -CO- and ester groups to the biochars, the adsorption increases with reduced leaching of metal(loid)s from that complex (Ghosh and Maiti, 2021). With a high pH range of around 8–11 and cation exchange capacity, this substance can also reduce the metal mobility from the soil complex and makes available to the plants for uptake. Hence, this biochar amendment is practically beneficial for the phytoremediation process to make it more enhanced and applicable at a large scale. For aquatic systems also, application of biochar can coagulate the pollutants and make available to the plant root system for uptake indicating the potential of biochar in both soil-water interphase to trigger the phytoremediation process (Shakoor et al., 2020). Different plant-based biochars are reported to be efficient in removal of HMs up to 42% without the assistance of phytoremediation (Ghosh and Maiti, 2021). Due to the immobilization of soil elements, biochar also helps in gathering the soil nutrients to promote plant growth, and with association of any suitable phytoremediator plant, this combination can efficiently remove a considerable amount of soil pollutants (Wiszniewska et al., 2016). Likewise, another exogenous substance is NPs of different origin to promote the phytoremediation technique (Zhou et al., 2021). With increasing industrialization, the contamination density and severity are also growing predominantly with HM contamination. Due to the effective surface area and minute size preference, the NPs are efficient in retaining a bulk content of contaminants through the surface adsorption process (Song et al., 2019). Among all the NPs, iron NP or zero-valent FeNP (or nZVI) has a diversified application with a high preference due to structural uniqueness and electronic properties. From terrestrial ornamental phytoremediating plants to aquatic macrophytes, all are combined with this nZVI to enhance the phytoremediation potential (Majumdar et al., 2021a; Zhu et al., 2018). Along with the accumulation of HMs to the surface, FeNP can also trigger the plant's oxidative defense system and minimize the ROS activities by increasing the SOD and GPX enzymatic performances (Zhou et al., 2021). Among other NPs, cerium dioxide (CeO_2), titanium dioxide (TiO_2), zinc oxide (ZnO), selenium and silicon (Se-Si) NPs are in use for similar goal with different degree of potentiality and target pollutants and associated plants for phytoremediation (Song et al., 2019).

4.4.4 *Microbial Amendment*

Microbial participations in soil pollutant remediation have been a long-term research agenda and many of the microbial genera like *Bacillus*, *Pseudomonas*, *Alcaligenes*, *Mycobacterium*, *Acidovorux* and *Rhodococcus* are mentioned in many reports for their potential influence during the course of phytoremediation (Sharma, 2021). Soil microbes can profusely modulate the plant-growth-promoting substances and generate organic acids and polysaccharides to confined contaminants in the soil system. Depending on the nature of the pollutants, microbes can change the redox status of the HMs while interacting under natural pH, temperature and moisture content (Majumdar et al., 2021b). *Bacillus*, *Escherichia* and *Mycobacterium* are reported frequently for their ability of breaking polycyclic aromatic hydrocarbons (PAH) and other HMs and can remove a considerable content of metals (about 17%–98%). Fungi like Basidiomycetes, Streptomycetes, Deuteromycetes and Actinomycetes can perform a wide range of PAH degradation (Yang et al., 2020). Negatively charged surface of microbial cells can adsorb cationic metal ions via ion exchange process whereas intracellular bindings of such metals are associated with protein molecules (Liu et al., 2017). Microbial cells are also reported to remove HMs by adsorbing to the extracellular polymeric substances (EPS) including proteinaceous, carbohydrates and complex organic molecules and accumulate at the rhizosphere (Pulsawat et al., 2003). Other than the EPS, microbes can produce variable amino acids and organic acids to dissolve the HM compounds to further alkylate, leach out and transport to either intracellular compartments or excrete extracellularly (Yang et al., 2020). Redox status alteration of HMs and other organic pollutants by soil microbes are frequently opted for the remediation process where polyvalent metal ions are converted to a lower valency state with lesser toxicity. *Bacillus*, *Pseudomonas*, *Arthrobacter*, *Rhizopus* and *Cymodocea* are well known for such redox transformation (Gadd, 2000). Microbial metabolites are secreted to chelate metal ions and further taken up by phytoremediator plants. Pyoverdine, pyochelin and desferrioxamine B-E are released by *Pseudomonas* and *Streptomyces* genera to chelate Cd, Cr and Pb and taken up by plants (Braud et al., 2009). Oxalate, citrate, tartarate, formate, acetate, gluconate and 2-ketogluconate are released by *Burkholderia*, *Gluconacetobacter* and *Pseudomonas* genera to immobilize Zn, Cd, Pb, Fe, Ni and Cu in the soil (Rajkumar et al., 2012). Such microbial secretions can trigger plant's growth while restricting the pollutant mobilization. In recent reports, genetically modified microbes are also studied for the remediation approach with further phytoremediation clean-up. Genetic modifications are performed to enhance the bacterial organic contaminant degradation capacity as well as high adsorption of metallic ions to the super-charged cell surfaces which allow greater accumulation rate of soil pollutants in a confined system (Sharma, 2021).

4.4.5 *Omics (Genomics, Metabolomics and Proteomics) Approaches*

Plants, that are associated with HM tolerance, are either genetically made up naturally or engineered to cope up with the toxicity. Several evidences are reported where the pivotal role of genomics and gene products has been confirmed. In the case of classical Mendelian genetics, hyperaccumulation and tolerance study toward Zn and Cd can be observed by crossing two different species of the same genus of *Arabidopsis helleri* and *Arabidopsis lyrate* where the progeny plants show segregated traits of hyperaccumulation and tolerance independently for both Cd and Zn (Bert et al., 2003). Mutants, on the other hand, which are experimentally created to boost some specific activity, can also give similar result as in the case of *A. thaliana*, T-DNA-inserted AtHMA4-mutant shows a lower accumulation of Zn and Cd in the leaves with higher accumulation in the roots compared to the wild-type variety, indicating an alternate way of resistance in plants (Verret et al., 2004). In the case of arsenic-contaminated soil, plants with dominant quantitative trait loci have shown greater tolerance than any wild-type variety progeny (Majumdar and Bose, 2018). Several genomic studies further investigated at the translated or protein level to find out the responsiveness of engineered plants toward HM pollutants (Maestri and Marmiroli, 2012; van de Mortel et al., 2008). Metagenomics analysis is also an important tool to understand the relation between plant-microbial interactions where plant-specific microbial community can be identified and related microbial inoculum can be used for plant-growth promotion and metal-uptake enhancement (Jaskulak, 2018). Recent advancement suggests that transcriptomics analysis can accurately indicate the plant gene/protein combinations effective against the pollutant transformation or reduction. Study provided evidence of mRNA abundance and atrazine-degrading activity by specific rhizospheric plant micribiome through transcriptomic analysis. (Bell et al., 2014). The same report also provided other substantial evidence of some antibiotic-resistant and biofilm-forming genes that were identified as hydrocarbon biodegradation activity in the willow root rhizosphere during a transcriptomic profiling indicating the interspecies interaction of microbial community. To identify a specific phytoremediator plant for any specific metal contaminated area, the plant transcriptomic profiling is also useful to understand the tolerance ability of that plant. In *Brassica* sp., such transcriptomics study helps to reveal the gene-specific products for sulfur- and chromium-contaminated environment along with the Cd-responsive genes in the hyperaccumulator *Sedum* sp. (Bell et al., 2014). Some plants can hydroxylate polychlorinated bi-phenyls, and using the transcriptomics analysis, those genes can be identified (Zhai et al., 2010). Proteomics can act as important device for deciphering the molecular mechanisms of interactions between HMs and differentcrop plants along with metabolite profiling For identification of physiological and molecular changes under Cd stress, proteomics study was used in *Populus* sp., and the result showed that the plant responses altered between tissues and leaves where different metabolite profiling was found under such stresses conditions (Durand et al., 2010). Hence, with gradual advancements, omics-techniques are becoming an inevitable part of the molecular studies of phytoremediation where soil-plant-microbial trilogy can be accurately understood.

4.5 Conclusions and Future Perspective

HMs mediated injury in crop plants possessing a significant threat. Further, their entry into the food chain may cause a negative impact on human health. Many studies have been conducted to understand the molecular mechanisms of HMs' uptake, translocation and detoxification in Arabidopsis and rice. The understanding of regulatory mechanisms of HM transport is still in developing phase. Interactions between HMs and nutrients and the substrate specificity of transporters are also paths for further research. Thus, it is important to identify and bind the sources of HMs pollution and also adapt new strategies which may help to mitigate the adverse effects of HMs in plants.

REFERENCES

Ali, H., Khan, E. and Sajad, M.A., 2013. Phytoremediation of heavy metals: Concepts and applications. *Chemosphere*, 91(7), pp. 869–881.

Ali, B., Song, W.J., Hu, W.Z., Luo, X.N., Gill, R.A., Wang, J. and Zhou, W.J., 2014. Hydrogen sulfide alleviates lead-induced photosynthetic and ultrastructural changes in oilseed rape. *Ecotoxicology and Environmental Safety*, 102, pp. 25–33.

Anjum, N.A., Singh, H.P., Khan, M.I.R., Masood, A., Per, T.S., Negi, A., Batish, D.R., Khan, N.A., Duarte, A.C., Pereira, E. and Ahmad, I., 2015. Too much is bad: An appraisal of phytotoxicity of elevated plant-beneficial heavy metal ions. *Environmental Science and Pollution Research*, 22(5), pp. 3361–3382.

Awa, S.H. and Hadibarata, T., 2020. Removal of heavy metals in contaminated soil by phytoremediation mechanism: A review. *Water, Air, & Soil Pollution*, 231(2), pp. 1–15.

Babula, P., Adam, V., Opatrilova, R., Zehnalek, J., Havel, L. and Kizek, R., 2009. Uncommon heavy metals, metalloids and their plant toxicity: A review. *Environmental Chemistry Letters*, pp. 275–317.

Bailey-Serres, J. and Mittler, R., 2006. The roles of reactive oxygen species in plant cells.

Basharat, Z., Novo, L.A. and Yasmin, A., 2018. Genome editing weds CRISPR: what is in it for phytoremediation?. Plants, 7(3), p.51.

Bell, T.H., Joly, S., Pitre, F.E. and Yergeau, E., 2014. Increasing phytoremediation efficiency and reliability using novel omics approaches. *TRENDS in Biotechnology*, 32(5), pp. 271–280.

Bert, V., Meerts, P., Saumitou-Laprade, P., Salis, P., Gruber, W. and Verbruggen, N., 2003. Genetic basis of Cd tolerance and hyperaccumulation in Arabidopsis halleri. *Plant and Soil*, 249(1), pp. 9–18.

Braud, A., Jézéquel, K., Bazot, S. and Lebeau, T., 2009. Enhanced phytoextraction of an agricultural Cr-and Pb-contaminated soil by bioaugmentation with siderophore-producing bacteria. Chemosphere, 74(2), pp.280-286.

Calderwood, A. and Kopriva, S., 2014. Hydrogen sulfide in plants: From dissipation of excess sulfur to signaling molecule. *Nitric Oxide*, 41, pp. 72–78.

Chen, Z., Sun, L., Liu, P., Liu, G., Tian, J. and Liao, H., 2015. Malate synthesis and secretion mediated by a manganese-enhanced malate dehydrogenase confers superior manganese tolerance in *Stylosanthes guianensis*. *Plant Physiology* 167, pp. 176–188.

Choudhary, S.P., Bhardwaj, R., Gupta, B.D., Dutt, P., Gupta, R.K., Biondi, S. and Kanwar, M., 2010. Epibrassinolide induces changes in indole-3-acetic acid, abscisic acid and polyamine concentrations and enhances antioxidant potential of radish seedlings under copper stress. *Physiologia Plantarum*, 140(3), pp. 280–296.

Clemens, S. and Ma, J.H., 2016. Toxic heavy metal and metalloid accumulation in crop plants and foods. *Annual Reviews of Plant Biology*, 67, pp. 489–512.

Correa-Aragunde, N., Foresi, N., Delledonne, M. and Lamattina, L., 2013. Auxin induces redox regulation of ascorbate peroxidase 1 activity by S-nitrosylation/denitrosylation balance resulting in changes of root growth pattern in Arabidopsis. *Journal of Experimental Botany*, 64(11), pp. 3339–3349.

Cristaldi, A., Conti, G.O., Jho, E.H., Zuccarello, P., Grasso, A., Copat, C. and Ferrante, M., 2017. Phytoremediation of contaminated soils by heavy metals and PAHs: A brief review. *Environmental Technology & Innovation*, 8, pp. 309–326.

Cui, J.X., Zhou, Y.H., Ding, J.G., Xia, X.J., Shi, K.A.I., Chen, S.C., Asami, T., Chen, Z. and Yu, J.Q., 2011. Role of nitric oxide in hydrogen peroxide-dependent induction of abiotic stress tolerance by brassinosteroids in cucumber. *Plant, Cell & Environment*, 34(2), pp. 347–358.

De La Luz Mora, M., Rosas, A., Ribera, A. and Rengel, Z., 2009. Differential tolerance to Mn toxicity in perennial ryegrass genotypes: Involvement of antioxidative enzymes and root exudation of carboxylates. *Plant and Soil*, 320, pp. 79–89.

Dietz, K.J., 2014. Redox regulation of transcription factors in plant stress acclimation and development. *Antioxidants & Redox Signaling*, 21(9), pp. 1356–1372.

Dixit, G., Singh, A.P., Kumar, A., Dwivedi, S., Deeba, F., Kumar, S., et al., 2015a. Sulfur alleviates arsenic toxicity by reducing its accumulation and modulating proteome, amino acids and thiol metabolism in rice leaves. *Scientific Reports*, 5, 16205.

Dixit, G., Singh, A.P., Kumar, A., Singh, P.K., Kumar, S., Dwivedi, S., et al., 2015b. Sulfur mediated reduction of arsenic toxicity involves efficient thiol metabolism and the antioxidant defense system in rice. *Journal of Hazardous Materials*, 298, 241–251.

Durand, T.C., Sergeant, K., Planchon, S., Carpin, S., Label, P., Morabito, D., Hausman, J.F. and Renaut, J., 2010. Acute metal stress in Populus tremula×P. alba (717-1B4 genotype): Leaf and cambial proteome changes induced by cadmium2+. *Proteomics*, 10(3), pp. 349–368.

Durner, J. and Klessig, D.F., 1996. Salicylic acid is a modulator of tobacco and mammalian catalases. *Journal of Biological Chemistry*, 271(45), pp. 28492–28501.

Fan, J., Xia, X., Hu, Z., Ziadi, N. and Liu, C., 2013. Excessive sulfur supply reduces arsenic accumulation in brown rice. *Plant, Soil and Environment*, 59(4), pp. 169–174.

Farooq, M.A., Ali, S., Hameed, A., Ishaque, W., Mahmood, K. and Iqbal, Z., 2013. Alleviation of cadmium toxicity by silicon is related to elevated photosynthesis, antioxidant enzymes; suppressed cadmium uptake and oxidative stress in cotton. *Ecotoxicology and Environmental Safety*, 96, pp. 242–249.

Gadd, G.M., 2000. Bioremedial potential of microbial mechanisms of metal mobilization and immobilization. *Current Opinion in Biotechnology*, 11(3), pp. 271–279.

Gang, L.I., ZHENG, M., Jianfeng, T.A.N.G., Hojae, S.H.I.M. and Chao, C.A.I., 2018. Effect of silicon on arsenic concentration and speciation in different rice tissues. Pedosphere, 28(3), pp.511-520.

Gholami, L., Rahimi, G. and Khademi Jolgeh Nezhad, A., 2020. Effect of thiourea-modified biochar on adsorption and fractionation of cadmium and lead in contaminated acidic soil. *International Journal of Phytoremediation*, 22(5), pp. 468–481.

Ghosh, D. and Maiti, S.K., 2021. Biochar assisted phytoremediation and biomass disposal in heavy metal contaminated mine soils: A review. *International Journal of Phytoremediation*, 23(6), pp. 559–576.

Gill, S.S., Hasanuzzaman, M., Nahar, K., Macovei, A. and Tuteja, N., 2013. Importance of nitric oxide in cadmium stress tolerance in crop plants. *Plant Physiology and Biochemistry*, 63, pp. 254–261.

Guerinot, M.L., 2000. The ZIP family of metal transporters. Biochimica et Biophysica Acta (BBA)-Biomembranes, 1465(1-2), pp.190-198.

Hayat, S., Maheshwari, P., Wani, A.S., Irfan, M., Alyemeni, M.N. and Ahmad, A., 2012. Comparative effect of 28 homobrassinolide and salicylic acid in the amelioration of NaCl stress in *Brassica juncea* L. *Plant Physiology and Biochemistry*, 53, pp. 61–68.

Hu, Z.Y., Zhu, Y.G., Li, M., Zhang, L.G., Cao, Z.H. and Smith, F.A., 2007. Sulfur (S)-induced enhancement of iron plaque formation in the rhizosphere reduces arsenic accumulation in rice (*Oryza sativa* L.) seedlings. *Environmental Pollution*, 147(2), pp. 387–393.

Hu, Y., Ge, Y., Zhang, C., Ju, T. and Cheng, W., 2009. Cadmium toxicity and translocation in rice seedlings are reduced by hydrogen peroxide pretreatment. *Plant Growth Regulation*, 59(1), pp. 51–61.

Huang, H., Li, M., Rizwan, M., Dai, Z., Yuan, Y., Hossain, M.M., Cao, M., Xiong, S. and Tu, S., 2021. Synergistic effect of silicon and selenium on the alleviation of cadmium toxicity in rice plants. *Journal of Hazardous Materials*, 401, p. 123393.

Hussain, D., Haydon, M.J., Wang, Y., Wong, E., Sherson, S.M., Young, J., Camakaris, J., Harper, J.F. and Cobbett, C.S., 2004. P-type ATPase heavy metal transporters with roles in essential zinc homeostasis in Arabidopsis. *The Plant Cell*, 16(5), pp. 1327–1339.

Hussain, M.M., Mohy-Ud-Din, W., Younas, F., Niazi, N.K., Bibi, I., Yang, X., Rasheed, F. and Farooqi, Z.U.R., 2022. Biochar: A game changer for sustainable agriculture. In E. Lichtfouse, M. Navarrete, P. Debaeke, S. Véronique and C. Alberola (eds), *Sustainable Agriculture* (pp. 143–157). Springer, Cham.

Inouhe, M., Huang, H., Chaudhary, S.K. and Gupta, D.K., 2012. Heavy metal bindings and their interactions with thiol peptides and other biological ligands in plant cells. In Metal toxicity in plants: perception, signaling and remediation (pp. 1-21). Springer, Berlin, Heidelberg.

Ishibashi, Y., Yamaguchi, H., Yuasa, T., Iwaya-Inoue, M., Arima, S. and Zheng, S.H., 2011. Hydrogen peroxide spraying alleviates drought stress in soybean plants. *Journal of Plant Physiology*, 168(13), pp. 1562–1567.

Jaskulak, M., 2018. Implementation of omics research to enhance phytoremediation efficiency-a review. *Inżynieria i Ochrona Środowiska,* 21.

Kong, J., Dong, Y., Xu, L., Liu, S. and Bai, X., 2014. Effects of foliar application of salicylic acid and nitric oxide in alleviating iron deficiency induced chlorosis of *Arachis hypogaea* L. *Botanical Studies*, 55(1), pp. 1–12.

Lajayer, B.A., Ghorbanpour, M. and Nikabadi, S., 2017. Heavy metals in contaminated environment: Destiny of secondary metabolite biosynthesis, oxidative status and phytoextraction in medicinal plants. *Ecotoxicology and Environmental Safety*, 145, pp. 377–390.

Lee, C.H., Huang, H.H., Syu, C.H., Lin, T.H. and Lee, D.Y., 2014. Increase of As release and phytotoxicity to rice seedlings in As-contaminated paddy soils by Si fertilizer application. Journal of hazardous materials, 276, pp.253-261.

Lin, H., Fang, C., Li, Y., Lin, W., He, J., Lin, R. and Lin, W., 2017. Cadmium-stress mitigation through gene expression of rice and silicon addition. *Plant Growth Regulation*, 81(1), pp. 91–101.

Li, J., Zhang, S. and Ding, X., 2022. Biochar combined with phosphate fertilizer application reduces soil cadmium availability and cadmium uptake of maize in Cd-contaminated soils. Environmental Science and Pollution Research, 29(17), pp.25925-25938.

Liu, S.H., Zeng, G.M., Niu, Q.Y., Liu, Y., Zhou, L., Jiang, L.H., Tan, X.F., Xu, P., Zhang, C. and Cheng, M., 2017. Bioremediation mechanisms of combined pollution of PAHs and heavy metals by bacteria and fungi: A mini review. *Bioresource Technology*, 224, pp. 25–33.

Liu, J., Hou, H., Zhao, L., Sun, Z. and Li, H., 2020. Protective Effect of foliar application of sulfur on photosynthesis and antioxidative defense system of rice under the stress of Cd. *Science of the Total Environment*, 710, p. 136230.

Ma, F., Wang, L., Li, J., Samma, M.K., Xie, Y., Wang, R., Wang, J., Zhang, J. and Shen, W., 2014. Interaction between HY1 and H_2O_2 in auxin-induced lateral root formation in Arabidopsis. *Plant Molecular Biology*, 85(1–2), pp. 49–61.

Maestri, E. and Marmiroli, M., 2012. Genetic and molecular aspects of metal tolerance and hyperaccumulation. In D.K. Gupta and L.M. Sandalio (eds), *Metal Toxicity in Plants: Perception, Signaling and Remediation* (pp. 41–63). Springer, Berlin, Heidelberg.

Majumdar, A. and Bose, S., 2017. Toxicogenesis and metabolism of arsenic in rice and wheat plants with probable mitigation strategies. In R. Knežević (ed.), *Arsenic: Risks of Exposure, Behavior in the Environment and Toxicology* (pp. 149–166). Nova Science Publishers, Hauppauge, New York.

Majumdar, A. and Bose, S., 2018. A glimpse on uptake kinetics and molecular responses of arsenic tolerance in Rice plants. In M. Hasanuzzaman, K. Nahar and M. Fujita (eds), *Mechanisms of Arsenic Toxicity and Tolerance in Plants* (pp. 299–315). Springer, Singapore.

Majumdar, A., Barla, A., Upadhyay, M.K., Ghosh, D., Chaudhuri, P., Srivastava, S. and Bose, S., 2018. Vermiremediation of metal (loid) s via Eichornia crassipes phytomass extraction: a sustainable technique for plant amelioration. *Journal of Environmental Management*, 220, pp. 118–125.

Majumdar, A., Upadhyay, M.K., Kumar, J.S., Barla, A., Srivastava, S., Jaiswal, M.K. and Bose, S., 2019. Ultra-structure alteration via enhanced silicon uptake in arsenic stressed rice cultivars under intermittent irrigation practices in Bengal delta basin. *Ecotoxicology and Environmental Safety*, 180, pp. 770–779.

Majumdar, A., Kumar, J.S. and Bose, S., 2020. Agricultural water management practices and environmental influences on arsenic dynamics in rice field. In S. Srivastava (ed.), *Arsenic in Drinking Water and Food* (pp. 425–443). Springer, Singapore.

Majumdar, A., Upadhyay, M.K., Giri, B., Srivastava, S., Srivastava, A.K., Jaiswal, M.K. and Bose, S., 2021b. Arsenic dynamics and flux assessment under drying-wetting irrigation and enhanced microbial diversity in paddy soils: A four year study in Bengal delta plain. *Journal of Hazardous Materials*, 409, p. 124443.

Majumdar, A., Upadhyay, M.K., Ojha, M., Afsal, F., Giri, B., Srivastava, S. and Bose, S., 2021a. Enhanced phytoremediation of Metal (loid) s via spiked ZVI nanoparticles: An urban clean-up strategy with ornamental plants. *Chemosphere,* p. 132588.

Marco, F., Alcázar, R., Altabella, T., Carrasco, P., Gill, S.S., Tuteja, N. and Tiburcio, A.F., 2012. Polyamines in developing stress-resistant crops. *Improving Crop Resistance to Abiotic Stress,* pp. 623–635.

McLaughlin, M.J., Smolders, E., Degryse, F. and Rietra, R., 2011. Uptake of metals from soil into vegetables. In Dealing with contaminated sites (pp. 325-367). Springer, Dordrecht.

Meiser, J. and Bauer, P., 2012. Looking for the hub in Fe signaling. *Plant Signaling & Behavior,* 7(6), pp. 688–690.

Miura, K. and Tada, Y., 2014. Regulation of water, salinity, and cold stress responses by salicylic acid. *Frontiers in Plant Science*, 5, p. 4.

Mora, M.D.L.L., Rosas, A., Ribera, A. and Rengel, Z., 2009. Differential tolerance to Mn toxicity in perennial ryegrass genotypes: Involvement of antioxidative enzymes and root exudation of carboxylates. *Plant and Soil*, 320(1), pp. 79–89.

Panchal, P., Miller, A.J. and Giri, J., 2021. Organic acids: Versatile stress-response roles in plants. *Journal of Experimental Botany*, 72(11), pp. 4038–4052.

Parvin, S., Lee, O.R., Sathiyaraj, G., Khorolragchaa, A., Kim, Y.J. and Yang, D.C., 2014. Spermidine alleviates the growth of saline-stressed ginseng seedlings through antioxidative defense system. *Gene*, 537(1), pp. 70–78.

Pulsawat, W., Leksawasdi, N., Rogers, P.L. and Foster, L.J.R., 2003. Anions effects on biosorption of Mn (II) by extracellular polymeric substance (EPS) from Rhizobium etli. *Biotechnology Letters*, 25(15), pp. 1267–1270.

Radhakrishnan, R. and Lee, I.J., 2013. Spermine promotes acclimation to osmotic stress by modifying antioxidant, abscisic acid, and jasmonic acid signals in soybean. *Journal of Plant Growth Regulation*, 32(1), pp. 22–30.

Rajkumar, M., Sandhya, S., Prasad, M.N.V. and Freitas, H., 2012. Perspectives of plant-associated microbes in heavy metal phytoremediation. *Biotechnology Advances*, 30(6), pp. 1562–1574.

Roychoudhury, A., Basu, S. and Sengupta, D.N., 2011. Amelioration of salinity stress by exogenously applied spermidine or spermine in three varieties of indica rice differing in their level of salt tolerance. *Journal of Plant Physiology*, 168(4), pp. 317–328.

Rubinovich, L. and Weiss, D., 2010. The Arabidopsis cysteine-rich protein GASA4 promotes GA responses and exhibits redox activity in bacteria and in planta. *The Plant Journal*, 64(6), pp. 1018–1027.

Sarkar, S.R., Majumdar, A., Barla, A., Pradhan, N., Singh, S., Ojha, N. and Bose, S., 2017. A conjugative study of Typha latifolia for expunge of phyto-available heavy metals in fly ash ameliorated soil. *Geoderma*, 305, pp. 354–362.

Schmidt, R. and Schippers, J.H., 2015. ROS-mediated redox signaling during cell differentiation in plants. *Biochimica et Biophysica Acta (BBA)-General Subjects*, 1850(8), pp. 1497–1508.

Seyfferth, A.L. and Fendorf, S., 2012. Silicate mineral impacts on the uptake and storage of arsenic and plant nutrients in rice (Oryza sativa L.). Environmental Science & Technology, 46(24), pp.13176-13183.

Shakoor, M.B., Ali, S., Rizwan, M., Abbas, F., Bibi, I., Riaz, M., Khalil, U., Niazi, N.K. and Rinklebe, J., 2020. A review of biochar-based sorbents for separation of heavy metals from water. *International Journal of Phytoremediation*, 22(2), pp. 111–126.

Sharma, P., 2021. Efficiency of bacteria and bacterial assisted phytoremediation of heavy metals: An update. *Bioresource Technology,* p. 124835.

Shu, S., Yuan, L.Y., Guo, S.R., Sun, J. and Yuan, Y.H., 2013. Effects of exogenous spermine on chlorophyll fluorescence, antioxidant system and ultrastructure of chloroplasts in *Cucumis sativus* L. under salt stress. *Plant Physiology and Biochemistry*, 63, pp. 209–216.

Siddiqui, M.H., Al-Whaibi, M.H. and Basalah, M.O., 2011. Interactive effect of calcium and gibberellin on nickel tolerance in relation to antioxidant systems in *Triticum aestivum* L. *Protoplasma*, 248(3), pp. 503–511.

Song, B., Xu, P., Chen, M., Tang, W., Zeng, G., Gong, J., Zhang, P. and Ye, S., 2019. Using nanomaterials to facilitate the phytoremediation of contaminated soil. *Critical Reviews in Environmental Science and Technology*, 49(9), pp. 791–824.

Song, W.Y., Yamaki, T., Yamaji, N., Ko, D., Jung, K.H., Fujii-Kashino, M., An, G., Martinoia, E., Lee, Y. and Ma, J.F., 2014. A rice ABC transporter, OsABCC1, reduces arsenic accumulation in the grain. *Proceedings of the National Academy of Sciences*, 111(44), pp. 15699–15704.

Srivastava, A.K., Suprasanna, P., Srivastava, S. and D'Souza, S.F., 2010. Thiourea mediated regulation in the expression profile of aquaporins and its impact on water homeostasis under salinity stress in Brassica juncea roots. *Plant Science*, 178(6), pp. 517–522.

Srivastava, A.K., Srivastava, S., Mishra, S., D'Souza, S.F. and Suprasanna, P., 2014. Identification of redox-regulated components of arsenate (AsV) tolerance through thiourea supplementation in rice. *Metallomics*, 6(9), pp. 1718–1730.

Srivastava, A.K., Pasala, R., Minhas, P.S. and Suprasanna, P., 2016. Plant bioregulators for sustainable agriculture: integrating redox signaling as a possible unifying mechanism. *Advances in Agronomy*, 137, pp. 237–278.

Srivastava, A.K., Pandey, M., Ghate, T., Kumar, V., Upadhyay, M.K., Majumdar, A., Sanjukta, A.K., Agrawal, A.K., Bose, S., Srivastava, S. and Suprasanna, P., 2021. Chemical intervention for enhancing growth and reducing grain arsenic accumulation in rice. *Environmental Pollution*, 276, p. 116719.

Suman, J., Uhlik, O., Viktorova, J. and Macek, T., 2018. Phytoextraction of heavy metals: a promising tool for clean-up of polluted environment? *Frontiers in Plant Science*, 9, p. 1476.

Sun, J., Wang, R., Zhang, X., Yu, Y., Zhao, R., Li, Z. and Chen, S., 2013. Hydrogen sulfide alleviates cadmium toxicity through regulations of cadmium transport across the plasma and vacuolar membranes in Populus euphratica cells. *Plant Physiology and Biochemistry*, 65, pp. 67–74.

Talukder, A.S.M.H.M., Meisner, C.A., Sarkar, M.A.R. and Islam, M.S., (2011) Effect of water management, tillage options and phosphorus status on arsenic uptake in rice. *Ecotox Environ Safety* 74, pp. 834–839.

Thakur, M., Praveen, S., Divte, P.R., Mitra, R., Kumar, M., Gupta, C.K., Kalidindi, U., Bansal, R., Roy, S., Anand, A. and Singh, B., 2022. Metal tolerance in plants: Molecular and physicochemical interface determines the "not so heavy effect" of heavy metals. *Chemosphere*, 287, p. 131957.

Upadhyay, M.K., Shukla, A., Yadav, P. and Srivastava, S., 2019. A review of arsenic in crops, vegetables, animals and food products. Food chemistry, 276, pp.608-618.

Upadhyay, M.K., Majumdar, A., Suresh Kumar, J. and Srivastava, S., 2020. Arsenic in rice agro-ecosystem: solutions for safe and sustainable rice production. *Frontiers in Sustainable Food Systems*, 4, p. 53.

Upadhyay, M.K., Majumdar, A., Barla, A., Bose, S. and Srivastava, S., 2021. Thiourea supplementation mediated reduction of grain arsenic in rice (*Oryza sativa* L.) cultivars: A two year field study. *Journal of Hazardous Materials*, 407, p. 124368.

Upadhyay, M.K., Majumdar, A., Srivastava, A.K., Bose, S., Suprasanna, P. and Srivastava, S., 2022. Antioxidant enzymes and transporter genes mediate arsenic stress reduction in rice (*Oryza sativa* L.) upon thiourea supplementation. *Chemosphere*, 292, p. 133482.

van de Mortel, J.E., Schat, H., Moerland, P.D., Van Themaat, E.V.L., Van Der Ent, S.J.O.E.R.D., Blankestijn, H., Ghandilyan, A., Tsiatsiani, S. and Aarts, M.G., 2008. Expression differences for genes involved in lignin, glutathione and sulphate metabolism in response to cadmium in Arabidopsis thaliana and the related Zn/Cd-hyperaccumulator Thlaspi caerulescens. *Plant, Cell & Environment*, 31(3), pp. 301–324.

Verret, F., Gravot, A., Auroy, P., Leonhardt, N., David, P., Nussaume, L., Vavasseur, A. and Richaud, P., 2004. Overexpression of AtHMA4 enhances root-to-shoot translocation of zinc and cadmium and plant metal tolerance. *FEBS Letters*, 576(3), pp. 306–312.

Vriet, C., Russinova, E. and Reuzeau, C., 2012. Boosting crop yields with plant steroids. *The Plant Cell*, 24(3), pp. 842–857.

Wiszniewska, A., Hanus-Fajerska, E., Muszyńska, E. and Ciarkowska, K., 2016. Natural organic amendments for improved phytoremediation of polluted soils: a review of recent progress. *Pedosphere*, 26(1), pp. 1–12.

Wu, X., Yao, X., Chen, J., Zhu, Z., Zhang, H. and Zha, D., 2014. Brassinosteroids protect photosynthesis and antioxidant system of eggplant seedlings from high-temperature stress. *Acta Physiologiae Plantarum*, 36(2), pp. 251–261.

Yadav, P., Srivastava, S., Patil, T., Raghuvanshi, R., Srivastava, A.K. and Suprasanna, P., 2021. Tracking the time-dependent and tissue-specific processes of arsenic accumulation and stress responses in rice (*Oryza sativa* L.). *Journal of Hazardous Materials*, 406, p. 124307.

Yamaji, N., Xia, J., Mitani-Ueno, N., Yokosho, K. and Feng Ma, J., 2013. Preferential delivery of zinc to developing tissues in rice is mediated by P-type heavy metal ATPase OsHMA2. *Plant Physiology*, 162(2), pp. 927–939.

Yang, Y., Liu, Y., Li, Z., Wang, Z., Li, C. and Wei, H., 2020. Significance of soil microbe in microbial-assisted phytoremediation: an effective way to enhance phytoremediation of contaminated soil. *International Journal of Environmental Science and Technology*, 17(4), pp. 2477–2484.

Yu, M., Lamattina, L., Spoel, S.H. and Loake, G.J., 2014. Nitric oxide function in plant biology: a redox cue in deconvolution. *New Phytologist*, 202(4), pp. 1142–1156.

Yun, B.W., Feechan, A., Yin, M., Saidi, N.B., Le Bihan, T., Yu, M., Moore, J.W., Kang, J.G., Kwon, E., Spoel, S.H. and Pallas, J.A., 2011. S-nitrosylation of NADPH oxidase regulates cell death in plant immunity. *Nature*, 478(7368), pp. 264–268.

Zhai, G., Lehmler, H.J. and Schnoor, J.L., 2010. Identification of hydroxylated metabolites of 3, 3′, 4, 4′-tetrachlorobiphenyl and metabolic pathway in whole poplar plants. *Chemosphere*, 81(4), pp. 523–528.

Zhang, J., Zhao, Q.Z., Duan, G.L. and Huang, Y.C., 2011. Influence of sulphur on arsenic accumulation and metabolism in rice seedlings. *Environmental and Experimental Botany*, 72(1), pp. 34–40.

Zhang, J., Zhao, C.Y., Liu, J., Song, R., Du, Y.X., Li, J.Z., Sun, H.Z., Duan, G.L. and Zhao, Q.Z., 2016. Influence of sulfur on transcription of genes involved in arsenic accumulation in rice grains. *Plant Molecular Biology Reporter*, 34(3), pp. 556–565.

Zheng, Y., Xu, X., Simmons, M., Zhang, C., Gao, F. and Li, Z., 2010. Responses of physiological parameters, grain yield, and grain quality to foliar application of potassium nitrate in two contrasting winter wheat cultivars under salinity stress. *Journal of Plant Nutrition and Soil Science*, 173(3), pp. 444–452.

Zhou, P., Adeel, M., Shakoor, N., Guo, M., Hao, Y., Azeem, I., Li, M., Liu, M. and Rui, Y., 2021. Application of nanoparticles alleviates heavy metals stress and promotes plant growth: An overview. *Nanomaterials*, 11(1), p. 26.

Zhu, S., Huang, X., Wang, D., Wang, L. and Ma, F., 2018. Enhanced hexavalent chromium removal performance and stabilization by magnetic iron nanoparticles assisted biochar in aqueous solution: mechanisms and application potential. *Chemosphere*, 207, pp. 50–59.

5

Chromium Dynamics in Soil-Plant System

Binaya Kumar Pattnaik
Symbiosis Institute of Geoinformatics (SIG), Symbiosis International (Deemed University) (SIU)

Chandan Sahu
Gangadhar Meher University

CONTENTS

5.1 Geochemistry of Chromium

Trivalent (Cr(III)) and hexavalent (Cr(VI)) forms of chromium (Cr) are the most stable form available in nature (Xia et al., 2019), although chromium varies from –2 to +6 valence state due to vacant d-orbital (Fendorf, 1995; Richard and Bourg, 1991). Intermediate redox reaction of chromium produces unstable Cr(IV) and Cr(V) (Wrobel et al., 2015). Speciation of chromium dictates the mobility, bioavailability and toxicity of chromium in the soil-plant system (Kožuh et al., 2000). Cr(III) acts as a glucose tolerance factor inside the human body which is required for proper functioning of insulin; on the contrary, Cr(VI) attracted the attention of scientists around the globe as a dangerous carcinogen. Anderson (1997) provided a detailed insight into the carcinogenic properties of Cr(VI). Danish et al. (2019) explained non-biological role of chromium in plant, which makes Cr a potentially toxic material.

Geochemistry of chromium is mostly governed by redox transformation mechanism, adsorption/desorption and precipitation/dissolution processes (Rai et al., 1989). Availability of redox couples can initiate the redox reaction in favourable pH conditions in soil. Hydrogen peroxide (H_2O_2), Fe(III) oxyhydroxide, Mn-oxides, dissolved oxygen and gaseous oxygen act as a probable electron donor for the oxidation of Cr(III) (Fantoni et al., 2002; Schroeder and Lee, 1975). There are uncertain evidences on the role of Fe(III) oxyhydroxide in the oxidation of Cr(III), which needs more in-depth investigation. Hydrogen peroxide oxidizes the Cr(III) at a very fast rate, as it is a strong oxidant (Pettine and Millero, 1990). Ball and Izbicki (2004) correlated high Cr(VI) concentration in groundwater of the western Mojave Desert, California with high concentration of dissolved oxygen. On the contrary, oxidation of Cr(III) to Cr(VI) by dissolved oxygen is a very slow process (Cranston and Murray, 1978). Mn oxide plays a crucial role in the oxidation of Cr(III) (Bartlett and James, 1979; Fendorf and Zasoski, 1992). In oxidizing condition of soil environment hexavalent chromium gets mobilized as CrO_4^{2-} and Mn(IV) get precipitated as MnO_2; but in reducing condition, trivalent chromium gets precipitated as $Cr(OH)_3$ and Mn gets mobilized as Mn^{2+}. Thus Mn and Cr demonstrate an inverse relationship in the context of solubility pattern in soil environment (Bartlett and James, 1979). Eary and Rai (1987) identified that oxidation of Cr(III) can take place on the surface of β-MnO_2 mineral within the pH range of 3–10.1; in addition to that, δ-MnO_2 (birnessite) and α-MnO_2 (cryptomelane) can also provide high surface energy for the oxidation of Cr(III). Trivalent chromium is found attached to the mineral surface which gets oxidized in the presence of Mn and gets converted to hexavalent form. During this process desorption of Cr(VI) and precipitation of Mn takes place (Schroeder and Lee, 1975). Richard and Bourg (1991) found traces of Mn mineral precipitated at site of oxidation of Cr(III) on mineral surface. Amacher and Baker (1979) explained the process of oxidation of Cr(III) by Mn through the following equation:

$$Cr^{3+} + 1.5\delta MnO_2(s) + H_2O(l) = HCrO_4^- + 1.5Mn^{2+} + H^+$$

$$Cr(OH)_2^+ + 3\beta\text{-}MnO_2(s) + 2H_2O(l) = HCrO_4^- + 3MnOOH(s) + 2H^+$$

Hexavalent chromium gets reduced to trivalent chromium in the presence of various reducing agents like hydrogen sulphide

DOI: 10.1201/9781003258063-5

(H_2S) (Smillie et al., 1981), organic matter (Schroeder and Lee, 1975), Fe^{2+} ion (Eary and Rai, 1989). Ding et al. (2016) highlighted that even in alkaline condition Cr(VI) can be reduced by organic matter and Fe(II). Agrawal et al. (2009) also pointed out a faster reduction of Cr(VI) in the presence of Fe(II)-DOM complex available in soil. Hug et al. (1996) described the photoreduction pathway of Cr(VI) reduction.

Ferrous-ion–containing mineral phases like magnetite, biotite, pyrite cause reduction of Cr(VI) (Deng et al., 1996; Eary and Rai, 1989; Peterson et al., 1997). The Cr(VI) can be reduced by Fe(II)-Fe(III) hydroxysalt green rust (Loyaux-Lawniczak et al., 1999). The following equation describes the reduction of hexavalent chromium by Fe(II) (Rai et al., 1988):

$$[3FeO]_{(s)} + 6H^+ + Cr^{6+} = Cr^{3+} + 3Fe^{3+} + 3H_2O \quad (5.1)$$

Kinetics of Cr(VI) reduction by Fe(II) is very fast which means all Cr(VI) will get reduced in excess of Fe(II) (Loyaux-Lawniczak et al., 2001). It is ten times faster than reoxidation rates (Saleh et al., 1989). Reduction of Cr(VI) can be possible by living tissue (Ishibashi et al., 1990) and residual organic matter (James and Bartlett, 1983a). Soluble low molecular weight organic molecules like gallic acid and ascorbic acid had shown prominent results in the effective reduction of Cr(VI) (James and Bartlett, 1983b). Sulphides generated from the reduction of dissolved sulphate, industrial waste and organic matter decomposition can reduce Cr(VI) (Schroeder and Lee, 1975; Richard and Bourg, 1991). Several reports cited the role of microorganism in the redox-chemistry of chromium (Dittert et al., 2014).

5.2 Chromium in Soil Environment

Chromium dynamic in the soil ecosystem is influenced by many properties of soil such as moisture content, microbial activities, soil Eh, soil pH, texture, organic matter, sulphide ions and Fe-Mn oxides (Banks et al., 2006; Dong et al., 2018; Xiao et al., 2021). Weathering of parent rock is the major natural source of chromium in soil. Chromium is present in the soil as adsorbed Cr(III) or insoluble $Cr(OH)_3$.aq form, which restricts its mobilization. Mobile ligands such as citric acid, diethylenetriaminepentaacetic acid (DTPA) and fulvic acid initiate complexation and followed by oxidation process of this immobile Cr(III) and resulted in the formation of mobile Cr(VI) (Wittbrodt and Palmer, 1995). The Cr(VI) exists in soluble form in neutral to alkaline condition, but in pH<6, it is available in soil as a dominant $HCrO_4^-$ form (Kotaś and Stasicka, 2000). The CrO_4^{2-} and $HCrO_4^-$ are the most soluble and bioavailable forms of chromium in soil. Binding of Cr(VI) in soil is dependent on soil mineralogical composition and pH (Richard and Bourg, 1991; Kotaś and Stasicka, 2000). The $HCrO_4^-$ remains in soluble form in acidic soil (Kotaś and Stasicka, 2000). Figure 5.1 describes processes of oxidation, reduction and complexation operating in the chromium cycle in soil environment. Xu et al. (2020) conducted a detailed investigation on the impact of soil pH on chromium chemistry in soil ecosystem. Redox potential (Eh) plays a crucial role in the speciation and concentration of chromium in soil. Zhang and Lin (2020) conducted toxicity characteristic leaching procedure to understand the impact of redox potential on chromium leachate concentration in soil, which explains the increase in redox potential in soil can enhance the chromium concentration in leachate (Figure 5.2).

At reduction site Cr(VI) gets reduced to Cr(III) by Fe(II) and S^{2-} (Bartlett and James, 1988; Palma et al., 2015). The Cr(III) generated due to the reduction of Cr(VI) is transported by mobile ligands to the oxidation site, whereas Fe(III) and S_{ox} are reduced back to Fe(II) and S^{2-} by organic reducer. At oxidation site, Cr(III) is oxidized to Cr(VI) by manganese oxide (Mn(IV)). The Mn(II) generated in this process is oxidized by atmospheric oxygen to Mn(IV) which can be reused at oxidation site.

The Cr(VI) present in the anionic forms gets adsorbed on the positively charged surface of aluminium and Fe hydroxides having inorganic hydroxyl group on their surface

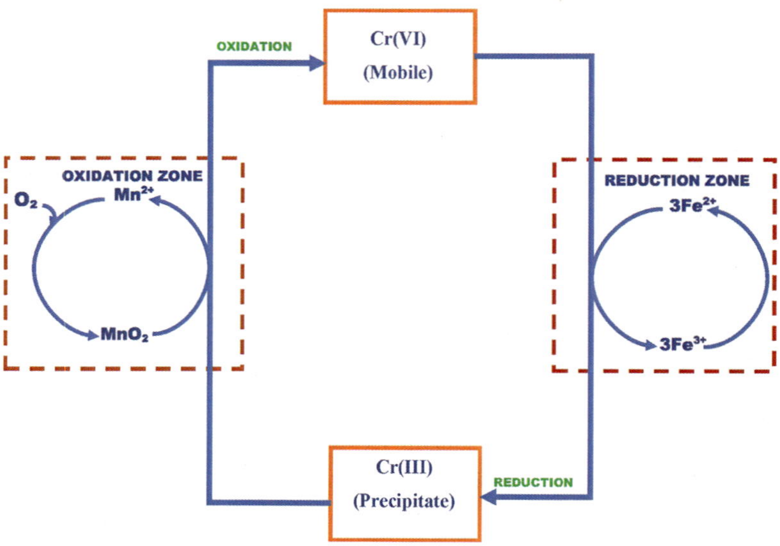

FIGURE 5.1 Redox reaction of chromium in soil ecosystem. (Modified from Bartlett and James (1988).)

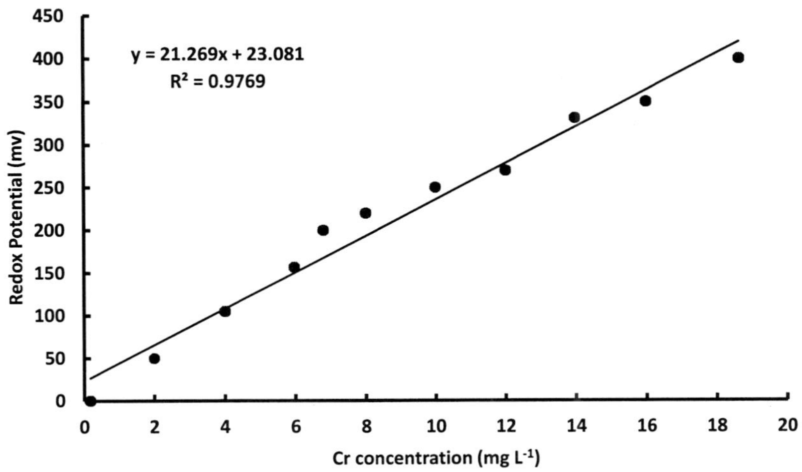

FIGURE 5.2 Relation of redox potential with chromium leachate concentration. (Modified from Zhang and Lin (2020).)

TABLE 5.1

World Mine Production and Reserves (USGS, 2021)

	Mine Production in Thousand Tons, Gross Weight, of Marketable Chromite Ore		Reserve (Shipping Grade)in Thousand Tons, Gross Weight, of Marketable Chromite Ore
	2019	2020	
United States	—	—	620
Finland	2,415	2,400	13,000
India	4,139	4,000	100,000
Kazakhstan	6,700	6,700	230,000
South Africa	16,395	16,000	200,000
Turkey	10,000	6,300	26,000
Other countries	5,110	4,800	NA

(Davis and Leckie, 1980; Benjamin and Bloom, 1981; Zachara et al., 1987). It also gets adsorbed on the surface of montmorillonite (Rai et al., 1987), natural soils and colloids (James and Bartlett, 1983b). It is a surface complexation process, which can be explained by the following equation (Rai et al., 1989):

$$SOH + H^+ + CrO_4^{2-} \Leftrightarrow SOH_2^+ - CrO_4^{2-}$$

Here SOH is surface inorganic hydroxyl and $SOH_2^+ - CrO_4^{2-}$ is adsorbed Cr(VI) surface complex.

In acidic condition, its adsorption to the mineral surface increases due to the protonation of a surface hydroxyl group and speciation of Cr(VI) (Zachara et al., 1987; Rai et al., 1989). Adsorption of chromium on the mineral surface is affected by several factors like the concentration of hydroxyl group on the surface, pH of the solution and equilibrium constant of adsorbents (Rai et al., 1989). Adsorbents such as aluminium oxide, iron oxide, montmorillonite and kaolinite form a coating or discrete particles on a solid matrix. Rai et al. (1988) observed the highest absorptivity of iron oxide for CrO_4^{2-} in comparison to another adsorbent in acidic to the neutral oxidized environment. Adsorbents having positive charge at low to neutral pH and low pH_{ZPC} values efficiently adsorb Cr(VI) (Richard and Bourg, 1991). Competitive dissolved anions and increased pH can reduce the adsorption of Cr(VI) at adsorption site (Richard and Bourg, 1991). Columbic attraction of Cr(VI) species on solid surfaces decreases due to the presence of competitive ions. Many authors had explained the competitive adsorption of hexavalent chromium with other ions (Stollenwerk and Grove, 1985; Music et al., 1986; Zachara et al., 1987). Chromate adsorption is influenced by electrostatic condition created by the major anion on the solid surface, which explains the least adsorption of Cr(VI) in groundwater and soil (Rai and Zachara, 1986). Xu et al. (2021) studied the oxidation process of Cr(III) by unsaturated oils which results in Cr(VI) in soil. Wang et al. (2020) investigated chromium migration in deep soil layer and its subsequent redox transformation at various depths. The authors also mentioned the impact of organic matter, soil permeability and amount of water passing through the soil layers on this vertical migration and transformation of chromium.

Mining activities such as mineral excavation and its processing, transportation, disposal of tailing and wastewater introduce the metal in soil environment (Dudka and Adriano, 1997; Navarro et al., 2008; Zhuang et al., 2009). Mining activities contribute a total of 33% of toxic Cr(VI) globally (Nayak et al., 2020). Table 5.1 provides details of worldwide chromite reserve and production, as per USGS January 2021 report. Discharge and dispersion of mine waste due to wind and/or water releases metal into the soil (Moore and Luoma, 1990).

Surface run-off from the overburden dump is severely contaminated with heavy metals. Contamination of mining soil is dependent on the type of deposit, extent of mineralization in mine tailings and physicochemical condition (Blowes and Jambor, 1990; Jung, 2001). García-Lorenzo et al. (2012) proposed three types of contamination transport mechanism in mining areas: (i) primary contaminants, which are deposited near the contamination source; (ii) secondary contaminants, which are produced by dispersion of trace element by water and wind and (iii) tertiary contaminants, which are transported due to trace element mobilization. Apart from mining, other industrial activities such as metal finishing, leather tanning and wood preservation also contribute a substantial amount of chromium in soil (Xu et al., 2020).

5.2.1 Natural Presence of Chromium in Rock and Soil

Ultramafic rock hosts a major share of naturally available chromium (Ao et al., 2022). Ultramafic rock consists of ferromagnesian minerals with <45% silica. It includes both peridotites and serpentinites. Peridotite contains magnesium-rich olivine and other minerals like chromite, orthopyroxene and clinopyroxene, whereas serpentinites are metamorphic rocks, formed by hydrothermal transformation of peridotite (O'Hanley, 1996; Kierczak et al., 2007). Soil formed by weathering of either peridotite or serpentinite is called serpentine soil, which contains high concentration of Mg and Fe but very low concentration of Ca and Al (Alexander, 2004). Ultramafic rocks most often contain world's famous chromite deposits. Chromite belongs to spinel group having general chemical composition AB_2O_4, where A represents Fe^{2+} and B may be Fe^{3+}, Al or Cr. It is subdivided into three groups: spinel series, magnetite series and chromite series where B is represented by Al, Fe^{3+} and Cr respectively (Hseu and Iizuka, 2013). Although chromite is resistant to alteration, having Cr(III) trapped in the lattice structure, there were evidences of its incongruent dissolution which may be a possible source of high level of Cr in soil (Garnier et al., 2008; Fandeur et al., 2009b). Mineralogy and geochemistry of serpentine soil are influenced by parent rock. In addition to these, climatic condition, topography, tectonic activities, biota and time also play an important role in rock weathering (Garnier et al., 2013; Hseu and Iizuka, 2013). In tropical climate complete hydrolysis of serpentine soil takes place which resulted in the highest leaching of Mg and insoluble elements like Al, Mn, Fe and Cr remains in the soil (Fandeur et al., 2009b). Thus, soil formed due to weathering of ultramafic rock is of special interest due to very high concentration chromium; however, the concentration of macronutrients is very low that causes low fertility in soil (Bonifacio et al., 1997). Minerological study using XANES was conducted by Fandeur et al. (2009a) which provides evidence of Cr(III) oxidation by Mn-oxides in natural system. Minerology of soil is a crucial factor for natural availability of metal in soil. Potential environmental contamination of toxic metal is dependent on its association with different mineral phases (Zelano et al., 2016). Major Cr-bearing mineral phases found in ultramafic tropical soil are chromite, Cr-bearing magnetite and goethite (Becquer et al., 2003, 2006; Oze et al., 2004a) and to some extent silicate phases like serpentine and pyroxene which possess Cr (Gasser et al., 1995; Quantin et al., 2008). A mineralogical study conducted by Garnier et al. (2006) on the ultramafic complex of Niquelandia in the Goias State, Brazil confirms the presence of iron oxide minerals (hematite and goethite), spinels (chromite and magnetite), quartz along with the presence of phyllosilicates and pyroxene. Oze et al. (2004b) identified chlorite, augite, enstatite along with other serpentinite phases in Franciscan complex of California.

5.2.2 Mobility and Bioavailability of Chromium in Soil

Heavy metal contamination has gotten worldwide attention due to its abundance, persistence, toxicity and accumulation in the environment (Dong et al., 2010). Metal contamination of soil ultimately affects the food chain due to its uptake by plant, which significantly impacts human health (Muchuweti et al., 2006; Sharma et al., 2007). The degree of soil contamination can be assessed by using different parameters such as geoaccumulation index, enrichment factor, pollution index and contamination index (Muller, 1969; Loska et al., 2004; Pattnaik and Equeenuddin, 2016). Trace element toxicity in soil is a complex phenomenon, which is influenced by pH, organic matter, redox condition and total concentration of metal (D'Amore et al., 2005; Burt et al., 2011). Total metal content in soil provides information on geological origin and anthropogenic impact on soil (Rao et al., 2008), but it is not effective to investigate bioavailability, mobility and fate of the contaminant (D'Amore et al., 2005). Potential toxicity of the metal present in the soil can be evaluated by assessing the process of metal distribution through geochemical partitioning which provides qualitative evidence of metal mobility and bioavailability (Li et al., 2009). Sequential extraction is the most promising tool that provides insight to the geochemical association of metals (Tessier et al., 1979; Sundaray et al., 2011; Ma et al., 2016). Extraction methods utilize different extractants to target the metal species in soil which are bound to or associated to different phases or compounds (Rao et al., 2008) such as residual fraction, Fe-Mn oxyhydroxide, organic fraction, carbonate fraction and exchangeable fraction. Metals associated with exchangeable fraction easily get dissolved in soil solution, whereas metals present in residual fraction resist the mobilization of metal as it is attached to the inner mineral lattice (Devesa-Rey et al., 2010; Rodríguez et al., 2009). Exchangeable and fraction of soil are associated with most soluble ions or metal complex with organic matter, which is the most mobile and available. Metals retain in their solid surface with weak electrostatic interaction, which gets mobilized by ion exchange process, reduction of pH, change in ionic composition and due to adsorption-desorption reaction (Arunachalam et al., 1996; Narwal et al., 1999). Metal associated with exchangeable fraction readily gets released into the environment. Metals associated with organic fraction of soil through adsorption, chelation and complexation mechanisms which can be mobilized by organic matter oxidation (Kennedy et al., 1997). Hence, the organic bound fraction is also known as oxidized faction.

The Fe-Mn oxyhydroxide phases are present in the soil as coating on mineral surfaces or fine discrete particles. These secondary oxides scavenge trace metal by the mechanism of surface complex formation, ion exchange, penetration of the lattice, co-precipitation and adsorption (Hall et al., 1996). Carbonate fraction metals are pH sensitive, whereas Fe-Mn oxide fraction metals are redox state sensitive. Metals associated with the residual fraction are attached to the crystal lattice structure of minerals which makes them most stable and immobile.

5.2.3 Plant Response to Cr(VI)

5.2.3.1 Plant Uptake and Translocation of Chromium

Since Cr is a non-essential element for plants, no specific uptake mechanism is involved. On most occasions, the uptake is facilitated by the carrier-mediated process involving anions like sulphate. Some species like Fe, S and P also showcase competing property with Cr for binding with the anions. Chromium uptake in plants can however be studied under two mechanisms: (i) passive mechanism involving the uptake of Cr(III) without the utilization of metabolic energy and (ii) active mechanism involving the uptake of Cr(VI) with the utilization of metabolic energy (Skeffington et al., 1976). Studies suggest that metabolic inhibitors tend to reduce the uptake of Cr(VI) but do not exhibit any effect on Cr(III) uptake. Further, crops grown independently with both the species of Cr have resulted in greater accumulation of Cr(VI) than in Cr(III).

The uptake of chromium from soil decreases with the progression of route from rhizosphere to aerial parts of plants. Maximum accumulation occurs in the root with regressive Cr concentration in the vegetative parts of plants. Tiwari et al. (2009) suggested through their experiments that Cr accumulation in pea plant part was in the order root>stem>leaves>seeds. However, the concentration of available Cr in different plant parts increases with the increase in concentration of Cr in the soil. Roots accumulate 10–100 folds more Cr than the aerial vegetative parts across plant species. Similarly, bean seeds tend to accumulate Cr as low as 0.1%. Species like *Eichhornia crassipes* and *Herniaria hirusta* have been a good accumulator of Cr in their roots. The probable reason for high rhizosphere Cr concentration may be attributed to the immobilization of chromium in the root cells. Other studies related to Cr toxicity suggest that some plants use roots as a storage organ for Cr while exhibiting Cr tolerance in their shoots (Fahad et al., 2015).

Several physicochemical factors viz. pH, temperature, electrical conductivity, moisture content, metal speciation etc. and plant traits like plant type, root system, leaf type etc. influence the uptake of chromium (Yamamoto and Kozlowski, 1987). Plant roots generate organic acids acting as a ligand for Cr binding. In this way, the organic acid converts the inorganic form to organic form of Cr making it readily available for plant uptake (Ertani et al., 2017). Chromium speciation in soil is influenced by the Eh factor that determines the biochemical behaviour of chromium (which exists in multiple oxidation states). Evidence in support of occurrence of Cr(VI) in oxygen-rich surrounding having neutral to alkaline pH (Ball and Izbicki, 2004), attest to the postulation of Cr behaviour in variable Eh and pH condition. Negative correlation between the soil pH and metal availability to plants has been documented previously (Shahid et al., 2016).

The translocation inside the plant is guided by the oxidation state, concentration of chromium and type of plant species involved. Hexavalent form of chromium exhibits greater accumulation inside plants as compared to its trivalent form (Cervantes et al., 2001). Chromium is mainly transported by the xylem as Cr(VI) moves faster in the xylem than the Cr(III). Electrostatic interaction of Cr(III) with the vessel walls prohibits the smooth movement of the Cr(III) (Skeffington et al., 1976). However, experiments have also shown the conversion of Cr(VI) to Cr(III) in both the root and shoot parts of the plants (Zayed et al., 1998). Sauerbeck (1991) in his study reported that of all the studied heavy metals (Cd, Zn, Cu, Ni), Cr showed the minimum transfer factor value.

5.2.3.2 Effects of Chromium on Plants

Chromium exhibits several morphological changes and biochemical alterations in plants. The early effect of Cr on plants is reduced germination potential and stunted root growth. The decrease in root length with increase in root diameter and root hairs is evidenced by the changes in proportions of the cortical and pith tissue layers. It is well documented that Cr tends to reduce the stem length; leaf number, biomass and area; number of flowers, grain size and biomass. While the reduced root length is expected due to the Cr-induced declining cell division, a secondary reason could be attributed to the collapse of root system due to direct contact of Cr species resulting in slow water absorption. This collapse is the main reason behind low nutrient and water transport to aerial plant parts consequential in low shoot growth (Barceló et al., 1986). Slow aerial development in plants can also occur due to interference of Cr in cellular metabolic activities in the aerial part mainly channelized during nutrient transport. Leaf chlorosis and leaf necrosis have been reported to be caused due to exposure of chromium at a concentration of 40–80 ppm (Jain et al., 2000). Chromium as compared to other heavy metals also has significant effects on young leaves (Pedreño et al., 1997).

Chromium has a serious effect in nutrient uptake capacity of plants through reduction in the uptake of Fe, Mg, P, Ca in plants by forming insoluble complexes that mask the sorption sites. Besides the nutrient uptake disruption, Cr also tends to cause imbalance in nutrient distribution and concentration in aerial parts of the plants. Studies confirming increased concentrations of Mn and P and decreased concentrations of S, Cu, Zn and Fe in *Citrullus* plant suggest the same (Dube et al., 2003). This may also be attributed to the decline in root growth and impairment of root penetration under Cr toxicity. Another reason could be declined nutrient translocation from physiologically active binding sites.

Chromium is also responsible for implicating major changes in the rate of photosynthesis, photophosphorylation and electron transport chain besides reducing the active enzyme activities in plants. Since chromium is utilized by the isolated chloroplasts as a Hill reagent, the observed effect is also significant in the concerned chloroplasts. Inhibitory effects on enzymatic activities and increased activity of reactive oxygen species (ROS) scavenging enzyme with changes in glutathione levels is another direct effect of chromium stress on plants. Protein degradation leading to reduction in chlorophyll-b synthesis can be best described due to Cr stress in plants.

Chromium has also shown its influence on the water content in plants. Some plants have been observed to wilt under Cr stress condition. The antagonistic relationship between the leaf water potential and Cr concentration in the case of bean has been reported by Barcelo (1985). On the other hand, a contradictory result was presented by Chatterjee and Chatterjee (2000) who, while working on cauliflower, reported that an increase in Cr concentration reduced the water potential and transpiration rates and increased the diffusive resistance and relative water content. Further, hexavalent chromium reduction to lower oxidation states is done by reducing agents like NADPH and $FADH_2$. This triggers the formation of complexes that further react with hydrogen peroxide to form ROS (particularly OH) which results in the alteration of DNA (Cervantes et al., 2001). Other intracellular agents that also show chromate-reducing property are cytochrome-P450, vitamin C etc. On the other hand, DNA shows Cr(III) sequestering tendency which may lead to oxidative DNA damage chiefly responsible for genotypic alterations. An overview of Cr(VI) toxicity in plants is presented in Figure 5.3.

Chromium also showcases hormetic effects on plants. Hormesis is a bi-phasic dose-response relationship between a metal and the plant. When metal induces stimulation at low doses but toxicity at high levels, it is known to cause hormesis in plants. Although chromium shows hormetic effects, there is no distinct reported value for Cr to cause stimulation or inhibitory effects in plants. Further, plant species and speciation exhibit variation in stimulation and toxicity levels in plants. Contradictory results related to Cr effects have been observed in previous studies. Some plants show hormetic results in terms of fertilization, development of toxicity and inhibitory effects, while other species of plants have been reported to show positive growth rate and plant productivity at low Cr doses (Pagano et al., 2008; Prasad et al., 2010).

Three routes of metabolic modifications can be pronounced in plants due to Cr: (i) changes in pigment production, (ii) over production of metabolites and (iii) production of biochemically new metabolites due to alteration in metabolic pool. While pigments like chlorophyll are responsible for the existence of plants, metabolites like glutathione damage the plants and another type of metabolites like phytochelatins can help plant counter the Cr stress condition. Nevertheless, some major effects on important enzymatic activities of plants are in direct connection to Cr stress. Few reported effects of Cr on enzymatic activities of plants are increase in nitrate reductase activity correlated with low root length, increase in iron reductase activity with conversion of ferric to ferrous ion (acceptable form of Fe to plant) and inactivation of superoxide dismutase at high concentration and depletion of catalase. However, protein-degrading ability of Cr in plants may lead to inhibition of nitrate reductase activity. Further, a decline in cysteine (amino acid that behaves as chelatin) may also cause degradation in sulphate-reducing enzymes and subsequently lead to toxicity in plants.

5.2.3.3 Bioremediation of Chromium

Poor translocation from rhizosphere to aerial vegetative plant parts is an obstacle for bioremediation of chromium. However, of all bioremediation measures, phytoaccumulation of

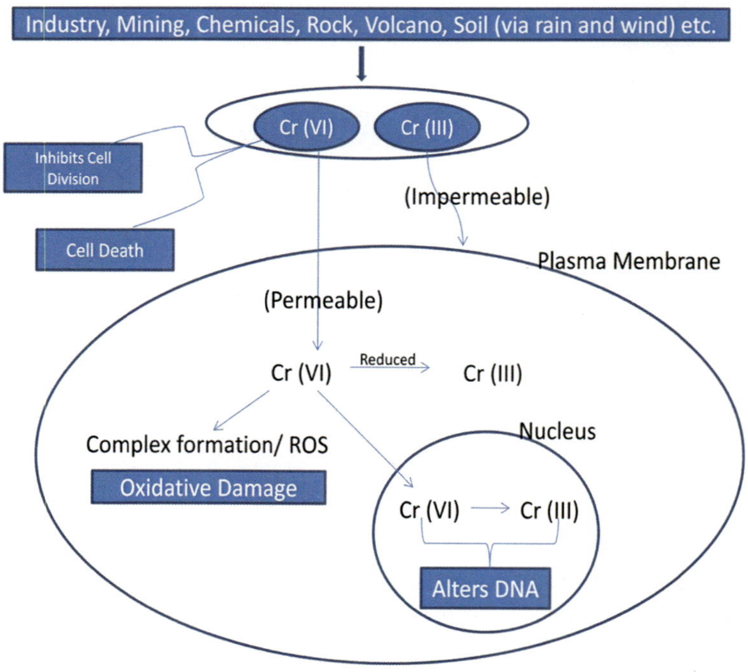

FIGURE 5.3 An overview of Cr(VI) toxicity in plants.

TABLE 5.2

Some Common Microorganisms and Plants Suitable for Cr Bioremediation

Algae	Fungi
Phormidium laminosum	*Aspergillus carbonarius*
Cyanidium caldarium	*Mucor mucedo*
Chlorella vulgaris	*Candida tropicalis*
Cladophora glomerata	*Rhizomucor miehei*
Enteromorpha intestinalis	*Penicillium chrysogenum*
Cyanidium caldarium	*Glomus mosseae*
Microspora amoena	*G. intraradices*
Bacteria	**Plant**
Alcaligenes eutrophus	*Conocephalum conicum*
Desulfovibrio vulgaris	*Laguncularia racemosa*
Pseudomonas aeruginosa	*Typha augustifolia*
P. putida	*Azolla pinnata*
Sporosarcina saromensis	*Cynodon dactylon*
Pannonibacter phragmitetus	*Panicum virgatum*
Vagococcus fluvialis	*Eichhornia crassipes*

chromium is the most preferred method to ameliorate chromium pollution. This might be due to the immobility of the chromium species in the root of plants. Although few in number, some of the noted hyper-accumulators of chromium are *Spartina argentinensis, Dyera costulata, Leersia hexandra, Pteris vittata* etc. Alyazouri et al. (2020) reported that *Portulaca oleracea* is an ideal species to uptake Cr at low soil organic matter content, alkaline pH range and with a sulphate concentration of around 300–600ppm. Kumar et al. (2019) from their work concluded that application of glycine betaine in *Sorghum bicolor* successfully ameliorated Cr(VI) up to 4ppm concentration. Table 5.2 presents an overview of various categories of organisms that are used for bioremediation of chromium.

Studies related to the nutrient culture suggest that DTPA, EDTA (Ethylenediamine tetraacetic acid) and EDDHA (ethylenediamine-N,N′-bis(2-hydroxyphenylacetic acid)) chelated Cr have the ability to be translocated by Phaseolus vulgaris (Athalye et al., 1995). Presence of organic acid also leads to enhanced Cr accumulation with a significant increase in soil-plant transfer ratio. In this connection phytostabilization of Cr has also been documented (Shanker et al., 2005). Although recent studies have advocated the use of microorganisms in bioremediation of Cr through bioaccumulation, biosorption, biotransformation etc. processes, bioreduction of toxic Cr species to non-toxic form or soluble to insoluble form or mobile to immobile form supersedes the other methods with multiple advantages (de Guillén-Jiménez et al., 2009). Below given are some processes employed in bioremediation of Cr.

5.2.3.3.1 Biosorption of Cr

A reversible physicochemical interaction between a heavy metal (Cr) and a biological material is known as biosorption. Being a passive process, the absorption of Cr does not involve energy and is performed by using both dead and live cells without the addition of nutrients. Microorganisms provide a high surface/volume ratio and also prove to be a cost-effective method of Cr removal. The main mechanism associated with the biosorption process is dependent on the presence of different functional groups of bio-molecules like lipids, polysaccharides, proteins etc. on the cell wall that effectively behaves as adsorption sites for metal binding. This adsorption may take place by physical or chemical means (Hamouda et al., 2020). Biosorption of Cr has been successfully achieved in cyanobacteria, algae, fungi, yeast, bacteria etc. with potentials of sludge, macrophytes, fruits, plants, vegetable residues to perform the same function.

Bioaccumulation of Cr Bioaccumulation of Cr by microorganisms is driven by the metallic concentration in the aqueous solution and the contact time. However, high metal concentration inhibits cell growth in plants (Dönmez and Aksu, 1999). Studies have revealed that Cr(VI) analogous to sulphate and phosphate ions enters the cells due to oxidative state-sensitive anion channel and non-selective mechanism facilitated via diffusion process (Ksheminska et al., 2008). Parameters like bio-concentration factor (BCF) and translocation factor (TF) have proved to be ideal measurements for evaluating the bioaccumulation potential. Hyper-accumulators of heavy metals have a BCF value in excess of one while the transducers have a value less than one. Similarly, the TF value determines the efficacy of a plant to translocate metals from root to shoot. A TF value greater than one indicates that the plant has a well-developed transport mechanism system actively engaged in translocating heavy metal from root to shoot (Usman et al., 2019).

Biotransformation of Cr Biotransformation of hexavalent chromium to trivalent form is the most environment-friendly bioremediation measure as the toxic form is converted to non-toxic form. The transformation process is generally facilitated by the enzymatic activity that is secreted by the plants (Jobby et al., 2019). Chromate reductase activity in the extracellular parts is evident in some microbial cultures. Bacteria and yeasts like *Pseudomonas, Bacillus, Arthrobacter, Pichia guilliermondii, Rhodotorula mucilaginosa, Cyberlindnera jadinii, Wickerhamomyces anomalus* etc. have proved to be efficient reducers of Cr(VI) to Cr(III).

REFERENCES

Agrawal, S.G., Fimmen, R.L., Chin, Y.-P., 2009. Reduction of Cr(VI) to Cr(III) by Fe(II) in the presence of fulvic acids and in lacustrine pore water. *Chem. Geol.* 262, 328–335. doi: 10.1016/j.chemgeo.2009.02.001.

Alexander, E.B., 2004. Serpentine soil redness, differences among peridotite and serpentinite materials, Klamath Mountains, California. *Int. Geol. Rev.* 46, 754–764.

Alyazouri, A., Jewsbury, R., Tayim, H., Humphreys, P., Al-Sayah, M.H., 2020. Uptake of chromium by portulaca oleracea from soil: Effects of organic content, pH, and sulphate concentration. *Appl. Environ. Soil Sci.* 2020, 3620726. doi: 10.1155/2020/3620726.

Amacher, M.C., Baker, D.E., 1979. Redox reactions involving chromium, plutonium, and manganese in soils.

Anderson, R.A., 1997. Chromium as an essential nutrient for humans. *Regul. Toxicol. Pharmacol.* 26, S35–S41. doi: 10.1006/rtph.1997.1136.

Ao, M., Chen, X., Deng, T., Sun, S., Tang, Y., Morel, J.L., Qiu, R., Wang, S., 2022. Chromium biogeochemical behaviour in soil-plant systems and remediation strategies: A critical review. *J. Hazard. Mater.* 424, 127233. doi: 10.1016/j.jhazmat.2021.127233.

Arunachalam, J., Emons, H., Krasnodebska, B., Mohl, C., 1996. Sequential extraction studies on homogenized forest soil samples. *Sci. Total Environ.* 181, 147–159.

Athalye, V. V., Ramachandran, V., D'Souza, T.J., 1995. Influence of chelating agents on plant uptake of 51Cr, 210Pb and 210Po. *Environ. Pollut.* 89, 47–53. doi: 10.1016/0269-7491(94)00047-H.

Ball, J.W., Izbicki, J.A., 2004. Occurrence of hexavalent chromium in ground water in the western Mojave Desert, California. *Appl. Geochem.* 19, 1123–1135. doi: 10.1016/j.apgeochem.2004.01.011.

Banks, M.K., Schwab, A.P., Henderson, C., 2006. Leaching and reduction of chromium in soil as affected by soil organic content and plants. *Chemosphere* 62, 255–264. doi: 10.1016/j.chemosphere.2005.05.020.

Barcelo, J., 1985. Leaf water potential in Cr (VI) treated bean plants (*Phaseolus vulgaris* L). *Plant Physiol. Suppl.* 77, 163–164.

Barceló, J., Poschenrieder, C., Gunsé, B., 1986. Water relations of chromium VI treated bush bean plants (*Phaseolus vulgaris* L. cv. Contender) under both normal and water stress conditions. *J. Exp. Bot.* 37, 178–187. doi: 10.1093/jxb/37.2.178.

Bartlett, R., James, B., 1979. Behavior of chromium in soils: III. Oxidation. *J. Environ. Qual.* 8, 31–35.

Bartlett, R.J., James, B.R., 1988. Mobility and bioavailability of chromium in soils. In: Nriagu, J.O., Nieboer, E. (Eds.), *Chromium in the Natural and Human Environments.* John Wiley & Sons, New York, pp. 267–571.

Becquer, T., Quantin, C., Rotte-Capet, S., Ghanbaja, J., Mustin, C., Herbillon, A.J., 2006. Sources of trace metals in Ferralsols in New Caledonia. *Eur. J. Soil Sci.* 57, 200–213. doi: 10.1111/j.1365-2389.2005.00730.x.

Becquer, T., Quantin, C., Sicot, M., Boudot, J.P., 2003. Chromium availability in ultramafic soils from New Caledonia. Sci. Total Environ. 301, 251–261. https://doi.org/10.1016/S0048-9697(02)00298-X

Benjamin, M.M., Bloom, N.S., 1981. Effects of strong binding of anionic adsorbates on adsorption of trace metals on amorphous iron oxyhydroxide. In: Tewari, P.H. (Ed.), *Adsorption from Aqueous Solutions.* Springer, Berlin, Germany, pp. 41–60.

Blowes, D.W., Jambor, J.L., 1990. The pore-water geochemistry and mineralogy of the vadose zone of the Waite Amulet mine tailings impoundment, Noranda, Quebec. *Appl. Geochem* 5, 327–346.

Bonifacio, E., Zanini, E., Boero, V., Franchini-Angela, M., 1997. Pedogenesis in a soil catena on serpentinite in north-western Italy. *Geoderma* 75, 33–51. doi: 10.1016/S0016-7061(96)00076-6.

Burt, R., Weber, T., Park, S., Yochum, S., Ferguson, R., 2011. Trace element concentration and speciation in selected mining-contaminated soils and water in Willow Creek Floodplain, Colorado. *Appl. Environ. Soil Sci.* 1–20. doi: 10.1155/2011/237071.

Cervantes, C., Campos-García, J., Devars, S., Gutiérrez-Corona, F., Loza-Tavera, H., Torres-Guzmán, J.C., Moreno-Sánchez, R., 2001. Interactions of chromium with microorganisms and plants. *FEMS Microbiol. Rev.* 25, 335–347. doi: 10.1016/S0168-6445(01)00057-2.

Chatterjee, J., Chatterjee, C., 2000. Phytotoxicity of cobalt, chromium and copper in cauliflower. *Environ. Pollut.* 109, 69–74. doi: 10.1016/S0269-7491(99)00238-9.

Cranston, R.E., Murray, J.W., 1978. The determination of chromium species in natural waters. *Anal. Chim. Acta* 99, 275–282.

D'Amore, J.J., Al-Abed, S.R., Scheckel, K.G., Ryan, J.A., 2005. Methods for speciation of metals in soils. *J. Environ. Qual.* 34, 1707. doi: 10.2134/jeq2004.0014.

Danish, S., Kiran, S., Fahad, S., Ahmad, N., Ali, M.A., Tahir, F.A., Rasheed, M.K., Shahzad, K., Li, X., Wang, D., Mubeen, M., Abbas, S., Munir, T.M., Hashmi, M.Z., Adnan, M., Saeed, B., Saud, S., Khan, M.N., Ullah, A., Nasim, W., 2019. Alleviation of chromium toxicity in maize by Fe fortification and chromium tolerant ACC deaminase producing plant growth promoting rhizobacteria. *Ecotoxicol. Environ. Saf.* 185, 109706. doi: 10.1016/j.ecoenv.2019.109706.

Davis, J.A., Leckie, J.O., 1980. Surface ionization and complexation at the oxide/water interface. 3. Adsorption of anions. *J. Colloid Interface Sci.* 74, 32–43.

de Guillén-Jiménez, F.M., Netzahuatl-Muñoz, A.R., Morales-Barrera, L., Cristiani-Urbina, E., 2009. Hexavalent chromium removal by Candida sp. in a concentric draft-tube airlift bioreactor. *Water Air Soil Pollut.* 204, 43. doi: 10.1007/s11270-009-0024-x.

Deng, Y., Stjernström, M., Banwart, S., 1996. Accumulation and remobilization of aqueous chromium(VI) at iron oxide surfaces: Application of a thin-film continuous flow-through reactor. *J. Contam. Hydrol.* 21, 141–151. doi: 10.1016/0169-7722(95)00039-9.

Devesa-Rey, R., Díaz-Fierros, F., Barral, M.T., 2010. Trace metals in river bed sediments: An assessment of their partitioning and bioavailability by using multivariate exploratory analysis. J. Environ. Manage. 91, 2471–2477. https://doi.org/10.1016/j.jenvman.2010.06.024

Ding, W., Stewart, D.I., Humphreys, P.N., Rout, S.P., Burke, I.T., 2016. Role of an organic carbon-rich soil and Fe(III) reduction in reducing the toxicity and environmental mobility of chromium(VI) at a {COPR} disposal site. *Sci. Total Environ.* 541, 1191–1199. doi: 10.1016/j.scitotenv.2015.09.150.

Dittert, I.M., de Lima Brandão, H., Pina, F., da Silva, E.A.B., de Souza, S.M.A.G.U., de Souza, A.A.U., Botelho, C.M.S., Boaventura, R.A.R., Vilar, V.J.P., 2014. Integrated reduction/oxidation reactions and sorption processes for Cr(VI) removal from aqueous solutions using Laminaria digitata macro-algae. *Chem. Eng. J.* 237, 443–454. doi: 10.1016/j.cej.2013.10.051.

Dong, H., Wei, G., Fan, W., Ma, S., Zhao, H., Zhang, W., Guan, X., Strathmann, T.J., 2018. Reinvestigating the role of reactive species in the oxidation of organic co-contaminants during Cr(VI) reactions with sulfite. *Chemosphere* 196, 593–597. doi: 10.1016/j.chemosphere.2017.12.194.

Dong, X., Li, C., Li, J., Wang, J., Liu, S., Ye, B., 2010. A novel approach for soil contamination assessment from heavy metal pollution: A linkage between discharge and adsorption. *J. Hazard. Mater.* 175, 1022–1030. doi: 10.1016/j.jhazmat.2009.10.112.

Dönmez, G., Aksu, Z., 1999. The effect of copper(II) ions on the growth and bioaccumulation properties of some yeasts. *Process Biochem.* 35, 135–142. doi: 10.1016/S0032-9592(99)00044-8.

Dube, B.K., Tewari, K., Chatterjee, J., Chatterjee, C., 2003. Excess chromium alters uptake and translocation of certain nutrients in citrullus. *Chemosphere* 53, 1147–1153.

Dudka, S., Adriano, D.C., 1997. Environmental impacts of metal ore mining and processing: A review. *J. Environ. Qual.* 26, 590–602.

Eary, L.E., Rai, D., 1987. Kinetics of chromium (III) oxidation to chromium (VI) by reaction with manganese dioxide. *Environ. Sci. Technol.* 21, 1187–1193.

Eary, L.E., Rai, D., 1989. Kinetics of chromate reduction by ferrous ions derived from hematite and biotite at 25 degrees C. *Am. J. Sci.* doi: 10.2475/ajs.289.2.180.

Ertani, A., Mietto, A., Borin, M., Nardi, S., 2017. Chromium in Agricultural Soils and Crops: A Review. *Water, Air, Soil Pollut.* 228, 190. doi: 10.1007/s11270-017-3356-y.

Fahad, S., Hussain, S., Saud, S., Hassan, S., Shan, D., Chen, Y., Deng, N., Khan, F., Wu, C., Wu, W., Shah, F., Ullah, B., Yousaf, M., Ali, S., Huang, J., 2015. Grain cadmium and zinc concentrations in maize influenced by genotypic variations and zinc fertilization. *Clean Soil Air Water* 43, 1433–1440. doi: 10.1002/clen.201400376.

Fandeur, D., Juillot, F., Morin, G., Livi, L., Cognigni, A., Webb, S.M., Ambrosi, J.P., Fritsch, E., Guyot, F., Brown, G.E., 2009a. XANES evidence for oxidation of Cr(III) to Cr(VI) by Mn-oxides in a lateritic regolith developed on serpentinized ultramafic rocks of New Caledonia. *Environ. Sci. Technol.* 43, 7384–7390. doi: 10.1021/es900498r.

Fandeur, D., Juillot, F., Morin, G., Olivi, L., Cognigni, A., Ambrosi, J.-P., Guyot, F., Fritsch, E., 2009b. Synchrotron-based speciation of chromium in an Oxisol from New Caledonia: Importance of secondary Fe-oxyhydroxides. *Am. Mineral.* 94, 710–719.

Fantoni, D., Brozzo, G., Canepa, M., Cipolli, F., Marini, L., Ottonello, G., Zuccolini, M., 2002. Natural hexavalent chromium in groundwaters interacting with ophiolitic rocks. *Environ. Geol.* 42, 871–882. doi: 10.1007/s00254-002-0605-0.

Fendorf, S.E., 1995. Surface reactions of chromium in soils and waters. *Geoderma* 67, 55–71. doi: 10.1016/0016-7061(94)00062-F.

Fendorf, S.E., Zasoski, R.J., 1992. Chromium(Iii) Oxidation By Delta-Mno2.1. characterization. *Environ. Sci. Technol.* 26, 79–85.

García-Lorenzo, M.L., Pérez-Sirvent, C., Martínez-Sánchez, M.J., Molina-Ruiz, J., 2012. Trace elements contamination in an abandoned mining site in a semiarid zone. *J. Geochem. Explor.* 113, 23–35.

Garnier, J., Quantin, C., Martins, E.S., Becquer, T., 2006. Solid speciation and availability of chromium in ultramafic soils from Niquelandia, Brazil. *J. Geochem. Explor.* 88, 206–209. doi: 10.1016/j.gexplo.2005.08.040.

Garnier, J., Quantin, C., Guimarães, E., Becquer, T., 2008. Can chromite weathering be a source of Cr in soils? *Mineral. Mag.* 72, 49–53. doi: 10.1180/minmag.2008.072.1.49.

Garnier, J., Quantin, C., Guimarães, E.M., Vantelon, D., Montargès-Pelletier, E., Becquer, T., 2013. Cr(VI) genesis and dynamics in Ferralsols developed from ultramafic rocks: The case of Niquelândia, Brazil. *Geoderma* 193–194, 256–264. doi: 10.1016/j.geoderma.2012.08.031.

Gasser, U.G., Juchler, S.J., Sticher, H., Hobson, W.A., 1995. The fate of chromium and nickel in subalpine soils derived from serpentinite. *Can. J. Soil Sci.* 75, 187–195.

Hall, G.E.M., Gauthier, G., Pelchat, J.-C., Pelchat, P., Vaive, J.E., 1996. Application of a sequential extraction scheme to ten geological certified reference materials for the determination of 20 elements. *J. Anal. At. Spectrom.* 11, 787–796.

Hamouda, R.A., El-Naggar, N.E.-A., Doleib, N.M., Saddiq, A.A., 2020. Bioprocessing strategies for cost-effective simultaneous removal of chromium and malachite green by marine alga Enteromorpha intestinalis. *Sci. Rep.* 10, 13479. doi: 10.1038/s41598-020-70251-3.

Hseu, Z.Y., Iizuka, Y., 2013. Pedogeochemical characteristics of chromite in a paddy soil derived from serpentinites. *Geoderma* 202–203, 126–133. doi: 10.1016/j.geoderma.2013.03.021.

Hug, S.J., Laubscher, H.-U., James, B.R., 1996. Iron (III) catalyzed photochemical reduction of chromium (VI) by oxalate and citrate in aqueous solutions. *Environ. Sci. Technol.* 31, 160–170.

Ishibashi, Y., Cervantes, C., Silver, S., 1990. Chromium reduction in Pseudomonas putida. *Appl. Environ. Microbiol.* 56, 2268–2270.

Jain, R., Srivastava, S., Madan, V.K., 2000. Influence of chromium on growth and cell division of sugarcane. *Indian J. Plant Physiol. (India).* 5, 228–231.

James, B.R., Bartlett, R.J., 1983a. Behavior of chromium in soils. VI. Interactions between oxidation-reduction and organic complexation. *J. Environ. Qual.* 12, 173–176.

James, B.R., Bartlett, R.J., 1983b. Behavior of chromium in soils: VII. Adsorption and reduction of hexavalent forms. *J. Environ. Qual.* 12, 177–181.

Jobby, R., Jha, P., Gupta, A., Gupte, A., Desai, N., 2019. Biotransformation of chromium by root nodule bacteria Sinorhizobium sp. SAR1. *PLoS One* 14, e0219387.

Jung, M.C., 2001. Heavy metal contamination of soils and waters in and around the Imcheon Au-Ag mine, Korea. Appl. Geochemistry 16, 1369–1375. https://doi.org/10.1016/S0883-2927(01)00040-3.

Kennedy, V.H., Sanchez, A.L., Oughton, D.H., Rowland, A.P., 1997. Use of single and sequential chemical extractants to assess radionuclide and heavy metal availability from soils for root uptake. *Analyst* 122, 89R–100R.

Kierczak, J., Neel, C., Bril, H., Puziewicz, J., 2007. Effect of mineralogy and pedoclimatic variations on Ni and Cr distribution in serpentine soils under temperate climate. *Geoderma* 142, 165–177. doi: 10.1016/j.geoderma.2007.08.009.

Kotaś, J., Stasicka, Z., 2000. Chromium occurrence in the environment and methods of its speciation. *Environ. Pollut.* 107, 263–283. doi: 10.1016/S0269-7491(99)00168-2.

Kožuh, N., Štupar, J., Gorenc, B., 2000. Reduction and oxidation processes of chromium in soils. *Environ. Sci. Technol.* 34, 112–119. doi: 10.1021/es981162m.

Ksheminska, H., Fedorovych, D., Honchar, T., Ivash, M., Gonchar, M., 2008. Yeast tolerance to chromium depends on extracellular chromate reduction and Cr (III) chelation. *Food Technol. Biotechnol.* 46, 419–426.

Kumar, P., Tokas, J., Singal, H.R., 2019. Amelioration of Chromium VI Toxicity in Sorghum (*Sorghum bicolor* L.) using Glycine Betaine. *Sci. Rep.* 9, 16020. doi: 10.1038/s41598-019-52479-w.

Li, Y.T., Becquer, T., Dai, J., Quantin, C., Benedetti, M.F., 2009. Ion activity and distribution of heavy metals in acid mine drainage polluted subtropical soils. Environ. Pollut. 157, 1249–1257. http://doi.org/10.1016/j.envpol.2008.11.050

Loska, K., Wiechulła, D., Korus, I., 2004. Metal contamination of farming soils affected by industry. Environ. Int. 30, 159–165. https://doi.org/10.1016/S0160-4120(03)00157-0

Loyaux-Lawniczak, S., Refait, P., Lecomte, P., Ehrhardt, J.-J., Génin, J.-M.R., 1999. The reduction of chromate ions by Fe(II) layered hydroxides. *Eur. Geosci. Union* 3, 593–599.

Loyaux-Lawniczak, S., Lecomte, P., Ehrhardt, J.J., 2001. Behavior of hexavalent chromium in a polluted groundwater: Redox processes and immobilization in soils. *Environ. Sci. Technol.* 35, 1350–1357. doi: 10.1021/es001073l.

Ma, X., Zuo, H., Tian, M., Zhang, L., Meng, J., Zhou, X., Min, N., Chang, X., Liu, Y., 2016. Assessment of heavy metals contamination in sediments from three adjacent regions of the Yellow River using metal chemical fractions and multivariate analysis techniques. Chemosphere 144, 264–272. https://doi.org/http://dx.doi.org/10.1016/j.chemosphere.2015.08.026

Moore, J.N., Luoma, S.N., 1990. Hazardous wastes from large-scale metal extraction: A case study. *Environ. Sci. Technol.* 24, 1278–1285.

Muchuweti, M., Birkett, J.W., Chinyanga, E., Zvauya, R., Scrimshaw, M.D., Lester, J.N., 2006. Heavy metal content of vegetables irrigated with mixtures of wastewater and sewage sludge in Zimbabwe: Implications for human health. *Agric. Ecosyst. Environ.* 112, 41–48.

Muller, G., 1969. Index of geoaccumulation in sediments of the Rhine River. Geo J. 2, 108–118.

Music, S., Ristic, M., Tonkovic, M., 1986. Sorption of chromium (VI) on hydrous iron oxides. *Zeitschrift f[ü]r Wasser-und Abwasser-forsch.* 19, 186–196.

Narwal, R.P., Singh, B.R., Salbu, B., 1999. Association of cadmium, zinc, copper, and nickel with components in naturally heavy metal-rich soils studied by parallel and sequential extractions. *Commun. Soil Sci. Plant Anal.* 30, 1209–1230.

Navarro, M.C., Pérez-Sirvent, C., Martínez-Sánchez, M.J., Vidal, J., Tovar, P.J., Bech, J., 2008. Abandoned mine sites as a source of contamination by heavy metals: A case study in a semi-arid zone. *J. Geochem. Explor.* 96, 183–193. doi: 10.1016/j.gexplo.2007.04.011.

Nayak, S., Rangabhashiyam, S., Balasubramani, P., Kale, P., 2020. A review of chromite mining in Sukinda Valley of India: Impact and potential remediation measures. *Int. J. Phytoremediation* 22, 804–818. doi: 10.1080/15226514.2020.1717432.

O'Hanley, D.S., 1996. *Serpentinites: Records of Tectonic and Petrological History.* Oxford Monographs on Geology and Geophysics. Oxford University Press on Demand, New York, USA, p. 277.

Oze, C., Fendorf, S., Bird, D.K., Coleman, R.G., 2004a. Chromium geochemistry of serpentine soils. *Int. Geol. Rev.* 46, 97–126. doi: 10.2747/0020-6814.46.2.97.

Oze, C., Fendorf, S., Bird, D.K., Coleman, R.G., 2004b. Chromium geochemistry in serpentinized ultramafic rocks and serpentine soils from the Franciscan complex of California. *Am. J. Sci.* 304, 67–101. doi: 10.2475/ajs.304.1.67.

Pagano, G., Castello, G., Gallo, M., Borriello, I., Guida, M., 2008. Complex mixture-associated hormesis and toxicity: The case of leather tanning industry. *Dose-Response* 6, 8–13.

Palma, L. Di, Gueye, M.T., Petrucci, E., 2015. Hexavalent chromium reduction in contaminated soil: A comparison between ferrous sulphate and nanoscale zero-valent iron. *J. Hazard. Mater.* 281, 70–76. doi: 10.1016/j.jhazmat.2014.07.058.

Pattnaik, B.K., Equeenuddin, S.M., 2016. Potentially toxic metal contamination and enzyme activities in soil around chromite mines at Sukinda Ultramafic Complex. *India. J. Geochem. Explor.* 168, 127–136. doi: 10.1016/j.gexplo.2016.06.011.

Pedreño, J.., Gómez, I., Moral, R., Palacios, G., Mataix, J., 1997. Heavy metals and plant nutrition and development. *Rec. Res. Dev. Phytochem.* 1, 173–179.

Peterson, M.L., White, A.F., Brown, G.E., Parks, G.A., 1997. Surface passivation of magnetite by reaction with aqueous Cr (VI): XAFS and TEM results. *Environ. Sci. Technol.* 31, 1573–1576.

Pettine, M., Millero, F.J., 1990. Chromium speciation in seawater: The probable role of hydrogen peroxide. *Limnol. Oceanogr.* 35, 730–736.

Prasad, A., Singh, A.K., Chand, S., Chanotiya, C.S., Patra, D.D., 2010. Effect of chromium and lead on yield, chemical composition of essential oil, and accumulation of heavy metals of mint species. *Commun. Soil Sci. Plant Anal.* 41, 2170–2186. doi: 10.1080/00103624.2010.504798.

Quantin, C., Ettler, V., Garnier, J., Šebek, O., 2008. Sources and extractibility of chromium and nickel in soil profiles developed on Czech serpentinites. *Comptes Rendus Geosci.* 340, 872–882. doi: 10.1016/j.crte.2008.07.013.

Rai, D., Zachara, J.M., 1986. Geochemical behavior of chromium species.

Rai, D., Sass, B.M., Moore, D.A., 1987. Chromium (III) hydrolysis constants and solubility of chromium(III) hydroxide. *Inorg. Chem.* 26, 345–349. doi: 10.1021/ic00250a002.

Rai, D., Zachara, J.M., Eary, L.E., 1988. Chromium reactions in geologic materials. Electric Power Research Institute, Palo Alto, California.

Rai, D., Eary, L.E., Zachara, J.M., 1989. Environmental chemistry of chromium. *Sci. Total Environ.* 86, 15–23.

Rao, C.R.M., Sahuquillo, A., Sanchez, J.F.L., 2008. A review of the different methods applied in environmental geochemistry for single and sequential extraction of trace elements in soils and related materials. *Water Air Soil Pollut.* 189, 291–333.

Rodríguez, L., Ruiz, E., Alonso-Azcárate, J., Rincón, J., 2009. Heavy metal distribution and chemical speciation in tailings and soils around a Pb–Zn mine in Spain. J. Environ. Manage. 90, 1106–1116. https://doi.org/10.1016/j.jenvman.2008.04.007

Richard, F.C., Bourg, A.C.M., 1991. Aqueous geochemistry of chromium: A review. *Water Res.* 25, 807–816. doi: 10.1016/0043-1354(91)90160-R.

Saleh, F.Y., Parkerton, T.F., Lewis, R. V, Huang, J.H., Dickson, K.L., 1989. Kinetics of chromium transformations in the environment. *Sci. Total Environ.* 86, 25–41. doi: 10.1016/0048-9697(89)90190-3.

Sauerbeck, D.R., 1991. Plant element and soil properties governing uptake and availability of heavy metals derived from sewage sludge. *Water Air Soil Pollut.* 57, 227–237. doi: 10.1007/BF00282886.

Schroeder, D.C, Lee, G.F., 1975. Potential transformations of chromium in natural waters. *Water Air Soil Pollut.* 4, 355–365. doi: 10.1007/BF00280721.

Shahid, M., Dumat, C., Khalid, S., Niazi, N.K., Antunes, P., 2016. Cadmium bioavailability, uptake, toxicity and detoxification in soil-plant system. In: Gunther, F.A., De Voogt, P. (Eds.), *Reviews of Environmental Contamination and Toxicology*, Volume 241. Springer, Berlin, Germany, pp. 73–137.

Shanker, A.K., Cervantes, C., Loza-Tavera, H., Avudainayagam, S., 2005. Chromium toxicity in plants. *Environ. Int.* 31, 739–753. doi: 10.1016/j.envint.2005.02.003.

Sharma, R.K., Agrawal, M., Marshall, F., 2007. Heavy metal contamination of soil and vegetables in suburban areas of Varanasi, India. *Ecotoxicol. Environ. Saf.* 66, 258–266.

Skeffington, R.A., Shewry, P.R., Peterson, P.J., 1976. Chromium uptake and transport in barley seedlings (*Hordeum vulgare* L.). *Planta* 132, 209–214. doi: 10.1007/BF00399719.

Smillie, R.H., Hunter, K., Loutit, M., 1981. Reduction of chromium (VI) by bacterially produced hydrogen sulphide in a marine environment. *Water Res.* 15, 1351–1354.

Stollenwerk, K.G., Grove, D.B., 1985. Adsorption and desorption of hexavalent chromium in an alluvial aquifer near Telluride, Colorado. *J. Environ. Qual.* 14, 150–155.

Sundaray, S.K., Nayak, B.B., Lin, S., Bhatta, D., 2011. Geochemical speciation and risk assessment of heavy metals in the river estuarine sediments—a case study: Mahanadi basin, India. *J. Hazard. Mater.* 186, 1837–1846.

Tessier, A., Campbell, P.G.C., Bisson, M., 1979. Sequential extraction procedure for the speciation of particulate trace metals. *Anal. Chem.* 51, 844–851. doi: 10.1021/ac50043a017.

Tiwari, K.K., Dwivedi, S., Singh, N.K., Rai, U.N., Tripathi, R.D., 2009. Chromium (VI) induced phytotoxicity and oxidative stress in pea (*Pisum sativum* L.): Biochemical changes and translocation of essential nutrients. *J. Environ. Biol.* 30, 389–394.

USGS, 2021. Mineral commodity summaries 2021.

Usman, K., Al-Ghouti, M.A., Abu-Dieyeh, M.H., 2019. The assessment of cadmium, chromium, copper, and nickel tolerance and bioaccumulation by shrub plant Tetraena qataranse. *Sci. Rep.* 9, 5658. doi: 10.1038/s41598-019-42029-9.

Wang, X., Li, L., Yan, X., Meng, X., Chen, Y., 2020. Processes of chromium (VI) migration and transformation in chromate production site: A case study from the middle of China. *Chemosphere* 257, 127282. doi: 10.1016/j.chemosphere.2020.127282.

Wittbrodt, P.R., Palmer, C.D., 1995. Reduction of Cr (VI) in the presence of excess soil fulvic acid. *Environ. Sci. Technol.* 29, 255–263.

Wrobel, K., Escobosa, A.R.C., Ibarra, A.A.G., Garcia, M.M., Barrientos, E.Y., 2015. Mechanistic insight into chromium(VI) reduction by oxalic acid in the presence of manganese(II). *J. Hazard. Mater.* 300, 144–152. doi: 10.1016/j.jhazmat.2015.06.066.

Xia, S., Song, Z., Jeyakumar, P., Shaheen, S.M., Rinklebe, J., Ok, Y.S., Bolan, N., Wang, H., 2019. A critical review on bioremediation technologies for Cr(VI)-contaminated soils and wastewater. *Crit. Rev. Environ. Sci. Technol.* 49, 1027–1078. doi: 10.1080/10643389.2018.1564526.

Xiao, W., Ye, X., Zhu, Z., Zhang, Q., Zhao, S., Chen, D., Gao, N., Hu, J., 2021. Continuous flooding stimulates root iron plaque formation and reduces chromium accumulation in rice (*Oryza sativa* L.). *Sci. Total Environ.* 788, 147786. doi: 10.1016/j.scitotenv.2021.147786.

Xu, T., Nan, F., Jiang, X., Tang, Y., Zeng, Y., Zhang, W., Shi, B., 2020. Effect of soil pH on the transport, fractionation, and oxidation of chromium(III). *Ecotoxicol. Environ. Saf.* 195, 110459. doi: 10.1016/j.ecoenv.2020.110459.

Xu, T., Jiang, X., Tang, Y., Zeng, Y., Zhang, W., Shi, B., 2021. Oxidation of trivalent chromium induced by unsaturated oils: A pathway for hexavalent chromium formation in soil. *J. Hazard. Mater.* 405, 124699. doi: 10.1016/j.jhazmat.2020.124699.

Yamamoto, F., Kozlowski, T.T., 1987. Effects of flooding, tilting of stems, and ethrel application on growth, stem anatomy, and ethylene production of acer platanoides seedlings. *Scand. J. For. Res.* 2, 141–156. doi: 10.1080/02827588709382453.

Zachara, J.M., Girvin, D.C., Schmidt, R.L., Resch, C.T., 1987. Chromate adsorption on amorphous iron oxyhydroxide in the presence of major groundwater ions. *Environ. Sci. Technol.* 21, 589–594. doi: 10.1021/es00160a010.

Zayed, A., Lytle, C.M., Qian, J.-H., Terry, N., 1998. Chromium accumulation, translocation and chemical speciation in vegetable crops. *Planta* 206, 293–299. doi: 10.1007/s004250050403.

Zelano, I., Sivry, Y., Quantin, C., Gélabert, A., Tharaud, M., Nowak, S., Garnier, J., Malandrino, M., Benedetti, M.F., 2016. Study of Ni exchangeable pool speciation in ultramafic and mining environments with isotopic exchange kinetic data and models. *Appl. Geochem.* 64, 146–156.

Zhang, W.-J., Lin, M.-F., 2020. Influence of redox potential on leaching behavior of a solidified chromium contaminated soil. *Sci. Total Environ.* 733, 139410. doi: 10.1016/j.scitotenv.2020.139410.

Zhuang, P., Zou, B., Li, N.Y., Li, Z.A., 2009. Heavy metal contamination in soils and food crops around Dabaoshan mine in Guangdong, China: Implication for human health. *Environ. Geochem. Health* 31, 707–715. doi: 10.1007/s10653-009-9248-3.

6

Overview on the Effects of Heavy Metals on the Biological Activities of Leafy Vegetables

Diyadyuti Das, Brahmarshi Mondal, Archita Dey, Deepanjan Mridha, and Tarit Roychowdhury
Jadavpur University

CONTENTS

6.1 Introduction

Green leafy vegetables are a hub of several nutritious contents provided to us. Now in today's world, the excessive accumulation of heavy metals is posing a problem for the basic metabolic activities of green leafy vegetables (Gupta et al., 2012). Most of the production of leafy vegetables is facing new challenges offered by biological and abiotic stresses. One of the catchiest things of concern in today's world is the effect of heavy metals on the biological activities of leafy vegetables. Mostly those metals pose a problem in their normal biological cycle and bring in them some negative as well as positive modifications. Therefore, the product which we procure from those leafy vegetables growing under an area having heavy-metal toxicity can be of mixed type. The accumulation of heavy metals like arsenic, mercury, iron, copper, manganese, nickel, cadmium, cobalt, zinc in soil has been occurring since long (Ghori et al., 2019). Almost all the developing countries in the world are deliberately consuming these veggies growing more or less in these heavy-metal–affected areas. The non-biodegradable nature of heavy metals comes up as the major problem in this arena. Their most dreaded characters are getting biomagnified as they are having a long biological half-life (Bagdatlioglu et al., 2010). The waste disposed of from several factories, disposal of sewage and application of fertilizers and smelting are accelerating the leaching of heavy toxic metals

DOI: 10.1201/9781003258063-6

into the groundwater or biomagnifying them into the surface of the soil (Aydinalp and Marinova, 2009). Leafy vegetables have been studied to primarily produce reactive oxygen species (ROS) in response to the accumulation of heavy metals in their surroundings (Yadav, 2010). Either this causes a havoc imbalance in the basic metabolism in the plants or else it totally disrupts the electron transport chain in these leafy vegetables (Qadir et al., 2004). There is no way we can purge them out biologically from our environment through some natural ways. Heavy metals can be categorized into two divisions: first one is the immobile one, i.e. it remains static in a single place where they are being accumulated and do not wander or get absorbed to some other places, the second one is the mobile one i.e. the dreadful one, those are the one which are affecting the leafy vegetables in a higher rate as they are being taken up by the roots of the plants through processes like endocytosis, diffusion, capillary action, transpiration etc. (Burakova et al., 2018). The positive effects generally include the heavy metals acting as the co-factors but the negative effects fall heavy over the positive ones. As soon as their concentration precedes the optimum limit, it becomes alarmingly toxic for the plant materials (Ali et al., 2011). Most of the plants that are growing under the influence of pesticides actually carry a large variety of heavy metals and also the plants that are growing in the coastal areas and are exposed to saline-rich soil turn up to be bio-accumulators (Li et al., 2010). Several significant enzymes and proteins are being ruined by the heavy-metal stress and they also replace the position of some beneficial metal ions from the leafy vegetables, which in turn sabotages the basic integrity of the metabolic system of the plants and the important processes like photosynthesis, homeostasis and the respiration are imbalanced (Hossain et al., 2012). Several countries have gone for analysis of the changes in the biochemical and physiological aspects of the plant. They have come up with some changes in their chlorophyll contents, peroxidase (POD) activity (Li et al., 2010) and there was oxidative stress leading to some pathways (Yadav, 2010). Their efforts have also suggested some effective alleviation techniques like chelation techniques using silicon (Ghori et al., 2019) and by genetically manipulating the genes for glutathiones and phytochelatins for making the plants tolerant to heavy-metal stress (Zhu et al., 1999).

6.2 Different Heavy Metals and Their Overall Reactions on Leafy Vegetables

6.2.1 Zinc

Zinc (Zn) is one of the most important metals required by leafy vegetables for their normal and proper development. They are also very essential in the synthesis mechanism of proteins, carbohydrates and auxin. It also serves as being the co-factor in several enzyme-dependent reactions in the plants. They are only helpful when present in trace amounts; however, when their amount crosses the optimum amount which is in between 150 and 300 mg/kg present in the soil (Vries et al., 2007), it comes out to be toxic and its over-accumulation can lead to the alteration of many enzymatic reactions and the co-accumulation of metals like copper and manganese in the roots and stems of leafy vegetables (Nagajyoti et al., 2010). Their over-accumulation can lead to the symptoms like purpling of leaves (Lee et al., 1996), thickening of the roots and a blockage in the cell division and elongation method. Zn toxicity has also emerged to be the sole cause for the loss of free calcium in plants crystallizing those (Todeschini et al., 2011) and has posed as a great hindrance in photosynthesis and ATP production by coherently inhibiting PS-I and PS-II (Assche and Clijsters, 1990). The toxic effect of zinc can also give rise to chlorosis in the young leaves (Ebbs and Kochian, 1997).

6.2.2 Copper

Copper (Cu) acts as a micronutrient in leafy vegetables (Thomas et al., 1998). There are evidences of interference of Cu in disrupting the mechanisms of ATP synthesis, photosystems and electron transport systems in which they are a part of proteins like plastocyanin and cytochrome oxidase, respectively (Demirevska-kepova et al., 2004), if it gets over-accumulated in the soil as an act of industrialization. The outcomes are the occurrence of chlorosis (Lewis et al., 2001), the production of ROS leading to the damage to macronutrients (Hegedüs et al., 2001) and an interruption to several metabolic pathways due to increased oxidative stress (Stadtman, 1991). Retardation in the root growth and a leakage in ion transport are some co-effects of copper toxicity in plants (Bouazizi et al., 2010). Finally, its toxic effects also include the delaying and degrading the plant germination and also affect the number of roots of that plant.

6.2.3 Chromium

Chromium (Cr) causes genuine defilement in groundwater, soil and dregs (Shankar et al., 2004) by the release of mechanical and sewage removal. The tanning industry has always been on the top for releasing Cr-rich sewage (Shankar et al., 2004). Chromium has a presence of 3,550 mt in the water bodies through discharge worldwide (Nriagu, 1990). The leading quality of chromium is that it can cause a very dreadful effect to the plants even in a very low concentration like 1.33 ppm (Gupta and Ali, 2013). Chromium toxicity causes chlorosis, development hindrance, withering of the apex and wounding of roots (Öztürk et al., 2015). The only positive effect of chromium is the generation of metabolites like phytochelatins and histidines to induce Cr stress tolerance in the plants (Schmfger, 2001). Major physiological activities of the plants like photosynthesis, plant-water relations etc. are hampered by Cr toxicity. There occurs an imbalance in the levels of mineral nutrition (Shankar et al., 2004). Its incessant production of the ROS (Yadav, 2010) causes alteration of the production of important photosynthetic pigments in plants (Boonyapookana et al., 2002) and antagonistically influences plants by increasing the generation of ascorbic acid and glutathiones (Shankar et al., 2004).

6.2.4 Lead

Lead (Pb) has come up as one of the most toxic heavy metals on earth that has over-accumulated in this nature due to the unforeseen uses of industrialization and due to the continuous weathering process of nature (Ashraf et al., 2016). Several plants have the capability to take up more than 50 mg of Pb which makes it even worse in terms of toxicity (Tanagahu et al., 2011). Soil productivity can be reversed by an absurd concentration of the Pb, whereas its scarcity can inhibit major processes like photosynthesis, absorption of water by plants and cell division. Its phytotoxicity includes reduced foliar structures, darkening of the green leaves, roots shortening and turning brown (Bhattacharyya et al., 2008). Lead also delays germination of seeds but it gets alleviated by some effective mechanisms and by the work of microorganisms (Ashraf and Ali, 2007). The toxicity further reduces the capability of uptaking nutrients by the plants and hence making them unable to perform nitrogen fixation (Guala et al., 2010). Blockage of Calvin cycle, electron transport chain and reducing stomatal pores are other toxic effects (Sharma and Dubey, 2005). Permeability of cell membrane, balance of mineral nutrients and the functions of hormones and some specific enzymes (namely sulphhydryl groups) are also reversed by this lead accumulation (Ghori et al., 2019).

6.2.5 Mercury

Mercury is one of those metals which has no positive effects on plants and can only act as an antagonist in the biochemical activities of the plant (Hameed et al., 2017). It is a kind of advantage due to its different existential forms but its phytotoxic effect is only visible when it exists as the ionic form (Hg^{2+}) which is re-methylated by an anaerobic bacterium (Tangahu et al., 2011) and is highly taken up by the plants (Kamal et al., 2004). According to the evidences collected till now, it is being reported that toxic level of Hg^{2+} has been really deadly for all the plants as they enter the plants through the water and immediately blocks the water channels inhibiting the proper flow of water to the plants. It also has a grave role in breaking down the lipid membranes by triggering the production of ROS to generate the extensive oxidative stress (Nagajyoti et al., 2010). It has also known to have a role in unbalancing the genetic stability of leafy vegetables (Malar et al., 2015). One of its deadliest effects is to tamper the electron transport chain in mitochondria and chloroplast changing or disabling their whole functions (Tangahu et al., 2011).

6.2.6 Cadmium

The pesticides used for the protection of the veggies against the pests actually house an enormous amount of cadmium in them. The optimum level for Cd is 100 mg/kg (Salt et al., 1995). Excessive Cd accumulation interferes in the uptake of other significant nutrients like Ca, K, Mg etc. that prohibits nitrate reductase which causes the blocking of nitrate absorption and translocation. It also disrupts the balance of water by actually dismantling the integration of the plasma membrane at the time of lipid peroxidation (Nagajyoti et al., 2010). Root tips browning, stunted growth and chlorosis are some of the common symptoms of cadmium toxicity (Wojcik and Tukiendorf, 2004). In some plants, cadmium replaces calcium in calcium-calmodulin pathway and inhibits phosphodiesterase activity which is highly dependent on calmodulin (Rivetta et al., 1997).

6.2.7 Arsenic

Arsenic has been considered toxic since long and there has been scientific journals alarming us regarding its toxicity. Though it is not being considered as a heavy metal, its toxic presence has a shocking measure of 3,470 mg/g in some of the plant species (Porter and Peterson, 1975). Arsenic, just like mercury, appears to occur in various methylated forms like arsenate and arsenite. Its over-exposure can cause malignancy (Das et al., 2021). These methylated forms are believed to cause phytotoxicity in plants. One of the beneficial effects of As is to lessen the toxic effects of another heavy-metal selenium in plants. Its inorganic forms serve to be the deadliest out of all. Sulphate application can somewhat reduce the effect of arsenic in *Oryza sativa* (Mridha et al., 2022). The main negative effect of As toxicity is that it is analogous to the phosphates and outcompetes then in the run for similar carriers persisting in plant plasmalemma. However, As influx can be reduced effectively by suppressing the carriers respective for As (Meharg and Macnair, 1992) (Table 6.1).

6.3 Physiological Responses by Plants

Plants do employ a lot of physiological responses in order to cope up with heavy-metal stress. They also undergo some major biochemical pathway changes. They impart two types of responses, either the tolerance-inducing one, else the one which chooses avoidance (Krzesłowska, 2011). They exhibit the rapid function of some transporter proteins, the change in chlorophyll content and POD activity (Li et al., 2010) as a way of response. They also perform the tolerance mechanisms with the help of some compounds like phytochelatins and the production of some secondary metabolites (Hossain et al., 2012).

6.3.1 The Response Exhibited by the Transport Proteins

Transport proteins play one of the key roles in heavy-metal stress and their homoeostasis. To get relieved from heavy-metal stress, ruling out the mechanisms for heavy-metal uptake is one of the options. Some transporter proteins (CPx-type) play a leading role in metal uptake in plants. The ATPase, ZIP Natural-Resistance–Linked Macrophage Protein Family (NRAMP), CTR transporter and the cation diffusion facilitator are the majors responsible for metal homeostasis and stress tolerance (Williams et al., 2000). The plants having the ability to transport divalent cations like Zn^{2+}, Fe^{2+}, Mn^{2+} and Cd^{+2} have shown to be one of the first members to house and utilize ZIP protein transporters (Eide et al., 1996). The ZIP family IRT1 gene is over-expressed in *Arabidopsis* and that leads to the over-accumulation of heavy metals like Fe, Cd and Zn (Connolly et al., 2002). ZIP 1 and ZIP 2 are the root stellar cells transporters that mobilize the Mn and Zn to cytoplasm from root stellar vacuoles, and they are further transferred to the parenchyma cells (Milner et al., 2013). The *Arabidopsis* roots and shoots possess ZIP4 which helps in Zn restriction and nutrition (Manara, 2012). NRAMP transporters are quite efficient in transporting a lot of metal ions (Nevo and

TABLE 6.1
The Effect of Different Heavy Metals on Different Plants

Heavy Metal	Affected Plant	Toxic Effects	References
Zinc	*Phaseolus vulgaris*	PS-I and PS-II are inhibited and ATP-synthesis is also affected. Zn displaces magnesium in Rubisco leading to blockage of carbon fixation	Assche and Clijsters (1990)
Copper	*Solanum melongena*	Inhibits or blocks seed germination, negatively affects root numbers and shoot length	Neelima and Reddy (2002)
Chromium	*Datura innoxia*	Alters the process of seed germination, inhibits photosynthesis and exposes it to oxidative stress	Vernay et al. (2008)
Mercury	*Digitaria sp.*	It replaces magnesium from chlorophyll, in turn prohibiting both PS-I and PS-II. Light and dark reactions are affected by its accumulation	Patra and Sharma (2002)
Lead	*Eicchornia crassipes*	Increases in antioxidant activity of the enzymes and increases in stress tolerance	Malar et al. (2016)
Cadmium	*Alternanthera bettzickiana*	Stunted growth, reduced biomass and inhibition of the photosynthetic pigment	Tauqeer et al. (2016)
Arsenic	*Brassica oleracea*	Defects in growth and nourishment of plants. Excessive oxidative stress due to ROS production	Abbas et al. (2017)
Combined heavy metals	*Helianthus annuus*	Production of promutagenic adducts	Chakravarty and Srivastava (1992)

TABLE 6.2
Different Protein Transporter Families and Their Roles

Transporter Family	Main Transporters	Defined Role	References
ZIP	ZIP 1, ZIP2, ZIP 4	Relocates Mn and Zn from root stellar vacuoles to cytoplasm and finally to parenchymatous cells. Induces zinc tolerance	Milner et al. (2013) and Manara (2012)
NRAMP	AtNRAMP, NRAMP1	Relocation of Fe and Cd. Moderating. MAPK pathway	Thomine et al. (2000) and Moisan et al. (2006)
ABC	AtPDR8, MRP, PDR, HMT1, AtABCC1, AtABCC2	Pumping out the metals, chelating and sequestering the heavy metals and translocating them into the vacuoles, transport of phytochelatin metalloids	Kim et al. (2007) and Ortiz et al. (1995)
CTR	CTR1	Serves as both membrane and lysosomal metal translocator protein	Puig and Thiele (2002)
ATPases	CPx, HMS and AtHMA4	Elimination of the soft cations and the induction of zinc tolerance and relocation	Ghori et al. (2019) and Mills et al. (2003)

Nelson, 2006). The NRAMP gene has been evidently reported to encode some metal transporters in plants and the Fe and the toxic Cd are being relocated by the AtNRAMP (Thomine et al., 2000). The NRAMP1 has a contribution to the process of the modification of the MAPK pathway (Moisan et al., 2006).

The eight sub-families of ABC transporters play a huge role in various cellular metabolisms related to heavy metals in plants. AtPDR8, an *Arabidopsis*, ABC transporter has emerged as one of the main cadmium removers by pumping it out among the metal transporters (Kim et al., 2007). MRP and PDR are two of the specific sub-families of ATP transporters that play a leading role in chelation and sequestration of the metals and their translocation to the vacuoles. The phytochelatin-cadmium complexes are transported into the vacuoles by HMT1, the very first ABC transporters of the vacuoles (Ortiz et al., 1995). AtABCC1 and AtABCC2 are two of the complex transporters known to transport the phytochelatin metalloids. CTR transporters are the sole reason for maintaining the homoeostasis of Cu. CTRs can both act as a cell membrane and lysosomal membrane proteins and translocates Cu either from extracellular spaces to cytosol or lysosome to cytosol, respectively (Puig and Thiele, 2002). ATPase is generally those types of transporters that utilize the energy released by ATP hydrolysis. The metal cations that are soft are being transported by the CPx or the HMS, by pumping them out of the cell (Ghori et al., 2019). The tolerance of zinc in plants and relocation of Zn from root to shoot is increased with the help of AtHMA4, a member of the HMA group (Mills et al., 2003) (Table 6.2).

6.3.2 Change in Chlorophyll Content

In an experiment conducted by Li et al. (2010), there was a comparison drawn between a plant Chinese Kale and another plant *Pakchoi*. Chinese Kale was somewhat tolerant to heavy-metal stress and the *Pakchoi* was a bit more susceptible to it. The result revealed that both the plants faced a reduction in their chlorophyll content than the control, the Chinese Kale being at more loss. It also brings a change in chlorophyll ratio that in turn alters the PS-I and PS-II ratios (Anderson, 1986). Chlorophyll contents do decrease due to the inhibition of enzymes associated with chlorophyll biosynthesis by heavy metals. Protochlorophyll reductase along with aminolevulonic acid synthesis is also hindered (Stobart et al., 1985).

6.4 The Heavy-Metal-Stress–Induced Signal Transduction Pathways

Many of the transcription factors get activated due to this heavy-metal stress which in turn activates the genes that are related to stress. These genes lead to the activation of some biochemical defence mechanisms that in turn alter some cascades of signalling pathways and modify their ways somehow.

6.4.1 The Primary and the Initial Line of Defence

The plant machinery causes the alteration of pH of the rhizosphere due to some root exudates that lead to the collection of the metal ions into the apoplast. Even the transporter proteins show least susceptibility to these metal ions and fails to carry them to the symplast and ultimately to the cell (Ghori et al., 2019). To make the plants more resistant, other useful metals have been reported to act antagonistically for the toxic metals and sequester them out with the help of the formation of plaques (Liu et al., 2008). Nitric oxide has been reported to reverse the ROS into hydrogen peroxide and oxygen (Saxena and Shekhawat, 2013). They also contribute to the production of glutathione and limit the free radicals by moving them out of the cell (Lamattina et al., 2003). HGA, one of the domains of the pectin, is chiefly responsible for trapping toxic heavy metals (Krzesłowska, 2011). Phytochelatins and metallochelatins use their Cys-residue to bind with the metal ions, help them accumulate in the vacuole and ultimately pump them out of the cell and protect the cell from intracellular injuries caused by oxidative stress (Hossain et al., 2012). Hyperaccumulators (a different kind of plant) provide protection to the plant proteins against ROS and protein degradation due to oxidative stress (Zhen et al., 2007). The defence mechanisms of the non-enzymatic antioxidants in plants like ascorbic acid, glutathiones and other antioxidants like glutathione reductase and superoxide dismutase (SOD) provide the plants a stable metal tolerance but it is not distributed properly over all plants (Viehweger, 2014).

6.4.2 Signalling Cascade

The toxic and prominent heavy metals like As, Cd, Pb and Hg have their devastating effects on the agronomical significant crops (Tchounwou et al., 2012). The consequence of their over-accumulation leads to the production of the lethal ROS inside the cells or their gathering ends up substituting some significant components required for the biochemical pathways of the cell cytoplasm (Stohs and Bagchi, 1995).

6.4.2.1 MAP Kinase Pathway

MAP kinase or mitogen-associated protein kinase is a kind of receptor tyrosine kinase that is activated in response to the specific ligand. This receptor kinase is localized in the plasma membrane of the cell and is involved in the regulation and proliferation of cells. For the prior phenomena, the pathway, the ERK-MAPK pathway comes into play. But in case of stress conditions, the well-accounted pathway which gets in action is the p38 MAPK pathway. It acts against the osmotic stress; therefore, the plant can acclimatize itself against the different stress-inducing calamities.

6.4.2.1.1 Biological Impact

It actually creates a havoc response against the toxic metal accumulation. Over-accumulation of cadmium chloride and copper chloride leads to the reposition of the hydrogen peroxide inside the cell by inducing the activation of the MAPK pathway via the signals offered by the ROS (Maksymiec and Krupa, 2006). The hydrogen peroxide molecules serve to be the signal of the stress then, by easily diffusing into the adjacent cells. This leads to a massive damage inside the cell by the displacement of the macromolecules (Hossain et al., 2015). Figure 6.1 gives a clear picture of the scenario.

6.4.2.2 Calcium-Calmodulin Pathway

It has been reported that during the stress response, the calcium-calmodulin pathway shows a major functional role. The presence of a high amount of Pb, Cd etc. can enhance the signalling cascade of the calcium-calmodulin pathway.

6.4.2.2.1 Biological Impact

The over-assemblage of cadmium leads to the increase in the production of Inositol-1,4,5-triphosphate (IP3) which is actually derived from phosphatidyl-4,5-bisphosphate that is present in the plasma membrane by the action of the phosphatidylinositol kinase. Then, the IP3 is transported to the cytoplasm from the cell membrane and in turn the calcium ion channel of the endoplasmic reticulum is opened. The increase in the amount of calcium ions then initiates an interaction with calmodulin to conciliate the shaping of calcium-CaM complex that facilitates the biosynthesis of SOD and GSH to retain the homeostasis of hydrogen peroxide. The calcium-CaM formation also takes place by the over-accumulation of Pb, Cr and La (Jalmi et al., 2018). Figure 6.2 gives an accurate picture of the pathway.

6.5 The Abrupt Change in the Biological Activities of the Leafy Vegetables and the Secondary Metabolites Produced to Cope Up with the Stress

Plants exposed to these heavy metals for a longer period of time lead to the production of ROS, that actually causes oxidative stress in the plants which ultimately leads to the activation and over-expression of various hormones, polyamines and polyphenols in order to deplete the oxidative stress within the leafy vegetables through their antioxidant role or the activation of the glutathione.

6.5.1 Role of Hormones Produced

The stress hormones like ethylene and jasmonic acid are produced mainly during heavy-metal stress along with the accumulation of salicylic acid (SA). The concentration of SA leads

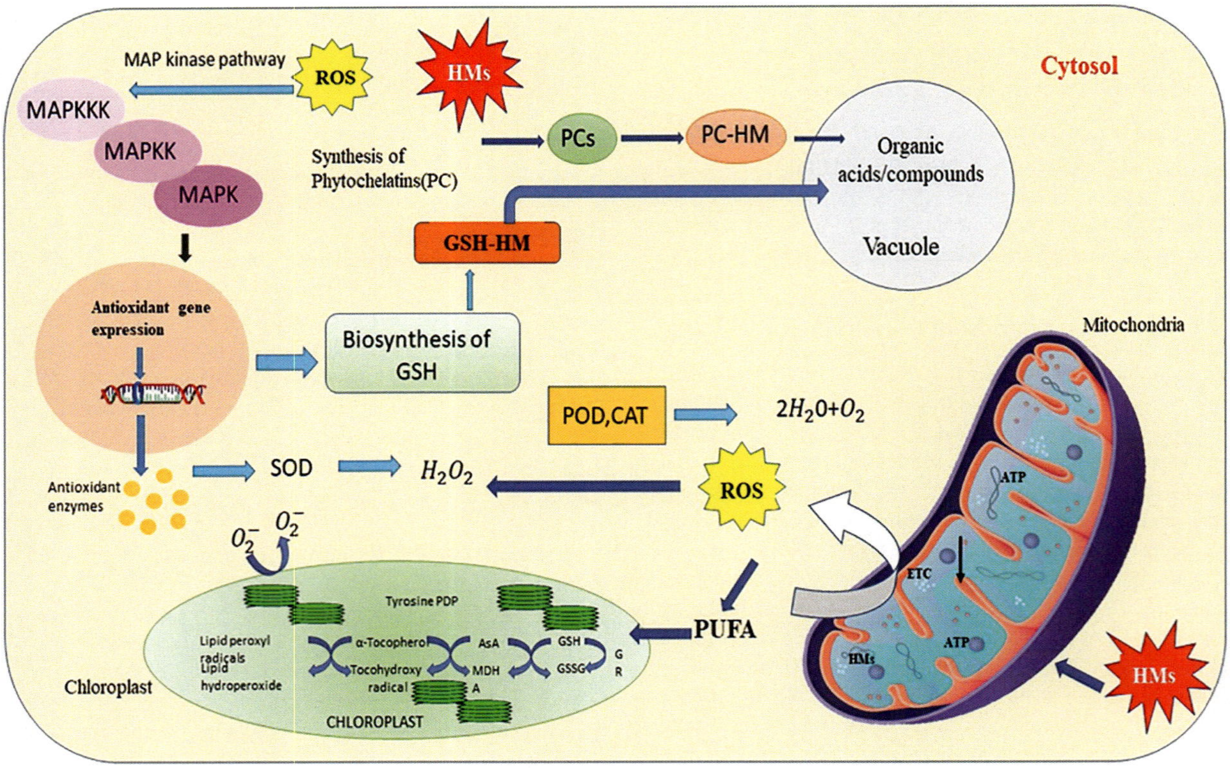

FIGURE 6.1 Mitogen-associated kinase pathway in response to the heavy-metal stress and consequences. (Modified from Sytar et al. (2013).)

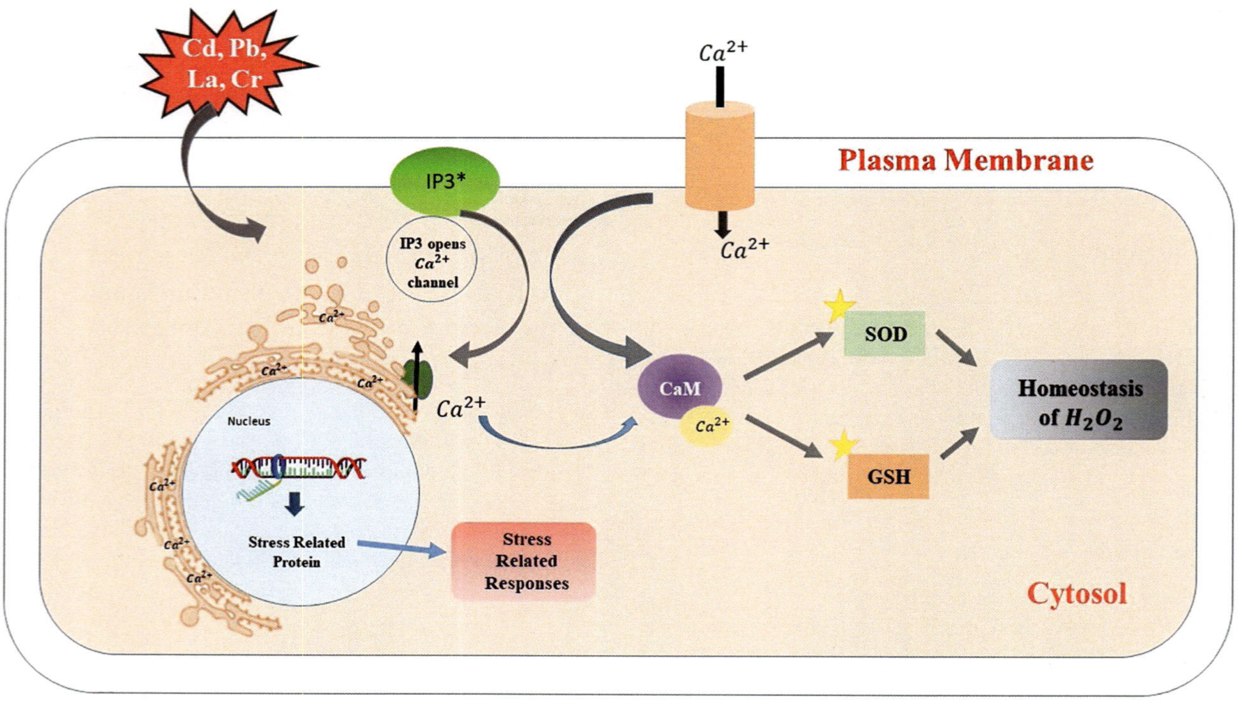

FIGURE 6.2 Representation of calcium-calmodulin-mediated signalling cascade during heavy-metal stress conditions. (Modified from Snedden and Fromm (1998).)

to the formation of the glutathione that in turn eliminates the ROS, maintains the antioxidant level and also helps various genes to induce their expression (Hossain et al., 2012). *Brassica juncea*, a plant treated by Cd, is reported to show mRNA that is responsive to auxin (Minglin et al., 2005). Mercury-treated *Arabidopsis* exhibited the activation of a protein named nitrilase and a gene called SAMT that leads to the biosynthesis of abscisic acid (ABA) and SA, respectively (Shao et al., 2010). Ethylene and ABA induce a cascade of events upon exposure to cadmium (Polle and Schützendübel, 2003). Proline that is found in the metal-tolerant plants works in alleviating the free radicals that are being produced due to ROS. They also produce glutathiones and glutathione-producing enzymes.

6.5.2 Role of Polyamines

On the other hand, polyamines (organic cations) like spermidine, spermine and putrescine play a pivotal role in both types of stresses by digesting the ROS. Proline and polyamines both come under "general adaptation syndrome" when plant produces a response against heavy-metal stress (Hossain et al., 2012).

6.5.3 Role of Polyphenols

Polyphenols are considered as one of the largest groups of metabolites that are mostly procured from the shikimic acid pathway. The level of polyphenols is generally found to rise in a leafy vegetable during heavy-metal stress which was estimated using Folin-Ciocalteu reagent (Jagadish et al., 2009). The polyphenols are divided into four sub-groups; however, two of them play a major role in the heavy-metal stress in leafy vegetables. Different kinds of polyphenols are listed below.

6.5.3.1 Phenolic Acids

They are basically the benzene ring-linked carboxyl group either derived from benzoic acid (hydroxybenzoic acid) or cinnamic acid (hydroxycinnamic acid). They have their intrinsic role during the heavy-metal stress and they are being produced even more to scavenge the ROS and eradicate the oxidative stress in plants. They are being produced by the activation of the phenylpropanoid biosynthetic pathway by the activity of the plant's antioxidant system (Ben-Abdallah et al., 2018). Chalcone synthase and phenylalanine ammonia lyase are the genes that induce the production of phenolic acids (Sharma et al., 2016). The p-coumaric acid acts as the precursor of the lignin (Goleniowski et al., 2013). Phenolic acids like gallic acid, caffeic acid and ferulic acid induce tolerance in leafy vegetables during heavy-metal stress (Wang et al., 2019). Vanillic acid plays an important role in providing stress tolerance to *Zea mays* from heavy metals like Cd, Pb and Cu (Kisa et al., 2016). Ellagic acid provides the same for *Amaranthus tricolor* (Sarker and Oba, 2018).

6.5.3.2 Flavonoids

Flavonoids are the most popular sub-group of the polyphenols and have six classes under them: flavonols, flavanons, flavon-3-ols, flavones, isoflavones and flavanonols. Flavonoids do have a very significant role in the eradication of the oxidative stress produced by the ROS and gearing up the leafy vegetables with a modest amount of stress tolerance. In tomatoes, heavy-metal stress increases the content of naringenin chalcone which chelates the heavy metals and sequesters them in compartments (Martinez et al., 2016). Luteolin and apigenin have a marvellous antioxidant activity during oxidative stress (Hodaei et al., 2018). The content of daidzin and genistein increases during heavy-metal stress and they help in scavenging free radicals (Gutierrez-Gonzalez et al., 2010). Similarly, epicatechin levels are increased during stress and upregulate the antioxidant system then (Cuong et al., 2020). In *Vitis vinifera,* quercetin and kaempferol help in engulfing the oxidative stress by getting their contents increased (Kőrösi et al., 2019). In *Arabidopsis*, different kinds of anthocyanins get accumulated during stress condition and help it cope up with the stress (Kovinich et al., 2014).

6.5.4 Role of Glutathione as an Antioxidant

Glutathione plays a leading role in heavy-metal sequestration and scavenging ROS (Foyer and Noctor, 2005). The nucleophilic nature of its thiol group builds up a marcaptide bond with the toxic metals. It also maintains the level of the hydrogen peroxide level by getting oxidized at the time of its degradation. Its reduced form also neutralizes the ROS-produced free radicals during stress (Foyer and Noctor, 2005). It actually binds with the heavy metals, translocates them to the cell vacuoles and detoxifies them there, protecting the cell. Their reduction can lead to a loss of balance in redox potential thus leading to the generation of a stress signal (Nocito et al., 2006). They also act as the precursor for the production of phytochelatins in heavy-metal stress. S-Nitrosoglutathiones that are induced under stress conditions bind to NO and accompany them in signalling. Reduced glutathiones are basically required for the regulation of glyoxylate pathways in heavy-metal stress (Yadav, 2010). A picture of the cycle has been represented in Figure 6.3.

6.6 Plant Adaptation Towards Heavy-Metal Stress

Due to this heavy-metal stress, plants have adapted some unique techniques of their own to cope up with the situation (Parmar et al., 2002). Several plants have adapted the use of another heavy-metal silicon to negate the effects of other heavy metals, which helps in prohibiting their uptake in plants, induces antioxidation and makes them precipitate along with it (Adrees et al., 2015). Plants also live in a symbiotic relationship with VAM to minimize the heavy-metal uptake (Emamverdian et al., 2015). The increase in proline and ethylene content also helps in the alleviation of the stress by its triple response (Thao et al., 2015). Sometimes, external application of plant hormones like 24-epibrassinolide (EBL) activates the enzymes that have antioxidant activity to eliminate the stress (Shahzad et al., 2018). Compartmentalizing the vacuoles and collecting

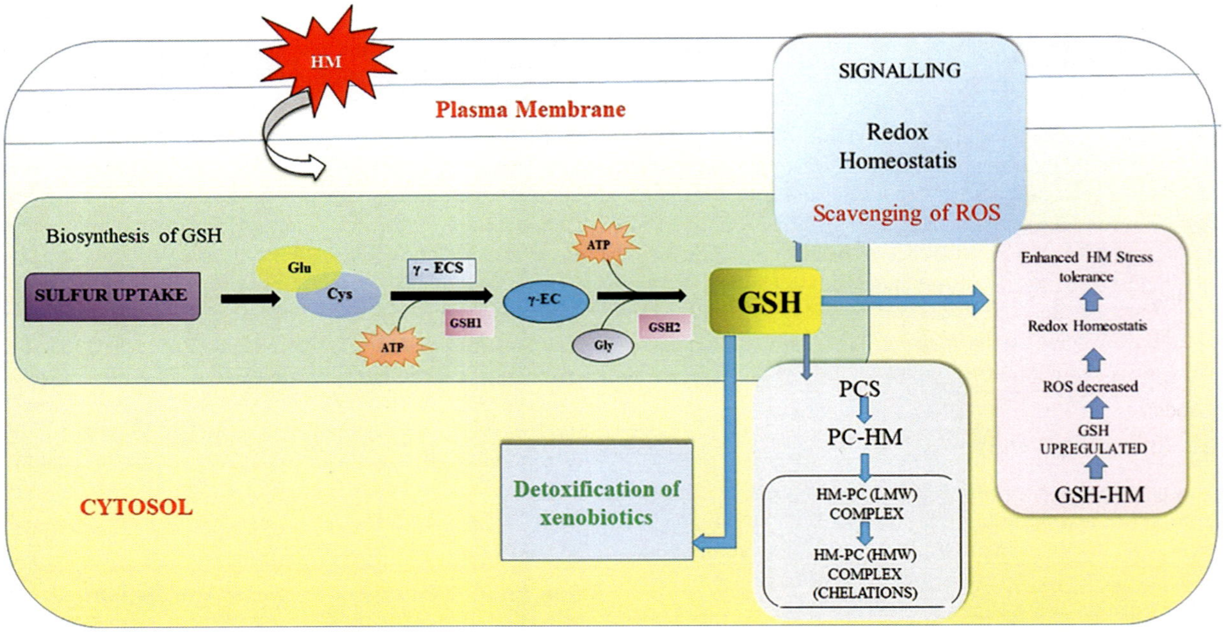

FIGURE 6.3 Contribution of glutathione in heavy-metal stress tolerance in plants. (Modified from Asgher et al. (2017).)

the heavy metals in different compartments to throw them out later is another alleviation technique (Sharma et al., 2016). The generation of plant phenolics in response to heavy-metal stress is another way of dealing with it (Kisa et al., 2016). The polyphenols many times play a significant role in scavenging the ROS and relieving the plants from the oxidative stress by the increase of the contents of phenolic acids and flavonoids which in turn chelates and sequesters the heavy metals to pump them out of the plant cells or else make the plants tolerant against these heavy metal stress.

6.7 Conclusion

The toxic effects of different heavy metals over the plants are posing great problems for the environment. Not only has it disrupted the normal metabolism of the plants but also has affected us indirectly as we are consuming these leafy vegetables in large amounts. Thus, they are being biomagnified in the ecosystem. From the list above, we have found mercury to be the most toxic one and copper being the least toxic one. We have also come to know some major physiological and biochemical changes that take place as soon as the plant is exposed to the heavy metals. These changes only ensure the increase in the heavy metal tolerance of the plants. Some of the proteins, phenolics and hormones are activated due to this alteration which in turn opposes the uptake of these toxic metals by the plants. Even though the plants end up uptaking them, these secondary metabolites and transport proteins do try to inhibit their movement inside the cell and use several techniques to sequester those metals into various compartments. There occur many changes in the signalling cascades and the production of ROS leading to the oxidative stress in the plants. At the end, we also come across the salvation techniques, where plants either by their innate techniques or by the external boosting up with the hormones like EBL, clutter out the toxic metals out of the cells. We must throw a light upon these stress alleviation methods for future use in the agronomic sector.

REFERENCES

Abbas, S.Z., Rafatullah, M., Ismail, N., Nastro, R.A., 2017. Enhanced bioremediation of toxic metals and harvesting electricity through sediment microbial fuel cell. *International Journal of Energy Research*, 41(14), 2345–2355.

Adrees, M., Ali, S., Rizwan, M., Zia-ur-Rehman, M., Ibrahim, M., Abbas, F., Farid, M., Qayyum, M.F., Irshad, M.K., 2015. Mechanisms of silicon-mediated alleviation of heavy metal toxicity in plants: A review. *Ecotoxicology and Environmental Safety*, 119, 186–197.

Ali, I., Khan, T.A., Asim, M., 2011. Removal of arsenic from water by electrocoagulation and electrodialysis techniques. *Separation & Purification Reviews*, 40(1), 25–42.

Anderson, J.M., 1986. Photoregulation of the composition, function, and structure of thylakoid membranes. *Annual Review of Plant Physiology*, 37(1), 93–136.

Asgher, M., Per, T.S., Anjum, S., Khan, M.I.R., Masood, A., Verma, S., Khan, N.A., 2017. Contribution of glutathione in heavy metal stress tolerance in plants. In: *Reactive Oxygen Species and Antioxidant Systems in Plants: Role and Regulation under Abiotic Stress*. Eds. Iqbal, M., Khan, R., Khan, N.A., Springer, Singapore, pp. 297–313.

Ashraf, R., Ali, T.A., 2007. Effect of heavy metals on soil microbial community and mung beans seed germination. *Pakistan Journal of Botany*, 39(2), 629.

Ashraf, M.Y., Roohi, M., Iqbal, Z., Ashraf, M., Öztürk, M., Gücel, S., 2016. Cadmium (Cd) and lead (Pb) induced changes in growth, some biochemical attributes, and mineral accumulation in two cultivars of mung bean [Vigna radiata (L.) Wilczek]. *Communications in Soil Science and Plant Analysis,* 47(4), 405–413.

Assche, F.V, Clijsters, H., 1990. Effects of metals on enzyme activity in plants. *Plant, Cell & Environment,* 13(3), 195–206.

Aydinalp, C., Marinova, S., 2009. The effects of heavy metals on seed germination and plant growth on alfalfa plant (Medicago sativa). *Bulgarian Journal of Agricultural Science,* 15(4), 347–350.

Bagdatlioglu, N., Nergiz, C., Ergonul, P.G., 2010. Heavy metal levels in leafy vegetables and some selected fruits. *Journal für Verbraucherschutz und Lebensmittelsicherheit,* 5(3), 421–428.

Ben-Abdallah, S., Zorrig, W., Amyot, L., Renaud, J., Hannoufa, A., Lachâal, M., Karray-Bouraoui, N., 2018. Potential production of polyphenols, carotenoids and glycoalkaloids in Solanum villosum Mill. under salt stress. *Biologia,* 74(3), 309–324.

Bhattacharyya, P., Chakrabarti, K., Chakraborty, A., Tripathy, S., & Powell, M. A. (2008). Fractionation and bioavailability of Pb in municipal solid waste compost and Pb uptake by rice straw and grain under submerged condition in amended soil. Geosciences Journal, 12(1), 41–45.

Boonyapookana, B., Upatham, E.S., Kruatrachue, M., Pokethitiyook, P., Singhakaew, S., 2002. Phytoaccumulation and phytotoxicity of cadmium and chromium in duckweed Wolffia globosa. *International Journal of Phytoremediation,* 4(2), 87–100.

Bouazizi, H., Jouili, H., Geitmann, A., El Ferjani, E., 2010. Copper toxicity in expanding leaves of *Phaseolus vulgaris* L.: Antioxidant enzyme response and nutrient element uptake. *Ecotoxicology and Environmental Safety,* 73(6), 1304–1308.

Burakova, E.A., Dyachkova, T.P., Rukhov, A.V., Tugolukov, E.N., Galunin, E.V., Tkachev, A.G., Ali, I., 2018. Novel and economic method of carbon nanotubes synthesis on a nickel magnesium oxide catalyst using microwave radiation. *Journal of Molecular Liquids,* 253, 340–346.

Chakravarty, B., Srivastava, S., 1992. Toxicity of some heavy metals in vivo and in vitro in Helianthus annuus. *Mutation Research Letters,* 283(4), 287–294.

Connolly, E.L., Fett, J.P., Guerinot, M.L., 2002. Expression of the IRT1 metal transporter is controlled by metals at the levels of transcript and protein accumulation. *The Plant Cell,* 14(6), 1347–1357.

Cuong, D.M., Kwon, S.J., Nguyen, B.V., Chun, S.W., Kim, J.K., Park, S.U., 2020. Effect of salinity stress on phenylpropanoid genes expression and related gene expression in wheat sprout. *Agronomy,* 10(3), 390.

Das, A., Joardar, M., Chowdhury, N.R., De, A., Mridha, D., Roychowdhury, T., 2021. Arsenic toxicity in livestock growing in arsenic endemic and control sites of West Bengal: Risk for human and environment. *Environmental Geochemistry and Health,* 43(8), 3005–3025.

Demirevska-Kepova, K., Simova-Stoilova, L., Stoyanova, Z., Hölzer, R., Feller, U., 2004. Biochemical changes in barley plants after excessive supply of copper and manganese. *Environmental and Experimental Botany,* 52(3), 253–266.

Ebbs, S.D., Kochian, L.V., 1997. Toxicity of zinc and copper to Brassica species: Implications for phytoremediation. *Journal of Environmental Quality,* 26(3), 776–781.

Eide, D., Broderius, M., Fett, J., Guerinot, M.L., 1996. A novel iron-regulated metal transporter from plants identified by functional expression in yeast. *Proceedings of the National Academy of Sciences,* 93(11), 5624–5628.

Emamverdian, A., Ding, Y., Mokhberdoran, F., Xie, Y., 2015. Heavy metal stress and some mechanisms of plant defense response. *The Scientific World Journal,* 2015, 756120.

Foyer, C.H., Noctor, G., 2005. Redox homeostasis and antioxidant signaling: A metabolic interface between stress perception and physiological responses. *The Plant Cell,* 17(7), 1866–1875.

Ghori, N.H., Ghori, T., Hayat, M.Q., Imadi, S.R., Gul, A., Altay, V., Ozturk, M., 2019. Heavy metal stress and responses in plants. *International Journal of Environmental Science and Technology,* 16(3), 1807–1828.

Goleniowski, M., Bonfill, M., Cusido, R., & Palazón, J. (2013). Phenolic acids. Natural products, 2013, 1951–1973.

Guala, S.D., Vega, F.A., Covelo, E.F., 2010. The dynamics of heavy metals in plant–soil interactions. *Ecological Modelling,* 221(8), 1148–1152.

Gupta, V.K., Ali, I., 2013. *Environmental Water: Advances in Treatment, Remediation and Recycling,* 1st Edition. Elsevier. eBook ISBN: 9780444594037.

Gupta, S., Nayek, S., Bhattacharya, P., 2012. Effect of air-borne heavy metals on the biochemical signature of tree species in an industrial region, with an emphasis on anticipated performance index. *Chemistry and Ecology,* 27(4), 381–392.

Gutierrez-Gonzalez, J.J., Guttikonda, S.K., Tran, L.S.P., Aldrich, D.L., Zhong, R., Yu, O., Nguyen, H.T., Sleper, D.A., 2010. Differential expression of isoflavone biosynthetic genes in soybean during water deficits. *Plant and Cell Physiology,* 51(6), 936–948.

Hameed, A., Qadri, T.N., Mahmooduzzafar, T.O.S., Ozturk, M., Altay, V., Ahmad, P., 2017. Physio-biochemical and nutritional responses of *Abelmoschus esculentus* (L.) Moench (Okra) under mercury contamination. *Fresenius Environmental Bulletin,* 26(10), 5814–5823.

Hegedüs, A., Erdei, S., Horváth, G., 2001. Comparative studies of H_2O_2 detoxifying enzymes in green and greening barley seedlings under cadmium stress. *Plant Science,* 160(6), 1085–1093.

Hodaei, M., Rahimmalek, M., Arzani, A., Talebi, M., 2018. The effect of water stress on phytochemical accumulation, bioactive compounds and expression of key genes involved in flavonoid biosynthesis in *Chrysanthemum morifolium* L. *Industrial Crops and Products,* 120, 295–304.

Hossain, M.A., Piyatida, P., da Silva, J.A.T., Fujita, M., 2012. Molecular mechanism of heavy metal toxicity and tolerance in plants: Central role of glutathione in detoxification of reactive oxygen species and methylglyoxal and in heavy metal chelation. *Journal of Botany,* 2012, 872875.

Hossain, M.A., Bhattacharjee, S., Armin, S.M., Qian, P., Xin, W., Li, H.Y., Burritt, D.J., Fujita, M., Tran, L.S.P., 2015. Hydrogen peroxide priming modulates abiotic oxidative stress tolerance: Insights from ROS detoxification and scavenging. *Frontiers in Plant Science,* 6, 420.

Jagadish, L.K., Krishnan, V.V., Shenbhagaraman, R., Kaviyarasan, V., 2009. Comparitive study on the antioxidant, anticancer and antimicrobial property of Agaricus bisporus (JE Lange) Imbach before and after boiling. *African Journal of Biotechnology,* 8(4), 654–661.

Jalmi, S.K., Bhagat, P.K., Verma, D., Noryang, S., Tayyeba, S., Singh, K., Sharma, D., Sinha, A.K., 2018. Traversing the links between heavy metal stress and plant signaling. *Frontiers in Plant Science,* 9, 12.

Kamal, M., Ghaly, A.E., Mahmoud, N., Cote, R., 2004. Phytoaccumulation of heavy metals by aquatic plants. *Environment International*, 29(8), 1029–1039.

Kim, D.Y., Bovet, L., Maeshima, M., Martinoia, E., Lee, Y., 2007. The ABC transporter AtPDR8 is a cadmium extrusion pump conferring heavy metal resistance. *The Plant Journal*, 50(2), 207–218.

Kısa, D., Elmastaş, M., Öztürk, L., Kayır, Ö., 2016. Responses of the phenolic compounds of Zea mays under heavy metal stress. *Applied Biological Chemistry*, 59(6), 813–820.

Kőrösi, L., Bouderias, S., Csepregi, K., Bognár, B., Teszlák, P., Scarpellini, A., Castelli, A., Hideg, É., Jakab, G., 2019. Nanostructured TiO_2-induced photocatalytic stress enhances the antioxidant capacity and phenolic content in the leaves of Vitis vinifera on a genotype-dependent manner. *Journal of Photochemistry and Photobiology B: Biology*, 190, 137–145.

Kovinich, N., Kayanja, G., Chanoca, A., Riedl, K., Otegui, M.S., Grotewold, E., 2014. Not all anthocyanins are born equal: Distinct patterns induced by stress in Arabidopsis. *Planta*, 240(5), 931–940.

Krzesłowska, M., 2011. The cell wall in plant cell response to trace metals: Polysaccharide remodeling and its role in defense strategy. *Acta Physiologiae Plantarum*, 33(1), 35–51.

Lamattina, L., García-Mata, C., Graziano, M., Pagnussat, G., 2003. Nitric oxide: The versatility of an extensive signal molecule. *Annual Review of Plant Biology*, 54(1), 109–136.

Lee, C.W., Choi, J.M., Pak, C.H., 1996. Micronutrient toxicity in seed geranium (Pelargonium× hortorum Bailey). *Journal of the American Society for Horticultural Science*, 121(1), 77–82.

Lewis, S., Donkin, M.E., Depledge, M.H., 2001. Hsp70 expression in Enteromorpha intestinalis (Chlorophyta) exposed to environmental stressors. *Aquatic Toxicology*, 51(3), 277–291.

Li, Q., Cai, S., Mo, C., Chu, B., Peng, L., Yang, F., 2010. Toxic effects of heavy metals and their accumulation in vegetables grown in a saline soil. *Ecotoxicology and Environmental Safety*, 73(1), 84–88.

Liu, H., Zhang, J., Christie, P., Zhang, F., 2008. Influence of iron plaque on uptake and accumulation of Cd by rice (*Oryza sativa* L.) seedlings grown in soil. *Science of the Total Environment*, 394(2–3), 361–368.

Maksymiec, W., Krupa, Z., 2006. The effects of short-term exposition to Cd, excess Cu ions and jasmonate on oxidative stress appearing in Arabidopsis thaliana. *Environmental and Experimental Botany*, 57(1–2), 187–194.

Malar, S., Sahi, S.V., Favas, P.J., Venkatachalam, P., 2015. Mercury heavy-metal-induced physiochemical changes and genotoxic alterations in water hyacinths [Eichhornia crassipes (Mart.)]. *Environmental Science and Pollution Research*, 22(6), 4597–4608.

Malar, S., Vikram, S.S., Favas, P.J., Perumal, V., 2016. Lead heavy metal toxicity induced changes on growth and antioxidative enzymes level in water hyacinths [Eichhornia crassipes (Mart.)]. *Botanical Studies*, 55(1), 1–11

Manara, A., 2012. Plant responses to heavy metal toxicity. In: *Plants and Heavy Metals*. Ed. Furini, A., Springer, Dordrecht, pp. 27–53.

Martinez, V., Mestre, T.C., Rubio, F., Girones-Vilaplana, A., Moreno, D.A., Mittler, R., Rivero, R.M., 2016. Accumulation of flavonols over hydroxycinnamic acids favors oxidative damage protection under abiotic stress. *Frontiers in Plant Science*, 7, 838.

Meharg, A.A., Macnair, M.R., 1992. Suppression of the high affinity phosphate uptake system: A mechanism of arsenate tolerance in *Holcus lanatus* L. *Journal of Experimental Botany*, 43(4), 519–524.

Mills, R.F., Krijger, G.C., Baccarini, P.J., Hall, J.L., Williams, L.E., 2003. Functional expression of AtHMA4, a P1B-type ATPase of the Zn/Co/Cd/Pb subclass. *The Plant Journal*, 35(2), 164–176.

Milner, M.J., Seamon, J., Craft, E., Kochian, L.V., 2013. Transport properties of members of the ZIP family in plants and their role in Zn and Mn homeostasis. *Journal of Experimental Botany*, 64(1), 369–381.

Minglin, L., Yuxiu, Z., Tuanyao, C., 2005. Identification of genes up-regulated in response to Cd exposure in *Brassica juncea* L. *Gene*, 363, 151–158.

Moisan, J., Thuraisingam, T., Henault, J., Sanctis, J.D., Radzioch, D., 2006. Role of SLC11A1 (formerly NRAMP1) in regulation of signal transduction induced by Toll-like receptor 7 ligands. *FEMS Immunology & Medical Microbiology*, 47(1), 138–147.

Mridha, D., Paul, I., De, A., Ray, I., Das, A., Joardar, M., Chowdhury, N.R., Bhadoria, P.B.S., Roychowdhury, T., 2022. Rice seed (IR64) priming with potassium humate for improvement of seed germination, seedling growth and antioxidant defense system under arsenic stress. *Ecotoxicology and Environmental Safety*, 219, 112313.

Nagajyoti, P.C., Lee, K.D., Sreekanth, T.V.M., 2010. Heavy metals, occurrence and toxicity for plants: A review. *Environmental Chemistry Letters*, 8(3), 199–216.

Neelima, P., Reddy, K.J., 2002. Interaction of copper and cadmium with seedling growth and biochemical responses in Solanum melongena. *Nature, Environment and Pollution Technology*, 1(3), 285–290.

Nevo, Y., Nelson, N., 2006. The NRAMP family of metal-ion transporters. *Biochimica et Biophysica Acta (BBA)-Molecular Cell Research*, 1763(7), 609–620.

Nocito, F.F., Lancilli, C., Crema, B., Fourcroy, P., Davidian, J.C., Sacchi, G.A., 2006. Heavy metal stress and sulfate uptake in maize roots. *Plant Physiology*, 141(3), 1138–1148.

Nriagu, J.O. (1990). Global metal pollution: poisoning the biosphere?. Environment: Science and Policy for Sustainable Development, 32(7), 7–33.

Ortiz, D.F., Ruscitti, T., McCue, K.F., Ow, D.W., 1995. Transport of Metal-binding Peptides by HMT1, A Fission Yeast ABC-type Vacuolar Membrane Protein (*). *Journal of Biological Chemistry*, 270(9), 4721–4728.

Öztürk, M., Ashraf, M., Aksoy, A., Ahmad, M.S.A., Hakeem, K.R. Eds., 2015. *Plants, Pollutants and Remediation*. Springer, Berlin/Heidelberg, Germany.

Parmar, N.G., Vithalani, S.D., Chanda, S.V., 2002. Alteration in growth and peroxidase activity by heavy metals in Phaseolus seedlings. *Acta Physiologiae Plantarum*, 24(1), 89–95.

Patra, M., Sharma, A., 2002. Mercury toxicity in plants. *The Botanical Review*, 66(3), 379–422.

Polle, A., Schützendübel, A., 2003. Heavy metal signalling in plants: Linking cellular and organismic responses. In: *Plant Responses to Abiotic Stress*, Eds. Hirt, H., Shinozaki, K., Springer, Berlin, Heidelberg, pp. 187–215.

Porter, E.K., Peterson, P.J., 1975. Arsenic accumulation by plants on mine waste (United Kingdom). *Science of the Total Environment*, 4(4), 365–371.

Puig, S., Thiele, D.J., 2002. Molecular mechanisms of copper uptake and distribution. *Current Opinion in Chemical Biology*, 6(2), 171–180.

Qadir, S., Qureshi, M.I., Javed, S., Abdin, M.Z., 2004. Genotypic variation in phytoremediation potential of Brassica juncea cultivars exposed to Cd stress. *Plant Science*, 167(5), 1171–1181.

Rivetta, A., Negrini, N., Cocucci, M., 1997. Involvement of Ca^{2+}-calmodulin in Cd^{2+} toxicity during the early phases of radish (*Raphanus sativus* L.) seed germination. *Plant, Cell & Environment*, 20(5), 600–608.

Salt, D.E., Prince, R.C., Pickering, I.J., Raskin, I., 1995. Mechanisms of cadmium mobility and accumulation in Indian mustard. *Plant Physiology*, 109(4), 1427–1433.

Sarker, U., Oba, S., 2018. Augmentation of leaf color parameters, pigments, vitamins, phenolic acids, flavonoids and antioxidant activity in selected Amaranthus tricolor under salinity stress. *Scientific Reports*, 8(1), 1–9.

Saxena, I., Shekhawat, G.S., 2013. Nitric oxide (NO) in alleviation of heavy metal induced phytotoxicity and its role in protein nitration. *Nitric Oxide*, 32, 13–20.

Schmfger, M.E.V., 2001. Phytochelatins: Complexation of metals and metalloids, studies on the phytochelatin synthase. Doctoral dissertation, PhD Thesis, Munich University of Technology (TUM), Munich.

Shahzad, B., Tanveer, M., Che, Z., Rehman, A., Cheema, S.A., Sharma, A., Song, H., Rehman, U.R., Zhaorong, D., 2018. Role of 24-epibrassinolide (EBL) in mediating heavy metal and pesticide induced oxidative stress in plants: A review. *Ecotoxicology and Environmental Safety*, 147, 935–944.

Shankar, A.K., Djanaguiraman, M., Pathmanabhan, G., Sudhagar, R., Avudainayagam, S., 2004. Uptake and phytoaccumulation of chromium by selected tree species. In: Wastewater treatment and waste management. Eds. Singh, V.P., Yadava, R.N., *Proceedings of the International Conference on Water and Environment (WE-2003)*, December 15–18, 2003, Bhopal, India.

Shao, H.B., Chu, L.Y., Ni, F.T., Guo, D.G., Li, H., Li, W.X., 2010. Perspective on phytoremediation for improving heavy metal-contaminated soils. In: *Plant Adaptation and Phytoremediation*, Eds. Ashraf, M., Ozturk, M., Ahmad, M.S.A., Springer, Berlin/Heidelberg, Germany, pp. 227–244.

Sharma, P., Dubey, R.S., 2005. Lead toxicity in plants. *Brazilian Journal of Plant Physiology*, 17(1), 35–52.

Sharma, S.S., Dietz, K.J., Mimura, T., 2016. Vacuolar compartmentalization as indispensable component of heavy metal detoxification in plants. *Plant, Cell & Environment*, 39(5), 1112–1126.

Snedden, W.A., Fromm, H., 1998. Calmodulin, calmodulin-related proteins and plant responses to the environment. *Trends in Plant Science*, 3(8), 299–304.

Stadtman, E.R., 1991. Metal-catalyzed oxidation of proteins. *The Journal of Biological Chemistry*, 266, 2005–2008.

Stobart, A.K., Griffiths, W.T., Ameen-Bukhari, I., Sherwood, R.P., 1985. The effect of Cd^{2+} on the biosynthesis of chlorophyll in leaves of barley. *Physiologia Plantarum*, 63(3), 293–298.

Stohs, S.J., Bagchi, D., 1995. Oxidative mechanisms in the toxicity of metal ions. *Free Radical Biology and Medicine*, 18(2), 321–336.

Sytar, O., Kumar, A., Latowski, D., Kuczynska, P., Strzałka, K., Prasad, M.N.V., 2013. Heavy metal-induced oxidative damage, defense reactions, and detoxification mechanisms in plants. *Acta Physiologiae Plantarum*, 35(4), 985–999.

Tangahu, B.V., Sheikh Abdullah, S.R., Basri, H., Idris, M., Anuar, N., Mukhlisin, M., 2011. A review on heavy metals (As, Pb, and Hg) uptake by plants through phytoremediation. *International Journal of Chemical Engineering*, 2011, 939161.

Tauqeer, H.M., Ali, S., Rizwan, M., Ali, Q., Saeed, R., Iftikhar, U., Ahmad, R., Farid, M., Abbasi, G.H., 2016. Phytoremediation of heavy metals by Alternanthera bettzickiana: Growth and physiological response. *Ecotoxicology and Environmental Safety*, 126, 138–146.

Tchounwou, P.B., Yedjou, C.G., Patlolla, A.K., Sutton, D.J., 2012. Heavy metals toxicity and the environment. *Molecular, Clinical and Environmental Toxicology*, 101, 133–164.

Thao, N.P., Khan, M.I.R., Thu, N.B.A., Hoang, X.L.T., Asgher, M., Khan, N.A., Trab, L.S.P., 2015. Role of ethylene and its cross talk with other signaling molecules in plant responses to heavy metal stress. *Plany Physiology*, 169(1), 73–84.

Thomas, J.C., Malick, F.K., Endreszl, C., Davies, E.C., Murray, K.S., 1998. Distinct responses to copper stress in the halophyte Mesembryanthemum crystallinum. *Physiologia Plantarum*, 102(3), 360–368.

Thomine, S., Wang, R., Ward, J.M., Crawford, N.M., Schroeder, J.I., 2000. Cadmium and iron transport by members of a plant metal transporter family in Arabidopsis with homology to Nramp genes. *Proceedings of the National Academy of Sciences*, 97(9), 4991–4996.

Todeschini, V., Lingua, G., D'agostino, G., Carniato, F., Roccotiello, E., Berta, G., 2011. Effects of high zinc concentration on poplar leaves: A morphological and biochemical study. *Environmental and Experimental Botany*, 71(1), 50–56.

Vernay, P., Gauthier-Moussard, C., Jean, L., Bordas, F., Faure, O., Ledoigt, G., Hitmi, A., 2008. Effect of chromium species on phytochemical and physiological parameters in Datura innoxia. *Chemosphere*, 72(5), 763–771.

Viehweger, K., 2014. How plants cope with heavy metals. *Botanical Studies*, 55(1), 1–12.

Vries, W.D., Lofts, S., Tipping, E., Meili, M., Groenenberg, J.E., Schütze, G., 2007. Impact of soil properties on critical concentrations of cadmium, lead, copper, zinc, and mercury in soil and soil solution in view of ecotoxicological effects. *Reviews of Environmental Contamination and Toxicology*, 191, 47–89.

Wang, Y., Du, H., Xie, M., Ma, G., Yang, W., Hu, Q., Pei, F., 2019. Characterization of the physical properties and biological activity of chitosan films grafted with gallic acid and caffeic acid: A comparison study. *Food Packaging and Shelf Life*, 22, 100401.

Williams, L.E., Pittman, J.K., Hall, J.L., 2000. Emerging mechanisms for heavy metal transport in plants. *Biochimica et Biophysica Acta (BBA)-Biomembranes*, 1465(1–2), 104–126.

Wojcik, M., Tukiendorf, A., 2004. Phytochelatin synthesis and cadmium localization in wild type of Arabidopsis thaliana. *Plant Growth Regulation*, 44(1), 71–80.

Yadav, S.K., 2010. Heavy metals toxicity in plants: An overview on the role of glutathione and phytochelatins in heavy metal stress tolerance of plants. *South African Journal of Botany*, 76(2), 167–179.

Zhen, Y., Qi, J.L., Wang, S.S., Su, J., Xu, G.H., Zhang, M.S., Miao, L.V., Peng, X.X., Tian, D., Yang, Y.H., 2007. Comparative proteome analysis of differentially expressed proteins induced by Al toxicity in soybean. *Physiologia Plantarum*, 131(4), 542–554.

Zhu, Y.L., Pilon-Smits, E.A., Jouanin, L., Terry, N., 1999. Overexpression of glutathione synthetase in Indian mustard enhances cadmium accumulation and tolerance. *Plant Physiology*, 119(1), 73–80.

7

Aluminum Toxicity and Ionic Homeostasis in Plants

Bedabrata Saha
Agricultural Research Organization

Bhaben Chowardhara
Arunachal University of Studies

Umakanta Chowra
Guwahati College

Chetan Kumar Panda
Odisha University of Agriculture and Technology

CONTENTS

7.1 Introduction

Aluminum (Al) toxicity in agricultural plants is a major problem that has been acknowledged at least since 1918. If concentrations of soluble and exchangeable Al in acid soils reach substantial level then it becomes toxic to plants (Saha et al., 2020). Acid soils are widespread and it limits plant production worldwide. They count for 30%–40% of arable landmass and excess of 70% of the landmass with potential to become arable. Acid soil has its impact felt on about 38% of farmland in Southeast Asia, 20% in East Asia, 56% in Sub-Saharan Africa, 31% in Latin America, and parts of North America. Acidity in soil occurs naturally in tropical and subtropical areas, but in temperate zones, it is a heightening trouble and is predominantly the consequence of acid rain in the industrial areas of USA, Canada, and Europe (Domingo, 2003; Vitorello et al., 2005; Bian et al., 2013). Acidity in soil is a primary hurdle for crop yield in the northeastern hilly regions (NE) of India. In this region estimated 65% of landmass is having pH below 5.5. Excessive deforestation coupled with shifting cultivation practices has resulted in tremendous soil loss and nutrient degradation. Aluminum (Al) and iron (Fe)-induced phosphorous (P) deficiencies majorly limit the production of rice (the staple food of the Indian subcontinent). Agricultural yield in the NE part of India has invariably been on the grimmer side when compared to the national norm, with soil acidity being the cause. Nearly 80% of agricultural landmass is acidic in NE rendered by high rainfall (Awasthi et al., 2017). Causes of soil acidity are leaching of excess nitrate, addition of some nitrogen-based fertilizers and buildup in mostly plant-based organic matter. Organic matter over the course of time decays to form carbonic acid and other weak organic acids rendering the soil acidic. Use of nitrogen fertilizers containing ammonium, which undergoes nitrification with the expulsion of hydrogen (H^+), causes more acidity. As there is increment in the proportion of ammonium in a given fertilizer, the acidification potency also increases, thus scaling down the pH (Saha et al., 2012). Al is observed in nature, mainly as aluminum silicates and aluminum oxides; a trace amount of the element is required to carry out the biological processes by living organisms. Al

DOI: 10.1201/9781003258063-7

subsists in +2 and +3 valency states. As pH subsides beneath 5.5, aluminosilicate clays and aluminum hydroxide minerals start to melt, expelling aluminum-hydroxy cations, $Al(H_2O)_6^{3+}$ and Al^{3+} (most phytotoxic form which retards root growth of plant as a primary consequence) (Roychowdhury et al., 2019). Comparatively fewer clay particles in sandy soils acidify speedily, imputable to their modest source of alkaline cations (buffering capacity) and more prominent leaching possibility (Excessive rainfall is an effective agent for removing basic cations) rendering the landmass acidic. Al toxicity decreases in the order of $Al^{3+} > Al(OH)^{2+} > Al(OH)_2^+$ (Kumar et al., 2012; Agrahari et al., 2020). Al toxicity is the causative agent for numerous physiological, morphological, and biochemical disruptions in plants due to stress, starting with root growth inhibition at the earliest instance (Awasthi et al., 2017, 2020, 2021). Al interferes with mitotic activity (decreases) and counters root elongation, accumulation of Al in nuclei of cells in the meristem of the root tip occurs within 30 minutes of exposure. Al^{3+} are extremely reactive cations (earmarked as Class A oxygen seeking) that preferentially bind to negative donor groups such as carboxyl and phosphate groups (Baluška & Mancuso, 2013). Al ions has its impact felt on multiple targets, including carboxyl residues of pectin in cell wall, glutamate/aspartate residue in several proteins, phospholipids in the plasma membrane, and phosphate residues in nucleotides/nucleic acids. This is a potent cell membrane destabilization process, Al in the early phase inhibits the uptake of sugar and water (disturbing cellular osmolality). And as a potent factor of reduction in energy sources, it hinders cell elongation. In the late phase, Al stimulates the synthesis of callose and reactive oxygen species (ROS) (Yamamoto, 2019). Al can cause cell death by (i) increment in Fe-arbitrated lipid peroxidation extending to plasma membrane (PM) integrity loss, (ii) mitochondrial dysfunction companioned by increase in ROS synthesis, and (iii) upregulation of vacuolar processing enzymes extending to vacuolar breakdown and PM integrity loss again (Silva, 2012). The transition zone of the root apex is where Al interacts and hinders root growth. This region is very sensitive to diverse environmental factors and also the region where changes in cyto-architectural physiology occurs (highly Al-sensitive portion is the distal transition zone of the root apex) (Sivaguru & Horst, 1998). The continuum basically the cell wall-plasma membrane-cytoskeleton is regarded the predominant/primary target of Al toxicity. Sensitivity of Al in a specific genotype with respect to different regions of root apical zone varies on the basis of proximity to the external exchangeable Al (Vitorello et al., 2005). In the apoplast, Al binds to pectins especially non-esterified pectins. Al^{3+}, when bound to the pectic matrix of the cell wall, renders its thickness and rigidity, which impact normal expansion and elongation. Pectin methyl esterase (PME) catalyze the de-methylsterification of homogalactouronates, releasing methanol and protons, creating negatively charged carboxyl groups (the extent of negative charge determines the binding of Al^{3+}) (Wolf et al., 2009). Strength/stiffness of the pectic gels is highly dependent upon its cross-link with Ca^{2+} ions which get dissociated when PMEs act, leading to cross-link with the intruding Al^{3+} ions. It's a known fact that Al adheres to cell wall polysaccharides and retards cell expansion (Yang et al., 2008). Lesser cell wall polysaccharide capacity with an increase in degree of methyl esterification leads to less number of carboxyl groups (cross-linked with Ca^{2+} ions), and this results in more Al exclusion as detected in Al-resistant "Nipponbare". The activity of PME was found to be constitutively in the higher side in Al-sensitive "Zhefu 802". The activity of Al-induced PME and subsequent pectin binding disturb the apoplastic Ca^{2+} equilibrium (these are major secondary messengers/signaling molecules which navigates itself/or their influx into cytoplasm decodes downstream stress response defense mechanism). The increased influx of Ca^{2+} in cytoplasm (disturbance in calcium homeostasis) rapidly triggers an event of root callose deposition (frequent markers of Al toxicity) as an Al-hindering mechanism and prevents cell death. Plasma membrane depolarization due to Al stress also triggers callose deposition. Depolarization is defined as increase of positive charge inside the cell (increment in functioning of cytosolic Ca^{2+} activates anion channels and, successively, the anion efflux leads to plasma membrane depolarization) (Thor, 2019). Under optimal levels of PME and in the absence of Al^{3+} ions, a physiological balance is maintained i.e., adequate number of methoxyl groups are being converted by PMEs on the polygalacturonic acid chain releasing both protons and methanol (PME remove methyl esters in a blockwise manner creating long contiguous chains of de-esterified pectin) (Wu et al., 2018). The negatively charged carboxyl group left exposed cross-link with Ca^{2+} ions. More the de-esterified pectins, stronger the interaction with Ca^{2+}. PMEs serve as stimulators of cell wall hydrolases such as polygalacturonases and pectate lyases which result in cell wall loosening and cell expansion. However, with Al exposure in question, an increased PME activity is observed leading to enhanced PME accumulation. The de-methylesterification of HG creates greater binding sites with Ca^{2+} but are effectively displaced by Al^{3+}. Al toxicity stress affects phosphoinositide-mediated signal transduction and inhibits phospholipase C enzyme activity significantly (Poot-Poot & Hernandez-Sotomayor, 2011). Al stress increases proline content, decreases carotenoid content in the leaves, amino acid capacity increment and decrease in protein holding capacity, increase in reducing and total sugar level. The Al-induced increment in total amino acid capacity is associated with the diminution in protein capacity as the increment in total amino acid is due to hydrolysis of protein which results in observable diminution of protein capacity. The increase in reducing and total sugar level may be attributed in maintaining the osmotic potential of cell sap when a plant experiences Al stress. Al toxicity reduces carbon dioxide assimilation rate, Al inhibits both Chl-a and Chl-b content in leaves, and Al^{3+} uptake in plant roots leads to an unwanted complexation. Soluble Al reacts with soluble phosphorous (P) and generates insoluble aluminum phosphate which is not available to plants. ROS production during Al stress causes depletion of adenosine triphosphate (ATP), inhibition of respiration, and considerable reduction in growth and yield (Silva, 2012). This suggests Al is capable of alteration in mitochondrial function and redox homeostasis in plants, and Al disrupts basipetal auxin transport (polar auxin transport-PAT) by severely interfering with the alignment of PIN2 proteins (which act as a major auxin efflux carrier). Polar auxin transport (basipetal auxin transport) into the transition zone of root apex is very essential to limit the phytotoxicity effect of Al^{3+} and enhance cell elongation but Al acts as an inhibitor of auxin flow just like chemical compounds like n-1-naphthylphthalamic acid and 2,3,5-triiodobenzoic acid (Shen et al., 2008; Barbez et al., 2017). Overexpression of genes encoding actin microfilaments conferring Al resistance in crops will be an efficient method to forefend the perturbation of actin microfilament when Al toxicity is experienced, which will pave the way for normal transport of auxin and root growth. Auxin-regulated (positively) cell elongation is widely known as the acid growth theory (Silva, 2012). The theory states that the plant hormone auxin actuates the initiation of plasma membrane-localized H^+-ATPases (proton pumps), ensuing in acidification of the apoplast. The diminution in pH of apoplast ensues cell wall relaxing enzymes, which, in agreement with turgor pressure, enables cellular expansion (Jaskowiak et al., 2019).

7.2 Role of Transporters in Ionic Homeostasis

Ion homeostasis is an essential physiological process to maintain proper growth and development in plants. Ions act as core metabolic regulatory substances that can improve plants performance under stress conditions as well. In fact, the main role of regulation is being performed by the transporters of ions. Several transporters are involved in ion acquisition, ion homeostasis, even during unfavorable environmental conditions for the plants. Many researchers have highlighted the role of transporters in abiotic stress responses in plants such as drought, salinity, metal, aluminum, etc. (Hirayama and Shinozaki, 2010). Although the need for proper and thorough understanding of ion transporters in ionic homeostasis and stress conditions is yet to be attained, so far, several types of transporters have been reported based on their localization and function; out of which, membrane-, cellular pore-, and cytoplasm-integrated transporters are mostly involved in regulatory and metabolic activity. Membrane- and cellular pore-integrated transporters are immobile and perform homeostatic action through exchange of ions between two sides of the membrane. Cytoplasmic transporters are mobile and relocate the ions through binding movement. The ionic imbalance signifies the dysfunctional activity of transporters which is crucial to upkeep plant's metabolic performance.

Studies have revealed that several transporters and signaling factors are tangled in plants abiotic stress responses. The molecular mechanism of their action during the stress condition is also getting unfurled gradually. Transporters and signaling factors mostly belong to the membrane-bound protein family. So basically, they are localized in membrane to maintain the ionic balance, to transport metabolites, toxic pollutants, and mineral nutrients by exchanging them from inside to outside or vice versa in the cell. Among these transporters, nutritional transporters are critically important for plants growth, nutrient acquirement, development, ionic homeostasis, and other stress responses (Bo et al., 2017). In recent reports, carbohydrate transporters were found to be involved in adjusting carbohydrate content to improve response to drought and waterlogging (Phukan et al., 2018; Ma et al., 2019). Mineral transporters regulate the macro and microelements to adjust the nutrient deficiencies in the cytoplasm, extracellular and vacuolar space (Li et al., 2019). Metal cation transporters are involved in ionic balance in the cell (Cai et al., 2019; Veraestrella et al., 2017).

Aluminum (Al) is generally phytotoxic at low pH and exists as Al^{3+}. The preliminary symptom of this phytotoxicity is inhibition root growth, which in long run reduces crop productivity and is visible within just few hours of exposure to Al^{3+}. Although plants have developed countermeasures for Al^{3+} toxicity which includes rhizopheric exudation and cytosolic synthesis of organic acids, especially citrate, malate, and oxalate, to scavenge the Al^{3+} moieties and maintain cellular homeostasis. This process is being primarily regulated by plasma membrane-localized transporter proteins for instance MATE (Multidrug And Toxic compound Extrusion) gene family and ALMT1 (aluminum-activated malate transporter) (Park et al. 2017). MATE and ALMT1 transporters help in scavenging Al^{3+} by secreting organic acids into the rhizosphere. Basically, there are two methods followed for the detoxification process, either the organic acids bind to Al^{3+} in the cytosol to form complexes which then undergo compartmentalization in vacuole or once exuded through root hair organic acids form these complexes in the rhizosphere, thus inactivating Al^{3+} (Upadhyay et al., 2019; Magalhaes et al., 2007). MATE and their homologs were found to be involved in enhancing Al tolerance in poplar and soybean, established from a Genome-Wide analysis of the MATE gene family (Li et al., 2017b; Liu et al., 2016; Garciaoliveira et al., 2018). In several plant species, MATE transporter was found to be the citrate transporter to enhance Al tolerance, for example, in rice bean, green gram, tobacco, black gram, *Brassica,* etc., on the other hand, malate has also been reported to be involved in Al tolerance in several crop plants (Saha et al., 2020), where ALMT1 is the malate transporter, first reported in *Arabidopsis* (Zhang et al., 2018a). In *Camelina sativa*, enhanced aluminum tolerance was detected with overexpression of the organic acid transporter gene, *Cs*ALMT1 (Park et al., 2017). Overexpression of *Sl*ALMT9, which is another homolog of ALMT1, found to be regulated by *Sl*WRKY42, enhanced malate accumulation in tomato fruit and aluminum tolerance (Ye et al., 2017). Two genes, namely STAR1 and STAR2 (for sensitivity to Al rhizotoxicityl), are also involved in Al^{3+} detoxification by forming a bacterial type ATP-binding cassette (ABC) transporter complex with UDP-glucose in rice (Huang et al., 2009). Another ABC transporter-like gene, Aluminum Sensitive 3 (ALS3), was found to have higher expression under Al^{3+} toxicity in buckwheat (Reynal-Lorens et al. 2015). *Os*NRAT1 (natural resistance and macrophage transporter protein) overexpression showed increased resistance to Al^{3+} toxicity in transgenic *Arabidopsis* by transporting in aluminum. *At*NRAMP3 is involved in metal transport into the vacuole, which initiates the gathering of excess Mn, Zn, and Fe in *Arabidopsis* (Thomine et al., 2003). *Zm*PGP1, a p-glycoprotein of the ABC transporter family, showed elevated Al toxicity in maize (Zhang et al., 2018b). Similarly, *Nt*SUT1, a plasma membrane-localized proton symporter, was found to be involved in enhanced aluminum tolerance and root growth promotion (Kariya et al., 2017). In recent studies, enhanced promoter activity and hence higher expression of a citrate efflux transporter, *Os*FRDL4, observed due to a retrotransposon insertion in the promoter region rendered better Al resistance in rice (Yokosho et al., 2016). Heterologous overexpression of NHX-type transporter, involved in K^+/H^+ fluxes, also promoted enhanced Al resistance in rice (Li et al., 2020).

Under an acidic environment, the concentration of Mn increases which can cause cellular damage to plants. NRAMP2 contains this irregularity, which is a divalent metal transporter of the NRAMP family. MTP8.2 (Mn-CDF metal tolerance protein), another tonoplast localized transporter, was found to be involved in Mn sequestration in the vacuoles of root and shoot (Takemoto et al., 2017). In contrast to Mn in acidic soil, phosphorous gets leached out and creates Pi-deficiency in plants. In *Arabidopsis*, ALS3 and *At*STAR1 (an ABC transporter complex) cordially act along with LOW PHOSPHATE ROOT1/2 (LPR1/2) to resolve the Pi-deficiencies by regulating Fe homeostasis (Dong et al., 2017).

Fe ion is another factor that influences acidic conditions. At the cellular level excess, iron can cause metabolic imbalance and cellular damage in plants. Fe excess and deficiency, both are regulated by several transporters in plants. Meliorated iron homeostasis in wheat was observed on overexpressing vacuolar iron transporter gene *Ta*VIT2 (Connorton et al., 2017). In rice, *Ap*KUP3 (*Alternanthera-philoxeroides*–based K^+ transporter gene) heterologous expression enhanced K^+ hoarding and improved drought resistance (Song et al., 2014). Under acidic soil conditions, Mg^{2+} ion concentration also becomes a major constraint for plant growth. In an acidic environment, Mg^{2+} deficiency occurs and leads to underdeveloped growth. CorA/MRS2/MGT-type Mg^{2+} transporters are crucial for ionic balance in the cell. In *Arabidopsis*, *At*MGT6 and *At*MGT7 are essential to maintain normal growth and development under low Mg^{2+} conditions or acid soil conditions (Gebert et al., 2009; Mao et al., 2014). *Zm*MGT10 and *Zm*MGT12 are found to be involved in Mg^{2+} transport to enhance tolerance to Mg^{2+} deficiency.

7.3 Role of Different Ionic Species in Aluminum Stress Response

Researchers have been investigating on a variety of techniques (modulating stress responses by twitching signaling molecules, use of exogenous ameliorants, switching on to superior resistant varieties, etc.) to reduce damage due to Al toxicity in crop plants. Different ionic species are crucial in aluminum-induced stress responses. In this segment, we dissertate on the function of ionic species in plants when exposed to aluminum stress.

7.3.1 Boron

Plants require boron (B) as a micronutrient. Boron is essential for the formation of cell walls. Boron (B^{3-}) being similar to the aluminum (Al^{3-}) in its ionic form, in many instances, has been found to replace Al, like in Al^{3+}-binding sites (carboxyl groups) present in the cell wall (Kochian, 1995). B has been reported to reduce Al toxicity in trifoliate oranges (Riaz et al., 2018). They showed that B prevents Al from entering the symplast by lowering Al immobilization. B also enhanced root ontogeny by minimizing root cell trauma and by diminution of ROS. Cell wall structure alteration due to B administration minimizes root oxidative damage (Yan et al., 2018a). In *Citrus grandis* roots, B boosts the scavenging ability of ROS and aldehydes, thus enhancing the manifestation of genes linked to lipid, amino acid, energy metabolisms, and cellular transport (Zhou et al., 2015). Yu et al. (2009) looked into the role played by B in reducing toxic effects of Al in pea plants (*Pisum sativum*). B reduces some of the effects of Al toxicity by restricting Al adhesion in cell walls, ensuing in diminished trauma to root and shoot tissues. B supplementation in Al-bearing hydroponic nutrient solution promotes root extension in pea (Li et al., 2017a), rape seedlings (Yan et al., 2018b). Supplementing with B prevents decrease in leaf chlorophyll concentration also (Yu et al., 2009). B also impacts pectin concentration and extent of methylation of pectin to influence the cell wall's capability to adsorb and desorb Al (Horst et al., 2010; Yan et al., 2018a). Furthermore, through modulating Al-related transporters in rice roots, B is able to improve Al transference from the cytoplasm to vacuoles (Zhu et al., 2019). B has also been reported to ameliorate Al toxicity in similar fashion in several other crop plants, like apple (Wojcik, 2003), sunflower (Ruiz et al., 2006), wheat (Hossain et al., 2004), French bean (Stass et al., 2007), and flax (Heidarabadi et al., 2011). The methods, through which B reduces toxic effects of Al in various plants, may be discrete and composite.

7.3.2 Magnesium

Magnesium (Mg) reduces the harmful effects of Al via improving carbon sectionalization from shoots to roots, boosting acid phosphatase activity, boosting the production and transudation of organic acid anions, preserving proton-ATPase activity, protecting against ROS, and inducing cytosolic calcium ascent (Bose et al., 2013). Al toxicity causes Mg deficiency, and exogenous treatment of Mg reduces the toxic effects of Al in wheat plants (Keltgens et al., 1999). Mg has been observed to minimize Al toxicity in soybean roots *via.* modification of citric acid production and exudation in the rhizosphere (Silva et al., 2001). Kong et al. (2020) reported that Mg helps in diminution of Al amass and imparts security from Al-stimulated oxidative stress through energization of antioxidative enzymes in maize. In wheat, increased malate and citrate exudation from the roots in the rhizosphere are the fundamental mechanics of heightened ontogeny falling out of foliar Mg spraying when exposed to acidic soil conditions (Kibria et al., 2021a, b). H^+-ATPase functioning was suppressed due to Al-induced toxicity in *Vigna umbellata* roots, but when 10 mM Mg was introduced into the same nutritional solution, the activity showed a marked increase and triggered citrate exudation (Yang et al., 2007). In *Vicia faba* also, when 20 mM Mg exogenic treatment was carried out after Al exposure, it boosted H^+-ATPase functioning and increased citrate transudation, reducing toxic effects of Al (Chen et al., 2015).

Mg supplementation not just merely reduces toxic effects caused by Al but also improves root ontogeny in a variety of plant species. Among legumes, Mg application has its effect felt upon on soybean (Glycine max) (Silva et al., 2001), rice bean (*V. umbellata*) (Yang et al., 2007), and broad bean (*Vicia faba*) (Chen et al., 2015). Furthermore, Mg transporter genes have been linked to the reduction of Al toxicity in plants. In *Nicotiana benthamiana*, overexpression of an Mg transporter gene (*At*MGT1) reduced toxic effects of Al by increasing Mg uptake (Deng et al., 2006)). Al treatment in nutritional solution (50M $AlCl_3$, pH 4.2) resulted in higher Mg absorption and free Mg content in the cytoplasm of Al-resistant *Arabidopsis* (Col-0 and alr104) than sensitive Arabidopsis (ALS-5 and ALS3) (Bose et al., 2013). This increase in cytosolic Mg feeding could be a significant indicator of a tolerant plant capable of reducing Al toxicity.

7.3.3 Calcium

Calcium (Ca) was found to reduce Al toxicity *via* electrostatic supplanting of Al from the cell membrane surface like boron. Ca reduces Al toxicity by regulating tricarboxylic acid cycle-related protein teemingness, enzyme functioning, vacuolar and mitochondrion ATP synthase, and cell wall constituent-related proteins, as well as inducing the transcription factor ethylene-insensitive 3 in *Arabidopsis* (Chen et al., 2019a, b). Ca-based rectificators, such as lime (e.g., calcite, dolomite), phosphogypsum (PG), and gypsum (G) with varied application programs on diverse crops to acid soils, have got favorable effects. According to Reyes-Díaz et al. (2011), gypsum ($CaSO_4$) effectively reduces Al toxicity in *Vaccinium corymbosum* L. by supplying a sufficient amount of Ca. G is a soft sulfate mineral made up of calcium sulfate dihydrate, which is commonly utilized as a soil amendment in acidic situations. Three high bush blueberries (*V. corymbosum* L.) cultivars were evaluated for the effects of gypsum in reducing Al toxicity (Alarcón-Poblete et al., 2019). It was observed that gypsum lowered Al levels, alleviated leaf morphological changes over and above photochemical and biochemical damage caused by Al. G is a good supply of calcium and sulfur, which helps plants' mineral profiles (N, P, K, Mn, and Zn) (Tuna et al., 2007). For example, a combined G (4.0 tha^{-1}) and NPK fertilizer (0.3 tha^{-1}) when applied to acid soil raised N, P, K, S, Mn, and Ca content in blueberry leaves, which improved stem elongation, bud survival, and flower number (Sanderson and Eaton, 2004). G can also impart Ca and S gains to nutrient-limited soils, resulting in increased soil fertility and alfalfa growth (Tirado-Corbalá et al., 2017). A variety of calcareous additions could be used to reduce Al toxicity and soil acidity (Toma and Saigusa, 1997). In acid sub-soils, where Al binds organic ligands (OL), the effect of SO_4 in reducing Al toxicity is critical. In Malaysian acid soils, complexing OL following G treatment, resulted in increased $AlSO^{4+}$ and decreased Al activity, resulting in greater corn production (Shamshuddin et al., 1991). Korcak (1988) discovered that G (0.60 tha^{-1}) when applied for 5 years raised soil pH and Ca levels, but that Ca levels in leaves and fruits were uneven.

7.3.4 Phosphorous

Phosphorous (P) is an all-important macro mineral that aids plant ontogeny in both normal and stressful environments (Razaq et al., 2017; Tariq et al., 2017). In acidic soils, Al/Fe fixes inorganic phosphate (Pi), which becomes crucial when the pH lowers, resulting in a severely restricted Pi (Chen and Liao, 2016). As a result, plants holding out in acid soils must contend with both toxic effects of Al and a lack of P. On acid soils, P when exogenously applied reduces toxic effects of Al in a variety of plants, including sorghum (Smith et al., 1994), buckwheat (Zheng et al., 2005), and wheat (Iqbal, 2013). Tan and Keltjens (1990) discovered that a modest dose of Al (0.4 mg/L) had no effect on plant biomass production, while a high Al dose (1.6 mg/L) drastically decreased plant ontogeny and dry matter yield (DMY) in sorghum. In this case, adding P reduced Al toxicity by improving root respiration and nutrient uptake, resulting in increased DMY. Al stress (150 mg/kg $AlCl_3$ soil) was found to reduce P capacity in wheat seedlings, whereas raised P level was observed after exogenous P (160 mg/kg P soil) was added to soil (Iqbal, 2013). A P threshold for reducing Al toxicity was established, stating that if the P/Al molar ratio in the root cells surpasses 5, Al toxicity can be alleviated (Chen et al., 2012). P treatment (8 g/plant) has recently been accounted to improve maple tree plant ontogeny in nursery settings (Razaq et al., 2017). In acidic soil, several successful strategies for crop yield under the toxic effects of Al and P deficit have been investigated. Increasing P accessibility through the application of organic amendments like, biochar, compost, etc., as organic amendments successfully fixed harmful Al and/or Fe in acid soil, increasing total P and Pi (Ch'ng et al., 2014). Besides this, numerous molecular breeding-related schemes for rectification of Al toxicity and P deficiency responses have been investigated. A 'root breeding' strategy was proposed for generating P-efficient plants by selecting plants that can use both native and exogenous P from acidic soils (Wang et al., 2010a). Genetic engineering for organic acid synthesis can improve crop yield for the sustenance of agriculture in acidic soil with Al toxicity can be perceived (Chen and Liao, 2016).

7.3.5 Sulfur

Sulfur (S) was largely ignored in soil research until the 1970s, despite its critical function in the formation of protein, vitamins, and chlorophyll in plants. S has recently gotten a lot of attention because of its ability to change metal toxicity while also playing an important role in plant growth and development (Guo et al., 2017; Dixit et al., 2016; Saifullah et al., 2016). Several investigations have shown that S-containing components in wheat (Zhang et al., 2010), barley (Dawood et al., 2012; Chen et al., 2013), oilseed rape (Qian et al., 2014; Ali et al., 2015), citrus plants (Guo et al., 2017), and soyabean (Wang et al., 2019) reduce Al toxicity. S protects against toxic effects of Al by: (i) enhancement of antioxidant activity and lowering ROS and lipid peroxidation; (ii) diminution of Al uptake; (iii) enhancement in intake of diverse nutrients such as P, Mg, and Ca; and (iv) increase in Al-activated exudation of organic acid anions into the rhizosphere. Exogenous S addition relieved the toxicity caused by numerous heavy metals in plants, albeit the effectiveness of the alleviation is largely dependent on the S-application tactics, dosages, and source (Zhang et al., 2016). S addition was recently shown to relieve Al toxicity in citrus by raising minerals (P, Mg, and Ca) and relative water content, lowering Al uptake and H_2O_2 synthesis (Guo et al., 2017). Furthermore, compared to an NPK fertilizer application,

S increased mineral components that supported the alleviation of numerous metal toxicities, including toxic effects of Al toxicity in plants (Guo et al., 2017; Dixit et al., 2016), and enhanced NO_3 and NH^{4+} levels in the soil (Skwierawska et al., 1997).

7.3.6 Silicon

In the earth's crust, silicon (Si) is the second most abundant. Plants absorb and amass Si varying from 1 to 100 g/kg Si dry weight in soil medium via a variety of absorption and transport routes (Farooq and Dietz, 2015). Rice is a distinct Si accumulator (up to 100 g/kg Si DW) (Farooq et al., 2019). Plants absorb Si in the form of orthosilicic acid (H_4SiO_4) despite the fact that the ingredient Si is not deemed necessary. The positive effects of Si on plant stress tolerance have long been known (Ma, 2004). Several investigations have shown that applying Si to plants grown hydroponically or in pot-based studies can diminish Al uptake (Singh et al., 2011; Qian et al., 2016; de Jesus et al., 2017). According to Singh et al. (2011), the application of Si exogenously relieved Al toxic effects by lowering Al buildup and avoiding Mg and Zn deficit in rice. In sorghum, Si demonstrated effectiveness via generating Al-Si composites in outer cellular tissues. As a result, the Al-Si combination prevented harmful Al from adhering to the cell wall (Kopittke et al., 2017). A unique technique has recently been used to decrease Al phytotoxicity in wheat by amending soil with Si-rich biochar (Qian et al., 2016). This reduced phytotoxicity is through a mechanism in which Al indulges in metal chelation with Si–biochar (Al-Si). Thus, Si particles lower the capacity of soil exchangeable Al and prevent harmful Al from migrating across plant tissues. Si-mediated alterations in SOD gene expression and activities of other antioxidant enzymes in rye grass and *Eucalyptus platyphylla* plant amended Al phytotoxicity (Pontigo et al., 2017; Lima et al., 2016). Si had an ameliorative impact on maize and rice seedlings exposed to Al toxicity (Corrales et al., 1997; Nhan and Hai, 2013). Silica nanoparticles reduced Al toxicity by boosting Al-detoxification mechanisms in roots via transudation of organic acids and polyphenols as well as prevented the plant from Al-activated oxidative stress by boosting the plant's antioxidative system (de Souza et al., 2019).

7.3.7 Nitric Oxide

Nitric oxide (NO) is a redox-active endogenous signaling chemical linked to plant ontogeny, seed germination, floral regulation, stomatal closure, the hypersensitive response, cell wall synthesis (Domingos et al., 2015), programmed cell death (Pan et al., 2017; Serrano et al., 2015), and heat stress (Rai et al., 2020). NO is involved in diverse physiological procedures, which involve facilitating rice seed germination and early root elongation in the presence of Al (Peres da Rocha Oliveiros Marciano et al., 2010), ROS and antioxidant enzymes in *Phaseolus vulgaris* (Wang et al., 2010b), and inhibiting plant programmed cell death (He et al., 2019; Pan et al., 2017). Exogenic NO defends wheat roots from oxidative stress caused by Al through modulating the AsA-GSH cycle (Sun et al., 2015).

7.3.8 Iron

Iron (Fe) is crucial ionic species for plant ontogeny. Very few studies have been reported on Fe and Al interaction. Bityutskii et al. (2017) mentioned that Fe plays an important role in plants under Al stress condition. Fe interacts negatively with Al, and this antagonistic relationship may impede Al movement through xylem in cucumber. The plant genotype that was most influenced by Fe's ameliorative impact was the more Fe-efficient genotype. The root content of serine, succinic, and fumaric acids were positively linked with xylem sap Fe concentrations, depicting a rudimentary mechanics of Al resistance that may entail the assistance of Fe transport to prevent Al intake via biosynthesis of these chelating chemicals.

7.3.9 Zinc

Zinc (Zn) is a vital nutrient for plants since it functions as a catalytic factor in enzymes as well as a structural constituent of proteins (Mattiello et al., 2015). Zn is found in high concentrations in plants and serves as a cofactor for over 300 enzymes. According to Su et al. (2020), under Al stress, the interaction of Zn boosted quantum yields and electron transfer rates of PSI and PSII. The increases in pmf, H^+-ATPase activity, and ATP synthase functioning, as well as the diminution in pH pmf between lumen and stroma, were mostly responsible for these beneficial effects. As such, more electrons were transported to NADPH production instead of ROS production, protecting photosystems from Al damage.

7.4 Conclusion

Thus, we conclude that ionic species of different elements have diverse roles to play in the alleviation of aluminum toxicity mediated through transporters which retain ionic homeostasis.

REFERENCES

Agrahari, R. K., Kobayashi, Y., Borgohain, P., Panda, S. K., & Koyama, H. (2020). Aluminum-specific upregulation of GmALS3 in the shoots of soybeans: A potential biomarker for managing soybean production in acidic soil regions. *Agronomy, 10*(9), 1228. doi: 10.3390/agronomy10091228.

Alarcón-Poblete, E., Inostroza-Blancheteau, C., Alberdi, M., Rengel, Z., & Reyes-Díaz, M. (2019). Molecular regulation of aluminum resistance and sulfur nutrition during root growth. *Planta, 247*(1), 27–39.

Alarcón-Poblete, E., Inostroza-Blancheteau, C., Latsague, M., Alberdi, M., de Oliveira Silva, F. M., Nunes-Nesi, A., ... & Reyes-Díaz, M. (2019). Gypsum application ameliorates morphological and photochemical damages provoked by Al toxicity in *Vaccinium corymbosum* L. cultivars. *Journal of Berry Research, 9*(4), 665–685.

Ali, S., Zeng, F., Qiu, L., & Zhang, G. (2011). The effect of chromium and aluminum on growth, root morphology, photosynthetic parameters and transpiration of the two barley cultivars. *Biologia Plantarum, 55*(2), 291–296.

Ali, B., Qian, P., Sun, R., Farooq, M. A., Gill, R. A., Wang, J., ... & Zhou, W. (2015). Hydrogen sulfide alleviates the aluminum-induced changes in Brassica napus as revealed by physiochemical and ultrastructural study of plant. *Environmental Science and Pollution Research, 22*(4), 3068–3081.

Awasthi, J. P., Saha, B., Regon, P., Sahoo, S., Chowra, U., Pradhan, A., ... & Panda, S. K. (2017). Morpho-physiological analysis of tolerance to aluminum toxicity in rice varieties of North East India. *PloS One*, *12*(4), e0176357. doi: 10.1371/journal.pone.0176357.

Awasthi, J. P., Saha, B., Panigrahi, J., Yanase, E., Koyama, H., & Panda, S. K. (2020). Redox balance, metabolic fingerprint and physiological characterization in contrasting North East Indian rice for Aluminum stress tolerance. *Scientific Reports*, *9*(1), 1–21. doi: 10.1038/s41598-019-45158-3.

Awasthi, J. P., Kusunoki, K., Saha, B., Kobayashi, Y., Koyama, H., & Panda, S. K. (2021). Comparative RNA-Seq analysis of the root revealed transcriptional regulation system for aluminum tolerance in contrasting indica rice of North East India. *Protoplasma*, *258*(3), 517–528. doi: 10.1007/s00709-020-01581-2.

Baluška, F., & Mancuso, S. (2013). Root apex transition zone as oscillatory zone. *Frontiers in Plant Science*, *4*, 354. doi: 10.3389/fpls.2013.00354.

Barbez, E., Dünser, K., Gaidora, A., Lendl, T., & Busch, W. (2017). Auxin steers root cell expansion via apoplastic pH regulation in Arabidopsis thaliana. *Proceedings of the National Academy of Sciences*, *114*(24), E4884–E4893. doi: 10.1073/pnas.1613499114.

Bian, M., Zhou, M., Sun, D., & Li, C. (2013). Molecular approaches unravel the mechanism of acid soil tolerance in plants. *The Crop Journal*, *1*(2), 91–104. doi: 10.1016/j.cj.2013.08.002.

Bityutskii, N. P., Yakkonen, K. L., Petrova, A. I., & Shavarda, A. L. 2017. Interactions between aluminium, iron and silicon in *Cucumber sativus* L. grown under acidic conditions. *Journal of Plant Physiology*, *218*, 100–108. doi: 10.1016/j.jplph.2017.08.003.

Bo, L., Xu, D., Ann, H., & Hassan, N. E. H. (2017). Advances in methods for identification and characterization of plant transporter function. *Journal of Experimental Botany*, *68*(15)445–4056. doi: 10.1093/jxb/erx140.

Bose, J., Babourina, O., Shabala, S., & Rengel, Z. (2013). Low-pH and aluminum resistance in Arabidopsis correlates with high cytosolic magnesium content and increased magnesium uptake by plant roots. *Plant and Cell Physiology*, *54*(7), 1093–1104. doi: 10.1093/pcp/pct064.

Cai, H., Huang, S., Che, J., Yamaji, N., & Ma, J. F. (2019). The tonoplast-localized transporter OsHMA3 plays an important role in maintaining Zn homeostasis in rice. *Journal of Experimental Botany*, *70*(10), 2717–2725. doi: 10.1093/jxb/erz091.

Ch'ng, H. Y., Ahmed, O. H., & Majid, N. M. A. (2014). Improving phosphorus availability in an acid soil using organic amendments produced from agroindustrial wastes. *The Scientific World Journal*, *2014*. doi: 10.1155/2014/506356.

Chen, Z. C., & Liao, H. (2016). Organic acid anions: An effective defensive weapon for plants against aluminum toxicity and phosphorus deficiency in acidic soils. *Journal of Genetics and Genomics*, *43*(11), 631–638. doi: 10.1016/j.jgg.2016.11.003.

Chen, Z. C., Zhao, X. Q., & Shen, R. F. (2010). The alleviating effect of ammonium on aluminum toxicity in Lespedeza bicolor results in decreased aluminum-induced malate secretion from roots compared with nitrate. *Plant and Soil*, *337*(1), 389–398.

Chen, R F., Zhang, F. L., Zhang, Q. M., Sun, Q. B., Dong, X. Y., & Shen, R. F. (2012). Aluminium–phosphorus interactions in plants growing on acid soils: Does phosphorus always alleviate aluminium toxicity? *Journal of the Science of Food and Agriculture*, *92*(5), 995–1000. doi: 10.1002/jsfa.4566.

Chen, J., Wang, W. H., Wu, F. H., You, C. Y., Liu, T. W., Dong, X. J., ... & Zheng, H. L. (2013). Hydrogen sulfide alleviates aluminum toxicity in barley seedlings. *Plant and Soil, 362*(1), 301–318.

Chen, Q., Kan, Q., Wang, P., Yu, W., Yu, Y., Zhao, Y., ... & Chen, L. (2015). Phosphorylation and interaction with the 14-3-3 protein of the plasma membrane H+-ATPase are involved in the regulation of magnesium-mediated increases in aluminum-induced citrate exudation in broad bean (*Vicia faba*. L). *Plant and Cell Physiology*, *56*(6), 1144–1153. doi: 10.1093/pcp/pcv038.

Chen, J., Duan, R. X., Hu, W. J., Zhang, N. N., Lin, X. Y., Zhang, J. H., & Zheng, H. L. (2019a). Unravelling calcium-alleviated aluminium toxicity in Arabidopsis thaliana: Insights into regulatory mechanisms using proteomics. *Journal of Proteomics*, *199*, 15–30. doi: 10.1016/j.jprot.2019.02.013.

Chen, Q., Wang, B., Ding, H., Zhang, J., & Li, S. (2019b). The role of NADP-malic enzyme in plants under stress. *Plant Science*, *281*, 206–212. doi: 10.1016/j.plantsci.2019.01.010.

Connorton, J. M., Jones, E. R., Rodriguezramiro, I., Fairweathertait, S. J., Uauy, C., & Balk, J. (2017). Wheat vacuolar iron transporter TaVIT2 transports Fe and Mn and is effective for biofortification. *Plant Physiology, 174*(4), 2434–2444. doi: 10.1104/pp.17.00672.

Corrales, I., Poschenrieder, C., & Barceló, J. (1997). Influence of silicon pretreatment on aluminium toxicity in maize roots. *Plant and Soil*, *190*(2), 203–209.

Dawood, M., Cao, F., Jahangir, M. M., Zhang, G., & Wu, F. (2012). Alleviation of aluminum toxicity by hydrogen sulfide is related to elevated ATPase, and suppressed aluminum uptake and oxidative stress in barley. *Journal of Hazardous Materials*, *209*, 121–128. doi: 10.1016/j.jhazmat.2011.12.076.

de Boer, W., & Kowalchuk, G. A. (2001). Nitrification in acid soils: Micro-organisms and mechanisms. *Soil Biology and Biochemistry*, *33*(7–8), 853–866.

de Jesus, L. R., Batista, B. L., & da Silva Lobato, A. K. (2017). Silicon reduces aluminum accumulation and mitigates toxic effects in cowpea plants. *Acta Physiologiae Plantarum, 39*(6), 138.

de Souza, L. C., Nogueira, D. C. S., Machado, L. C., Costa, T. C., da Silva Martins, J. T., Mendes, C. A. P., ... & de Araújo Brito, A. E. (2016). Nitrogen compounds, proteins and amino acids in corn subjected to doses of aluminum. *African Journal of Agricultural Research*, *11*(17), 1519–1524. doi: 10.5897/AJAR2015.10758.

Deng, W., Luo, K., Li, D., Zheng, X., Wei, X., Smith, W., ... & Pei, Y. (2006). Overexpression of an Arabidopsis magnesium transport gene, AtMGT1, in Nicotiana benthamiana confers Al tolerance. *Journal of Experimental Botany*, *57*(15), 4235–4243. doi: 10.1093/jxb/erl201.

Dixit, G., Singh, A. P., Kumar, A., Mishra, S., Dwivedi, S., Kumar, S., ... & Tripathi, R. D. (2016). Reduced arsenic accumulation in rice (*Oryza sativa* L.) shoot involves sulfur mediated improved thiol metabolism, antioxidant system and altered arsenic transporters. *Plant Physiology and Biochemistry*, *99*, 86–96. doi: 10.1016/j.plaphy.2015.11.005.

Domingo, J. L. (2003), December 6. *Aluminum (Aluminium): Toxicology*. Encyclopedia of Food Sciences and Nutrition (Second Edition).

Domingos, P., Prado, A. M., Wong, A., Gehring, C., & Feijo, J. A. (2015). Nitric oxide: A multitasked signaling gas in plants. *Molecular Plant*, *8*(4), 506–520. doi: 10.1016/j.molp.2014.12.010.

Dong, J., Pineros, M.A., Li, X., Yang, H., Liu, Y., Murphy, A. S., Kochian, L. V., & Liu, D. (2017). An Arabidopsis ABC transporter mediates phosphate deficiency-induced remodeling of root architecture by modulating iron homeostasis in roots. *Molecular Plant, 10*(2), 244–259. doi: 10.1016/j.molp.2016.11.001.

Epstein, E. (1994). The anomaly of silicon in plant biology. *Proceedings of the National Academy of Sciences, 91*(1), 11–17.

Farooq, M. A., & Dietz, K. J. (2015). Silicon as versatile player in plant and human biology: Overlooked and poorly understood. *Frontiers in Plant Science*, *6*, 994. doi: 10.3389/fpls.2015.00994.

Farooq, M. A., Saqib, Z. A., Akhtar, J., Bakhat, H. F., Pasala, R. K., & Dietz, K. J. (2019). Protective role of silicon (Si) against combined stress of salinity and boron (B) toxicity by improving antioxidant enzymes activity in rice. *Silicon, 11*(4), 2193–2197. doi: 10.1007/s12633-015-9346-z.

Garciaoliveira, A. L., Benito, C., Guedespinto, H., & Martinslopes, P. (2018). Molecular cloning of TaMATE2 homoeologues potentially related to aluminium tolerance in bread wheat (*Triticum aestivum* L.). *Plant Biology, 20*(5), 817–824. doi: 10.1111/plb.12864.

Gebert, M., Meschenmoser. K., Svidova, S., Weghuber, J., Schweyen, R. J., Eifler, K., Lenz, H., Weyand, K., & Knoop, V. P. C. (2009). A rootexpressed magnesium transporter of the MRS2/MGT gene family in Arabidopsis thaliana allows for growth in low-Mg^{2+} environments. *Plant Cell, 21*(12), 4018–4030. doi: 10.1105/tpc.109.070557.

Guo, T., Zhang, G., Zhou, M., Wu, F., & Chen, J. (2004). Effects of aluminum and cadmium toxicity on growth and antioxidant enzyme activities of two barley genotypes with different Al resistance. *Plant and Soil, 258*(1), 241–248.

Guo, J. H., Liu, X. J., Zhang, Y., Shen, J. L., Han, W. X., Zhang, W. F., ... & Zhang, F. S. (2010). Significant acidification in major Chinese croplands. *Science, 327*(5968), 1008–1010.

Guo, P., Li, Q., Qi, Y. P., Yang, L. T., Ye, X., Chen, H. H., & Chen, L. S. (2017). Sulfur-mediated-alleviation of aluminum-toxicity in Citrus grandis seedlings. *International Journal of Molecular Sciences*, *18*(12), 2570. doi: 10.3390/ijms18122570.

Habibi, G., & Shahinfar, M. (2021). The role of exogenous silicon to mitigate Al_2O_3 nanoparticle-induced toxicity in barley (*Hordeum vulgare* L.). *Acta Agriculturae Slovenica*, *117*(1), 1–9.

He, H. Y., He, L. F., Gu, M. H., & Li, X. F. (2012). Nitric oxide improves aluminum tolerance by regulating hormonal equilibrium in the root apices of rye and wheat. *Plant Science, 183*, 123–130.

He, H., Li, Y., & He, L. F. (2019). Role of nitric oxide and hydrogen sulfide in plant aluminum tolerance. *Biometals*, *32*(1), 1–9. doi: 10.1007/s10534-018-0156-9.

Heidarabadi, M. D., Ghanati, F., & Fujiwara, T. (2011). Interaction between boron and aluminum and their effects on phenolic metabolism of *Linum usitatissimum* L. roots. *Plant Physiology and Biochemistry*, *49*(12), 1377–1383. doi: 10.1016/j.plaphy.2011.09.008.

Hirayama, T., & Shinozaki, K. (2010). Research on plant abiotic stress responses in the post-genome era: Past, present and future. *The Plant Journal*, *61*(6), 1041–1052. doi: 10.1111/j.1365-313X.2010.04124.x.

Horst, W. J., Wang, Y., & Eticha, D. (2010). The role of the root apoplast in aluminium-induced inhibition of root elongation and in aluminium resistance of plants: A review. *Annals of Botany*, *106*(1), 185–197. doi: 10.1093/aob/mcq053.

Hossain, A. Z., Hossain, M. A., Koyama, H., & Hara, T. (2004). Effects of aluminum and boron supply on growth of seedlings among 15 cultivars of wheat (*Triticum aestivum* L.) grown in Bangladesh. *Soil Science and Plant Nutrition*, *50*(2), 189–195. doi: 10.1080/00380768.2004.10409271.

Hossain, M. A., Ashrafuzzaman, M., Hossain, A. K. M., Ismail, M., & Koyama, H. (2014). Role of accumulated calcium in alleviating aluminum injury in wheat plants. *The Scientific World Journal*, 1-6.

Huang, C., Yamaji, N., Mitani. N., Yano, M., Nagamura, Y., Ma, J. F. (2009). A bacterial-type ABC transporter is involved in aluminum tolerance in rice. *Plant Cell, 21*(2), 655–667. doi: 10.1105/tpc.108.064543.

Iqbal, M. T. (2013). Phosphorus enhances aluminium tolerance in both aluminium-tolerant and aluminium-sensitive wheat seedlings. *South African Journal of Plant and Soil*, *30*(1), 13–21. https://hdl.handle.net/10520/EJC133584.

Jamal, S. N., Iqbal, M. Z., & Athar, M. (2006). Effect of aluminum and chromium on the germination and growth of two Vigna species. *International Journal of Environmental Science & Technology, 3*(1), 53–58.

Jaskowiak, J., Kwasniewska, J., Milewska-Hendel, A., Kurczynska, E. U., Szurman-Zubrzycka, M., & Szarejko, I. (2019). Aluminum alters the histology and pectin cell wall composition of barley roots. *International Journal of Molecular Sciences*, *20*(12), 3039. doi: 10.3390/ijms20123039.

Kariya, K., Sameeullah. M., Sasaki. T., Yamamoto. Y. (2017). Overexpression of the sucrose transporter gene NtSUT1 alleviates aluminum-induced inhibition of root elongation in tobacco (*Nicotiana tabacum* L.). *Soil Science and Plant Nutrition, 63*(1), 45–54. doi: 10.1080/00380768.2017.1283646.

Kibria, M. G., Barton, L., & Rengel, Z. (2021a). Applying foliar magnesium enhances wheat growth in acidic soil by stimulating exudation of malate and citrate. *Plant and Soil*, 1–14. doi: 10.1007/s11104-021-04984-z.

Kibria, M. G., Barton, L., & Rengel, Z. (2021b). Genetic aluminium resistance coupled with foliar magnesium application enhances wheat growth in acidic soil. *Journal of the Science of Food and Agriculture*, *101*(11), 4643–4652. doi: 10.1002/jsfa.11107.

Kinraide, T. B. (1998). Three mechanisms for the calcium alleviation of mineral toxicities. *Plant Physiology*, *118*(2), 513–520. doi: 10.1104/pp.118.2.513.

Kinraide, T. B., Pedler, J. F., & Parker, D. R. (2004). Relative effectiveness of calcium and magnesium in the alleviation of rhizotoxicity in wheat induced by copper, zinc, aluminum, sodium, and low pH. *Plant and Soil, 259*(1), 201–208. doi: 10.1023/B:PLSO.0000020972.18777.99.

Klotz, F., & Horst, W. J. (1988). Effect of ammonium-and nitrate-nitrogen nutrition on aluminium tolerance of soybean (*Glycine max* L.). *Plant and Soil, 111*(1), 59–65.

Kochian, L. V. (1995). Cellular mechanisms of aluminum toxicity and resistance in plants. *Annual Review of Plant Biology, 46*(1), 237–260.

Kong, X., Peng, Z., Li, D., Ma, W., An, R., Khan, D., ... & Chen, Q. (2020). Magnesium decreases aluminum accumulation and plays a role in protecting maize from aluminum-induced oxidative stress. *Plant and Soil, 457*(1), 71–81. doi: 10.1007/s11104-020-04605-1.

Kopittke, P. M., Gianoncelli, A., Kourousias, G., Green, K., & McKenna, B. A. (2017). Alleviation of Al toxicity by Si is associated with the formation of Al–Si complexes in root tissues of sorghum. *Frontiers in Plant Science, 8*, 2189. doi: 10.3389/fpls.2017.02189.

Korcak, R. F. (1988). Fluidized bed material applied at disposal levels: Effects on an apple orchard. *American Society of Agronomy, Crop Science Society of America, and Soil Science Society of America, 17*(3), 469–473. doi: 10.2134/jeq1988.00472425001700030021x.

Kumar, M., Hazarika, S., Choudhury, B. U., Ramesh, T., Verma, B. C., & Bordoloi, L. J. (2012). Liming and integrated nutrient management for enhancing maize productivity on acidic soils of northeast India. *Indian Journal of Hill Farming, 25*(1), 36–38.

Li, X. W., Liu, J. Y., Fang, J., Tao, L., Shen, R. F., Li, Y. L., ... & Yu, M. (2017a). Boron supply enhances aluminum tolerance in root border cells of pea (Pisum sativum) by interacting with cell wall pectins. *Frontiers in Plant Science, 8*, 742. doi: 10.3389/fpls.2017.00742.

Li, N., Meng, H., Xing, H., Liang, L, Zhao, X., & Luo, K. (2017b). Genome-wide analysis of MATE transporters and molecular characterization of aluminum resistance in Populus. *Journal of Experimental Botany, 68*(20), 5669–5683.

Li, J., Wang, Y., Zheng, L., Li, Y., Zhou, X., Li, J., ... & Zhang, W. (2019). The intracellular transporter AtNRAMP6 is involved in Fe homeostasis in Arabidopsis. *Frontiers in Plant Science*, 1124. doi: 10.3389/fpls.2019.01124.

Li, W., Du, J., Feng, H., Wu, Q., Xu, G., Shabala, S., & Yu, L. (2020). Function of NHX-type transporters in improving rice tolerance to aluminum stress and soil acidity. *Planta, 251*(3), 1–13. doi: 10.1007/s00425-020-03361-x.

Lima, M. D. R., Barros, J. U., Barbosa, M. A. M., Segura, F. R., Silva, F. F., Batista, B. L., & Lobato, A. K. S. (2016). Silicon mitigates oxidative stress and has positive effects in Eucalyptus platyphylla under aluminium toxicity. *Plant, Soil and Environment, 62*(4), 164–170. doi: 10.17221/85/2016-PSE.

Liu J, Li Y, Wang W, Gai J, Li Y (2016) Genome-wide analysis of MATE transporters and expression patterns of a subgroup of MATE genes in response to aluminum toxicity in soybean. *BMC Genomics, 17*(1), 1–15.

Ma, J. F. (2004). Role of silicon in enhancing the resistance of plants to biotic and abiotic stresses. *Soil Science and Plant Nutrition, 50*(1), 11–18. doi: 10.1080/00380768.2004.10408447.

Ma, Q. J., Sun, M. H., Lu, J., Kang, H., You, C. X., & Hao, Y. J. (2019). An apple sucrose transporter MdSUT2. 2 is a phosphorylation target for protein kinase MdCIPK22 in response to drought. *Plant Biotechnology Journal, 17*(3), 625–637. doi: 10.1111/pbi.13003.

Magalhaes, J. V., Liu, J., Guimaraes, C. T., Lana, U. G., Alves, V., Wang, Y. H., ... & Kochian, L. V. (2007). A gene in the multidrug and toxic compound extrusion (MATE) family confers aluminum tolerance in sorghum. *Nature Genetics, 39*(9), 1156–1161. doi: 10.1038/ng2074.

Mao, D., Chen, J., Tian, L., Liu, Z., Yang, L., Tang, R., ... & Luan, S. (2014). Arabidopsis transporter MGT6 mediates magnesium uptake and is required for growth under magnesium limitation. *The Plant Cell, 26*(5), 2234–2248. doi: 10.1105/tpc.114.124628.

Mattiello, E. M., Ruiz, H. A., Neves, J. C., Ventrella, M. C., & Araújo, W. L. (2015). Zinc deficiency affects physiological and anatomical characteristics in maize leaves. *Journal of Plant Physiology, 183*, 138–143. doi: 10.1016/j.jplph.2015.05.014.

McCain, S., & Davies, M. S. (1983). The influence of background solution on root responses to aluminium in *Holcus lanatus* L. *Plant and Soil, 73*(3), 425–430.

Nhan, P. P., & Hai, N. T. (2013). Amelioration of aluminum toxicity on OM4900 rice seedlings by sodium silicate. *African Journal of Plant Science, 7*(6), 208–212. doi: 10.5897/AJPS11.306.

Noble, A. D., & Sumner, M. E. (1988). Calcium and Al interactions and soybean growth in nutrient solutions. *Communications in Soil Science and Plant Analysis, 19*(7–12), 1119–1131.

Okem, A., Southway, C., Stirk, W. A., Street, R. A., Finnie, J. F., & Van Staden, J. (2015). Effect of cadmium and aluminum on growth, metabolite content and biological activity in Drimia elata (Jacq.) Hyacinthaceae. *South African Journal of Botany, 98*, 142–147.

Ozyigit, I. I., Vardar, F., Yasar, U., & Akinci, S. (2013). Long-term effects of aluminum and cadmium on growth, leaf anatomy, and photosynthetic pigments of cotton. *Communications in Soil Science and Plant Analysis, 44*(21), 3076–3091.

Pan, C. L., Yao, S. C., Xiong, W. J., Luo, S. Z., Wang, Y. L., Wang, A. Q., ... & He, L. F. (2017). Nitric oxide inhibits Al-induced programmed cell death in root tips of peanut (*Arachis hypogaea* L.) by affecting physiological properties of antioxidants systems and cell wall. *Frontiers in Physiology, 8*, 1037. doi: 10.3389/fphys.2017.01037.

Park, W., Kim, H. S., Park, T. W., Lee, Y. H., & Ahn, S. J. (2017). Functional characterization of plasma membrane-localized organic acid transporter (CsALMT1) involved in aluminum tolerance in *Camelina sativa* L. *Plant Biotechnology Reports, 11*(3), 181–192. doi: 10.1007/s11816-017-0441-z.

Peres da Rocha Oliveiros Marciano, D., Toledo Ramos, F., Neiva Alvim, M., Ronaldo Magalhaes, J., & Giovanni Costa França, M. (2010). Nitric oxide reduces the stress effects of aluminum on the process of germination and early root growth of rice. *Journal of Plant Nutrition and Soil Science*, *173*(6), 885–891. doi: 10.1002/jpln.200900312.

Phukan, U. J., Jeena, G. S., Tripathi, V., & Shukla, R. K. (2018). Ma RAP 2-4, a waterlogging-responsive ERF from Mentha, regulates bidirectional sugar transporter At SWEET 10 to modulate stress response in Arabidopsis. *Plant Biotechnology Journal*, *16*(1), 221–233. doi: 10.1111/pbi.12762.

Pontigo, S., Godoy, K., Jiménez, H., Gutiérrez-Moraga, A., Mora, M. D. L. L., & Cartes, P. (2017). Silicon-mediated alleviation of aluminum toxicity by modulation of Al/Si uptake and antioxidant performance in ryegrass plants. *Frontiers in Plant Science*, *8*, 642. doi: 10.3389/fpls.2017.00642.

Poot-Poot, W., & Teresa Hernandez-Sotomayor, S. M. (2011). Aluminum stress and its role in the phospholipid signaling pathway in plants and possible biotechnological applications. *IUBMB Life*, *63*(10), 864–872. doi: 10.1002/iub.550.

Prabagar, S., Hodson, M. J., & Evans, D. E. (2011). Silicon amelioration of aluminium toxicity and cell death in suspension cultures of Norway spruce (*Picea abies* (L.) Karst.). *Environmental and Experimental Botany*, *70*(2–3), 266–276.

Qian, P., Sun, R., Ali, B., Gill, R. A., Xu, L., & Zhou, W. (2014). Effects of hydrogen sulfide on growth, antioxidative capacity, and ultrastructural changes in oilseed rape seedlings under aluminum toxicity. *Journal of Plant Growth Regulation*, *33*(3), 526–538. doi: 10.1007/s00344-013-9402-0.

Qian, L., Chen, B., & Chen, M. (2016). Novel alleviation mechanisms of aluminum phytotoxicity via released biosilicon from rice straw-derived biochars. *Scientific Reports*, *6*(1), 1–11. doi: 10.1038/srep29346.

Rai, K. K., Pandey, N., & Rai, S. P. (2020). Salicylic acid and nitric oxide signaling in plant heat stress. *Physiologia Plantarum*, *168*(2), 241–255.doi: 10.1111/ppl.12958.

Razaq, M., Zhang, P., & Shen, H. L. (2017). Influence of nitrogen and phosphorous on the growth and root morphology of Acer mono. *PLoS One*, *12*(2), e0171321. doi: 10.1371/journal.pone.0171321.

Reyes-Díaz, M., Meriño-Gergichevich, C., Alarcón, E., Alberdi, M., & Horst, W. J. (2011). Calcium sulfate ameliorates the effect of aluminum toxicity differentially in genotypes of highbush blueberry (*Vaccinium corymbosum* L.). *Journal of Soil Science and Plant Nutrition*, *11*(4), 59–78. doi: 10.4067/S0718-95162011000400005.

Reyna-Llorens, I., Corrales, I., Poschenrieder, C., Barcelo, J., & Cruz-Ortega, R. (2015). Both aluminum and ABA induce the expression of an ABC-like transporter gene (FeALS3) in the Al-tolerant species Fagopyrum esculentum. *Environmental and Experimental Botany*, *111*, 74–82. doi: 10.1016/j.envexpbot.2014.11.005.

Riaz, M., Yan, L., Wu, X., Hussain, S., Aziz, O., Wang, Y., ... & Jiang, C. (2018). Boron alleviates the aluminum toxicity in trifoliate orange by regulating antioxidant defense system and reducing root cell injury. *Journal of Environmental Management*, 208, 149–158. doi: 10.1016/j.jenvman.2017.12.008.

Roychowdhury, R., Khan, M. H., & Choudhury, S. (2019). Physiological and molecular responses for metalloid stress in rice: A comprehensive overview. *Advances in Rice Research for Abiotic Stress Tolerance*, 341–369. doi: 10.1016/B978-0-12-814332-2.00016-2.

Ruiz, J. M., Rivero, R. M., & Romero, L. (2006). Boron increases synthesis of glutathione in sunflower plants subjected to aluminum stress. *Plant and Soil*, *279*(1), 25–30. doi: 10.1007/s11104-005-7931-4.

Saha, R., Chaudhary, R. S., & Somasundaram, J. (2012). Soil health management under hill agroecosystem of North East India. *Applied and Environmental Soil Science*, *2012*. doi: 10.1155/2012/696174.

Saha, B., Swain, D., Borgohain, P., Rout, G. R., Koyama, H., & Panda, S. K. (2020). Enhanced exudation of malate in the rhizosphere due to AtALMT1 overexpression in blackgram (*Vigna mungo* L.) confers increased aluminium tolerance. *Plant Biology*, *22*(4), 701–708. doi: 10.1111/plb.13112.

Saifullah, K. M. N., Iqbal, M., Naeem, A., Bibi, S., Waraich, E. A., & Dahlawi, S. (2016). Elemental sulfur improves growth and phytoremediative ability of wheat grown in lead-contaminated calcareous soil. *International Journal of Phytoremediation*, *18*(10), 1022–1028. doi: 10.1080/15226514.2016.1146226.

Sanderson, K. R., & Eaton, L. J. (2004). Gypsum: An alternative to chemical fertilizers in lowbush blueberry production. *Small Fruits Review*, *3*(1–2), 57–71. doi: 10.1300/J301v03n01_07.

Sasaki, T., Egami, Y., Ono, T., & Hirose, K. (2004). First-principles calculation of transport properties of single-row aluminium nanowires suspended between semi-infinite crystalline electrodes. *Nanotechnology*, *15*(12), 1882.

Schroder, J. L., Zhang, H., Girma, K., Raun, W. R., Penn, C. J., & Payton, M. E. (2011). Soil acidification from long-term use of nitrogen fertilizers on winter wheat. *Soil Science Society of America Journal*, *75*(3), 957–964.

Serrano, I., Romero-Puertas, M. C., Sandalio, L. M., & Olmedilla, A. (2015). The role of reactive oxygen species and nitric oxide in programmed cell death associated with self-incompatibility. *Journal of Experimental Botany*, *66*(10), 2869–2876.

Shamshuddin, J., Fauziah, I. C., & Sharifuddin, H. A. H. (1991). Effects of limestone and gypsum application to a Malaysian Ultisol on soil solution composition and yields of maize and groundnut. *Plant and Soil*, *134*(1), 45–52.

Shamsi, I. H., Wei, K., Jilani, G., & Zhang, G. P. (2007). Interactions of cadmium and aluminum toxicity in their effect on growth and physiological parameters in soybean. *Journal of Zhejiang University Science B*, *8*(3), 181–188.

Shen, H., Hou, N., Schlicht, M., Wan, Y., Mancuso, S., & Baluska, F. (2008). Aluminium toxicity targets PIN2 in Arabidopsis root apices: Effects on PIN2 endocytosis, vesicular recycling, and polar auxin transport. *Chinese Science Bulletin*, *53*(16), 2480–2487.

Silva, S. (2012). Aluminium Toxicity Targets in Plants. *Journal of Botany*. 2012, 1–8. https://doi.org/10.1155/2012/219462

Silva, I. R., Smyth, T. J., Israel, D. W., Raper, C. D., & Rufty, T. W. (2001). Magnesium ameliorates aluminum rhizotoxicity in soybean by increasing citric acid production and exudation by roots. *Plant and Cell Physiology*, *42*(5), 546–554.

Singh, V. P., Tripathi, D. K., Kumar, D., & Chauhan, D. K. (2011). Influence of exogenous silicon addition on aluminium tolerance in rice seedlings. *Biological Trace Element Research*, *144*(1), 1260–1274.

Sivaguru, M., & Horst, W. J. (1998). The distal part of the transition zone is the most aluminum-sensitive apical root zone of maize. *Plant Physiology*, *116*(1), 155–163.

Skwierawska, M., Zawartka, L., & Zawadzki, B. (1997). The effect of different rates and forms of sulphur applied on changes of soil agrochemical properties. *Extraction, 1*(50).

Smith, C. J., Peoples, M. B., Keerthisinghe, G., James, T. R., Garden, D. L., & Tuomi, S. S. (1994). Effect of surface applications of lime, gypsum and phosphogypsum on the alleviating of surface and subsurface acidity in a soil under pasture. *Soil Research, 32*(5), 995–1008.

Song, Z. Z., Yang, S. Y., Zuo, J., & Su, Y. H. (2014). Over-expression of ApKUP3 enhances potassium nutrition and drought tolerance in transgenic rice. *Biol Plant, 58*(4), 649–658.

Stass, A., Kotur, Z., & Horst, W. J. (2007). Effect of boron on the expression of aluminium toxicity in Phaseolus vulgaris. *Physiologia Plantarum, 131*(2), 283–290.

Su, L., Xie, J., Wen, W., Li, J., Zhou, P., & An, Y. (2020). Interaction of zinc and IAA alleviate aluminum-induced damage on photosystems via promoting proton motive force and reducing proton gradient in alfalfa. *BMC Plant Biology, 20*(1), 1–17.

Sun, L., Liang, C., Chen, Z., Liu, P., Tian, J., Liu, G., & Liao, H. (2015). Superior aluminium (Al) tolerance of Stylosanthes is achieved mainly by malate synthesis through an Al-enhanced malic enzyme, Sg ME 1. *New Phytologist, 202*(1), 209–219.

Takemoto, Y., Tsunemitsu, Y., Fujiikashino, M., Mitanіueno, N., Yamaji, N., Ma, J. F., Kato, S., Iwasaki, K., & Ueno, D. (2017). The tonoplast localized transporter MTP82 contributes to manganese detoxification in the shoots and roots of *Oryza sativa* L. *Plant Cell Physiology, 58*(9), 1573–15802.

Tan, K., & Keltjens, W. G. (1990). Interaction between aluminium and phosphorus in sorghum plants. *Plant and Soil, 124*(1), 15–23.

Tariq, A., Pan, K., Olatunji, O. A., Graciano, C., Li, Z., Sun, F., ... & Liu, C. (2017). Phosphorous application improves drought tolerance of Phoebe zhennan. *Frontiers in Plant Science, 8*, 1561.

Thomine, S., Lelievre, F., Debarbieux, E., Schroeder, J. I., & Barbierbrygoo, H. (2003). AtNRAMP3, a multispecific vacuolar metal transporter involved in plant responses to iron deficiency. *The Plant Journal, 34*(5), 685–695.

Thor, K. (2019). Calcium—Nutrient and messenger. *Frontiers in Plant Science, 10*, 440.

Tirado-Corbalá, R., Slater, B. K., Dick, W. A., & Barker, D. (2017). Alfalfa responses to gypsum application measured using undisturbed soil columns. *Plants, 6*(3), 29.

Toma, M., & Saigusa, M. (1997). Effects of phosphogypsum on amelioration of strongly acid nonallophanic andosols. *Plant and Soil, 192*(1), 49–55.

Tuna, A. L., Kaya, C., Ashraf, M., Altunlu, H., Yokas, I., & Yagmur, B. (2007). The effects of calcium sulphate on growth, membrane stability and nutrient uptake of tomato plants grown under salt stress. *Environmental and Experimental Botany, 59*(2), 173–178.

Upadhyay, N., Kar, D., Mahajan, B. D., Nanda, S., Rahiman, R., Panchakshari, N., Bhagavatula, L., & Datta, S. (2019). The multitasking abilities of MATE transporters in plants. *Journal of Experimental Botany, 70*(18), 4643–4656.

Veraestrella, R., Gomezmendez, M. F., Amezcuaromero, J. C., Barkla, B. J., Rosassantiago, P., & Pantoja, O. (2017). Cadmium and zinc activate adaptive mechanisms in Nicotiana tabacum similar to those observed in metal tolerant plants. *Planta, 246*(3), 433–451.

Vitorello, V. A., Capaldi, F. R., & Stefanuto, V. A. (2005). Recent advances in aluminum toxicity and resistance in higher plants. *Brazilian Journal of Plant Physiology, 17*, 129–143.

Wang, Y., Stass, A., & Horst, W. J. (2004). Apoplastic binding of aluminum is involved in silicon-induced amelioration of aluminum toxicity in maize. *Plant Physiology, 136*(3), 3762–3770.

Wang, W., Zhao, X. Q., Chen, R. F., Dong, X. Y., Lan, P., Ma, J. F., & Shen, R. F. (2015). Altered cell wall properties are responsible for ammonium-reduced aluminium accumulation in rice roots. *Plant, Cell & Environment, 38*(7), 1382–1390.

Wang, H., Ji, F., Zhang, Y., Hou, J., Liu, W., Huang, J., & Liang, W. (2019). Interactions between hydrogen sulphide and nitric oxide regulate two soybean citrate transporters during the alleviation of aluminium toxicity. *Plant, Cell & Environment, 42*(8), 2340–2356.

Wang, X., Yan, X., & Liao, H. (2010a). Genetic improvement for phosphorus efficiency in soybean: A radical approach. *Annals of Botany, 106*(1), 215–222.

Wang, H. H., Huang, J. J., & Bi, Y. R. (2010b). Nitrate reductase-dependent nitric oxide production is involved in aluminum tolerance in red kidney bean roots. *Plant Science, 179*(3), 281–288.

Wojcik, P. (2003). Impact of boron on biomass production and nutrition of aluminum-stressed apple rootstocks. *Journal of Plant Nutrition, 26*(12), 2439–2451.

Wolf, S., Mouille, G., & Pelloux, J. (2009). Homogalacturonan methyl-esterification and plant development. *Molecular Plant, 2*(5), 851–860.

Wu, H. C., Bulgakov, V. P., & Jinn, T. L. (2018). Pectin methylesterases: Cell wall remodeling proteins are required for plant response to heat stress. *Frontiers in Plant Science, 9*, 1612.

Xiao, Z., Yan, G., Ye, M., & Liang, Y. (2021). Silicon relieves aluminum-induced inhibition of cell elongation in rice root apex by reducing the deposition of aluminum in the cell wall. *Plant and Soil*, 1–17.

Yamamoto, Y. (2019). Aluminum toxicity in plant cells: Mechanisms of cell death and inhibition of cell elongation. *Soil Science and Plant Nutrition, 65*(1), 41–55.

Yan, L., Riaz, M., Wu, X., Du, C., Liu, Y., & Jiang, C. (2018a). Ameliorative effects of boron on aluminum induced variations of cell wall cellulose and pectin components in trifoliate orange (*Poncirus trifoliate* (L.) Raf.) rootstock. *Environmental Pollution, 240*, 764–774.

Yan, L., Riaz, M., Wu, X., Wang, Y., Du, C., & Jiang, C. (2018b). Interaction of boron and aluminum on the physiological characteristics of rape (*Brassica napus* L.) seedlings. *Acta Physiologiae Plantarum, 40*(2), 1–11.

Yang, J. L., You, J. F., Li, Y. Y., Wu, P., & Zheng, S. J. (2007). Magnesium enhances aluminum-induced citrate secretion in rice bean roots (Vigna umbellata) by restoring plasma membrane H+-ATPase activity. *Plant and Cell Physiology, 48*(1), 66–73.

Yang, J. L., Li, Y. Y., Zhang, Y. J., Zhang, S. S., Wu, Y. R., Wu, P., & Zheng, S. J. (2008). Cell wall polysaccharides are specifically involved in the exclusion of aluminum from the rice root apex. *Plant Physiology, 146*(2), 602–611.

Ye, J., Wang, X., Hu, T., Zhang, F., Wang, B., Li, C., Yang, T., Li, H., Lu, Y., & Giovannoni, J. J. (2017). An InDel in the promoter of Al-ACTIVATED MALATE TRANSPORTER9 selected during tomato domestication determines fruit malate contents and aluminum tolerance. *Plant Cell, 29*(9), 2249–2268.

Yokosho, K., Yamaji, N., Fujiikashino, M., & Ma, J. F. (2016). Retrotransposon mediated aluminum tolerance through enhanced expression of the citrate transporter OsFRDL4. *Plant Physiology, 172*(4), 2327–2336.

Yu, M., Shen, R., Xiao, H., Xu, M., Wang, H., Wang, H., ... & Bian, J. (2009). Boron alleviates aluminum toxicity in pea (*Pisum sativum*). *Plant and Soil, 314*(1), 87–98.

Zhang, H., Tan, Z. Q., Hu, L. Y., Wang, S. H., Luo, J. P., & Jones, R. L. (2010). Hydrogen sulfide alleviates aluminum toxicity in germinating wheat seedlings. *Journal of Integrative Plant Biology, 52*(6), 556–567.

Zhang, J., Zhao, C. Y., Liu, J., Song, R., Du, Y. X., Li, J. Z., ... & Zhao, Q. Z. (2016). Influence of sulfur on transcription of genes involved in arsenic accumulation in rice grains. *Plant Molecular Biology Reporter, 34*(3), 556–565.

Zhang, L., Wu, X., Wang, J., Qi, C., Wang, X., Wang, G., Li, M., Li, X., & Guo, Y. (2018a). BoALMT1, an Al-induced malate transporter in cabbage, enhances aluminum tolerance in Arabidopsis thaliana. *Frontiers Plant Science, 8*, 2156.

Zhang, M., Lu, X., Li, C., Zhang, B., Zhang, C., Zhang, X. S., & Ding, Z. (2018b) Auxin efflux carrier ZmPGP1 mediates root growth inhibition under aluminum stress. *Plant Physiology, 177*(2), 819–832.

Zhao, X. Q., Guo, S. W., Shinmachi, F., Sunairi, M., Noguchi, A., Hasegawa, I., & Shen, R. F. (2013). Aluminium tolerance in rice is antagonistic with nitrate preference and synergistic with ammonium preference. *Annals of Botany, 111*(1), 69–77.

Zheng, S. J., Yang, J. L., He, Y. F., Yu, X. H., Zhang, L., You, J. F., ... & Matsumoto, H. (2005). Immobilization of aluminum with phosphorus in roots is associated with high aluminum resistance in buckwheat. *Plant Physiology, 138*(1), 297–303.

Zhou, X. X., Yang, L. T., Qi, Y. P., Guo, P., & Chen, L. S. (2015). Mechanisms on boron-induced alleviation of aluminum-toxicity in Citrus grandis seedlings at a transcriptional level revealed by cDNA-AFLP analysis. *PLoS One, 10*(3), e0115485.

Zhou, J., Li, Z., Zhou, T., Xin, Z., Wu, L., Luo, Y., & Christie, P. (2020). Aluminum toxicity decreases the phytoextraction capability by cadmium/zinc hyperaccumulator Sedum plumbizincicola in acid soils. *Science of the Total Environment, 711*, 134591.

Zhu, C. Q., Zhang, J. H., Sun, L. M., Zhu, L. F., Abliz, B., Hu, W. J., ... & Jin, Q. Y. (2018). Hydrogen sulfide alleviates aluminum toxicity via decreasing apoplast and symplast Al contents in rice. *Frontiers in Plant Science, 9*, 294.

Zhu, C. Q., Cao, X. C., Zhu, L. F., Hu, W. J., Hu, A. Y., Abliz, B., ... & Zhang, J. H. (2019). Boron reduces cell wall aluminum content in rice (*Oryza sativa*) roots by decreasing H_2O_2 accumulation. *Plant Physiology and Biochemistry, 138*, 80–90.

8

Drought and Heat Stress Tolerance in Field Crops: Consequences and Adaptation Strategies

Lalichetti Sagar and Subhashisa Praharaj
Centurion University of Technology and Management

Sultan Singh
Sri Karan Narendra Agriculture University

Meenakshi Attri
Sher-e-Kashmir University of Technology and Management

Biswajit Pramanick
Dr. Rajendra Prasad Central Agricultural University

Sagar Maitra
Centurion University of Technology and Management

Akbar Hossain
Bangladesh Wheat and Maize Research Institute

Tanmoy Shankar, Jnana Bharati Palai, and Upasana Sahoo
Centurion University of Technology and Management

CONTENTS

8.1 Introduction

Plants require an optimal environment in terms of moisture availability, temperature, relative humidity, solar radiation etc. for achieving maximum productivity. Any significant deviations from the optimal value of these parameters affect the plant productivity and the plants are said to be stressed. Moisture and temperature are the two important factors that affect plant growth, development and yield (Guajardo et al. 2016; Chen et al. 2018; Kumar et al. 2012). As high temperature leads to excessive evapotranspiration, hence the plants are exposed to severe moisture stress. For this reason, plants express many common symptoms to both heat and moisture stresses. In this chapter, plant stress tolerance mechanisms for both these stresses have been systematically presented.

Under changing climate, there are frequent weather aberrations such as insufficient rainfall, untimely rainfall, longer dry spell and frequent heatwaves. Such frequent anomalies have

DOI: 10.1201/9781003258063-8

made it imperative to study both moisture and heat stress, for their better scientific understanding. Moisture stress leads to reduced cell turgidity which results in a lower leaf area that in turn reduces the photosynthetic activity. Moisture stress also impacts the physiological and biochemical levels of the plants, ultimately resulting in lower crop performance and poor yield. Heat stress affects the plant and the harshness of it varies with the level of strain occurred. Just like moisture, high-temperature stress influences the plants at their morphological, physiological and biochemical levels.

To ensure optimal productivity under these stresses, the plant needs to have stress tolerance mechanisms. For drought stress tolerance, the role of signal transduction pathway (Sharma et al. 2019), osmotic adjustment (Imadi et al. 2016) and glyoxylate pathway (Zhang et al. 2016) as well as stress proteins are worth mentioning. Like heat stress, the stress tolerance mechanisms for heat stress involves the role of antioxidants (Tripathy 2016), phytohormones (Li et al. 2018a), heat shock proteins (HSPs) (Usman et al. 2014) and osmolytes. Each of these tolerance mechanisms has been thoroughly discussed in the coming sections. As the resultant, impacts of limited soil moisture and high-temperature effects are similar to some extent, especially at the physiological and biological levels. Hence, some tolerance mechanisms suitable for one may work for the other also.

In low-income and developing countries, agriculture is mostly done in open fields, where it is difficult to control weather parameters. Therefore, crops are commonly exposed to stresses. Moreover, lower irrigation coverage also increases the risk of crop failure under drought stress. Considering these facts, it can be assumed that the negative outcome of the stresses appears more disastrous in developing countries, with poor infrastructure to counteract the stress environment. In this regard, developing stress-tolerant varieties, having better adaptation to drought and heat stress, can be an excellent option. In this chapter, different plant tolerance mechanisms to deficit moisture- and heat-related burdens have been discussed.

8.2 Drought Stress Tolerance in Field Crops

Drought or soil moisture deficit causes stress in crops resulting in a reduction in cell turgidity. Further, it affects important physiological activities such as photosynthesis, respiration and transpiration. To manage the harsh influences of drought, the plants develop some tolerance mechanisms as they can survive even under stressful environment. The drought stress tolerance mechanisms have been explained in the following sections.

8.2.1 Signal Transduction in Drought Tolerance

In general, the signal transduction pathway includes the perception of drought signals by primary receptors. Further, these perceived signals are amplified and transmitted for the synthesis of drought-resistant proteins inducing stress tolerance (Sharma et al. 2019). Under drought, the biochemical behaviour of the plant changes and generates a signal which was primarily perceived by the signal receptors and then transmitted to target sites, mainly through specific protein kinase molecules (Farooq et al. 2009). Receptor-like kinases perceive the external drought stimulus and lead to bring changes in the cytoplasmic environment and help in converting the perceived external stimulus to internal stimuli (Wang et al. 2018a). Similarly, histidine kinase is another important signal cascade molecule that modifies the signalling hormonal responses to biotic or abiotic stress (Verma et al. 2016). Consequently, the perceived signal gets amplified by the mediation of various secondary messengers [calcium, reactive oxygen species (ROS) etc.] or plant hormones (Xia et al. 2015). Henceforth, due to the existence of many cross-talk networks for each signal, it activates multiple cascade mechanisms to tolerate stress.

The role of calcium is widely popular. The rise in calcium concentration is perceived mainly by three calcium receptors, *viz.*, calmodulin, calmodulin-related proteins and calcineurin B-like proteins that trigger activation of drought adaptive mechanisms, respectively (Kachenmeister et al. 2014). In calcium-mediated signalling, the signals transmitted from the receptors were transmitted through the calcium-mediated pathway such that the external drought stimulus was first perceived by phospholipase C receptor, which produces inositol triphosphate and diacylglycerol, leading to an accelerated increase in cytoplasmic calcium (Jing et al. 2019). Further, this enhanced calcium concentration improves the signal strength and activates protein phosphorylation (Ardito et al. 2017). Further, the ROS-mediated signal amplification was mainly attributed to long-distance signalling of drought stress (Sachdev et al. 2021). Perception of drought by plants impairs cellular homeostasis and thus results in the production of ROS from cell organelles, *viz.*, chloroplast, mitochondria and peroxisomes (Das and Roychoudhury 2014). The enhanced concentration of ROS molecules is detrimental to plant metabolic activity (Hasanuzzaman et al. 2020). However, their role in the signal transduction pathways was widely reported (Ortega et al. 2020). Moreover, every plant usually has an innate scavenging mechanism to tolerate oxidative stress (Tewari et al. 2021). Abscisic acid (ABA)-mediated ROS due to drought led to complete closure of stomata and reported a significant reduction of water due to transpiration (Rajab et al. 2019).

Drought signal transduction is highly complex in nature; interorganellar signalling is also reported between genomes of nucleus and cell organelles influencing their phenotypic expression (Nair et al. 2019). This signalling is of two types, *viz.*, anterograde and retrograde (Wagner et al. 2018). In anterograde signalling, the nucleus regulates the functioning of all other organelles, while in retrograde signalling, cell organelles regulate the functioning of the nucleus (Liu and Li 2019).

8.2.2 Phytohormones in Drought Stress Tolerance

Recent advances in soil moisture stress evidenced the role of phytohormones in adapting the stress. Plant hormones act as secondary messengers and trigger various drought adaptive physiological functions upon perception. Among various phytohormones, the ABA is more important in tolerating stress (Guajardo et al. 2016). In specific, ABA concentration in the plant cell determines the stomatal movement as a response to adapt soil moisture stress (Bertolino et al. 2019). However, the research studies revealed that ABA signalling is operated by two protein receptors, *viz.*, pyrabactin and a regulatory component of the ABA receptor. SNF1-related protein kinase 2s (SnRK2s) and protein phosphatase 2Cs (PP2Cs) regulate the ABA signalling in the plant (Wang et al. 2018b). During the drought, the ABA concentrations surge which inhibit PP2C and allow to activate SnRK2s that facilitate drought-responsive synthesis of transcriptional factors (Lee et al. 2015). ABA-mediated synthesis of transcriptional factors plays an important role in activating enzymes responsible for the synthesis of ROS scavengers, attributing to tolerate drought-induced oxidative stress (Kalia et al. 2017).

Cytokinins are another important hormone that has a marked influence on regulating leaf senescence (Hönig et al. 2018). Drought-induced plants downregulate cytokinin synthesis and adenosine phosphate isopentyl transferase enzymes which initiates leaf senescence, while under optimum conditions, leaf senescence was delayed by the influence of adequate cytokinin concentration (Rubia et al. 2014). Additionally, cytokinin content mediates the production of antioxidants that help in tolerating drought-induced oxidative stresses (Gujjar and Supaibulwatana 2019).

Auxin and gibberellic acids were also taking part in different metabolic pathways and were involved to tolerate stress. Root-to-shoot ratio is another important concern (Rastogi et al. 2013). The crops under drought were observed to enhance the auxin concentration and consequently reported increased lateral root development (Farooq et al. 2009). Further, the enhanced root development and proliferation promoted the total water intake that was supported in the maintenance of turgidity (Sharp et al. 2004). Similarly, exogenous application of salicylic acid and jasmonic acid took part in preventing lipid peroxidation through stimulation of ROS-detoxifying enzymes (Kalachova et al. 2012). Further, collectively they helped in retaining more water in the plants.

8.2.3 Osmotic Adjustment in Drought Stress Tolerance

Drought or soil moisture stress occurs when the plant is unable to balance the rate of water lost with its uptake (Imadi et al. 2016). Water balance within the plant system can be hampered in two ways, *viz.*, reduced water uptake or by increased transpiration losses (Chen et al. 2018). This water deficit triggers innate multifaceted drought tolerance responses (Chourasia 2017). Decreasing the root water potential by gathering the osmotically active substances to improve plant water uptake is a common response in tolerating drought (Kamiloğlu et al. 2014). Moreover, the osmoregulatory potential is species and environmental specific (Cao et al. 2018).

Accumulation of osmotically active compounds contributes in maintaining the turgor and regulating the stomatal movement of the plant (Lee 2010). Stomata are highly sensitive to water availability; they usually close at reduced turgor and vice versa. Stomatal opening is a key adaptive mechanism to tolerate stress (Rezae et al. 2008). However, the duration of stomatal opening determines the yield of the crop (Pujiwati 2014). Osmotic adjustment is highly significant in roots compared to other plant parts (Mittal et al. 2015). In roots, reduced root water potential helps to establish the continuance of water uptake by plants. However, the length, as well as depth, of root also determines the moisture extraction (Henry et al. 2012).

The most common osmotically active compounds include amino acids, sugars, polyols and quaternary ammonium salts (Khan et al. 2020). Drought stress slows down the photosynthesis and the utilization of metabolites (Ghotbi-Ravandi et al. 2014). In wheat, increased carbohydrate accumulation was reported under drought stress (Kandoliya et al. 2015). This accumulation plays an important role in improving cell membrane stability and scavenging of ROS. Proline is an important osmolyte that imparts stress resistance (Karuppanapandian et al. 2011). In general, water stress triggers upregulation of proline and its concentration peters out upon stress withdrawal (Solanki and Sarangi 2014). When a plant perceives drought, the glutamic acid present in the cytosol or plastids were phosphorylated into glutamate semialdehyde in the presence of an enzyme, glutamyl kinase. Further, glutamate semialdehyde was reduced to proline with pyrroline-5-carboxylate as an intermediate (Skidmore et al. 2011). High proline concentration confers improved membrane stabilization, minimization of protein degradation and osmotic regulation (Godard et al. 2020). However, when normal conditions are restored, the proline enters mitochondria through specific uniporters where it oxidizes back to glutamic acid (Zulkifli et al. 2020). For instance, (Nikolaeva et al. 2015) observed an increased adaptation to drought through osmotic adjustments under elevated proline concentration.

Similarly, the protective role of Glycine betaine was widely investigated and reported to impart dehydration tolerance in leaf chloroplast (Hasan et al. 2021). This osmolyte was reported to be synthesized from precursor choline and an intermediate betaine aldehyde, respectively (Bhuiyan et al. 2007). Besides, Glycine betaine synthesis was usually facilitated in the presence of two enzymes, *viz.*, Choline monooxygenase and betaine aldehyde dehydrogenase (BADH) (Khan et al. 2009). Further, reported to reduce ROS production thus attributing towards enhanced tolerance to drought stress (Cruz de Carvalho 2008).

8.2.4 Glyoxalase Pathway in Drought Tolerance

In the recent scenario of climate change, the plants are frequently exposed to drought stress. The sessile habit of plants evolved them to adapt to stress through several gene-regulated mechanisms (Ahuja et al. 2010). In general, plants when exposed to abiotic stress elevate the concentration of harmful secondary metabolites, *viz.*, ROS, methylglyoxal (Xie et al. 2019). However, methylglyoxal accumulation in plants was deleterious for their growth and development. Excessive accumulation of methylglyoxal limits the functioning of glutathione (Upadhyaya et al. 2011). In plants, glutathione was an important antioxidant, popularly recognized for its role in ROS detoxification (Mulata 2017). However, a high concentration of methylglyoxal is reported to deplete glutathione into hydroxyacyl glutathione (Proshad et al. 2018). In addition, methylglyoxal also inhibits the production of essential proteins, nucleic acids and lipids leading to cellular dysfunction (Li et al. 2018b).

Recent investigations reported the potential of the innate glyoxalase pathway in the detoxification of harmful methylglyoxal (Zhang et al. 2016). This mechanism is divided into two steps which are catalysed by *glyoxalase* I and glyoxalase II enzymes, respectively (Goo et al. 2015). In step 1, the methylglyoxal is converted to lactoylglutathione which is subsequently reduced to lactic acid in step 2, henceforth conferring drought stress tolerance (Paoli et al. 2010). However, overexpression of genes involved in the production of glyoxalase enzymes determines the innate ability of the plant to tolerate drought stress (El-Esawi and Alayafi 2019).

Besides all these negative impacts of the accumulation of methylglyoxal, it is also reported to be associated with certain beneficial roles in signalling pathways too (Delle Monache et al. 2021). The role of methylglyoxal in the triggering of ROS synthesis is attributed to regulating stomatal movement (Hoque et al. 2016), when the contraction of methylglyoxal increases the stomata closes and prevents the water loss through transpiration and vice versa (Lau et al. 2021).

8.2.5 Stress Induced Proteins in Drought Stress Tolerance

The drought tolerance in plants is highly complex in nature. Unravelling the gene expression and function under drought paves a way to engineer plants that produce optimum yield even under stress (Kumar 2020). The most popular drought-responsive genes were namely late embryogenesis abundant (LEA) proteins, aquaporins and HSPs (Song et al. 2017).

The LEA proteins were induced with a significant rise in ABA concentration under drought stress (Maleki et al. 2011). These proteins were primarily discovered during the late embryonic stages and helped in the proper sequestration of ions (Kosecka-Strojek et al. 2021). Under drought-tolerant plants, it was reported that LEA proteins were heavily accumulated (Butt et al. 2017). When the role of LEA proteins during water deficit stress was assessed, it was reported to act as a molecular chaperone and help in ROS sequestration (Artur et al. 2019). For instance, the rice plants overexpressed with OsLEA reported to produce optimum yields even under drought stress (Verma et al. 2019).

Aquaporins are another important group of proteins that belong to the family of major intrinsic proteins (Park et al. 2010). These proteins play a key role in facilitating the intercellular transport of water and solutes (Wanat 2020). In general, they are found on the cell membranes and conferred to regulate the stomatal movement (Ye et al. 2020). Aquaporins are categorized into five subgroups based on their existing location, *viz.*, Tonoplast intrinsic proteins (vacuole), plasma membrane intrinsic proteins (plasma membrane), X intrinsic proteins, nodulin -26-like intrinsic proteins and small basic intrinsic proteins (Kadam et al. 2017). In an experiment, it was found that the drought tolerance ability of sorghum was comparatively more than maize because of overexpression of PIP 1;5 and PIP 2;3 in the leaves and roots of sorghum (Hasan et al. 2017).

HSPs were widely popular for their role in treating high-temperature stress (Usman et al. 2014). However, their overexpression also identified severe drought conditions (Mbambalala et al. 2020). They usually act as molecular chaperones and prevent misfolding of proteins even under drought (Chatterjee et al. 2020). Consequently, attributes produce normal yield even under stress conditions, thus inducing drought stress tolerance (Sallam et al. 2019).

8.3 High-Temperature Stress Tolerance in Field Crops

Exposure of plants to high temperatures affects their growth and development by influencing different physiological as well as biochemical activities. Plants develop a tolerance mechanism against high temperature by virtue of which they survive even under high-temperature stress conditions.

8.3.1 Antioxidants in High-Temperature Stress Tolerance

ROS are atoms or molecules with unpaired valence electrons involved in signalling the cascade of the plants (Tripathy 2016). The production of ROS under ambient temperature is usually neutralized by antioxidants (Das and Roychoudhury 2014). However, extreme temperatures elicit an unbalance between production and scavenging of ROS and thus resulting in their accumulation within the cell (Karuppanapandian et al. 2011). Cellular ROS is mainly of two types, *viz.*, radicals and non-radicals (Noreen et al. 2018). The free radical form of ROS includes superoxide, hydroxyl, peroxyl, semiquinone, alkoxyl, carbonate etc. and non-radicals include singlet oxygen, hydrogen peroxide, ozone, hypochlorous acid etc. (Vatansever et al. 2013).

Perception of high-temperature signals by plant triggers the ROS production (Boguszewska and Zagdańska 2012). Overproduction of ROS has a detrimental effect on membrane stability resulting in leaky membranes due to lipid peroxidation (Del Río 2015). However, heat stress-tolerant plants will acclimatize to the negative impact of ROS upregulation through the synthesis of ROS scavenging molecules/antioxidants and try to maintain the balance between both antioxidants and ROS (Medina et al. 2021). The rate of antioxidant metabolite synthesis also influences the tolerance of crops which was governed by the many enzymes, *viz.*, catalase, superoxide dismutase, ascorbate peroxidase, glutathione reductase, vitamin C etc. (Zhu et al. 2020). Therefore, both rate of neutralization and concentration of antioxidants in the cellular system determine the lenience of plants to oxidative stress persuaded by high temperature.

Antioxidants produced in the heat-tolerant species detoxify the ROS either by donating electrons in the intercellular cell space to the free radical or by scavenging the ROS from the cell system. Any delay in ROS neutralization has a destructive effect on the physiological process of the field crops resulting in complete cell death (Sharma et al. 2012). The establishment of equilibrium between ROS produced and antioxidants is a key to withstanding stress (Ozougwu 2016). Exposure to high-temperature stress disrupts this balance and increases the ROS concentration which initially causes damage to biologically active compounds, *viz.*, lipids, amino acids, proteins, nucleic acids and proteins and is the key in tolerating oxidative stress (Tripathy and Mohanty 2017). It further results in cell organelle dysfunction, membrane leakage, reduced photosynthesis etc. and consequently resulting in the reduction of crop growth and yield (Hajihashemi et al. 2020).

In tobacco, ROS scavenging enzymes are overexpressed when exposed to high-temperature stress and increase the concentration of antioxidant metabolites within the cell, thus promoting heat stress tolerance (Li et al. 2018c). Similarly, in rice, exposure to high temperatures increased the malondialdehyde and ROS concentration. The response of different cultivars of rice in tolerating heat stress varies upon increased expression of Cu/Zn–SODa in rice anther (Zhao et al. 2018a). Another experiment in rice indicated that low catalase activity in response to heat stress-induced ROS accumulation resulted in pollen infertility at the meiosis stage (Zhao et al. 2018b).

8.3.2 Role of Phytohormones in High-Temperature Stress Tolerance

As plants cannot move away from unfavourable conditions, they are naturally facilitated with inherent defence mechanisms to tolerate both biotic and abiotic stress (Devi et al. 2017). However, timely perception of stress and activation of defence system is the key to surviving extremities of the environment (Dragoş and Tanasescu 2010). Phytohormones are important members of the signal transduction pathway and act as middlemen in the activation of plant defence to stress (Hossain et al. 2021a) (Table 8.1). They are triggered when stress stimuli are perceived and activate the defence mechanism for plant tolerance (Sehgal et al. 2016). However, the cytosolic calcium/calmodulin was not increased significantly upon encountering heat stress but observed to show increased elevation during recovery from heat stress (Priya et al. 2019).

The role of ABA is highly significant to cope with extreme temperatures. Elevated temperatures induce oxidative stress apart from their direct role on plant growth and development (Kumar et al. 2012). ABA's role in instigating antioxidants attributes in establishing an equilibrium between ROS and antioxidants and neutralizing the excess ROS induced due to high-temperature stress (Arif et al. 2020).

In an investigation on *Arabidopsis*, it was observed that exogenous application of ABA significantly protected against heat stress-induced oxidative damage (Larkindale and Knight, 2002). Besides, ABA signal induces the production of molecular chaperones to tolerate heat stress and also physiologically take part in maintaining optimum water balance by regulating stomatal opening and closing (Farooq et al. 2009). This role of ABA attributed thermo-tolerance to wild-type rice genotype (Li et al. 2020a). The role of auxin in cell division and enlargement was indirectly responsible for heat stress tolerance. In heat-sensitive plants, the auxin production was inhibited upon exposure to abnormally high temperatures and thereby resulting in inhibition of growth and development of the crop. However, it was observed that with exogenous application of auxin stress-induced growth response was nullified and cell division and enlargement were resumed to normal (Majdaand Robert 2018).

Cytokinin is another important phytohormone that attributes to thermo-tolerance within the field crops (Tiwari et al. 2020). Cytokinin plays a significant role in achieving heat stress tolerance through leaf cooling. This leaf cooling was accomplished by increased transpiration which is correlated with the opening of stomata as facilitated by cytokinin upregulation (Macková et al. 2013). Besides, cytokinin also induces several HSPs upon perceiving heat stress signals and helps in the protection of protein misfolding (Park et al. 2005). However, salicylic acids play a crucial role in recovery in different field crops, *viz.*, legumes, mustard, tobacco etc. from heat stress. Apart from this exogenous application, salicylic acids impart thermo-tolerance in wheat preventing protein degradation (Sita et al. 2017). Brassinosteroids are another important group of phytohormones that attribute thermo-tolerance due to improved antioxidant activity accompanied with improved stomatal conductance and net photosynthetic rate (Ogweno et al. 2008).

8.3.3 Role of HSPs in High-Temperature Stress Tolerance

In general, HSPs play a major role in tolerating high-temperature stress. As soon as the plant senses heat stress, it upregulates the production of monomeric heat shock factors present in the cytoplasm of the plant cell (Usman et al. 2014; Maitra et al. 2021a). These monomeric heat shock factors move from the cytoplasm into the nucleus and induce the activity of several heat shock genes, by binding the promoter region of respective genes (Boltaña et al. 2017). Further, these activated heat shock genes encode the synthesis of HSPs. HSPs are otherwise called chaperones, mainly due to their function in protecting the degradation of heat-sensitive proteins in the cell

TABLE 8.1

Phytohormones and Their Role in Different Crops to Tolerate Heat Stress

Phytohormones	Role in Heat Stress Response	Crop	References
Abscisic acid	Signalling to induce HSPs	Rice	Li et al. (2017)
	Regulates plant water balance	Wheat	Travaglia et al. (2010)
	Antioxidant production	Tobacco	Dai et al. (2020)
Auxin	Improves plasticity	Rice	Zhang et al. (2018)
Cytokinin	Regulates stomatal movement	Tobacco	Macková et al. (2013)
Salicylic acid	Protein stabilization	Wheat	Munir and Shabbir (2018)
Brassinosteroids	Stomatal conductance	Rice	Kothari and Lachowiec (2021)

TABLE 8.2
Classification of Major Classes of HSP Family Based on Their Function

Major Classes of HSP Family	Energy Requirement	Function	References
HSP 100	ATP dependent	Disaggregation of proteins	Mogk et al. (2015)
HSP 90	ATP dependent	Mediates signal transduction	Xu et al. (2012)
HSP 70	ATP dependent	Stabilizes new proteins	Usman et al. (2017)
HSP 60	ATP dependent	Recovers denatured proteins	Usman et al. (2014)
sHSP	ATP independent	Stabilizes protein structure	Al-Whaibi (2011)

TABLE 8.3
Heat Stress Proteins Involved in Tolerating High-Temperature Stress in Field Crops

Crop	Protein Heat Stress	References
Maize	bzip60	Li et al. (2020b)
	HSFTF13	Li et al. (2018c)
	HSP26	Li et al. (2020b)
	HSP101	Young et al. (2001)
	bZIP60	Li et al. (2020b)
Wheat	HSP101	Young et al. (2001)
	HSP 100	Sumesh et al. (2008)
	HSP101	Campbell et al. (2001)
	HSP16.9	Garg et al. (2012)
	HSP20	Muthusamy et al. (2017)
	sHSP26	Chauhan et al. (2012)
Rice	ROS	Zhao et al. (2018a)
	OsHSFA2d	Cheng et al. (2015)
	HSP17.7	Xu et al. (2021)
	HSP18.2	
	HSP21	
	HSP83.1	
	HSP101	
	bZIP50	Hayashi et al. (2012)
	OsHsfB4b	Singh et al. (2012)
	OsHsfA2c	
	OsRGB1	Biswas et al. (2019)

through protein folding in turn preserving their stability (Sung and MacRae 2011). The HSP family is dynamic in nature. However, the majority of HSPs are classified under five major classes of HSP family, *viz.*, HSP 100, HSP 90, HSP 70, HSP 60, sHSP (Tkáčová and Angelovičová 2012).

HSP 100, being an important member of AAA ATPase superfamily, plays a key role in removing all the polypeptides that were toxic and non-functional to the plants (Tables 8.2 and 8.3) which were produced as a result of misfolding of proteins (Usman et al. 2014). Besides, HSP 100 helps in the solubilization of aggregated proteins and in association with HSP 70 and all the solubilized proteins were refolded into their previous conformation under normal conditions (Żwirowski et al. 2017). Consequently, this helped in normalizing the plant condition under a post stress scenario. The second major class of HSP constituted in the cytoplasm of the plant species is HSP 90 (Bita and Gerats 2013). Its chaperone activity is targetted to certain substrates such as steroid receptor proteins, kinases, involved in the cell cycle, etc. leading to mediate the plant signal transduction pathway (Tripathy et al. 2019). The third and most abundant class of HSP is HSP 70 (Gade et al. 2010). This class of proteins helps in the stabilization of newly synthesized proteins from aggregation and misfolding (Yadav et al. 2019). Besides, it also plays a significant role in signal transduction due to its marked role in the modulation of protein A, C and phosphatases (Proud 2007). The last major class of HSP includes HSP (Tóth et al. 2015). These sHSPs range between 15 and 42 kilo Daltons and are induced upon exposure to any abiotic stress (Khan et al. 2019 and Hossain et al. 2021b). However, these sHSPs usually impart tolerance by binding partially folded proteins and play a key role in improving protein stability (Pegoraro et al. 2011).

Maize upon exposure to a high temperature of 45°C for 4 hours doubled the manifested protein spots during 2D electrophoresis and on the other hand registered overexpression of four HSP, *viz.*, HSP 26, HSP 16.9, ATPase beta subunit and an unknown HSP imparting increased tolerance to high temperature (Abou-Deif et al. 2019). Similarly, when garden pea is subjected to high temperature during its early seedling stage upregulated the expression of sHSP, *viz.*, Pshsp 22.7 (Endoplasmic Reticulum), Pshsp 26.2 (Chloroplastic) and Pshsp 22.9 (Mitochondrial), respectively (Talalaiev and Korduym 2014).

8.3.4 Omics Approach in Tolerating High-Temperature Stress

This advanced approach is based on the changes that were observed in genome, microme, transcriptome, proteome and metabolome in relation to elevated temperatures for an extended period of time. This correlation often improved the scope of imparting stress tolerance to plants. Omics approach includes five levels, *viz.*, genomics, transcriptomics, micromics, proteomics and metabolomics (Bhardwaj et al. 2021; Ibraheem et al. 2018). High-temperature stress tolerance as a response to the presence and expression of several genes at the genome level of a crop that has the role of withstanding stress is covered under genomics (Fahad et al. 2019; Maitra et al. 2021b). Transcriptome includes a set of transcripts expressed in a cell and it is the second level of understanding stress tolerance in an omics approach, which is otherwise termed as transcriptomics (Wang et al. 2019). Similarly, if a micro RNA helps in understanding the tolerance of crops to stress that is termed as micromics (Ferdous et al. 2015). In the next level of omics, the stress tolerance was studied at a protein level, depending on their presence and expression in plants when imposed with stress is termed proteomics (Sahoo et al. 2020). Plants in response to tolerate stress were observed to produce certain stress-specific metabolites. An omics approach that involves the study of metabolites that helps in the maintenance of normal functioning of the cell is termed metabolomics (Isah 2019; Gupta et al. 2013).

The proteomics approach during high temperatures enabled the researchers to understand the biochemical tolerance of plants to heat stress (Jegadeesan et al. 2018). This approach compares the status of protein profiling when exposed to stress which is directly related to the various biological processes imparting stress tolerance in plants (Priya et al. 2019). Several proteins that are highly sensitive to heat stress include mainly HSPs, Rubisco, ATP synthase etc. (Usman et al. 2014). Protein analysis in wheat, under high-temperature stress, identified that the activity of Rubisco increased by 40 times in an interval of 4 hours (Degen et al. 2021). The transcriptomics approach in heat stress shows how different sets of transcriptomes at a particular growth stage vary in concentration on exposure to high-temperature extremes (Sun et al. 2020). Transcriptome analysis of gobhi sarson revealed that around 442 differentially expressed genes were observed in the seed, especially at 40°C, 60°C and 40/60°C, and this modulation of transcriptomes plays a key role in tolerating heat stress (Gao et al. 2021). Similarly, several metabolites, like amino acids, were released in certain heat-tolerant plants when exposed to extreme temperatures in turn imparting high-temperature stress tolerance in the species (Hasanuzzaman et al. 2013). Further, some metabolites involved in osmotic adjustment, *viz.*, Glycine betaine, proline etc. also play a role in imparting high-temperature tolerance in crops by maintaining crop water balance and ensuring membrane stability (Ali et al. 2020). In a study, when sugarcane sprouts are grown at high temperatures (40°C C), instead of its ambient temperature (28°C C), reported a significant increase in the synthesis of primary metabolites, *viz.*, Glycine betaine proline, soluble sugars etc., and secondary metabolites, *viz.*, anthocyanin, carotenoids, phenolics etc. (Wahid 2007).

8.4 Conclusion

Moisture and heat stress are the two most frequently occurring stresses in agriculture. Both genetic and physiological approaches in addition to agronomic alteration may be useful in managing these stresses. Many physiological and biochemical activities are found to impart stress tolerance in plants. The role of antioxidants, phytohormones, HSP, osmolytes is worth mentioning for heat stress tolerance in plants. Plants also follow many strategies to adapt and tolerate moisture stress. With climate change, such stresses are expected to be even more frequent in the upcoming years. In the developing countries where agriculture is largely rainfed and done under uncontrolled environments or in open field conditions, the occurrence of moisture and heat stress severely limits crop productivity. Unless such stresses are carefully managed they may put food security at risk. It must be kept in mind that stress has not only agricultural implications but also affects the economy of a country. Hence, care must be taken to develop an integrated strategy involving physiological, biochemical, genetic and agronomic strategies to manage these stresses for ensuring food security.

REFERENCES

Abou-Deif, M.H., Rashed, M.A.S., Khalil, K.M. and F.E.S. Mahmoud. 2019. Proteomic analysis of heat shock proteins in maize (*Zea mays* L.). *Bulletin of the National Research Centre* 43:1–9.

Ahuja, I., de Vos, R.C., Bones, A.M. and R.D. Hall. 2010. Plant molecular stress responses face climate change. *Trends in Plant Science* 15:664–674.

Ali, S., Abbas, Z., Seleiman, M.F., Rizwan, M., Yavas, I., Alhammad, B.A. and D. Kalderis. 2020. Glycine betaine accumulation, significance and interests for heavy metal tolerance in plants. *Plants* 9:896.

Al-Whaibi, M.H. 2011. Plant heat-shock proteins: A mini-review. *Journal of King Saud University-Science* 23:139–150.

Ardito, F., Giuliani, M., Perrone, D., Troiano, G. and Lo Muzio. 2017. The crucial role of protein phosphorylation in cell signaling and its use as targeted therapy. *International Journal of Molecular Medicine* 40:271–280.

Arif, Y., Singh, P., Siddiqui, H., Bajguz, A. and S. Hayat. 2020. Salinity induced physiological and biochemical changes in plants: An omic approach towards salt stress tolerance. *Plant Physiology and Biochemistry* 156:64–77.

Artur, M.A.S., Rienstra, J., Dennis, T.J., Farrant, J.M., Ligterink, W. and H. Hilhorst. 2019. Structural plasticity of intrinsically disordered LEA proteins from Xerophytaschlechteri provides protection in vitro and in vivo. *Frontiers in Plant Science* 10:1272.

Bertolino, L.T., Caine, R.S., and J.E. Gray. 2019. Impact of stomatal density and morphology on water-use efficiency in a changing world. *Frontiers in Plant Science* 10:225.

Bhardwaj, A., Devi, P., Chaudhary, S., Rani, A., Jha, U.C., Kumar, S. and H. Nayyar. 2021. 'Omics' approaches in developing combined drought and heat tolerance in food crops. *Plant Cell Reports* 41:699–739.

Bhuiyan, N.H., Hamada, A., Yamada, N., Rai, V., Hibino, T. and T. Takabe. 2007. Regulation of betaine synthesis by precursor supply and choline monooxygenase expression in Amaranthus tricolor. *Journal of Experimental Botany* 58:4203–4212.

Biswas, S., Islam, M.N., Sarker, S., Tuteja, N. and Z.I. Seraj. 2019. Overexpression of heterotrimeric G protein beta subunit gene (OsRGB1) confers both heat and salinity stress tolerance in rice. *Plant Physiology Biochemistry* 144:334–344.

Bita, C. and T. Gerats. 2013. Plant tolerance to high temperature in a changing environment: Scientific fundamentals and production of heat stress-tolerant crops. *Frontiers in Plant Science* 4:273.

Boguszewska, D. and B. Zagdańska. 2012. ROS as signaling molecules and enzymes of plant response to unfavorable environmental conditions. In: V. Lushchak and H. Semchyshyn (Eds), *Oxidative Stress–Molecular Mechanisms and Biological Effects*. IntechOpen: London, pp. 341–362.

Boltaña, S., Sanhueza, N., Aguilar, A., Gallardo-Escarate, C., Arriagada, G., Valdes, J.A. and R.A. Quinones. 2017. Influences of thermal environment on fish growth. *Ecology and Evolution* 7:6814–6825.

Butt, Y.N., Fatima, Q., Nasar, S., Ikram, J. and S. Akram. 2017. Drought tolerance in plants: A review. *Research & Reviews: Journal of Ecology and Environmental Sciences* 5:20–28.

Campbell, J.L., Yklueva, N., Zheng, H., Nieto-Sotelo, J., Henry, T.N. and H.T. Guyen. 2001. Cloning of new members of heat shock protein HSP101 gene family in wheat (*Triticum aestivum* (L.) Moench) inducible by heat, dehydration, and ABA[1]Biochimicaet Biophysica Acta (BBA). *Gene Structure and Expression* 1517:270–277.

Cao, Q., Gu, J., Wang, D., Liang, F., Zhang, H., Li, X. and S. Yin. 2018. Physiological mechanism of osmoregulatory adaptation in anguillid eels. *Fish Physiology and Biochemistry* 2:423–433.

Chatterjee, A., Rajarshi, K., Ghosh, H., Singh, M.K., Roy, O.P. and S. Ray. 2020. Molecular chaperones in protein folding and stress management in cyanobacteria. In: P.K. Singh, A. Kumar, A.K. Shrivistava, and V.K. Singh (Eds), *Advances in Cyanobacterial Biology*. Academic Press: Cambridge, MA, pp. 119–128.

Chauhan, H., Khurana, N., Nijhavan, A., Khurana, J.P. and P. Khurana. 2012. The wheat chloroplastic small heat shock protein (sHSP26) is involved in seed maturation and germination and imparts tolerance to heat stress. *Plant Cell and Environment* 35:1879–2059.

Chen, D., Wang, S., Yin, L. and X. Deng. 2018. How does silicon mediate plant water uptake and loss under water deficiency. *Front Plant Science* 9:281.

Cheng, Q., Zhou, Y., Liu, Z., Zhang, L., Song, G., Guo, Z., Wang, W., Qu, X., Zhu, Y., Yang, D. and E. Flemetakis. 2015. An alternatively spliced heat shock transcription factor, OsHSFA2dI, functions in the heat stress-induced unfolded protein response in rice. *Plant Biology* 17:419–429.

Chourasia, K.N. 2017. Resistance/Tolerance mechanism under water deficit (Drought) condition in plants. *International Journal of Current Microbiology and Applied Sciences* 6:66–78.

Cruz de Carvalho, M.H. 2008. Drought stress and reactive oxygen species: Production, scavenging and signaling. *Plant Signaling and Behavior* 3:156–165.

Dai, X., Wang, Y., Chen, Y., Li, H., Xu, S., Yang, T. and Z. Xia. 2020. Overexpression of NtDOG1L-T improves heat stress tolerance by modulation of antioxidant capability and defense, heat and ABA-related gene expression in tobacco. *Frontiers in Plant Science* 11:1676.

Das, K. and A. Roychoudhury. 2014. Reactive oxygen species (ROS) and response of antioxidants as ROS-scavengers during environmental stress in plants. *Frontiers in Environmental Science* 2:53.

Degen, G.E., Orr, D.J. and E. Carmo-Silva. 2021. Heat-induced changes in the abundance of wheat Rubisco activase isoforms. *New Phytologist* 229:1298–1311.

Del Río, L.A. 2015. ROS and RNS in plant physiology: An overview. *Journal of Experimental Botany* 10:2827–2837.

Delle Monache, S., Pulcini, F., Frosini, R., Mattei, V., Talesa, V.N. and C. Antognelli. 2021. Methylglyoxal-dependent glycative stress is prevented by the natural antioxidant oleuropein in human dental pulp stem cells through Nrf2/Glo1 pathway. *Antioxidants* 10:716.

Devi, E.L., Kumar, S., Singh, T.B., Sharma, S.K., Beemrote, A., Devi, C.P. and S.H. Wani. 2017. Adaptation strategies and defence mechanisms of plants during environmental stress. In: M. Ghorbanpour and A. Varma (Eds), *Medicinal Plants and Environmental Challenges*. Springer: Cham, pp. 359–413.

Dragos, D. and M.D. Tanasescu. 2010. The effect of stress on the defense systems. *Journal of Medicine and Life* 1:10.

El-Esawi, M.A., and A.A. Alayafi. 2019. Overexpression of StDREB2 transcription factor enhances drought stress tolerance in cotton (*Gossypium barbadense* L.). *Genes* 10:142.

Fahad, S., Adnan, M., Hassan, S., Saud, S., Hussain, S., Wu, C. and J. Huang. 2019. Rice responses and tolerance to high temperature. In: M. Hasanuzzaman, M. Fujita, K. Nahar and J.K. Biswas (Eds), *Advances in Rice Research for Abiotic Stress Tolerance*. Woodhead Publishing: Sawston, Cambridge, pp. 201–224.

Farooq, M., Wahid, A., Kobayashi, N., Fujita, D. and S.M.A. Basra. 2009. Plant drought stress: Effects, mechanisms and management. *Agronomy for Sustainable Development* 29:185–212.

Ferdous, J., Hussain, S.S. and B.J. Shi. 2015. Role of micro RNA s in plant drought tolerance. *Plant Biotechnology Journal* 13:293–305.

Gade, N., Mahapatra, R.K., Sonawane, A., Singh, V.K., Doreswamy, R. and M. Saini. 2010. Molecular characterization of heat shock protein 70–1 gene of goat (*Capra hircus*). *Molecular Biology International* 2010:7.

Gao, G., Hu, J., Zhang, X., Zhang, F., Li, M. and X. Wu. 2021. Transcriptome analysis reveals genes expression pattern of seed response to heat stress in *Brassica napus* L. *Oil Crop Science* 6:87–96.

Garg, D., Sareen, S., Dalal, S., Tiwari, R. and R. Singh. 2012. Heat shock protein-based SNP marker for terminal heat stress in wheat (*Triticum aestivum* L.). *Australian Journal of Crop Science* 6:1516–1521.

Ghotbi-Ravandi, A.A., Shahbazi, M., Shariati, M. and P. Mulo. 2014. Effects of mild and severe drought stress on photosynthetic efficiency in tolerant and susceptible barley (*Hordeum vulgare* L.) genotypes. *Journal of Agronomy and Crop Science* 200:403–415.

Godard, T., Zuhlke, D., Richter, G., Wall, M., Rohde, M., Riedel, K. and R. Biedendieck. 2020. Metabolic rearrangements causing elevated proline and polyhydroxybutyrate accumulation during the osmotic adaptation response of Bacillus megaterium. *Frontiers in Bioengineering and Biotechnology* 8:47.

Goo, Y.K., Yamagishi, J., Ueno, A., Terkawi, M.A., Aboge, G.O., Kwak, D. and X. Xuan. 2015. Characterization of Toxoplasma gondii glyoxalase 1 and evaluation of inhibitory effects of curcumin on the enzyme and parasite cultures. *Parasites and Vectors* 8:1–8.

Guajardo, E., Correa, J.A. and L. Contreras-Porcia. 2016 Role of abscisic acid (ABA) in activating antioxidant tolerance responses to desiccation stress in intertidal seaweed species. *Planta* 243:767–781.

Gujjar, R.S. and K. Supaibulwatana. 2019. The mode of cytokinin functions assisting plant adaptations to osmotic stresses. *Plants* 8:542.

Gupta, B., Sengupta, A., Saha, J. and K. Gupta. 2013. Plant abiotic stress: 'Omics' approach. *Journal of Plant Biochemistry and Physiology* 1:1000e108.

Hajihashemi, S., Mbarki, S., Skalicky, M., Noedoost, F., Raeisi, M. and M. Brestic. 2020. Effect of wastewater irrigation on photosynthesis, growth, and anatomical features of two wheat cultivars (*Triticum aestivum* L.). *Water* 12:607.

Hasan, S.A., Rabei, S.H., Nada, R.M. and G.M. Abogadallah. 2017. Water use efficiency in the drought-stressed sorghum and maize in relation to expression of aquaporin genes. *Biologia Plantarum* 61:127–137.

Hasan, M., Skalicky, M., Jahan, M.S., Hossain, M., Anwar, Z., Nie, Z.F. and X.W. Fang. 2021. Spermine: Its emerging role in regulating drought stress responses in plants. *Cells* 10:261.

Hasanuzzaman, M., Nahar, K., Alam, M., Roychowdhury, R. and M. Fujita. 2013. Physiological, biochemical, and molecular mechanisms of heat stress tolerance in plants. *International Journal of Molecular Sciences* 14:9643–9684.

Hasanuzzaman, M., Bhuyan, M.H.M., Parvin, K., Bhuiyan, T.F, Anee, T.I., Nahar, K. and M. Fujita. 2020. Regulation of ROS metabolism in plants under environmental stress: A review of recent experimental evidence. *International Journal of Molecular Sciences* 21:8695.

Hayashi, S., Wakasa, Y., Takahashi, H., Kawakatsu, T. and F. Takaiwa. 2012. Signal transduction by IRE1-mediated splicing of bZIP50 and other stress sensors in the endoplasmic reticulum stress response of rice. *The Plant Journal,* 69:946–956.

Henry, A., Cal, A.J., Batoto, T.C., Torres, R.O. and R. Serraj. 2012. Root attributes affecting water uptake of rice (*Oryza sativa*) under drought. *Journal of Experimental Botany* 63:4751–4763.

Hönig, M., Plihalova, L., Husickova, A., Nisler, J. and K. Dolezal. 2018. Role of cytokinins in senescence, antioxidant defence and photosynthesis. *International Journal of Molecular Sciences* 19:4045.

Hoque, T.S., Hossain, M.A., Mostofa, M.G., Burritt, D.J., Fujita, M. and L.S.P. Tran. 2016. Methylglyoxal: An emerging signaling molecule in plant abiotic stress responses and tolerance. *Frontiers in Plant Science* 7:1341.

Hossain, A., Raza, A., Maitra, S., Asaduzzaman, M., Islam, M.R., Hossain, M.J. and T. Aftab. 2021a. Strigolactones: A novel carotenoid-derived phytohormone-biosynthesis, transporters, signalling, and mechanisms in abiotic stress. In: T. Aftab and K.R. Hakeem (Eds), *Plant Growth Regulators: Signalling Under Stress Conditions.* Springer Nature: Switzerland AG, pp. 275–303.

Hossain, A., Skalicky, M., Brestic, M., Maitra, S., Ashraful Alam, M., Syed, M.A. and T. Islam. 2021b. Consequences and mitigation strategies of abiotic stresses in wheat (*Triticum aestivum* L.) under the changing climate. *Agronomy* 11:241.

Ibraheem, O., Adigun, R.O. and I.T. Olatunji. 2018. Omics technologies in unraveling plant stress responses; using sorghum as a model crop, how far have we gone. *Vegetos* 31:2.

Imadi, S.R., Gul, A., Dikilitas, M., Karakas, S., Sharma, I. and P. Ahmad. 2016. Water stress: Types, causes, and impact on plant growth and development. In: Parvaiz Ahmad (Ed.). *Water Stress and Crop Plants.* John Wiley & Sons: Chichester, UK, pp. 343–355.

Isah, T. 2019. Stress and defense responses in plant secondary metabolites production. *Biological Research* 52:39.

Jegadeesan, S., Chaturvedi, P., Ghatak, A., Pressman, E., Meir, S., Faigenboim, A. and N. Firon. 2018. Proteomics of heat-stress and ethylene-mediated thermotolerance mechanisms in tomato pollen grains. *Frontiers in Plant Science* 9:1558.

Jing, X., Cai, C., Fan, S., Wang, L. and X. Zeng. 2019. Spatial and temporal calcium signaling and its physiological effects in Moso Bamboo under drought stress. *Forests* 10:224.

Kachenmeister, F., Kachenmeister, K., Kayser, M., Wrage-MAlnnig, N. and J. Isselstein. 2014. Effects of drought stress and sward botanical composition on the nutritive value of grassland herbage. *International Journal of Agriculture and Biology* 16:1560–8530.

Kadam, S., Abril, A., Dhanapal, A.P., Koester, R.P., Vermerris, W., Jose, S. and F.B. Fritschi. 2017. Characterization and regulation of aquaporin genes of sorghum [*Sorghum bicolor* (L.) Moench] in response to waterlogging stress. *Frontiers in Plant Science* 8:862.

Kalachova, T.A., Iakovenko, O.M., Kretinin, S.V. and V.S. Kravets. 2012. Effects of salicylic and jasmonic acid on phospholipase D activity and the level of active oxygen species in soybean seedlings. *Biochemistry (Moscow) Supplement Series A: Membrane and Cell Biology* 6:243–248.

Kalia, R., Sareen, S., Nagpal, A., Katnoria, J. and R. Bhardwaj. 2017. ROS-induced transcription factors during oxidative stress in plants: A tabulated review. In: M. Iqbal, R. Khan and N.A. Khan (Eds), *Reactive Oxygen Species and Antioxidant Systems in Plants: Role and Regulation under Abiotic Stress.* Springer: Singapore, pp. 129–158.

Kamiloğlu, O., Sivritepe, N., Onder, S. and H. Daghan. 2014. Effects of water stress on plant growth and physiological characteristics of some grape varieties. *Fresenius Environmental Bulletin* 23:2155–2163.

Kandoliya, U.K., Marviya, G.V., Patel, N.J., Vakharia, D.N. and B.A. Golakiya. 2015. Effect of drought at different growth stage on carbohydrates and lipids composition of groundnut (*Arachis hypogaea* L.) pod. *International Journal of Current Research and Academic Review* 3:281–287.

Karuppanapandian, T., Moon, J.C., Kim, C., Manoharan, K. and W. Kim. 2011. Reactive oxygen species in plants: Their generation, signal transduction, and scavenging mechanisms. *Australian Journal of Crop Science* 5:709–725.

Khan, M.S., Yu, X., Kikuchi, A., Asahina, M. and K.N. Watanabe. 2009. Genetic engineering of glycine betaine biosynthesis to enhance abiotic stress tolerance in plants. *Plant Biotechnology* 26:125–134.

Khan, A., Ali, M., Khattak, A.M., Gai, W.X., Zhang, H.X., Wei, A.M. and Z.H. Gong. 2019. Heat shock proteins: Dynamic biomolecules to counter plant biotic and abiotic stresses. *International Journal of Molecular Sciences* 20:5321.

Khan, N., Ali, S., Zandi, P., Mehmood, A., Ullah, S., Ikram, M. and M.B. Babar. 2020. Role of sugars, amino acids and organic acids in improving plant abiotic stress tolerance. *Pakistan Journal of Botany* 52:355–363.

Kosecka-Strojek, M., Trzeciak, J., Homa, J., Trzeciak, K., Władyka, B., Trela, M. and M.W. Lis. 2021. Effect of Staphylococcus aureus infection on the heat stress protein 70 (HSP70) level in chicken embryo tissues. *Poultry Science* 100:101119.

Kothari, A. and J. Lachowiec. 2021. Roles of brassinosteroids in mitigating heat stress damage in cereal crops. *International Journal of Molecular Sciences* 22:2706.

Kumar, S. 2020. Abiotic stresses and their effects on plant growth, yield and nutritional quality of agricultural produce. *International Journal of Research in Agriculture and Food Sciences* 4:367–378.

Kumar, S., Kaushal, N., Nayyar, H. and P. Gaur. 2012. Abscisic acid induces heat tolerance in chickpea (*Cicer arietinum* L.) seedlings by facilitated accumulation of osmoprotectants. *Acta Physiologiae Plantarum* 34:1651–1658.

Larkindale, J. and M.R. Knight. 2002. Protection against heat stress-induced oxidative damage in Arabidopsis involves calcium, abscisic acid, ethylene, and salicylic acid. *Plant Physiology* 128:682–695.

Lau, S.E., Hamdan, M.F., Pua, T.L., Saidi, N.B. and B.C. Tan. 2021. Plant nitric oxide signaling under drought stress. *Plants* 10:360.

Lee, J.S. 2010. Stomatal opening mechanism of CAM plants. *Journal of Plant Biology* 1:19–23.

Lee, H.Y., Jang, G., Um, T., Kim, J.K., Lee, J.S., Do and Y. Choi. 2015. The soluble ABA receptor PYL8 regulates drought resistance by controlling ABA signaling in *Arabidopsis. Plant Biotechnology Reports* 9:319–330.

Li, J., Li, Y., Yin, Z., Jiang, J., Zhang, M., Guo, X. and Z. Li. 2017. Os ASR 5 enhances drought tolerance through a stomatal closure pathway associated with ABA and H_2O_2 signalling in rice. *Plant Biotechnology Journal* 15:183–196.

Li, Q., Wang, W., Wang, W., Zhang, G., Liu, Y., Wang, Y. and W. Wang. 2018a. Wheat F-box protein gene TaFBA1 is involved in plant tolerance to heat stress. *Frontiers in Plant Science* 9:521.

Li, S., Liu, S. and C.T. Ho. 2018b. Safety issues of methylglyoxal and potential scavengers. *Frontiers of Agricultural Science and Engineering* 5:312–320.

Li, Z. Srivastava, R. Tang, J. Zheng, Z. and S.H. Howell. 2018c. Cis-effects condition the induction of a major unfolded protein response factor, ZmbZIP60, in response to heat stress in maize. *Frontiers in Plant Science* 9:1–11.

Li, G., Zhang, C., Zhang, G., Fu, W., Feng, B., Chen, T. and G. Fu. 2020a. Abscisic acid negatively modulates heat tolerance in rolled leaf rice by increasing leaf temperature and regulating energy homeostasis. *Rice* 13:1–16.

Li, Z., Tang, J., Srivastava, R. and D.C. Bassham. 2020b. Howell, the transcription factor bZIP60 links the unfolded protein response (UPR) to the heat stress response (HSR) in maize. *Plant Cell* 32:3559–3575.

Liu, L. and J. Li. 2019. Communications between the endoplasmic reticulum and other organelles during abiotic stress response in plants. *Frontiers in Plant Science* 10:749.

Macková, H., Hronkova, M., Dobra, J., Turečková, V., Novak, O., Lubovska, Z. and R. Vankova. 2013. Enhanced drought and heat stress tolerance of tobacco plants with ectopically enhanced cytokinin oxidase/dehydrogenase gene expression. *Journal of Experimental Botany* 64:2805–2815.

Maitra, S., Bhadra, P., Yadav, A.N., Palai, J.B., Jena, J. and T. Shankar. 2021a. The omics strategies for abiotic stress responses and microbe-mediated mitigation in plants. In: A.N. Yadav (Ed.), *Soil Microbiomes for Sustainable Agriculture*. Springer: Cham, pp. 315–377.

Maitra, S., Pramanick, B., Dey, P., Bhadra, P., Shankar, T. and K. Anand. 2021b. Thermotolerant Soil Microbes and their role in mitigation of heat stress in plants. In: A.N. Yadav (Ed.), *Soil Microbiomes for Sustainable Agriculture*. Springer: Cham, pp. 203–242.

Majda, M. and S. Robert. 2018. The role of auxin in cell wall expansion. *International Journal of Molecular Sciences* 19:951.

Maleki, M., Ebrahimzade, H., Gholami, M. and V. Niknam. 2011. The effect of drought stress and exogenous abscisic acid on growth, protein content and antioxidative enzyme activity in saffron (*Crocus sativus* L.). *African Journal of Biotechnology* 10:9068–9075.

Mbambalala, N., Panda, S.K. and C. van der Vyver. 2020. Overexpression of AtBBX29 improves drought tolerance by maintaining photosynthesis and enhancing the antioxidant and osmolyte capacity of sugarcane plants. *Plant Molecular Biology Reporter* 39:1–15.

Medina, E., Kim, S.H., Yun, M. and W.G. Choi. 2021. Recapitulation of the function and role of ROS generated in response to heat stress in plants. *Plants* 10:371.

Mittal, V., Singh, N., Narwal, S., Mamrutha, H., Tiwari, V. and I. Sharma. 2015. Effect of osmotic stress on root architecture and defensive system in wheat genotypes at seedling stage. *Journal of Wheat Research* 7:52–59.

Mogk, A., Kummer, E. and B. Bukau. 2015. Cooperation of Hsp70 and Hsp100 chaperone machines in protein disaggregation. *Frontiers in Molecular Biosciences* 2:22.

Mulata, H.N. 2017. Role of endogenous and exogenous glutathione in the detoxification of free radicals. *Research & Reviews: A Journal of Toxicology* 7:1–14.

Munir, M. and G. Shabbir. 2018. Salicylic acid mediated heat stress tolerance in selected bread wheat genotypes of Pakistan. *Pakistan Journal of Botany* 50:2141–2146.

Muthusamy, K., Dalal, M., Chinnusamy, V. and K.C. Bansal. 2017. Genome-wide identification and analysis of biotic and abiotic stress regulation of small heat shock protein (HSP20) family genes in bread wheat. *Journal of Plant Physiology* 211:100–113.

Nair, A., Chauhan, P., Saha, B. and K.F. Kubatzky. 2019. Conceptual evolution of cell signaling. *International Journal of Molecular Sciences* 20:3292.

Nikolaeva, M.K., Maevskaya, S.N. and P.Y. Voronin. 2015. Activities of antioxidant and osmoprotective systems and photosynthetic gas exchange in maize seedlings under drought conditions. *Russian Journal of Plant Physiology* 62:314–21.

Noreen, A., Bukhari, D.A. and A. Rehman. 2018. Reactive oxygen species: Synthesis and their relationship with cancer-a review. *Pakistan Journal of Zoology* 50:1951–1963.

Ogweno, J.O., Song, X.S., Shi, K., Hu, W.H., Mao, W.H., Zhou, Y.H. and S. Nogues. 2008. Brassinosteroids alleviate heat-induced inhibition of photosynthesis by increasing carboxylation efficiency and enhancing antioxidant systems in Lycopersicon esculentum. *Journal of Plant Growth Regulation* 1:49–57.

Ortega, M.A., Fraile-Martinez, O., Asunsolo, A., Bujan, J., Garcia-Honduvilla, N. and S. Coca. 2020. Signal transduction pathways in breast cancer: The important role of PI3K/Akt/mTOR. *Journal of Oncology* 1:1–11.

Ozougwu, J.C. 2016. The role of reactive oxygen species and antioxidants in oxidative stress. *International Journal of Research* 3:1–8.

Paoli, T., Faulkner, J., Okennedy, R. and E. Keshavarz-Moore. 2010. A study of D-lactate and extracellular methylglyoxal production in lactate Re-Utilizing CHO cultures. *Biotechnology and Bioengineering* 107:182–189.

Park, H.G., Han, S.I., Oh, S.Y. and H.S. Kang. 2005. Cellular responses to mild heat stress. *Cellular and Molecular Life Sciences* 62:10–23.

Park, W., Scheffler, B.E., Bauer, P.J. and B.T. Campbell. 2010. Identification of the family of aquaporin genes and their expression in upland cotton (*Gossypium hirsutum* L.). *BMC Plant Biology* 10:1–17.

Pegoraro, C., Mertz, L.M., da Maia, L.C., Rombaldi, C.V. and A.C. de Oliveira. 2011. Importance of heat shock proteins in maize. *Journal of Crop Science and Biotechnology* 14:85–95.

Priya, M., Dhanker, O.P., Siddique, K.H., HanumanthaRao, B., Nair, R.M., Pandey, S. and H. Nayyar. 2019. Drought and heat stress-related proteins: An update about their functional relevance in imparting stress tolerance in agricultural crops. *Theoretical and Applied Genetics* 132:1607–1638.

Proshad, R., Kormoker, T., Mursheed, N., Islam, M.M., Bhuyan, M.I., Islam, M.S. and T.N. Mithu. 2018. Heavy metal toxicity in agricultural soil due to rapid industrialization in Bangladesh: A review. *International Journal of Advanced Geosciences* 6:83–88.

Proud, C.G. 2007. Signalling to translation: How signal transduction pathways control the protein synthetic machinery. *Biochemical Journal* 2:217–234.

Pujiwati, I. 2014. The pattern of stomatal opening through the exposure of high-frequency sound wave with the different duration and age of soybeans (*Glycine max* (L.) Merril). *Agriculture Science* 2:69–77.

Rajab, H., Khan, M.S., Malagoli, M., Hell, R. and M. Wirtz. 2019. Sulfate-induced stomata closure requires the canonical ABA signal transduction machinery. *Plants* 1:21.

Rastogi, A., Siddiqui, A., Mishra, B.K., Srivastava, M., Pandey, R., Misra, P. and S. Shukla. 2013. Effect of auxin and gibberellic acid on growth and yield components of linseed (*Linumusitatissimum* L.). *Crop Breeding and Applied Biotechnology* 13:136–143.

Rezaei Nejad, A. and U. van Meeteren. 2008. Dynamics of adaptation of stomatal behaviour to moderate or high relative air humidity in Tradescantia virginiana. *Journal of Experimental Botany* 59:289–301.

Rubia, L., Rangan, L., Choudhury, R.R., Kaminek, M., Dobrev, P., Malbeck, J. and M. Elliott. 2014. Changes in the chlorophyll content and cytokinin levels in the top three leaves of new plant type rice during grain filling. *Journal of Plant Growth Regulation* 33:66–76.

Sachdev, S., Ansari, S.A., Ansari, M.I., Fujita, M. and M. Hasanuzzaman. 2021. Abiotic stress and reactive oxygen species: Generation, signaling, and defense mechanisms. *Antioxidants* 10:277.

Sahoo, J.P., Behera, L., Sharma, S.S., Praveena, J., Nayak, S.K. and K.C. Samal. 2020. Omics studies and systems biology perspective towards abiotic stress response in plants. *American Journal of Plant Sciences* 11:2172.

Sallam, A., Alqudah, A.M., Dawood, M.F., Baenziger, P.S. and A. Borner. 2019. Drought stress tolerance in wheat and barley: Advances in physiology, breeding and genetics research. *International Journal of Molecular Sciences* 20:3137.

Sehgal, A., Sita, K. and H. Nayyar. 2016. Heat stress in plants: Sensing and defense mechanisms. *The Journal of Plant Science Research* 2:195.

Sharma, V., Goel, P., Kumar, S. and A.K. Singh. 2019. An apple transcription factor, MdDREB76, confers salt and drought tolerance in transgenic tobacco by activating the expression of stress-responsive genes. *Plant Cell Reports* 2:221–241.

Sharma, P., Jha, A.B., Dubey, R.S. and M. Pessarakli. 2012. Reactive oxygen species, oxidative damage, and antioxidative defense mechanism in plants under stressful conditions. *Journal of Botany* 2012:26.217037.

Sharp, R.E., Poroyko, V., Hejlek, L.G., Spollen, W.G., Springer, G.K., Bohnert, H.J. and H.T. Nguyen. 2004. Root growth maintenance during water deficits: Physiology to functional genomics. *Journal of Experimental Botany* 55:2343–2351.

Singh, A., Mittal, D., Lavania, D., Agarwal, M., Mishra, R.C. and A. Grover. 2012. OsHsfA2c and OsHsfB4b are involved in the transcriptional regulation of cytoplasmic OsClpB (Hsp100) gene in rice (*Oryza sativa* L.). *Cell Stress and Chaperones* 17:243–254.

Sita, K., Sehgal, A., Hanumantha Rao, B., Nair, R.M., Vara Prasad, P.V., Kumar, S. and H. Nayyar. 2017. Food legumes and rising temperatures: Effects, adaptive functional mechanisms specific to reproductive growth stage and strategies to improve heat tolerance. *Frontiers in Plant Science* 8:1658.

Skidmore, D.L., Chitayat, D., Morgan, T., Hinek, A., Fischer, B., Dimopoulou, A. and S.P. Robertson. 2011. Further expansion of the phenotypic spectrum associated with mutations in ALDH18A1, encoding Δ1-pyrroline-5-carboxylate synthase (P5CS). *American Journal of Medical Genetics Part A* 155:1848–1856.

Solanki, J.K. and S.K. Sarangi. 2014. Effect of drought stress on proline accumulation in peanut genotypes. *International Journal of Advanced Research* 2:301–309.

Song, K., Yim, W.C. and B.M. Lee. 2017. Expression of heat shock proteins by heat stress in Soybean. *Plant Breeding and Biotechnology* 5:344–353.

Sumesh, K.V., Sharma-Natu, P. and M.C. Ghildiyal. 2008. Starch synthase activity and heat shock protein in relation to thermal tolerance of developing wheat grains. *Biologia Plantarum* 52:749–753.

Sun, M., Huang, D., Zhang, A., Khan, I., Yan, H., Wang, X. and L. Huang. 2020. Transcriptome analysis of heat stress and drought stress in pearl millet based on Pacbio full-length transcriptome sequencing. *BMC Plant Biology* 20:1–15.

Sung, Y.Y. and T.H. MacRae. 2011. Heat shock proteins and disease control in aquatic organisms. *Journal of Aquaculture Research & Development* S:2.

Talalaiev, O. and E. Korduym. 2014. Expression of small heat shock protein (sHSP) genes in the garden pea (*Pisum sativum*) under slow horizontal clinorotation. *Plant Signaling and Behavior* 9:e29035.

Tewari, R.K., Yadav, N., Gupta, R. and P. Kumar. 2021. Oxidative stress under macronutrient deficiency in plants. *Journal of Soil Science and Plant Nutrition* 21:1–28.

Tiwari, P., Bajpai, M., Singh, L.K., Mishra, S. and A.N. Yadav. 2020. Phytohormones producing fungal communities: Metabolic engineering for abiotic stress tolerance in crops. In: A.N. Yadav, S. Mishra, D. Kour, N. Yadav and A. Kumar (Eds), *Agriculturally Important Fungi for Sustainable Agriculture*. Springer: Cham, pp. 171–197.

Tkáčová, J. and M. Angelovičová. 2012. Heat shock proteins (HSPs): A review. *Scientific Papers Animal Science and Biotechnologies* 1:349–353.

Tóth, M.E., Gombos, I. and M. Santha. 2015. Heat shock proteins and their role in human diseases. *Acta Biologica Szegediensis* 59:121–141.

Travaglia, C., Reinoso, H., Cohen, A., Luna, C., Tommasino, E., Castillo, C. and R. Bottini. 2010. Exogenous ABA increases yield in field-grown wheat with moderate water restriction. *Journal of Plant Growth Regulation* 29:366–374.

Tripathy, A. 2016. Oxidative stress, reactive oxygen species (ROS) and antioxidative defense system, with special reference to fish. *International Journal of Current Research in Biosciences and Plant Biology* 3:79–89.

Tripathy, S. and P.K. Mohanty. 2017. Reactive oxygen species (ROS) are boon or bane. *International Journal of Pharmaceutical Sciences and Research* 8:1.

Tripathy, S.K., Nayak, G., Naik, J., Patnaik, M., Dash, A.P., Sahoo, D.B. and A.M. Prusti. 2019. Signal transduction in plants under drought and salt stress-an overview. *International Journal of Current Microbiology and Applied Sciences* 8:318–325.

Upadhyaya, C.P., Venkatesh, J., Gururani, M.A., Asnin, L., Sharma, K., Ajappala, H. and S.W. Park. 2011. Transgenic potato overproducing L-ascorbic acid resisted an increase in methylglyoxal under salinity stress via maintaining higher reduced glutathione level and glyoxalase enzyme activity. *Biotechnology Letters* 33:2297–2307.

Usman, M.G., Rafii, M.Y., Ismail, M.R., Malek, M.A., Latif, M.A. and Y. Oladosu. 2014. Heat shock proteins: Functions and response against heat stress in plants. *International Journal of Science Technology Research* 3:204–218.

Usman, M.G., Rafii, M.Y., Martini, M.Y., Yusuff, O.A., Ismail, M.R. and G. Miah. 2017. Molecular analysis of Hsp70 mechanisms in plants and their function in response to stress. *Biotechnology and Genetic Engineering Reviews* 33:26–39.

Vatansever, F., de Melo, W.C., Avci, P., Vecchio, D., Sadasivam, M., Gupta, A. and M.R. Hamblin. 2013. Antimicrobial strategies centered around reactive oxygen species-bactericidal antibiotics, photodynamic therapy, and beyond. *FEMS Microbiology Reviews* 37:955–989.

Verma, V., Ravindran, P. and P.P. Kumar. 2016. Plant hormone-mediated regulation of stress responses. *BMC Plant Biology* 16:1–10.

Verma, R.K., Santosh Kumar, V.V., Yadav, S.K., Pushkar, S., Rao, M.V. and V. Chinnusamy. 2019. Overexpression of ABA receptor PYL10 gene confers drought and cold tolerance to indica rice. *Frontiers in Plant Science* 10:1488.

Wagner, S., Van Aken, O., Elsasser, M. and M. Schwarzlander. 2018. Mitochondrial energy signaling and its role in the low-oxygen stress response of plants. *Plant Physiology* 176:1156–1170.

Wahid, A. 2007. Physiological implications of metabolite biosynthesis for net assimilation and heat-stress tolerance of sugarcane (*Saccharum officinarum*) sprouts. *Journal of Plant Research* 120:219–228.

Wanat, K. 2020. Biological barriers, and the influence of protein binding on the passage of drugs across them. *Molecular Biology Reports* 47:3221–3231.

Wang, B., Liu, C., Zhang, D., He, C., Zhang, J. and Z. Li. 2019. Effects of maize organ-specific drought stress response on yields from transcriptome analysis. *BMC Plant Biology* 19:1–19.

Wang, J., Li, C., Yao, X., Liu, S., Zhang, P. and K. Chen. 2018a. The Antarctic moss leucine-rich repeat receptor-like kinase (PnLRR-RLK2) functions in salinity and drought stress adaptation. *Polar Biology* 41:353–364.

Wang, Y.G., Fu, F.L., Yu, H.Q., Hu, T., Zhang, Y.Y., Tao, Y. and W.C. Li. 2018b. Interaction network of core ABA signaling components in maize. *Plant Molecular Biology* 96:245–263.

Xia, X.J., Zhou, Y.H., Shi, K., Zhou, J., Foyer, C.H. and J.Q. Yu. 2015. Interplay between reactive oxygen species and hormones in the control of plant development and stress tolerance. *Journal of Experimental Botany* 66:2839–2856.

Xie, X., He, Z., Chen, N., Tang, Z., Wang, Q. and Y. Cai. 2019. The roles of environmental factors in regulation of oxidative stress in plant. *BioMed Research International* 2019,4:1–11.

Xu, Z.S., Li, Z.Y., Chen, Y., Chen, M., Li, L.C. and Y.Z. Ma. 2012. Heat shock protein 90 in plants: Molecular mechanisms and roles in stress responses. *International Journal of Molecular Sciences* 13,12:15706–15723.

Xu, Y., Chu, C. and S. Yao. 2021. The impact of high-temperature stress on rice: Challenges and solutions. *The Crop Journal* 9:963–976.

Yadav, K., Yadav, A., Vashistha, P., Pandey, V.P. and U.N. Dwivedi. 2019. Protein misfolding diseases and therapeutic approaches. *Current Protein and Peptide Science* 20:1226–1245.

Ye, W., Ando, E., Rhaman, M.S., Tahjib-Ul-Arif, M., Okuma, E., Nakamura, Y.Y. Murata. 2020. Inhibition of light-induced stomatal opening by allyl isothiocyanate does not require guard cell cytosolic Ca^{2+} signaling. *Journal of Experimental Botany* 71:2922–2932.

Young, T.E., Ling, J., Geisler-Lee, C.J., Tanguay, R.L., Caldwell, C. and D.R. Gallie. 2001. Developmental and thermal regulation of the maize heat shock protein, HSP101. *Plant Physiology* 127:777–791.

Zhang, M.M., Ong, C.L.Y., Walker, M.J. and A.G. McEwan. 2016. Defence against methylglyoxal in group a streptococcus: A role for glyoxylase I in bacterial virulence and survival in neutrophils. *Pathogens and Disease* 74:1–8.

Zhang, C., Li, G., Chen, T., Feng, B., Fu, W., Yan, J. and G. Fu. 2018. Heat stress induces spikelet sterility in rice at anthesis through inhibition of pollen tube elongation interfering with auxin homeostasis in pollinated pistils. *Rice* 11:1–12.

Zhao, Q., Zhou, L., Liu, J., Du, X., Huang, F., Pan, G. and F. Cheng. 2018a. Relationship of ROS accumulation and superoxide dismutase isozymes in developing anther with floret fertility of rice under heat stress. *Plant Physiology and Biochemistry* 122:90–101.

Zhao, Q., Zhou, L., Liu, J., Cao, Z., Du, X., Huang, F. and F. Cheng. 2018b. Involvement of CAT in the detoxification of HT-induced ROS burst in rice anther and its relation to pollen fertility. *Plant Cell Reports* 37:741–757.

Zhu, Y., Luo, X., Wei, M., Khan, A., Munsif, F., Huang, T. and Z. Shan. 2020. Antioxidant enzymatic activity and its related genes expression in cassava leaves at different growth stages play key roles in sustaining yield and drought tolerance under moisture stress. *Journal of Plant Growth Regulation* 39:594–607.

Zulkifli, M., Neff, J.K., Timbalia, S.A., Garza, N.M., Chen, Y., Watrous, J.D. and V.M. Gohil. 2020. Yeast homologs of human MCUR1 regulate mitochondrial proline metabolism. *Nature Communications* 11:1–15.

Żwirowski, S., Klosowska, A., Obuchowski, I., Nillegoda, N.B., Pirog, A., Zietkiewicz, S. and K. Liberek. 2017. Hsp70 displaces small heat shock proteins from aggregates to initiate protein refolding. *The EMBO Journal* 36:783–796.

9

Drought and High-Temperature Stress Tolerance in Field Crops

Goutam Kumar Dash
Centurion University of Technology and Management

Soumya Kumar Sahoo and Jagdish Jena
Siksha O Anusandhan University

Madhusmita Barik, Selukash Parida, Mirza Jaynul Baig, and Padmini Swain
ICAR-National Rice Research Institute

CONTENTS

9.1 Introduction

Global food security relies mainly on crop production and its productivity, which are highly affected by various abiotic stresses. Among these stresses, drought and heat stress have a huge impact on the physiology of plant growth, development and crop yield. Irregular rainfall patterns and reductions in precipitation cause drought (Lobell et al. 2015), similarly, atmospheric temperature rise above the threshold limit causes high-temperature stress which hampers plant growth within a very short duration of time. The average global temperature in the last 140 years increased by 0.85°C (IPCC 2013) mainly due to an increase in the level of carbon dioxide and other greenhouse gases. An increase in global temperature develops drought-prone areas that impact the productivity of major crops (Zhao et al. 2017). Drought and heat stress are generally studied individually, though they occur combinedly in the field condition. The temperature and water status of plants are highly correlated. High temperature increases evapotranspiration rate ultimately causing drought stress (Lamaoui et al. 2018) under field conditions. The combined effect of drought and high-temperature stresses caused significant damage to plant growth compared to the effect of individual stresses (De Boeck et al. 2016). Moreover, in plants, cross-synergism effect is also observed in which plant faces multiple stresses at a particular time and has a more deleterious effect, considerably higher than that of single stress (Prasad, Staggenborg, and Ristic 2015).

Plants respond differently to individual and combined stresses, making it difficult to understand the influence of these stresses on the plant and the plant's response to these stresses due to the complex phenomena controlled by several genes and interconnected signaling pathways. The impact of these stresses on the plants is influenced by environmental factors (e.g. rainfall patterns, temperature), edaphic factors (e.g. soil temperature, moisture, soil nutrient content) and plant developmental stages (vegetative or reproductive stages) along with its duration and intensity. Recent studies revealed that plants exhibit more or less similar physiological responses to both drought and high temperature (Zandalinas et al. 2018; Lamaoui et al. 2018). Drought and high-temperature stress have a negative impact on transpiration efficiency, leaf area, photosynthetic efficiency and stomatal conductance (Zandalinas et al. 2018; Farooq et al. 2019). The photosynthesis rate decreases under drought and heat stresses due to stomata closure (Farooq et al. 2019). High temperature destabilizes protein structure, hampers protein synthesis, inactivates the enzymes and damages biomembranes by generating reactive oxygen species (ROS) that ultimately affect cell division (Wahid et al. 2007; Hossain et al. 2016).

Plants have several stress adaptive and avoiding mechanisms to cope with the stresses. The plants exhibit specific strategies to deal with combined stresses, which are quite different from those exhibited by the individual stresses (Pandey, Ramegowda, and Senthil-Kumar 2015). Understanding the different mechanisms involved in this tolerance response might help in mitigating the negative effect of stresses and development of stress-tolerant varieties. Advanced genomic approaches such as mapping of genomic regions for complex traits, discovery of candidate genes, use of markers for genetic

DOI: 10.1201/9781003258063-9

dissection and image-based phenotyping tools such as visible and near-infrared reflectance techniques, infrared imaging can be efficiently used to unravel the stress tolerance mechanisms. This chapter provides a clear-cut insight regarding the impact of the combined and individual stresses and the approaches employed to improve stress tolerance in different crops.

9.2 Drought and Heat Stress Response on Field Crops

The fast increase in human population and drastic climate change is globally affecting the food security (Lesk, Rowhani, and Ramankutty 2016). Among different abiotic stresses, drought and heat stresses have a severe impact on crop productivity, yields and ultimately the food security leading to lower farmers' income. In crops, significant progress in research has been achieved on drought or high-temperature stress responses, and, in recent times, the simultaneous effect of both stresses is attaining additional importance. The two main causes responsible for the regular onset of drought stress worldwide are reduced precipitation and irregular rainfall patterns (Lobell, Schlenker, and Costa-Roberts 2011).

Plant growth and development are usually affected by high-temperature stress (Wahid et al. 2007). Presently, the frequent occurrence of world heat waves due to the increasing atmospheric temperatures is affecting severely the global crop productivity. It may cause alteration in the developmental stages of the agricultural crop plants (Porter 2005). Here, we presented the effect of drought and high-temperature stress on field crops along with plant growth, architecture, biomass partitioning, canopy temperature (CT), evapotranspiration, regulation of gas exchange and regulation in hydraulic conductance under both the stress conditions.

9.2.1 Plant Growth, Architecture and Biomass Partitioning under Drought and High Temperature

Stress can occur at any stage of the plant life cycle. It causes poor seed germination and impaired seedlings establishment when occurs at the initial stage of plant life cycle. Growth of the plant is mostly accomplished by cell division, cell enlargement and cell differentiation. Drought stress retards the process of mitosis and elongation of cells due to turgor loss which results in poor growth of the plant (Hussain et al. 2008; Taiz and Zeiger 2006). Turgor pressure and assimilate supply have vital roles in the leaf expansion. Under drought, reduced turgor pressure and slow photosynthetic rate mostly limit the expansion of the leaf (Rucker et al. 1995) along with plant height, size of the leaf, girth of the stem and biomass accumulation at various stages (Khan et al. 2015; Dash et al. 2017).

High temperatures in plants may cause growth inhibition, leaf senescence and discoloration of leaves and fruits (Vollenweider and Günthardt-Goerg 2005). It can also decrease the seed germination potential and poor seedling establishment. High-temperature effects vary with the duration, developmental stage and intensity of the stress on different cereal crops (Fahad et al. 2016). In rice, heat stress reduces spikes, whereas floret numbers and rate of seed-setting in sorghum are severely affected by high temperature (Fahad et al. 2016). Anthers and pollens inside a floret are highly sensitive to high temperatures than ovules. In sugarcane (*Saccharum officinarum* L.) and maize (*Zea mays* L.), heat stress significantly reduces growth and net assimilation rate (Wahid and Close 2007). Significant reduction in biomass accumulation and internodal length along with early leaf senescence were observed under high-temperature stress in sugarcane (Ebrahim et al. 1998).

9.2.2 CT and Evapotranspiration under Drought and High Temperature

Though high-temperature stress increases precipitation but in turn accelerates evaporation from land, water and from the leaf surface. The process of transpiration regulates the CT of plants which has a direct association with stomatal density and stomatal conductance (Rebetzke et al. 2012) and is a key determinant of drought tolerance.

Increment in atmospheric temperature increases the Vapour Pressure Deficit (VPD) which ultimately increases the transpiration rate. However, during higher evaporative demand, stomata closure for a longer period of time checks transpiration and increases the plant CT (Mir et al. 2012; Munjal and Rana 2003). CT differences correlate perfectly with the transpiration rate in various crops such as potatoes, wheat, rice and sugar beet. Temperature difference between the air and plant canopies is measured as CT depression (CTD), which is a direct indicator of whole-plant water status (Balota et al. 2008) and helps in evaluating the response of plants to high temperature (Reynolds et al. 2007) and drought (Royo et al. 2002). When the canopy is cooler than the air, CTD is positive which is directly correlated with an increase in the crop yield (Yang and Zhang 2006; Fischer et al. 1998).

To measure CT, a thermal imagery system is generally used which is a powerful tool that can rapidly measure the temperature gradient between different plant canopies. Usually, two approaches, namely thermal infrared imaging and infrared thermography, are used to measure CT which are identical methods that measure the extent of evaporative cooling of a crop canopy and provide information about the plant water status. In plants, CT is one of the integrative traits that reflect the stomatal conductance as well as the balance that exists between the root-water uptake and shoot transpiration (Berger et al. 2010). For these reasons, CT has been used effectively as a selection criterion for identifying drought tolerant genotypes (Jones et al. 2009). Assessing evapotranspiration in crops under drought and high-temperature stress has been in practice in farmland irrigation and has gained worldwide attention (Shi et al. 2015). Drought stress has become a constraint to the ecological development of agriculture in the basin. Thus, it is essential to understand and estimate crop evapotranspiration which can improve the irrigation program development, increase the water-use efficiency (WUE), and guarantee high and stable crop yields.

Several methods are used to determine crop evapotranspiration. One of them employed a computational method that determines the reference crop evapotranspiration and adopts the Penman formula, comprising the computation of solar radiation. This method mainly includes aerodynamics, the

Bowen ratio-energy balance and remote sensing. The Food and Agriculture Organization has recommended another method named as crop coefficient method which has extensive applicability (Girona et al. 2004) and is characterized by reliable precision, easy operation and practicability. This method has been adopted globally and is able to distinguish the transpiration of crop from soil evaporation (De Jonge et al. 2012). However, this method is mostly used to estimate evapotranspiration of crops that are not grown under drought stress, and only a few investigations have been performed on food crops under water stress conditions (Zhao and Ji 2011).

9.2.3 Regulation of Gas Exchange

Carbon dioxide exchange is hampered both during drought and heat stress (Yan, Zhong, and Shangguan 2016). Drought reduces leaf water potential which induces stomata closure and reduces entry of CO_2 into the leaf reducing carbon dioxide fixation (Wang et al. 2018). Also, high temperatures damage photosynthetic machinery and decrease CO_2 concentration inside chloroplast and inactivate enzymes (Perdomo et al. 2017; Piveta et al. 2021; Flexas et al. 2007; Nayak et al. 2022). The reduction in photosynthesis is also ascribed to inactivation of photosystem (PSII) reaction center, disruption in electron transport rate and decrease in quantum efficiency of PSII photochemistry (Fv/Fm) (Parida et al. 2021). The damage of oxygen-evolving complexes is also responsible, which may hamper the flow of electrons to the PSII acceptor site (De Ronde et al. 2004). Photosynthetic rate is also inhibited due to diminution in the activity of RuBisCO (Morales et al. 2003).

The genotypes maintaining a higher photosynthesis rate under both stresses indicate greater tolerance ability (Wang et al. 2012). Stomatal conductance is a crucial trait that regulates photosynthetic rate under stress conditions. Partial closure of stomata reduces transpiration rate (T_r) to maintain internal plant water status and enhance adaptive tolerance mechanism under stress conditions (Park et al. 2020). Additionally, WUE of the crops is mainly regulated by g_s (Chen et al. 2016). Many researchers had proposed that a reduction in stomatal conductance under drought stress is well related with an increase in WUE (Ruggiero et al. 2017).

9.2.4 Regulation in Hydraulic Conductance under Drought and High Temperature

Plant water status is regulated by stomata by controlling water loss through transpiration such that it matches with the ability of root-leaf hydraulic system to supply water to leaves for the process of transpiration and photosynthesis (Meinzer 2002). According to the earlier studies, leaf and root hydraulic conductance can account for >70% of the entire hydraulic conductivity; hence, hydraulic conductance can be used to measure the water transportability (Domec et al. 2009). Some plant species, i.e., anisohydric species, can gain carbon at the cost of plant hydraulic dysfunction due to the open stomata under mild drought; however, strong stomatal regulation occurs under severe drought in these plants to limit hydraulic failure (Creek et al. 2018). In isohydric species, rapid stomatal closure avoids hydraulic dysfunction by minimizing water loss in response to drought stress (Martin-StPaul, Delzon, and Cochard 2017). In plants, various stress response mechanisms are functioned by synchronization between stomata and hydraulic traits. Under stress conditions, plants avoid hydraulic failure through an adjustment at structural and physiological levels that regulate water transport. Stomatal closure at short periods is one of the mechanisms that maintain water potential at tolerable limits by minimizing water loss. With the severity of drought stress, both shoot development and root hydraulic conductivity (Lpr) are reduced strongly. It was found that the root hydraulic conductivity in maize plants grown in hydroponic culture was inhibited by drought stress induced by polyethylene glycol (Hossain et al. 2021). So, root hydraulic conductivity is one of the major factors that reflects root water uptake ability and is closely correlated with water relations in plants.

9.3 Approaches to Improve Tolerance to the Combined Effect of Drought and High Temperature

Studies on combined heat and drought stress response on crops are very limited though a large number of reports are available on individual stress responses (Rizhsky, Liang, and Mittler 2002; Rizhsky et al. 2004; Lawas et al. 2019; Rang et al. 2011). Previously, conventional breeding mostly focused on increasing grain yield under normal conditions. As the climate change is getting more severe day by day, the grain yield is plateauing due to the inadaptability of the varieties to the adverse condition (Grassini et al. 2013). Moreover, during the variety development process, the genes responsible for stress tolerance are gradually lost causing these varieties more vulnerable to different stresses. A large number of tolerant genotypes have been developed through conventional breeding program; however, very few high-temperature tolerant varieties have been produced to date. Advancement of new techniques such as genome sequencing and availability of genome sequence information of a huge number of varieties made it possible to identify genes and Quantitative trait locus (QTLs) for the combined stress tolerance. From most of the studies conducted for high-temperature tolerance, N22 came out to be a highly tolerant genotype (Jagadish, Craufurd, and Wheeler 2007; Rang et al. 2011). Several attempts are also taken to identify the QTLs linked with high-temperature tolerance (Ye et al. 2015). A study of meta-QTL analysis revealed colocalization of yield QTLs with flowering date, spike weight, CTD and plant height (Acuña-Galindo et al. 2015). Directional selection has been used now to select varieties, by evaluating the varieties in alternating low to high stressful regimes, that has stable yield in a varied range of stressful environment having relatively low GxE interactions (Chapman et al. 2012; Lopes et al. 2012). Also, the selection of traits with higher plasticity and their related QTLs may help in enhancing grain yield in stressful environments (Nicotra et al. 2010; Aspinwall et al. 2015). Identifying genes and QTLs for the combined stresses, robust phenotyping methods are required. Researchers are now using high-throughput phenotyping platforms to quantify the spatio-temporal variation in different traits under combined stress. With the advancement in omics technologies, the genes responsible for individual stresses and combined stress have been identified. A transcriptomic study of Moroberakan and N22 revealed that heat tolerance exhibited by N22 is the manifestation of the heat shock proteins and transcription factors that provide cellular tolerance through their chaperon activities. In addition to these, several genes involved in sugar metabolism such as INV2 (cell wall invertase) and MST8 (sugar transporter) were differentially expressed. The genes identified through omics approaches are now engineered through transgenic technology to enhance individual tolerance in rice (Yu et al. 2016; Latha et al. 2017). Introgression of such genes has been proven to show individual stress tolerance in rice. SNAC overexpression enhanced drought and heat tolerance individually through ROS homeostasis (Fang et al. 2015). Similarly, OsRab7 overexpression confers drought and heat tolerance by changing the expression pattern of genes responsible for ROS, osmolyte and stress-responsive genes (El-Esawi and Alayafi 2019). To date, only one gene OsWRKY11 has been overexpressed through a transgenic approach that conferred combined stress tolerance at vegetative stage in rice (Wu et al. 2009). Recently with the advancement of high-throughput phenotyping technology, a large number of genotypes are evaluated for combined stresses and most probably in the recent future the characterized genes can be edited through genome editing technology to develop new tolerant varieties.

9.4 Strategic Consideration and Future Prospects

Unpredictable weather abnormalities and global climate change due to different anthropogenic activities have imposed a challenge for the crops to use the limited resource base efficiently and to produce to their full potential. Also, these weather abnormalities cause certain collateral damages like more insect pest infestation, loss of soil fertility, frequent occurrence of heat and drought stress etc., which limit crop productivity. In response to these phenomena, plants have several complex stress adaptation mechanisms that help plants to mollify diverse abiotic and biotic stresses. These complex adaptive mechanisms are always being tried to understand by the scientific community and are now possible through new high-throughput genotyping and phenotyping technologies. The interactions of plants with soil microbiomes and gene x environmental interaction are yet to be studied as an opportunity to understand their complexity, so that different models can be developed to mitigate the abiotic stresses for achieving food security. Microbes surrounding the rhizosphere significantly influence the physiological, biochemical, molecular and developmental changes in plants to escape drought and heat stress. The opportunity to understand this microbiome mechanism and developing novel technology can help novel plant type to sustain in different stress environments.

Gene identification, characterization and editing tools and techniques helped the scientific community to find out the genes responsible for various stress responses. Finding the suitable gene and their insertion to generate new plant type has now become the most opportunistic field for plant physiologists and biotechnologists. All of these advanced omics technologies, genome sequencing and genetic engineering could be used along with the conventional breeding approach in a holistic manner to develop new plant type that can cope with the various biotic and abiotic stresses and can be suitably adapted to the changing environmental conditions.

Crop modeling approach is the convergence of all these research activities whether we are dealing with crop improvement, crop production or crop protection approach. Crop modeling approach to find the suitable line for a specific environmental condition simulating different morphological and physiological attributes with the changing phyllosphere and rhizosphere environment creates a huge opportunity for the scientific community to achieve the desired target i.e., draught and heat stress management and improving the food security. Future research must emphasize developing new plant types that can tolerate and adapt to different stress conditions using advanced sequencing, genomics and transcriptomics techniques along with a crop modeling approach, which can be more holistic and fruitful.

REFERENCES

Acuña-Galindo, M.A., R. Esten Mason, N.K. Subramanian, and D.B. Hays. 2015. "Meta-Analysis of Wheat QTL Regions Associated with Adaptation to Drought and Heat Stress." *Crop Science* 55(2). John Wiley & Sons, Ltd: 477–92. doi:10.2135/CROPSCI2013.11.0793.

Aspinwall, M.J., M.E. Loik, V.R. de Dios, M.G. Tjoelker, P.R. Payton, and D.T. Tissue. 2015. "Utilizing Intraspecific Variation in Phenotypic Plasticity to Bolster Agricultural and Forest Productivity under Climate Change." *Plant, Cell & Environment* 38 (9): 1752–64. doi:10.1111/PCE.12424.

Balota, M., W.A. Payne, S.R. Evett, and T.R. Peters. 2008. "Morphological and Physiological Traits Associated with Canopy Temperature Depression in Three Closely Related Wheat Lines." *Crop Science* 48 (5). John Wiley & Sons, Ltd: 1897–1910. doi:10.2135/CROPSCI2007.06.0317.

Berger B, Parent B, Tester M. High-throughput shoot imaging to study drought responses. Journal of Experimental Botany. 2010;61(13):3519-3528.

Chapman, S.C., S. Chakraborty, M. Fernanda Dreccer, and S. Mark Howden. 2012. "Plant Adaptation to Climate Change-Opportunities and Priorities in Breeding." *Crop and Pasture Science* 63 (3): 251–68. doi:10.1071/CP11303.

Chen, Y., L. Liu, Q. Guo, Z. Zhu, and L. Zhang. 2016. "Effects of Different Water Management Options and Fertilizer Supply on Photosynthesis, Fluorescence Parameters and Water Use Efficiency of Prunella Vulgaris Seedlings." *Biological Research* 49 (1): 1–9. doi:10.1186/s40659-016-0069-4.

Creek, D., C.J. Blackman, T.J. Brodribb, B. Choat, and D.T. Tissue. 2018. "Coordination between Leaf, Stem, and Root Hydraulics and Gas Exchange in Three Arid-Zone Angiosperms during Severe Drought and Recovery." *Plant, Cell & Environment* 41 (12). John Wiley & Sons, Ltd: 2869–81. doi:10.1111/PCE.13418.

Dash, G.K., M. Barik, A.K. Debata, M.J. Baig, and P. Swain. 2017. "Identification of Most Important Rice Root Morphological Markers in Response to Contrasting Moisture Regimes under Vegetative Stage Drought." *Acta Physiologiae Plantarum* 39 (1): 1–13. doi:10.1007/S11738-016-2297-1/FIGURES/2.

De Boeck, H.J., S. Bassin, M. Verlinden, M. Zeiter, and E. Hiltbrunner. 2016. "Simulated Heat Waves Affected Alpine Grassland Only in Combination with Drought." *The New Phytologist* 209(2): 531–41. doi:10.1111/NPH.13601.

De Ronde, J.A., W.A. Cress, G.H.J. Krüger, R.J. Strasser, and J. Van Staden. 2004. "Photosynthetic Response of Transgenic Soybean Plants, Containing an Arabidopsis P5CR Gene, during Heat and Drought Stress." *Journal of Plant Physiology* 161 (11): 1211–24. doi:10.1016/j.jplph.2004.01.014.

De Jonge, K.C., J.C. Ascough, A.A. Andales, N.C. Hansen, L.A. Garcia, and M. Arabi. 2012. "Improving Evapotranspiration Simulations in the CERES-Maize Model under Limited Irrigation." *Agricultural Water Management* 115 (December). Elsevier: 92–103. doi:10.1016/J.AGWAT.2012.08.013.

Domec, J.C., S. Palmroth, E. Ward, C.A. Maier, M.Th Érézien, and R. Oren. 2009. "Acclimation of Leaf Hydraulic Conductance and Stomatal Conductance of Pinus Taeda (Loblolly Pine) to Long-Term Growth in Elevated CO(2) (Free-Air CO(2) Enrichment) and N-Fertilization." *Plant, Cell & Environment* 32 (11): 1500–1512. doi:10.1111/J.1365-3040.2009.02014.X.

Ebrahim, M.K., O. Zingsheim, M.N. El-Shourbagy, P.H. Moore, and E. Komor. 1998. "Growth and Sugar Storage in Sugarcane Grown at Temperatures below and above Optimum." *Journal of Plant Physiology* 153 (5–6). Urban & Fischer: 593–602. doi:10.1016/S0176-1617(98)80209-5.

El-Esawi, M.A., and A.A. Alayafi. 2019. "Overexpression of Rice Rab7 Gene Improves Drought and Heat Tolerance and Increases Grain Yield in Rice (Oryza Sativa L.)." *Genes* 10 (1). Multidisciplinary Digital Publishing Institute (MDPI). doi:10.3390/GENES10010056.

Fahad, S., S. Hussain, S. Saud, F. Khan, S. Hassan, A.W. Nasim, M. Arif, F. Wang, and J. Huang. 2016. "Exogenously Applied Plant Growth Regulators Affect Heat-Stressed Rice Pollens." *Journal of Agronomy and Crop Science* 202 (2). John Wiley & Sons, Ltd: 139–50. doi:10.1111/JAC.12148.

Fang, Y., K. Liao, H. Du, Y. Xu, H. Song, X. Li, and L. Xiong. 2015. "A Stress-Responsive NAC Transcription Factor SNAC3 Confers Heat and Drought Tolerance through Modulation of Reactive Oxygen Species in Rice." *Journal of Experimental Botany* 66 (21). J Exp Bot: 6803–17. doi:10.1093/JXB/ERV386.

Farooq, M., M. Hussain, S. Ul-Allah, and K.H.M. Siddique. 2019. "Physiological and Agronomic Approaches for Improving Water-Use Efficiency in Crop Plants." *Agricultural Water Management* 219 (June). Elsevier: 95–108. doi:10.1016/J.AGWAT.2019.04.010.

Fischer, R.A., D. Rees, K.D. Sayre, Z.M. Lu, A.G. Condon, and A. Larque Saavedra. 1998. "Wheat Yield Progress Associated with Higher Stomatal Conductance and Photosynthetic Rate, and Cooler Canopies." *Crop Science* 38 (6). John Wiley & Sons, Ltd: 1467–75. doi:10.2135/CROPSCI1998.0011183X003800060011X.

Flexas, J., M.F. Ortuño, M. Ribas-Carbo, A. Diaz-Espejo, I.D. Flórez-Sarasa, and H. Medrano. 2007. "Mesophyll Conductance to CO_2 in Arabidopsis Thaliana." *New Phytologist* 175 (3): 501–11. doi:10.1111/j.1469-8137.2007.02111.x.

Girona, J., J. Marsal, M. Mata, and J. Del Campo. 2004. "Pear Crop Coefficients Obtained in a Large Weighing Lysimeter." *Acta Horticulturae* 664. International Society for Horticultural Science: 277–81. doi:10.17660/ACTAHORTIC.2004.664.33.

Grassini, P., K.M. Eskridge, and K.G. Cassman. 2013. "Distinguishing between Yield Advances and Yield Plateaus in Historical Crop Production Trends." *Nature Communications* 4 (1). Nature Publishing Group: 1–11. doi:10.1038/ncomms3918.

Hossain, A., M. Skalicky, M. Brestic, S. Maitra, M.A. Alam, M.A. Syed, J. Hossain, et al. 2021. "Consequences and Mitigation Strategies of Abiotic Stresses in Wheat (*Triticum Aestivum* L.) under the Changing Climate." *Agronomy* 11 (2). Multidisciplinary Digital Publishing Institute: 241. doi:10.3390/AGRONOMY11020241.

Hossain, M.A., S.H. Wani, S. Bhattacharjee, D.J. Burritl, and L.S.P. Tran. 2016. Drought Stress Tolerance in Plants, Vol 1: Physiology and Biochemistry. *Drought Stress Tolerance in Plants, Vol 1: Physiology and Biochemistry*. Vol. 1. doi:10.1007/978-3-319-28899-4.

Hussain, M., M.A. Malik, M. Farooq, M.Y. Ashraf, and M.A. Cheema. 2008. "Improving Drought Tolerance by Exogenous Application of Glycinebetaine and Salicylic Acid in Sunflower." *Journal of Agronomy and Crop Science* 194 (3). John Wiley & Sons, Ltd: 193–99. doi:10.1111/J.1439-037X.2008.00305.X.

IPCC. 2013. "Summary for Policymakers. In: *Climate Change 2013: The Physical Science Basis. Contribution of Working Group I to the Fifth Assessment Report of the Intergovernmental Panel on Climate Change.*" Cambridge University Press, Cambridge, U.K. and New York, NY. https://www.ipcc.ch/report/ar5/wg1/.

Jagadish, S.V.K., P.Q. Craufurd, and T.R. Wheeler. 2007. "High Temperature Stress and Spikelet Fertility in Rice (*Oryza Sativa* L.)." *Journal of Experimental Botany* 58 (7). Oxford Academic: 1627–35. doi:10.1093/JXB/ERM003.

Jones, H.G., R. Serraj, B.R. Loveys, L. Xiong, A. Wheaton, and A.H. Price. 2009. "Hermal Infrared Imaging of Crop Canopies for the Remote Diagnosis and Quantification of Plant Responses to Water Stress in the FieldT." *Functional Plant Biology : FPB* 36 (11): 978–89. doi:10.1071/FP09123.

Khan, M.B., M. Hussain, A. Raza, S. Farooq, and K. Jabran. 2015. "Seed Priming with CaCl 2 and Ridge Planting for Improved Drought Resistance in Maize." *Turkish Journal of Agriculture and Forestry* 39: 193–203. doi:10.3906/tar-1405-39.

Lamaoui, M., M. Jemo, R. Datla, and F. Bekkaoui. 2018. "Heat and Drought Stresses in Crops and Approaches for Their Mitigation." *Frontiers in Chemistry* 6 (February). Frontiers Media S.A.: 26. doi:10.3389/FCHEM.2018.00026/BIBTEX.

Latha, G.M., T. Mohapatra, A.S. Geetanjali, and K.R.S. Sambasiva Rao. 2017. "Engineering Rice for Abiotic Stress Tolerance: A Review-Indian Journals." *Engineering Rice for Abiotic Stress Tolerance* 11 (4): 396–413. https://www.indianjournals.com/ijor.aspx?target=ijor:ctbp&volume=11&issue=4&article=008.

Lawas, L.M.F., X. Li, A. Erban, J. Kopka, S.V. Krishna Jagadish, E. Zuther, and D.K. Hincha. 2019. "Metabolic Responses of Rice Cultivars with Different Tolerance to Combined Drought and Heat Stress under Field Conditions." *GigaScience* 8 (5). doi:10.1093/GIGASCIENCE/GIZ050.

Lesk, C., P. Rowhani, and N. Ramankutty. 2016. "Influence of Extreme Weather Disasters on Global Crop Production." *Nature* 529 (7584). Nature Publishing Group: 84–87. doi:10.1038/nature16467.

Lobell, D.B., G.L. Hammer, K. Chenu, B. Zheng, G. Mclean, and S.C. Chapman. 2015. "The Shifting Influence of Drought and Heat Stress for Crops in Northeast Australia." *Global Change Biology* 21 (11): 4115–27. doi:10.1111/GCB.13022.

Lobell, D.B., W. Schlenker, and J. Costa-Roberts. 2011. "Climate Trends and Global Crop Production since 1980." *Science (New York, N.Y.)* 333 (6042): 616–20. doi:10.1126/SCIENCE.1204531.

Lopes, M.S., M.P. Reynolds, M.R. Jalal-Kamali, M. Moussa, Y. Feltaous, I.S.A. Tahir, N. Barma, M. Vargas, Y. Mannes, and M. Baum. 2012. "The Yield Correlations of Selectable Physiological Traits in a Population of Advanced Spring Wheat Lines Grown in Warm and Drought Environments." *Field Crops Research* 128 (March). Elsevier: 129–36. doi:10.1016/J.FCR.2011.12.017.

Martin-StPaul, N., S. Delzon, and H. Cochard. 2017. "Plant Resistance to Drought Depends on Timely Stomatal Closure." *Ecology Letters* 20 (11). John Wiley & Sons, Ltd: 1437–47. doi:10.1111/ELE.12851.

Meinzer, F.C. 2002. "Co-Ordination of Vapour and Liquid Phase Water Transport Properties in Plants." *Plant, Cell & Environment* 25 (2). John Wiley & Sons, Ltd: 265–74. doi:10.1046/J.1365-3040.2002.00781.X.

Mir, R.R., M. Zaman-Allah, N. Sreenivasulu, R. Trethowan, and R.K. Varshney. 2012. "Integrated Genomics, Physiology and Breeding Approaches for Improving Drought Tolerance in Crops." *Theoretical and Applied Genetics* 125 (4): 625–45. doi:10.1007/s00122-012-1904-9.

Morales, D., P. Rodriguez, J. Dell' Amico, E. Nicolas, A. Torrecillas, and M.J. Sanchez-Blanco. 2003. "High-Temperature Preconditioning and Thermal Shock Imposition Affects Water Relations, Gas Exchange and Root Hydraulic Conductivity in Tomato."

Munjal, R., and R. Rana. 2003. "Evaluation of Physiological Traits in Wheat (*Triticum Aestivum* L.)." *Terminal High Temperature Tolerance Proceedings of the Tenth International Wheat Genetics Symposium*, Poestum, Italy, Class | Request PDF. https://www.researchgate.net/publication/285310499_Evaluation_of_Physiological_traits_in_wheat_Triticum_aestivum_L_for_terminal_high_temperature_tolerance_Proceedings_of_the_Tenth_International_Wheat_Genetics_Symposium_Poestum_Italy_Class.

Nayak, L., D. Panda, G.K. Dash, M.K. Lal, P. Swain, M.J. Baig, and A. Kumar. 2022. "A Chloroplast Glycolate Catabolic Pathway Bypassing the Endogenous Photorespiratory Cycle Enhances Photosynthesis, Biomass and Yield in Rice (*Oryza Sativa* L.)." *Plant Science : An International Journal of Experimental Plant Biology* 314. doi:10.1016/J.PLANTSCI.2021.111103.

Nicotra, A.B., O.K. Atkin, S.P. Bonser, A.M. Davidson, E.J. Finnegan, U. Mathesius, P. Poot, et al. 2010. "Plant Phenotypic Plasticity in a Changing Climate." *Trends in Plant Science* 15 (12). Elsevier Current Trends: 684–92. doi:10.1016/J.TPLANTS.2010.09.008.

Pandey, P., V. Ramegowda, and M. Senthil-Kumar. 2015. "Shared and Unique Responses of Plants to Multiple Individual Stresses and Stress Combinations: Physiological and Molecular Mechanisms." *Frontiers in Plant Science* 6. doi:10.3389/FPLS.2015.00723.

Parida, S., G.K. Dash, S.K. Sahoo, M. Barik, M.J. Baig, R.K. Panda, and P. Swain. 2021. "Photosystem II Photochemistry and Chlorophyll Intensity in Rice (*Oryza Sativa* L.) under Drought at Flowering Stage." *International Journal of Current Microbiology and Applied Sciences* 10 (2): 746–56. doi:10.20546/ijcmas.2021.1002.089.

Park, S.I., J.J. Kim, S.Y. Shin, Y.S. Kim, and H.S. Yoon. 2020. "ASR Enhances Environmental Stress Tolerance and Improves Grain Yield by Modulating Stomatal Closure in Rice." *Frontiers in Plant Science* 10: 1–17. doi:10.3389/fpls.2019.01752.

Perdomo, J.A., S. Capó-Bauçà, E. Carmo-Silva, and J. Galmés. 2017. "Rubisco and Rubisco Activase Play an Important Role in the Biochemical Limitations of Photosynthesis in Rice, Wheat, and Maize under High Temperature and Water Deficit." *Frontiers in Plant Science* 8: 1–15. doi:10.3389/fpls.2017.00490.

Piveta, L.B, N. Roma-Burgos, J.A. Noldin, V.E. Viana, C. de Oliveira, F.P. Lamego, and L.A. de Avila. 2021. "Molecular and Physiological Responses of Rice and Weedy Rice to Heat and Drought Stress †." *Agriculture (Switzerland)* 11 (1): 1–23. doi:10.3390/agriculture11010009.

Porter, J.R. 2005. "Rising Temperatures Are Likely to Reduce Crop Yields." *Nature* 436 (7048). Nature Publishing Group: 174–174. doi:10.1038/436174b.

Prasad, P.V.V., S.A. Staggenborg, and Z. Ristic. 2015. "Impacts of Drought and/or Heat Stress on Physiological, Developmental, Growth, and Yield Processes of Crop Plants." *Response of Crops to Limited Water: Understanding and Modeling Water Stress Effects on Plant Growth Processes*. John Wiley & Sons, Ltd, 301–55. doi:10.2134/ADVAGRICSYSTMODEL1.C11.

Rang, Z.W., S.V.K. Jagadish, Q.M. Zhou, P.Q. Craufurd, and S. Heuer. 2011. "Effect of High Temperature and Water Stress on Pollen Germination and Spikelet Fertility in Rice." *Environmental and Experimental Botany 70* (1). Elsevier: 58–65. doi:10.1016/J.ENVEXPBOT.2010.08.009.

Rebetzke, G.J., A.R. Rattey, G.D. Farquhar, R.A. Richards, A.G. Condon, G.J. Rebetzke, A.R. Rattey, G.D. Farquhar, and R.A. Richards. 2012. "Genomic Regions for Canopy Temperature and Their Genetic Association with Stomatal Conductance and Grain Yield in Wheat." *Functional Plant Biology* 40 (1). CSIRO PUBLISHING: 14–33. doi:10.1071/FP12184.

Reynolds, M.P., C.S. Pierre, A.S.I. Saad, M. Vargas, and A.G. Condon. 2007. "Evaluating Potential Genetic Gains in Wheat Associated with Stress-Adaptive Trait Expression in Elite Genetic Resources under Drought and Heat Stress." *Crop Science* 47. doi:10.2135/cropsci2007.10.0022IPBS.

Rizhsky, L., H. Liang, and R. Mittler. 2002. "The Combined Effect of Drought Stress and Heat Shock on Gene Expression in Tobacco." *Plant Physiology* 130 (3). Oxford Academic: 1143–51. doi:10.1104/PP.006858.

Rizhsky, L., H. Liang, J. Shuman, V. Shulaev, S. Davletova, and R. Mittler. 2004. "When Defense Pathways Collide. The Response of Arabidopsis to a Combination of Drought and Heat Stress." *Plant Physiology* 134 (4): 1683–96. doi:10.1104/PP.103.033431.

Royo, C., D. Villegas, L.F. García del Moral, S. Elhani, N. Aparicio, Y. Rharrabti, and J.L. Araus. 2002. "Comparative Performance of Carbon Isotope Discrimination and Canopy Temperature Depression as Predictors of Genotype Differences in Durum Wheat Yield in Spain." *Australian Journal of Agricultural Research* 53 (5). CSIRO PUBLISHING: 561–69. doi:10.1071/AR01016.

Rucker, K.S., C.K. Kvien, C.C. Holbrook, and J.E. Hook. 1995. "Identification of Peanut Genotypes with Improved Drought Avoidance Traits." *Peanut Science* 22 (1). Allen Press: 14–18. doi:10.3146/PNUT.22.1.0003.

Ruggiero, A., P. Punzo, S. Landi, A. Costa, M.J. Van Oosten, and S. Grillo. 2017. "Improving Plant Water Use Efficiency through Molecular Genetics." *Horticulturae* 3 (2): 1–22. doi:10.3390/horticulturae3020031.

Shi, X., C. Huanjie, Z. Lili, Y. Pei, and W. Zishen. 2015. "Estimation of Greenhouse Tomato Evapotranspiration under Deficit Irrigation Based on SIMDualKc Model." *Transactions of the Chinese Society of Agricultural Engineering* 31(8): 131–38.

Taiz, L., and E. Zeiger. 2006. *Plant Physiology*, 4th Edition. Sinauer Associates, Inc.: Sunderland. https://www.scirp.org/(S(351jmbntvnsjt1aadkozje))/reference/referencespapers.aspx?referenceid=717115.

Vollenweider, P., and M.S. Günthardt-Goerg. 2005. "Diagnosis of Abiotic and Biotic Stress Factors Using the Visible Symptoms in Foliage." *Environmental Pollution (Barking, Essex : 1987)* 137 (3): 455–65. doi:10.1016/J.ENVPOL.2005.01.032.

Wahid, A., and T.J. Close. 2007. "Expression of Dehydrins under Heat Stress and Their Relationship with Water Relations of Sugarcane Leaves." *Biologia Plantarum* 51 (1). Springer: 104–9. doi:10.1007/S10535-007-0021-0.

Wahid, A., S. Gelani, M. Ashraf, and M.R. Foolad. 2007. "Heat Tolerance in Plants: An Overview." *Environmental and Experimental Botany 61* (3). Elsevier: 199–223. doi:10.1016/J.ENVEXPBOT.2007.05.011.

Wang, W., C. Wang, D. Pan, Y. Zhang, B. Luo, and J. Ji. 2018. "Effects of Drought Stress on Photosynthesis and Chlorophyll Fluorescence Images of Soybean (Glycine Max) Seedlings." *International Journal of Agricultural and Biological Engineering* 11 (2): 196–201. doi:10.25165/j.ijabe.20181102.3390.

Wang, Z.X., L. Chen, J. Ai, H.Y. Qin, Y.X. Liu, P.L. Xu, Z.Q. Jiao, Y. Zhao, and Q.T. Zhang. 2012. "Photosynthesis and Activity of Photosystem II in Response to Drought Stress in Amur Grape (Vitis Amurensis Rupr.)." *Photosynthetica* 50 (2): 189–96. doi:10.1007/s11099-012-0023-9.

Wu, X., Y. Shiroto, S. Kishitani, Y. Ito, and K. Toriyama. 2009. "Enhanced Heat and Drought Tolerance in Transgenic Rice Seedlings Overexpressing OsWRKY11 under the Control of HSP101 Promoter." *Plant Cell Reports* 28 (1): 21–30. doi:10.1007/S00299-008-0614-X.

Yan, W., Y. Zhong, and Z. Shangguan. 2016. "A Meta-Analysis of Leaf Gas Exchange and Water Status Responses to Drought." *Scientific Reports* 6 (January). Nature Publishing Group: 1–9. doi:10.1038/srep20917.

Yang, J., and J. Zhang. 2006. "Grain Filling of Cereals under Soil Drying." *New Phytologist* 169 (2): 223–36. doi:10.1111/j.1469-8137.2005.01597.x.

Ye, C., F.A. Tenorio, M.A. Argayoso, M.A. Laza, H.J. Koh, E.D. Redoña, K.S.V. Jagadish, and G.B. Gregorio. 2015. "Identifying and Confirming Quantitative Trait Loci Associated with Heat Tolerance at Flowering Stage in Different Rice Populations." *BMC Genetics* 16 (1). BioMed Central Ltd.: 1–10. doi:10.1186/S12863-015-0199-7/FIGURES/4.

Yu, J., Y. Lai, X. Wu, G. Wu, and C. Guo. 2016. "Overexpression of OsEm1 Encoding a Group I LEA Protein Confers Enhanced Drought Tolerance in Rice." *Biochemical and Biophysical Research Communications* 478 (2): 703–9. doi:10.1016/J.BBRC.2016.08.010.

Zandalinas, S.I., R. Mittler, D. Balfagón, V. Arbona, and A. Gómez-Cadenas. 2018. "Plant Adaptations to the Combination of Drought and High Temperatures." *Physiologia Plantarum* 162 (1): 2–12. doi:10.1111/PPL.12540.

Zhao, L., and X. Ji. 2011. "Quantification of Transpiration and Evaporation over Agricultural Field Using the Fao-56 Dual Crop Coefficient Approach: A Case Study of the Maize Field in an Oasis in the Middlestream of the Heihe River Basin in Northwest China." *AGRIS* 43 (19): 4016–26. https://agris.fao.org/agris-search/search.do?recordID=CN2011001584.

Zhao, C., B. Liu, S. Piao, X. Wang, D.B. Lobell, Y. Huang, M. Huang, et al. 2017. "Temperature Increase Reduces Global Yields of Major Crops in Four Independent Estimates." *Proceedings of the National Academy of Sciences of the United States of America* 114 (35): 9326–31. doi:10.1073/PNAS.1701762114.

10

Improving Submergence Tolerance in Rice: Recent Progress and Future Perspectives

Debabrata Panda, Jijnasa Barik, and Prafulla K. Behera
Central University of Odisha

CONTENTS

10.1 Introduction

Flooding of rice fields in lowland and deep water rice regions particularly in South and South-East Asia causes several hurdles in agriculture sector and results in huge crop loss. It adversely affects above 22 million hectares of rain-fed lowland rice areas which demands proper action plans to find out remedial measures to overcome the same (Panda and Barik 2021). In India, more than 5 million hectares of lowlands depending on rain fed are drastically damaged by submergence leading to a loss of 4 million tonnes of rice every year (Bal and Adhya 2021; Ikmal et al. 2021). Rice is a staple crop feeding >3.5 billion people globally (Stallworth et al. 2021), and about 90% of rice is cultivated in Asia. Moreover, Asian people use rice in daily diet. Rice is a semiaquatic species which is grown under a wide range of agro-climatic conditions, ranging from coastal belt to higher hilly

DOI: 10.1201/9781003258063-10

areas. Submergence may be either partial or complete. In complete submergence, the plants fully remain under water, while in the other, the plant remains partially under water. Complete submergence of rice plant due to continuous flooding can adversely affect plant growth and its development. Though some traditional rice varieties can withstand submergence to some extent, there are many other rice varieties (mostly improved varieties) which are highly susceptible to complete flooding (Fukao et al. 2011; Mickelbart et al. 2015). More than 50 years ago, low-yielding submergence–tolerant rice landraces were found in the eastern India and a study was made on the aspect of physiological as well as molecular approaches of resistance to complete flooding. Subsequently, *SUBMERGENCE-1* (*'SUB1'*) controlling submergence tolerance was identified (Xu and Mackill 1996). This Quantitative trait locus (QTL) acts as a suppressing factor on 'ethylene-activated shoot elongation' under water, resulting in low consumption of carbohydrate and increasing survival rates. *SUB1* QTL was then incorporated successfully in several high-yielding mega varieties in Asia like Swarna, Savitri, BR11, IR64, Ciherang etc. (Septiningsih et al. 2015). Since then, millions of rice cultivators adapted the *SUB1*-introgressed high-yielding varieties throughout the globe (Ismail et al. 2013).

Global climate is rapidly and unexpectedly changing and needless to say that such global climatic change brings untimely heavy rains. The same leads to water-logging and flooding of different depths and duration in vulnerable flood-prone regions (Hirabayashi et al. 2013; Panda et al. 2021). This adversely affects rice production mainly in rain-fed agro-ecosystem. It is also reported that many of the available submergence tolerant donors have less combining capacity and agronomic characters (Ram et al. 2002). Further, the *Sub-1*-incorporated high-yielding varieties due to their short height are amenable to vagaries of flooding in variable different depths. For this reason, farmers are greatly discouraged from adapting these improved varieties (Goswami et al. 2017; Afrin et al. 2018). Hence, rice genotypes that can withstand submergence are greatly required and need to develop genotype with higher capacity to withstand the stress due to longer duration of submergence for its survival. To overcome the situation, it is necessary to explore and identify noble genes and proper understanding of complexity involved therein. In this chapter, responses and adaptation strategies of rice plant under submergence are summarized from morphological level to molecular level. The objective of the chapter is to improve the understanding of responses and tolerance mechanism of rice to submergence stress, which is subsequently helpful in developing new rice varieties with improved tolerance.

10.2 Types of Flooding

Flooding can be classified into three classes based on the plant growth stage in relation to variable flooding condition and duration.

10.2.1 Anaerobic Germination (AG)

The first of its kind is AG. It is commonly found in rain-fed lowland areas where farmers follow direct seeding procedure and encounter flooding due to heavy rainfall. In such a situation, the seed completely remains submerged. They suffer from hypoxia or anoxia resulting in poor or no germination, least crop growth and death of seedlings (Ismail et al. 2009; Barik et al. 2019).

10.2.2 Short-Term Flooding/Flash Flooding

Short-term or flash flooding is the second of its kind, where submergence takes place during seedling or tillering stage. In this type of flooding category, generally plants completely inundate under flood water. Such flooding remains for a short duration up to 2 weeks (Sasidharan et al. 2018). Generally, this happens in river basin areas and causes total yield loss.

10.2.3 Long-Term Flooding

The third type is on the account of 'long-term flooding' where water gets stagnated for months together with variance in water depth and duration. Long-term flooding is of different types i.e. partial/stagnant (water depth 30–50 cm), semi-deep (water depth >100 cm) and very deep water (water depth up to 3 or 4 m) (Kuanar et al. 2017; Oladosu et al. 2020). So, irrespective of the crop growth stages, duration and depth of flood, always remains as the critical constraint to rice production in lowland and deep water areas.

10.3 Environmental Characterization of Flood Water

The consequences of submergence stress is multiple, as well as complex. Plants are stressed by low light intensity, limitation in gas movement, efflux of soil nutrients and high vulnerability to pests and diseases when they are under the constraints of flooding (Figure 10.1). The survival of plant under water is not only determined by the depth and duration of flood water but also by the physicochemical characteristics of flood water i.e. oxygen and CO_2 concentration, pH, turbidity, light intensity and temperature (Setter et al. 1997; Oladosu et al. 2020). Flooding has a negative influence on rice because inadequate gas exchange hinders biochemical approaches like photosynthesis and respiration (Panda et al. 2008). Inadequate gas diffusion is a big issue subjected to flooding because as compared to air the gas diffusion under water is about 10,000 times slower (Armstrong and Drew 2002). The main source of impairment to plants under flooded soil is due to depletion of oxygen (O_2), which creates a condition of hypoxia (less concentration of O_2) or anoxia (absence of O_2). The soil microorganism also reacts to lower O_2 by generating enzyme with high affinity towards oxygen, thus rapidly consuming the available oxygen in the soil. The following processes will further change the physicochemical properties of soil such as pH, redox potential and formation of harmful soil phytotoxins. These reductions in redox potential of soil also affect the balance of charge condition of various soil minerals and toxic ions and substance generation like H_2S, Fe^{2+} and Mn^{2+} and impair severe injury to plants subjected to waterlogged constraints (Setter et al. 2009).

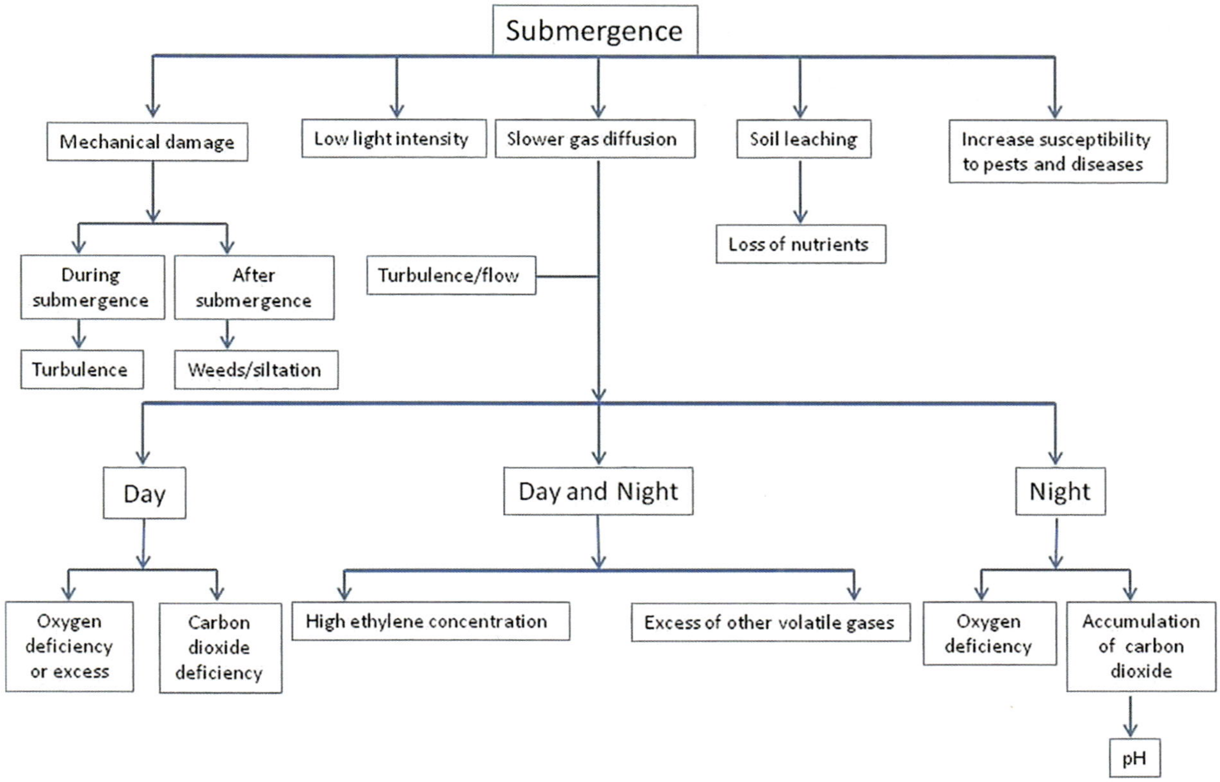

FIGURE 10.1 Environmental parameters affecting submergence tolerance in rice.

As a result of this unfavourable soil condition, fewer nutrients would be available in soils for plants and cause ethylene and CO_2 accumulation in the flooded parts of plants (Bailey-Serres and Voesenek 2008). In addition, the deposits of volatile organic acids like propionic acids in waterlogged soils damage the plant roots. High concentration of carbon dioxide in flood water together with these organic acids creates an impact of acid loads on the root cells. The turbidity in flood water also lowers light transmission and sediments silt on the underwater plant which subsequently limits the capacity of plants to photosynthesize through chloroplast impairment (Panda et al. 2006). Higher temperature (30°C) speeds up plant mortality and lower temperature (20°C) enhances plant survival under water. Higher temperature reduces the solubility of oxygen and carbon dioxide in flood water and speeds up anaerobic respiration, causing the plant to starve and die sooner. Flooding is considered to be the most significant abiotic stress which impacts global species distribution and agricultural output (Ram et al. 2002; Jackson et al. 2009).

10.4 Impact of Submergence Stress on Rice Plant

10.4.1 Stress during Submergence

Flooding-induced submergence is an important environmental factor reducing plant productivity to a greater extent. Plant growth and development can be highly disturbed and restricted under flooding. The metabolic disturbances due to flooding result in tissue injury and subsequently cause plant death (Fukao et al. 2019). In sensitive rice varieties, flooding causes chlorophyll degradation, reduction in rubisco activity as well as damage to photosynthetic apparatus which eventually leads to a strong decrement of photosynthetic rate (Panda et al. 2008). Under flash flooding situations, excessive elongation of shoot is considered to be injurious and harmful for the plant, as it enhances energy depletion and increases plant mortality. During de-submergence situation, the tall plants tend to lodge and cause yield loss with declined grain quality (Ella et al. 2003). It has also been noticed that the survival of the rice seedling has negatively correlated with under water extension growth in short-term flooding condition (Panda and Barik 2021).

10.4.2 Stress after De-Submergence/Re-Aeration

Anoxic plants experience an instant rise in light intensity and oxygen concentration upon de-submergence when plants become re-exposed to air. The visual symptoms of injuries in plants develop gradually as the flood water recedes. One of the major stresses which plant encounters after post submergence is 'photoinhibition' to the photosynthetic apparatus. Panda et al. (2006) confirmed the same as they observed the enhanced minimum fluorescence in rice indicates photoinhibition. So, photoprotection along with recovery from photoinhibition is the foremost challenge faced by plants after de-submergence. Re-aeration can lead to excessive production of reactive

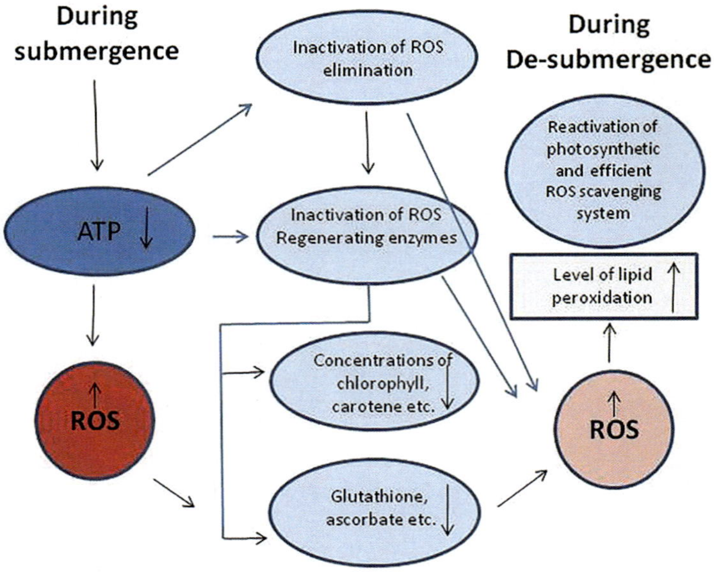

FIGURE 10.2 Changes in the metabolism of plant under submergence and during de-submergence.

oxygen species (ROS) and toxic oxidative substances which are highly reactive and harmful to plants. The excessive excitation of chlorophyll molecules in high light intensity can lead to the production of singlet oxygen which induces lipid peroxidation in photosynthetic membranes. Higher carbohydrate levels and efficient ROS scavenging systems in plants after post submergence are key factors for determining the capability for quick regeneration and also recovery growth when water recedes (Ram et al. 2002; Singh et al. 2014; Upadhyay 2018). Figure 10.2 represents the changes in metabolism in plants during and after submergence.

10.5 Adaptive Mechanisms to Submergence Stress

Flood tolerant crops are able to withstand the adverse impacts of flooding and subsequent post-flooding stress by changing the anatomy and morphology of plant parts, adjusting metabolic pathways and also through alteration in gene expression (Agarwal and Grover 2006). Some of the adaptive traits help rice to tolerate prolonged soil flooding or complete submergence such as organization of longitudinal interconnected air cavities or aerenchyma, accumulation of non-structural carbohydrates, maintenance of higher chlorophyll content and better antioxidant defence mechanism, the ability to maintain active fermentative metabolism and under water photosynthesis (Panda et al. 2021).

10.5.1 Morphological Adaptation

10.5.1.1 Aerenchyma Formation

Aerenchyma formation plays an essential role in plant's survival and functioning by supplying oxygen from the aerial portion of plants to roots under oxygen-deficient environment (Nishiuchi et al. 2012). 'Aerenchyma' is a specialized tissue of plant with longitudinal interconnection of large gas spaces which provide a pathway for long-distance movement of gases along the plant organ and contributes to the ventilation of volatile compounds, for example carbon dioxide, methane, ethanol, ethylene and acetaldehyde (Sou et al. 2021). The aerenchyma tissue in rice plant is well established in internodes, sheath, roots as well as midrib of leaves (Pradhan et al. 2017). Mainly two types of aerenchyma are present depending upon its formation: schizogenous and lysigenous. The first one is formed by cell separation during development of tissues, as found in the case of *Rumex palustris*. The formation of the second type of aerenchyma occurs through death and cell lysis, for example in rice, maize, wheat and barley. In rice, aerenchyma is believed as a key anatomical adaptation for seedling viability under submergence (Yamauchi et al. 2020).

10.5.1.2 Adventitious Root Formation

Adventitious root formation in plants ensures fast supply of water and nutrients to the growing parts. In the case of rice, flooding activates the apical meristem tissue of adventitious root primordia which leads to assemblage of cell cycle gene mRNA (Kuroha and Ashikari 2020). During flooding, just before root arises, the epidermal tissue encounters programmed cell death and initiate adventitious root growth (Steffens et al. 2012). The growth of adventitious root is stimulated by ethylene under submergence (Kuroha and Ashikari 2020) and can be detected within few hours, for example after 10 hours in deep water rice (Dawood et al. 2014). According to the study by Rich et al. (2012), aerenchyma number in adventitious root is more than leaves.

10.5.1.3 Leaf Gas Film (LGF) Formation

At the time of flooding, the surface of the leaf becomes hydrophobic and a thin layer of air is trapped among surrounding flood water and leaf surface, which is referred to as LGF. It helps in internal aeration (by exchange of oxygen and carbon dioxide) while under water and hence increases the submergence resistance ability in rice (Kuroha and Ashikari 2020). The increase in gaseous exchange due to LGFs stimulates internal aeration in plants under submergence. This increase in aeration inside plant leads to an increase in net under water photosynthesis by supplying carbon dioxide in the day time and promoting oxygen uptake at night for respiration process (Colmer et al. 2011). As a result of this, LGFs help in production of sugar in leaf through underwater photosynthesis and in turn plant biomass (Pedersen et al. 2009). The *LGF1* gene involved in surface hydrophobicity and retention of LGFs under submergence was identified (Kurokawa et al. 2018) which contributes greatly to enhance submergence resistance in rice.

10.5.1.4 Shoot Elongation

Plant height is an important factor which influences crop yield as well as the survival of seedlings under submergence. Quiescence and prolongation are two opposite approaches in rice which provide adaptation towards flood according to flooding types (Panda et al. 2021). Both these approaches are caused by ethylene response factors (ERFs) (Xu et al. 2006).

10.5.1.4.1 Quiescence Strategy

Complete submergence which occurs for a short duration of 10–14 days after heavy rain or overflowing of rivers is known as 'flash flooding'. Rice genotypes tolerant of flash flood condition adapt a slow growth strategy where submerged stem under goes minimal or no elongation (Barik et al. 2020). It helps plants in maintaining adequate carbohydrate reserves to keep metabolism going during submergence as well as for recovery and regeneration, when flood water recedes (Perata and Voesenek 2007; Singh et al. 2017). In the complete flooding condition, a negative correlation can be observed between the elongation of vegetative parts and survival (Azarin et al. 2017). This is also confirmed by utilizing 903 genotypes of rice collected from IRRI (gene bank). Thus, maintenance of a higher amount of carbohydrates in plant before flooding along with restricted (less) shoot extension and retention of chlorophyll content are essential traits to survive under complete submergence (Panda and Barik 2021). The flooding tolerance based on the quiescence strategy is accorded with important QTL i.e. '*SUBMERGENCE-1*' on chromosome number 9 of rice. The *SUB-1* gene was successfully incorporated into many mega high-yielding varieties and the *Sub-1* incorporated into popular varieties were then released in Asia and Africa for commercial cultivation (Mackill et al. 2012).

10.5.1.4.2 Escape Strategy

In deep water areas, stem elongation with rising flood water to escape complete submergence is one of the vital characteristics of adapted genotypes. This escape strategy is associated with high carbohydrate consumption (Vergara et al. 2014). The fast elongation due to flooding is mainly stimulated by oxygen deficit and accumulation of ethylene hormone (Singh et al. 2017). Rice under deep water can elongate nearly 25 cm of height per day to maintain its leaves over the water surfaces, to maintain photosynthesis process and oxygen delivery, resulting in improved survival (Bailey-Serres and Voesenek 2008). This type of rice can grow up to 5 m tall while retaining its growing tips above the water's surface (Panda and Barik 2021). The major QTL responsible for internode elongation under deep water flooding was located on chromosome number 12 and contains two ERF family genes i.e. *SNORKEL1 (SK1)* and *SNORKEL2 (SK2)*. In complete submerged/ flooding condition, these two genes promote internode elongation by increasing gibberellic acid (GA) production (Hattori et al. 2009; Nagai et al. 2014; Minami et al. 2018).

10.5.1.5 Root Traits

Rice has the capacity to sustain soil water-logging by generating aerenchyma and an impediment in the root peripheral cells to prevent radial oxygen loss (ROL) (Yamauchi et al. 2018). This is an adaptive character which facilitates longitudinal oxygen movement via aerenchyma towards root tip by reducing oxygen loss; as a result, enabling rice roots elongation in the anoxic environment (Armstrong 1979). This barrier to ROL not only aerates the rhizosphere but also causes impediment in entry of soil-derived toxic compounds and gases into the roots (Colmer and Voesenek 2009). Both upland and lowland rice varieties use the traits under submerged conditions. This hurdle to ROL is treated to be an adaptive characteristic in plants growing in submerged environment (Neubauer et al. 2007).

10.5.1.6 Coleoptile Elongation during AG

In spite of being well adapted to varied flooding conditions, majority of rice cultivars are sensitive to anaerobiosis at the time of germination and early stages of embryo development (Angaji et al. 2010). Rice seed has the ability to germinate under anaerobic environment up to some extent i.e. coleoptiles elongation only (Ismail et al. 2012). The elongation of coleoptiles plays an important role in promoting the growing seedlings exposed to atmosphere aids crop establishment. The development of root occurs followed by the leaves getting in contact with air (Panda et al. 2021). The coleoptiles elongate due to cell expansion rather than cell division, because cell division process requires high energy than cell expansion (Magneschi and Perata 2009). The tolerant rice varieties are capable of faster coleoptiles emergence, high germination rate, gene expression of 'expansin genes', upregulation of amylase activity and anaerobic respiration which is reverse in susceptible genotypes (Barik et al. 2019; Mondal et al. 2020).

10.5.2 Physiological Adaptation

10.5.2.1 Photosynthesis

The process 'photosynthesis' is the initiating point for carbon metabolism and becomes the entry point for energy in plants. Submergence seriously affects the most crucial process for plant growth i.e. photosynthesis and causes a significant reduction in energy supply. Inundation of aerial parts of plant under complete submergence severely reduces under water gas exchange and inhibits oxygen uptake, thereby decreasing photosynthesis as well as transpiration, and as a result of this, stored carbohydrate in plants gets exhausted leading to plant death (Das et al. 2009). Moreover, photosynthesis is limited by the turbidity of flooding water which can check the light accessibility. Along with low light intensity, various other factors such as degradation of chlorophyll, lipid peroxidation, stomata closure and poor intracellular CO_2 concentration also lead to decrease in photosynthesis (Yang et al. 2017). Underwater photosynthesis is not only reduced by low light intensity or less carbon dioxide availability but also due to structural and functional impairment of chloroplast, majorly due to inhibition of photosystem II (PSII) photochemistry (Panda et al. 2006). The tolerant varieties have the ability to maintain an elevated level of photosynthetic rate on account of photosynthetic apparatus's protection, retention of more chlorophyll content as well as better stomatal conductance while under submergence and during post submergence as compared to sensitive genotypes (Panda and Sarkar 2013; Panda and Barik 2021). Chlorophyll fluorescence is a highly versatile tool and technique that gives information about relationships between PSII reaction centres and antenna complexes in rice crops under submergence stress (Falqueto et al. 2010). Submergence adversely affected photosynthetic system further which was confirmed from decreased values of fluorescence parameters in both resistance and sensitive varieties (Barik et al. 2019). Hence, different key parameters could be quantified through chlorophyll fluorescence technique which is helpful for identification of the tolerant and intolerant genotypes in a short duration.

10.5.2.2 Chlorophyll

Chlorophyll has an important role in photosynthesis and the total content of chlorophyll is a key parameter to find out the potentiality of photosynthesis process in plants. Submergence accelerates chlorophyll degradation in the leaves of submerged seedlings. This is caused by ethylene, accumulated in plant tissue by stimulating the gene expression as well as activity of 'chlorophyllase' enzyme. This is the first enzyme associated with the breakdown of chlorophyll molecules. Due to this, the capacity for carbon dioxide fixation is reduced during submergence and after re-exposure to atmosphere (Ella et al. 2003). Decrease in chlorophyll under submergence is a frequent phenomenon in rice seedlings and can be used as an index of submergence tolerance in rice (Panda and Barik 2021). Retention of chlorophyll content by inhibiting ethylene biosynthesis may enhance under water photosynthesis (Pradhan et al. 2017). Further, chlorophyll retention in plant while remaining under water and after submergence becomes very crucial for plant viability as it supports photosynthesis process during flooding and is helpful in quick recovery and regeneration following de-submergence (Singh et al. 2014). Thus, tolerance of submergence stress is associated with the ability to maintain higher chlorophyll content in plant tissues during and after submergence.

10.5.3 Biochemical Adaptation

10.5.3.1 Hormonal Regulation

The regulation of elongation in plant is mostly carried out by three different phytohormones like ethylene, abscisic acid and GA. In rice, ethylene hormone is a prime regulator in relation to submergence. Due to physical entrapment and biosynthesis under stress, ethylene production is enhanced in submerged plants, stimulating a variety of acclimation responses such as adventitious root generation, shoot enlargement and metabolic activities of carbohydrate. According to Hattori et al. (2009), ethylene maintains foliage to escape flooding via speed stem elongation regulated by *SK1* and *SK2* in deep water rice system. Also in the case of flash-flood–tolerant rice genotypes, ethylene inhibits stem elongation amid the quiescence response via AP2/ERF transcription factor, SUB1A-1 (Kuroha and Ashikari 2020). The SUB1A checks ethylene-induced elongation growth through two mechanisms: (i) elevating GA repressors i.e. SLR1 and SLR2, restricting responsiveness of GA and (ii) increasing GA catabolic activities by genes involved in brassinosteroid synthesis, including one that catabolizes GA (Schmitz et al. 2013). Thus both mechanisms restrict GA-induced growth and are helpful in preserving energy and carbohydrate metabolism subjected to flooding and also recovery after de-submergence.

10.5.3.2 Carbohydrate Reserves and Energy Metabolism

In plants, presence of soluble sugars and starch are significant packets of energy for the function of cell and regulation during submersion, also subsequent tissue regeneration after de-submergence. During submergence constraints, metabolic activities of carbohydrates appear to be a key element subjected to flash flood resistance and the 'quiescence strategy' is defined by moderate growth, which is thought to preserve energy (Panda and Barik 2021). A strong correlation and substantial link between carbohydrate levels before submergence and submergence tolerance is consistently observed through earlier studies. Also, the carbohydrate concentration present during submergence is related to flash flood tolerance and is resolved by the balance of consumption and supply; thus, quick shoot enlargement promotes carbohydrate consumption (Fukao and Bailey-Serres 2004). Accordingly, the tolerant genotypes consume starch reserves slowly as compared to genotypes growing speedily upon complete submergence. Previous investigation with submergence tolerant and susceptible rice genotypes revealed about tolerant variety consisting of 30%–50% more non-structural carbohydrate than susceptible variety and these non-structural carbohydrates are being used by plants as their primary energy source during submergence to keep their metabolism running (Sarkar et al. 2006).

10.5.3.3 Activation of Alcoholic Fermentation

As carbohydrate metabolism is strongly inhibited under water, plants activate alternative pathways to sustain under oxygen-limiting environment. During AG, the alcoholic fermentation in rice is strongly activated which is indicated from higher activities of alcohol dehydrogenase, aldehyde dehydrogenase and pyruvate decarboxylase enzymes (Ram et al. 2002; Panda and Barik 2021). The enzymes such as alpha-amylases, sucrose synthase and aldolase are highly expressed in hypoxia tolerant genotypes during germination whereas all these enzymes are inhibited in sensitive genotypes. Though anaerobic fermentation produces only two adenosine triphosphate molecules (ATP molecules), it is considered to be highly essential for energy production, resulting in plant survival and tolerance to low oxygen conditions (Agarwal and Grover 2006; Panda and Barik 2021).

10.5.3.4 Induction of Anaerobic Proteins

Submergence adversely affects the synthesis of proteins. The oxygen deprivation leads to the repression of synthesis of most anaerobic proteins. At the same time, it induces the production of many anaerobic proteins, majority are being enzymes related to glycolysis and fermentation process (Sarkar et al. 2006). The number as well as level of anaerobic protein production may vary in anoxia tolerant and sensitive species. The role of anaerobic proteins in induction of anoxia tolerance was confirmed through experimental investigation with ruthenium red, which hinders anaerobic gene expression, leading to lowering in tolerance to anoxic condition (Ram et al. 2002). Different proteins/enzymes induced upon oxygen deprivation condition/submergence condition are summarized in Table 10.1.

10.5.3.5 Oxidative Metabolism and Antioxidant Defence System

In plants' green tissues, chloroplasts and peroxisomes are the main sources and production sites of ROS. ROS like hydroxyl radical ($OH^•$), $O_2^{•-}$, $^1O_2$ and H_2O_2 are being considerably accumulated subjected to flooding stress constraints, which causes oxidative effect leading to cell death. Excess generation of these ROS due to stress impacts in plant cells, is very reactive and can oxidize lipids, DNA and proteins which leads to cellular destruction and plant death. On the other hand, these increased generations of ROS during stress situation act as activator of various intracellular stress response pathways (You and Chan 2015). These free radicals are efficiently detoxified by different defence systems (enzymatic and non-enzymatic) of plants and thereby plants protect themselves against oxidative damages. Different antioxidant enzymes like ascorbate peroxidase, monodehydroascorbate reductase and dehydroascorbate reductase and many non-enzymatic like glutathione, carotenoids, flavonoids and tocopherols work together to maintain ROS homeostasis in plant (Upadhyay 2018). Further, the upregulation of antioxidant enzyme is intimately linked to the flood tolerance of rice crop. Rice varieties are capable of enhancing anti-oxidative defence systems under water and upon de-submergence have better tolerance capability as observed from higher survival rates (Panda and Sarkar 2013; Barik et al. 2019).

10.5.4 Molecular Adaptation

The genetic foundation for rice flooding/submergence tolerance has been extensively researched with the development of molecular techniques and associated primers that allow for the identification of distinct QTL and detailed mapping of the submergence tolerance genes in rice. The QTLs associated with AG, flash flooding tolerance and long-term flooding stress have been identified by various researchers.

TABLE 10.1

Proteins and Enzymes Found to Be Linked with Submergence Stress in Rice Plant

Sl. No.	Pathway	Proteins/Enzymes	Observations	References
1	Carbohydrate metabolism	Sucrose synthase	Higher enzyme activity under anoxia in rice roots	Kato-Noguchi (2004)
		α-amylase	Upregulation of Amy3 in rice embryos	Hwang et al. (1999)
		Aldehyde dehydrogenase	Transcripts prompted during flooding stress	Tsuji et al. (2003)
2	Glycolysis	Glyceraldehyde 3-phosphate dehydrogenase	Flooding causes a significant increase in transcript level	Umeda and Uchimiya (1994)
		Fructose 1,6-bisphosphate aldolase	Transcript levels in rice roots increased upon flooding but did decline under typical aerobic circumstances	
		Phosphofructokinase	Activity prompted in studied rice cultivars but lower in sensitive cultivar	Gibbs et al. (2000)
			Enzyme activity induced in hypoxic and anoxic conditions and noticed to be induced in higher amounts in case of tolerant genotype than the sensitive cultivar	
			pdc1 transcript was significantly induced anoxic condition whereas marginal induction was shown by *pdc2*	Hossain et al. (1996)
			pdc1 transcript accumulated in submergence condition; faster accumulation is observed in sensitive variety than the tolerant variety	Umeda and Uchimiya (1994)
		Alcohol dehydrogenase	*adh1* transcript assembled during flooding stress; faster accumulation is noticed in sensitive variety as compared to tolerant variety	Umeda and Uchimiya (1994)
			adh genes induced in anaerobic condition	Xie and Wu (1989)
			adh activity increased upon anoxic condition	Gibbs et al. (2000)
4	Transport	ATPase	Higher enzyme assay (two fold) in anoxic condition	Carystinos et al. (1995)
		Vacuolar H^+-pyrophosphatase	Enhancement of enzyme activity, proteins and mRNA under anoxic condition	

11.3.1 Mechanical Approach

When excess salt got accumulated in soil surface (in the form salt deposition or salt crust) scraping of salt is the one viable option (Figure 11.3). In most cases, salt is deposited on the soil's surface through evapotranspiration when the water table is very high, practically at the soil's surface. Flooding with good-quality irrigation water is highly effective when salt is dominated by NaCl and Na_2SO_4. This salt poses a very higher degree of solubility, therefore easily gets dissolved in water and gradually moves down below the crop rootzone with percolated water. Ditching or trenching in crop fields with fixed spacing is helpful to collect excess soluble salt from cropping area, followed by draining out the salt water from crop field. Subsurface drainage is functioning similar principle with ditching, to collect excess salt water from crop fields and dump the salt water into a large drainage outlet so that salt may permanently flushed out from the area. The details of the working principle of subsurface drainage are depicted in Figure 11.4. In subsurface drainage, perforated pipes are installed below 2- to 3-foot depth, so that salt present in the crop rooting zone is washed away with rain or drainage water.

Adoption of subsurface drainage technology in the salt-affected belt of North-Western India had shown it effectiveness through the reduction in soil salinity associated with yield increment in wheat, maize, and pear millet crops. Experiment conducted on 23 blocks showed 9.7%–66% reduction in EC_e (from 3.8–11.4 dS/m to 2.1–7.4 dS/m) within 5 years of experimentation, which results in an increase in yield by 17.7%, 21.4%, and 39.8% under wheat, rice, and pearl millet respectively in those areas (Datta et al., 2004). Later, another study conducted in Haryana and the Andaman and Nicobar Islands, India reported that in the long run subsurface drainage technology minimized overall 44% soil salinity and 49.5% drain water salinity and increased cropping intensity by 20.32%. Further, yield of rice, wheat, cotton, and mustard crop was increased by 20.46%, 16.26%, 19.75%, and 15.01% respectively over undrained control plot (Raju et al., 2016).

11.3.2 Soil and Crop Management Practices

Efficient agronomic intervention can suitably manage salt stress in crop plant for which below agronomic management practices need to be followed.

Growing salt-tolerant crops: The easiest way is the growing salt-tolerant crops, which are capable of growing in salt-affected soil by virtue of their own mechanism. Various field crops, like rice, maize, cotton, sorghum, red gram, green gram, sunflower,

FIGURE 11.3 Deposition of excess salt in soil surface (a) and in the form surface crust (b) (this excess surface deposited salt in (a) and salt crust formed in (b).

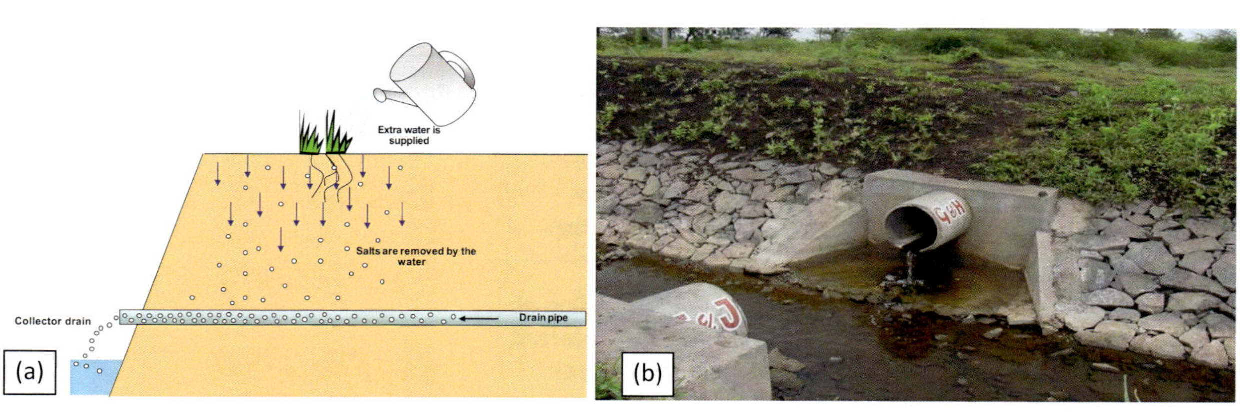

FIGURE 11.4 Removal of excess salt through subsurface drainage system (a) and (b).

linseed, sugar beet, pumpkin, bitter gourd, beet roots, asparagus, and spinach, among vegetables, were reported to grow well in salt-affected condition (Gupta and Abrol, 1990).

Alteration in planting geometry: Research work showed that (i) planting of crops in paired rows on broad sloping ridges and (ii) single crop rows on asymmetric ridges poses a low risk of salt injury compared to planting in top of ridges or bottom of furrows. In ridge and furrow methods, maximum accumulation of salt occurs in peak of the ridges; hence, these peaks of ridges are called zone of salt accumulation (Figure 11.5). This method is more suitable for salt-sensitive to moderate salt-tolerant crops, especially in a semi-arid environment. For coastal areas, where seawater inundation problem is very frequent due to shallow water table, broad bed furrow (BBF) or three-tier farming is more suitable. In this, three wide beds (at least 10–20 m wide) are constructed at three different heights. Soil from first bed excavated and piled up in third bed, therefore an difference in elevation was created which helps the cultivation of different type crops.

On the lowest bed, which is filled with water, fish is cultivated; in the middle, rice is grown and at the upper bed, vegetable was cultivated. These three-tier BBF cultivation system is sustainable for coastal regions where paddy and fish are grown together (Figure 11.6b).

Seed priming: Salt injury to crops is more prominent during the seed germination stage. Seed priming aims to allow crop seeds to germinate under a salt-stress-free environment. Although the viability of seed priming technology is short, still this technology can offer the best low-cost tool to manage salt injury to crop plants. Seed priming with some chemicals is found to overcome the negative effect of salt stress by increasing the lag phase during seed germination. Many seed priming methodologies, such as seed priming with chemicals, osmoregulatory substances, hormones, and nutrient elements, still exist. In chemical seed priming methods, various organic or inorganic chemical substances like hydrogen peroxide, ascorbic acid, salicylic acid, glutathione, sodium nitroprusside, selenium, chitosan, urea, etc. are used. Seed priming is done with substances

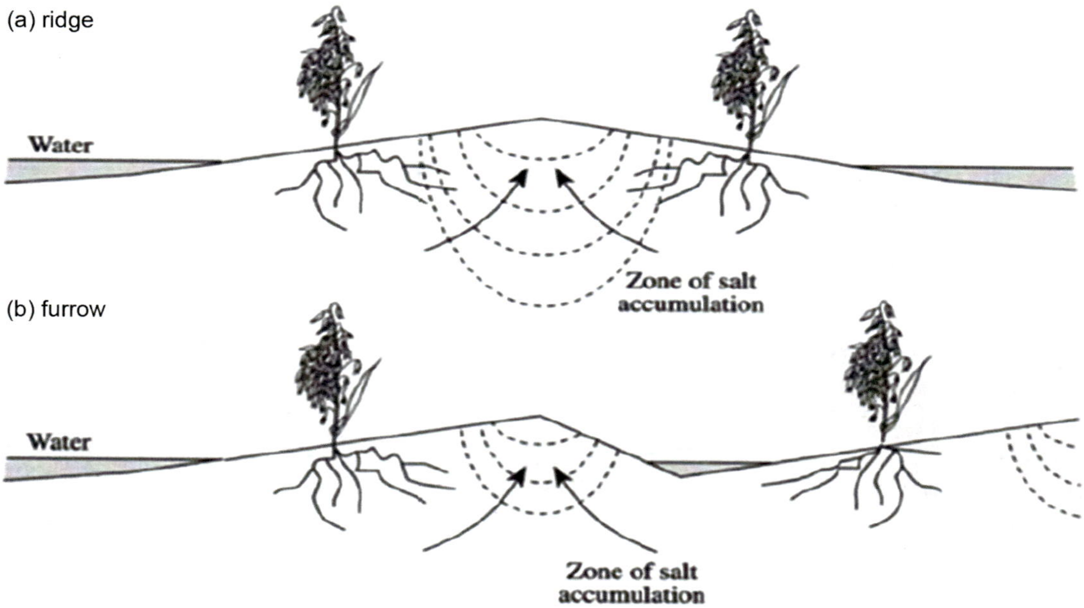

FIGURE 11.5 Avoidance of salt stress by creation of microtopography of ridge and furrow systems. (Adapted from Berstein and Fireman (1957).)

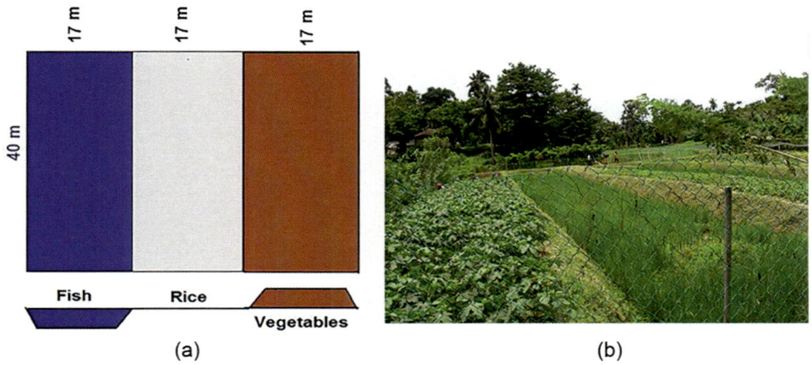

FIGURE 11.6 Pictorial representation of BBF or three-tier farming, (a) concept of three-tier farming, (b) vegetable-paddy-fish farming in coastal region.

TABLE 11.2

Seed Priming for Enhanced Germination through Various Agents in Field Crops

Name of the crop	Seed priming agent	References
Wheat	Ascorbic acid, indole acetic acid (IAA), potassium silicate (K_2SiO_3), proline	Feghhenabi et al. (2020) and Wajid et al. (2019)
Maize	Gibberellins (GA3), potassium silicate (K_2SiO_3)	Tsegay and Andargie (2018) and Khan et al. (2015)
Barley	H_2O_2	Kilic and Kahraman (2016)
Oat	Gibberellin (GA3)	Chauhan et al. (2019)
Sorghum	Gibberellin (GA3) $CaCl_2$, NaCl, PEG, KCl	Chen et al. (2021)
Lentil (*Lathyrus sativus*)	Salicylic acid, H_2O_2, GA3	Bouallègue et al. (2019)
Tomato	Epigallocatechin-3-gallate (EGCG)	Ahammed et al. (2018)
Melon	Malanotin	Castañares et al. (2019)
Cucumber	Silicon (NaSi)	Gou et al. (2020)
Field pea	Gibberellins (GA3)	Tsegay and Andargie (2018)

which modify seed water potential or regulate osmotic potential called osmopriming. Sorbitol, poly-ethylene glycol (PEG), and mannitol are commonly used for osmopriming. Seed priming with these substances reduces water potential outer seed layer, thus restricting entry to excess water inside the seed which, in turn, deactivates ROS species formation. In some cases, inorganic salts, called halopriming, like KCl, NaCl, $NaNO_3$, K_3PO_4, KNO_3, $CaCl_2$, and $MgSO_4$ are used. Hormonal priming is a seed treatment with hormones such as auxins, gibberellins (GA_3), ethylene, etc. The beneficial findings of seed priming to mitigate salt stress are presented in Table 11.2.

Nutrient management: It was observed that soil salinity sometimes gets aggravated due to faulty nutrient management practices. The application of organic manures along with straight fertilizers is found more beneficial over straight fertilizer alone. On the decomposition of organic manures, CO_2 and organic acid are released in soil solution which enhances the dissolution of salt due to a rise in partial pressure of CO_2 and lower soil pH. Applying additional P fertilizer to saline soil may help reduce the toxicity of Cl to plants; the effect was more so with single super phosphate (SSP) over di-ammonium phosphate (DAP). SSP contains a greater proportion of $CaSO_4$ which allows formation of H_2SO_4 and provides Ca^{2+} ion which minimizes the toxic effect of Na^+. Applying FYM and green manure crops, such as planting Sesbania rostrata in the field before transplanting rice, has increased yields in salt affected soils (Jat et al., 2020). Improvement in rice and wheat productivity under balanced fertilization, integrated nutrient management with FYM 10 t/ha was reported from North West India (Yaduvanshi and Sharma, 2016). Integrated approach (NPK+FYM and NPK+Pressmud) increased the yield of rice and wheat by 104% and 159% over control. Similarly, the application of commercial humic acid at 24 and 48 kg/ha and FYM 10 t/ha reportedly reduced EC_e and sodium adsorption ratio vis-à-vis increment in root length and yield (up to 136%) of rice under saline-alkali soil (Shaaban et al., 2013).

11.3.3 Biological Approach

Various biological agents like plant-growth-promoting microbes (PGPMs) and green plants were utilized.

11.3.3.1 Microbial intervention

The ever-deteriorating climatic conditions and various anthropogenic activities have led to the rise in the magnitude of global warming. It is a very serious condition from the plant development point of view as it will hinder their growth due to the accumulation of salt in soil and reducing water availability for plant uptake. Plants' rhizosphere harbors diverse regime of microorganisms that plays an important role in nutrient acquisition for plants and also aids in their defense system against biotic and abiotic stresses. Microorganisms, being an integral component of essential biological processes and ultimately soil health, could be taken upon as an effective measure to alleviate and assist plants in withstanding salinity stress. Moreover, salinity also affects microbial diversity and related biochemical properties, thus deteriorating soil health.

Be it rhizosphere or phyllosphere, there is an intimate relation between plants and the associated microbial communities. Plant-microbe interaction is vital for the sustainability of agriculture, as it regulates the biological processes. Therefore, the microbes are considered ecological engineers to encounter the biotic and abiotic stresses in the environment. The group of microbes, including bacteria, fungi, actinomycetes, etc., and arbuscules mycorrhiza associated with plants improve their growth, commonly known as PGPMs. These PGPMs provide additional protection to plants under stress conditions. PGPMs do so by regulating the phytohormones, enhancing antioxidant complex, and improving nutrient availability and siderophore production. The 1-aminocyclopropane-1-carboxylic acid (ACC) deaminase enzyme degrades ACC, therefore preventing accumulation of ethylene. Several soil microbes produce this enzyme which when inoculated under salt stress helps the plant to alleviate the stress (Orozco-Mosqueda et al., 2020).

Plants upon encountering salinity stress undergo secondary stress like ionic and oxidative stresses. Due to this, the plant suffers biochemical and physiological alterations leading to loss in crop productivity. However, when plants are inoculated with PGPRs, they alleviate the negative effects induced due to stress. Morphological and biochemical modifications have been observed in plants upon inoculation

TABLE 11.3
Salt-Tolerant Crop Varieties Developed in India

Crops	Varieties Developed	Adaptability	Potential Yield
Rice	CSR10, CSR11, CSR12, CSR13, CSR19, CSR23, CSR27, CSR30, and CSR36	Up to EC_e 11 dS/m	Up to 6 t/ha under salt-free condition, >(2–4) t/ha under salt-stress condition
Wheat	KRL 1-4, KRL 19, KRL 210, KRL 213	Up to EC_e 7 dS/m	Up to 5.5 t/ha under salt-free condition 2.5–3.5 t/ha under salt-stress situation
Indian mustard	Pusa bold, Varuna, Kranti, CS52, CS54, CS56	EC_e up to 9 dS/m	>1.6 t/ha with in EC_e 6–9 dS/m
Chickpea	Karnal Chana1, CSG 88101	EC_e up to 6 dS/m	Up to 2.1 t/ha under salt-free condition, >1.6 t/ha under salt-stress situation
Sugarcane	Co-453, Co1341	EC_e up to 9 dS/m	–

with PGPM, resulting in improved tolerance and resistance to abiotic stresses. PGPM inoculation caused upregulation of drought-related genes such as Early Response to Dehydration 15 and ABA-responsive genes, heat shock protein, dehydrin-like protein, and dehydration-responsive element–binding protein genes, etc. (Kaushal and Wani, 2016). In another study, *Triticum aestivum* exposed to salt stress (200 mM) was inoculated with eight halotolerant and halophilic bacteria. The inoculation resulted in a significant increase in the root-shoot lengths and total biomass of plants (Orhan, 2016). Being able to tolerate salt stress, these bacteria with PGP properties could be a very good alternative for salt-affected crops. Similarly, several rhizosphere-competent bacteria synthesize osmoprotectants that help the plant withstand or result in salt stress. *Bacillus amyloliquefaciens*, a plant-growth-promoting bacteria, was tested in the rice crop under hydroponic and soil conditions. The inoculation resulted in the upregulation of NADP-Me2, SOSI, EREBP, SERK1, and BADH genes. This further caused a surge of osmoprotectants utilizing the bacterial population. Similarly, a study was conducted to evaluate the efficacy of a PGPR strain upon inoculation in salt-affected *Avena sativa*. Physiological parameters like relative water content, shoot-root length, and plant dry weight were significantly higher, and biochemical parameters like antioxidant enzymes, proline content, and MDA were significantly lower in IG3 inoculated plants. Inoculation of PGPR alleviated the negative effects of salt stress as indicated by lowered stress response (Sapre et al., 2018).

11.3.3.2 Phytoremediation or bio-saline agriculture

This method employs green plants to remove excess salt water from soil through transpiration by trees using solar radiation. Some plants like eucalyptus, *Acacia arabica*, *Prosopis juliflora*, *Albizzia*, etc. transpire huge quantities of water compared to others and therefore helps to removal of salt containing water from soil. This tree-based bio-saline agriculture gives an option to prevent the development of waterlogged and saline soils, especially in areas where the conventional system of drainage is not possible. This tree-based remediation process only lowers the water table. It does not remove salts but minimizes their accumulation in the root zone, so that crop can be grown in micro-salt-free environment.

11.3.3.3 Breeding of salt-tolerant crop through conventional methods

In India, various salt-tolerant crops such as rice, wheat, chickpea, mustard, barley, sugar beet, and sugarcane had been released through conventional breeding approach (Singh and Sharma, 2006; Sharma, 2010). Various salt-tolerant aromatic rice varieties like CSR10, CSR13, CSR19, CSR27, CSR30, and CSR36 had been released which can produce grain yield >2–4 t/ha under salt-stress condition and up to 6 t/ha under salt-free condition, with salt tolerance (EC value) limit up to $EC_e < 11$ dS/m. A detailed list of salt-tolerant crops developed in India through breeding approach is given in Table 11.3.

11.3.3.4 Breeding of salt tolerance in crops through genetic engineering and Marker-Assisted Selection (MAS)

Traditional breeding procedures for developing abiotic-stress–resistant genotypes have been used for a long time to introduce useful traits into selected cultivars (Ashraf, 2010; Varshney et al., 2011). Salt-stress-tolerant rice and wheat cultivars have been generated by using the traditional breeding approach (Munns et al., 2006; Singh et al., 2010). Although resistance to salt is governed by multiple genes, such approaches have provided limited accomplishment, which explains the lack of availability of commercial salt-tolerant cultivars. Furthermore, the process of conventional plant breeding is lengthy and labor demanding, depends on the availability of well-characterized genetic resources, and can also involve linkage drag responsible for the introduction of unwanted characteristics together with the selected one. As a result, biotechnological approaches like genetic engineering and molecular breeding appear to be a viable option.

During the last three decades, incredible developments have been accomplished toward cloning and molecular characterization of various genes governing salt-stress tolerance. Such genes are characterized into two main categories: genes that act as regulators or effectors. The effector genes mainly play an important role in ion transport, osmolyte accumulation, ROS scavenging and maintenance of protein structure and function, etc., whereas regulatory genes play a vital role in

TABLE 11.4

Example of Genes Leading to Improvement of Salt-Stress Tolerance of Crop Plants through Genetic Engineering

Transgene	Donor	Transgenic Plant	References
AtNHX	*Arabidopsis*	Tomato, *B. napus*, wheat, cotton	Zhang and Blumwald (2001), Zhang et al. (2001), Xue et al. (2004), and He et al. (2005)
AtNHX+AVP1	*Arabidopsis*	Cotton, barley, cotton, tomato	Pasapula et al. (2011), Schilling et al. (2014), Shen et al. (2015), Gouiaa and Khoudi (2015)
Ncl	Soybean	Soybean	Do et al. (2016)
mt1D	*Escherichia coli*	Tobacco	Tarczynski et al. (1992)
codA	*E. coli*	Rice	Sakamoto et al. (1998)
P5CS	*Arabidopsis*	Tobacco	Kishor et al. (1995)
AtAPX	*Arabidopsis*	Tobacco	Badawi et al. (2004)
HVA7	Barley	Rice	Xu et al. (1996)
DREB1B/CBF1	*Arabidopsis*	Rice	Oh et al. (2005)
DREB1A/CBF3	*Arabidopsis*	Rice	Hu et al. (2006)
SNAC1	Rice	Rice	Hu et al. (2008)
SlAREB1	Tomato	Tomato	Orellana et al. (2010)
ZFP179	Rice	Rice	Sun et al. (2010)
OsCDPK21	Rice	Rice	Asano et al. (2011)
GhMPK2	Cotton	Tobacco	Zhang et al. (2011)

Source: Reprinted from Hanin et al. (2016).

the regulation of gene expression and signaling transduction. Such genes include various TFs and various enzyme-encoding genes like proteases, protein kinases, and phosphatases. Several studies have described that enhanced expression of such genes has improved resistance to different abiotic stress conditions in addition to salt stress (Ashraf, 2009; Türkan and Demiral, 2009; Cominelli et al., 2013; Roy et al., 2014; Fita et al., 2015; Table 11.4).

11.3.3.4.1 Incorporation of Ion Transporter Genes

The significance of ion transporters is characterized due to their vital function in the detoxification of higher salt concentrations and ion homeostasis and are, therefore, evident candidates of genes amidst effectors. The overexpression of *A. thaliana* vacuolar Na^+/H^+ antiporter (*AtNHX1*) improved salt-stress resistance in *A. thaliana* (Apse et al., 1999). The transgenic events were observed to display enhanced ability of sequestration of Na^+ into vacuoles, thereby avoiding its toxic buildup into the cell cytoplasm. The overexpression of *AtNHX1* and related NHX proteins in tomato, *Brassica napus*, wheat and cotton, and in other plant species has also been revealed to improve salt-stress tolerance (Zhang and Blumwald, 2001; Pardo et al., 2006; Munns and Tester, 2008). It was also observed that in the *AtNHX1* overexpressing plants in tomato can withstand higher salt stress, with the buildup of Na^+ in their leaves. Apart from this, the improved salt tolerance may be because of its capability to the ability of K^+ influx in the vacuolar membrane (Leidi et al., 2010). Overexpression of vacuolar cation/H^+ antiporter along with its H^+-PPase also shows improved tolerance to salt stress. This was evident from the overexpression of wheat *TNHX1* and *TVP1* genes, which resulted in enhanced tolerance to salinity stress resistance in *A. thaliana* (Brini et al., 2007a), *Nicotiana tabacum* (Gouiaa et al., 2012), and *Solanum lycopersicum* (Gouiaa and Khoudi, 2015). Overexpression of only *AtAVP1* displayed increased salt-stress resistance in barley (Schilling et al. 2014). Similarly, the HKT (high-affinity K^+ transporter) family and the SOS (Salt Overlay Sensitive) pathway proteins also play an important role in regulating Na^+ transport and have been utilized in transgenic strategies for improving tolerance to salt stress in various crops. Improved salt-stress tolerance utilizing HKTs has been observed to be doing well only when it is expressed in the stele tissues or its expression was under the control of tissue-specific and salt-inducible expression (Møller et al., 2009; Roy et al., 2014). Increased expression of the *Ncl* gene in a salt-sensitive soybean genotype, Kariyutaka, resulted in improved salinity resistance in the genetically modified soybean events.

11.3.3.4.2 Incorporation of Genes Involved in Osmolyte Accumulation

Plants exposed to salt stress survive by Na^+ exclusion from the cell cytoplasm, as well as by buildup of a variety of osmolytes (compatible solutes) to counter the vacuolar osmotic stress. Proline, sucrose, and glycine betaine are among the most thoroughly investigated compatible solutes synthesized in the instances of salt stress exposed in plants (Flowers et al., 1977). Accordingly, engineering crops with the increased buildup of these osmolytes is believed to be a solution for improving tolerance to salt stress in crop plants. The *mtlD* gene from bacteria, coding for MtlD, playing role in the synthesis of mannitol improves salinity tolerance upon its transfer in *N. tabacum* (Tarczynski et al., 1992, 1993). Likewise, improved tolerance to salt and cold stress was observed in *Oryza sativa* upon overexpression of *choline oxidase* (Sakamoto et al., 1998). In addition, transgenic tobacco lines overexpressing the *P5CS* gene showed increased proline quantities and enhanced resistance to salt and drought (Kishor et al., 1995). Likewise, in genetically modified rice lines overexpressing the moth bean *P5CS* gene improved resistance to salinity stress (Su and Wu, 2004).

11.3.3.4.3 Incorporation of Antioxidant Systems and Protective Proteins

Apart from its significance as signal transduction elements modulating intra and intercellular interactions to different abiotic stress factors (Apel and Hirt, 2004), ROS can also cause injury to plant cells when plants are exposed to salinity. For detoxification of ROS produced during salinity stress, plants synthesize a higher amount of antioxidative ROS-scavenging enzymes like CAT, APX, SOD, and GPX. On overexpression of ROS-scavenging systems in various crops, tolerance to diverse stresses like salt stress has been observed (Rodriguez-Rosales et al., 1997; Roxas et al., 1997; McKersie et al., 1999; Badawi et al., 2004; Miller et al., 2008). The increased expression of APX in *N. tabacum* chloroplasts improved resistance to salt, and drought stress has been observed (Badawi et al., 2004). Additionally, osmolites like LEA proteins and osmotin help in improving salinity tolerance by defending macromolecules from injuries salinity and drought. LEA (late embryogenesis abundant) protein, HVA7 gene, cloned from barley, after transforming into rice, gives tolerance to drought and salinity stress (Xu et al., 1996). Especially, two LEA proteins also called dehydrin proteins are mainly exciting and have imparted improved tolerance to various abiotic stress resistance (Hanin et al., 2011). Brini et al. (2007) have observed improved resistance to salinity and osmotic stress in transgenic *A. thaliana* expressing wheat dehydrin DHN-5.

11.3.3.4.4 Incorporation of Transcription Factors and Signaling Proteins

The reaction of plants to salt stress is a very intricate phenomenon that engages several genes engaged in different or overlapping regulatory mechanisms. Regulatory proteins like signaling proteins and TFs are gaining increasing attention because they have the ability to control the transcription of various downstream genes. TFs belonging to the families of NAC (NAM, ATAF, and CUC), DREB, MYB, Cys2/His2 zinc finger, MYC, bZIP, AP2/ERF (AP2/EREBP), and WRKY hold a crucial function in salinity resistance (Lata and Prasad, 2011; Lata et al., 2011; Turan et al., 2012; Golldack et al., 2011, 2014; Nuruzzaman et al., 2013). Transgenic *A. thaliana* and rice plants overexpressing CBF1/DREB1B or CBF3/DREB1A under constitutive promoter (35S) drastically improve drought, freezing, and high salt-stress resistance (Jaglo-Ottosen et al., 1998; Liu et al., 1998; Kasuga et al., 1999; Oh et al., 2005). Similarly, increased expression of stress-responsive NAC1 (SNAC1) and SNAC2 gives tolerance to salinity stress (Hu et al., 2006, 2008). Overexpression of SlAREB1 TF in *S. lycopersicum* imparts drought and salinity stress tolerance (Orellana et al., 2010). Similarly, rice lines overexpressing the ZFP179 showed increased tolerance to salt stress (Sun et al., 2010).

In addition, the overexpressing genes encoding signaling proteins like phosphatases and kinases for salinity tolerance have also been studied. The ectopic overexpression of *Gossypium hirsutum GhMPK2* enhances salt and water stress tolerance in *N. tabacum* (Zhang et al., 2011). CDPKs (cyclin-dependent protein kinases) play an important role in salinity stress response. Transgenic rice lines overexpressing *OsCDPK21* showed improved tolerance to salt stress (Asano et al., 2011). Apart from these, regulators of MAPKs (mitogen-activated protein kinases), the MAPK phosphatases, also play an important role in abiotic stress responses.

11.3.3.4.5 MAS for Salt-Stress Tolerance

The employment of biological techniques focusing on the utilization or creation of crop plants capable of withstanding extremely high salt concentrations is one method for creating salt-tolerant crops. This strategy incorporates (i) diversification of the cropping systems to incorporate crops which are having the ability to tolerant high salt concentration; (ii) exploration of wild species of crops that are acclimatized to saline conditions; and/or (iii) genetic modification of crops by breeding programs and selection to generate varieties with the improved level of salt tolerance. Breeding for genotypes that are tolerant to salinity and grow more efficiently than the traditional/usual cultivars at high salt concentration is a basic strategy and is believed to be economical also (Blum, 1988; Ashraf et al., 2008). The major drawbacks in the application of usual breeding programs for enhancing salinity resistance in crops are its sluggishness and the polygenic characteristics of salt resistance attributes. Plant breeders, therefore, use the strategy of coupling QTL analysis with MAS, is a strategy of identifying a marker/s linked to a quantitative/qualitative characteristic that shows polymorphism among the diverse parent genotypes (Ashraf and Foolad, 2013). Therefore, future research in salinity stress tolerance demands the development of efficient molecular markers. Genome sequences of crops like rice, sorghum, soybean, barley, and maize are available, and the recent progress made in NGS (next-generation sequencing) technologies can aid in developing high-resolution genetic maps for generating crops tolerant to salinity stress. In rice, 17 "Saltol" QTLs have been delimited and can be a potential QTL for salt resistance (Thomson et al., 2010). Furthermore, "Saltol" QTL also helps in the maintenance of Na^+/K^+ shoot ratio (Thomson et al., 2010). Nax1 QTL from wheat was mapped to the *TaHKT1* gene that helps in the removal of Na^+ from leaf sheath xylem thereby evading buildup in leaves (Huang et al., 2006). On chromosome 3, the QTL (*Ncl*) controlling salt tolerance in soybean has been identified (Lee et al., 2004; Hamwieh et al., 2011). Furthermore, utilizing MAS, the Ncl allele was successfully incorporated into soybean, resulting in a salinity-tolerant genotype (Do et al., 2016). Regardless of these examples, MAS-supported breeding strategies are still restricted due to the undesirable transfer of linked traits along with the QTLs. However, recent advances in NGS tools can radically contribute to discovery of molecular markers for obtaining high-density genetic maps, which is a requirement for accurate genomic location and more rapid cloning of new QTLs/genes governing salinity tolerance.

11.3.4 Chemical Approach

In chemical approach, various chemical substances were used to ameliorate excess salt present in soil or foliar spray of some selective chemicals to improve salt tolerance in plants.

Application of soil amendments: Various chemical substances like gypsum ($CaSO_4$, $2H_2O$) and acid-forming chemical substances like iron sulfate, aluminum sulfate, iron pyrites (FeS_2), etc. had been used to remove excess salt from soil. The application of gypsum in salt-affected alkali had been reported to be beneficial by several researchers (Bajwa and Josan, 1989; Gupta and Abrol, 1990; Minhas and Sharma, 2003; Khan et al., 2006; Sharma and Arora, 2008). Gypsum application increased rice grain yield by 22% when applied at 1 t/ha and the effect was further enlarged (55% more grain yield) when gypsum was applied at 2 t/ha (Khan et al., 2006). The application increased the yield of rice as high as 132% over control associated with cumulative decrease in EC_e by 66% over initial (Niazi et al., 2000). In mustard, gypsum application increased the mustard yield by 29.8% over control (Sharma and Arora, 2008).

Application of nanosubstances: Recent studies conducted on nanoparticles highlighted that nanoparticles used in the form of nanofertilizer or nanopesticide are highly successful in the domain of agricultural studies. It was observed that both seed priming and foliar spray with some nanosubstances may positively affect alleviating salt stress (Aazami et al., 2021; Wazid et al., 2020; Elshoky et al., 2021; Haripriya et al., 2018). Seed priming with nanoparticles allows them enter inside seed tissues through aquaporins or new pore space production and after entering inside seed tissue nanoparticles regulate various biochemical processes which directly or indirectly improve seed germination (Aazami et al., 2021). Application of various nanosubstances like nano-ZnO, nano-Fe_2O_3, nano-Ag, nano-Se, nano-TiO_2, nano-chitosan, etc. had shown to counter-balance salt-stress effect in crop plants. The application of nanoparticles elevated the production of anti-stress enzymes, influences certain protein synthesis under salt stress, and prevents cell damage by counterbalancing excess Na^+ ion toxicity in plant tissues and suppression of ROS substances. In wheat, foliar application of Ag nanoparticles (2–10 mM) results an increase in chlorophyll content in leaf tissues followed by overall photosynthesis, and regulated ABA production under salt-stress situation (Mohammed et al., 2017; Wazid et al., 2020). Elsewhere foliar spray with nano-ZnO and Fe_2O_3 in wheat had shown to be beneficial in terms of leaf Fe and Zn concentration, total leaf area, plant height and shoot dry weight under salt stress (Fathi et al., 2017). Similarly, Fathi et al. (2017) reported that foliar spray with nano-ZnO and Fe_2O_3 improved growth, photosynthesis, and chlorophyll content of maize under salt stress. In another study, it was documented that application nano-cerium oxide (CeO) improves maize yield under salt stress (Liu et al., 2022). In barley, soil application of nano-TiO_2 (500–2,000 mg/kg) results in better growth, physiological parameters (such as elevated level of SOD, CAT, and APX), and photosynthesis rate of barley in seedling stage under salt-stress condition, whereas in control (no soil application of TiO2) stunted growth of barley seedlings was recorded (Karami and Sepehri, 2018). Another study suggests that the application of selenium (Se) nanoparticles at seedling stage counters the negative effect of salt stress and increases the total phenolic content and MDA activity of treated barley seedlings (Habibi and Aleyasin, 2020). Elsewhere, foliar spray of nano-ZnO improved salt tolerance in field pea and finger millet respectively; that was the same for nano-chitosan application on French bean (Elshoky et al., 2021; Haripriya et al., 2018; Zayed et al., 2017).

11.4 Conclusion and Path Ahead

It is quite obvious that soil salinity is going to be a serious threat to crop production in future. Gradual land degradation and desertification under the context of climate change may further aggravate land salinization threat. Managing salt stress using agronomic management and chemical methods may not be sufficient in the upcoming days. The development of salt-tolerant crop varieties is the most effective and economically sound management option. More focus has to be given to the development of salt-tolerant crop varieties through the molecular crop breeding approach. Over the years with the advancement of biotechnology and molecular breeding approach, transfer of salt-tolerant character from one variety or other plant or microbes to desired crop species become quite easier than in previous decades. Further advancement of microbiology salt-tolerant PGPR may play an important role in future crop production. Although nanotechnology has already proven effective in alleviating salt stress in various plants, future research on large-scale field demonstration of nanotechnology is to be carried out in salt-affected areas globally. With these advancements, we may expect that all these approaches together may provide solutions for salt stress in future crop production.

Acknowledgment: The authors are thankful to ICAR-Directorate of Weed Research, Jabalpur, India and ICAR-Indian Institute of Water Management, Bhubaneswar, India for providing the necessary facilities.

REFERENCES

Aazami, Mohammad Ali, Farzad Rasouli, and Asghar Ebrahimzadeh. "Oxidative damage, antioxidant mechanism and gene expression in tomato responding to salinity stress under in vitro conditions and application of iron and zinc oxide nanoparticles on callus induction and plant regeneration." *BMC Plant Biology* 21, no. 1 (2021): 1–23.

Abdelgadir, E. M., M. Oka, and H. Fujiyama. "Characteristics of nitrate uptake by plants under salinity." *Journal of Plant Nutrition* 28, no. 1 (2005): 33–46.

Ahammed, Golam Jalal, Yang Li, Xin Li, Wen-Yan Han, and Shuangchen Chen. "Epigallocatechin-3-gallate alleviates salinity-retarded seed germination and oxidative stress in tomato." *Journal of Plant Growth Regulation* 37, no. 4 (2018): 1349–1356.

Albacete, Alfonso A., Cristina Martínez-Andújar, and Francisco Pérez-Alfocea. "Hormonal and metabolic regulation of source–sink relations under salinity and drought: from plant survival to crop yield stability." *Biotechnology Advances* 32, no. 1 (2014): 12–30.

Amirjani, M. R. "Effect of salinity stress on growth, mineral composition, proline content, antioxidant enzymes of soybean." American Journal of Plant Physiology 5, no. 6 (2010): 350–360.

Apel, Klaus, and Heribert Hirt. "Reactive oxygen species: metabolism, oxidative stress, and signaling transduction." Annual review of plant biology 55 (2004): 373–399.

Apse, Maris P., Gilad S. Aharon, Wayne A. Snedden, and Eduardo Blumwald. "Salt tolerance conferred by overexpression of a vacuolar Na+/H+ antiport in Arabidopsis." Science 285, no. 5431 (1999): 1256–1258.

Asano, Takayuki, Makoto Hakata, Hidemitsu Nakamura, Naohiro Aoki, Setsuko Komatsu, Hiroaki Ichikawa, Hirohiko Hirochika, and Ryu Ohsugi. "Functional characterisation of OsCPK21, a calcium-dependent protein kinase that confers salt tolerance in rice." Plant molecular biology 75, no. 1 (2011): 179–191.

Ashraf, M. "Biotechnological approach of improving plant salt tolerance using antioxidants as markers." Biotechnology advances 27, no. 1 (2009): 84–93.

Ashraf, M. "Inducing drought tolerance in plants: recent advances." Biotechnology advances 28, no. 1 (2010): 169–183.

Ashraf, Ml, H. R. Athar, P. J. C. Harris, and T. R. Kwon. "Some prospective strategies for improving crop salt tolerance." Advances in agronomy 97 (2008): 45–110.

Ashraf, Muhammad, and Majid R. Foolad. "Crop breeding for salt tolerance in the era of molecular markers and marker-assisted selection." Plant Breeding 132, no. 1 (2013): 10–20.

AYDIN, Adil, K. A. N. T. Canan, and Metin TURAN. "Hydrogel substrate alleviates salt stress with increase antioxidant enzymes activity of bean (Phaseolus vulgaris L.) under salinity stress." African Journal of Agricultural Research 6, no. 3 (2011): 715–724.

Badawi, Ghazi Hamid, Yasuo Yamauchi, Emi Shimada, Ryozo Sasaki, Naoyoshi Kawano, Kunisuke Tanaka, and Kiyoshi Tanaka. "Enhanced tolerance to salt stress and water deficit by overexpressing superoxide dismutase in tobacco (Nicotiana tabacum) chloroplasts." *Plant Science* 166, no. 4 (2004): 919–928.

Bagwasi, Gaesejwe, Gert A. Agenbag, and Pieter A. Swanepoel. "Effect of salinity on the germination of wheat and barley in South Africa." *Crop, Forage & Turfgrass Management* 6, no. 1 (2020): e20069.

Begara-Morales, Juan C., Beatriz Sánchez-Calvo, Mounira Chaki, Raquel Valderrama, Capilla Mata-Pérez, Javier López-Jaramillo, María N. Padilla, Alfonso Carreras, Francisco J. Corpas, and Juan B. Barroso. "Dual regulation of cytosolic ascorbate peroxidase (APX) by tyrosine nitration and S-nitrosylation." *Journal of Experimental Botany* 65, no. 2 (2014): 527–538.

Bose, Jayakumar, Ana Rodrigo-Moreno, and Sergey Shabala. "ROS homeostasis in halophytes in the context of salinity stress tolerance." *Journal of Experimental Botany* 65, no. 5 (2014): 1241–1257.

Bouallègue, Amal, Fatma Souissi, Issam Nouairi, Monia Souibgui, Zouhaier Abbes, and Haythem Mhadhbi. "Physiological and biochemicals changes modulated by seeds' priming of lentil (*Lens culinaris* L.) under salt stress at germination stage." *Acta Scientiarum Polonorum Hortorum Cultus* 18, no. 5 (2019): 27–38.

Bouaziz, Donia, Rania Jbir, Safa Charfeddine, Mohamed Najib Saidi, and Radhia Gargouri-Bouzid. "The StDREB1 transcription factor is involved in oxidative stress response and enhances tolerance to salt stress." *Plant Cell, Tissue and Organ Culture (PCTOC)* 121, no. 1 (2015): 237–248.

Brini, Faïçal, Moez Hanin, Imed Mezghani, Gerald A. Berkowitz, and Khaled Masmoudi. "Overexpression of wheat Na^+/H^+ antiporter TNHX1 and H^+-pyrophosphatase TVP1 improve salt-and drought-stress tolerance in Arabidopsis thaliana plants." *Journal of Experimental Botany* 58, no. 2 (2007): 301–308.

Brini, Marisa, and Ernesto Carafoli. "The plasma membrane Ca^{2+} ATPase and the plasma membrane sodium calcium exchanger cooperate in the regulation of cell calcium." *Cold Spring Harbor Perspectives in Biology* 3, no. 2 (2011): a004168.

Carpýcý, E. B., N. Celýk, and Gjajob Bayram. "Effects of salt stress on germination of some maize (*Zea mays* L.) cultivars." *African Journal of Biotechnology* 8, no. 19 (2009).

Castañares, José Luis, and Carlos Alberto Bouzo. "Effect of exogenous melatonin on seed germination and seedling growth in melon (*Cucumis melo* L.) under salt stress." *Horticultural Plant Journal* 5, no. 2 (2019): 79–87.

Chakraborty, Koushik, Sujit K. Bishi, Nisha Goswami, Amrit L. Singh, Debarati Bhaduri, and Pratap V. Zala. "Salinity-induced changes in seed germination and the expression profile of antioxidant enzymes in peanut as early and late responses in emerging radicles." *Acta Physiologiae Plantarum* 41, no. 8 (2019): 1–16.

Chauhan, Anju, Bassam A. AbuAmarah, Atul Kumar, J. S. Verma, Hamed A. Ghramh, Khalid Ali Khan, and Mohammad Javed Ansari. "Influence of gibberellic acid and different salt concentrations on germination percentage and physiological parameters of oat cultivars." *Saudi Journal of Biological Sciences* 26, no. 6 (2019): 1298–1304.

Chen, Xiaofei, Ruidong Zhang, Yifan Xing, Bing Jiang, Bang Li, Xiaoxue Xu, and Yufei Zhou. "The efficacy of different seed priming agents for promoting sorghum germination under salt stress." *PLoS One* 16, no. 1 (2021): e0245505.

Datta, K. K., Laxmi Tewari, and P. K. Joshi. "Impact of subsurface drainage on improvement of crop production and farm income in north-west India." *Irrigation and Drainage Systems* 18, no. 1 (2004): 43–56.

Drira, Marwa, Walid Saibi, Imen Amara, Khaled Masmoudi, Moez Hanin, and Faiçal Brini. "Wheat dehydrin K-segments ensure bacterial stress tolerance, antiaggregation and antimicrobial effects." *Applied Biochemistry and Biotechnology* 175, no. 7 (2015): 3310–3321.

Duan, Jiuju, Juan Li, Shirong Guo, and Yunyan Kang. "Exogenous spermidine affects polyamine metabolism in salinity-stressed Cucumis sativus roots and enhances short-term salinity tolerance." *Journal of Plant Physiology* 165, no. 15 (2008): 1620–1635.

El-Hendawy, Salah, Adel Elshafei, Nasser Al-Suhaibani, Majed Alotabi, Wael Hassan, Yaser Hassan Dewir, and Kamel Abdella. "Assessment of the salt tolerance of wheat genotypes during the germination stage based on germination ability parameters and associated SSR markers." *Journal of Plant Interactions* 14, no. 1 (2019): 151–163.

Elshoky, Hisham A., Ekaterina Yotsova, Mohamed A. Farghali, Khaled Y. Farroh, Kh El-Sayed, Heba Elsayed Elzorkany, George Rashkov et al. "Impact of foliar spray of zinc oxide nanoparticles on the photosynthesis of *Pisum sativum* L. under salt stress." *Plant Physiology and Biochemistry* 167 (2021): 607–618.

Etehadnia, Masoomeh, Doug R. Waterer, and Karen K. Tanino. "The method of ABA application affects salt stress responses in resistant and sensitive potato lines." *Journal of Plant Growth Regulation* 27, no. 4 (2008): 331–341.

Fan, Huaifu, Shirong Guo, Yansheng Jiao, Runhua Zhang, and Juan Li. "Effects of exogenous nitric oxide on growth, active oxygen species metabolism, and photosynthetic characteristics in cucumber seedlings under NaCl stress." *Frontiers of Agriculture in China* 1, no. 3 (2007): 308–314.

Fariduddin, Q., T. A. Khan, M. Yusuf, S. T. Aafaqee, and R. R. A. E. Khalil. "Ameliorative role of salicylic acid and spermidine in the presence of excess salt in Lycopersicon esculentum." *Photosynthetica* 56, no. 3 (2018): 750–762.

Farooq, Muhammad, Mubshar Hussain, Abdul Wakeel, and Kadambot HM Siddique. "Salt stress in maize: effects, resistance mechanisms, and management. A review." *Agronomy for Sustainable Development* 35, no. 2 (2015): 461–481.

Fathi, Alireza, Morteza Zahedi, and Shahram Torabian. "Effect of interaction between salinity and nanoparticles (Fe_2O_3 and ZnO) on physiological parameters of *Zea mays* L." *Journal of Plant Nutrition* 40, no. 19 (2017): 2745–2755.

Fathi, Alireza, Morteza Zahedi, Shahram Torabian, and Amirhossein Khoshgoftar. "Response of wheat genotypes to foliar spray of ZnO and Fe_2O_3 nanoparticles under salt stress." *Journal of Plant Nutrition* 40, no. 10 (2017): 1376–1385.

Feghhenabi, Faride, Hashem Hadi, Habib Khodaverdiloo, and Martinus Th Van Genuchten. "Seed priming alleviated salinity stress during germination and emergence of wheat (*Triticum aestivum* L.)." *Agricultural Water Management* 231 (2020): 106022.

Flowers, T. J., P. F. Troke, and A. R. Yeo. "The mechanism of salt tolerance in halophytes." *Annual Review of Plant Physiology* 28, no. 1 (1977): 89–121.

Gou, Tianyun, Xinhang Chen, Rong Han, Jiaqi Liu, Yongxing Zhu, and Haijun Gong. "Silicon can improve seed germination and ameliorate oxidative damage of bud seedlings in cucumber under salt stress." *Acta Physiologiae Plantarum* 42, no. 1 (2020): 1–11.

Guo, Ying-Hui, Yue-Ping Yu, Dong Wang, Chang-Ai Wu, Guo-Dong Yang, Jin-Guang Huang, and Cheng-Chao Zheng. "GhZFP1, a novel CCCH-type zinc finger protein from cotton, enhances salt stress tolerance and fungal disease resistance in transgenic tobacco by interacting with GZIRD21A and GZIPR5." *New Phytologist* 183, no. 1 (2009): 62–75.

Gupta, Bhaskar, and Bingru Huang. "Mechanism of salinity tolerance in plants: physiological, biochemical, and molecular characterization." *International Journal of Genomics* 2014 (2014).

Gupta, Raj K., and I. P. Abrol. "Salt-affected soils: Their reclamation and management for crop production." In *Advances in Soil Science*, pp. 223–288. Springer, New York, 1990.

Habibi, Ghader, and Yahya Aleyasin. "Green synthesis of Se nanoparticles and its effect on salt tolerance of barley plants." *International Journal of Nano Dimension* 11, no. 2 (2020): 145–157.

Hanin, Moez, Faïçal Brini, Chantal Ebel, Yosuke Toda, Shin Takeda, and Khaled Masmoudi. "Plant dehydrins and stress tolerance: versatile proteins for complex mechanisms." *Plant Signaling & Behavior* 6, no. 10 (2011): 1503–1509.

Haripriya, P., P. M. Stella, and S. Anusuya. "Foliar spray of zinc oxide nanoparticles improves salt tolerance in finger millet crops under glass-house condition." *SCIOL Biotechnology* 1 (2018): 20–29.

Hasegawa, Paul M. "Sodium (Na^+) homeostasis and salt tolerance of plants." *Environmental and Experimental Botany* 92 (2013): 19–31.

Hossain, M. Sazzad, Marcus Persicke, Abdelaleim Ismail ElSayed, Jörn Kalinowski, and Karl-Josef Dietz. "Metabolite profiling at the cellular and subcellular level reveals metabolites associated with salinity tolerance in sugar beet." *Journal of Experimental Botany* 68, no. 21–22 (2017): 5961–5976.

Hu, Xiaohui, Yi Zhang, Yu Shi, Zhi Zhang, Zhirong Zou, Hao Zhang, and Jiuzhou Zhao. "Effect of exogenous spermidine on polyamine content and metabolism in tomato exposed to salinity–alkalinity mixed stress." *Plant Physiology and Biochemistry* 57 (2012): 200–209.

James, Richard A., Carol Blake, Caitlin S. Byrt, and Rana Munns. "Major genes for Na^+ exclusion, Nax1 and Nax2 (wheat HKT1; 4 and HKT1; 5), decrease Na^+ accumulation in bread wheat leaves under saline and waterlogged conditions." *Journal of Experimental Botany* 62, no. 8 (2011): 2939–2947.

Jiang, Aiming, Lu Gan, Yi Tu, Hongxia Ma, Jinming Zhang, Zhaojian Song, Yuchi He, Detian Cai, and Xiaoqiao Xue. "The effect of genome duplication on seed germination and seedling growth of rice under salt stress." *Australian Journal of Crop Science* 7, no. 12 (2013): 1814–1821.

Karami, Afshin, and Ali Sepehri. "Nano titanium dioxide and nitric oxide alleviate salt induced changes in seedling growth, physiological and photosynthesis attributes of barley." *Zemdirbyste-Agriculture* 105, no. 2 (2018).

Khan, R., A. R. Gurmani, M. S. Khan, and A. H. Gurmani. "Effect of gypsum application on rice yield under wheat rice system." *International Journal of Agriculture and Biology* 8, no. 4 (2006): 536–538.

Khan, Waqas-ud-Din, Tariq Aziz, Ejaz Ahmad Waraich, and Muhammad Khalid. "Silicon application improves germination and vegetative growth in maize grown under salt stress." *Pakistan Journal of Agricultural Sciences* 52, no. 4 (2015).

Khan, M. A. H., MA Baset Mia, J. U. Ahmed, M. A. Karim, and M. M. H. Saikat. "Morpho-anatomical appraisal of some pulse crops under salinity stress." *Bangladesh Journal of Agricultural Research* 44, no. 3 (2019): 439–451.

Kilic, Semra, and Ayten Kahraman. "The mitigation effects of exogenous hydrogen peroxide when alleviating seed germination and seedling growth inhibition on salinity-induced stress in barley." *Polish Journal of Environmental Studies* 25, no. 3 (2016).

Kim, June-Sik, Junya Mizoi, Takuya Yoshida, Yasunari Fujita, Jun Nakajima, Teppei Ohori, Daisuke Todaka et al. "An ABRE promoter sequence is involved in osmotic stress-responsive expression of the DREB2A gene, which encodes a transcription factor regulating drought-inducible genes in Arabidopsis." *Plant and Cell Physiology* 52, no. 12 (2011): 2136–2146.

Kishor, P. B. Kavi, Zonglie Hong, Guo-Hua Miao, Chein-An A. Hu, and Desh Pal S. Verma. "Overexpression of [delta]-pyrroline-5-carboxylate synthetase increases proline production and confers osmotolerance in transgenic plants." *Plant Physiology* 108, no. 4 (1995): 1387–1394.

Kopyra, Małgorzata, and Edward A. Gwóźdź. "Nitric oxide stimulates seed germination and counteracts the inhibitory effect of heavy metals and salinity on root growth of Lupinus luteus." *Plant Physiology and Biochemistry* 41, no. 11–12 (2003): 1011–1017.

Liu, Ji-Hong, Wei Wang, Hao Wu, Xiaoqing Gong, and Takaya Moriguchi. "Polyamines function in stress tolerance: from synthesis to regulation." *Frontiers in Plant Science* 6 (2015a): 827.

Liu, Ming, Meijie Chu, Yanfeng Ding, Shaohua Wang, Zhenghui Liu, She Tang, Chengqiang Ding, and Ganghua Li. "Exogenous spermidine alleviates oxidative damage and reduce yield loss in rice submerged at tillering stage." *Frontiers in Plant Science* 6 (2015b): 919.

Liu, Li, Weili Xia, Haixia Li, Hanlai Zeng, Benhui Wei, Suoyi Han, and Changxi Yin. "Salinity inhibits rice seed germination by reducing α-amylase activity via decreased bioactive gibberellin content." *Frontiers in Plant Science* 9 (2018): 275.

Liu, Yinglin, Xuesong Cao, Le Yue, Chanxi Wang, Mengna Tao, Zhenyu Wang, and Baoshan Xing. "Foliar-applied cerium oxide nanomaterials improve maize yield under salinity stress: Reactive oxygen species homeostasis and rhizobacteria regulation." *Environmental Pollution* (2022): 118900.

Manai, Jamel, Houda Gouia, and Francisco J. Corpas. "Redox and nitric oxide homeostasis are affected in tomato (Solanum lycopersicum) roots under salinity-induced oxidative stress." *Journal of Plant Physiology* 171, no. 12 (2014): 1028–1035.

Mandal, Chiranjib, Nirmalya Ghosh, Malay Kumar Adak, and Narottam Dey. "Interaction of polyamine on oxidative stress induced by exogenously applied hydrogen peroxide in Salvinia natans Linn." *Theoretical and Experimental Plant Physiology* 25 (2013): 223–230.

McKersie, Bryan D., Stephen R. Bowley, and Kim S. Jones. "Winter survival of transgenic alfalfa overexpressing superoxide dismutase." *Plant Physiology* 119, no. 3 (1999): 839–848.

Miller, Gad, Vladimir Shulaev, and Ron Mittler. "Reactive oxygen signaling and abiotic stress." *Physiologia Plantarum* 133, no. 3 (2008): 481–489.

Minhas, P. S., and O. P. Sharma. "Management of soil salinity and alkalinity problems in India." *Journal of Crop Production* 7, no. 1–2 (2003): 181–230.

Mittler, Ron, Sandy Vanderauwera, Martin Gollery, and Frank Van Breusegem. "Reactive oxygen gene network of plants." *Trends in Plant Science* 9, no. 10 (2004): 490–498.

Mittler, Ron. "ROS are good." *Trends in Plant Science* 22, no. 1 (2017): 11–19.

Mohamed, Abdel Kareem S.H,. Muhammad Farooq Qayyum, Ahmed Mustafa Abdel-Hadi, Rabia Abdur Rehman, Shafaqat Ali, and Muhammad Rizwan. "Interactive effect of salinity and silver nanoparticles on photosynthetic and biochemical parameters of wheat." *Archives of Agronomy and Soil Science* 63, no. 12 (2017): 1736–1747.

Mostofa, Mohammad Golam, Masayuki Fujita, and Lam-Son Phan Tran. "Nitric oxide mediates hydrogen peroxide-and salicylic acid-induced salt tolerance in rice (*Oryza sativa* L.) seedlings." *Plant Growth Regulation* 77, no. 3 (2015): 265–277.

Németh, Mónika, Tibor Janda, Eszter Horváth, Emil Páldi, and Gabriella Szalai. "Exogenous salicylic acid increases polyamine content but may decrease drought tolerance in maize." *Plant Science* 162, no. 4 (2002): 569–574.

Niazi, B. H., N. Hussain, M. Salim, G. Hussain, and A. Ali. "Effect of split application of gypsum on wheat yield and properties of a saline-sodic soil irrigated with brackish water." *International Journal of Agriculture and Biology* 3 (2000): 234–236.

Niu, Xiaomu, Ray A. Bressan, Paul M. Hasegawa, and José M. Pardo. "Ion homeostasis in NaCl stress environments." *Plant Physiology* 109, no. 3 (1995): 735.

Oliveira, Halley C., Bruna C. R. Gomes, Milena T. Pelegrino, and Amedea B. Seabra. "Nitric oxide-releasing chitosan nanoparticles alleviate the effects of salt stress in maize plants." *Nitric Oxide* 61 (2016): 10–19.

Rajabi Dehnavi, Ahmad, Morteza Zahedi, Agnieszka Ludwiczak, Stefany Cardenas Perez, and Agnieszka Piernik. "Effect of salinity on seed germination and seedling development of sorghum (*Sorghum bicolor* (L.) Moench) genotypes." *Agronomy* 10, no. 6 (2020): 859.

Raju, R., K. Thimmappa, Parveen Kumar, Satyendra Kumar, and R.S. Tripathi. "Reclamation of saline soils through subsurface drainage technology in Haryana–an economic impact analysis." (2016).

Roxas, Virginia P., Roger K. Smith, Eric R. Allen, and Randy D. Allen. "Overexpression of glutathione S-transferase/glutathioneperoxidase enhances the growth of transgenic tobacco seedlings during stress." *Nature Biotechnology* 15, no. 10 (1997): 988–991.

Sabagh, Ayman EL, Mohammad Sohidul Islam, Milan Skalicky, M. Ali Raza, Kulvir Singh, M. Anwar Hossain, Akbar Hossain et al. "Salinity stress in wheat (*Triticum aestivum* L.) in the changing climate: adaptation and management strategies." *Frontiers in Agronomy* 3(2021): 661932.

Sakamoto, Atsushi, and Alia Norio Murata. "Metabolic engineering of rice leading to biosynthesis of glycinebetaine and tolerance to salt and cold." *Plant Molecular Biology* 38, no. 6 (1998): 1011–1019.

Sang, Q. Q., S. Shu, X. Shan, S. R. Guo, and J. Sun. "Effects of exogenous spermidine on antioxidant system of tomato seedlings exposed to high temperature stress." *Russian Journal of Plant Physiology* 63, no. 5 (2016): 645–655.

Sehrawat, Nirmala, Mukesh Yadav, Kangila V. Bhat, Raj K. Sairam, and Pawan K. Jaiwal. "Effect of salinity stress on mungbean [*Vigna radiata* (L.) Wilczek] during consecutive summer and spring seasons." *Journal of Agricultural Sciences, Belgrade* 60, no. 1 (2015): 23–32.

Serrano, Ramón, Jose M. Mulet, Gabino Rios, Jose A. Marquez, Iñigo F. De Larrinoa, Martin P. Leube, Iratxe Mendizabal et al. "A glimpse of the mechanisms of ion homeostasis during salt stress." *Journal of Experimental Botany* (1999): 1023–1036.

Shaaban, M., M. Abid, and R. A. I. Abou-Shanab. "Amelioration of salt affected soils in rice paddy system by application of organic and inorganic amendments." *Plant, Soil and Environment* 59, no. 5 (2013): 227–233.

Sharma, S. K. "Salt tolerant crop varieties developed at CSSRI Karnal, India: A success story." *Cereal Research Communications* 14 (2010): 474–483.

Sharma, K. R., and Sanjay Arora. "Effectiveness of gypsum and pyrite for sulfur fertilization on yield and quality of Indian mustard under rainfed conditions of north India." *Communications in Soil Science and Plant Analysis* 39, no. 15–16 (2008): 2431–2439.

Sharma, Pushp, Virender Sardana, and S. S. Banga. "Salt tolerance of Indian mustard (Brassica juncea) at germination and early seedling growth." *Environmental and Experimental Biology* 11 (2013): 39–46.

Siddiqui, Manzer H., Mohamed H. Al-Whaibi, and Mohammed O. Basalah. "Role of nitric oxide in tolerance of plants to abiotic stress." *Protoplasma* 248, no. 3 (2011): 447–455.

Su, Jin, and Ray Wu. "Stress-inducible synthesis of proline in transgenic rice confers faster growth under stress conditions than that with constitutive synthesis." *Plant Science* 166, no. 4 (2004): 941–948.

Sun, Xinbo, Xin Li, Junfei Zhu, Ning Huang, Xiuju Bian, Huibin Li, Lihong Wang, and Liebao Han. "Polyamines and ethylene metabolism during cold acclimation in zoysiagrass (Zoysia Japonica Steud.)." *Acta Physiologiae Plantarum* 42, no. 8 (2020): 1–10.

Suzuki, Nobuhiro, Amith R. Devireddy, Madhuri A. Inupakutika, Aaron Baxter, Gad Miller, Luhua Song, Elena Shulaev, Rajeev K. Azad, Vladimir Shulaev, and Ron Mittler. "Ultra-fast alterations in mRNA levels uncover multiple players in light stress acclimation in plants." *The Plant Journal* 84, no. 4 (2015): 760–772.

Tang, Ruimin, Suyan Niu, Guodong Zhang, Guanshui Chen, Muhammad Haroon, Qing Yang, Om P. Rajora, and Xiu-Qing Li. "Physiological and growth responses of potato cultivars to heat stress." *Botany* 96, no. 12 (2018): 897–912.

Tarczynski, Mitchell C., Richard G. Jensen, and Hans J. Bohnert. "Expression of a bacterial mtlD gene in transgenic tobacco leads to production and accumulation of mannitol." *Proceedings of the National Academy of Sciences* 89, no. 7 (1992): 2600–2604.

Tarczynski, Mitchell C., and William H. Outlaw Jr. "The interactive effects of pH, L-malate, and glucose-6-phosphate on guard-cell phosphoenolpyruvate carboxylase." *Plant Physiology* 103, no. 4 (1993): 1189–1194.

Tiburcio, Antonio F., Teresa Altabella, Marta Bitrián, and Rubén Alcázar. "The roles of polyamines during the lifespan of plants: From development to stress." *Planta* 240, no. 1 (2014): 1–18.

Tsegay, Berhanu Abraha, and Melkamu Andargie. "Seed priming with gibberellic acid (GA3) alleviates salinity induced inhibition of germination and seedling growth of *Zea mays* L., *Pisum sativum* var. abyssinicum A. Braun and *Lathyrus sativus* L." *Journal of Crop Science and Biotechnology* 21, no. 3 (2018): 261–267.

Uchida, Akio, Andre T. Jagendorf, Takashi Hibino, Teruhiro Takabe, and Tetsuko Takabe. "Effects of hydrogen peroxide and nitric oxide on both salt and heat stress tolerance in rice." *Plant Science* 163, no. 3 (2002): 515–523.

Vibhuti, Charu Shahi, Kiran Bargali, and S.S. Bargali. "Seed germination and seedling growth parameters of rice (*Oryza sativa* L.) varieties as affected by salt and water stress." *Indian Journal of Agricultural Sciences* 85, no. 1 (2015): 102–108.

Wahid, Iram, Sarika Kumari, Rafiq Ahmad, Sofi J. Hussain, Saud Alamri, Manzer H. Siddiqui, and M. Iqbal R. Khan. "Silver nanoparticle regulates salt tolerance in wheat through changes in ABA concentration, ion homeostasis, and defense systems." *Biomolecules* 10, no. 11 (2020): 1506.

Wajid, Mahboob, Muhammad Athar Khan, Muhammad Ubaidullah Shirazi, and Faisal Summiya. "Seed priming modulates germination potential, osmoprotectants accumulation and ionic uptake in wheat seedlings under salt stress." *International Journal of Agriculture and Biology* 22, no. 3 (2019): 594–600.

Wang, Jian-Ying, Jian Wang, Vera A. Golovina, Li Li, Oleksandr Platoshyn, and Jason Xiao-Jian Yuan. "Role of K+ channel expression in polyamine-dependent intestinal epithelial cell migration." *American Journal of Physiology-Cell Physiology* 278, no. 2 (2000): C303–C314.

Wang, Feng, Pei Liu, Tingting Nie, Huixian Wei, and Zhenggang Cui. "Characterization of a polyamine microsphere and its adsorption for protein." *International Journal of Molecular Sciences* 14, no. 1 (2013): 17–29.

Xu, Deping, Xiaolan Duan, Baiyang Wang, Bimei Hong, Tuan-Hua David Ho, and Ray Wu. "Expression of a late embryogenesis abundant protein gene, HVA1, from barley confers tolerance to water deficit and salt stress in transgenic rice." *Plant Physiology* 110, no. 1 (1996): 249–257.

Yaduvanshi, N. P. S., and D. R. Sharma. "Utilization of organics, amendment and fertilizers with sodic water irrigation: Long-term effect on soil properties and rice-wheat productivity." *Journal of the Indian Society of Soil Science* 64, no. 3 (2016): 255–260.

Yang, Huqing, Fenghua Wu, and Jiyu Cheng. "Reduced chilling injury in cucumber by nitric oxide and the antioxidant response." *Food Chemistry* 127, no. 3 (2011): 1237–1242.

Zayed, M. M., S. H. Elkafafi, Amina M. G. Zedan, and Sherifa F. M. Dawoud. "Effect of nano chitosan on growth, physiological and biochemical parameters of Phaseolus vulgaris under salt stress." *Journal of Plant Production* 8, no. 5 (2017): 577–585.

Zhang, Yanyan, Liling Wang, Youliang Liu, Qun Zhang, Qiuping Wei, and Wenhua Zhang. "Nitric oxide enhances salt tolerance in maize seedlings through increasing activities of proton-pump and Na+/H+ antiport in the tonoplast." *Planta* 224, no. 3 (2006): 545–555.

Zhang, Y., Q. Yao, Y. Shi, X. Li, L. Hou, G. Xing, and G. J. Ahammed. Elevated CO_2 improves antioxidant capacity, ion homeostasis, and polyamine metabolism in tomato seedlings under Ca (NO_3) 2-induced salt stress. *Scientia Horticulturae* 273 (2020): 109644.

Zhou, Xiaoting, Hailiang Zhao, Kai Cao, Lipan Hu, Tianhao Du, František Baluška, and Zhirong Zou. "Beneficial roles of melatonin on redox regulation of photosynthetic electron transport and synthesis of D1 protein in tomato seedlings under salt stress." *Frontiers in Plant Science* 7 (2016): 1823.

Zhu, Jian-Kang. "Regulation of ion homeostasis under salt stress." *Current Opinion in Plant Biology* 6, no. 5 (2003): 441–445.

Zhu, Jian-Kang. "Abiotic stress signaling and responses in plants." *Cell* 167, no. 2 (2016): 313–324.

12

Compatible Solutes Engineering to Balance Salt (Na^+) and ROS-Induced Changes in Potassium Homeostasis

Dipankar Barman, Lekshmy Sathee, Birendra K. Padhan, and Archana Watts
ICAR-Indian Agricultural Research Institute

CONTENTS

12.1 Introduction

Salinity stress imposes severe damaging effects on the growth and development of crop plants. In an attempt to combat salinity and to complete their lifecycle effectively, plants employ different adaptive strategies such as sodium (Na^+) ion exclusion and potassium (K^+) homeostasis, compatible solute accumulation in the cytosol and upregulation of reactive oxygen species (ROS) scavenging systems (Shabala & Cuin, 2008). A higher than the optimum concentration of salt (NaCl) in soil reduces the plant's ability to take up water and hinders cellular metabolism by accumulating toxic chloride (Cl^-) and Na^+ ions. Sodium ions are particularly detrimental to the uptake of potassium (K^+) ions, plant metabolism, enzyme activities and photosynthesis. Conversely, salt accumulation impairs the homeostasis of other essential nutrients ions like calcium (Ca^{2+}), magnesium (Mg^{2+}) and nitrate (NO_3^-) (Lekshmy et al., 2013).

In a salinity-stressed plant, the restriction of cytosolic Na^+ influx, active Na^+ efflux and compartmentalisation of Na^+ in the vacuole effectively reduce toxic effects. In *Arabidopsis,* salt overly sensitive (SOS) signalling and the activation of component protein Na^+/H^+ antiporter (SOS1) minimises the impact of salinity. *Arabidopsis thaliana* mutants defective in SOS1 impart salt sensitivity and defects in the long-distance movement of Na^+ (Qiu et al., 2002). The cytosolic K^+/Na^+ ratio falls dramatically under saline conditions. NaCl-induced membrane depolarisation leads to the activation of outward-rectifying K^+ channels (Zhu, 2002, 2003). In salinity-stressed *Arabidopsis*, *Eucalyptus* and wheat plants, the high-affinity K^+ transporter (*HKT1*) functions as a low-affinity Na^+ transporter (Chinnusamy et al., 2005).

Salt stress induces the accumulation of ROS, which causes oxidative damage to macromolecules, including lipids, proteins and DNA. Non-enzymatic antioxidants like ascorbate, glutathione, α-tocopherol and carotenoids and enzymatic antioxidants such as superoxide dismutase (SOD), catalase (CAT), ascorbate peroxidase (APX) and glutathione reductase (GR) combat ROS and maintain cellular integrity (Affenzeller et al., 2009; Mittova et al., 2003; Sairam & Srivastava, 2002). The cross talk of ion exclusion module component SOS1 with ROS scavenging through interaction with radical-induced cell death 1 is reported in *Arabidopsis* (Katiyar-Agarwal et al., 2006; Shapiguzov et al., 2019). ROS also transduce stress signals and regulates plasma membrane Ca^{2+} and K^+ permeable channels in roots.

Accumulation of organic compounds that do not interfere with cellular enzymes and functions is an important mechanism attributed to plant survival under salinity. Accumulation of osmolytes in the cytoplasm for osmotic adjustment is a vital salt tolerance mechanism found in halophytes (Flowers, Munns, & Colmer, 2015). Glycine betaine (GB), proline, polyols/sugar alcohols and quaternary amines function as compatible solutes (Munns et al., 2020; Munns & Tester, 2008).

DOI: 10.1201/9781003258063-12

In addition to their conventional roles in plant metabolism, these solutes play an essential role in osmotic adjustment and maintaining ionic balance, reducing salt influx or modulating its exclusion (Shabala & Cuin, 2008). Accumulation of these compatible solutes under salinity conditions resulted in a gradual net increase in cellular solute concentration. The accumulation of ions is a strategy for osmotic adjustment in vacuoles and not in cytosol. As the vacuolar build-up of ions progresses, cytosol creates a counterbalance by accumulating non-ionic solutes to sustain the cellular water potential equilibrium. Metabolic profiling in *Arabidopsis*, rice and *Lotus* suggested a change in balance among amino acids and organic acids in response to salt treatment (Acosta-Motos et al., 2017; Sanchez et al., 2008). In halophytic plants, proline or GB was found at sufficiently high leaf concentrations to contribute to the osmotic pressure in the cell (Leach, et al., 1990). In glycophytes, though compatible solute accumulation is comparatively low, if deployed ultimately to the cytoplasm, it can exert an osmotic effect sufficient for survival. In Plantago, an accumulation of sorbitol and sucrose reduced the rate of phloem sap exudation after NaCl treatment (Kartashov et al., 2008; Koyro, 2006).

Trehalose imparted salt tolerance by maintaining sodium and potassium balance, increasing K level in shoots and reducing the Na accumulation (Garg et al., 2002). Trehalose is also speculated to involve in scavenging ROS and conferring protection to the machinery of protein synthesis (Zhu, 2002). GB plays a crucial role in facilitating osmotic adjustment and safeguarding subcellular structures in stressed plants, as well as the transcriptional and translational machinery by acting as a molecular chaperone in enzyme refolding during salinity stress (Ashraf, 2007). Compatible solutes including mannitol, fructans, trehalose, sorbitol, mannitol and proline help reduce oxidative stress produced by salt (Chang et al., 2014; Zhang et al., 2019). Melatonin in plants is considered a growth regulator and enhances antioxidant capacity and confers stress tolerance (Zhao et al., 2018). Functions of melatonin have been studied in various crops. It is found that exogenous melatonin application enhances the antioxidant enzymes and improves the activities of SOD, POD and CAT in cucumber under salt stress. Exogenous addition of salicylic acid (SA) has been shown in numerous investigations to alleviate toxicity symptoms caused by salt stress in various plant species (Gunes et al., 2007). SA is a signalling molecule that has been linked to salinity defence responses. However, little is known about SA signalling and salt tolerance mechanisms in plants. The accumulation of sorbitol in the leaves of control plants was augmented by salt stress induced by increasing NaCl levels, and the sorbitol concentration after SA pretreatment remained elevated during salt stress (Jayakannan et al., 2013, 2015; Smirnoff and Cumbes, 1989; Sharma et al., 2017). Polyamines are well-known growth regulators of plants and are involved in developing stress responses. Polyamines are always regarded as ROS scavengers and activators of the antioxidant system under stress conditions. Polyamine-induced plant salt tolerance is closely related to their regulation of intracellular ion homeostasis (Shabala & Pottosin, 2014) and ROS scavenging.

12.2 Compatible Solutes in Relation to K^+ Balance and ROS Homeostasis under Salinity

Maintenance of optimum K^+/Na^+ ratio in the cytosol is considered an indicator of salinity tolerance. Potassium is an essential primary macro nutrient that plays a crucial role in ionic homeostasis in plant cells, and K^+ depletion and imbalance significantly affect plant growth and development. While Na^+ is a beneficial nutrient that can undertake some function of K^+ due to their chemical similarities, at higher concentrations severely affects K^+ homeostasis under salt stress (Adams & Shin, 2014). Salt stress increases excessive Na^+ accumulation, which competes with K^+ and inhibits its uptake resulting in cytoplasmic K^+ deficiency. The excessive Na^+ accumulated in cytosol due to salt stress is sequestrated into the vacuole, which is achieved by increased K^+ concentration to a subtoxic level and the accumulation of proline and other osmolytes, creating increased osmotic pressure in the subcellular components (Isayenkov & Maathuis, 2019). Under stress, proline acts as a potent ROS scavenger and directly or indirectly maintains a desired ROS homeostasis (Hayat et al., 2012). K^+/Na^+ ratio is maintained by signalling molecules like Ca^{2+} and H_2O_2, which activate different K^+ and Na^+ transporters. Increasing Ca^{2+} accumulation inhibits outward-rectifying K^+ channels (KORs), and non-selective cation channels, the main contributor to Na^+ influx, are also involved in K^+ efflux, thus maintaining a higher K^+/Na^+ ratio under salt stress (Shabala & Cuin, 2008). At the same time, H_2O_2 reduces the Na^+/K^+ balance and further induces salt resistance (Ma et al., 2012; Xie, Wan, Li, & Zhang, 2014). Another study observed that ROS activated KOR, thus directly responsible for salt-induced K^+ loss in the cell leading to programmed cell division (PCD) (Demidchik, 2018). These results indicate that an adequate level of ROS is required for better salt tolerance. In *Arabidopsis*, *AtrbohD* and *AtrbohF*, the NADPH oxidases, are involved in ROS production and increase the K^+/Na^+ ratio by upregulating inward K^+ current in *Arabidopsis* (Ma et al., 2012).

12.2.1 Proline

Proline is a multifunctional proteogenic amino acid known for its diverse role in plant growth, development and overall stress tolerance. Its physiochemical properties like neutral pH, zwitterionic nature and high solubility of 6.5 mol/L water (at 25°C) make it a perfect candidate as a potent compatible solute in plant system (Hare & Cress, 1998). Apart from its major role as an osmoprotectant, it helps in membrane and protein stability by acting as a protein-compatible hydrotrope by redox homeostasis, free radical scavenging under different environmental stresses and maintaining optimum $NADP^+/NADPH$ ratio (Hare & Cress, 1997). The biosynthesis and accumulation of proline are induced under drought stress, temperature stress, salinity and heavy metal toxicity etc. It is determined by its biosynthesis, catabolism and transport (Hare & Cress, 1999).

Under salinity stress, proline gets accumulated in the cytosol and regulates cytoplasmic osmotic balance (Ketchum, Warren, Klima, Lopez-Gutiérrez, & Nabors, 1991; Leigh, Ahmad, &

Jones, 1981). Proline accumulation under salt stress may involve induction or/and activation of proline biosynthesis genes, relaxation of proline feedback mechanism and downregulation of proline catabolic genes. Proline is synthesised in the plant by the most dominant glutamate pathway and less active ornithine pathway under osmotic stress (Surender Reddy et al., 2015). Proline biosynthesis through the glutamate pathway is regulated by the rate-limiting enzyme Δ^1-pyrroline-5-carboxylate synthase (P5CS) (Silva-Ortega et al., 2008). In *Arabidopsis,* ABA1, ABI1 and AXR2 regulate the expression of *AtP5CS* under osmotic stress. During recovery from water/osmotic stress, the transcript abundance of *P5CS* is reduced while that of proline dehydrogenase (*PDH*) is increased (Kovács et al., 2019). Similarly, conformational change in P5CS protein by replacing Phe with Ala at position 129 eliminates feedback inhibition with a higher proportion of proline synthesis. Increased proline content is also related to fewer free radicals and more osmotic stress tolerance. Activation of R2R3-MYB protein gene GmMYB68 confers salt tolerance by regulating the P5CR gene along with other osmolyte and other stress relative MYB transcription factors (He et al., 2020). P5CS predominates the role of proline biosynthesis over P5CR protein in the glutamate pathway (Dudziak et al., 2019). Some other studies show that proline biosynthesis is also regulated by TsPIP1;1 aquaporin and an inorganic pyrophosphorylase gene, *ThPIP1* associated with upregulation of vacuolar cation/proton exchanger 1 (*CAX1*), cation/H^+ antiporter and L-type LecRK, genes in rice conferring salt and alkali stress, respectively, but their exact mechanism in proline biosynthesis is not well known (He et al., 2017). In maize, *WRKY114,* which has a DNA-binding activity, negatively regulates *P5CS* and *TPP1*in rice, thus showing a negative role in salt stress tolerance (Bo et al., 2020). In medicago, the *MtP5CS3* gene has a direct role in proline build-up during salinity stress (Kim & Nam, 2013). Overall, proline biosynthesis under salt stress is mainly regulated by *P5CS1* gene activation, while *P5CS2* is only weekly induced. Proline biosynthesis is also influenced by light-regulated transcription factor *ELONGATED HYPOCOTYL 5* (*HY5*) and inhibited by brassinosteroid (Kovács et al., 2019). Proline is catabolised in mitochondria by PDH and P5C dehydrogenase (P5CDH), which convert proline to glutamate (Verslues & Sharma, 2010); PDH is a rate-limiting enzyme among the two (Liang, Zhang, Natarajan, & Becker, 2013). Mutation of the *OsMPT3* gene is linked to lower proline accumulation by altering the transcript level of *ProDH* and *P5CS* mRNA under salt stress (Huang et al., 2020). NaCl stress increases H_2O_2 level, which is a signalling molecule to upregulate proline synthesis and downregulate degradation pathway, thus accumulating more proline under salt stress (Liu, Huang, Lin, & Sun, 2020). Finding demonstrates that maintaining a balance between proline synthesis and degradation and an efficient way of achieving adequate redox balance under stressful situations is critical for plant growth and development (Miller et al., 2009).

A proper ROS homeostasis and proline accumulation is maintained by siRNA under salt stress which is regulated by the *P5CDH-SRO5* gene pair. Salt stress induces ROS production, which in turn causes *SRO5* expression. This causes nat-siRNA processing and downregulating the P5CDH gene related to the proline catalytic pathway by forming the SRO5-P5CDH dsRNA complex and initiating the DCL2-dependent formation of 24-nt nat-siRNA, ultimately degrading *P5CDH* mRNA. At the same time, SRO5 protein inhibits P5C-mediated excessive ROS accumulation, which is toxic to the cellular mechanisms, thus fine-tuning the regulatory loop of ROS-proline accumulation and stress response under salt stress (Borsani et al., 2005). Furthermore, proline induces transcription of salt-stress–responsive genes with proline-responsive elements in the promoter region, which opens a broad way to find out proline-responsive genes under different physiological conditions (Satoh, et al., 2002). Proline homeostasis is also essential for maintaining cell division in the context of abiotic stress. Ethylene plays an essential role in K^+/Na^+ homeostasis through RBOHF-dependent ROS accumulation and reduced root Na influx and HAK5-mediated K uptake in *Arabidopsis* under salt stress which might be in turn regulated by ROS-induced ABA signalling and proline accumulation (Liang et al., 2013; Wei et al., 2017). Na^+/H^+ exchanger (*AtNHX1*) enhances salt tolerance by vascular sequestration of K^+ and increases intercellular K^+ retention along with the accumulation of proline and sugar in the cytosol (Assaha et al., 2017; Leidi et al., 2010).

12.2.2 Trehalose

Trehalose (-D-glucopyranosyl-1,1-D-glucopyranoside) is a non-reducing disaccharide and functions as an energy source and as a compatible solute (Tsai & Gazzarrini, 2014). Earlier trehalose was thought to be restricted to resurrection plants, but new findings show it is distributed across plant species, including the model plant *Arabidopsis* and crop plants. The trehalose concentration ranges from ~10 µg to 10 mg/g FW in different plant species (Chen & Murata, 2002). Some basic properties like glycosidic bond formation, water replacement and glass formation mechanism make it a stable solute under diverse conditions and appropriate membrane and molecule stabiliser (Hottiger et al., 1987). Trehalose plays an essential role in alleviating abiotic stress tolerance of many organisms through its ability to scavenge ROS and acts as a versatile chaperone that protects protein synthesis machinery, enzymes, proteins and lipid membranes (Paul, 2007). The external application increases endogenous trehalose level and provides salinity tolerance (Chen & Murata, 2002). Trehalose is also important for *Arabidopsis* embryo development, abscisic acid signalling, glucose signalling and abiotic stress signalling (Coello & Martínez-Barajas, 2014; Van Dijken, Schluepmann, & Smeekens, 2004).

Biosynthesis of trehalose occurs in five different pathways in bacteria, whereas animals, fungi and plants use only one pathway (Ilhan et al., 2015). Two enzymes govern trehalose biosynthesis: trehalose-6-phosphate synthase (TPS) and trehalose phosphatase (TPP), where TPS enzymes catalyse the production of trehalose-6-phosphate (T6P), and TPP enzyme catalyses the production of trehalose (Van Dijken et al., 2004). *TPS* genes have two subclasses, type-I and type-II, based on amino acid sequence variation and the ability to complement different TPS yeast (Ge et al., 2008). Whole-genome sequencing and analysis identified 11 TPSs and 10 TPPs in *Arabidopsis*, nine TPSs and nine TPPs in the rice genome

(Iordachescu & Imai, 2008). Out of the total, TPS and TPP genes in Arabidopsis *AtTPS1* and *AtTPS6* are catalytically (Vogel et al., 2001) efficient. In rice, trehalose accumulation was accelerated under salt stress by induced expression of different biosynthesis genes (Iordachescu & Imai, 2008). TPP is a magnesium-dependent phosphatase/phosphotransferase that belongs to the L-2-haloacid dehalogenase (HAD) superfamily (Ge et al., 2008). This superfamily has three conserved motifs viz. motif 1 (DXDX (T/V) sequence), motif 2 ((S/T) GX sequence) and motif 3 (K(X) 16–30(G/S)(D/S)XXX(D/N) sequence) which is a component of the enzyme's active site. *Arabidopsis* has two functional genes, TPPA and TPPB, encoding TPP protein, and rice has two TPP genes. *OsTPP1* and *OsTPP2,* while maize has *RAMOSA3* (*RA3*) (Vogel et al., 2001). The trehalose degradation pathway involves the trehalase enzyme, which might be an essential regulatory point for trehalose homeostasis under stress conditions (Elbein et al., 2003). Research shows that *Arabidopsis TPS1* homozygous mutants are lethal for embryo development, while induced expression of *AtTPS1* has proved its role in vegetative and reproductive growth (Eastmond et al., 2002). In rice, the transgenic overexpression of *Escherichia coli* trehalose biosynthetic genes (*otsA* and *otsB*) increases trehalose biosynthesis and provides salt stress tolerance. These transgenic plants maintained a higher K^+ ion content over Na^+ ion and thus have a lower K^+/Na^+ ratio both in shoot and root, which helps in ion homeostasis under salt stress (Garg, 2002). These plants also show a better photosynthetic rate than non-transgenic plants. Overexpression of the *Arabidopsis* trehalose-6-phosphate synthase gene (*AtTPS1*) produces a higher level of trehalose-6-phosphate accumulation and has a role in sugar signalling during plant development. A transgenic TPSP, a bifunctional fusion of *Escherichia coli* TPS and TPP, shows a better salt stress tolerance with an elevated transcript level of *Lip5* while maintaining average growth in rice and *Arabidopsis* (Jang et al., 2003; Miranda et al., 2007). Similar results were observed in trehalose-6-phosphate synthase/phosphatase fusion transgenic tomato with an accelerated tolerance to slat and other abiotic stresses (Lyu et al., 2013). Overexpression of *OsTPS1* confers salt stress tolerance and induces expression of *HSP70* (heat shock protein), *CRP* (cold-regulated protein), *ELIP* (early light-inducible protein), *RAB16C* (responsive to ABA), *DHN6* (dehydrin), LEA14A (late embryogenesis abundant protein) and *LEA14A* (late embryogenesis abundant protein) related to multiple stress tolerance (Luo et al., 2010). TPP is the terminal enzyme in trehalose synthesis, and overexpression of *OsTPP1* in rice enhances salt and cold stress tolerance via inducing different stress-inducible genes like *Lip5, Lip9, OsAsr, DREB1B* and mitogen-activated protein kinases (*MPK*) (Ge et al., 2008). A chloroplast-localised *AtTPPD* overexpression line was more tolerant to salinity, whereas the *tppd* mutants were more sensitive and is redox-regulated under stress. A higher expression of trehalase in stomatal guard cells indicates it plays a crucial function in abiotic stress tolerance and needs further characterisation (Van Houtte et al., 2013). *AtTRE1* is involved in stomatal closure by inducing HXK and ABA-dependent mechanism.

Application of trehalose has a positive effect of improving the K^+/Na^+ ratio, reducing ROS and MDA content while enhancing the antioxidative capacity, including enzymatic and non-enzymatic antioxidants under salt stress in maize, wheat, rice, *Arabidopsis* and *Catharanthus roseus* (Rohman et al., 2019; Sadak, 2019; Weng et al., 2014; Yang et al., 2014). In wheat, trehalose maintains K^+/Na^+ homeostasis by inducing *NHX1*and *SOS1* expression while increased alternative oxidase (*AOX)* expression aids ROS homeostasis under salinity stress (Nemat Alla, Badran, & Mohammed, 2019). Exogenous application of trehalose promotes flowering under salinity due to reduced Na^+ toxicity in inflorescence resulting from local Na^+ transport from leaf to inflorescence (Yang et al., 2014). Application of exogenous trehalose was also helpful under salinity stress to increase other compatible solutes proline, trehalose, soluble sugars (fructose, glucose and sucrose), amino acids (proline, arginine, threonine and glutamate) and alkaloids (vindoline, catharanthine and vinblastine) (Chang et al., 2014). Another study on rice suggests that both proline and trehalose have a role in salt stress tolerance but play a prominent beneficial role during the recovery period (Nounjan et al., 2012). However, little is known about the trehalose transporter system in the plant. Trehalose plays an important role in multiple abiotic stress tolerance, including salinity stress in different crop plants through regulating ROS, antioxidative system, K^+/Na^+ ratio, osmolyte level etc. Still, the proper signalling mechanism under salinity stress and other abiotic stresses are yet to be discovered.

12.2.3 Glycine Betaine

Plants rely on a different cellular mechanism to tolerate salinity stress, and one of the adaptations to improve cellular activities is to accumulate organic and inorganic molecules. GB or N, N, N-trimethylglycine betaine is a quaternary amino acid derivatives-cum-compatible solute that is widely distributed among different kingdoms that regulate osmotic adjustment and provide protection to cellular components under salinity stress (Mansour & Ali, 2017). Studies show that salt-tolerant genotypes amass higher GB accumulation under salinity and tolerant genotypes generally accumulate a higher GB content than the sensitive genotypes (Sairam et al., 2002). While exogenous application of GB improves salt stress tolerance in many plant species, even lower levels of GB confers salt stress tolerance which might indicate its role in the stress signalling mechanism. Few studies show a negative role of GB in plant salt stress tolerance (Szabados et al., 2011). Two independent researches suggest that GB does not act as an antioxidant while indirectly activating the ROS defence system and antioxidative system to tolerate salinity stress (Chen & Murata, 2008; Malekzadeh, 2015). These contrasting results make this research area more interesting to generate robust evidence on the molecular mechanism underlying GB-mediated salinity tolerance.

From a genetic engineering point of view, GB homeostasis is essential, including its biosynthesis and degradation pathway. GB is synthesised in two pathways one is through choline-dependent, and another is glycine-dependent pathway. The choline-dependant pathway is predominant in higher plants and follows the two-step oxidation of choline to GB with the help of two enzymes, i.e. ferredoxin-dependent choline monooxygenase (CMO), which oxidises choline to betaine aldehyde (BA) and betain aldehyde dehydrogenase (BADH) converts

BA to GB with the help of $NAD(P)^+$ coenzyme (Weigel et al., 1986). Under abiotic stress, GB is accumulated in a high amount in plants in an ABA-dependent pathway (Storey et al., 1977). A transgenic approach to increasing GB by manipulating its biosynthesis pathway has been used in many crops to achieve salinity tolerance (Mansour & Ali, 2017). Transgenic tobacco with the CMO gene from *Salicornia europaea* confers tolerance to salt stress (Zhang et al., 2009). In cotton, *AhCMO* transgenic lines increase GB biosynthesis and provide tolerance to salinity. While expressing the CMO, gene in plastid improves salt and drought stress through GB accumulation in tobacco. Using the bacterial *codA* gene, which codes for choline oxidase or CHO that oxidises choline to GB, many crops like *Arabidopsis*, rice, tomato and *Brassica juncia* have been transformed to accumulate more GB under salt stress to counteract the adverse effects of salinity stress, including other osmotic stress at various growth stages (Prasad et al., 2000). Fusion of an *A. thaliana* heat shock protein (HSP) terminator increases the transcript level of the *codA* gene and improves GB production and salinity stress tolerance of *Eucalyptus camaldulensis* (Tran et al., 2018). Overexpression of the *BADH* increases GB accumulation to alleviate salinity stress in tobacco, rice, tomato, sweet potato etc. (Fan et al., 2012; Kishitani et al., 2000; Li et al., 2003; Zhou et al., 2007). In *BADH*-overexpressed wheat, a high accumulation of GB improves photosynthesis by better functioning of the thylakoid membrane and higher Ca^{2+}-ATPase activity under salt stress (Tian et al., 2017). Environmental factors like light and nitric oxide regulate *BADH* transcript level and protein activity, thus maintaining proper homeostasis in sunflower under salt stress and providing new information on GB-NO cross talk where NO acts as a long-distance elicitor (Kumari et al., 2019). RNAi-mediated downregulation of *OsBADH1* in rice resulted in ROS accumulation while there without affecting GB accumulation (Yang et al., 2014). Glycine betaine translocation is maintained by a phloem-localised BvBet/ProT1 transporter system, which transports GB and proline at a pH-dependent manner (Yamada et al., 2011). Degradation of GB is carried out by betaine homocysteine methyl transferase (BHMT) in cyanobacteria *Aphanothece halophytica. In contrast,* in higher plants, its degradation path is not known (Valenzuela-Soto & Figueroa-Soto, 2019).

GB is one of the main compatible compounds which significantly increase osmotic pressure and balance vascular osmotic potential in plants (Raza et al., 2007). Under different abiotic stress conditions, GB acts as a zwitter ion to stabilise protein complexes and maintain functional and structural integrity of cellular organelles from the deleterious effect of ROS. GB plays a vital role to reduce oxidative stress by activating antioxidant enzymes (SOD, CAT, DHAR APX, MDHAR, GPX, GR etc.) and increase the contents of AsA, GSH and GSH/ GSSG ratio to scavenge more ROS generated during salinity stress (Fan et al., 2012). Under stress, PSII activity gets severely affected by ROS, which is stabilised by GB by preventing salt (Na^+ and Cl^-) induced damage to PSII proteins (the 18, 23 and 33kDa proteins) (Papageorgiou & Murata, 1995). In *Arabidopsis*, OH·- causes K^+ leakage in the root, which is alleviated by exogenous application of GB (Cuin & Shabala, 2007). In *Beta vulgaris,* GB maintains an intracellular osmotic balance of Na^+ between cytoplasm and vacuole (Subbarao et al., 2001). Salt-stress-induced increase in Na^+/K^+ ratio is reduced by GB in rice signifying that GB suppressed the apoplastic Na^+ flow that reduced Na^+ uptake in rice plants and higher K^+ concentration (Sobahan et al., 2009). Higher expression of BADH is well correlated with transcript accumulation of different antioxidant enzymes (Kumar et al., 2004; Tian et al., 2017). Under various abiotic stresses, BADH function is mainly regulated by MAPK and ABA signalling pathways. Furthermore, the exogenous application of 100 mM GB increases its endogenous level in wheat varieties under salinity stress, which improves yield attributes. This increase in GB content was correlated with an enhanced antioxidative system, reduced Na^+/K^+ and reduced Na^+/Ca^{2+} ratio in leaves and roots (Raza et al., 2007). In genetically engineered tomato with *codA* gene alleviates K^+ efflux in salt stress by upregulating K^+ transporter *LKT1,* which is a low-affinity influx protein. Na^+/H^+ antiporter *SOS1,* Na^+ transporter *NHX1* and *NHX2* help to accumulate extra Na^+ to the vacuole with the help of plasma membrane H+-ATPases, LHA1 and LHA2, and thus regulate K^+ and Na^+ homeostasis under salinity stress (Wei et al., 2017). Similarly, a recent study on maize seedlings shows GB's wide range of regulation under salt stress. These GB-mediated salt stress tolerance mechanisms involve different transcription factor families (MYB, MYB-related, NAC, bZIP, WRKY), signal transduction mechanisms (CDPK, CBL-interacting protein kinase, CBL-interacting protein kinase 14-like), ROS-scavenging mechanism (Apx1-cytosolic ascorbate peroxidase, peroxidase 2-like), secondary metabolism (Ent-cassadiene C2-murhydroxylase-like, Cyanidin 3-O-rutinoside 5-O-glucosyltransferase) and ion transport mechanism (metal ion-binding protein, potassium transporter 21-like and phosphate transporter protein) (Chen et al., 2020). GB was able to modulate a Ca^{2+} -CaM signalling cascade under NaCl stress by increasing Ca^{2+} influx and enhancing the expression of heat shock transcription factor (HSF) genes resulting in higher levels of HSPs like HSP70 (Li et al., 2014). However, foolproof evidence and exact mechanism for different correlative data on GB and salt stress are lacking even though many studies on exogenous application and molecular modulation of GB metabolism in different plant systems are reported.

12.3 Role of Plant Growth Regulators in Modulating Compatible Solute-Mediated K^+ Balance

12.3.1 Polyamines

Polyamines are aliphatic compounds ubiquitously present across all living organisms and play an essential role in plant biotic and abiotic stress tolerance (Hussain et al., 2011). The most abundant polyamines are putrescine (Put, 1, 4- diaminobutane), spermidine (SPD, N -3-aminopropyl-1, 4-diaminobutane) and spermine (SPM, bis (N -3-aminopropyl)-1,4-diaminobutane). Free and conjugative polyamines scavenge ROS under stress and also activate production of different antioxidative enzymes (Quinet et al., 2010). Exogenous application of polyamines like SPM improves redox homeostasis and provides tolerance to salinity in different crops (Saha et al., 2015). In *Cucumis sativus*, exogenous application of SPD regulates polyamine metabolism and alleviates salinity stress by increasing proline accumulation which might be correlated with an increase in K^+/Na^+ (Duan et al., 2008; Liu et al., 2007). According to (Larher et al., 1998), polyamine and proline synthesis work as a substrate-product mechanism. The polyamine catabolism pathway indirectly generates a substrate for proline metabolism. Proline is degraded by diamine oxidase (DAO) and polyamine oxidase (PAO), which produces Δ1-pyrroline and is converted to γ-aminobutyric acid. This γ-amino butyric acid produces succinate by transamination and oxidation reaction, used in the Krebs cycle to produce proline. Other reports suggest that oxidation of polyamine by H_2O_2 formed due to oxidative stress can lead to proline biosynthesis (Santa-Cruz et al., 1999). In spinach, the application of putrescine increases proline content under control and salt stress condition (Öztürk & Demir, 2003). Different other studies have proved the correlation of polyamine with proline accumulation under salt stress in rice, wheat, tomato, maize, sorghum and bean with higher performance in salt-tolerant genotypes (Campestre et al., 2011; Chai et al., 2010; Saha et al., 2015). While increased proline content was also observed under salinity stress with a concomitant decrease of polyamine content, polyamine degradation leads to proline biosynthesis (Larher et al., 1998). Application of 1mM SPD increases proline content in M-1–48, Nona bokra, and to the highest level in aromatic rice Gobindobhog (GB) under 200mM NaCl stress (Roychoudhury & Tripathi, 2020). Studies from OE and RNAi lines of the *P5CS* gene show that proline biosynthesis is closely related to increased polyamine biosynthesis and metabolism genes under salt stress which is well correlated with K^+ and Na^+ ion homeostasis (Guan et al., 2020). Another research shows that *P5CS1* is an S-nitrosylation target under SPD treatment, and SPD strongly induces *P5CS1* gene expression (Tanou et al., 2014). In *spd* and *spm* mutants *PAO* were observed to produce more proline and exhibit higher salt stress tolerance while maintaining a lower Na^+ concentration (Zarza et al., 2017). Under salt stress, proline metabolism-related gene expression was similar in SPM-deficient mutants and wild type (WT). Even though a wide range of literature shows the correlation of polyamine with proline accumulation, the exact cross talk between proline metabolism and polyamine signalling under abiotic stress tolerance is not deciphered. Similarly, the role of polyamine on K^+ and ROS homeostasis through modulating trehalose and GB metabolism is not well known under salinity stress.

12.3.2 Salicylic Acid

SA is a hormone-cum-signalling molecule present ubiquitously in plants that plays a significant role in salinity stress (Jayakannan et al., 2015). Exogenous application and mutant studies of SA demonstrated its ameliorative role in salinity stress (Hayat et al., 2010). Application of NaCl and SA in lentil plants in combination with NaCl provided higher salt tolerance by increasing compatible solute concentration leading to better turgor maintenance. Treatment of 0.5mM SA induces a significant increase in free proline content in the shoot but not in the root, correlated with increased P5CR and reduced proline oxidase activity. Similarly, SA application also induces accumulation of GB in both shoot and root compared to control, thus providing tolerance to salt stress (Misra & Saxena, 2009). In another study, SA application improved photosynthesis by increasing GB accumulation and suppressing ethylene production under salt stress in *Vigna radiata L.* (Khan, Asgher, & Khan, 2014).

Furthermore, *Tanacetum parthenium L.* and *Phaseolus vulgaris* SA affect proline accumulation and reduce ROS levels by improving enzymatic antioxidant mechanisms (like POX, APX and MDHAR), thus acclimatise adverse effects of salt stress (Mallahi et al., 2018; Palma et al., 2009). (Shakirova et al., 2003) suggested a possible role of SA-induced ABA in proline accumulation in wheat seedlings under salinity. But the key regulatory mechanism in terms of molecular signalling of salicylic-acid–mediated ROS and K^+ homeostasis through regulating compatible solute metabolism is unclear and needs further evidence.

12.3.3 Melatonin

Melatonin is an indole amine molecule ubiquitously distributed across animal and plant kingdoms. It plays a significant role as a hormone in animals and a multi-regulatory signalling molecule in plants under different environmental conditions (Zhao et al., 2018). Melatonin is an ancient molecule that principally acts as an antioxidant while it also regulates different physiological processes ranging from growth and development to biotic and abiotic stress tolerance. For the last decade, many studies have proved the role of melatonin in adaptation to abiotic stresses, including salinity (Zhan et al., 2019). Under salinity stress, melatonin works mainly in three different ways to optimise ROS homeostasis. Firstly, it acts as an antioxidant to directly scavenge ROS and RNS; secondly, it improves antioxidative enzymes activity like APX, CAT, SOD, POD, GR and GPX; and thirdly, it activates regulatory mechanism to induce antioxidants (AsA and GSH) level under salinity stress (de Campos et al., 2012; Chen et al., 2018). Recent studies on melatonin-mediated salinity tolerance show its role in ion homeostasis. Under salinity stress, melatonin increases the K^+/Na^+ ratio in different crops by regulating multiple ion transporters like NHX, SOS and AKT and mediated by NO signalling (Jiang et al., 2021; Shah et al., 2021; Wang et al., 2018; Zhan et al., 2019; Zhao et al., 2018). Melatonin-mediated polyamine metabolism is an integral part of the salinity tolerance mechanism (Gong, Shi, Dou, Song, & Ma, 2017; Gong et al., 2017; Ke et al., 2018). Under salt stress, melatonin upregulates *SAMDC1, -3, -4*, and *SPDS1, -3, -5, and -6* genes and suppresses PAO and DAO activity, thus improving salinity tolerance (Ji et al., 2017). The melatonin receptor *CAND2/PMTR1* is directly involved in the H_2O_2 and Ca^{2+} signalling mechanism, which indicates a possible involvement in ROS homeostasis under salinity tolerance and needs further study to prove this hypothesis. (Chen et al. 2018; Siddiqui et al., 2019) observed direct correlation of melatonin application and proline accumulation through induction of proline biosynthesis protein P5CS activity under NaCl stress. Similar results were observed in *Avena sativa,* where both the biosynthesis gene *P5CS* and *PDH1* were unregulated under salinity stress with melatonin treatment which improves photosynthetic machinery and antioxidative system (Varghese et al., 2019). Melatonin-induced proline content is an essential factor for increased K^+/Na^+ ratio under salinity via seed priming (Dawood & El-Awadi, 2015). Even though few studies prove melatonin's role in regulation of proline-mediated salinity tolerance, a broader understanding on the regulation of different other compatible solutes like trehalose and GB under salinity stress and their crosstalk to regulate ROS and K^+/Na^+ homeostasis are lacking.

12.4 Compatible Solutes Engineering to Modulate K^+ Balance and ROS Homeostasis under Salt Stress

The build-up of sugars and other suitable solutes that function as osmoprotectants is a frequent response of organisms to drought, salt and low-temperature stresses. Even though using ions for osmotic adjustment may be more energy efficient than producing organic osmolytes under osmotic pressure, many plants accumulate organic osmolytes to cope with osmotic stress. Under salt stress, genes involved in osmoprotectant production are upregulated. Accumulation of proline, GB, polyols, sugar alcohols and soluble sugars are regulated transcriptionally by salt stress (Zhu, 2002). Transgenic rice plants overproducing GB (Kishitani et al., 2000) and trehalose (Garg et al., 2002) absorbed less Na+ ions while maintaining K^+ uptake. GB and trehalose serve as osmoprotectants by preserving protein quaternary structures and highly organised membrane states. Proline is a carbon and nitrogen sink as well as a free-radical scavenger. It also helps to keep subcellular structures (membranes and proteins) stable and the cellular redox potential balanced. Wheat cultivars acquired trehalose during salt and drought stress. Under stress, the activity of the trehalose metabolising enzyme trehalose-6-phosphate synthase (TPS) rose dramatically (El-Bashiti, et al., 2005). Differences in endogenous levels of different osmoprotectants might also be contributing to salinity tolerance in crops.

To overcome the adverse effects of salt stress and enhance the salt tolerance of plants, various strategic approaches such as molecular breeding of salt-tolerant genotypes/plants and genetic engineering of multiple osmolytes can be a better alternative and need of the hour (Parida & Das, 2005). Transgenic plants with an improved accumulation of compatible solutes such as GB (Lilius et al., 1996; Romero et al., 1997), mannitol (Tarczynski et al., 1993), ononitol (Sheveleva, Chmara, Bohnert, & Jensen, 1997), proline (Kavi Kishor et al., 1995) and fructan (Pilon-Smits et al., 1995) showed promising results in hyperosmotic tolerance.

Wheat (*Triticum aestivum L.*) transformed with a betA gene-encoding choline dehydrogenase from *Escherichia coli* (*E. coli*) showed varying levels of expression of the betA gene in different transgenic lines. Transgenic lines L2 and L3 under salt stress led to a higher accumulation of GB, and the K^+/Na^+ ratio was significantly higher than the WT. The results indicated that a higher concentration of GB effectively protected membrane integrity and macromolecules, essential to retain K^+ in cells under salinity. This finding was supported by lower electrolytes leakage and MDA content in both transgenic lines L2 and L3 under salt stress. The transgenic lines also had higher germination percentages, more tillers and higher yield than WT under salt treatment (He et al., 2010).

Arthrobacter globiformis codA gene encoding choline oxidase gene-altered for targeting in the chloroplast and cytosol was transformed into tomato (*Solanum lycopersicum cv.* 'Moneymaker'). Transformed plants showed a higher accumulation of GB compared to WT. Moreover, under stress, photosynthetic rate and antioxidant enzyme activities were significantly improved, and low ROS accumulation was observed in the transformed plants. Transgenic lines also showed decreased K^+ efflux and increased Na^+ efflux in the roots under salt stress resulting in a lower Na^+/K^+ ratio than WT plants. Furthermore, exogenous application of GB in WT plants showed a reduction in NaCl-induced K^+ efflux and an increase in Na^+ efflux. GB also enhanced the activity of K^+ transporters and Na^+/H^+ antiporter, and H^+-ATPase genes were all activated by NaCl. These findings imply that the higher salt tolerance imparted by codA in transgenic tomato plants may be related to GB's control of ion channels and transporters, allowing transgenic plants to retain high K levels and low Na levels under salt stress conditions (Wei et al., 2017).

Protoplasts isolated from barley seedling root cells were applied with patch-clamp techniques, and the effects of polyamines on ion-channel activities were investigated. Exogenous application of polyamines blocked Na^+ and K^+ influx in root epidermal and cortical cells. K^+ efflux increased while K^+ influx was blocked in root xylem parenchyma of barley seedlings on the application of extracellular spermidine. Polyamines were shown to maintain Na^+/K^+ homeostasis by curbing root level Na^+ influx and thwarting shoot K^+ efflux (Aldesuquy et al., 2014; Chai et al., 2020; Quinet et al., 2010).

A transgenic plant harbouring *E. coli* TPSP fusion gene in IR64 background was tolerant to salinity by maintaining a higher K^+/Na^+ and stomatal conductance and improved photosynthetic rate. These transgenic plants were able to tolerate a high pH of 9.9, EC 10 d/Sm and drought of 30%–35% soil moisture content while maintaining a higher yield than WT (Joshi et al., 2020). This indicates that trehalose might be playing a regulatory role in selective ion (Na^+ and K^+) movement and proper homeostasis under salt stress which needs further research to explain this assumption at the molecular level. Apart from ionic balance, trehalose acts as a signalling molecule to protect cellular mechanisms by activating genes related to sugar sensing and carbon metabolism genes under different abiotic stresses (Chang et al., 2014; Joshi et al., 2020; Nounjan et al., 2012).

Kiani et al. (2017) found that the barley mutant genotype (73-M4–30) showed an optimum K^+/Na^+ ratio in roots and comparatively enhanced K^+/Na^+ ratio in shoots under salt stress conditions. Osmolytes play an essential role in ion homeostasis, thus imparting salt tolerance to plants. To better understand the role of osmolyte-related genes in maintaining K+ homeostasis under salt stress, it is indispensable to identify the signalling pathways and genes involved in their synthesis and distribution in plants, which can be utilised for transgenic overexpression or genome editing.

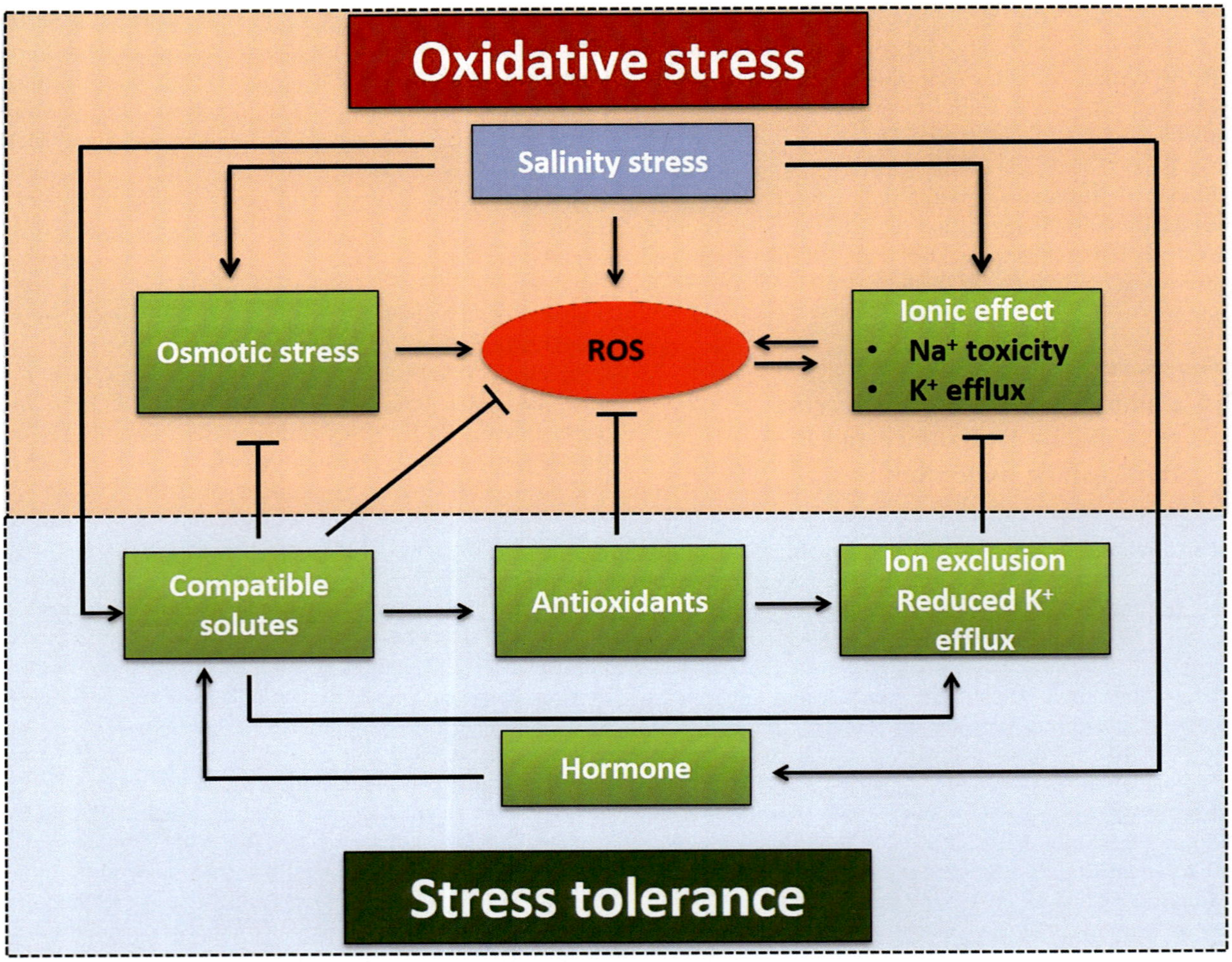

FIGURE 12.1 Schematic model representing impact of salt stress on plants and the adaptive mechanisms employed by plants to cope with stress.

12.5 Summary and Conclusions

The build-up of compatible solutes in response to salinity stress is common in organisms ranging from bacteria to plants and mammals. Under pressure, increased levels of osmolytes typically contribute to maintaining turgor and facilitating continued water uptake. Osmolyte accumulation is achieved by increased activity and expression of genes that code for enzymes involved in osmolyte production, which leads to osmolyte build-up. Stress acclimation occurred due to osmotic adjustment, resulting in a substantial positive connection between osmolyte accumulation and salinity tolerance. In this write-up, we have attempted to compile information on the role of compatible solutes in scavenging stress-induced ROS, induction of hormone signalling and maintaining optimum K^+/Na^+ ratio by reducing K^+ efflux (Figure 12.1). A brief account of the use of candidate genes associated with compatible solute accumulation for engineering for salt tolerance is also discussed. Engineering plants for stress-induced differential accumulation of compatible solutes either by conventional/molecular breeding or transgenic manipulation is a possible path to accomplish salinity tolerance.

REFERENCES

Acosta-Motos, J. R., Ortuño, M. F., Bernal-Vicente, A., Diaz-Vivancos, P., Sanchez-Blanco, M. J., & Hernandez, J. A. (2017). Plant responses to salt stress: Adaptive mechanisms. *Agronomy*, *7*, 18. doi: 10.3390/agronomy7010018.

Adams, E., & Shin, R. (2014). Transport, signaling, and homeostasis of potassium and sodium in plants. *Journal of Integrative Plant Biology*, *56*(3), 231–249. doi: 10.1111/JIPB.12159.

Affenzeller, M. J., Darehshouri, A., Andosch, A., Lütz, C., & Lütz-Meindl, U. (2009). Salt stress-induced cell death in the unicellular green alga Micrasterias denticulata. *Journal of Experimental Botany*, *60*(3), 939–954. doi: 10.1093/jxb/ern348.

Aldesuquy, H., Baka, Z., & Mickky, B. (2014). Kinetin and spermine mediated induction of salt tolerance in wheat plants: Leaf area, photosynthesis and chloroplast ultrastructure of flag leaf at ear emergence. *Egyptian Journal of Basic and Applied Sciences*, *1*(2), 77–87. doi: 10.1016/j.ejbas.2014.03.002.

Ashraf, M. F. (2007). Roles of glycine betaine and proline in improving plant abiotic stress resistance. *Environmental and Experimental Botany*, *59*(2), 206–216. doi: 10.1016/j.envexpbot.2005.12.006.

Assaha, D. V. M., Ueda, A., Saneoka, H., Al-Yahyai, R., & Yaish, M. W. (2017). The role of Na^+ and K^+ transporters in salt stress adaptation in glycophytes. *Frontiers in Physiology, 8*. doi: 10.3389/fphys.2017.00509.

Bo, C., Chen, H., Luo, G., Li, W., Zhang, X., Ma, Q., … Cai, R. (2020). Maize WRKY114 gene negatively regulates salt-stress tolerance in transgenic rice. *Plant Cell Reports, 39*(1), 135–148. doi: 10.1007/s00299-019-02481-3.

Borsani, O., Zhu, J., Verslues, P. E., Sunkar, R., & Zhu, J.-K. (2005). Endogenous siRNAs derived from a pair of natural cis-antisense transcripts regulate salt tolerance in Arabidopsis. *Cell, 123*(7), 1279–1291. doi: 10.1016/j.cell.2005.11.035.

Campestre, M. P., Bordenave, C. D., Origone, A. C., Menéndez, A. B., Ruiz, O. A., Rodríguez, A. A., & Maiale, S. J. (2011). Polyamine catabolism is involved in response to salt stress in soybean hypocotyls. *Journal of Plant Physiology, 168*(11), 1234–1240. doi: 10.1016/J.JPLPH.2011.01.007.

Chai, Y. Y., Jiang, C. D., Shi, L., Shi, T. S., & Gu, W. B. (2010). Effects of exogenous spermine on sweet sorghum during germination under salinity. *Biologia Plantarum, 54*(1), 145–148. doi: 10.1007/S10535-010-0023-1.

Chai, H., Guo, J., Zhong, Y., Hsu, C. C., Zou, C., Wang, P., … Shi, H. (2020). The plasma-membrane polyamine transporter PUT3 is regulated by the Na^+/H^+ antiporter SOS1 and protein kinase SOS2. *New Phytologist, 226*(3), 785–797. doi: 10.1111/nph.16407.

Chang, B., Yang, L., Cong, W., Zu, Y., & Tang, Z. (2014). The improved resistance to high salinity induced by trehalose is associated with ionic regulation and osmotic adjustment in Catharanthus roseus. *Plant Physiology and Biochemistry, 77*, 140–148. doi: 10.1016/j.plaphy.2014.02.001.

Chen, T. H. H., & Murata, N. (2002). Enhancement of tolerance of abiotic stress by metabolic engineering of betaines and other compatible solutes. *Current Opinion in Plant Biology, 5*, 250–257. doi: 10.1016/S1369-5266(02)00255-8.

Chen, T. H. H., & Murata, N. (2008). Glycinebetaine: An effective protectant against abiotic stress in plants. *Trends in Plant Science, 13*(9), 499–505. doi: 10.1016/J.TPLANTS.2008.06.007.

Chen, Y. E., Mao, J. J., Sun, L. Q., Huang, B., Ding, C. B., Gu, Y., … Yuan, M. (2018). Exogenous melatonin enhances salt stress tolerance in maize seedlings by improving antioxidant and photosynthetic capacity. *Physiologia Plantarum, 164*(3), 349–363. doi: 10.1111/PPL.12737.

Chen, F., Fang, P., Zeng, W., Ding, Y., Zhuang, Z., & Peng, Y. (2020). Comparing transcriptome expression profiles to reveal the mechanisms of salt tolerance and exogenous glycine betaine mitigation in maize seedlings. *PLoS One, 15*(5), e0233616. doi: 10.1371/JOURNAL.PONE.0233616.

Chinnusamy, V., Jagendorf, A., & Zhu, J. K. (2005). Understanding and improving salt tolerance in plants. *Crop Science, 45*(2), 437–448. doi: 10.2135/cropsci2005.0437.

Coello, P., & Martínez-Barajas, E. (2014). SnRK1 is differentially regulated in the cotyledon and embryo axe of bean (*Phaseolus vulgaris* L) seeds. *Plant Physiology and Biochemistry, 80*, 153–159. doi: 10.1016/j.plaphy.2014.03.033.

Cuin, T. A., & Shabala, S. (2007). Compatible solutes reduce ROS-induced potassium efflux in Arabidopsis roots. *Plant, Cell & Environment, 30*(7), 875–885. doi: 10.1111/J.1365-3040.2007.01674.X.

Dawood, M. G., & El-Awadi, M. E. (2015). Alleviation of salinity stress on *Vicia faba* L. plants via seed priming with melatonin. *Acta Biológica Colombiana, 20*(2), 223–235. doi: 10.15446/ABC.V20N2.43291.

de Campos, M. L. O., de Hsie, B. S., de Granja, J. A. A., Correia, R. M., de Almeida-Cortez, J. S., & Pompelli, M. F. (2012). Photosynthesis and antioxidant activity in *Jatropha curcas* L. under salt stress. *Brazilian Journal of Plant Physiology, 24*(1), 55–67. doi: 10.1590/S1677-04202012000100008.

Demidchik, V. (2018). ROS-activated ion channels in plants: Biophysical characteristics, physiological functions and molecular nature. *International Journal of Molecular Sciences, 19*(4), 1263.

Duan, J. J., Li, J., Guo, S., & Kang, Y. (2008). Exogenous spermidine affects polyamine metabolism in salinity-stressed Cucumis sativus roots and enhances short-term salinity tolerance. *Journal of Plant Physiology, 165*(15), 1620–1635. doi: 10.1016/J.JPLPH.2007.11.006.

Dudziak, K., Zapalska, M., Börner, A., Szczerba, H., Kowalczyk, K., & Nowak, M. (2019). Analysis of wheat gene expression related to the oxidative stress response and signal transduction under short-term osmotic stress. *Scientific Reports, 9*(1). doi: 10.1038/s41598-019-39154-w.

Eastmond, P. J., Van Dijken, A. J. H., Spielman, M., Kerr, A., Tissier, A. F., Dickinson, H. G., … Graham, I. A. (2002). Trehalose-6-phosphate synthase 1, which catalyses the first step in trehalose synthesis, is essential for Arabidopsis embryo maturation. *The Plant Journal : For Cell and Molecular Biology, 29*(2), 225–235. doi: 10.1046/J.1365-313X.2002.01220.X.

El-Bashiti, T., Hamamci, H., Öktem, H. A., & Yücel, M. (2005). Biochemical analysis of trehalose and its metabolizing enzymes in wheat under abiotic stress conditions. *Plant Science, 169*(1), 47–54. doi: 10.1016/j.plantsci.2005.02.024.

Elbein, A. D., Pan, Y. T., Pastuszak, I., & Carroll, D. (2003). New insights on trehalose: A multifunctional molecule. *Glycobiology, 13*(4). doi: 10.1093/GLYCOB/CWG047.

Fan, W., Zhang, M., Zhang, H., & Zhang, P. (2012). Improved tolerance to various abiotic stresses in transgenic sweet potato (ipomoea batatas) expressing spinach betaine aldehyde dehydrogenase. *PLoS One, 7*(5), e37344. doi: 10.1371/JOURNAL.PONE.0037344.

Flowers, T. J., Munns, R., & Colmer, T. D. (2015). Sodium chloride toxicity and the cellular basis of salt tolerance in halophytes. *Annals of Botany, 115*(3), 419–431.

Garg, A. K., Kim, J. K., Owens, T. G., Ranwala, A. P., Do Choi, Y., Kochian, L. V., & Wu, R. J. (2002). Trehalose accumulation in rice plants confers high tolerance levels to different abiotic stresses. *Proceedings of the National Academy of Sciences of the United States of America, 99*(25), 15898–15903. doi: 10.1073/pnas.252637799.

Ge, L. F., Chao, D. Y., Shi, M., Zhu, M. Z., Gao, J. P., & Lin, H. X. (2008). Overexpression of the trehalose-6-phosphate phosphatase gene OsTPP1 confers stress tolerance in rice and results in the activation of stress responsive genes. *Planta, 228*(1), 191–201. doi: 10.1007/S00425-008-0729-X.

Gong, X., Shi, S., Dou, F., Song, Y., & Ma, F. (2017). Exogenous melatonin alleviates alkaline stress in malus hupehensis rehd by regulating the biosynthesis of polyamines. *Molecules : A Journal of Synthetic Chemistry and Natural Product Chemistry, 22*(9). doi: 10.3390/MOLECULES22091542.

Guan, C., Cui, X., Liu, H. Y., Li, X., Li, M. Q., & Zhang, Y. W. (2020). Proline biosynthesis enzyme genes confer salt tolerance to switchgrass (*Panicum virgatum* L.) in cooperation with polyamines metabolism. *Frontiers in Plant Science*, *11*, 46. doi: 10.3389/FPLS.2020.00046/BIBTEX.

Gunes, A., Inal, A., Alpaslan, M., Eraslan, F., Bagci, E. G., & Cicek, N. (2007). Salicylic acid induced changes on some physiological parameters symptomatic for oxidative stress and mineral nutrition in maize (*Zea mays* L.) grown under salinity. *Journal of Plant Physiology, 164*(6), 728–736. doi: 10.1016/j.jplph.2005.12.009.

Hare, P. D. & Cress, W. C. (1997). Metabolic implications of stress-induced proline accumulation in plants. *Plant Growth Regulation*, *21*, 79–102.

Hare, P. D. & Cress, W. C. (1998). Dissecting the roles of osmolyte accumulation during stress. *Plant Cell & Environment*, *21*, 535–553.

Hare, P. D. & Cress, W. C. (1999). Proline synthesis and degradation: A model system for elucidating stress-related signal transduction. *Journal of Experimental Botany*, *50*(333), 413–434. doi: 10.1093/jexbot/50.333.413.

Hayat, Q., Hayat, S., Irfan, M., & Ahmad, A. (2010). Effect of exogenous salicylic acid under changing environment: A review. *Environmental and Experimental Botany*, *68*, 14–25. doi: 10.1016/j.envexpbot.2009.08.005.

Hayat, S., Hayat, Q., Alyemeni, M. N., Wani, A. S., Pichtel, J., & Ahmad, A. (2012). Role of proline under changing environments: A review. *Plant Signaling and Behavior*, *7*. doi: 10.4161/psb.21949.

He, C., Yang, A., Zhang, W., Gao, Q., & Zhang, J. (2010). Improved salt tolerance of transgenic wheat by introducing betA gene for glycine betaine synthesis. *Plant Cell, Tissue and Organ Culture (PCTOC)*, *101*(1), 65–78. doi: 10.1007/s11240-009-9665-0.

He, R., Yu, G., Han, X., Han, J., Li, W., Wang, B., … Cheng, X. (2017). ThPP1 gene, encodes an inorganic pyrophosphatase in Thellungiella halophila, enhanced the tolerance of the transgenic rice to alkali stress. *Plant Cell Reports*, *36*(12), 1929–1942. doi: 10.1007/S00299-017-2208-Y.

He, Y., Dong, Y., Yang, X., Guo, D., Qian, X., Yan, F., … Wang, Q. (2020). Functional activation of a novel R2R3-MYB protein gene, GmMYB68, confers salt-alkali resistance in soybean (*Glycine max* L.). *Genome*, *63*(1), 13–26. doi: 10.1139/gen-2018-0132.

Hottiger, T., Boller, T., & Wiemken, A. (1987). Rapid changes of heat and desiccation tolerance correlated with changes of trehalose content in Saccharomyces cerevisiae cells subjected to temperature shifts. *FEBS Letters*, *220*(1), 113–115. doi: 10.1016/0014-5793(87)80886-4.

Huang, S., Xin, S., Xie, G., Han, J., Liu, Z., Wang, B., … Cheng, X. (2020). Mutagenesis reveals that the rice OsMPT3 gene is an important osmotic regulatory factor. *The Crop Journal*, *8*(3), 465–479. doi: 10.1016/J.CJ.2020.02.001.

Hussain, S. S., Ali, M., Ahmad, M., & Siddique, K. H. M. (2011). Polyamines: Natural and engineered abiotic and biotic stress tolerance in plants. *Biotechnology Advances*, *29*, 300–311.

Ilhan, S., Ozdemir, F., & Bor, M. (2015). Contribution of trehalose biosynthetic pathway to drought stress tolerance of Capparis ovata Desf. *Plant Biology*, *17*(2), 402–407. doi: 10.1111/PLB.12227.

Iordachescu, M., & Imai, R. (2008). Trehalose biosynthesis in response to abiotic stresses. *Journal of Integrative Plant Biology*, *50*(10), 1223–1229. doi: 10.1111/J.1744-7909.2008.00736.X.

Isayenkov, S. V., & Maathuis, F. J. M. (2019). Plant salinity stress: Many unanswered questions remain. *Frontiers in Plant Science*, *10*. doi: 10.3389/fpls.2019.00080.

Jang, I. C., Oh, S. J., Seo, J. S., Choi, W. Bin, Song, S. I., Kim, C. H., … Kim, J. K. (2003). Expression of a bifunctional fusion of the Escherichia coli genes for trehalose-6-phosphate synthase and trehalose-6-phosphate phosphatase in transgenic rice plants increases trehalose accumulation and abiotic stress tolerance without stunting growth. *Plant Physiology*, *131*(2), 516–524. doi: 10.1104/PP.007237.

Jayakannan, M., Bose, J., Babourina, O., Rengel, Z., & Shabala, S. (2013). Salicylic acid improves salinity tolerance in Arabidopsis by restoring membrane potential and preventing salt-induced K^+ loss via a GORK channel. *Journal of Experimental Botany*, *4*, 1–14.

Jayakannan, M., Bose, J., Babourina, O., Rengel, Z., & Shabala, S. (2015). Salicylic acid in plant salinity stress signalling and tolerance. *Plant Growth Regulation*, *76*(1), 25–40. doi: 10.1007/s10725-015-0028-z.

Ji, T., Li, S., Huang, M., Di, Q., Wang, X., Wei, M., … Yang, F. (2017). Overexpression of cucumber phospholipase D alpha gene (CsPLDα) in tobacco enhanced salinity stress tolerance by regulating Na^+-K+ balance and lipid peroxidation. *Frontiers in Plant Science*, *8*. doi: 10.3389/fpls.2017.00499.

Jiang, D., Lu, B., Liu, L., Duan, W., Meng, Y., Li, J., … Li, C. (2021). Exogenous melatonin improves the salt tolerance of cotton by removing active oxygen and protecting photosynthetic organs. *BMC Plant Biology*, *21*(1). doi: 10.1186/s12870-021-03082-7.

Joshi, R., Sahoo, K. K., Singh, A. K., Anwar, K., Pundir, P., Gautam, R. K., … Singla-Pareek, S. L. (2020). Enhancing trehalose biosynthesis improves yield potential in marker-free transgenic rice under drought, saline, and sodic conditions. *Journal of Experimental Botany*, *71*(2), 653–668. doi: 10.1093/JXB/ERZ462.

Kartashov, A. V., Radyukina, N. L., Ivanov, Y. V., Pashkovskii, P. P., Shevyakova, N. I., & Kuznetsov, V. V. (2008). Role of antioxidant systems in wild plant adaptation to salt stress. *Russian Journal of Plant Physiology*, *55*(4), 463–468. doi: 10.1134/S1021443708040055.

Katiyar-Agarwal, S., Zhu, J., Kim, K., Agarwal, M., Fu, X., Huang, A., & Zhu, J. K. (2006). The plasma membrane Na^+/H^+ antiporter SOS1 interacts with RCD1 and functions in oxidative stress tolerance in Arabidopsis. *Proceedings of the National Academy of Sciences of the United States of America*, *103*(49), 18816–18821. doi: 10.1073/pnas.0604711103.

Kavi Kishor, P. B., Hong, Z., Miao, G. H., Hu, C. A. A., & Verma, D. P. S. (1995). Overexpression of [delta]-Pyrroline-5-Carboxylate synthetase increases proline production and confers osmotolerance in transgenic plants. *Plant Physiology*, *108*(4), 1387–1394. doi: 10.1104/PP.108.4.1387.

Ke, Q., Ye, J., Wang, B., Ren, J., Yin, L., Deng, X., & Wang, S. (2018). Melatonin mitigates salt stress in wheat seedlings by modulating polyamine metabolism. *Frontiers in Plant Science*, *9*. doi: 10.3389/FPLS.2018.00914.

Ketchum, R. E. B., Warren, R. S., Klima, L. J., Lopez-Gutiérrez, F., & Nabors, M. W. (1991). The mechanism and regulation of proline accumulation in suspension cell cultures of the halophytic grass *Distichlis spicata L. Journal of Plant Physiology, 137*(3), 368–374. doi: 10.1016/S0176-1617(11)80147-1.

Khan, M. I. R., Asgher, M., & Khan, N. A. (2014). Alleviation of salt-induced photosynthesis and growth inhibition by salicylic acid involves glycinebetaine and ethylene in mungbean (*Vigna radiata* L.). *Plant Physiology and Biochemistry, 80*, 67–74. doi: 10.1016/j.plaphy.2014.03.026.

Kiani, D., Soltanloo, H., Ramezanpour, S. S., Nasrolahnezhad Qumi, A. A., Yamchi, A., Zaynali Nezhad, K., & Tavakol, E. (2017). A barley mutant with improved salt tolerance through ion homeostasis and ROS scavenging under salt stress. *Acta Physiologiae Plantarum, 39*(3), 1–14. doi: 10.1007/S11738-017-2359-Z/FIGURES/10.

Kim, G. B., & Nam, Y. W. (2013). A novel Δ(1)-pyrroline-5-carboxylate synthetase gene of Medicago truncatula plays a predominant role in stress-induced proline accumulation during symbiotic nitrogen fixation. *Journal of Plant Physiology, 170*(3), 291–302. doi: 10.1016/J.JPLPH.2012.10.004.

Kishitani, S., Takanami, T., Suzuki, M., Oikawa, M., Yokoi, S., Ishitani, M., … Takabe, T. (2000). Compatibility of glycinebetaine in rice plants: Evaluation using transgenic rice plants with a gene for peroxisomal betaine aldehyde dehydrogenase from barley. *Plant, Cell and Environment, 23*(1), 107–114. doi: 10.1046/J.1365-3040.2000.00527.X.

Kovács, H., Aleksza, D., Baba, A. I., Hajdu, A., Király, A. M., Zsigmond, L., … Szabados, L. (2019). Light control of salt-induced proline accumulation is mediated by ELONGATED HYPOCOTYL 5 in Arabidopsis. *Frontiers in Plant Science, 10*, 1584. doi: 10.3389/fpls.2019.01584.

Koyro, H.-W. (2006). Effect of salinity on growth, photosynthesis, water relations and solute composition of the potential cash crop halophyte *Plantago coronopus* (L.). *Environmental and Experimental Botany, 56*(2), 136–146.

Kumar, S., Dhingra, A., & Daniell, H. (2004). Plastid-expressed betaine aldehyde dehydrogenase gene in carrot cultured cells, roots, and leaves confers enhanced salt tolerance. *Plant Physiology, 136*(1), 2843–2854. doi: 10.1104/PP.104.045187.

Kumari, A., Kapoor, R., & Bhatla, S. C. (2019). Nitric oxide and light co-regulate glycine betaine homeostasis in sunflower seedling cotyledons by modulating betaine aldehyde dehydrogenase transcript levels and activity. *Plant Signaling & Behavior, 14*(11). doi: 10.1080/15592324.2019.1666656.

Larher, F., Aziz, A., Deleu, C., Lemesle, P., Ghaffar, A., Bouchard, F., & Plasman, M. (1998). Suppression of the osmoinduced proline response of rapeseed leaf discs by polyamines. *Physiologia Plantarum, 102*(1), 139–147. doi: 10.1034/j.1399-3054.1998.1020118.x.

Leach, R. P., Rogers, J., Wheeler, K. P., Flowers, T. J., & Yeo, A. R. (1990). Molecular markers for ion compartmentation in cells of higher plants. *Journal of Experimental Botany, 41*(9), 1079–1087. doi: 10.1093/JXB/41.9.1079.

Leidi, E. O., Barragán, V., Rubio, L., El-Hamdaoui, A., Ruiz, M. T., Cubero, B., … Pardo, J. M. (2010). The AtNHX1 exchanger mediates potassium compartmentation in vacuoles of transgenic tomato. *The Plant Journal : For Cell and Molecular Biology, 61*(3), 495–506. doi: 10.1111/j.1365-313X.2009.04073.x.

Leigh, R. A., Ahmad, N., & Jones, R. G. W. (1981). Assessment of glycinebetaine and proline compartmentation by analysis of isolated beet vacuoles. *Planta, 153*(1), 34–41. doi: 10.1007/BF00385315.

Lekshmy, S., Sairam, R. K., & Kushwaha, S. R. (2013). Effect of long-term salinity stress on growth and nutrient uptake in contrasting wheat genotypes. *Indian Journal of Plant Physiology, 18*(4), 344–353. doi: 10.1007/s40502-014-0059-x.

Li, M., Guo, S., Xu, Y., Meng, Q., Li, G., & Yang, X. (2014). Glycine betaine-mediated potentiation of HSP gene expression involves calcium signaling pathways in tobacco exposed to NaCl stress. *Physiologia Plantarum, 150*(1), 63–75. doi: 10.1111/PPL.12067.

Li, Q. L., Gao, X. R., Yu, X. H., Wang, X. Z., & An, L. J. (2003). Molecular cloning and characterization of betaine aldehyde dehydrogenase gene from Suaeda liaotungensis and its use in improved tolerance to salinity in transgenic tobacco. *Biotechnology Letters, 25*(17), 1431–1436. doi: 10.1023/A:1025003628446.

Liang, X., Zhang, L., Natarajan, S. K., & Becker, D. F. (2013). Proline mechanisms of stress survival. *Antioxidants & Redox Signaling, 19*(9), 998–1011. doi: 10.1089/ARS.2012.5074.

Lilius, C., Holmberg, N., & Bülow, L. (1996). Enhanced NaCl stress tolerance in transgenic tobacco expressing bacterial choline dehydrogenase. *Bio/Technology 14*(2), 177–180. doi: 10.1038/nbt0296-177.

Liu, J. H., Kitashiba, H., Wang, J., Ban, Y., & Moriguchi, T. (2007). Polyamines and their ability to provide environmental stress tolerance to plants. *Plant Biotechnol, 24*, 117–126.

Liu, L., Huang, L., Lin, X., & Sun, C. (2020). Hydrogen peroxide alleviates salinity-induced damage through enhancing proline accumulation in wheat seedlings. *Plant Cell Reports, 39*(5), 567–575. doi: 10.1007/s00299-020-02513-3.

Luo, Y., Li, F., Wang, G. P., Yang, X. H., & Wang, W. (2010). Exogenous supplied trehalose protects thylakoid membranes of winter wheat from heat-induced damage. *Biologia Plantarum, 54*, 495–501.

Lyu, J. Il, Min, S. R., Lee, J. H., Lim, Y. H., Kim, J. K., Bae, C. H., & Liu, J. R. (2013). Overexpression of a trehalose-6-phosphate synthase/phosphatase fusion gene enhances tolerance and photosynthesis during drought and salt stress without growth aberrations in tomato. *Plant Cell, Tissue and Organ Culture, 112*(2), 257–262. doi: 10.1007/S11240-012-0225-7/FIGURES/5.

Ma, L., Zhang, H., Sun, L., Jiao, Y., Zhang, G., Miao, C., & Hao, F. (2012). NADPH oxidase AtrbohD and AtrbohF function in ROS-dependent regulation of Na^+/K^+homeostasis in Arabidopsis under salt stress. *Journal of Experimental Botany, 63*(1), 305–317. doi: 10.1093/jxb/err280.

Malekzadeh, P. (2015). Influence of exogenous application of glycinebetaine on antioxidative system and growth of salt-stressed soybean seedlings (*Glycine max* L.). *Physiology and Molecular Biology of Plants : An International Journal of Functional Plant Biology, 21*(2), 225–232. doi: 10.1007/S12298-015-0292-4.

Mallahi, T., Saharkhiz, M. J., & Javanmardi, J. (2018). Salicylic acid changes morpho-physiological attributes of feverfew (*Tanacetum parthenium* L.) under salinity stress. *Acta Ecologica Sinica, 38*(5), 351–355. doi: 10.1016/J.CHNAES.2018.02.003.

Mansour, M. M. F., & Ali, E. F. (2017). Glycinebetaine in saline conditions: An assessment of the current state of knowledge. *Acta Physiologiae Plantarum, 39*(2), 1–17. doi: 10.1007/S11738-017-2357-1.

Miller, G., Schlauch, K., Tam, R., Cortes, D., Torres, M. A., Shulaev, V., ... Mittler, R. (2009). The plant NADPH oxidase RBOHD mediates rapid systemic signaling in response to diverse stimuli. *Science Signaling*, *2*(84), ra45–ra45. doi: 10.1126/scisignal.2000448.

Miranda, J. A., Avonce, N., Suárez, R., Thevelein, J. M., Van Dijck, P., & Iturriaga, G. (2007). A bifunctional TPS-TPP enzyme from yeast confers tolerance to multiple and extreme abiotic-stress conditions in transgenic Arabidopsis. *Planta*, *226*(6), 1411–1421. doi: 10.1007/S00425-007-0579-Y.

Misra, N., & Saxena, P. (2009). Effect of salicylic acid on proline metabolism in lentil grown under salinity stress. *Plant Science*, *177*(3), 181–189. doi: 10.1016/j.plantsci.2009.05.007.

Mittova, V., Tal, M., Volokita, M., & Guy, M. (2003). Up-regulation of the leaf mitochondrial and peroxisomal antioxidative systems in response to salt-induced oxidative stress in the wild salt-tolerant tomato species *Lycopersicon pennellii*. *Plant, Cell and Environment*, *26*(6), 845–856. doi: 10.1046/j.1365-3040.2003.01016.x.

Munns, R., & Tester, M. (2008). Mechanisms of salinity tolerance. *Annual Review of Plant Biology*, *59*, 651–681. doi: 10.1146/ANNUREV.ARPLANT.59.032607.092911.

Munns, R., Day, D. A., Fricke, W., Watt, M., Arsova, B., Barkla, B. J., ... Tyerman, S. D. (2020). Energy costs of salt tolerance in crop plants. *New Phytologist*, *225*, 1072–1090. doi: 10.1111/nph.15864.

Nemat Alla, M., Badran, E., & Mohammed, F. (2019). Exogenous trehalose alleviates the adverse effects of salinity stress in wheat. *Turkish Journal of Botany*, *43*(1), 48–57. doi: 10.3906/BOT-1803-36.

Nounjan, N., Nghia, P. T., & Theerakulpisut, P. (2012). Exogenous proline and trehalose promote recovery of rice seedlings from salt-stress and differentially modulate antioxidant enzymes and expression of related genes. *Journal of Plant Physiology*, *6*(169), 596–604. doi: 10.1016/J.JPLPH.2012.01.004.

Öztürk, L., & Demir, Y. (2003). Effects of putrescine and ethephon on some oxidative stress enzyme activities and proline content in salt stressed spinach leaves. *Plant Growth Regulation*, *40*(1), 89–95. doi: 10.1023/A:1023078819935.

Palma, F., Lluch, C., Iribarne, C., García-Garrido, J. M., & Tejera García, N. A. (2009). Combined effect of salicylic acid and salinity on some antioxidant activities, oxidative stress and metabolite accumulation in Phaseolus vulgaris. *Plant Growth Regulation*, *58*(3), 307–316. doi: 10.1007/S10725-009-9380-1.

Papageorgiou, G. C., & Murata, N. (1995). The unusually strong stabilizing effects of glycine betaine on the structure and function of the oxygen-evolving Photosystem II complex. *Photosynthesis Research*, *44*(3), 243–252. doi: 10.1007/BF00048597.

Parida, A. K., & Das, A. B. (2005). Salt tolerance and salinity effects on plants: A review. *Ecotoxicology and Environmental Safety*, *60*(3), 324–349. doi: 10.1016/j.ecoenv.2004.06.010.

Paul, M. (2007). Trehalose 6-phosphate. *Current Opinion in Plant Biology*, *10*(3), 303–309. doi: 10.1016/j.pbi.2007.04.001.

Pilon-Smits, E. A. H., Ebskamp, M. J. M., Paul, M. J., Jeuken, M. J. W., Weisbeek, P. J., & Smeekens, S. C. M. (1995). Improved performance of transgenic fructan-accumulating tobacco under drought stress. *Plant Physiology*, *107*(1), 125–130. doi: 10.1104/PP.107.1.125.

Prasad, K. V. S. K., Sharmila, P., Kumar, P. A., & Saradhi, P. P. (2000). Transformation of *Brassica juncea* (L.) Czern with bacterial codA gene enhances its tolerance to salt stress. *Molecular Breeding*, *6*(5), 489–499. doi: 10.1023/A:1026542109965.

Qiu, Q. S., Guo, Y., Dietrich, M. A., Schumaker, K. S., & Zhu, J. K. (2002). Regulation of SOS1, a plasma membrane Na^+/H^+ exchanger in Arabidopsis thaliana, by SOS2 and SOS3. *Proceedings of the National Academy of Sciences of the United States of America*, *99*(12), 8436–8441. doi: 10.1073/pnas.122224699.

Quinet, M., Ndayiragije, A., Lefèvre, I., Lambillotte, B., Dupont-Gillain, C. C., & Lutts, S. (2010). Putrescine differently influences the effect of salt stress on polyamine metabolism and ethylene synthesis in rice cultivars differing in salt resistance. *Journal of Experimental Botany*, *61*(10), 2719–2733. doi: 10.1093/jxb/erq118.

Raza, S. H., Athar, H. R., Ashraf, M., & Hameed, A. (2007). Glycinebetaine-induced modulation of antioxidant enzymes activities and ion accumulation in two wheat cultivars differing in salt tolerance. *Environmental and Experimental Botany*, *60*(3), 368–376. doi: 10.1016/J.ENVEXPBOT.2006.12.009.

Rohman, M. M., Islam, M. R., Monsur, M. B., Amiruzzaman, M., Fujita, M., & Hasanuzzaman, M. (2019). Trehalose protects maize plants from salt stress and phosphorus deficiency. *Plants*, *8*(12), 568. doi: 10.3390/PLANTS8120568.

Romero, C., Bellés, J. M., Vayá, J. L., Serrano, R., & Culiáñez-Macià, F. A. (1997). Expression of the yeast trehalose-6-phosphate synthase gene in transgenic tobacco plants: Pleiotropic phenotypes include drought tolerance. *Planta*, *201*(3), 293–297. doi: 10.1007/S004250050069.

Roychoudhury, A., & Tripathi, D. K. (Eds.) (2020). *Protective Chemical Agents in the Amelioration of Plant Abiotic Stress*. doi: 10.1002/9781119552154.

Sadak, M. S. (2019). Physiological role of trehalose on enhancing salinity tolerance of wheat plant. *Bulletin of the National Research Centre*, *43*(1), 1–10. doi: 10.1186/S42269-019-0098-6.

Saha, J., Brauer, E. K., Sengupta, A., Popescu, S. C., Gupta, K., & Gupta, B. (2015). Polyamines as redox homeostasis regulators during salt stress in plants. *Frontiers in Environmental Science*, *3*(APR), 21. doi: 10.3389/FENVS.2015.00021/BIBTEX.

Sairam, R. K., & Srivastava, G. C. (2002). Changes in antioxidant activity in sub-cellular fractions of tolerant and susceptible wheat genotypes in response to long term salt stress. *Plant Science*, *162*(6), 897–904. doi: 10.1016/S0168-9452(02)00037-7.

Sairam, R. K., Rao, K., & Srivastava, G. C. (2002). Differential response of wheat genotypes to long term salinity stress in relation to oxidative stress, antioxidant activity and osmolyte concentration. *Plant Science*, *163*(5), 1037–1046. doi: 10.1016/S0168-9452(02)00278-9.

Sanchez, D. H., Lippold, F., Redestig, H., Hannah, M. A., Erban, A., Krämer, U., ... Udvardi, M. K. (2008). Integrative functional genomics of salt acclimatization in the model legume Lotus japonicus. *Plant Journal*, *53*(6), 973–987. doi: 10.1111/j.1365-313X.2007.03381.x.

Santa-Cruz, A., Acosta, M., Rus, A., & Bolarin, M. C. (1999). Short-term salt tolerance mechanisms in differentially salt tolerant tomato species. *Plant Physiology and Biochemistry*, *37*(1), 65–71. doi: 10.1016/S0981-9428(99)80068-0.

Satoh, R., Nakashima, K., Seki, M., Shinozaki, K., & Yamaguchi-Shinozaki, K. (2002). ACTCAT, a novel cis-acting element for proline- and hypoosmolarity-responsive expression of the ProDH gene encoding proline dehydrogenase in Arabidopsis. *Plant Physiology, 130*(2), 709–719. doi: 10.1104/PP.009993.

Shabala, S., & Cuin, T. A. (2008). Potassium transport and plant salt tolerance. *Physiologia Plantarum, 133*(4), 651–669. doi: 10.1111/J.1399-3054.2007.01008.X.

Shabala, S., & Pottosin, I. (2014). Regulation of potassium transport in plants under hostile conditions: Implications for abiotic and biotic stress tolerance. *Physiologia Plantarum, 151*, 257–279. doi: 10.1111/ppl.12165.

Shah, A. N., Tanveer, M., Abbas, A., Fahad, S., Baloch, M. S., Ahmad, M. I., ... Song, Y. (2021). Targeting salt stress coping mechanisms for stress tolerance in Brassica: A research perspective. *Plant Physiology and Biochemistry, 158*, 53–64. doi: 10.1016/j.plaphy.2020.11.044.

Shakirova, F. M., Sakhabutdinova, A. R., Bezrukova, M. V., Fatkhutdinova, R. A., & Fatkhutdinova, D. R. (2003). Changes in the hormonal status of wheat seedlings induced by salicylic acid and salinity. *Plant Science, 164*(3), 317–322. doi: 10.1016/S0168-9452(02)00415-6.

Shapiguzov, A., Vainonen, J. P., Hunter, K., Tossavainen, H., Tiwari, A., Järvi, S., ... Kangasjärvi, J. (2019). Arabidopsis RCD1 coordinates chloroplast and mitochondrial functions through interaction with ANAC transcription factors. *ELife, 8*, e43284. doi: 10.7554/eLife.43284.

Sharma, M., Gupta, S. K., Majumder, B., Maurya, V. K., Deeba, F., Alam, A., & Pandey, V. (2017). Salicylic acid mediated growth, physiological and proteomic responses in two wheat varieties under drought stress. *Journal of Proteomics, 163*, 28–51. doi: 10.1016/j.jprot.2017.05.011.

Sheveleva, E., Chmara, W., Bohnert, H. J., & Jensen, R. G. (1997). Increased salt and drought tolerance by d-ononitol production in transgenic *Nicotiana tabacum* L. *Plant Physiology, 115*(3), 1211. doi: 10.1104/PP.115.3.1211.

Siddiqui, M. H., Alamri, S., Al-Khaishany, M. Y., Khan, M. N., Al-Amri, A., Ali, H. M., ... Alsahli, A. A. (2019). Exogenous melatonin counteracts NaCl-induced damage by regulating the antioxidant system, proline and carbohydrates metabolism in tomato seedlings. *International Journal of Molecular Sciences, 20*(2). doi: 10.3390/IJMS20020353.

Silva-Ortega, C. O., Ochoa-Alfaro, A. E., Reyes-Agüero, J. A., Aguado-Santacruz, G. A., & Jiménez-Bremont, J. F. (2008). Salt stress increases the expression of p5cs gene and induces proline accumulation in cactus pear. *Plant Physiology and Biochemistry, 46*(1), 82–92. doi: 10.1016/j.plaphy.2007.10.011.

Smirnoff, N. & Cumbes, Q. C. (1989). Hydroxyl radical scavenging activity of compatible solutes. *Phytochemistry, 28*(4), 1057–1060. doi: 10.1016/0031-9422(89)80182-7.

Sobahan, M. A., Arias, C. R., Okuma, E., Shimoishi, Y., Nakamura, Y., Hirai, Y., ... Murata, Y. (2009). Exogenous proline and glycinebetaine suppress apoplastic flow to reduce Na^+ uptake in rice seedlings. *Bioscience, Biotechnology, and Biochemistry, 73*(9), 2037–2042. doi: 10.1271/BBB.90244.

Storey, R., Ahmad, N., & Wyn Jones, R. G. (1977). Taxonomic and ecological aspects of the distribution of glycinebetaine and related compounds in plants. *Oecologia, 27*(4), 319–332. doi: 10.1007/BF00345565.

Subbarao, G. V., Wheeler, R. M., Levine, L. H., & Stutte, G. W. (2001). Glycine betaine accumulation, ionic and water relations of red-beet at contrasting levels of sodium supply. *Journal of Plant Physiology, 158*(6), 767–776. doi: 10.1078/0176-1617-00309.

Surender Reddy, P., Jogeswar, G., Rasineni, G. K., Maheswari, M., Reddy, A. R., Varshney, R. K., & Kavi Kishor, P. B. (2015). Proline over-accumulation alleviates salt stress and protects photosynthetic and antioxidant enzyme activities in transgenic sorghum [*Sorghum bicolor* (L.) Moench]. *Plant Physiology and Biochemistry, 94*, 104–113. doi: 10.1016/j.plaphy.2015.05.014.

Szabados, L., Kovács, H., Zilberstein, A., & Bouchereau, A. (2011). Plants in extreme environments: Importance of protective compounds in stress tolerance. *Advances in Botanical Research, 57*, 105–150. doi: 10.1016/B978-0-12-387692-8.00004-7.

Tanou, G., Ziogas, V., Belghazi, M., Christou, A., Filippou, P., Job, D., ... Molassiotis, A. (2014). Polyamines reprogram oxidative and nitrosative status and the proteome of citrus plants exposed to salinity stress. *Plant, Cell & Environment, 37*(4), 864–885. doi: 10.1111/PCE.12204.

Tarczynski, M. C., Jensen, R. G., & Bohnert, H. J. (1993). Stress protection of transgenic tobacco by production of the osmolyte mannitol. *Science (New York, N.Y.), 259*(5094), 508–510. doi: 10.1126/SCIENCE.259.5094.508.

Tian, F., Wang, W., Liang, C., Wang, X., Wang, G., & Wang, W. (2017). Overaccumulation of glycine betaine makes the function of the thylakoid membrane better in wheat under salt stress. *The Crop Journal, 5*(1), 73–82. doi: 10.1016/J.CJ.2016.05.008.

Tran, N. H. T., Oguchi, T., Matsunaga, E., Kawaoka, A., Watanabe, K. N., & Kikuchi, A. (2018). Environmental risk assessment of impacts of transgenic Eucalyptus camaldulensis events highly expressing bacterial Choline Oxidase A gene. *Plant Biotechnology, 35*(4), 393. doi: 10.5511/PLANTBIOTECHNOLOGY.18.0831A.

Tsai, A. Y. L., & Gazzarrini, S. (2014, April 1). Trehalose-6-phosphate and SnRK1 kinases in plant development and signaling: The emerging picture. *Frontiers in Plant Science, 5*. doi: 10.3389/fpls.2014.00119.

Valenzuela-Soto, E. M., & Figueroa-Soto, C. G. (2019). Biosynthesis and degradation of glycine betaine and its potential to control plant growth and development. *Osmoprotectant-Mediated Abiotic Stress Tolerance in Plants*, 123–140. doi: 10.1007/978-3-030-27423-8_5.

Van Dijken, A. J. H., Schluepmann, H., & Smeekens, S. C. M. (2004). Arabidopsis trehalose-6-phosphate synthase 1 is essential for normal vegetative growth and transition to flowering. *Plant Physiology, 135*(2), 969–977. doi: 10.1104/pp.104.039743.

Van Houtte, H., Vandesteene, L., López-Galvis, L., Lemmens, L., Kissel, E., Carpentier, S., ... Van Dijck, P. (2013). Overexpression of the trehalase gene AtTRE1 leads to increased drought stress tolerance in arabidopsis and is involved in abscisic acid-induced stomatal closure. *Plant Physiology, 161*(3), 1158–1171. doi: 10.1104/PP.112.211391.

Varghese, N., Alyammahi, O., Nasreddine, S., Alhassani, A., & Gururani, M. A. (2019). Melatonin positively influences the photosynthetic machinery and antioxidant system of avena sativa during salinity stress. *Plants (Basel, Switzerland), 8*(12). doi: 10.3390/PLANTS8120610.

Verslues, P. E., & Sharma, S. (2010). Proline metabolism and its implications for plant-environment interaction. *The Arabidopsis Book*, *8*, e0140. doi: 10.1199/TAB.0140.

Vogel, G., Fiehn, O., Jean-Richard-Dit-Bressel, L., Boller, T., Wiemken, A., Aeschbacher, R. A., & Wingler, A. (2001). Trehalose metabolism in Arabidopsis: Occurrence of trehalose and molecular cloning and characterization of trehalose-6-phosphate synthase homologues. *Journal of Experimental Botany*, *52*(362), 1817–1826. doi: 10.1093/JEXBOT/52.362.1817.

Wang, Y., Reiter, R. J., & Chan, Z. (2018). Phytomelatonin: A universal abiotic stress regulator. *Journal of Experimental Botany*, *69*(5), 963–974. doi: 10.1093/jxb/erx473.

Wei, D., Zhang, W., Wang, C., Meng, Q., Li, G., Chen, T. H. H., & Yang, X. (2017). Genetic engineering of the biosynthesis of glycinebetaine leads to alleviate salt-induced potassium efflux and enhances salt tolerance in tomato plants. *Plant Science : An International Journal of Experimental Plant Biology*, *257*, 74–83. doi: 10.1016/j.plantsci.2017.01.012.

Weigel, P., Weretilnyk, E. A., & Hanson, A. D. (1986). Betaine aldehyde oxidation by spinach chloroplasts. *Plant Physiology*, *82*(3), 753. doi: 10.1104/PP.82.3.753.

Weng, C. M., Lu, J. X., Wan, H. F., Wang, S. W., Wang, Z., Lu, K., & Liang, Y. (2014). Over-expression of BnMAPK1 in brassica napus enhances tolerance to drought stress. *Journal of Integrative Agriculture*, *13*(11), 2407–2415. doi: 10.1016/S2095-3119(13)60696-6.

Xie, H. T., Wan, Z. Y., Li, S., & Zhang, Y. (2014). Spatiotemporal production of reactive oxygen species by nadph oxidase is critical for tapetal programmed cell death and pollen development in Arabidopsis. *Plant Cell*, *26*(5), 2007–2023. doi: 10.1105/tpc.114.125427.

Yamada, N., Sakakibara, S., Tsutsumi, K., Waditee, R., Tanaka, Y., & Takabe, T. (2011). Expression and substrate specificity of betaine/proline transporters suggest a novel choline transport mechanism in sugar beet. *Journal of Plant Physiology*, *168*(14), 1609–1616. doi: 10.1016/J.JPLPH.2011.03.007.

Yang, L., Zhao, X., Zhu, H., Paul, M., Zu, Y., & Tang, Z. (2014). Exogenous trehalose largely alleviates ionic unbalance, ROS burst, and PCD occurrence induced by high salinity in arabidopsis seedlings. *Frontiers in Plant Science*, *5*, 570. doi: 10.3389/FPLS.2014.00570/BIBTEX.

Zarza, X., Atanasov, K. E., Marco, F., Arbona, V., Carrasco, P., Kopka, J., … Alcázar, R. (2017). Polyamine oxidase 5 loss-of-function mutations in Arabidopsis thaliana trigger metabolic and transcriptional reprogramming and promote salt stress tolerance. *Plant, Cell & Environment*, *40*(4), 527–542. doi: 10.1111/PCE.12714.

Zhan, H., Nie, X., Zhang, T., Li, S., Wang, X., Du, X., … Song, W. (2019). Melatonin: A small molecule but important for salt stress tolerance in plants. *International Journal of Molecular Sciences*, *20*(3). doi: 10.3390/IJMS20030709.

Zhang, H., Dong, H., Li, W., Sun, Y., Chen, S., & Kong, X. (2009). Increased glycine betaine synthesis and salinity tolerance in AhCMO transgenic cotton lines. *Molecular Breeding*, *23*(2), 289–298. doi: 10.1007/S11032-008-9233-Z/FIGURES/5.

Zhang, A., Liu, Y., Wang, F., Li, T., Chen, Z., Kong, D., … Luo, L. (2019). Enhanced rice salinity tolerance via CRISPR/Cas9-targeted mutagenesis of the OsRR22 gene. *Molecular Breeding*, *39*(3), 1–10. doi: 10.1007/s11032-019-0954-y.

Zhao, G., Zhao, Y., Yu, X., Kiprotich, F., Han, H., Guan, R., … Shen, W. (2018). Nitric oxide is required for melatonin-enhanced tolerance against salinity stress in rapeseed (*Brassica napus* L.) seedlings. *International Journal of Molecular Sciences*, *19*(7). doi: 10.3390/ijms19071912.

Zhou, S. F., Chen, X. Y., Xue, X. N., Zhang, X. G., & Li, Y. X. (2007). Physiological and growth responses of tomato progenies harboring the betaine alhyde dehydrogenase gene to salt stress. *Journal of Integrative Plant Biology*, *49*(5), 628–637. doi: 10.1111/J.1744-7909.2007.00464.X.

Zhu, J. K. (2002). Salt and drought stress signal transduction in plants. *Annual Review of Plant Biology*, *53*, 247–273. doi: 10.1146/annurev.arplant.53.091401.143329.

Zhu, J. K. (2003). Regulation of ion homeostasis under salt stress. *Current Opinion in Plant Biology*, 6, 441–445. doi: 10.1016/S1369-5266(03)00085-2.

13

Metabolomics and Molecular Physiology Perspective for Drought and Salinity Stress Tolerance

Sagar Sudam Jadhav, Renu Kumari, Sanjeet Kumar Mahtha, Ravi Kiran Purama, Vinita Lamba, and Gitanjali Yadav
National Institute of Plant Genome Research

CONTENTS

13.1 Introduction

Industrialization, deforestation and continuous use of fossil fuels have led to steady increase in atmospheric CO_2 concentrations from 280 µmol CO_2/mol air to a present level of 400 µmol CO_2/mol air (Vaughan et al. 2019). This has resulted in greenhouse effects, warmer temperatures and increased abiotic stress events. Abiotic stresses such as salinity, flood, cold, ultra violet radiations, drought, heavy metal and heat severely affect crop productivity (Hatfield et al. 2011; Boyer et al. 2013). Recently, it is reported that due to dry European summer of 2018, cereal production declined by 8% (https://ec.europa.eu/info/sites/info/files/foodfarmingfisheries/farming/documents/short-term-outlook-autumn-2018_en.pdf). Though plants are sessile, they have evolved the remarkable ability to combat abiotic stresses, and this evolution has taken place at physiological, biochemical, cellular as well as at molecular levels. For example, during evolution, decreasing atmospheric CO_2 concentrations resulted in the development of C4 and crassulacean acid metabolism pathways of carbon fixation in plants (Sage 2001). Hence, C4 plants possess specialized cells named bundle sheath cells and therefore demonstrate a nice example of molecular as well as cellular evolution. C4 plants (e.g. Sorghum) have higher carbon fixation efficiency because of the minimized effect of photorespiration, a wasteful biochemical process that primarily occurs in C3 plants, such as rice. Moreover, most C4 plants are native to hot, dry environments and tropics and are tolerant to abiotic stresses. Many such favorable traits of C4 plants intrigued international scientists and, therefore, attempts to transfer C4 traits into rice have started (Ermakova et al. 2020). In this pursuit, many key metabolic enzymes of the C4 pathway and enzymes responsible for the formation of bundle sheath cells are being investigated. With the successful transfer of C4 traits into rice, high-yielding rice varieties with increased photosynthetic efficiency by 50%, improved nitrogen use efficiency and doubled water use efficiency can be developed (https://c4rice.com/). This will minimize challenges posed by abiotic stresses and the issues related to food security.

Among all abiotic stresses, drought and salinity are mainly affecting agriculture production. Therefore, many drought- and salt-tolerance-related breeding programs or quantitative trait loci (QTL) mapping studies are prospecting alleles from wild species or cultivars that have traits such as high drought/salt tolerance, high leaf chlorophyll content, enhanced photosynthetic capacity, lower cuticular transpiration, efficient root system and stomatal density (Shashidhar et al. 2013). The final objective of such programs is to introgress these favorable traits into elite crop cultivars without having negative epistatic effects. Recently, one of the studies has shown that multiple copies of phosphatase and inositol-3-phosphate synthase

DOI: 10.1201/9781003258063-13

(genes in the myo-inositol biosynthesis pathway) impart salt tolerance to wild tomato *Solanum pimpinellifolium* (Razali et al. 2018). Myo-inositol acts as a precursor for the synthesis of InsP6 (myo-Inositol hexakisphosphate) and ascorbic acid. InsP6 triggers Ca^{2+} release from guard cells while ascorbic acid can scavenge reactive oxygen species (ROS) under stress (Lemtiri-Chlieh et al. 2003). Myo-inositol and other plant metabolic pathways are under genetic, metabolic and hormonal regulation. Some of them are still hidden and are induced upon abiotic/biotic stress and treatment with epigenetic modifiers (Vasanthakumari et al. 2015). Various studies showed drought stress-related increase in primary as well as species-specific secondary metabolite contents. Here, we will mainly review the progress that has been made in the drought and salinity stress research with metabolomics and molecular perspective.

13.2 Metabolomics under Abiotic Stress

With the advancements in mass spectrometry and data analysis, plant metabolomics field is being updated and progressing speedily. Plant metabolome is the ultimate product of genome, transcriptome, proteome, post-translational modifications (PTMs), epigenome and environmental attacks. During most abiotic stresses, ROS are formed and can severely damage cellular components, such as membrane, ATP synthase and other key enzymes (Bowler et al. 1992). As a result, cellular antioxidant machinery synthesizes ROS scavengers, such as peroxidase (POX), catalase and superoxide dismutase (SOD) as well as antioxidant metabolites e.g. glutathione, ascorbic acid, α-tocopherol, anthocyanins, phenols, flavonoids, terpenes etc. in higher amounts (Rice-Evans et al. 1997; Hayat et al. 2012). These secondary metabolites contain protonated amino and imino groups that stabilize cellular substructures by facilitating electrostatic interactions with them (Edreva et al. 2007). Additionally, plants also produce ROS-detoxifying osmoregulatory compatible solutes like trehalose, pipecolate betaine, sucrose, polyols, proline betaine, glycine betaine, alanine betaine and proline (Ashraf and Foolad 2007). Therefore, to meet the high demand for these key metabolites and enzymes or substrates involved thereof, metabolic networks get reconfigured instantaneously upon imposition of abiotic stress (Fàbregas and Fernie 2019). This metabolic reconfiguration is also species, stress intensity and type of abiotic stress-specific.

Both drought and salt stress induce water stress in the plants. Plants respond to drought stress by reduction in stomatal aperture and root elongation. Contrary, in the case of salt stress, in order to accumulate higher amounts of chloride, heavier roots are formed. These physiological changes are accompanied by some of the common metabolomic changes. For example, during salinity stress, osmosensing mechanisms lead to an increase in ABA content while that happens in drought to trigger osmotic ion efflux, loss of guard cell turgor and closure of stomata (Tombesi et al. 2015). Amino acids glycine and proline reported as active drought and salt-stress-responsive metabolites. Abiotic stress-responsive metabolically linked pathways, such as pathways of lignin, cell wall, phenylpropanoid, photorespiration, fatty acid and hormone, are shown in Figure 13.1.

13.2.1 Plant Metabolomics under Drought Stress

Genes responsible for different cellular functions such as transport, photosynthesis, protein processing, and primary and secondary metabolism were mainly represented in the chloroplast proteome of tomato plants that were subjected to drought stress (Tamburino et al. 2017). These five processes shape metabolome by affecting each other for metabolic homeostasis under stress conditions. The imposition of drought stress causes an early and rapid increase in sugars. For example, raffinose family oligosaccharides, such as galactinol, myo-inositol and raffinose, were rapidly increased under dehydration stress in Arabidopsis (Fàbregas and Fernie 2019). Numbers of studies showed that these changes in sugar levels are followed by enhanced levels of amino acids (phenylalanine GABA, tryptophan, proline and branched chain amino acids) and various secondary metabolites (Fàbregas and Fernie 2019). Published reports implicated polyamine derivatives hordatines, terpenoid blumenol C derivatives, lignin, ascorbate as well as dehydroascorbate in drought stress tolerance (Hu et al. 2009; Piasecka et al. 2016). Under water stress condition, an increase in p-coumarate, proline, caffeate and malate while the decrease in ferulate content was found in the maize xylem sap (Alvarez et al. 2008). Drought responsiveness and affected metabolome are typically a species-specific mechanism and can be seen in Table 13.1. Besides above mentioned metabolomic changes, Vitamin E was also reported to be protecting plants against oxidative stress (Maeda et al. 2006). Various studies proved that isoprenes alter the expression of some genes in abiotic stress signaling pathways. For example, isoprenes regulate ROS synthesis, and their lipophilic properties stabilize thylakoid membranes at high temperatures (Velikova et al. 2015). Isoprene emission also enhances the synthesis of chlorophylls and carotenoids as well as induces gibberellic acid (GA) and jasmonic acid (JA)-mediated signaling (Zuo et al. 2019). Many metabolic pathways are interlinked because of the shared common precursors or key enzymes.

It is well documented that under drought stress, polyphenols, acetate and flavonoid pathways are induced (Kim et al. 2017). In this connection, the work done by Perez de Souza et al. (2020) in Arabidopsis is very interesting and enlightening. This work supports the hypothesis that the flavonoid regulatory network affects carbon flux through shikimate, phenylpropanoid and acetate pathways. Hence, identifying key transcription factors regulating such metabolic linkage will increase our understanding of metabolic regulation of drought-responsive pathways.

13.2.2 Plant Metabolomics under Salinity Stress

During salinity stress, osmotic adjustments are hindered because of lower water uptake and increased cytotoxicity due to excess Na^+ and Cl^- ions. Osmotic stress severely affects thylakoid membrane stability and photosynthetic pigment synthesis, disturbing metabolic processes such as photophosphorylation, photosynthesis, respiration, ROS scavenging mechanisms and

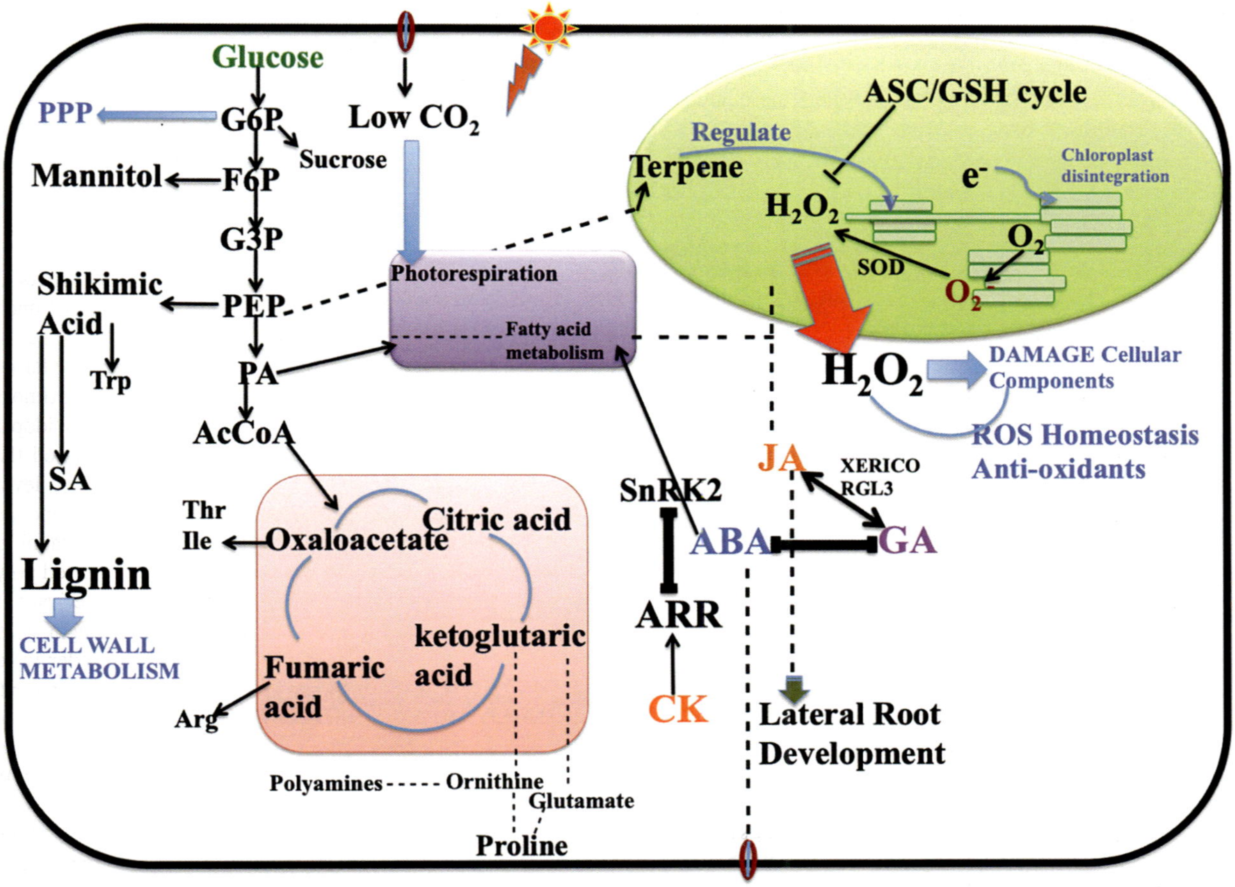

FIGURE 13.1 Linked metabolic and hormonal network underlying abiotic stress responses. Chloroplast, peroxisome and mitochondria are indicated by light green, purple and pink colors. G6P, Glucose-6-phosphate; F6P, Fructose-6-phosphate; G3P, Glucose-3-phosphate; PEP, Phosphoenolpyruvate; PA, Pyruvic acid; Trp, Tryptophan; Thr, Threonine; Ile, Isoleucin; Arg, Arginine.

hormone metabolism (especially ABA and JA) (Veiga et al. 2007). These salt-responsive metabolic perturbations induce metabolic homeostasis pathways and hence plant synthesizes osmoregulatory metabolites, for example, proline, sucrose, malate and fumarate. Table 13.2 lists recent studies reporting salt stress-responsive metabolic changes in different plant species. Additionally, plant species-specific increase in artemisinin, stevioside, capsaicin, 20-hydroxyecdysone, sorbitol, anthocyanins, polyamines, polyols and polyphenols under salinity stress has also been reported (Arrowsmith et al. 2012; Shahverdi et al. 2019; Muchate et al. 2019).

Exploiting microorganisms for developing abiotic stress tolerance is presently an active area of research. Auge and Stodola (1990) interestingly found that mycorrhiza maintains root turgor during drought. This work was then followed by many studies that showed the role of halo-tolerant plant-growth-promoting rhizobacteria (HT-PGPR) in conferring salt tolerance to plants. HT-PGPR produces metabolites, such as osmoprotectants, exopolysaccharides and volatile organic compounds (VOCs), and also helps in maintaining plant ion homeostasis (Upadhyay and Singh, 2015). For example, arbuscular mycorrhizal infections affect soluble sugars, proline and trehalose metabolism to regulate plant growth, yield and adaptation to salinity stress (Kumar et al. 2010). Mukherjee et al. (2019) reported that *Halomonas* sp. Exo1 produces exopolysaccharides, helps in osmotic stress tolerance and promotes the growth of rice crop. The integrative use of exopolysaccharides and *Alcaligenes* sp. showed promising results by enhancing plant growth under saline conditions (Fatima et al. 2020). Additionally, VOCs producing *Bacillus subtilis* GB03 were found to be imparting salt tolerance in *Arabidopsis* because of its tissue-specific activity of *HKT1* (High-affinity K^+ Transporter) (Zhang et al. 2008).

13.2.3 Hormonal Control of Drought and Salinity Tolerance

Drought and salinity stress response mechanisms are under the tight control of coordinated signaling between different plant hormones and their components as shown in Figure 13.1. For example, hormone ABA receptors PYR1 and PYLs positively regulate drought and salinity stress responses by inhibiting PP2C phosphatase activity and controlling ABA signaling (Fujii et al. 2011). However, Achard et al. (2006) showed that DELLA proteins are negative regulators in salinity stress-responsive signaling. It is well known that GAs regulate the rapid reduction of DELLA proteins. Therefore, both of these studies emphasize the importance of GA signaling (in addition

TABLE 13.1
Drought Stress-Responsive Leaf Metabolomics in Different Plant Species

Plant Species	Metabolites Increased	Metabolites Decreased	References
Arabidopsis thaliana Col-0	Five different flavonols, five different anthocyanins, galactinol, raffinose and proline		Nakabayashi et al. (2014)
A. thaliana Col-0	Fatty acid unsaturation, DGDG:MGDG ratio		Gigon et al. (2004)
Arachis hypogaea cv. TAG-24	D-xylopyranose, stearic acid, D-ribose, agmatine, cadaverine, syringic acid, vanillic acid		Gundraniya et al. (2020)
Glycine max cv. W82	Tryptophan, proline, glycine, 1,3-dihydroxyacetone	Tyrosine, alanine, succinate, coumestrol	Das et al. (2017)
G. max cv. BR16	Proline, sugars, putrescine, spermidine, ornithine		Lima et al. (2019)
Hordeum vulgare RILs	Hordatines, terpenoid blumenol C derivatives, ferulic and sinapic acid derivatives, acylated glycosides of flavones		Piasecka et al. (2016)
Nicotiana tabacum cv. Burley 21	Galactinol, raffinose, dehydroascorbate	Oxoglutaric acid, mannitol, trehalose, inosine, GABA, GSSG	Rabara et al. (2015)
Sesamum indicum ZZM3330	ABA, proline, arginine, 2-aminoadipate, aromatic and branched chain amino acids, GABA, saccharopine, allantoin, lysine	Nucleosides, nucleotides, stachyose, galactose	You et al. (2019)
Triticum aestivum cv. ***Kedar, Gayetri***	Proline, phenol, ascorbate	Total chlorophyll	Chakraborty and Pradhan (2012)
T. aestivum cv. AK, CS	Phenolics, flavonoids, anthocyanin, schaftoside		Ma et al. (2014)
T. aestivum cv. Bahar	Proline, methionine, arginine, lysine, aromatic and branched chain amino acids	Glutamate	Michaletti et al. (2018)
T. aestivum cv. LA754	Tryptophan, valine, citric acid, fumaric acid, malic acid, proline, α-linolenic acid/γ-linolenic acid, phosphoric acid, glucose, fructose	Aspartic acid, glutamic acid, serine, alanine, glycine, putrescine	Kang et al. (2019)
Zea mays	Tryptophan, asparagine, β-alanine, phenylalanine, isoleucine	Erythritol, maltose, malate, fumarate	Obata et al. (2015)

TABLE 13.2
Salt-Stress-Responsive Leaf Metabolomics in Different Plant Species

Plant Species	Metabolites Increased	Metabolites Decreased	References
A. thaliana Col-0	Raffinose, galactinol		Nishizawa et al. (2008)
A. hypogaea cv. LH14	Urea, leucine, uridine, histidine, sedoheptulose, caffeic acid	Malate, 4-hydroxybutyrate, quinic acid	Cui et al. (2018b)
Gossypium hirsutum cv. ***SCRC 28, K638***	Tannin, gossypol, flavonoid		Wang et al. (2015b)
Lonicera japonica	Quinic acid, caffeic acid, ferulic acid, isochlorogenic acid A, luteolin-5-O-β-D-glucopyranoside, lonicerin, flavoyadorinin-B, isoquercitrin, aastragalin	Iridoids	Cai et al. (2020)
Oryza sativa cv. Sujala, MTU 7029, Bhutnath, Nonabokra	(Intolerant – trehalose, amino acids, ferulic acid, serotonin, gentisic acid, vanillic acid) (Insensitive – Cinnamic acid derivatives)	(Intolerant – Hydroxybenzoic acid, adenosine) (Insensitive – Adenine, galactinol, trehalose)	Gupta and De (2017)
Salicornia herbacea	Shikimic acid, vitamin K1, indole-3-carboxylic acid, steroids		Lee et al. (2016)
S. indicum WZM3063	13 Free amino acids, 8 sugars, ABA, aspartate, glutamate	N-acetylglutamic acid	Zhang et al. (2019a)
S. lycopersicum cv. MM	Proline, maltose, rhamnose, tyrosine, serine	Methionine, asparagine, fumarate, glutamine, arginine, alanine, phenylalanine	Devkar et al. (2020)

to ABA) during salt stress response and hint that DELLA proteins might be linking ABA-mediated abiotic stress signaling with GA-regulated signaling. In this connection, RING-H2 zinc finger factor *XERICO*, which was shown to be acting downstream of DELLAs, was also found to be acting at the interface of ABA and GA signaling pathways (Ko et al. 2006). Moreover, RGL3, a DELLA protein, was also reported to interact with JAZ proteins (Wild et al. 2012). Furthermore, drought-induced expression of many genes in JA biosynthesis and signaling pathways was studied in rice (Du et al. 2013).

Recently, *SnRK2s* (sucrose nonfermenting1-related kinases 2s) reported to phosphorylate *ARR5*, a well-known negative regulator of cytokinin signaling (Huang et al. 2018). This study showed antagonistic signaling between cytokinin and ABA during drought stress response.

13.3 Molecular Physiology under Abiotic Stress

Drought and salinity stress-responsive genes and transcription factors have been listed in Table 13.3. Among these studies, the identification of drought, cold and salinity stress-responsive novel *cis*-acting elements in rd29A and rd29B promoter was an important finding (Yamaguchi-Shinozaki and Shinozaki 1994). The same group went on further and successfully identified drought and cold stress-responsive transcription factors (TFs) *DREB1* (dehydration-responsive element-binding 1) and *DREB2* (Liu et al. 1998). Tahmasebi et al. (2019) processed publically available gene expression datasets that were available for 85 samples of cotton at different abiotic stresses. Mainly TF families, such as bZIP, C2H2, NAC, G2 like, WRKY, MYB and ERF, were upregulated, while TF families GRAS, ARF and S1Fa-like were downregulated in stress conditions. By using publically available microarray data, Chen et al. (2014) analyzed all soybean HD-Zip genes and found 2 HD-Zip III, 4 HD-Zip II and 12 HD-Zip I genes to be regulating salinity stress response. In our recently published collaborative work, we analyzed publically available transcriptome metadata from a total of 226 arrays, and 89 susceptible and 137 tolerant arrays, for their differentially expressed genes. This meta-transcriptomics study identified 6,956 "abiotic stress-tolerant" genes and stress-responsive networks (Smita et al. 2020). Moreover, by co-expression analysis, we found complex regulatory network containing genes related to the pathways of signal transduction, ubiquinone and osmotic adjustment.

13.3.1 Drought-Responsive Molecular Physiology

Previous experiments in maize showed that drought caused a considerable decline in water use efficiency (50.87%), transpiration rate (37.84%), net photosynthesis (33.22%), stomatal conductance (25.54%), intercellular CO_2 (5.86%) and intrinsic water use efficiency (11.58%) over that of well water control (Anjum et al. 2011). Such drastic physiological responses are regulated through complex networks of TFs, hormones and epigenetic mechanisms, and some of them have been listed in Table 13.3. By studying *yy1* mutant in Arabidopsis, Li et al. (2016) found that TF *AtYY1* negatively regulates ABA-responsive gene expression by interacting with TFs *ABI4* and *ABR1*. Various studies supported that ABA accumulation is positively correlated with drought tolerance (Yamazaki et al. 1995). In light of this, ABA-dependent and independent signaling cascades have been elucidated. For example, abscisic acid-responsive element binding protein (*AREB)/ABF* was identified by microarray analysis in Arabidopsis, rice and soybean (Maruyama et al. 2012). Transcription factor families, such as Zinc finger proteins, NAC, bZIP, AP2/ERF, HD-ZIP and MYB, have also been implicated in this signaling cascade. Basing on these findings, ABA-dependent and independent signaling network has been updated and presented in Figure 13.2. In addition to the TFs and genes mentioned in Figure 13.2, genes *PhyB*, *OsTOC1* and *Ghd7* were also found to be acting in an ABA-dependent manner while *OsGI, OsELF3, OsPRR37* and *OsMADS50* were

TABLE 13.3
Genes and Transcription Factors Involved in Drought and Salt-Stress Response

Gene/TF	Abiotic Stress	References
DREB	Drought	Liu et al. (1998)
Adc		Capell et al. (2004)
SOD, APX		Lu et al. (2010)
ERD1		Tran et al. (2006)
GmNAC8		Yang et al. (2020)
XERICO		Ko et al. (2006)
ZPT2-3		Sugano et al. (2003)
Med25 mediator		Elfving et al. (2011)
AREB/ABFs		Maruyama et al. (2012)
SnRK2		Fujita et al. (2013)
TaNAC2, TaNAC69		Mao et al. (2012)
		Xue et al. (2011)
GmNAC20		Hao et al. (2011)
GmbZIP1		Gao et al. (2011)
HvDRF1		Xue and Loveridge (2004)
		Chen et al. (2012)
OsbZIP16		Zhao et al. (2007)
miR-169g, miR393		
NHX1	Salt	Gaxiola et al. (2001)
AVP1		Pasapula et al. (2011)
SlATAF1		Devkar et al. (2020)
JUB1, ANAC019,		Wu et al. (2012)
ANAC055 and ANAC072		Tran et al. (2004)
ANAC092, ANAC016		Balazadeh et al. (2010)
		Kim et al. (2013)
SNAC1		Hu et al. (2006)
OsNAC6		Nakashima et al. (2007)
OsNAC045		Zheng et al. (2009)
OsNAC5		Takasaki et al. (2010)
OsNAC106		Sakuraba et al. (2015)
ONAC022		Hong et al. (2016)
OsNAC2, OsAP37, OsCCX11		Mao et al. (2018)
SlSOS1		Olias et al. (2009)
LeNHX2		Huertas et al. (2013)
SlAREB1		Orellana et al. (2010)
SlARS1		Campos et al. (2016)
SlDREB2		Hichri et al. (2016)
SlbZIP1		Zhu et al. (2018)
SlNAC4		Zhu et al. (2014)
SlNAC35		Wang et al. (2016)
SlNAC11		Wang et al. (2017)
OsCDPK21		Asano et al. (2011)
OsbZIP23	Salt, drought	Xiang et al. (2008)
AtbHLH17, AtWRKY28		Babitha et al. (2013)

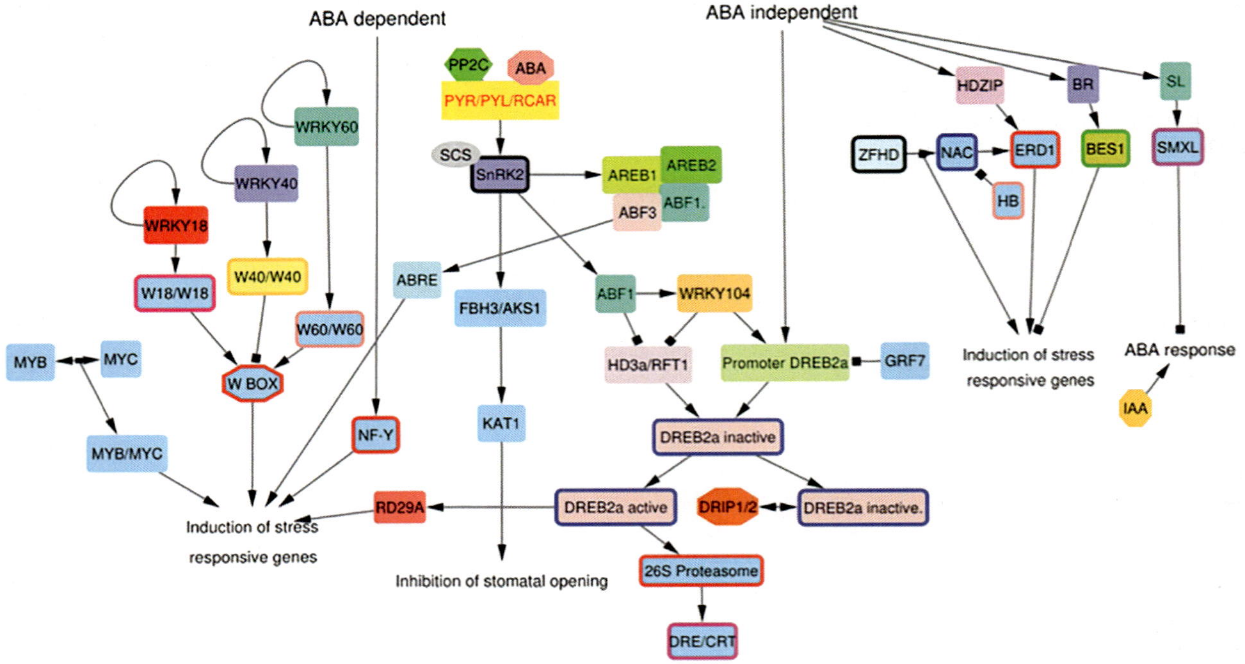

FIGURE 13.2 Generalized network of drought- and salinity-stress-responsive genes, transcription factors and hormonal signaling (Singh and Laxmi 2015). BR, SL and IAA are brassinosteroids, strigolactones and indole acetic acid, respectively.

found to be acting independent of ABA (Du et al. 2018). Wang et al. (2018) reported the role of flowering repressor *SVP* and ABA catabolism pathway genes (*CYP707A1/3* and *AtBG1*) in drought tolerance. Moreover, sulfate transporter subfamily 3 was also implicated in ABA biosynthesis and drought stress response (Chen et al. 2019).

Studying natural variation may unravel novel genetic machinery underlying stress tolerance (Grandillo et al. 2011). With next generation sequencing techniques, such as nanopore sequencing, scientists are now sequencing many wild species and are unraveling stress tolerance mechanisms in those species. Two such studies have been carried out in wild tomatoes *S. pimpinellifolium* (discussed earlier) and *Solanum pennellii*. Bolger et al. (2014) sequenced *S. pennellii* and identified that stress-responsive genes from *S. pennellii* show a significant enrichment of Copia transposable elements within their 5 kb. This study confirms the vital role played by transposable elements in the evolution of abiotic stress response mechanisms. In our research work, we are studying the molecular architecture and expression profile of 3 *HMGR* (3-hydroxy-3-methylglutaryl-coenzyme A reductase) genes of mevalonate pathway in chickpea. Interestingly, it was found that among 3 *HMGR* genes, *HMGR2* and *HMGR3* are specifically highly expressing under abiotic stresses, which will be further pursued with a transgenic approach. Simultaneously, assembling transcriptome of *Lantana camara* is underway. *Lantana camara* reported showing drought-avoidance strategies because of its deep root system and high water-holding capacity (Castillo et al. 2007). We hope that transcriptomics of *Lantana camara* will unravel many key novel metabolic enzymes responsible for its drought-avoidance physiology.

Previous reports showed that drought stress-responsive genes get induced upon the interaction between TFs such as *TaNF-YA10–1* and *NF-YA/NF-YB* (Ma et al. 2015). Among 14 and 17 HD-Zip subfamily I genes in rice and arabidopsis, respectively, some are also reported to be regulated by drought and salt stresses (Bray 2004; Agalou et al. 2008). Therefore, publically available transcriptome datasets could be exploited to perform gene co-expression network (GCN) analysis to annotate stress-responsive and poorly characterized TF genes (Reményi et al. 2004). In such GCNs, TF candidates are key regulatory target because alterations in functions of TFs cause major metabolic diseases. Therefore, various studies have been focused on studying genome-wide analysis of abiotic stress-responsive TF families. By using the SMART (http://smart.embl-heidelberg.de/), Pfam (http://pfam.janelia.org/) and BLASTP search, Zhao et al. (2016) identified genome-wide *TIFY* TF gene family and found *GaJAZ5* to be regulating drought stress response in cotton. By performing conserved motif analysis, phylogenetic construction and interaction network analysis for 33 HD-Zip TFs in tea plants, Shen et al. (2019) found that gene *Cshdz8* had a high identity to *HB6* (well-known ABA-dependent growth regulator) and therefore might regulate drought stress response and ABA biosynthesis. In maize, 15 important TF families (containing 1,436 genes) were characterized by analyzing TF-specific domains, protein-protein interactions, gene structure, *cis*-regulatory elements and motifs, and expression assays (Mittal et al. 2018). Doing this, the authors found leaf-specific bZIP TF, *GRMZM2G140355*, as a major drought response regulator.

13.3.2 Salinity Stress-Responsive Molecular Physiology

Dubouzet et al. (2003) identified *JUB1* regulating drought, salinity and osmotic stress tolerance by directly controlling the expression of *DREB2A* a key TF, while Achard et al.

(2006) found Arabidopsis quadruple-DELLA mutant (lacking DELLAs such as *GAI*, *RGA*, *RGL1* and *RGL2*) seedlings to be salt tolerant. Therefore, DELLAs and *JUB1* could be one of the main targets for post-transcriptional/-translational modifications during salinity stress. Various TFs acting downstream of DELLAs and JUB1 have been listed in Table 13.3. Let us have a look at how molecular physiology regulates ROS scavenging, lipid profile and kinase pathway in response to salinity. Arabidopsis TF *WRKY33* was found to be targeting genes such as *glutathione S-transferase (GSTU11)*, *POX* and *lipoxygenase* (*LOX1*) under salt stress (Jiang and Deyholos 2009). Furthermore, *ThWRKY4* from *Tamarix hispida* also reported to control the expression and activity of *SOD* and *POX* and regulate ROS biosynthesis (Zheng et al. 2013).

The number of genes involved in modulating salinity-stress-responsive lipid profiles has been identified. It was shown that ABA-mediated regulation of stomatal aperture requires phospholipase C (Hunt et al. 2003). Next, Darwish et al. (2009) found enhanced levels of PA (phosphatidic acid), DGPP (diacylglycerol pyrophosphate) and PIP2 (phosphatidylinositol bisphosphate) upon salinity stress in rice and implicated diacylglycerol kinase and phospholipase C as important enzymes in salt-stress-induced signaling. Here, PA influences the membrane integrity and activity of ion channels (Falkenburger et al. 2010). Besides this, role of phospholipase Dα1 in the biosynthesis of ABA and PA has also been elucidated (Jia et al. 2013).

During salt stress response, SOS (salt overly sensitive) pathway performs a central role in maintaining and regulating ion homeostasis. Interestingly, activation of plasma membrane-localized SOS1 (Na^+/H^+ antiporter) by the interaction of SnRK3 SOS2-like protein kinases with SOS3 has been documented (Chinnusamy et al. 2004; Du et al. 2011). Additionally, sodium chloride and PA-mediated regulation of *MPK6* via *SOS1* phosphorylation have been studied (Yu et al. 2010). Moreover, Liu et al. (2013) validated the interaction of C_2H_2-type zinc finger protein ZAT6 with *MPK6* and its role in regulating salt and osmotic stress responses in Arabidopsis. Nowadays, research is focused on elucidating signaling cascades that happen during the interaction of ROS signaling with MAP kinase pathway. In this attempt, Schmidt et al. (2013) identified TF *SERF1* (*salt-responsive ERF1)* to be linking both pathways in rice.

13.4 Kinases and RNA Metabolism

Kinases play a multifaceted role in the physiological functioning of the plants. Members of *SnRK2* (a multiple gene family with ten redundant members belonging to Ser/Thr kinases) are integral part of ABA-dependent signaling pathways of plant developmental processes. However, multiple *SnRK2* members work either in an ABA-dependent or independent manner to alter the drought and osmotic stress tolerance (Kulik et al. 2011; Saruhashi et al. 2015). For example, ABA activate *SnRK2.6* (*OST1*) which then phosphorylate plasma membrane intrinsic protein 2;1 (PIP2;1) aquaporin and cause closure of stomata (Grondin et al. 2015). Contrary, *SnRK2.4* and *SnRK2.10* reported modifying the root system architecture during salt stress in an ABA independent manner (McLoughlin et al. 2012). Additionally, *SnRK2.3* and *SnRK2.8* proteins have been implicated in the phosphorylation and import of drought stress-responsive TFs *HAT1* and *NTL6*, respectively (Kim et al. 2012; Tan et al. 2018). Through transcriptomic profiling of the RNAi and overexpression lines of *OsSAPK9* (stress-activated protein kinase 9), Zhang et al. (2019b) demonstrated its dual functioning in alleviating bacterial blight resistance and salt stress tolerance in rice. Furthermore, overexpression of *OsSAPK* members in rice confirmed tolerance to oxidative stress (*OsSAPK4*), salt stress (all 10 *OsSAPKs*) and drought stress (*OsSAPK9*) (Kobayashi et al. 2004; Diédhiou et al. 2008; Dey et al. 2016).

Gene expression comprises transcription and translation, and among them, RNA metabolism is the sensitive and key regulatory pathway that is mostly affected during abiotic stress response. Recently, it is reported that ABA insensitive subclass I *SnRK2s* and its substrate *VARICOSE* (an mRNA-decapping activator) control salinity stress response by regulating mRNA decay (Soma et al. 2017). In addition to this, published reports confirm the role of RISC (RNA-induced silencing complex) complex proteins in the stress response of plants. Published studies reported induction of DEAD-box DNA helicases in pea (*PDH45* and *PDH47*) and rice (*OsSUV3*) under abiotic stress and ABA treatment (Vashisht et al. 2005; Sanan-Mishra et al. 2005). Transgenic rice overexpressing *OsSUV3* showed that this protein exhibits DNA/RNA helicase and ATPase activity and improves antioxidant and photosynthetic capacity and imparts salt stress tolerance (Tuteja et al. 2013). The same group implicated *Pisum sativum* PsP68 DEAD-box helicases in cold, salinity and heat stress response (Banu et al. 2015). Transgenic rice lines overexpressing *PsP68* showed lower loss of chlorophyll, improved salinity tolerance, increased proline content, lower electrolyte leakage, decreased ROS accumulation and better water retention. Authors also confirmed *in vitro* interaction of PsP68 with many other proteins, including argonaute protein, (*AGO1*) and demonstrated that *PsP68* acts as a key regulator in molecular switching of the stress-influenced responses in transgenic lines.

Plants have to respond promptly to environmental changes for their survival, and their response varies from transcriptional regulation to post-transcriptional regulation. Various studies proved crucial roles of AGO proteins and small RNAs in regulating abiotic stress-influenced gene expression. For example, Bai et al. (2012) found upregulation of four dicer-like proteins (SlDCL1, 2a, 2c and 2d) under different abiotic stresses. Here, three *RDR* (RNA-dependent RNA polymerase) genes (*SlRDR2, 6a* and *3a*) and five *AGO* genes (*SlAGO1a, 1b, 4c, 4b* and *5*) showed high expression levels under abiotic stresses. *SlAGO10a* and *10b* were expressed strongly under heat and salt stress conditions, respectively. AGO proteins are RNA-binding proteins that play a primary role in the small RNA-guided gene-silencing pathways (Bohmert et al. 1998). Based on phylogenetic relationship and capacity to bind to small RNAs, plant AGO proteins can be classified into three major groups: AGO1/5/10, AGO2/3/7 and AGO4/6/8/9 (Vaucheret 2008). Published studies reported 19, 17 and 10 different AGOs in rice, maize and arabidopsis, respectively (Morel et al. 2002; Kapoor et al. 2008; Zhai et al. 2019). Recently, Zhai et al. (2019) reported gene expression study of AGOs in maize seedlings subjected to different abiotic stresses and found upregulation of 9 and 2 *ZmAGOs* in response to drought and heat, respectively. Here, the authors implicated *AGO18a/b* in regulating drought-stress-responsive gene expression.

In addition to the helicases and argonautes discussed above, microRNAs (miRNAs) are the key post-transcriptional regulators under salt stress conditions. Li et al. (2012) demonstrated that ABA regulates transcriptional levels of miRNA *MIR168a*, which then maintains stable *AGO1* transcript levels during drought stress response in Arabidopsis. Furthermore, Wang et al. (2015a) reported that *HOS1* (*High Expression of Osmotically Responsive Genes1*) regulates *AGO1* and *miR168a/b* levels by transcriptional regulation of *MIR168b*. Furthermore, *HOS1* mutant was found to be a ubiquitin E3 ligase-regulating cold-stress responses and mRNA export across nuclear pores. Dolata et al. (2016) proved that AGO1 played a key regulatory role in salt stress by controlling the levels of *miR161* and *miR173*. By using *snrk2.2/3/6* mutants and *in vitro* assays, Yan et al. (2017) confirmed the role of *SnRK2s* in the modulation of miRNA levels. miRNA-processing complex consists of zinc finger protein SE (SERRATE) and double-stranded RNA-binding protein HYL1 (HYPONASTIC LEAVES). These two proteins phosphorylate SnRK2 kinases *in vitro* by directly interacting with them. Recently, abiotic stress-responsive *Raf-like MAPKKKs* functioning upstream of the SnRK2 kinases have been identified in Arabidopsis (Soma et al. 2020).

13.5 Prospecting of Drought and Salt Stress-Responsive Metabolic QTL (mQTLs)/Alleles

In order to screen stress-responsive genotypes or identify candidate genes responsible for quantitative trait variation, various strategies have been employed, for example, the use of $\delta^{13}C$ and QTL studies. QTL mapping studies help in identifying quantitative trait regulating candidate genes and their evolutionary perspective. In this attempt, F2 and recombinant inbred lines (RILs) populations were developed and studied. Various studies have reported the development of introgression lines (ILs) and backcross inbred lines for several different species of tomato, melon and rice (Eshed and Zamir, 1994; Kumar et al. 2020). *S. pennellii* have been reported to be extreme stress tolerant and therefore panel of 76 *S. pennellii* ILs was developed and exploited to prospect favorable alleles required for abiotic stress tolerance. For example, Gur and Zamir (2004) found QTLs on ILs 7–5–5, 8–3 and 9–2–5 to be regulating yield under drought. Additionally, Gong et al. (2010) found ILs 2–5 and 9–1 as drought-tolerant ILs and have identified 400 drought-responsive genes. A recent study subjected a population of 100 barley RILs to drought and by secondary metabolite profiling found cold, heat and oxidative stress-responsive secondary metabolite mQTLs and genes (Piasecka et al. 2016).

Several grain yield QTLs for reproductive stage drought stress have been identified and validated in various popular lowland and upland rice varieties (Dixit et al. 2017). Recently, an innovative green super rice technology was implemented to develop salt- and drought-tolerant ILs (Ali et al. 2017). Cui et al. (2018a) reported the development of higher yielding drought stress-responsive 72 ILs. In addition to these studies, a panel consisting of 25,000 ILs in 11 elite rice genetic backgrounds was developed (Kumar et al. 2020). This study developed high-quality ILs that gave high yield even under severe reproductive stage drought. The same research group used marker-assisted selection approach to integrate biotic and abiotic stress-responsive QTLs from five different donor parents in the Swarna with drought background. Doing these, they developed seven ILs with high-yielding trait under various stresses.

Marker-assisted selection has not been exploited much for salinity-stress-responsive QTL identification. Rivandi et al. (2010) reported *HvNax4* as the locus harboring candidate genes for salt tolerance in barley. Further, fine mapping of this locus identified 34 salt-stress-responsive candidate genes. Shavrukov et al. (2010) reported *HvNax3* as the major locus for salinity stress tolerance.

13.6 Conclusion

We discuss this section at three different levels as amino acid metabolism, hormone metabolism and epigenetic reprogramming. Amino acid metabolic network is complex, not only because of its linkage with the tri-carboxylic acid cycle and its central role in nitrogen metabolism but also because amino acids are precursors for various secondary metabolites, hormones and cell wall components. For example, methionine is the precursor for chlorophyll, ethylene and polyamine biosynthesis. Glycine and proline are major constituents of plant cell wall structural proteins while phenylalanine and tyrosine are precursors of flavonoids as well as lignins (Fraser and Chapple, 2011). Metabolic linkage between amino acid and secondary metabolic pathways is shown in Figure 13.1. Chlorophylls, lignins, flavonoids and ethylene are metabolites playing key roles during abiotic stress tolerance (Hu et al. 2009). Plant metabolism is under the control of transcriptional regulation, allosteric control of the activity of a key enzyme and PTMs. Therefore, understanding genetic or metabolic regulation of amino acid metabolism is the key for abiotic stress research. In this pursuit, by studying publically available Arabidopsis microarray data, Less and Galili (2008) proposed a novel regulatory program for abiotic stress-responsive amino acid metabolic regulation. Here, authors found that the allosteric biosynthetic enzyme is regulated post-translationally and in feedback inhibition, while the first catabolic enzyme is regulated at transcriptional level. Recently, the genetic regulation of leucine metabolism was studied (Ning et al. 2015). This study found that *Solanum lycopersicum* gene *SlIPMS3* encodes a leucine inhibition-insensitive version of isopropylmalate synthase, missing the carboxyl-terminal 160 amino acids. Novel molecular architecture of this *SlIPMS3*, therefore, resulted in enhanced production of acylsugars (biotic stress-responsive secondary metabolites). Comparative metabolomics and transcriptomics of wild tomato species with that of domesticated have led to such interesting finding. Day-by-day genomes and transcriptomes of various wild species are becoming available and therefore there is much more scope for such findings to unravel abiotic stress-specific key metabolic bottlenecks and novel enzymes.

During abiotic stress, maintaining cellular metabolic/ion homeostasis becomes priority, and this happens at the cost of global reprogramming of primary metabolism. Upon imposition of abiotic stress, rapid activation of sugars and lipid signaling cascades happen (Hou et al. 2016). However, such metabolic reprogrammings are tightly regulated by the interplay of hormones and epigenetic machinery. Karpac and Jasper (2011) have reported that hormonal regulation of lipid and glucose metabolism is pivotal for metabolic homeostasis and energy balance; and histone deacetylases (HDACs) play important role in this metabolic signaling. In line with this, HDACs from several plant species such as barley, soybean, rice, maize, tomato, arabidopsis, common bean etc. have been reported to be playing role in several abiotic stress responses (Jadko 2015; Hayford et al. 2017; Liu et al. 2017). Besides HDACs, various abiotic stress-responsive novel epigenetic components are being identified. Review by Asensi-Fabado et al. (2017) listed such epigenetic studies that have been carried out on several different plant species subjected to different abiotic stresses. Growing evidence also suggests that nuclear cap-binding complex-guided RNA metabolism plays a primary role in drought and salt stress response by modulating ethylene and ABA (Daszkowska-Golec 2018). Future studies involving stable isotope approaches, co-expression network analysis, transcription factor signaling and its correlation with HDACs might increase our understanding of factors that play key roles in metabolic homeostasis under abiotic stress. Exploiting metabolomics and molecular physiology of abiotic stress-responsive archaebacteria can also be helpful in this endeavor.

Conflict of Interest

The authors declare that they have no conflict of interests.

REFERENCES

Achard, P., Cheng, H., De Grauwe, L. et al. 2006. Integration of plant responses to environmentally activated phytohormonal signals. *Science (New York, N.Y.)*, 311, 91–94.

Agalou, A., Purwantomo, S., Overnäs E. et al. 2008. A genome-wide survey of *HD-Zip* genes in rice and analysis of drought-responsive family members. *Plant Molecular Biology*, 66, 87–103.

Ali, J., Xu, J.L., Gao, Y.M., et al. 2017. Harnessing the hidden genetic diversity for improving multiple abiotic stress tolerance in rice (*Oryza sativa* L.). *PLoS One*, 12, e0172515.

Alvarez, S., Marsh, E.L., Schroeder, S.G. and Schachtman, D.P. 2008. Metabolomic and proteomic changes in the xylem sap of maize under drought. *Plant, Cell & Environment*, 31, 325–340.

Anjum, S., Xie, X., Wang, L., Saleem, M., Man, C. and Lei, W. 2011. Morphological, physiological and biochemical responses of plants to drought stress. *African Journal of Agricultural Research,* 6.

Arrowsmith, S., Egan, T.P., Meekins, J.F., Powers, D. and Metcalfe, M. 2012. Effects of salt stress on capsaicin content, growth, and fluorescence in a Jalapeño cultivar of *Capsicum annuum* (Solanaceae). *Bios*, 83, 1–7.

Asano, T., Hakata, M., Nakamura, H., et al. 2011. Functional characterisation of OsCPK21, a calcium-dependent protein kinase that confers salt tolerance in rice. *Plant Molecular Biology*, 75, 179–191.

Asensi-Fabado, M.A., Amtmann, A. and Perrella, G. 2017. Plant responses to abiotic stress: The chromatin context of transcriptional regulation. Biochimica et biophysica acta. *Gene Regulatory Mechanisms*, 1860, 106–122.

Ashraf, M. and Foolad, M.R. 2007. Roles of glycine betaine and proline in improving plant abiotic stress resistance. *Environmental and Experimental Botany*, 59, 206–216.

Augé, R. M. and Stodola, J. W. 1990. Apparent increase in symplastic water contributes to greater turgor in mycorrhizal roots of droughted Rosa plants. *New Phytol.* 115: 285–295.

Babitha, K.C., Ramu, S.V., Pruthvi, V. et al. 2013. Co-expression of *AtbHLH17* and *AtWRKY28* confers resistance to abiotic stress in Arabidopsis. *Transgenic Research*, 22, 327–341.

Bai, M., Yang, G.S., Chen, W.T. et al. 2012. Genome-wide identification of *Dicer-like, Argonaute* and *RNA-dependent RNA polymerase* gene families and their expression analyses in response to viral infection and abiotic stresses in *Solanum lycopersicum*. *Gene*, 501, 52–62.

Balazadeh, S., Siddiqui, H., Allu A.D. et al. 2010. A gene regulatory network controlled by the NAC transcription factor ANAC092/AtNAC2/ORE1 during salt-promoted senescence. *The Plant Journal*, 62, 250–264.

Banu, M.S.A., Huda, K.M.K., Sahoo, R.K. et al. 2015. Pea p68 imparts salinity stress tolerance in rice by scavenging of ROS-mediated H_2O_2 and interacts with Argonaute. *Plant Molecular Biology Reporter*, 33, 221–238.

Bohmert, K., Camus, I., Bellini, C., Bouchez, D., Caboche, M. and Benning, C. 1998. AGO1 defines a novel locus of Arabidopsis controlling leaf development. *The EMBO Journal*, 17, 170–180.

Bolger, A., Scossa, F., Bolger M.E. et al. 2014. The genome of the stress-tolerant wild tomato species *Solanum pennellii. Nature Genetics*, 46, 1034–1038.

Bowler, C., Montagu, M.V., and Inze, D. 1992. Superoxide dismutase and stress tolerance. *Annual Review of Plant Physiology and Plant Molecular Biology*, 43, 83–116.

Boyer, J.S., Byrne, P., Cassman, K.G. et al. 2013. The U.S. drought of 2012 in perspective: A call to action. *Global Food Security*, 2, 139–143.

Bray, E.A. 2004. Genes commonly regulated by water-deficit stress in *Arabidopsis thaliana*. *Journal of Experimental Botany*, 55, 2331–2341.

Cai, Z., Chen, H., Chen, J., et al. 2020. Metabolomics characterizes the metabolic changes of Lonicerae Japonicae Flos under different salt stresses. *PLoS One,* 15, e0243111.

Campos, J.F., Cara, B., Pérez-Martín F. et al. 2016. The tomato mutant *ars1* (*altered response to salt stress 1*) identifies an R1-type MYB transcription factor involved in stomatal closure under salt acclimation. *Plant Biotechnology Journal*, 14, 1345–1356.

Capell, T., Bassie, L. and Christou, P. 2004. Modulation of the polyamine biosynthetic pathway in transgenic rice confers tolerance to drought stress. *Proceedings of the National Academy of Sciences of the United States of America*, 101, 9909–9914.

Castillo, J.M., Leira-Doce, P., Carrión-Tacuri J. et al. 2007. Contrasting strategies to cope with drought by invasive and endemic species of Lantana in Galapagos. *Biodiversity and Conservation*, 16, 2123–2136.

Chakraborty, U. and Pradhan, B. 2012. Drought stress-induced oxidative stress and antioxidative responses in four wheat (*Triticum aestivum* L.) varieties. *Archives of Agronomy and Soil Science*, 58, 617–630.

Chen, H., Chen, W., Zhou, J. et al. 2012. Basic leucine zipper transcription factor OsbZIP16 positively regulates drought resistance in rice. *Plant Science: An International Journal of Experimental Plant Biology*, 193–194, 8–17.

Chen, X., Chen, Z., Zhao, H. et al. 2014. Genome-wide analysis of soybean *HD-Zip* gene family and expression profiling under salinity and drought treatments. *PLoS One*, 9, e87156.

Chen, Z., Zhao, P.X., Miao Z. et al. 2019. SULTR3s function in chloroplast sulfate uptake and affect ABA biosynthesis and the stress response. *Plant Physiology*, 180, 593–604.

Chinnusamy, V., Schumaker, K. and Zhu, J.K. 2004. Molecular genetic perspectives on cross-talk and specificity in abiotic stress signalling in plants. *Journal of Experimental Botany*, 55, 225–236.

Cui, F., Sui, N., Duan, G. et al. 2018a. Identification of metabolites and transcripts involved in salt stress and recovery in peanut. *Frontiers in Plant Science*, 9, 217.

Cui, Y., Zhang, W., Lin, X., Xu, S., Xu, J. and Li, Z. 2018b. Simultaneous improvement and genetic dissection of drought tolerance using selected breeding populations of rice. *Frontiers in Plant Science*, 9:320.

Darwish, E., Testerink, C., Khalil, M., El-Shihy, O. and Munnik, T. 2009. Phospholipid signaling responses in salt-stressed rice leaves. *Plant & Cell Physiology*, 50, 986–997.

Das, A., Rushton, P.J. and Rohila, J.S. 2017. Metabolomic profiling of soybeans (*Glycine max* L.) reveals the importance of sugar and nitrogen metabolism under drought and heat stress. *Plants (Basel, Switzerland)*, 6.

Daszkowska-Golec, A. 2018. Emerging roles of the nuclear cap-binding complex in abiotic stress responses. *Plant Physiology*, 176, 242–253.

de Souza, L.P., Garbowicz, K., Brotman, Y., et al. 2020. The acetate pathway supports flavonoid and lipid biosynthesis in Arabidopsis. *Plant Physiology*, 182, 857–869.

Devkar, V., Thirumalaikumar, V.P., Xue G. et al. 2020. Multifaceted regulatory function of tomato SlTAF1 in the response to salinity stress. *New Phytologist*, 225, 1681–1698.

Dey, A., Samanta, M.K., Gayen, S. and Maiti, M.K. 2016. The sucrose non-fermenting 1-related kinase 2 gene *SAPK9* improves drought tolerance and grain yield in rice by modulating cellular osmotic potential, stomatal closure and stress-responsive gene expression. *BMC Plant Biology*, 16, 158.

Diédhiou, C.J., Popova, O.V., Dietz, K.-J. and Golldack, D. 2008. The SNF1-type serine-threonine protein kinase SAPK4 regulates stress-responsive gene expression in rice. *BMC Plant Biology*, 8, 49.

Dixit, S., Singh, A., Sandhu, N. et al. 2017. Combining drought and submergence tolerance in rice: Marker-assisted breeding and QTL combination effects. *Molecular Breeding*, 37, 143–143.

Dolata, J., Bajczyk, M., Bielewicz, D. et al. 2016. Salt stress reveals a new role for *ARGONAUTE1* in miRNA biogenesis at the transcriptional and posttranscriptional levels. *Plant Physiology*, 172, 297–312.

Du, W., Lin, H., Chen, S. et al. 2011. Phosphorylation of SOS3-like calcium-binding proteins by their interacting SOS2-like protein kinases is a common regulatory mechanism in Arabidopsis. *Plant Physiology*, 156, 2235–2243.

Du, H., Liu, H. and Xiong, L. 2013. Endogenous auxin and jasmonic acid levels are differentially modulated by abiotic stresses in rice. *Frontiers in Plant Science*, 4, 397–397.

Du, H., Huang, F., Wu, N. et al. 2018. Integrative regulation of drought escape through ABA-dependent and -independent pathways in rice. *Molecular Plant*, 11, 584–597.

Dubouzet, J.G., Sakuma, Y., Ito, Y., et al. 2003. *OsDREB* genes in rice, *Oryza sativa* L., encode transcription activators that function in drought-, high-salt- and cold-responsive gene expression. *The Plant Journal*, 33, 751–763.

Edreva, A., Velikova, V., Tsonev, T. et al. 2007. Stress-protective role of secondary metabolites: Diversity of functions and mechanisms. *General and Applied Plant Physiology* 34.

Elfving, N., Davoine, C., Benlloch, R. et al. 2011. The *Arabidopsis thaliana* Med25 mediator subunit integrates environmental cues to control plant development. *Proceedings of the National Academy of Sciences*, 108, 8245–8250.

Ermakova, M., Danila, F.R., Furbank, R.T. and von Caemmerer, S. 2020. On the road to C4 rice: Advances and perspectives. *The Plant Journal*, 101, 940–950.

Eshed, Y. and Zamir, D. 1994. A genomic library of *Lycopersicon pennellii in L. esculentum*: A tool for fine mapping of genes. *Euphytica*, 79, 175–179.

Fàbregas, N. and Fernie, A.R. 2019. The metabolic response to drought. *Journal of Experimental Botany*, 70, 1077–1085.

Falkenburger, B.H., Jensen, J.B. and Hille, B. 2010. Kinetics of PIP2 metabolism and KCNQ2/3 channel regulation studied with a voltage-sensitive phosphatase in living cells. *The Journal of General Physiology*, 135, 99–114.

Fatima, T., Mishra, I., Verma, R. and Arora, N.K. 2020. Mechanisms of halotolerant plant growth promoting Alcaligenes sp. involved in salt tolerance and enhancement of the growth of rice under salinity stress. *3 Biotech*, 10, 361.

Fraser, C.M. and Chapple, C. 2011. The phenylpropanoid pathway in Arabidopsis. *Arabidopsis Book*, 9, e0152–e0152.

Fujii H, Verslues PE, Zhu J-K. Arabidopsis decuple mutant reveals the importance of SnRK2 kinases in osmotic stress responses in vivo. *Proc Natl Acad Sci USA*. 2011;108:1717–22.

Fujita, Y., Yoshida, T. and Yamaguchi-Shinozaki, K. 2013. Pivotal role of the AREB/ABF-SnRK2 pathway in ABRE-mediated transcription in response to osmotic stress in plants. *Physiologia Plantarum*, 147, 15–27.

Gao, S.Q., Chen, M., Xu, Z.S. et al. 2011. The soybean GmbZIP1 transcription factor enhances multiple abiotic stress tolerances in transgenic plants. *Plant Molecular Biology*, 75, 537–553.

Gaxiola, R.A., Li, J., Undurraga, S. et al. (2001) Drought- and salt-tolerant plants result from overexpression of the AVP1 H+-pump. *Proceedings of the National Academy of Sciences*, 98, 11444–11449.

Gigon, A., Matos, A.R., Laffray, D., Zuily-Fodil, Y. and Pham-Thi, A.T. 2004. Effect of drought stress on lipid metabolism in the leaves of *Arabidopsis thaliana* (ecotype Columbia). *Annals of Botany*, 94, 345–351.

Gong, P., Zhang, J., Li, H. et al. 2010. Transcriptional profiles of drought-responsive genes in modulating transcription signal transduction, and biochemical pathways in tomato. *Journal of Experimental Botany*, 61, 3563–3575.

Grandillo, S., Chetelat, R., Knapp S. et al. 2011. Solanum sect. Lycopersicon. In: *Wild Crop Relatives: Genomic and Breeding Resources: Vegetables*, Kole, C. (ed.), Berlin, Heidelberg: Springer Berlin Heidelberg, pp. 129–215.

Grondin, A., Rodrigues, O., Verdoucq, L. et al. 2015. Aquaporins contribute to ABA-triggered stomatal closure through OST1-mediated phosphorylation. *The Plant Cell*, 27, 1945–1954.

Gundaraniya S.A,, Ambalam P.S., Tomar R.S. Metabolomic Profiling of Drought-Tolerant and Susceptible Peanut (Arachis hypogaea L.) Genotypes in Response to Drought Stress. *ACS Omega*. 2020 Nov 20;5(48):31209–31219.

Gupta, P. and De, B. 2017. Metabolomics analysis of rice responses to salinity stress revealed elevation of serotonin, and gentisic acid levels in leaves of tolerant varieties. *Plant Signal Behav*, 12, e1335845.

Gur, A. and Zamir, D. 2004. Unused natural variation can lift yield barriers in plant breeding. *PLoS Biology*, 2, e245.

Hao, Y.J., Wei, W., Song Q. et al. 2011. Soybean NAC transcription factors promote abiotic stress tolerance and lateral root formation in transgenic plants. *The Plant Journal: For Cell and Molecular Biology*, 68, 302–313.

Hatfield, J.L., Boote, K.J., Kimball, B.A., et al. 2011. Climate impacts on agriculture: Implications for crop production. *Agronomy Journal*, 103, 351–370.

Hayat, S., Hayat, Q., Alyemeni, M. et al. 2012. Role of proline under changing environment: A review. *Plant Signaling & Behavior,* 7 (11): 1456–66.

Hayford, R.K.A., Ligaba-Osena, A., Subramani M. et al. 2017. Characterization and expression analysis of common bean histone deacetylase 6 during development and cold stress response. *International Journal of Genomics*, 2017, 2502691.

Hichri, I., Muhovski, Y., Clippe, A. et al. 2016. SlDREB2, a tomato dehydration-responsive element-binding 2 transcription factor, mediates salt stress tolerance in tomato and Arabidopsis. *Plant, Cell & Environment*, 39, 62–79.

Hong, Y., Zhang, H., Huang, L., Li, D. and Song, F. 2016. Overexpression of a stress-responsive NAC transcription factor Gene *ONAC022* improves drought and salt tolerance in rice. *Frontiers in Plant Science*, 7, 4–4.

Hou, Q., Ufer, G. and Bartels, D. 2016. Lipid signalling in plant responses to abiotic stress. *Plant, Cell & Environment*, 39, 1029–1048.

Hu, H., Dai, M., Yao, J. et al. 2006. Overexpressing a NAM, ATAF, and CUC (NAC) transcription factor enhances drought resistance and salt tolerance in rice. *Proceedings of the National Academy of Sciences of the United States of America*, 103, 12987–12992.

Hu, Y., Li, W., Xu, Y. et al. 2009. Differential expression of candidate genes for lignin biosynthesis under drought stress in maize leaves. *Journal of Applied Genetics*, 50, 213–223.

Huang, X., Hou, L., Meng, J. et al. 2018. The antagonistic action of abscisic acid and cytokinin signaling mediates drought stress response in Arabidopsis. *Molecular Plant*, 11, 970–982.

Huertas, R., Rubio, L., Cagnac, O. et al. 2013. The K+/H+ antiporter LeNHX2 increases salt tolerance by improving K+ homeostasis in transgenic tomato. *Plant, Cell & Environment*, 36, 2135–2149.

Hunt, L., Mills, L., Pical, C. et al. 2003. Phospholipase C is required for the control of stomatal aperture by ABA. *The Plant Journal: For Cell and Molecular Biology*, 34, 47–55.

Jadko, S.I. 2015. Histone deacetylase activity and reactive oxygen species content in the tissue culture of *Arabidopsis thaliana* under normal conditions and development of acute osmotic stress. *The Ukrainian Biochemical Journal*, 87, 57–62.

Jia, Y., Tao, F. and Li, W. 2013. Lipid profiling demonstrates that suppressing Arabidopsis phospholipase Dδ retards ABA-promoted leaf senescence by attenuating lipid degradation. *PLoS One*, 8, e65687.

Jiang, Y. and Deyholos, M.K. 2009. Functional characterization of Arabidopsis NaCl-inducible WRKY25 and WRKY33 transcription factors in abiotic stresses. *Plant Molecular Biology*, 69, 91–105.

Kang Z., Babar M.A, Khan N., et al. (2019) Comparative metabolomic profiling in the roots and leaves in contrasting genotypes reveals complex mechanisms involved in post-anthesis drought tolerance in wheat. *PLOS ONE* 14(3): e0213502.

Kapoor, M., Arora, R., Lama, T. et al. 2008. Genome-wide identification, organization and phylogenetic analysis of *Dicer-like*, *Argonaute* and *RNA-dependent RNA Polymerase* gene families and their expression analysis during reproductive development and stress in rice. *BMC Genomics*, 9, 451.

Karpac, J. and Jasper, H. 2011. Metabolic homeostasis: HDACs take center stage. *Cell*, 145, 497–499.

Kim, M.J., Park, M.J., Seo, P.J. et al. 2012. Controlled nuclear import of the transcription factor NTL6 reveals a cytoplasmic role of SnRK2.8 in the drought-stress response. *The Biochemical Journal*, 448, 353–363.

Kim, Y.S., Sakuraba, Y., Han, S.H., Yoo, S.C. and Paek, N.C. 2013. Mutation of the Arabidopsis NAC016 transcription factor delays leaf senescence. *Plant & Cell Physiology*, 54, 1660–1672.

Kim, J.M., To, T.K., et al. 2017. Acetate-mediated novel survival strategy against drought in plants. *Nature Plants*, 3, 17097.

Ko, J.H., Yang, S.H. and Han, K.H. 2006. Upregulation of an Arabidopsis RING-H2 gene, XERICO, confers drought tolerance through increased abscisic acid biosynthesis. *The Plant Journal*, 47, 343–355.

Kobayashi, Y., Yamamoto, S., Minami, H., Kagaya, Y. and Hattori, T. 2004. Differential activation of the rice sucrose nonfermenting1-related protein kinase2 family by hyperosmotic stress and abscisic acid. *Plant Cell*, 16, 1163–1177.

Kulik, A., Wawer, I., Krzywińska, E., Bucholc, M. and Dobrowolska, G. 2011. SnRK2 protein kinases--key regulators of plant response to abiotic stresses. *OMICS*, 15, 859–872.

Kumar, A., Sharma, S. and Mishra, S. 2010. Influence of arbuscular mycorrhizal (AM) fungi and salinity on seedling growth, solute accumulation and mycorrhizal dependency of *Jatropha curcas* L. *Journal of Plant Growth Regulation*, 29, 297–306.

Kumar, A., Sandhu, N., Venkateshwarlu, C., et al. 2020. Development of introgression lines in high yielding, semi-dwarf genetic backgrounds to enable improvement of modern rice varieties for tolerance to multiple abiotic stresses free from undesirable linkage drag. *Scientific Reports,* 10, 13073.

Lee, S.J., Jeong, E.M., Ki, A.Y. et al. 2016. Oxidative defense metabolites induced by salinity stress in roots of *Salicornia herbacea*. *Journal of Plant Physiology*, 206, 133–142.

Lemtiri-Chlieh, F., MacRobbie, E.A.C., Webb A. et al. 2003. Inositol hexakisphosphate mobilizes an endomembrane store of calcium in guard cells. *Proceedings of the National Academy of Sciences*, 100, 10091–10095.

Less, H. and Galili, G. 2008. Principal transcriptional programs regulating plant amino acid metabolism in response to abiotic stresses. *Plant Physiology*, 147, 316–330.

Li, W., Cui, X., Meng, Z. et al. 2012. Transcriptional regulation of Arabidopsis MIR168a and argonautel homeostasis in abscisic acid and abiotic stress responses. *Plant Physiology*, 158, 1279–1292.

Li, T., Wu, X.Y., Li, H., Song, J.H. and Liu, J.Y. 2016. A dual-function transcription factor, AtYY1, is a novel negative regulator of the Arabidopsis ABA response network. *Molecular Plant*, 9, 650–661.

Lima, L.L., Balbi, B.P., Mesquita, R.O., et al. 2019. Proteomic and metabolomic analysis of a drought tolerant soybean cultivar from Brazilian Savanna. *Crop Breeding, Genetics and Genomics*, 1, e190022.

Liu, Q., Kasuga, M., Sakuma, Y. et al. 1998. Two transcription factors, DREB1 and DREB2, with an EREBP/AP2 DNA binding domain separate two cellular signal transduction pathways in drought- and low-temperature-responsive gene expression, respectively, in Arabidopsis. *Plant Cell*, 10, 1391–1406.

Liu, X.M., Nguyen, X.C., Kim, K.E. et al. 2013. Phosphorylation of the zinc finger transcriptional regulator ZAT6 by MPK6 regulates Arabidopsis seed germination under salt and osmotic stress. *Biochemical and Biophysical Research Communications*, 430, 1054–1059.

Liu, X., Wei, W., Zhu, W. et al. 2017. Histone deacetylase AtSRT1 links metabolic flux and stress response in Arabidopsis. *Molecular Plant*, 10, 1510–1522.

Lu, Y., Deng, X. and Kwak, S. 2010. Over expression of CuZn superoxide dismutase (CuZn SOD) and ascorbate peroxidase (APX) in transgenic sweet potato enhances tolerance and recovery from drought stress. *African Journal of Biotechnology*, 9, 8378–8391.

Ma, D., Sun, D., Wang, C., Li, Y. and Guo, T. 2014. Expression of flavonoid biosynthesis genes and accumulation of flavonoid in wheat leaves in response to drought stress. *Plant Physiology and Biochemistry: PPB*, 80, 60–66.

Ma, X., Li, C. and Wang, M. 2015. Wheat NF-YA10 functions independently in salinity and drought stress. *Bioengineered*, 6, 245–247.

Maeda, H., Song, W., Sage, T.L. and DellaPenna, D. 2006. Tocopherols play a crucial role in low-temperature adaptation and phloem loading in Arabidopsis. *The Plant Cell*, 18, 2710–2732.

Mao, X., Zhang, H., Qian, X. et al. 2012. TaNAC2, a NAC-type wheat transcription factor conferring enhanced multiple abiotic stress tolerances in Arabidopsis. *Journal of Experimental Botany*, 63, 2933–2946.

Mao, C., Ding, J., Zhang, B., Xi, D. and Ming, F. 2018. OsNAC2 positively affects salt-induced cell death and binds to the OsAP37 and OsCOX11 promoters. *The Plant Journal*, 94, 454–468.

Maruyama, K., Todaka, D., Mizoi J. et al. 2012. Identification of *cis*-acting promoter elements in cold- and dehydration-induced transcriptional pathways in Arabidopsis, rice, and soybean. *DNA Research*, 19, 37–49.

McLoughlin, F., Galvan-Ampudia, C.S., Julkowska M.M. et al. 2012. The Snf1-related protein kinases SnRK2.4 and SnRK2.10 are involved in maintenance of root system architecture during salt stress. *The Plant Journal: For Cell and Molecular Biology*, 72, 436–449.

Mittal, S., Banduni, P., Mallikarjuna, M.G. et al. 2018. Structural, functional, and evolutionary characterization of major drought transcription factors families in maize. *Frontiers in Chemistry*, 6, 177–177.

Morel, J.B., Godon, C., Mourrain, P. et al. 2002. Fertile hypomorphic ARGONAUTE (ago1) mutants impaired in post-transcriptional gene silencing and virus resistance. *The Plant Cell*, 14, 629–639.

Muchate, N.S., Rajurkar, N.S., Suprasanna, P. and Nikam, T.D. 2019. NaCl induced salt adaptive changes and enhanced accumulation of 20-hydroxyecdysone in the in vitro shoot cultures of *Spinacia oleracea* (L.). *Scientific Reports*, 9, 12522.

Mukherjee, P., Mitra, A. and Roy, M. 2019. Halomonas rhizobacteria of avicennia marina of Indian Sundarbans promote rice growth under saline and heavy metal stresses through exopolysaccharide production. *Frontiers in Microbiology*, 10, 1207.

Nakabayashi, R., Mori, T. and Saito, K. 2014. Alternation of flavonoid accumulation under drought stress in *Arabidopsis thaliana*. *Plant Signaling & Behavior*, 9.e29518.

Nakashima, K., Tran, L.S., Van Nguyen, D. et al. 2007. Functional analysis of a NAC-type transcription factor OsNAC6 involved in abiotic and biotic stress-responsive gene expression in rice. *The Plant Journal: For Cell and Molecular Biology*, 51, 617–630.

Ning, J., Moghe, G.D., Leong, B. et al. 2015. A feedback-insensitive isopropylmalate synthase affects acylsugar composition in cultivated and wild tomato. *Plant Physiology*, 169, 1821–1835.

Nishizawa, A., Yabuta, Y. and Shigeoka, S. 2008. Galactinol and raffinose constitute a novel function to protect plants from oxidative damage. *Plant Physiology*, 147, 1251–1263.

Obata, T., Witt, S., Lisec, J. et al. 2015. Metabolite profiles of maize leaves in drought, heat, and combined stress field trials reveal the relationship between metabolism and grain yield. *Plant Physiology*, 169, 2665–2683.

Olias, R., Eljakaoui, Z., Li, J. et al. 2009. The plasma membrane Na+/H+ antiporter SOS1 is essential for salt tolerance in tomato and affects the partitioning of Na+ between plant organs. *Plant, Cell & Environment*, 32, 904–916.

Orellana, S., Yañez, M., Espinoza, A. et al. 2010. The transcription factor SlAREB1 confers drought, salt stress tolerance and regulates biotic and abiotic stress-related genes in tomato. *Plant, Cell & Environment*, 33, 2191–2208.

Pasapula, V., Shen, G., Kuppu S. et al. 2011. Expression of an Arabidopsis vacuolar H+-pyrophosphatase gene (AVP1) in cotton improves drought- and salt tolerance and increases fibre yield in the field conditions. *Plant Biotechnology Journal*, 9, 88–99.

Piasecka, A., Sawikowska, A., Kuczyńska A. et al. 2016. Drought related secondary metabolites of barley (*Hordeum vulgare L.*) leaves and their mQTLs. *The Plant Journal*, 89(5), 898–913.

Rabara, R.C., Tripathi, P., Reese, R.N. et al. 2015. Tobacco drought stress responses reveal new targets for Solanaceae crop improvement. *BMC Genomics*, 16, 484.

Razali, R., Bougouffa, S., Morton, M.J.L. et al. 2018. The genome sequence of the wild tomato *Solanum pimpinellifolium* provides insights into salinity tolerance. *Frontiers in Plant Science*, 9, 1402.

Reményi, A., Scholer, H.R. and Wilmanns, M. 2004. Combinatorial control of gene expression. *Nature Structural and Molecular Biology*, 11, 812.

Rice-Evans, C., Miller, N. and Paganga, G. 1997. Antioxidant properties of phenolic compounds. *Trends in Plant Science*, 2, 152–159.

Rivandi, J., Miyazaki, J., Hrmova, M. et al. 2010. A SOS3 homologue maps to HvNax4, a barley locus controlling an environmentally sensitive Na+ exclusion trait. *Journal of Experimental Botany*, 62, 1201–1216.

Sage, R.F. 2001. C4 Plants. In *Encyclopedia of Biodiversity* Levin, S.A. (ed.) New York: Elsevier, pp. 575–598.

Sakuraba, Y., Kim, Y.-S., Han, S.-H., Lee, B.-D. and Paek, N.-C. 2015. The Arabidopsis transcription factor NAC016 promotes drought stress responses by repressing AREB1 transcription through a trifurcate feed-forward regulatory loop involving NAP. *The Plant Cell*, 27, 1771–1787.

Sanan-Mishra, N., Pham, X.H., Sopory, S.K. and Tuteja, N. 2005. Pea DNA helicase 45 overexpression in tobacco confers high salinity tolerance without affecting yield. *Proceedings of the National Academy of Sciences of the United States of America*, 102, 509–514.

Saruhashi, M., Kumar Ghosh, T., Arai, K. et al. 2015. Plant Raf-like kinase integrates abscisic acid and hyperosmotic stress signaling upstream of SNF1-related protein kinase2. *Proceedings of the National Academy of Sciences*, 112, E6388–E6396.

Schmidt, R., Mieulet, D., Hubberten, H.-M., et al. 2013. Salt-responsive ERF1 regulates reactive oxygen species-dependent signaling during the initial response to salt stress in rice. *The Plant Cell*, 25, 2115–2131.

Shahverdi A.M., Omidi, H. and Tabatabaei, S.J. 2019. Stevia (Stevia rebaudiana Bertoni) responses to NaCl stress: Growth, photosynthetic pigments, diterpene glycosides and ion content in root and shoot. *Journal of the Saudi Society of Agricultural Sciences*, 18, 355–360.

Shashidhar, H.E., Kanbar, A., Toorchi, M. et al. 2013. Breeding for drought resistance using whole plant architecture: Conventional and molecular approach. In *Plant Breeding from Laboratories to Fields*, Andersen, S. B. (ed.), Norderstedt, Germany: Bod – Books on Demand, pp. 122–151.

Shavrukov, Y., Gupta, N.K., Miyazaki, J. et al. 2010. HvNax3--a locus controlling shoot sodium exclusion derived from wild barley (*Hordeum vulgare* ssp. spontaneum). *Functional & Integrative Genomics*, 10, 277–291.

Shen, W., Li, H., Teng, R. et al. 2019. Genomic and transcriptomic analyses of HD-Zip family transcription factors and their responses to abiotic stress in tea plant (*Camellia sinensis*). *Genomics*, 111, 1142–1151.

Singh, D. and Laxmi, A. 2015. Transcriptional regulation of drought response: A tortuous network of transcriptional factors. *Frontiers in Plant Science*, 6, 895.

Smita, S., Katiyar, A., Lenka, S.K. et al. 2020. Gene network modules associated with abiotic stress response in tolerant rice genotypes identified by transcriptome meta-analysis. *Functional & Integrative Genomics*, 20, 29–49.

Soma, F., Mogami, J., Yoshida, T. et al. 2017. ABA-unresponsive SnRK2 protein kinases regulate mRNA decay under osmotic stress in plants. *Nature Plants*, 3, 16204.

Soma, F., Takahashi, F., Suzuki, T., Shinozaki, K. and Yamaguchi-Shinozaki, K. 2020. Plant Raf-like kinases regulate the mRNA population upstream of ABA-unresponsive SnRK2 kinases under drought stress. *Nature Communications*, 11, 1373.

Sugano, S., Kaminaka, H., Rybka, Z. et al. 2003. Stress-responsive zinc finger gene ZPT2–3 plays a role in drought tolerance in petunia. *The Plant Journal*, 36, 830–841.

Tahmasebi, A., Ashrafi-Dehkordi, E., Shahriari, A.G., Mazloomi, S.M. and Ebrahimie, E. 2019. Integrative meta-analysis of transcriptomic responses to abiotic stress in cotton. *Progress in Biophysics and Molecular Biology*, 146, 112–122.

Takasaki, H., Maruyama, K., Kidokoro, S. et al. 2010. The abiotic stress-responsive NAC-type transcription factor OsNAC5 regulates stress-inducible genes and stress tolerance in rice. *Molecular Genetics and Genomics: MGG*, 284, 173–183.

Tamburino, R., Vitale, M., Ruggiero, A. et al. 2017. Chloroplast proteome response to drought stress and recovery in tomato (*Solanum lycopersicum* L.). *BMC Plant Biology*, 17, 40.

Tan, W., Zhang, D., Zhou, H. et al. 2018. Transcription factor HAT1 is a substrate of SnRK2.3 kinase and negatively regulates ABA synthesis and signaling in Arabidopsis responding to drought. *PLoS Genetics*, 14, e1007336.

Tombesi, S., Nardini, A., Frioni, T. et al. 2015. Stomatal closure is induced by hydraulic signals and maintained by ABA in drought-stressed grapevine. *Scientific Reports*, 5, 12449.

Tran, L.-S.P., Nakashima, K., Sakuma, Y. et al. 2004. Isolation and functional analysis of Arabidopsis stress-inducible NAC transcription factors that bind to a drought-responsive *cis*-element in the early responsive to dehydration stress 1 promoter. *The Plant Cell*, 16, 2481–2498.

Tran, L.-S.P., Nakashima, K., Sakuma, Y. et al. 2006. Role of the ZFHD1 and NAC transcription factors in drought-inducible expression of the early responsive to dehydration stress 1 (ERD1) gene of Arabidopsis. *Plant and Cell Physiology Supplement*, 2006, 807–807.

Tuteja, N., Sahoo, R.K., Garg, B. and Tuteja, R. 2013. OsSUV3 dual helicase functions in salinity stress tolerance by maintaining photosynthesis and antioxidant machinery in rice (*Oryza sativa* L. cv. IR64). *The Plant Journal*, 76, 115–127.

Upadhyay, S.K. and Singh, D.P. 2015. Effect of salt-tolerant plant growth-promoting rhizobacteria on wheat plants and soil health in a saline environment. *Plant Biology*, 17, 288–293.

Vasanthakumari, M.M., Jadhav, S.S., Sachin, N. et al. 2015. Restoration of camptothecine production in attenuated endophytic fungus on re-inoculation into host plant and treatment with DNA methyltransferase inhibitor. *World Journal of Microbiology and Biotechnology*, 31, 1629–1639.

Vashisht, A.A., Pradhan, A., Tuteja, R. and Tuteja, N. 2005. Cold- and salinity stress-induced bipolar pea DNA helicase 47 is involved in protein synthesis and stimulated by phosphorylation with protein kinase C. *The Plant Journal*, 44, 76–87.

Vaucheret, H. 2008. Plant Argonautes. *Trends in Plant Science*, 13, 350–358.

Vaughan, C., Hansen, J., Roudier, P., Watkiss, P. and Carr, E. 2019. Evaluating agricultural weather and climate services in Africa: Evidence, methods, and a learning agenda. *WIREs Climate Change*, 10, e586.

Veiga, T.A., Silva, S.C., Francisco, A.C. et al. 2007. Inhibition of photophosphorylation and electron transport chain in thylakoids by lasiodiplodin, a natural product from *Botryosphaeria rhodina*. *Journal of Agricultural and Food Chemistry*, 55, 4217–4221.

Velikova, V., Müller, C., Ghirardo, A. et al. 2015. Knocking down of isoprene emission modifies the lipid matrix of thylakoid membranes and influences the chloroplast ultrastructure in poplar. *Plant Physiology*, 168, 859–870.

Wang, B., Duan, C.G., Wang, X. et al. 2015a. HOS1 regulates Argonautel by promoting the transcription of the microRNA gene MIR168b in Arabidopsis. *The Plant Journal*, 81(6), 861–870.

Wang, G., Zhang, S., Ma, X. et al. 2016. A stress-associated NAC transcription factor (SlNAC35) from tomato plays a positive role in biotic and abiotic stresses. *Physiologia Plantarum*, 158, 45–64.

Wang, L., Hu, Z., Zhu, M. et al. 2017. The abiotic stress-responsive NAC transcription factor SlNAC11 is involved in drought and salt response in tomato (*Solanum lycopersicum* L.). *Plant Cell, Tissue and Organ Culture (PCTOC)*, 129, 161–174.

Wang, Z., Wang, F., Hong, Y. et al. 2018. The flowering repressor SVP confers drought resistance in Arabidopsis by regulating abscisic acid catabolism. *Molecular Plant*, 11, 1184–1197.

Wild, M., Davière, J.M., Cheminant, S. et al. 2012. The Arabidopsis DELLA RGALIKE3 is a direct target of MYC2 and modulates jasmonate signaling responses. *The Plant Cell*, 24, 3307–3319.

Wu, A., Allu, A.D., Garapati, P. et al. 2012. JUNGBRUNNEN1, a reactive oxygen species-responsive NAC transcription factor, regulates longevity in Arabidopsis. *The Plant Cell*, 24, 482–506.

Xiang, Y., Tang, N., Du, H., Ye, H. and Xiong, L. 2008. Characterization of OsbZIP23 as a key player of the basic leucine zipper transcription factor family for conferring abscisic acid sensitivity and salinity and drought tolerance in rice. *Plant Physiology*, 148, 1938–1952.

Xue, G.P. and Loveridge, C.W. 2004. HvDRF1 is involved in abscisic acid-mediated gene regulation in barley and produces two forms of AP2 transcriptional activators, interacting preferably with a CT-rich element. *The Plant Journal: For Cell and Molecular Biology*, 37, 326–339.

Xue, G.P., Way, H.M., Richardson, T. et al. 2011. Overexpression of TaNAC69 leads to enhanced transcript levels of stress up-regulated genes and dehydration tolerance in bread wheat. *Molecular Plant*, 4, 697–712.

Yamaguchi-Shinozaki, K. and Shinozaki, K. 1994. A novel *cis*-acting element in an Arabidopsis gene is involved in responsiveness to drought, low-temperature, or high-salt stress. *Plant Cell*, 6, 251–264.

Yamazaki H., Takaaki, N. and Masaji, K. 1995. Effects of (+)-S-Abscisic acid on the quality of stored cucumber and tomato seedlings. *HortScience*, 30, 80–82.

Yan, J., Wang, P., Wang, B. et al. 2017. The SnRK2 kinases modulate miRNA accumulation in Arabidopsis. *PLoS Genetics*, 13, e1006753.

Yang, C., Huang, Y., Lv, W. et al. 2020. GmNAC8 acts as a positive regulator in soybean drought stress. *Plant Science*, 293, 110442.

You, J., Zhang, Y., Liu, A. et al. 2019. Transcriptomic and metabolomic profiling of drought-tolerant and susceptible sesame genotypes in response to drought stress. *BMC Plant Biology*, 19, 267.

Yu, L., Nie, J., Cao, C. et al. 2010. Phosphatidic acid mediates salt stress response by regulation of MPK6 in *Arabidopsis thaliana*. *The New Phytologist*, 188, 762–773.

Zhai, L., Teng, F., Zheng, K., Xiao, J., Deng, W. and Sun, W. 2019. Expression analysis of Argonaute genes in maize (*Zea mays* L.) in response to abiotic stress. *Hereditas*, 156, 27.

Zhang, H., Kim, M.S., Sun, Y. et al. 2008. Soil bacteria confer plant salt tolerance by tissue-specific regulation of the sodium transporter HKT1. *Molecular Plant-Microbe Interactions:MPMI*, 21, 737–744.

Zhang, Y., Li, D., Zhou, R., et al. 2019a. Transcriptome and metabolome analyses of two contrasting sesame genotypes reveal the crucial biological pathways involved in rapid adaptive response to salt stress. *BMC Plant Biology*, 19, 66.

Zhang, F., Zeng, D., Huang, L. et al. 2019b. Stress-activated protein kinase OsSAPK9 regulates tolerance to salt stress and resistance to bacterial blight in Rice. *Rice*, 12, 80.

Zhao, B., Liang, R., Ge, L., et al. 2007. Identification of drought-induced microRNAs in rice. *Biochemical and Biophysical Research Communications*, 354, 585–590.

Zhao, G., Song, Y., Wang, C., et al. 2016. Genome-wide identification and functional analysis of the TIFY gene family in response to drought in cotton. *Molecular Genetics and Genomics*. 291.

Zheng, L., Liu, G., Meng, X., et al. 2013. A WRKY gene from *Tamarix hispida*, ThWRKY4, mediates abiotic stress responses by modulating reactive oxygen species and expression of stress-responsive genes. *Plant Molecular Biology*, 82, 303–320.

Zheng, X., Chen, B., Lu, G. and Han, B. 2009. Overexpression of a NAC transcription factor enhances rice drought and salt tolerance. *Biochemical and Biophysical Research Communications*, 379, 985–989.

Zhu, M., Chen, G., Zhang, J., et al. 2014. The abiotic stress-responsive NAC-type transcription factor SlNAC4 regulates salt and drought tolerance and stress-related genes in tomato (*Solanum lycopersicum*). *Plant Cell Reports*, 33, 1851–1863.

Zhu, M., Meng, X., Cai, J., et al. 2018. Basic leucine zipper transcription factor SlbZIP1 mediates salt and drought stress tolerance in tomato. *BMC Plant Biology*, 18, 83.

Zuo, Z., Weraduwage, S.M., Lantz, A.T., et al. 2019. Isoprene acts as a signaling molecule in gene networks important for stress responses and plant growth. *Plant Physiology*, 180, 124–152.

14

UV Stress in Plants: A Curse in Plant Productivity and Blessing in Food Security

Jayjit Majumdar
University of Kalyani

CONTENTS

14.1 Introduction

Physiological response to UV radiation as well as blue light is among the oldest but least understood phenomena in plant photobiology (Senger and Schmidt 1994). The UV (Ultra Violet) radiation refers to the electromagnetic radiation having a wavelength of 100–400 nm, and it is divided into three separate bands according to their wavelengths, such as UV-A (315–400 nm), UV-B (280–315 nm), and UV-C (100–280 nm). As the atmosphere filter out the UV radiation below 290 nm and does not reach plant cover, our major concern here is regarding UV-B and UV-A radiation (Holzinger and Lütz 2006). Responses to radiation are generally divided on the amount of light needed, whether the response requires continuous exposure or irradiation, and efficiency elicited by pulses of light (Kami et al. 2010). Distribution and internal permeation of UV-B radiation vary greatly from species to species depending on numerous factors such as morphology, anatomy, pigments, and other physicochemical changes in leaves, resulting from UV radiation. Less epidermal penetration was found in the case of conifer needles, whereas herbaceous dicots are more prone to deeper penetration. Woody dicots and grasses have intermediate penetration of UV-B radiation (Day, Vogelmann, and DeLucia 1992). Day et al. (1992) found a weak correlation between epidermal thickness and UV-B penetration. The majority of UV-B screening is done by flavonoids present on the outer membrane of the chloroplast and screening compounds present in the vacuoles of the epidermal cells, and epicuticle wax does not play a vital role in screening the UV-B radiation. UV radiation in plants induces expression of stress signal employing synthesis of different types of a secondary metabolite which not only acts as a defense against radiation but also some other biotic and abiotic stress. This chapter is dedicated to understanding the defense mechanism developed by plants after environmental UV radiation and application of UV dose under experimental and field conditions and how UV radiation can be utilized for crop protection against pests and pathogens and crop improvement in the pre- and post-harvest phase.

14.2 UV Reception in Plants

Plants have a legion of sensory mechanisms to sense stress and transduce signals from environmental stimulants to their own physiological and biochemical responses. These proteins are very sensitive toward changes in wavelength, duration, intensity direction, etc. UV resistance proteins are known receptors of UV radiation in plant cells, such as U-B resistance 8 (UVR8). UVR8 can sense radiation of wavelength ranging 280–315 nm (mostly sensitive at 285 nm) and triggers the plant's stress

DOI: 10.1201/9781003258063-14

signaling mechanisms. UV-B acclimation is directly influenced under UVR8-mediated gene activation, DNA repair, photoinhibition, phototropism, stomatal opening, stomatal opening, thermomorphogenesis, flavonoid synthesis, changes in circadian rhythm, stress tolerance, etc. (Yin and Ulm 2017; Moriconi et al. 2018). Initial research conducted on the identification and isolation of UV-R8 (*Arabidopsis thaliana* mutant of UV resistance locus8 (UVR8) was conducted by Kliebenstein et al. (2002); recently several other authors have identified the structure and molecular basis of the physiological role of UVR8. UVR8 isolated from Arabidopsis isolated comprises 400 amino acids. This protein consists of a seven-bladed b-propeller core with an N-terminal extension and a flexible C-terminal region comprising ~60 amino acids (Wu et al. 2012; Christie et al. 2012; Jenkins 2014a, b). UVR8 is rich in tryptophan (14 tryptophans). Ultrastructure has revealed that apart from one molecule of tryptophan (W400), which is located in the C-terminus, the rest of the 13 tryptophan molecules are organized into two groups. Six tryptophan molecules are located in the core, whereas the rest of them form a dimeric interface (Christie et al. 2012; O'Hara and Jenkins 2012). *A. thaliana* is one of the vastly studied plants in photobiology. In the case of *A. thaliana*, this species uses 13 photoreceptors, apart from UVR8, there are five red/infrared regions phytochromes (phyA-E), phototropins (photl and phot2), cryptochromes (cryl and cry2), Zeitlupes (ZTL, FKF1, and LKP2) to recognize blue lights. These photoreceptors function differently at a different light intensity as well as wavelength (Kami et al. 2010; Heijde and Ulm 2012; Tilbrook et al. 2013). Photoreceptors such as cryl and cry2 at low continuous exposure of UV-A and blue light trigger anthocyanin accumulation, cotyledon opening, modulation of phototropism, and hypocotyl growth inhibition but in the case of high continuous exposure accelerates flowering, inhibition of petiole growth, stomata opening, stomatal development, and cotyledon expansion (Casal and Boccalandro 1995; Jackson and Jenkins 1995; Bagnall, King, and Hangarter 1996; Karsch et al. 1998; Lin et al. 1998; Ahmad et al. 2002; Whippo and Hangarter 2003; Ohgishi et al. 2004; Mao et al. 2005; Jenkins 2009; Kang et al. 2009;). Receptor-like phytochrome A and B (phyA and phyb) at high continuous exposure of UVA and blue light promote cotyledon expansion (Neff and Van Volkenburgh 1994; Neff and Chory 1998), whereas phytochrome B and C at low continuous exposure can inhibit hypocotyl growth (Casal and Boccalandro 1995; Franklin et al. 2003).

14.3 Plant Responses to UV Radiation

14.3.1 Germination

Phytochromes are the responsible photoreceptors for seed germination. They trigger the synthesis of gibberellins (GA) as well as signaling. Several commercially grown crops such as kale (red Prussian), cabbage (early golden acre), radish (cherry bell), and agave (century plant) seeds have shown higher germination (after 24 hours and 15 days) and root growth under UVB light against fluorescent light as a control (Noble 2002). Sugimoto (2013) has suggested that buckwheat (var Shinano Oosoba), cabbage (var Okina), Chinese cabbage 'Kyoto 3', mizuna (*Brassica rapa* L. Japonica Group), parsley 'curly paramount', and turnip 'Syogoin Ooma-rukabu' have a positive influence of UV light on germination of seeds, whereas several other crops like eggplant 'Chikuyo', lettuce 'Great Lake', onion 'OL Ki', radish 'Minowase', parsley 'Curly paramount', pea 'Hyogo Kinusaya', perilla 'Chirimen Aoshiso', spinach 'Alright', Swiss chard 'Umaina', and Welsh onion 'Kujo Futonegi' had germinated early under the influence of light and dark condition rather than UVB radiation.

14.4 Physiological Responses UV Radiation Mediated by Photoreceptors

14.4.1 Photosynthesis

Different species have different sensitivity toward UV radiation and their productivity directly depends on their UV perceptions well as their sensitivity toward it. It has been found that C3 species are more sensitive to UV radiation. UVB effect is primarily on PSII, but there is a little significant effect on PSI itself. This sensitivity is primarily associated with hill activity (Teramura 1983). In earlier studies, it has been found that there is a significant reduction of chlorophyll concentration upon UV-B radiation in crops like collard, oats, soybean, etc. Low PAR (photosynthetically active radiation) also found to be a causal agent in significant chlorophyll reduction in chlorophyll in crops like bean, cabbage, pea, barley, and corn (Garrard et al. 1976; Basiouny et al. 1978; Tevini et al. 1981; Teramura 1983; Vu et al. 1981). Basiouny et al. (1978) found that hill activity was unaffected by UVB in peanut (*Arachis hypogaea* L.) as well as in resistant C4 plants. This is basically due to ultrastructural changes in the chloroplast. Disruption of the structural integrity of membrane due to higher UV radiance found to damage machinery responsible for light as well as dark reaction, mesophyll tissues also become resistant to carbon dioxide assimilation and limit photosynthesis in numerous plants. This is due to the effect on the electron transport chain and carboxylation reaction (Teramura 1983; Bornman and Teramura 1993). From several studies conducted in past reveal that enhanced UV radiation directly can cause disintegration of the thylakoid membrane, damage to PSII, reduction in carbon fixation and oxygen production, decrease in rubisco activity, decreased chlorophyll, and starch content, and reduction in dry weight. Indirect effects may include changes in leaf thickness, anatomy, changes in canopy morphology, and induction of stomatal closure (Tevini, Iwanzik, and Thoma 1981; Teramura 1983; Bornman and Terramura 1993; Teramura and Sullivan 1994; Allen, Nogués, and Baker 1998; Tevini 1999; Bolink 2003). Mature leaves of pea (*Pisum sativum* L. cv. Meteor) on irradiation with high-intensity UV (40 kJ/m^2d) were found to be synchronized with active photon flux density (PPFD> 450 $mmol/m^2s$) and found to reduce light-saturated CO_2 assimilation rate (Caldwell 1971). UV inactivation of photosynthesis is believed to be the inactivation of Mn in water oxidizing complexes (redox-active tyrosines (Tyr-z, Tyr-D)). Some researchers believe that the main reason for photosynthetic inhibition is due to the absorption of UV light by Mn ions as Mn III, and Mn IV oxidation states (Lidon et al. 2012; Rai et al. 2018). It has been found that UV radiation changes the gene expression of the production of photosynthetic proteins including rubisco. Genes of rbcs – small subunit of Rubisco, atpc-g subunit of ATP synthase, cab – chl a/b binding protein of the light-harvesting antenna of PS II are more sensitive toward UV than genes encoded in the chloroplast, and this varies according to the developmental stages of the plant (Jordan et al. 1992, 1994; Jordan 2002; Mpoloka 2008; Mishra et al. 2008).

14.4.2 Stomatal Opening and Closure

Stomatal adjustment is greatly influenced by Abscisic acid (ABA) and electrolyte fluxes in the guard cells and mainly by K^+ ion fluxes. A large number of environmental conditions/factors govern the electrolyte mobility in guard cells. It has been found that UV-A stimulates the opening of the stomata, whereas UV-C closes the stomata, this might affect ultrastructure and cause DNA damage (Wright and Murphy 1982; Negash and Bjom 1986; MacRobbie 1998; Jansen and Van Den Noort 2000; He et al. 2005). In filed and experimental conditions revealed UV-B reduces stomatal conductance/or aperture (Teramura 1983; Middleton and Teramura 1993; Musil and Wand 1993; Day and Vogelmann 1995; Nogués and Baker 1995; Ambasht and Agrawal 1998; Jansen and Van Den Noort 2000). In a greenhouse experiment on pea, it was found that stomatal limitation toward CO_2 uptake and lower amount of water usage as a result of a 65% decrease in adaxial stomatal conductance on supplemented UV-B radiation in greenhouse conditions (Nogues and Baker 1995), whereas in the well-watered condition in droughted pea plants, similar effects have been reduced to 50% (Allen et al. 1999). In the case of cucumber, which is found to be a very UV-responsive plant, initial UV exposure resulted in decreased stomatal conductance and prolonged exposure significantly increased stomatal aperture (Teramura 1983), whereas in the case of some members of Ericaceae increase in UV-induced stomatal conductance was observed and rest of the members did not reveal significant results (Musil and Wand 1993).

14.5 UV and Plant Ultrastructure

During the developmental phases of tissues and cells, UV radiation can alter the ultrastructure of cells and their organelles. The effect of UV radiation is multifarious in higher plants. The steps may include perceptivity of signals, transduction of signals, a remolding of the photosynthetic machinery and other organelles, and changes in gene expression (Holzinger and Lütz 2006). As in the case of other stresses, UV-B radiation also reduces the leaf area. Some of the most vulnerable crops for leaf area reduction are soybean, bean, pea, cowpea, cucumber, watermelon, Brussels sprout, etc. Prolonged UV radiation can reduce leaf area up to 60%, which in turn will reduce the overall productivity of the crops (Biggs and Kossuth; 1978; González et al. 1981). But moderate UV radiation with PAR can increase plant productivity in various crops such as potato and mustard (Teramura 1983). UV radiation also causes browning, scorching, glazing, chlorosis, necrosis, and desiccation in the case of susceptible crops or plants (Teramura 1983; Bornman and Teramura 1993; Rai et al. 2018). As UV damages membrane integrity, thus it may affect thylakoid integrity and ultimately affect photosynthesis. Plants try to cope up with this by increasing leaf thickness by adding bulk to mesophyll tissue (Kakani et al. 2003; Rai et al. 2020). UV radiation and reduction in chlorophyll content are in positive correlation. Chlorophyll reduction in crop plants ranges from low 10% to as high as 70%, and this reduction is higher in dicot plants (10%–78%) than monocot species (0%–33%) (Bornman and Teramura 1993; Tevini 1999).

14.6 Mechanism of UV Inhibition and Damage Control

One defense mechanism against UV radiation in plants is the increase of flavonoids (such as isovitexin, phenylpropanoids, cinnamoyl esters, flavones, flavonols, and anthocyanins esterified with cinnamic acid). Flavonoids are thought to protect photosynthetic pigments by absorbing UV radiation. Flavonoid concentration built up in the upper epidermal region of the leaves. Their best absorbance is found in the wavelength of 300 nm. An increase in flavonoid content with UV irradiation is well reported (Tevini, Braun, and Fieser 1991; Bornman and Terramura 1993; Middleton and Teramura 1993; Tevini 1999). Increase in content and accumulation of furanocoumarins, polyketides, and terpenoids such as cannabinoids (shikimic acid pathway) in epidermal tissues have been with enhanced UV radiation. This type of phenolic compound diminishes UV radiation and allows PAR to palisade and mesophyll tissues allowing photosynthesis, whereas sinapic esters (lignin precursor) act as UV screens in *Arabidopsis* (Stapleton 1992; Landry, Chapple, and Last 1995; Mpoloka 2008). This increase of phenolics and flavonoids is the result of the activity of enzyme phenylalanine ammonium lyase, enzyme chalcone synthase (CHS), etc. (Tevini and Teramura 1989; Mpoloka 2008). Photoinhibition takes place in higher illumination. During this situation, the plastoquinone pool gets completely reduced and the radical pair P680 forms a triplet P680 which in reaction with oxygen releases oxygen radical which degenerates D1 and D2 protein (key subunit of PSII). Full reduction of electron acceptors occurs when reduction equivalents are not being fully consumed during the dark reaction as stomata are closed. In high photon fluxes, the xanthophyll cycle gets activated, which acts as a sink for excitation energy in PSII. Another defense mechanism to prevent damage caused by UV radiation is the afresh synthesis of D1 and D2 proteins in PSII (Campbell et al. 1998; Tevini 2004).

Mitigation of DNA damage due to UV radiation in different organisms has a different DNA repair or damage control strategy. This is a well-studied topic in the case of microbes and mammals. One of the defense mechanisms is the formation of pyrimidine dimers to inhibit transcription of UV-induced mutation. DNA double-strand break in chromosomes is one of the detrimental effects of UV radiation. The repair mechanism involves excision repair (nucleotide and base excision), recombinational, photoreactivation, and post-replicational repair (Caldwell 1981; Mpoloka 2008). Double-strand break may be repaired by post-replication recombination, such as nonhomologous end joining (illegitimate recombination), homologous recombination. cDNAs such as DRT111 and DRT112 have been isolated from *Arabidopsis*; these are postulated to play a vital role in homologous recombination. Rearrangement of somatic homologous DNA has been reported in Arabidopsis as well as in tobacco. This phenomenon of somatic homologous rearrangement of DNA is aggravated in progenies exposed to high UV radiation than their predecessors. Germ cells exposed to UV radiation may experience permanent changes into progeny population (Ries et al. 2000a,b).

14.7 Use of UV in Crop Improvement

Although UV radiation causes a reduction in photosynthetic activities, such as photosynthetic rate, stomatal conductance, transpiration, internal CO_2 concentration, and photochemical efficiency, some cases are reported where UV can be used for the improvement of crop biochemical quality and yield. In colder climates, lots of different vegetables are cultivated in greenhouse conditions, which significantly reduce the direct UV irradiation during crop growth. Sometimes, these crops were supplemented with UV radiation to induce positive oxidative stress conditions and simulate antioxidant defense mechanisms in the crop. Tomato (*Solanum lycopersicum*) cultivar such as MicroTom was supplemented with UV-A (1 and 4 hours) and UV-B radiation (1 and 5 minutes) during their flowering and fruiting period (Mariz-Ponte et al. 2018). Within a month of growth, it was found that flowering and fruit ripening was synchronized. The receptor genes for UV-A/Blue and UV-B such as *cry1a* and *uvr8* were upregulated with increased expression of *hy5* and *uvr8*. UV radiation increased SOD (superoxide dismutase), an increase of H_2O_2 content along with minimalization of membrane disruption and lipid peroxidation. Phenolic content increase in response to antioxidant mechanism was found which acts UV screen. This was done by upregulation of *chs* and *fls* genes responsible for chalcone synthase and flavonol synthase enzymes which in turn are responsible for the synthesis of phenylpropanoid. Mariz-Ponte et al. (2018) in their study found that UV-A and UV-B in all doses increase flowering and fruiting, and lower doses of UV-B induced lower redox disorder and flowering/fruiting synchronization. With elevated UV-B radiation in the growth of two Curcuma species, *Curcuma longa* (turmeric) and *Curcuma caesia* (black turmeric), in field condition increased essential oil content 16% and 9%, respectively, irrespective of reduction of photosynthetic rate, stomatal conductance, carbon assimilation, etc. In the case of *C. caesia*, the amount of monoterpenes was increased rather than sesquiterpenes, whereas in the case of *C. longa*, all types of terpenes content increased with elevated UV-B radiation. Elevated UV-B inducted aromatic compounds such as epiglobulol, germacrene, 4-terpineol, and major active compound (1, 8-cineole) in *C. caesia*, whereas *C. longa* revealed an increased concentration of anti-cancer compounds such as alpha-terpinolene (61%), beta-caryophyllene (60%), and beta-sesquiphellandrene (32%). Overall, most of the compounds were increased in the case of *C. longa* upon elevated UV radiation (Jaiswal and Agrawal 2021). Enhanced germination rate and increase in dry mass have been reported in the case of *Vigna radiata* (Mung beans) seeds primed/pretreated with UV-A and UV-C radiation over their non-treated counterparts (Hamid and Jawaid 2011).

14.8 Application of UV in Disease Control

UV-induced biochemical signatures such as uvr8 can be very beneficial in disease and pest control as well as the post-harvest quality of crops and other plants. As a result of UV-induced stress flavonoids, hydroxycinnamic acid, SOD, catalase, peroxidase, several other antioxidants, phenolics, tannins, and stress-induced defense mechanisms spike in leaf as well as in other parts of the plant tissues (Tevini, Braun, and Fieser 1991; Bornman and Teramura 1993; Middleton and Teramura 1993; Tevini 1999). The phenylpropanoid pathways lead to the production of several different groups of flavonoids (such as flavones, flavonols, isoflavonols, isoflavones, anthocyanidins, and dehydroflavones) as well as nonflavonoids such as hydroxyl cinnamic acid, chalcones, dehydroxychalcones, jasmonic acid, abscisic acid, and ethylene (Meyer, Van de Poel, and De Coninck 2021). Most of the nonflavonoid compounds such as lignans, chalcones, suberin, and coumarin are formed from the derivatives of hydroxycinnamic acid. These compounds act as a scavenger for ROS. Many of these phenolic compounds pose antifungal properties. They either directly act as growth inhibitors, or increase microbial cell membrane permeability by altering membrane protein structure and composition. Derivatives of hydroxycinnamic acid, like p-coumaroylagmatine, feruloylagmatine, p-coumaroylputrescine and feruloylputrescine, agmatine, sesquiterpenoids, and putrescine, generated in host *A. thaliana* proved to be excellent in defense against *Alternaria brassicicola* and *Botrytis cinerea* (Ton et al. 2002; Van Der Ent et al. 2008; Muroi et al. 2009). Zealexin derived in maize is found to be very much effective against *Aspergillus flavus*, *Rhizopus microspores*, and *Fusarium graminearum* (Huffaker et al. 2011). Several secondary metabolites such as tannins, salicylic acid, azulenes, sesquiterpenes, coumarins, and quinones which accumulate in leaves and other parts of plants as a result of UV defense mechanism can efficiently be used as antifungal compounds. Some of them are presently being commercially utilized for the natural treatment of fungal infections of humans also (Arif et al. 2009). Camalexin, a broad-spectrum antimicrobial phytoalexin alkaloid commonly found in *Arabidopsis* and several members of Brassicaceae, can be induced upon UV radiation/exposure. Camalexin can damage the cell membrane of several genera of fungi and bacteria, this is also known to induce programmed cell death in *Botrytis cineraria* (Rogers, Glazebrook, and Ausubel 1996; Stotz et al. 2011; Ahuja, Kissen, and Bones 2012; Meyer, Van de Poel, and De Coninck 2021). Glucosinolates are sulfur-rich tryptophan derivatives abundantly present in the members of Brassicaceae

TABLE 14.1

UV Dose and Increase of Disease Resistance in Commonly Grown Crops

Crop	Pathogen/Pest	UV-B Daily dose (kJ/m^2)	Increase Specialized Metabolites, Proteins, and Hormones	References
Rice	*Magnaporthe oryzae*	2.5–7.5	Flavonoids, phenols, PAL, LOX, β-1,3-glucanase	Li et al. (2018)
Rice	*Mythimna separata, Chilo suppressalis*	1 kJ m^{-2}	JA and JA-Ile, p-coumaroylputrescine, feruloylputrescine, caffeoylputrescine	Qi et al. (2018)
Tobacco	*Spodoptera litura*	3 kJ m^{-2}	JA and JA-Ile, chlorogenic acid, rutin, di-caffeoyl spermidine	Qi et al. (2018)
Maize	*S. litura, M. separata*	3 kJ/m	JA and JA-Ile, DIMBOA-Glc, HDMBOA-Glc	Qi et al. (2018)
Soybean	*Spodoptera frugiperd*		Faster phenolic accumulation, isorhamnetin- and quercetin-based flavonoid	Winter and Rostás (2008)
Soybean	*Nezara viridula, Piezodorus guildinii*		Isoflavonoids daizin and genistin	Zavala et al. (2015)
Broccoli	*Myzus persicae, Pieris brassicae*	0.6–1	Total aliphatic glucosinolates, indolyl glucosinolates, aliphatic glucosinolates, compositional change of indolyl glucosinolates, flavonoids (esp. kaempferol), glucosinolates CHS, CHI, PBS3, PR-1, PR-2, PR-4, BG3, osmotin34	Mewis et al. (2012)
Cucumber	*Podosphaera xanthii*	03–0.9	Flavonols – Anthocyanins, both after inoculation with pathogen	Suthaparan et al. (2014)
Strawberry	*Podosphaera aphanis*	0.2–6.5	β-1,3-glucanase, osmotin-like protein, CHI, PAL, CHS, CHI	Kanto et al. (2014)
Strawberry	*Sphaerotheca aphanis*	5–32		Kanto et al., (2014, 2009)
Arabidopsis	*B. cinerea*	5.5	Sinapates	Demkura and Ballaré (2012)
Arabidopsis	*Pieris rapae, Trichoplusia ni*	1.5	Flavonoids	Grant-Petersson and Renwick (1996)
Arabidopsis	*S. litura*	1	JA and JA-Ile	Qi et al. (2018)

BG3-β-1, 3-glucanase 3; CHS, chalcone synthase; CHI, chalcone isomerase; DIMBOA-Glc, DIMBOA glucoside; FLS, flavonol synthase; HDMBOA-Glc, 4,7-dimethoxy-2-[3,4,5-trihydroxy-6-(hydroxymethyl)oxan-2-yl]oxy-1,4-benzoxazin-3-one; JA, jasmonic acid; JA-Ile, jasmonic acid-Isoleucine; LOX, lipoxygenase; PAL, phenylalanine; PR-1, PR-2, PR-4, pathogen-related protein.

(such as *Arabidopsis*, cabbage, mustard, and Broccoli) is found to be a very good defense mechanism against different insect pests along with other pathogens (Ahuja, Kissen, and Bones 2012; Buxdorf et al. 2013; Pusztahelyi, Holb, and Pócsi 2015; Meyer, Van de Poel, and De Coninck 2021). It can be said that UV dose can be effectively utilized for disease prevention and protection against a variety of pathogens and pests. For pathogen-specific disease prevention, UV application needs to vary (duration, dose, intensity, nocturnal, and diurnal application). Low-dose UV application is found to increase plant defense mechanism where higher dose can cause a reduction in photosynthesis, reactive oxidative stress, drought, growth retardation, and other detrimental and morphological changes (Meyer, Van de Poel, and De Coninck 2021). Seed priming is a well-known technique for better germination; it also helps to cope up with stress (environmental, chemical, and biotic) induced hindrances like toxicity, growth inhibition, and vulnerability toward pathogen (Moulick, Santra, and Ghosh 2018b, a). The application of UV dose for the minimization of pest and pathogens for some common commercially grown crops is summarized in Table 14.1. Calculated low UV dose priming in crops and medicinal plants can make them resistant to environmental and abiotic stress and subsequent higher adaptation and yield. But this needs extensive research and field trial. It has been suggested that night-time application of low-dose UV in strawberry, cucumber, rosemary, and several other crop cultivations and applications in floriculture can reduce the risk of contamination of *Pseudomonas xanthii, Pseudomonas aphanis, Golovinomyces biocellatus*, etc. and suppress their symptoms (Meyer, Van de Poel, and De Coninck 2021).

14.9 Application of UV in Pre and Post-Harvest for Crop Management and Improvement

Application of UV radiation in pre- and post-harvest of crops and vegetables reduces microbial growth and extends their shelf life. Low-dose application of UV-C during preharvest was found to increase antioxidant content in leafy vegetables like spinach (*Spinacia oleracea*), which is being cultivated around the world in huge quantities. (Martínez-Sánchez et al. 2019) have applied two different UV-C doses (1.5 and 3 kJ/m^2) in three different life stages of spinach (immature, baby, and mature commercial stage) in field conditions. It was found that immature leaves are more vulnerable to UV regarding their growth and development. In the case of the second-growth stage, application of UV-C dose at 3 kJ/m^2 reduced Vitamin C content. But in other cases, overall total phenolic content increased, and a reduction in microbial load (bacterial and yeast) in the second and third stages of development was observed. UV-C radiation in fresh-cut baby spinach at 0, 2.4, 7.2, 12, and 24 kJ/m^2 on both surfaces was found to reduce the growth of *Listeria monocytogenes* and *Salmonella enterica*. A significant decrease in psychrotrophic counts of

Enterobacteriaceae was found during the first 4 days of storage along with *Pseudomonas marginalis,* without showing discoloration, morphological changes, and cell shape (Escalona et al. 2010). In fresh-cut kalian-hybrid broccoli, UV-C dose (≤2.5 kJ/m^2) inactivates the growth and metabolism of *Escherichia coli, Salmonella enteritidis*, and *L. monocytogenes* and keeps these microbial counts stable during shelf life at 5°C–10°C (Martínez-Hernández et al. 2015). Prestorage treatment of strawberries with a hormetic dose of UV-C can extend their shelf life along with the deactivation of pathogens like *B. cinerea*. Enzymatic activity of phenylalanine ammonia-lyase and higher polyphenol oxidase concentration increased after 4 and 24 hours of storage, whereas β-1,3-glucanase and peroxidase revealed a biphasic pattern compared to the control set (Pombo et al. 2011). It has been suggested that abiotic stress, like wounding, altered gaseous composition in the environment, and water stress along with PAR and UV can enhance nutraceutical content in freshly harvested fruits and vegetables (Cisneros-Zevallos 2003).

The content of several commercially applied pesticides and fungicides (Esfenverate, Pirimicarb, Dimethoate, Imidacloprid, Buprofezin, Pyraclostrobin, Boscalid, Fludioxonil, Iprodione, and several others) significantly gets reduced within 12–24 hours of UV radiation (Mishra et al. 2008; Lassalle et al. 2014; Choi et al. 2020). Photodegradation by UV has revealed a significant reduction in the content of trichlorfon, diuron, dimethoate, simazine, terbuthylazine, chlorpyrifos methyl, parathion methyl, pirimiphos methyl, fenitrothion, terbutryn, malathion, chlorpyrifos, parathion ethyl, metidathion, and endosulfan from virgin olive oil (Nieto, Hodaifa, and Casanova 2009).

14.10 Conclusion

With the thinning of the ozone layer, UV radiation has gradually increased over the last few decades. Enhanced environmental radiation of UV along with water scarcity and increase in global temperature became a threat to plant productivity around the world. Plants' response to UV radiation depends on wavelength, intensity, in vivo or in vitro condition of growth and development, dose, duration, day or night application, and several other factors. Research on the impact of UV radiation in the last few decades has revealed that UV poses a great threat toward food security, morphological changes, yield, productivity, growth retardation, delay in maturation, retardation of reproductive organ development, flowering, fruiting, visual appeal of the harvest, sustenance, extinction threat of different commercially grown crops and wild species. Several groups of plants have naturally developed a defense mechanism against this threat via attenuation by physical methods, nullifying radiation-induced damage, and biochemical defense mechanisms to cope up with this stress. Plants' biochemical defense mechanism mostly relies on their UV receptor proteins such as UVR8, which triggers alternative of the customary pathway for the synthesis of different groups secondary metabolites, stress enzymes and proteins, essential oils, and their derivatives. These secondary metabolites help the crops/plants toward their disease resistance to varieties of pathogens and pests and to cope up with other environmental, toxicological, and abiotic stress. UV application in different stages in crop growth and development can persuade reproductive growth, pest, and pathogen resistance. The post-harvest application can resist or limit pathogens and pests through direct UV action or through a spike of a certain secondary metabolite which has proved to restrict pathogen proliferation and pest occurrence. UV-B and UV-C radiation itself and along with other stress can enhance nutraceutical content in harvested fruit and vegetables which is ultimately beneficial for harvest quality, economy and also beneficial for human health. Altogether UV-B radiation has an enormous potential toward harvest quality improvement and plants defense mechanisms, and this phenomenon has opened a broad multidimensional research scope of UV application in crop protection.

REFERENCES

Ahmad, M., N. Grancher, M. Heil, R. C. Black, B. Giovani, P. Galland, and D. Lardemer. 2002. "Action Spectrum for Cryptochrome-Dependent Hypocotyl Growth Inhibition in Arabidopsis." *Plant Physiology* 129(2): 774–85. doi: 10.1104/pp.010969.

Ahuja, I., R. Kissen, and A. M. Bones. 2012. "Phytoalexins in Defense against Pathogens." *Trends in Plant Science* 17(2): 73–90. doi: 10.1016/j.tplants.2011.11.002.

Allen, D. J., S. Nogués, and N. R. Baker. 1998. "Ozone Depletion and Increased UV-B Radiation: Is There a Real Threat to Photosynthesis?" *Journal of Experimental Botany* 49(328): 1775–88. doi: 10.1093/jxb/49.328.1775.

Allen, D. J., S. Nogués, J. I. L. Morison, P. D. Greenslade, A. R. McLeod, and N. R. Baker. 1999. "A Thirty Percent Increase in UV-B Has No Impact on Photosynthesis in Well-Watered and Droughted Pea Plants in the Field." *Global Change Biology* 5(2): 235–44. doi: 10.1046/j.1365-2486.1999.00227.x.

Ambasht, N. K., and M. Agrawal. 1998. "Physiological and Biochemical Responses of Sorghum Vulgare Plants to Supplemental Ultraviolet-B Radiation." *Canadian Journal of Botany* 76 (7): 1290–94. https://doi.org/10.1139/b98-137.

Arif, T., J. D. Bhosale, N. Kumar, T. K. Mandal, R. S. Bendre, G. S. Lavekar, and R. Dabur. 2009. "Natural products: Antifungal agents derived from plants." *Journal of Asian Natural Products Research* 11(7): 621–38. doi: 10.1080/10286020902942350.

Bagnall, D. J., R. W. King, and R. P. Hangarter. 1996. "Blue-Light Promotion of Flowering Is Absent in Hy4 Mutants of Arabidopsis." *Planta* 200(2): 278–80. doi: 10.1007/BF00208319.

Basiouny, F. M., T. K Van, and R. H. Biggs. 1978. Some Morphological and Biochemical Characteristics of C3 and C4 Plants Irradiated with UV-B. *Physiologia Plantarum*, 42(1), 29–32. doi: 10.1111/j.1399-3054.1978.tb01533.x.

Biggs, R. H. and Kossuth, S. V. 1978. Effects of ultraviolet-B radiation enhancement under field conditions on potatoes, tomatoes, corn, rice, southern peas, peanuts, squash, mustard and radish. UV-B Biological and Climatic Effects Research (BACER), Final Report, EPA, Washington, DC.

Bolink, E. 2003. "Causes of Leaf Area Reduction and Implications of Acclimation to UV-B Radiation in Pisum Sativum L."

Bornman, J. F., and A. H. Teramura. 1993. "Effects of Ultraviolet-B Radiation on Terrestrial Plants." In: L. O. Björn, J. Moan, W. Nultsch, and A. R. Young (Eds), *Environmental UV Photobiology*. Springer US. doi: 10.1007/978-1-4899-2406-3_14.

Buxdorf, K., H. Yaffe, O. Barda, and M. Levy. 2013. "The Effects of Glucosinolates and Their Breakdown Products on Necrotrophic Fungi." *PLoS One* 8(8). doi: 10.1371/journal.pone.0070771.

Caldwell, M. M. 1971. Solar UV Irradiation and the Growth and Development of Higher Plants. In: A. C. Giese (Ed.) *Photophysiology*, 131–177. Academic Press. doi: 10.1016/B978-0-12-282606-1.50010-6.

Caldwell, M. M. 1981. "Plant Response to Solar Ultraviolet Radiation." *Physiological Plant Ecology I*, 169–97. doi: 10.1007/978-3-642-68090-8_7.

Campbell, D., M. J. Eriksson, G. Öquist, P. Gustafsson, and A. K. Clarke. 1998. "The Cyanobacterium Synechococcus Resists UV-B by Exchanging Photosystem II Reaction-Center D1 Proteins." *Proceedings of the National Academy of Sciences of the United States of America* 95(1): 364–69. doi: 10.1073/pnas.95.1.364.

Casal, J. J., and H. Boccalandro. 1995. "Co-action between phytochrome B and HY4 in arabidopsis Thaliana." *Planta* 197(2): 213–18. doi: 10.1007/BF00202639.

Choi, S. W., H. M. Shahbaz, J. U. Kim, D. H. Kim, S. Yoon, S. H. Jeong, J. Park, and D. U. Lee. 2020. "Photolysis and TiO_2 photocatalytic treatment under UVC/VUV irradiation for simultaneous degradation of pesticides and microorganisms." *Applied Sciences (Switzerland)* 10(13). doi: 10.3390/app10134493.

Christie, J. M., A. S. Arvai, K. J. Baxter, M. Heilmann, A. J. Pratt, A. O'Hara, S. M. Kelly, M. Hothornbrian, B. O. Smith, K. Hitomi, G. I. Jenkins, and E. D. Getzoff., 2012. "Plant UVR8 photoreceptor senses UV-B by tryptophan-mediated disruption of cross-dimer salt bridges." Science 335(6075): 1492-1496. doi: 10.1126/science.1218091

Cisneros-Zevallos, L. 2003. "The Use of Controlled Postharvest Abiotic Stresses as a Tool for Enhancing the Nutraceutical Content and Adding-Value." *Concise Reviews and Hypotheses in Food Science* 68(5): 1560–65.

Day, T. A., and T. C. Vogelmann. 1995. "Alterations in Photosynthesis and Pigment Distributions in Pea Leaves Following UV-B Exposure." *Physiologia Plantarum* 94(3): 433–40. doi: 10.1111/j.1399-3054.1995.tb00950.x.

Day, T. A., T. C. Vogelmann, and E. H. DeLucia. 1992. "Are Some Plant Life Forms More Effective than Others in Screening out Ultraviolet-B Radiation?" *Oecologia* 92(4): 513–19. doi: 10.1007/BF00317843.

Demkura, P. V., and C. L. Ballaré. 2012. "UVR8 Mediates UV-B-Induced *Arabidopsis* Defense Responses against *Botrytis Cinerea* by Controlling Sinapate Accumulation." *Molecular Plant* 5 (3): 642–52. https://doi.org/10.1093/mp/sss025.

Escalona, V. H., E. Aguayo, G. B. Martínez-Hernández, and F. Artés. 2010. "UV-C Doses to Reduce Pathogen and Spoilage Bacterial Growth in Vitro and in Baby Spinach." *Postharvest Biology and Technology* 56(3): 223–31. doi: 10.1016/j.postharvbio.2010.01.008.

Franklin, K. A., U. Praekelt, W. M. Stoddart, O. E. Billingham, K. J. Halliday, and G. C. Whitelam. 2003. "Phytochromes B, D, and E Act Redundantly to Control Multiple Physiological Responses in Arabidopsis." *Plant Physiology* 131(3): 1340–46. doi: 10.1104/pp.102.015487.

Garrard, L. A., Van, T. K. and West, S. H. 1976. Plant Response to Middle Ultraviolet (UV-B) Radiation: Carbohydrate Levels and Chloroplast Reactions. *Proceedings Soil and Crop Science Society of Florida* 36: 184–188.

González, R., R. Mepsted, A. R. Wellburn, N. D. Paul, M. Tevini, W. Iwanzik, U. Thoma, and E. Bolink. 1981. "Non-Photosynthetic Mechanisms of Growth Reduction in Pea (*Pisum Sativum* L.) Exposed to UV-B Radiation." *Planta* 153(1): 388–94. doi: 10.1046/j.1365-3040.1998.00243.x.

Grant-Petersson, J., and J. A. A. Renwick. 1996. "Effects of Ultraviolet-B Exposure of *Arabidopsis Thaliana* on Herbivory by Two Crucifer-Feeding Insects (Lepidoptera)." *Environmental Entomology* 25 (1): 135–42. https://doi.org/10.1093/ee/25.1.135.

Hamid, N., and F. Jawaid. 2011. "Influence of Seed Pre-Treatment by UV-A and UV-C Radiation on Germination and Growth of Mung Beans." *Pakistan Journal of Chemistry* 1(4): 164–67. doi: 10.15228/2011.v01.i04.p04.

He, J. M., H. Xu, X. P. She, X. G. Song, and W. M. Zhao. 2005. "The Role and the Interrelationship of Hydrogen Peroxide and Nitric Oxide in the UV-B-Induced Stomatal Closure in Broad Bean." *Functional Plant Biology* 32(3): 237–47. doi: 10.1071/FP04185.

Heijde, M., and R. Ulm. 2012. "UV-B Photoreceptor-Mediated Signalling in Plants." *Trends in Plant Science* 17(4): 230–37. doi: 10.1016/j.tplants.2012.01.007.

Holzinger, A., and C. Lütz. 2006. "Algae and UV Irradiation: Effects on Ultrastructure and Related Metabolic Functions." *Micron* 37(3): 190–207. doi: 10.1016/j.micron.2005.10.015.

Huffaker, A., F. Kaplan, M. M. Vaughan, N. J. Dafoe, X. Ni, J. R. Rocca, H. T. Alborn, P. E.A. Teal, and E. A. Schmelz. 2011. "Novel Acidic Sesquiterpenoids Constitute a Dominant Class of Pathogen-Induced Phytoalexins in Maize." *Plant Physiology* 156(4): 2082–97. doi: 10.1104/pp.111.179457.

Jackson, J. A., and G. I. Jenkins. 1995. "Extension-Growth Responses and Expression of Flavonoid Biosynthesis Genes in the Arabidopsis Hy4 Mutant." *Planta* 197(2): 233–39. doi: 10.1007/BF00202642.

Jaiswal, D., and S. B. Agrawal. 2021. "Ultraviolet-B Induced Changes in Physiology, Phenylpropanoid Pathway, and Essential Oil Composition in Two Curcuma Species (*C. Caesia Roxb.* and *C. Longa* L.)." *Ecotoxicology and Environmental Safety* 208. doi: 10.1016/j.ecoenv.2020.111739.

Jansen, M. A. K., and R. E. Van Den Noort. 2000. "Ultraviolet-B Radiation Induces Complex Alterations in Stomatal Behaviour." *Physiologia Plantarum* 110(2): 189–94. doi: 10.1034/j.1399-3054.2000.110207.x.

Jenkins, G. I. 2009. "Signal Transduction in Responses to UV-B Radiation." *Annual Review of Plant Biology* 60: 407–31. doi: 10.1146/annurev.arplant.59.032607.092953.

Jenkins, G. I. 2014a. "Structure and Function of the UV-B Photoreceptor UVR8." *Current Opinion in Structural Biology* 29: 52–57. doi: 10.1016/j.sbi.2014.09.004.

Jenkins, G. I. 2014b. "The UV-B Photoreceptor UVR8: From Structure to Physiology." *Plant Cell* 26(1): 21–37. doi: 10.1105/tpc.113.119446.

Jordan, B. R. 2002. "Molecular Response of Plant Cells to UV-B Stress." *Functional Plant Biology* 29(8): 909–16. doi: 10.1071/FP02062.

Jordan, B. R., J. He, W. S. Chow, and J. M. Anderson. 1992. "Changes in MRNA Levels and Polypeptide Subunits of Ribulose 1,5-Bisphosphate Carboxylase in Response to Supplementary Ultraviolet-B Radiation." *Plant, Cell & Environment* 15(1): 91–98. doi: 10.1111/j.1365-3040.1992.tb01461.x.

Jordan, B. R., P. E. James, A. Strid, and R. G. Anthony. 1994. "The Effect of Ultraviolet-B Radiation on Gene Expression and Pigment Composition in Etiolated and Green Pea Leaf Tissue: UV-B-induced Changes Are Gene-specific and Dependent upon the Developmental Stage." *Plant, Cell & Environment* 17(1): 45–54. doi: 10.1111/j.1365-3040.1994.tb00264.x.

Kakani, V. G., K. R. Reddy, D. Zhao, and K. Sailaja. 2003. "Field Crop Responses to Ultraviolet-B Radiation: A Review." *Agricultural and Forest Meteorology*, 120: 191–218 Elsevier. doi: 10.1016/j.agrformet.2003.08.015.

Kami, C., S. Lorrain, P. Hornitschek, and C. Fankhauser. 2010. "Light-Regulated Plant Growth and Development." *Current Topics in Developmental Biology* 91(C): 29–66. doi: 10.1016/S0070-2153(10)91002-8.

Kang, C. Y., H. L. Lian, F. F. Wang, J. R. Huang, and H. Q. Yang. 2009. "Cryptochromes, Phytochromes, and COP1 Regulate Light-Controlled Stomatal Development in Arabidopsis." *Plant Cell* 21(9): 2624–41. doi: 10.1105/tpc.109.069765.

Kanto, T., K. Matsuura, M. Yamada, T. Usami, and Y. Amemiya. 2009. "UV-B Radiation for Control of Strawberry Powdery Mildew." *Acta Horticulturae* 842: 359–62. doi: 10.17660/ActaHortic.2009.842.68.

Kanto, T., Matsuura, K., Ogawa, T., Yamada, M., Ishiwata, M., Usami, T., and Amemiya, Y. 2014. A new UV-B lighting system controls powdery mildew of strawberry. *Acta Horticulturae* 1049, 655–660.

Karsch, F. J., C. J. I. Woodfiil, B. Malpaux, N. The, and C. D. C. Klein. 1998. "References and Notes 1." *Science*, 80, 319–438.

Kliebenstein, D. J., J. E. Lim, L. G. Landry, and R. L. Last. 2002. "Arabidopsis UVR8 Regulates Ultraviolet-B Signal Transduction and Tolerance and Contains Sequence Similarity to Human Regulator of Chromatin Condensation 1." *Plant Physiology* 130 (1): 234–43. https://doi.org/10.1104/pp.005041.

Landry, L. G., C. C. S. Chapple, and R. L. Last. 1995. "Arabidopsis Mutants Lacking Phenolic Sunscreens Exhibit Enhanced Ultraviolet-B Injury and Oxidative Damage." *Plant Physiology* 109(4): 1159–66. doi: 10.1104/pp.109.4.1159.

Lassalle, Y., H. Jellouli, L. Ballerini, Y. Souissi, É. Nicol, S. Bourcier, and S. Bouchonnet. 2014. "Ultraviolet-Vis Degradation of Iprodione and Estimation of the Acute Toxicity of Its Photodegradation Products." *Journal of Chromatography A* 1371: 146–53. https://doi.org/10.1016/j.chroma.2014.10.051.

Li, Xiang, Y. He, C. Xie, Y. Zu, F. Zhan, X. Mei, Y. Xia, and Y. Li. 2018. "Effects of UV-B Radiation on the Infectivity of: Magnaporthe Oryzae and Rice Disease-Resistant Physiology in Yuanyang Terraces." *Photochemical and Photobiological Sciences* 17(1): 8–17. doi: 10.1039/c7pp00139h.

Lidon, F. J. C., F. H. Reboredo, A. E. Leitão, M. M. A. Silva, M. P. Duarte, and J. C. Ramalho. 2012. "Impact of UV-B Radiation on Photosynthesis: An Overview." *Emirates Journal of Food and Agriculture* 24(6): 546–56. doi: 10.9755/ejfa.v24i6.14673.

Lin, C., H. Yang, H. Guo, T. Mockler, J. Chen, and A. R. Cashmore. 1998. "Enhancement of Blue-Light Sensitivity of Arabidopsis Seedlings by a Blue Light Receptor Cryptochrome 2." *Proceedings of the National Academy of Sciences of the United States of America* 95(5): 2686–90. doi: 10.1073/pnas.95.5.2686.

MacRobbie, E. A.G. 1998. "Signal Transduction and Ion Channels in Guard Cells." *Philosophical Transactions of the Royal Society B: Biological Sciences* 353(1374): 1475–88. doi: 10.1098/rstb.1998.0303.

Mao, J., Y. C. Zhang, Y. Sang, Q. H. Li, and H. Q. Yang. 2005. "A role for arabidopsis cryptochromes and COP1 in the regulation of stomatal opening." *Proceedings of the National Academy of Sciences of the United States of America* 102(34): 12270–75. doi: 10.1073/pnas.0501011102.

Mariz-Ponte, N., R. J. Mendes, S. Sario, J. M.P. Ferreira de Oliveira, P. Melo, and C. Santos. 2018. "Tomato Plants Use Non-Enzymatic Antioxidant Pathways to Cope with Moderate UV-A/B Irradiation: A Contribution to the Use of UV-A/B in Horticulture." *Journal of Plant Physiology* 221(November 2017): 32–42. doi: 10.1016/j.jplph.2017.11.013.

Martínez-Hernández, G. B., J. P. Huertas, J. Navarro-Rico, P. A. Gómez, F. Artés, A. Palop, and F. Artés-Hernández. 2015. "Inactivation Kinetics of Foodborne Pathogens by UV-C Radiation and Its Subsequent Growth in Fresh-Cut Kailan-Hybrid Broccoli." *Food Microbiology* 46: 263–71. doi: 10.1016/j.fm.2014.08.008.

Martínez-Sánchez, A., P. Lozano-Pastor, F. Artés-Hernández, F. Artés, and E. Aguayo. 2019. "Preharvest UV-C Treatment Improves the Quality of Spinach Primary Production and Postharvest Storage." *Postharvest Biology and Technology* 155(February): 130–39. doi: 10.1016/j.postharvbio.2019.05.021.

Mewis, I., M. Schreiner, C. N. Nguyen, A. Krumbein, C. Ulrichs, M. Lohse, and R. Zrenner. 2012. "UV-B Irradiation Changes Specifically the Secondary Metabolite Profile in Broccoli Sprouts: Induced Signaling Overlaps with Defense Response to Biotic Stressors." *Plant and Cell Physiology* 53(9): 1546–60. doi: 10.1093/pcp/pcs096.

Meyer, P., B. Van de Poel, and B. De Coninck. 2021. "UV-B Light and Its Application Potential to Reduce Disease and Pest Incidence in Crops." *Horticulture Research* 8(1). doi: 10.1038/s41438-021-00629-5.

Middleton, E. M., and A. H. Teramura. 1993. "The Role of Flavonol Glycosides and Carotenoids in Protecting Soybean from Ultraviolet-B Damage." *Plant Physiology* 103(3): 741–52. doi: 10.1104/pp.103.3.741.

Mishra, V., G. Srivastava, S. M. Prasad, and G. Abraham. 2008. "Growth, Photosynthetic Pigments and Photosynthetic Activity during Seedling Stage of Cowpea (Vigna Unguiculata) in Response to UV-B and Dimethoate." *Pesticide Biochemistry and Physiology* 92(1): 30–37. doi: 10.1016/j.pestbp.2008.05.003.

Moriconi, V., M. Binkert, C. Costigliolo, R. Sellaro, R. Ulm, and J. J. Casal. 2018. "Perception of Sunflecks by the UV-B Photoreceptor UV RESISTANCE LOCUS8." *Plant Physiology* 177(1): 75–81. doi: 10.1104/PP.18.00048.

Moulick, D., S. C. Santra, and D. Ghosh. 2018a. "Effect of Selenium Induced Seed Priming on Arsenic Accumulation in Rice Plant and Subsequent Transmission in Human Food Chain." *Ecotoxicology and Environmental Safety* 152(January): 67–77. doi: 10.1016/j.ecoenv.2018.01.037.

Moulick, D., S. C. Santra, and D. Ghosh. 2018b. "Rice Seed Priming with Se: A Novel Approach to Mitigate As Induced Adverse Consequences on Growth, Yield and As Load in Brown Rice." *Journal of Hazardous Materials* 355: 187–96. doi: 10.1016/j.jhazmat.2018.05.017.

Mpoloka, S. W. 2008. "Effects of Prolonged UV-B Exposure in Plants." *African Journal of Biotechnology* 7(25): 4874–83. doi: 10.4314/ajb.v7i25.59692.

Muroi, A., A. Ishihara, C. Tanaka, A. Ishizuka, J. Takabayashi, H. Miyoshi, and T. Nishioka. 2009. "Accumulation of Hydroxycinnamic Acid Amides Induced by Pathogen Infection and Identification of Agmatine Coumaroyltransferase in Arabidopsis Thaliana." *Planta* 230(3): 517–27. doi: 10.1007/s00425-009-0960-0.

Musil, C. F., and S. J. E. Wand. 1993. "Responses of Sclerophyllous Ericaceae to Enhanced Levels of Ultraviolet-B Radiation." *Environmental and Experimental Botany* 33(2): 233–42. doi: 10.1016/0098-8472(93)90069-R.

Neff, M. M., and E. Van Volkenburgh. 1994. "Light-Stimulated Cotyledon Expansion in Arabidopsis Seedlings: The Role of Phytochrome B." *Plant Physiology* 104(3): 1027–32. doi: 10.1104/pp.104.3.1027.

Neff, M. and J. Chory 1998. "Genetic Interactions between Phytochrome A, Phytochrome B, and Cryptochrome 1 during *Arabidopsis* Development." *Plant Physiology* 118 (1): 27–36. https://doi.org/10.1104/pp.118.1.27.

Negash, L., and L. O. Bjom. 1986. "Stomata Closure by Ultrayiolet Radiation." *Physiologia Plantarum* 66: 360–64.

Nieto, L. M., G. Hodaifa, and M. S. Casanova. 2009. "Elimination of Pesticide Residues from Virgin Olive Oil by Ultraviolet Light: Preliminary Results." *Journal of Hazardous Materials* 168(1): 555–59. doi: 10.1016/j.jhazmat.2009.02.030.

Noble, R. E. 2002. "Effects of UV-Irradiation on Seed Germination" 299(March): 173–76.

Nogués, S., and N. R. Baker. 1995. "Evaluation of the Role of Damage to Photosystem II in the Inhibition of CO_2 Assimilation in Pea Leaves on Exposure to UV-B Radiation." *Plant, Cell & Environment* 18(7): 781–87. doi: 10.1111/j.1365-3040.1995.tb00581.x.

O'Hara, A., and G. I. Jenkins. 2012. "In Vivo Function of Tryptophans in the Arabidopsis UV-B Photoreceptor UVR8." *Plant Cell* 24(9): 3755–66. doi: 10.1105/tpc.112.101451.

Ohgishi, M., K. Saji, K. Okada, and T. Sakai. 2004. Functional Analysis of each Blue Light Receptor, cryl, cry2, photl, and phot2, by Using Combinatorial Multiple Mutants in Arabidopsis. *Proceedings of the National Academy of Sciences* 101(8): 2223–2228.

Pombo, M. A., H. G. Rosli, G. A. Martínez, and P. M. Civello. 2011. "UV-C Treatment Affects the Expression and Activity of Defense Genes in Strawberry Fruit (Fragaria×ananassa, Duch.)." *Postharvest Biology and Technology* 59(1): 94–102. doi: 10.1016/j.postharvbio.2010.08.003.

Pusztahelyi, T., I. J. Holb, and I. Pócsi. 2015. "Secondary Metabolites in Fungus-Plant Interactions." *Frontiers in Plant Science* 6: 1–23. doi: 10.3389/fpls.2015.00573.

Qi, J., M. Zhang, C. Lu, C. Hettenhausen, Q. Tan, G. Cao, X. Zhu, G. Wu, and J. Wu. 2018. "Ultraviolet-B Enhances the Resistance of Multiple Plant Species to Lepidopteran Insect Herbivory through the Jasmonic Acid Pathway." *Scientific Reports* 8(1): 1–9. doi: 10.1038/s41598-017-18600–7.

Rai, R., S. Singh, S. Yadav, A. Chatterjee, S. Rai, A. Shankar, and L. Rai. 2018. "Impact of UV-B Radiation on Photosynthesis and Productivity of Crop." In *Environment and Photosynthesis a Future Prospect*, ed V. P. Singh, S. Singh, R. Singh, S. M. Prasadno, 336–346. India: Studium Press India.

Rai, N., A. O'Hara, D. Farkas, O. Safronov, K. Ratanasopa, F. Wang, A. V. Lindfors, et al. 2020. The Photoreceptor UVR8 Mediates the Perception of Both UV-B and UV-A Wavelengths up to 350 Nm of Sunlight with Responsivity Moderated by Cryptochromes. *Plant Cell and Environment* 43. doi: 10.1111/pce.13752.

Ries, G., G. Buchholz, H. Frohnmeyer, and B. Hohn. 2000a. "UV-Damage-Mediated Induction of Homologous Recombination in Arabidopsis Is Dependent on Photosynthetically Active Radiation." *Proceedings of the National Academy of Sciences of the United States of America* 97(24): 13425–29. doi: 10.1073/pnas.230251897.

Ries, G., W. Heller, H. Puchta, H. Sandermann, H. K. Seldlitz, and B. Hohn. 2000b. "Elevated UV-B Radiation Reduces Genome Stability in Plants." *Nature* 406(6791): 98–101. doi: 10.1038/35017595.

Rogers, E. E., J. Glazebrook, and F. M. Ausubel. 1996. "Mode of Action of the Arabidopsis Thaliana Phytoalexin Camalexin and Its Role in Arabidopsis-Pathogen Interactions." *Molecular Plant-Microbe Interactions*. doi: 10.1094/MPMI-9-0748.

Senger, H., & Schmidt, W. (1994). Diversity of photoreceptors. In *Photomorphogenesis in plants* (pp. 301-325). Springer, Dordrecht.

Stapleton, A. E. 1992. "Ultraviolet Radiation and Plants: Burning Questions." The Plant Cell 4(11): 1353-1358. doi:10.1105/tpc.4.11.1353

Stotz, H. U., Y. Sawada, Y. Shimada, M. Y. Hirai, E. Sasaki, M. Krischke, P. D. Brown, K. Saito, and Y. Kamiya. 2011. "Role of Camalexin, Indole Glucosinolates, and Side Chain Modification of Glucosinolate-Derived Isothiocyanates in Defense of Arabidopsis against Sclerotinia Sclerotiorum." *Plant Journal* 67(1): 81–93. doi: 10.1111/j.1365-313X.2011.04578.x.

Sugimoto, K. 2013. "Seed germination under UV-B irradiation." *Bulletin Minamikyushu University* 43(A):1-9.

Suthaparan, A., A. Stensvand, K. A. Solhaug, S. Torre, K. H. Telfer, A. K. Ruud, L. M. Mortensen, D. M. Gadoury, R. C. Seem, and H. R. Gislerød. 2014. "Suppression of Cucumber Powdery Mildew by Supplemental UV-B Radiation in Greenhouses Can Be Augmented or Reduced by Background Radiation Quality." *Plant Disease* 98(10): 1349–57. doi: 10.1094/PDIS-03-13-0222-RE.

Teramura, A. H. 1983. "Effects of Ultraviolet-B Radiation on the Growth and Yield of Crop Plants."

Teramura, A. H, and J. H. Sullivan. 1994. "Effects of UV-B Radiation on Photosynthesis and Growth of Terrestrial Plants." *Photosynthesis Research* 39.

Tevini, M. 1999. UV-Effects on Plants. In G.S. Singhal, G. Renger, S.K. Sopory, K.-D. Irrgang and Govindjee (eds), *Concepts in Photobiology: Photosynthesis and Photomorphogenesis Concepts in Photobiology*, 588–613. Narosa Publishing House, New Delhi, India. doi: 10.1007/978-94-011-4832-0_19.

Tevini, M. 2004. Plant Responses to Ultraviolet Radiation Stress. In: G. C. Papageorgiou and Govindjee (eds), *Chlorophyll a Fluorescence: Advances in Photosynthesis and Respiration*, eds. 605–621. Springer, Dordrecht. doi: 10.1007/978-1-4020-3218-9_23

Tevini, M., and A. H. Teramura. 1989. "Uv-B Effects on Terrestrial Plants." *Photochemistry and Photobiology* 50(4): 479–87. doi: 10.1111/j.1751-1097.1989.tb05552.x.

Tevini, M., W. Iwanzik, and U. Thoma. 1981. "Some Effects of Enhanced UV-B Irradiation on the Growth and Composition of Plants." *Planta* 153(4): 388–94. doi: 10.1007/BF00384258.

Tevini, M., J. Braun, and G. Fieser. 1991. "The Protective Function of the Epidermal Layer of Rye Seedlings Against Ultraviolet-B Radiation." *Photochemistry and Photobiology* 53(3): 329–33. doi: 10.1111/j.1751-1097.1991.tb03636.x.

Tilbrook, K., A. B. Arongaus, M. Binkert, M. Heijde, R. Yin, and R. Ulm. 2013. "The UVR8 UV-B Photoreceptor: Perception, Signaling and Response." *The Arabidopsis Book* 11(11): e0164. doi: 10.1199/tab.0164.

Ton, J., J. A. Van Pelt, L. C. Van Loon, and C. M.J. Pieterse. 2002. "Differential Effectiveness of Salicylate-Dependent and Jasmonate/Ethylene-Dependent Induced Resistance in Arabidopsis." *Molecular Plant-Microbe Interactions* 15(1): 27–34. doi: 10.1094/MPMI.2002.15.1.27.

Van Der Ent, S., B. W. M. Verhagen, R. Van Doorn, D. Bakker, M. G. Verlaan, M. J. C. Pel, R. G. Joosten, et al. 2008. "MYB72 Is Required in Early Signaling Steps of Rhizobacteria-Induced Systemic Resistance in Arabidopsis." *Plant Physiology* 146(3): 1293–1304. doi: 10.1104/pp.107.113829.

Vu, CV, L. H. Allen, Jr., and L. A. Garrard. 1982. Effects of supplemental UV-B radiation on primary photosynthetic carboxylating enzymes and soluble proteins in leaves of C3 and C4 crop plants. *Physiologia Plantarum*, 55(1), 11-16. https://onlinelibrary.wiley.com/doi/abs/10.1111/j.1399-3054.1982.tb00277.x.

Whippo, C. W., and R. P. Hangarter. 2003. "Second Positive Phototropism Results from Coordinated Co-Action of the Phototropins and Cryptochromes." *Plant Physiology* 132(3): 1499–1507. doi: 10.1104/pp.102.018481.

Winter, T. R., and M. Rostás. 2008. "Ambient Ultraviolet Radiation Induces Protective Responses in Soybean but Does Not Attenuate Indirect Defense." *Environmental Pollution* 155(2): 290–97. doi: 10.1016/j.envpol.2007.11.018.

Wright, L. A., and T. M. Murphy. 1982. "Short-Wave Ultraviolet Light Closes Leaf Stomata." *American Journal of Botany* 69(7): 1196–99. doi: 10.1002/j.1537-2197.1982.tb13364.x.

Wu, D., Q. Hu, Z. Yan, W. Chen, C. Yan, X. Huang, J. Zhang, et al. 2012. "Structural Basis of Ultraviolet-B Perception by UVR8." *Nature* 484(7393): 214–19. doi: 10.1038/NATURE10931.

Yin, R., and R. Ulm. 2017. "How Plants Cope with UV-B: From Perception to Response." *Current Opinion in Plant Biology* 37: 42–48. doi: 10.1016/j.pbi.2017.03.013.

Zavala, J. A., C. A. Mazza, F. M. Dillon, H. D. Chludil, and C. L. Ballaré. 2015. "Soybean Resistance to Stink Bugs (Nezara Viridula and Piezodorus Guildinii) Increases with Exposure to Solar UV-B Radiation and Correlates with Isoflavonoid Content in Pods under Field Conditions." *Plant Cell and Environment* 38(5): 920–28. doi: 10.1111/pce.12368.

15

Impact of Elevated CO_2 and O_3 on Field Crops and Adaptive Strategies through Agro-Technology

Arkabanee Mukherjee
Indian Institute of Tropical Meteorology

Swati Hazra
Sharda University

CONTENTS

15.1 Introduction

Food security of the steadily growing human population on earth is the biggest challenge in the 21st century. The global population has already crossed the 7 billion mark in 2021 and has been projected to be about 11 billion by 2100 (UN, 2021). This rising population has put the agricultural production system worldwide under an immense pressure leading to the risk of food insecurity and rise in global hunger which is evident in recent years as per the report by FAO et al. (2018). At present 821 million, which is approximately one out of every nine people on earth, are estimated to be undernourished and the situation is worsening especially in Africa and South America (FAO et al., 2018). Climate has a profound impact on the agricultural productivity and distribution of crops grown around the world (O'Brien et al., 2021). Severe food crises of recent times are mostly linked to the major environmental disasters, of which 80% are directly related to climate variability and extreme events (Molotoks et al., 2021). Therefore, climate change plays a key role in increasing the risk of food insecurity and global hunger (FAO et al., 2018).

Anthropogenic activities since the era of industrialization have been causing the emission of greenhouse gases (GHGs) and other non-GHG pollutants which are the major drivers of climate change (IPCC, 2013). These changes include global warming and the related phenomena, such as rise in the sea level, alteration in precipitation distribution and its pattern, climate variability, and occurrence of frequent extreme climatic events. Additionally, there is a rising concern over the tropospheric ozone accumulation leading towards severe injuries in plant, animal, and human health and thus becomes lethal to all living organisms (Zhang et al., 2019; Unger et al., 2020). Therefore, changes in climate related to the emission of GHG and non-GHG pollutants leading to global warming, climatic variabilities, and surface ozone accumulation, all conjointly would have a major impact on the food production systems and gradually make the Earth uninhabitable for humans. Thus, assessment of the impact of climate change on agriculture is the primary focus of many research groups all over the world to find out the suitable adaptation measures for sustainable crop production system. Innumerable experimental and modelling studies have already been conducted around the world to check particularly the effects of rising atmospheric CO_2

DOI: 10.1201/9781003258063-15

and tropospheric O_3 on the field crops. In this chapter, we attempt to give a consolidated account of such research focussing on the strategic agro-technological adaptations for the production of field crops.

15.2 Trends of Atmospheric CO_2 and Tropospheric O_3

There have been continuous emission of GHGs and other non-GHG pollutants since the industrial flourishment took place in the 18th century. The historical and future trends as projected by analysing the data of CO_2 and O_3 accumulation on the Earth's surface over time have been discussed in this section.

15.2.1 Carbon Dioxide Emission and Impact on Global Climate

Carbon dioxide is one of the major greenhouse gases causing the gradual rise in global surface temperatures. Natural processes such as decomposition, respiration, and ocean release can be the sources of CO_2 emission. However, the concentration of atmospheric CO_2 has been on the rise since 1750 mainly due to various anthropogenic activities, for example, deforestation, cement production, and burning of fossil fuels. In 2011, 40% increase was noted in atmospheric [CO_2] which had been about 280 ppm during the pre-industrial period (IPCC, 2013). Nevertheless, there was about 80% rise in the annual emission of CO_2 during the period of 1970–2004 (IPCC, 2007). Also, the rate of annual increase in global [CO_2] in recent years has been doubled since the pre-1970 rate, i.e., <1 ppm/year (Hofmann et al., 2009). The trend of atmospheric CO_2 over the last 40 years has been depicted in Figure 15.1.

This rising atmospheric CO_2 level has the largest contribution towards the total radiative forcing (RF), i.e., changes in energy flux caused by the drivers of climate change (IPCC, 2013). Consequently, the global mean surface temperature has been about 1°C higher in 2017 than the pre-industrial level, and it is currently increasing at the rate of about 0.2°C per decade because of the past and ongoing anthropogenic RF (IPCC, 2018). Therefore, it is expected to cross the 2°C mark by the end of 21st century. There are also regional and temporal variations in the extent of warming and actual rise in the atmospheric temperature is much higher than the global annual average, especially in the Arctic region, leading to the melting of ice cover and rise in the sea level at much quicker rate than estimated on the basis of mean global temperature rise. However, an international treaty on the climate change, known as The Paris Agreement, has been adopted in 2015, to limit the global warming to 1.5°C–2°C as compared to the pre-industrial levels (UNFCCC, 2015). Accordingly, emission pathways based on the trajectories of global anthropogenic emission models over the 21st century have been recognized and categorized (IPCC, 2018) as:

A. **No overshoot**: limiting global warming to below 1.5°C
B. **1.5°C limited-overshoot**: limiting warming to below 1.6°C and returning to 1.5°C by 2100
C. **Higher-overshoot**: warming exceeding 1.6°C but still returning to 1.5°C by 2100

In the special report published by the IPCC (2018), five major Reasons For Concern (RFCs) in relation to the changes in climate have been identified as "RFC1: Unique and threatened systems, RFC2: Extreme weather events, RFC3: Distribution of impacts, RFC4: Global aggregate impacts, and RFC5: Large-scale singular events". Among these, higher risk is associated with RFC1, 2, and 3 which would also have significant impact on the agricultural sector.

15.2.2 Ozone: The Background

Ozone when present at the stratospheric level, is the saviour of the biosphere from the harmful UV radiations (Seinfeld

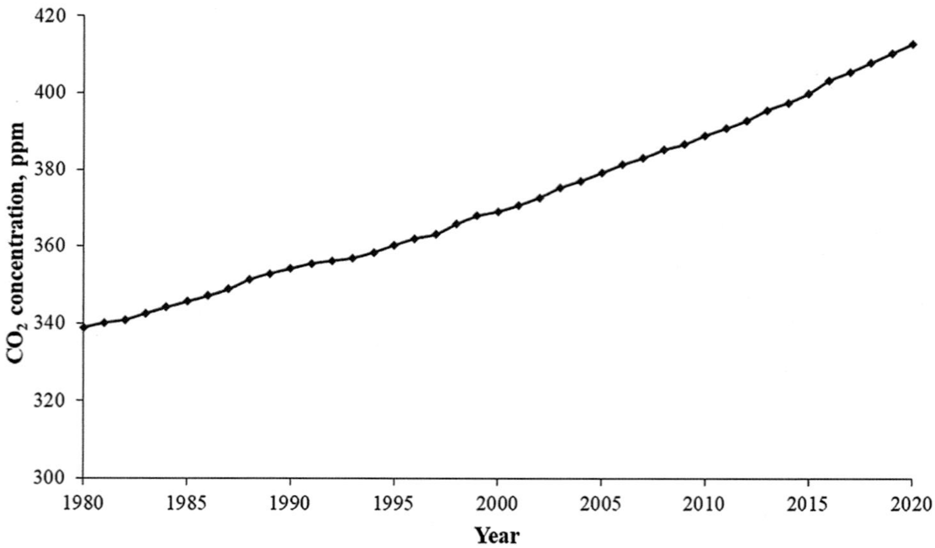

FIGURE 15.1 Changes in global marine surface CO_2 concentration over the last 40 years (1980–2020). (Data source: NOAA, ESRL (2021).)

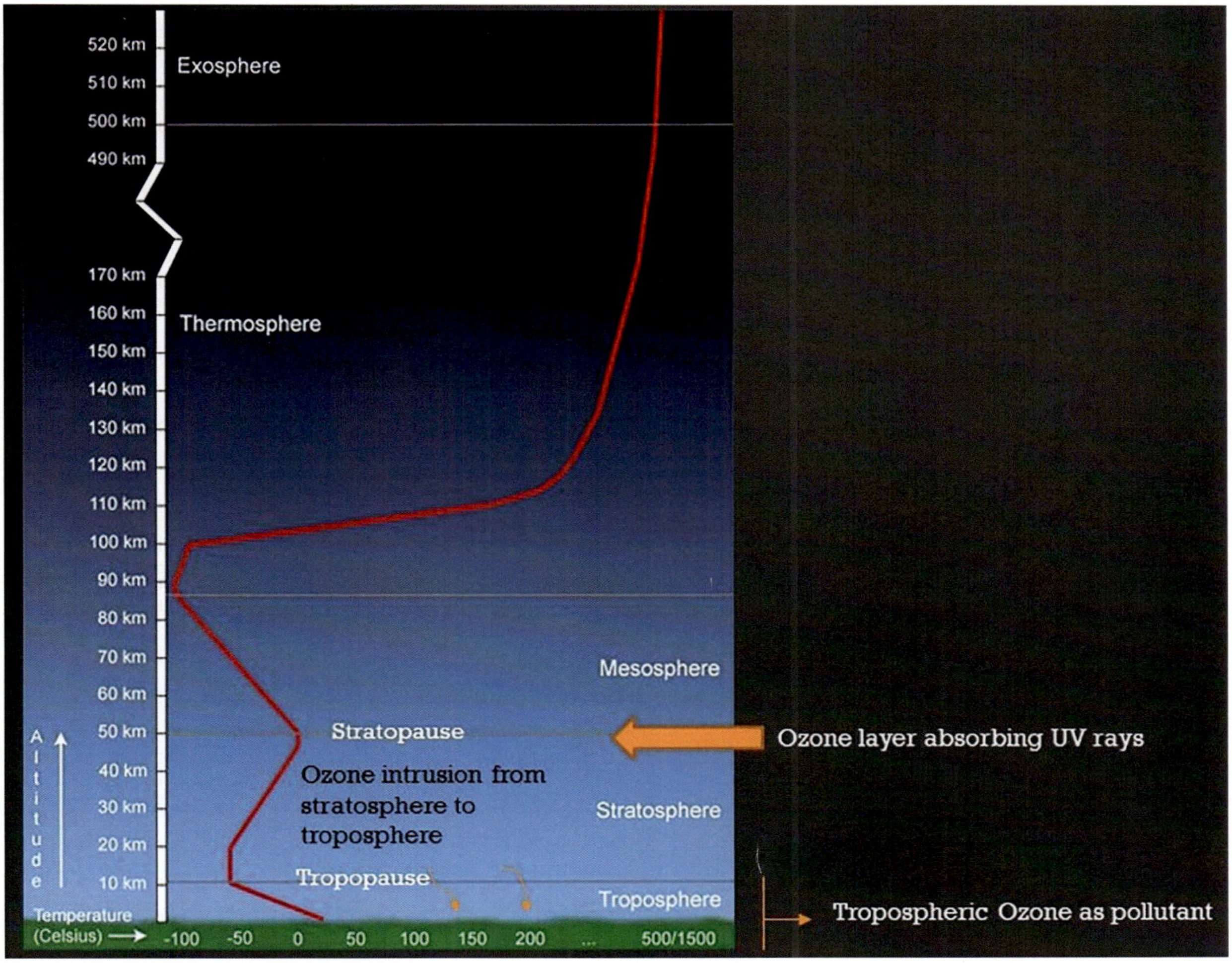

FIGURE 15.2 Temperature profile and ozone dynamics in different atmospheric strata.

and Pandis, 2016). However, the same is considered as a GHG in tropospheric level due to its ability to increase the RF (Stevenson et al., 2013; Simpson et al., 2014). Gradually, the gas has been enlisted as the "criteria air pollutant" by the United States Environmental Protection Agency (US EPA) due to the harmful effects of ozone imparted on the biological organisms. Further, the permissible concentration of tropospheric O_3 is revised by the US EPA to 0.07 part per million (ppm) in 8-hour exposure (NAAQS, 2015). Climatic variability, frequent extreme weather events, and rising global temperature decide the fate of the tropospheric ozone formation. Significant global and regional climatic conditions like El Nino and Southern osscilation interject with tropospheric ozone precursor availibility, and thus surface [O_3] increases (Lu et al., 2019). Influence of temperature, radiation, and availibility of O_3 precursors also diversify ozone formation pathway in stratosphere and troposphere. Considering the vertical distribution of temperature, a steady temperature reduction occurs with the increase in altitude in the troposphere, resulting in a temperature minimum of −50°C at the tropopause. The temperature again rises in the stratosphere region and reaches up to 2°C at stratopause (Figure 15.2). The temperature along

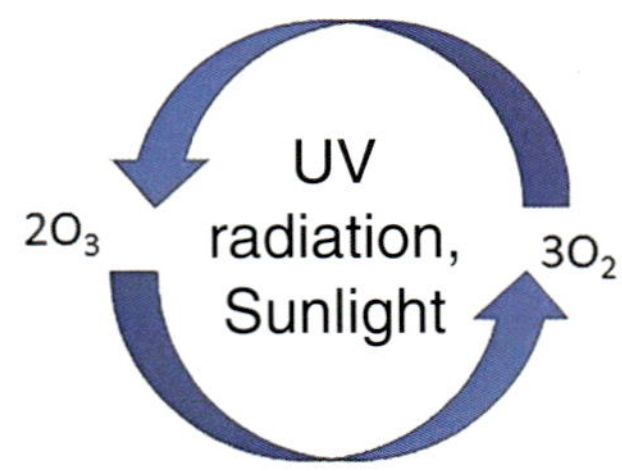

FIGURE 15.3 Ozone formation and destruction at equilibrium in stratosphere.

with the sunlight penetration and UV radiation have been proved to be a favourable condition for the continuous formation and breakdown of O_3 from O_2 in the atmospheric region known as the "ozone layer" (Figure 15.3).

15.2.2.1 O_3 Chemistry

In stratosphere, the formation and breaking down of ozone follows a straight pattern which persists in equilibrium. As already mentioned, the UV radiation acts as a fuel to maintain this process in continuity. The simplified chemical equation is mentioned in Figure 15.3, but the process occurs in the following steps of reactions.

$$O_2 + \text{UV radiation} \rightarrow O + O \quad \text{(R1)}$$

$$2O + 2O_2 + N_2 \rightarrow 2O_3 + 2N_2 \quad \text{(R2)}$$

N_2 is not taking any active participation in the chemical reaction (R2) and acting merely as a catalyst.

$$O_3 + \text{UV radiation} \rightarrow O_2 + O \quad \text{(R3)}$$

$$O + O_3 \rightarrow O_2 + O \quad \text{(R4)}$$

The loss of ozone atom occurs due to braking down of the same in to nascent oxygen and the procedure continues (R4).

The 90% of total [O_3] in the atmosphere exists within 15–35 km altitude (ozone layer). Hence, the massive portion of harmful UV rays (<240 nm) gets absorbed within this layer (Chapman 1930) causing a significant temperature rise in this stratum. Unlike the O_3 from the ozone layer, tropospheric O_3 (within 0–10 km from the Earth's surface) is already designated to be harmful for human and other living organisms of this region. O_3 chemistry in the troposphere is strongly non-linear and much more complex than in the stratosphere. Precursor gases such as oxides of nitrogen (NO_x), carbon monoxide (CO), and volatile organic compounds (VOCs) are the major drivers of tropospheric O_3 formation.

Mainly NO_2, both from natural and anthropogenic sources, initiates the photolytic reaction at <424 nm radiation.

$$NO_2 \rightarrow NO + [O] \quad \text{(R5)}$$

The atomic oxygen reacts with existing O_2 molecules in the atmosphere to form O_3 in the presence of a third component.

$$[O] + O_2 + M \rightarrow O_3 + M \quad \text{(R6)}$$

Presence of NO in the atmosphere again reacts with the existing O_3. Thus, NO_x in different forms acts as both the source and the sink for O_3.

$$O_3 + NO \rightarrow NO_2 + O_2 \quad \text{(R7)}$$

Alkyl hydride (RH) and alkoxy radicals follow a different pathway in radical formation which finally leads to ozone formation.

$$RO_2 + O_2 \rightarrow HO_2 + \text{Carbonyl} \quad \text{(R8)}$$

$$RH + O_2 + OH \rightarrow RO_2 + H_2O \quad \text{(R9)}$$

$$HCHO \rightarrow H + CO \quad \text{(R10)}$$

Photolysis of non-methyl hydrocarbon (NMHC) with O_2 results in O_3 formation.

From R8 and R9, the two products HO_2 and RO_2 oxidize NO into NO_2 which again acts as a source for O_3 formation.

$$HO_2 + NO \rightarrow NO_2 + HO \quad \text{(R11)}$$

$$RO_2 + NO \rightarrow RO + NO_2 \quad \text{(R12)}$$

$$CO + OH \rightarrow CO_2 + HO_2 \quad \text{(R13)}$$

Also, oxidation of CO, hydrocarbons (RH), or alkoxy radical (RO), form HO_2 and RO_2. Absence of OH radical, however, can terminate this reaction chain.

Tropospheric ozone has a longer persistancy especially in the higher altitudes. The scavenging by NO_x and dry deposition occur nearby the surface level. Thus, ozone molecules formed in the lower altitudes can move upwards and be transported to a different location, and thereby enriching the latter without even any precursor (Van Dop et al., 1987; Comrie and Yarnal, 1992).

Apart from the anthropogenic activities, the biogenic VOCs (isoprene, monoterpene, and sesquiterpenes) originated from the terrestrial vegetation can be as high as 600 Tg C/year (Guenther et al., 2012). Unlike the complex tropospheric O_3 chemistry, these biogenic VOCs, such as isoprenes, already have been reported to impart a major influence on ozone chemistry due to its oxidation caused by OH^-, NO_x, and O_3 (Squire et al., 2015). In case of low nitrate availability in the atmosphere, PAN (peroxy acyl nitrates) are produced from the reaction between VOCs and the existing ozone molecules. The PAN act as another precursor for the surface O_3 (Atkinson et al., 2000). Hence, formation of O_3 also depends on the precursors (NO_x, VOC, and PAN) chemistry.

15.2.2.2 Surface Ozone Pollution: Past, Present, and Future

Majorly, O_3 precursors are anthropogenic in nature. The very first observation of increasing [O_3] in the troposphere was reported in 1800, measured by the first 300 air quality monitoring stations in the world distributed over Europe and USA. At a later period, a report on the measurement of [O_3] in Athens and Greece during the first half of the last century (1901–1940) mentioned the highest magnitude of the same to be 20 ppb (Varotsos and Cartalis, 1991). A simulation study using the baseline data obtained from the period of 1970–1995 indicated a triple increment of [O_3] over the Asian countries by 2010 (Jacob et al., 1999). This was due to the higher population growth which consequently led to more energy consumption. This study is one of the first fews stating that a long-range transport of [O_3] from the other continents could add up to 20–40 ppb [O_3] in the surface atmosphere without the scavenging NO_x available in the regional air. Nevertheless, the increasing anthropogenic activities and urbanization are directly proportional to the formation of ozone precursors (Mittal et al., 2007; Cho et al., 2011; Rai et al., 2012) which may lead to a significantly high [O_3] (100 ppb) as already reported by Colbeck (1985) in the USA in the last century. The air quality monitoring stations positioned in the mediterrean regions indicated an yearly 0.4% average reduction in the rural areas but an increment occurred at semi-urban/urban areas of the same percentage magnitude between 2000 and 2011 (Saxena et al., 2019). During summer, Beijing, China reported up to ~200 ppb surface [O_3] (Wang et al., 2017). Long-range transportation is responsible for the countries with low anthropogenic activities but higher surface [O_3] (Fiore et al., 2009). Both seasonal and diurnal variations in [O_3] also have been noticed during the last century. Stratospheric intrusion of ozone is another source of the gas in the surface atmosphere during November to April (Colbeck, 1985). In the starting of this century, 50–100 ppb O_3 has already been reported in the tropospheric region in various episodic moments in the Northern hemisphere which is more prone to the tropospheric ozone exposure (Morgan et al., 2006). An investigation by Permadi and Oanh (2008) manifested a strong influence of the meteorological parameters (solar radiation, wind speed, and temperature) on the favourable condition for surface ozone formation. They also reported the highest value of surface ozone concentration (>200 ppbv) in Jakarta in 2002.

A close monitoring of background surface [O_3] magnitude recorded 1–4 ppb/annum increment during the period of 2013–2019 (Liu and Wang, 2020a, b; Li et al., 2020). The current situation in O_3 pollution scenario has taken a leap due to the corona virus (COVID-19) pandemic worldwide. This pandemic caused worldwide lockdown and thus putting a hold to the anthropogenic emission of pollutants which provides the rarest opportunity to study the background pollution level of each pollutants. However, in case of surface ozone concentration, the magnitude reaches to the maximum level during the lockdown period worldwide (Siciliano et al., 2020; Wang et al., 2021; Grange et al., 2021). Similar time period during the pre-lockdown years recorded a maximum 30% rise in [O_3] under the favourable meteorological condition (Sicard et al., 2020). Effect of the same on the global food security is also taking its grasp as studies on the crop model simulations based on the previous and current parameters are incorporated for future prediction. A study using the partial derivative-linear regression model considering the ozone and temperature projections by the Community Earth System Model clearly reflects the food insecurity in US alone. The model implicates 13%, 43%, and 28% loss in US wheat, maize, and soybean production up to 2050 (Tai and Martin, 2017). Another model simulation based on current data to project the tropospheric ozone situation in 2100 considering climate change issues like prolonged summer and extreme weather insinuate bad air quality areas would get worse and less polluted will be in better condition (Schnell et al., 2016).

15.3 Effect of Elevated CO_2 and O_3 on Field Crops

Elevated levels of atmospheric CO_2 and surface O_3, independently or conjointly, have the potential to influence the growth and development of crops. The same has been elaborated in the following sub-sections.

15.3.1 Crop Responses to Elevated CO_2

Crop growth, development, and yield depend largely on photosynthesis, the process in which carbon from the atmospheric CO_2 is assimilated into the organic molecules inside the green tissues of plants. Therefore, rising atmospheric [CO_2] has an immense potential to influence the agricultural production. However, there are several other factors, such as temperature, water and nutrients, and physiological processes (respiration, transpiration etc.), which concurrently regulate the final growth and yield of plants.

15.3.1.1 Crop Growth and Yield under Elevated CO_2

More than 100 years ago, in 1804, Nicholas-Théodore de Saussure first demonstrated that the atmospheric CO_2 is the primary source of carbon for the assimilation process (photosynthesis) going inside the plant bodies which finally led to the accumulation of biomass and thus crop yield (Hill, 2012). Since then, thousands of plant growth experiments have been carried out under the elevated CO_2, involving both C_3 and C_4 crops such as cereals (wheat, rice, barley, oats, maize, and sorghum), pulses (beans, peas, and soybean), vegetables (potatoes, sweet pepper, tomato, cucumber, and lettuce), fruits (strawberry, grapes, and citrus), fibre-yielding (cotton), oil-yielding (sunflower), flowers (roses, carnations, petunias, etc.), and high-value crops (tobacco, sugarbeet, and sugarcane) (Wittwer, 1970; Kimball, 1983; Jablonski et al., 2002; Ainsworth and Long, 2005; Singh and Jasrai, 2012; Wang et al., 2013; Challinor et al., 2014; Abebe et al., 2016; Kimball, 2016; Chaturvedi et al., 2017; Kumar et al., 2017). In most of these studies, CO_2-enriched microclimate was created artificially inside the glasshouses, chambers, or greenhouses. Open-top chambers (OTCs), which have been proved to be a cost-effective technology to maintain a high CO_2 level at the near-natural field conditions, were widely used in more recent researches. Fewer experiments were conducted in the open field under the elevated CO_2 level using the more expensive free air CO_2 enrichment (FACE) technology. The trend of increasing dry matter accumulation and crop yield under the elevated [CO_2] is noted in almost all cases and thus has established the "fertilization effect" of CO_2, especially in C_3 crops. Model simulation studies on diverse crops also supported same results (Tubiello and Ewert, 2002), especially under the adequate supply of nitrogen and water (Asseng et al., 2004; O'Leary et al., 2015). However, the wide range of elevated CO_2 level (450–1,250 ppm) set for various crops in these experiments, conducted over the past decades, has led to a greater variability in the extent of the biomass and yield increments. Also, the biomass and yield gain of the same crop varied considerably due to genotypes, growth conditions, and the CO_2 enrichment technology used. For instances, various rice cultivars showed differential responses to the elevated [CO_2] resulting in 3%–36% increase in grain yield over the ambient level in a FACE experiment conducted in Japan (Hasegawa et al., 2013). Similarly, yield enhancement of major cereals in FACE trials was found to be ~50% less than that reported in the enclosure studies (Long et al., 2006). Findings of some major

TABLE 15.1
Some Experiments Conducted on Major Field Crops under Elevated [CO_2] Varying from 490 to 700 ppm

Crop	Elevated [CO_2] (ppm)	Cultivation and CO_2 Enrichment Technique	Change under Elevated [CO_2]	References
Sorghum, C_4	795	Rhizotron box, closed enrichment chambers	+53% biomass, +59% grain yield	Chaudhuri et al. (1986)
Cotton, C_3	650	Field plot, OTC	+60% above ground biomass; +63% seed cotton yield (lint plus seed)	Kimball and Mauney (1993)
Spring wheat, C_3	520 and 680	Pot culture, OTC	Average +36% shoot biomass; +34% grain yield under elevated [CO_2]	Fangmeier et al. (1996)
Winter wheat, C_3	700	Polyethylene-covered tunnels	+6%–31% biomass; +7%–44% seed dry weight	Wheeler et al. (1996)
Rice, C_3	586–645	Field plot, FACE	Average +13% plant biomass; +11% grain yield	Kim et al. (2003)
Spring wheat, C_3	537–572	Field plot, FACE	Average +11.8% above ground biomass; +10.4% grain yield	Högy et al. (2009a)
Maize, C_4	550	Field plot, FACE	+24% biomass; +41% grain yield under limited water supply	Manderscheid et al. (2014)
Rice, C_3	490	Field plot, OTC	Average +8% above ground biomass; –4% grain yield	Satapathy et al. (2015)
Soybean, C_3	550	Pot culture, OTC	Average +46% shoot biomass; +40% seed yield	Li et al. (2017)

research conducted around the world to assess the effect of elevated CO_2 on the growth and yield of field crops are listed in Table 15.1.

The biomass and yield enhancement in response to the elevated CO_2 level is further regulated by the presence of other growth factors, especially water, nitrogen, and temperature (Kimball, 2016). Wide variation (–17 to +79%) in wheat yield increment due to the varying level of supplemental irrigation under an elevated [CO_2] in the semi-arid condition indicated a strong interaction between CO_2 and water availability (Fitzgerald et al., 2016). Generally, better biomass and yield responses to the elevated [CO_2] were noted under dry or water-restricted conditions for both C_3 (Bishop et al., 2014) and C_4 (Manderscheid et al., 2014) crops. Furthermore, a meta-analysis study of the previous FACE results showed that the elevated [CO_2] led to about 19% yield gain in C_3 cereals (wheat, rice, and barley) under the abundant nitrogen and water supply, whereas the increment was comparatively lower under the limited N treatments (Kimball, 2016). Nevertheless, the fertilization effect of elevated [CO_2] has been more or less evident in the results of all these experiments. Concomitant rise in atmospheric temperature, however, may counteract with the growth and yield stimulations imparted by the CO_2-enriched environment (Lobell et al., 2011). Global warming poses a serious threat to the crops which are particularly sensitive to heat stress, such as wheat (Slafer and Rawson, 1994; Prasad and Djanaguiraman, 2014; Asseng et al, 2015). It has been found that ~0.4°C rise in annual global temperature during the period of 1981–2002 reduced the global wheat yield by 3%, which actually counteracted about 3.5% yield gain due to ~35 ppm increase in global CO_2 level (Lobell et al., 2011). Simulation study on global wheat production has projected ~6% reduction for every 1°C further increase in global atmospheric temperature (Asseng et al, 2015).

15.3.1.2 Crop Quality under Elevated CO_2

Field crops such as cereals (wheat, rice, maize, and barley) and pulses constitute a major portion of the human diet. Therefore, any change in nutritional composition of these food crops would have a great impact on the human health. Environmental and growth factors, such as temperature, water, and nutrients availability, make a significant contribution towards the nutrients present in the consumable parts of the crops (Altenbach et al., 2002). For instance, elevated [CO_2], alone or in combination with other factors, such as temperature, water, and fertilizer application, is able to alter the grain nutrients and minerals composition of rice (Chaturvedi et al., 2017; Wang et al., 2020), wheat (Manderscheid et al., 1995; Högy and Fangmeier, 2008), barley (Manderscheid et al., 1995), and other crops (Uprety et al., 2010). Elevated [CO_2] promotes the photosynthetic carbon assimilation resulting in a higher accumulation of carbohydrates (starch) in the sink tissues (stem, storage, and reproductive organs) of C_3 crops, especially in the cereal grains. On the contrary, concentration of other nutrients decreases due to the "dilution by starch" (Kuehny et al., 1991; Xu et al., 2015). In addition to this, a downregulated rate of transpiration under the elevated [CO_2] (Conroy, 1992) can lower down the nutrient uptake by plants and thereby decrease the grain mineral nutrients content. There are reports of such reduction of Fe and Zn content in wheat, rice, and barley due to the [CO_2] elevation (Manderscheid et al., 1995; Fangmeier et al., 1997; Pleijel et al., 2000). In the separate CO_2-enrichment studies on sorghum, wheat, and barley, concentration of Mn, Cd, Si, and S decreased significantly (Prior et al., 2008; Högy et al., 2009a; Erbs et al., 2010). However, there are a few instances where elevated [CO_2] increased some macronutrients (Ca and Mg) content in grains (Prior et al., 2008). In addition to carbohydrates and minerals, wheat grains also contain a substantial amount of storage proteins (gluten) which ultimately regulate the dough-making properties of flour. Therefore, a reduction in the protein content and alterations of gluten components (glutenins and gliadins) under the elevated [CO_2] (Högy et al., 2009b) could destabilize the whole bakery industries around the world.

15.3.2 Surface O_3 Exposure and Plant Health

Trees act as a huge sink due to the dry deposition of O_3 under certain meteorological conditions. The 45% of total O_3 deposition across the ecosystem occurs through the stomata (Clifton et al., 2020) which is majorly the entry point of ozone. Also, minor openings like cuticles contribute to some extent. The entrance of ozone in the plant system through the stomatal aperture breaks the molecule down into a more reactive oxygen species (ROS) and peroxide (H_2O_2) by "ozonolysis". Oxidants generated by the process influence the guard cells and affect stomatal conductance (g_s). Peroxides originated from ozonolysis are responsible for gene alterations and premature cell death or senescence (Miller et al., 1999). In the higher plants, ozonolysis has deleterious effect on light and dark reaction of photosynthesis. Additionally, a delayed stomatal aperture movement, known as "stomatal sluggishness", was observed due to the ROS interaction with guard cells. However, the sensitivity of stomata towards O_3 varies greatly among crops. Cereal crops like wheat are more sensitive than rice. Similarly, fruits like watermelon and legumes are more prone to be affected by higher [O_3] exposure (Table 15.2).

Similar to the studies on the effects of elevated [CO_2] on crops, OTC and Free air [O_3] elevation (FACE) experiments were done to determine the crop responses to ozone exposure which were much indicated by the stomatal conductance (Feng et al., 2019). Leaves are the targetted entry point for O_3 into the plant system; water content and [O_3] concentration in this part are inversely proportional to each other (Massman, 1998; Hoshika et al., 2015). Hence, sluggishness in stomata causes partial closure of the aperture, thereby reducing the water content in the leaves and leading to the higher ozone exposure of the plant (Paoletti and Grulke, 2010). Apart from the loss of stomatal conductance, reduction in photosynthetic activities, premature senescence, and changes in the antioxidant systems are the major factors affecting crop health, and thus yield. Like stomatal conductance, effect of ozone on growth also differs among various crops and cultivars (Ainsworth, 2008; Ashmore, 2005; Biswas et al., 2008a,b). In the scenario of accelarating population growth, compromising yield is an unacceptable option, hence outcome of a crop simulation model considering the current ozone elevation scenarios predicting ~17–35 billion USD worth of loss in food production industry in 2030 is an alarming circumstance (Mills and Harmens, 2011). Although rice, among all the major crops, is less sensitive to the O_3 exposure (Mills et al., 2007; Wang et al., 2007), susceptibility of rice towards the higher background [O_3] is well established (Pandey et al. 2015, 2018; Singh and Agrawal 2017). Therefore, O_3 concentration, already recorded as high as 100 ppb during the winter season over the Indian subcontinent, can put rice cultivation under a greater risk. Indian subcontinent being one of the major rice cultivating areas in the world, 8% drop in the production due to the ozone exposure during khariff (June- September) and rabi (December- February) seasons results in a loss of ~1.5 billion USD (Sharma et al., 2019).

TABLE 15.2

Sensitivity of Different Field Crops and Vegetables Towards [O_3] Exposure

O_3-Sensitive Crops	O_3 Moderately Sensitive Crops	O_3-Resistant or Less Sensitive Crop
Wheat	Rice	Barley
Pulses	Maize	Strawberry
Turnip	Potato	Plum
Cotton	Sugarbeet	
Onion	Rapeseed	
Tomato	Broccoli	

Previous reports on the hourly magnitude of [O_3] during the higher radiation period or sudden high ozone episodes may reach up to ~90 ppb in Mexico (Borja-Aburto et al., 1997), Cairo (Khoder, 2009), and India (Jain et al., 2005; Sharma et al., 2013). Although rice is moderately sensitive, this level of [O_3] is putting rice cultivation under a threat. Visible changes in rice plants, such as leaf bronzing and leaf injury, indicate a high O_3 exposure (Chaudhary and Agrawal 2013; Frei, 2015). Wheat, on the other hand, showed significant yield loss with this [O_3], mainly because of the lower grain weight (1,000 grain weight). Weak correlation between the starch content and an increased micronutrient level in the wheat grains manifests into poor nutritional yield per capita due to the ozone exposure (Broberg et al., 2015). Though a short-term exposure to ozone is unable to impart any visible leaf injury in the plant, the same can affect the Calvin cycle, and thus reduce the light receiving potency of the plant as well as the carbon assimilation (Goumenaki et al., 2010; Sarkar et al., 2010). Altered metabolisms within the cell impair activities of Ribulose-1,5-Bisphosphate Carboxylase/Oxygenase (RuBisCO) (Agarwal et al., 2002; Cho et al., 2008). In case of wheat, the formation of RuBisCO activase protein also declines as a result of O_3 exposure to a higher concentration. However, a rise in the protein spots homologous to that of RuBisCO occured under such exposure. Rice, on the other hand, already reported similar biochemical changes under both O_3 and H_2O_2 stress condition. Carbon assimilation process requires the aforesaid protein to catalyse the activation of RuBisCO (Agarwal et al., 2002; Wan and Liu, 2008). Another reason for crop health detoriation in high [O_3] environment is the ROS formation due to ozonolysis in the apoplast region of a cell, modifying both secondary and tertiary structure of a protein to make those more suceptible towards proteolysis (Stadtman and Oliver, 1991; Sarkar et al., 2010).

In addition to the C assimilation, anomaly and alteration in protein structures due to the ozone exposure also affect the yield greatly. A meta-analysis including six major field crops, such as potato (*Solanum tuberosum* L.), barley (*Hordeum vulgare* L.), wheat (*Triticum aestivum* L.), rice (*Oryza sativa* L.), bean (*Phaseolus vulgaris* L.), and soybean (*Glycine max* L.), showed 5%–19% reduction in average yield of all crops under the tropospheric [O_3] magnitude of 31–50 ppb as compared to the outcome with a base [O_3] (≥26 ppb) (Feng and Kobayashi, 2009; Osborne et al., 2016). Furthermore,

10%–20% yield loss was projected by the crop simulation models for all six crops if exposed to <50–75 ppb level of [O_3]. This meta-analysis based on the linear dose model output from the study enlightens the comparative yield performance among all the mentioned crops. At AOT 40 level of exposure, 5% yield reduction was noticed in the case of soyabean, whereas no significant change was observed in potato in terms of the number of tuber or the weight of single tuber. However, cereal crops like barley and wheat showed an altered scenario where individual grain weight (3.5%), grains per panicle (3.1%), and panicles per plant (8.9%) significantly decreased when compared between the base and elevated levels of ozone exposure (Feng and Kobayashi, 2009). The OTC and O_3-FACE are also two widely approved methods for studying the effects of ozone in crop physiology. Among these two, O_3-FACE is preferred for the aforementioned purpose. Comparative analysis among these two methods showed a significant difference in crop yield at AOT 40 level. Cereals, such as rice and wheat, had higher sensitivity to O_3-FACE but legumes like soybean were more sensitive to O_3 when exposed in OTCs (Feng et al., 2019). In spite of the moderate sensitivity towards the ozone exposure, rice showed significant yield reduction in O_3-FACE.

15.3.3 Effect of Elevated CO_2 and Tropospheric O_3 Interactions on Crops

Several enrichment studies have been carried out to compare the effects of elevated [CO_2] and [O_3] on physiological processes, growth, and yield of crops. Past study on wheat reported that net CO_2 assimilation rate was enhanced by 47% when [CO_2] was elevated up to 700 ppm which was in contrast with about 20% decrease in net CO_2 assimilation rate associated with 19% reduced stomatal conductance under elevated [O_3] of ~75 ppb (Balaguer et al., 1995). Thus, plant growth was depressed under the ozone exposure, whereas the high [CO_2] promoted the accumulation of dry matter and formation of new sinks in the form of increased number of tillers. However, plants exposed to the combination of elevated [CO_2] and [O_3] exhibited growth characteristics equivalent to those of the control plants maintained under ambient [CO_2] and [O_3] due to the counterbalancing effects of both the gases on the leaf photosynthesis rate and carbohydrate availability. Similar findings were reported in another experiment where wheat plants were concomitantly exposed to elevated CO_2 and O_3 of ~500 ppm and 40 ppb, respectively (Rudorff et al., 1996). In this experiment, CO_2 enrichment was found to ameliorate the O_3-induced stress effects on wheat, i.e. reduction in biomass, grain yield, and radiation use efficiency. However, in the same study, maize remained unresponsive to the elevated [CO_2] and a significant loss of grain yield was incurred due to the O_3-induced damages during the flowering process. Interestingly, CO_2 and O_3 both enter in the plant system through the stomatal openings. Combined effects of CO_2 and O_3 reduce g_s (stomatal conductance unit=mol H_2O/m^2s) value (Ainsworth and Long, 2005, Feng et al., 2019). Vapour pressure deficit (VPD) in plant cells regulates the stomatal conductance, transpiration, and water loss (Grossiord et al., 2020). Higher VPD can be correlated with the stress conditions, such as heatwaves or drought. The elevated [O_3] also can corroborate with this climatic condition (Johansson et al., 2020).

15.4 Adaptive Strategies to Elevated CO_2 and Tropospheric O_3

Gradually rising global population has put a major emphasis on the food security. However, there are several constraints on the agricultural sector, and climate change is a major one. Though increasing atmospheric [CO_2] has some potential to increase the crop yield, availability of water and nutrients could be the limiting factors. Degradation of grain quality under elevated [CO_2] as elaborated in sub-section 15.3.1.2, is another concern of the researchers in the field of agriculture. Zinc deficiency in crops imposed by the increasing atmospheric [CO_2] and its impact on human health is already gaining attention among the scientists, doctors, and common people (Myers et al., 2015). Moreover, rising temperature and surface [O_3] have a detrimental effect on human health and also have deleterious effect on agricultural produce. According to a report, 4.2 billion people on earth will be at risk of hunger and undernourishment by 2050 in the absence of any adaptation or innovation in the field of agriculture (Dawson et al., 2016). Therefore, suitable agrotechnological adaptations are the need of the hour to sustain the global agricultural production and crop quality.

Simulation studies on wheat, rice, and maize in the tropical and temperate regions under the scenario of 2°C local warming projected yield loss of more than 10% likely in the second half of the 21st century which could be avoided to some extent with the help of proper adaptive strategies (Challinor et al., 2014). These adaptations might be crop-level management at the existing cropping systems such as alternative varieties, planting dates, and irrigation. Adaptive strategies at broader level might include changes in crop species, integration of grazing, crop relocation, and complete transformation of the farming systems, such as from irrigated to dryland. Heat-stress–induced floret sterility and reduction in harvest index could be avoided by incorporating heat-tolerance trait of wild wheat (*Aegilops* spp.) or heat-escaping mechanism of wild rice (*Oryza officinalis*) into the novel field crops through breeding (Prasad et al., 2017).

Grain quality of any crop is influenced by G×E×M, i.e., the genetic constituents, environmental factors, and management practices. Therefore, suitable adaptation strategy against the elevated CO_2-induced degradation of grain nutrients composition could be the manipulation of any one or more of these factors. Studies have demonstrated that future breeding approach to improve grain quality of wheat could be based on the varietal differences in the responses to elevated [CO_2] in regards of minerals, storage proteins content and composition (Fernando et al., 2014). Performance of modern technologies of crop production, such as organic farming and precision agriculture, should be analysed thoroughly under the elevated [CO_2] to come up with a solution through the nutrient management practices.

In the case of CO_2, increasing pattern covers evenly within the atmosphere, whereas regional and temporal variations are stronger in O_3 exposure among the crops. Significant yield loss was noted in the case of crops having higher sensitivity towards the ozone exposure. For example, studies on European spring wheat varieties showed that few varieties, for example cv. Minaret (Fangmeier et al., 1996), were more resistant to the O_3-induced damages than the more susceptible ones, i.e. cv. Albis (Fuhrer et al., 1989) and cv. Turbo (Fangmeier et al., 1994) when exposed to the equivalent level of [O_3]. Therefore, allocation of crops based on the spatial and temporal variations of O_3 exposure could be proved to be a good adaptive strategy. Changes in irrigation pattern and genotypic characteristics also can help under the rising tropospheric [O_3]. A study by Teixeira et al. (2011) showed that shifting of the crop calendar by delaying the sowing date or selecting crops with different maturity duration until harvest could be a contributing approach. Chemical antioxidants can also ameliorate ozone injury in the vegetation. Studies to identify and quantify the optimum concentration and combination of such artificial antioxidants (fungicides, insecticides, herbicides, growth inducers, etc.) on ozone sensitive crops were conducted (Archambault et al., 2000). Nevertheless, other persistant side effects from these antioxidants proved the ozone injury mitigation to be less effective. In this course of action, a systematic antioxidant EDU (ethylene diurea) was brought in the market as foliar spray and soil drench for vegetation (Carnahan et al., 1978; Archambault et al., 2000; Manning, 2011). The EDU proved to be useful to protect plants from the higher [O_3] exposure. Application of EDU has been a cost-effective method for bulk screening in the rural areas (Pandey et al., 2015) due to resource limitation. Furthermore, organic alternatives of EDU, such as (i) ascorbic acid, (ii) phytohormones, (iii) flavonoids, and (iv) polyamines are extensively being studied as the preferred options for amelioration of ozone injury in the crop at different stages of its cultivation without compromising the plant health (Didyk and Blum, 2011).

15.5 Conclusion

Effects of elevated [CO_2] and [O_3] on field crops are two major concerns of climate change adaptation research and have been widely studied over the past few decades. Many research outcomes predicted a drastic impact on the agricultural production with quite high degree of certainty. Quality of crop is also compromised due to climate change. Therefore, search for sustainable agro-technological adaptations have begun since the projections about the future climate partly became the reality. The concept of climate-smart agriculture has emerged as a consolidated approach to ensure the food security of present and future generations. However, proper implementation of the climate-smart agricultural practices is possible only by the close coordination among the farmers, researchers, private sector, civil society, and policymakers (Lipper et al., 2014).

List of Abbreviations

1. **AOT 40**- Accumulated hourly O_3 exposure above threshold concentration of 40 ppb
2. **FACE** - Free air CO_2/O_3 enrichment
3. **OTC** – Open-top chamber

REFERENCES

Abebe A, Pathak H, Singh SD, Bhatia A, Harit RC, and Kumar V. 2016. Growth, yield and quality of maize with elevated atmospheric carbon dioxide and temperature in north-west India. *Agriculture, Ecosystems and Environment.* 218:66–72 doi: 10.1016/j.agee.2015.11.014.

Agarwal GK, Rakwal R, Yonekura M, Kubo A, and Saji H. 2002. Proteome analysis of differentially displayed proteins as a tool for investigating ozone stress in rice (*Oryza sativa* L.) seedlings. *Proteomics.* 2(8): 947–959. doi: 10.1002/1615-9861(200208)2:8<947::AID-PROT947>3.0.CO;2-J.

Ainsworth EA. 2008. Rice production in a changing climate: a meta-analysis of responses to elevated carbon dioxide and elevated ozone concentration. *Global Change Biology.* 14(7):1642–1650. doi: 10.1111/j.1365-2486.2008.01594.x.

Ainsworth EA, and Long SP. 2005. What have we learned from 15 years of free-air CO_2 enrichment (FACE)? A meta-analytic review of the responses of photosynthesis, canopy properties and plant production to rising CO_2. *New Phytologist.* 165(2):351–372. doi: 10.1111/j.1469-8137.2004.01224.x.

Altenbach SB, Kothari KM and Lieu D. 2002. Environmental conditions during wheat grain development alter temporal regulation of major gluten protein genes. *Cereal Chemistry.* 79(2):279–285. doi: 10.1094/CCHEM.2002.79.2.279.

Archambault D, Slaski DJ, and Li JJ. 2000. Ozone protection in plants. The potential use of chemical protectants to measure oxidant damage in Alberta crops. Report Prepared for the Air Research Users Group, Alberta Environment, Edmonton, Canada. https://open.alberta.ca/publications/0778512274 (accessed April 10, 2022).

Ashmore MR. 2005. Assessing the future global impacts of ozone on vegetation. *Plant, Cell and Environment.* 28(8):949–964. doi: 10.1111/j.1365-3040.2005.01341.x.

Asseng S, Jamieson PD, Kimball B, et al. 2004. Simulated wheat growth affected by rising temperature, increased water deficit and elevated atmospheric CO_2. *Field Crops Research.* 85(2–3):85–102. doi: 10.1016/S0378-4290(03)00154-0.

Asseng S, Ewert F, Martre P, et al. 2015. Rising temperatures reduce global wheat production. *Nature Climate Change.* 5(2):143–147.doi: 10.1038/nclimate2470.

Atkinson R. 2000. Atmospheric chemistry of VOCs and NO*x*. *Atmospheric Environment.* 34(12–14):2063–2101. doi: 10.1016/S1352-2310(99)00460-4.

Balaguer L, Barnes JD, Panicucci A, and Borland AM. 1995. Production and utilization of assimilates in wheat (*Triticum aestivum* L.) leaves exposed to elevated O_3 and/or CO_2. *New Phytologist.* 129(4):557–568. doi: 10.1111/j.1469-8137.1995.tb03023.x.

Bishop KA, Leakey ADB, and Ainsworth EA. 2014. How seasonal temperature or water inputs affect the relative response of C_3 crops to elevated [CO_2]: A global analysis of open top chamber and free air CO_2 enrichment studies. *Food and Energy Security.* 3(1):33–45. doi: 10.1002/fes3.44.

Biswas DK, Xu H, Li YG, et al. 2008a. Assessing the genetic relatedness of higher ozone sensitivity of modern wheat to its wild and cultivated progenitors/relatives. *Journal of Experimental Botany.* 59(4):951–963. doi: 10.1093/jxb/ern022.

Biswas DK, Xu H, Li YG, et al. 2008b. Genotypic differences in leaf biochemical, physiological and growth responses to ozone in 20 winter wheat cultivars released over the past 60 years. *Global Change Biology.* 14(1):46–59. doi: 10.1111/j.1365-2486.2007.01477.x.

Borja-Aburto VH, Loomis DP, Bangdiwala SI., Shy CM, and Rascon-Pacheco RA. 1997. Ozone, suspended particulates, and daily mortality in Mexico City. *American Journal of Epidemiology.* 145(3):258–268. doi: 10.1093/oxfordjournals.aje.a009099.

Broberg MC, Feng Z, Xin Y, and Pleijel H. 2015. Ozone effects on wheat grain quality: A summary. *Environmental Pollution.* 197:203–213 doi: 10.1016/j.envpol.2014.12.009.

Carnahan JE, Jenner EL, and Wat EKW. 1978. Prevention of ozone injury to plants by a new protectant chemical. *Phytopathology.* 68(122):1229.

Challinor AJ, Watson J, Lobell DB., Howden SM, Smith DR, and Chhetri N. 2014. A meta-analysis of crop yield under climate change and adaptation. *Nature Climate Change.* 4(4):287–291. doi: 10.1038/nclimate2153.

Chapman, S. 1930. a theory of upper-atmospheric ozone: Edward Stanford. *Memoirs of the Royal Meteorological Society.* 3:103–125.

Chaturvedi AK, Bahuguna RN, Pal M, Shah D, Maurya S, and Jagadish KSV. 2017. Elevated CO_2 and heat stress interactions affect grain yield, quality and mineral nutrient composition in rice under field conditions. *Field Crops Research.* 206:149–157 doi: 10.1016/j.fcr.2017.02.018.

Chaudhary N, and Agrawal SB. 2013. Intraspecific responses of six Indian clover cultivars under ambient and elevated levels of ozone. *Environmental Science and Pollution Research.* 20(8):5318–5329. doi: 10.1007/s11356-013-1517-0.

Chaudhuri UN, Burnett RB, Kirkham MB, and Kanemasu ET. 1986. Effect of carbon dioxide on sorghum yield, root growth, and water use. *Agricultural and Forest Meteorology.* 37(2):109–122. doi: 10.1016/0168-1923(86)90002-X.

Cho K, Shibato J, Agrawal GK, et al. 2008. Integrated transcriptomics, proteomics, and metabolomics analyses to survey ozone responses in the leaves of rice seedling. *Journal of Proteome Research.* 7(7):2980–2998. doi: 10.1021/pr800128q.

Cho K, Tiwari S, Agrawal SB, Torres NL, Agrawal M, Sarkar A, Shibato J, Agrawal GK, Kubo A, and Rakwal R. 2011. Tropospheric ozone and plants: absorption, responses, and consequences. In: Whitacre D. (ed) *Reviews of Environmental Contamination and Toxicology.* 212:61-111, Springer, New York, NY. doi: 10.1007/978-1-4419-8453-1_3

Clifton OE, Fiore AM, Massman WJ, et al. 2020. Dry deposition of ozone over land: processes, measurement, and modeling. *Reviews of Geophysics.* 58(1):e2019RG000670. doi: 10.1029/2019RG000670.

Colbeck I. 1985. Photochemical ozone pollution. *Weather.* 40(8):241–245.

Comrie AC, and Yarnal B. 1992. Relationships between synoptic-scale atmospheric circulation and ozone concentrations in metropolitan Pittsburgh, Pennsylvania. *Atmospheric Environment. Part B. Urban Atmosphere.* 26(3):301–312. doi: 10.1016/0957-1272(92)90006-E.

Conroy JP. 1992. Influence of elevated atmospheric CO_2 concentrations on plant nutrition. *Australian Journal of Botany.* 40(5):445–456. doi: 10.1071/BT9920445.

Dawson TP, Perryman AH, and Osborne TM. 2016. Modelling impacts of climate change on global food security. *Climatic Change.* 134(3):429–440. doi: 10.1007/s10584-014-1277-y.

Didyk NP, and Blum OB. 2011. Natural antioxidants of plant origin against ozone damage of sensitive crops. *Acta Physiologiae Plantarum.* 33(1):25–34. doi: 10.1007/s11738-010-0527-5.

Erbs M, Manderscheid R, Jansen G, Seddig S, Pacholski A, and Weigel HJ. 2010. Effects of free-air CO_2 enrichment and nitrogen supply on grain quality parameters and elemental composition of wheat and barley grown in a crop rotation. *Agriculture, Ecosystems and Environment.* 136(1–2):59–68. doi: 10.1016/j.agee.2009.11.009.

Fangmeier A, Brockerhoff U, Grüters U, and Jäger HJ. 1994. Growth and yield responses of spring wheat (*Triticum aestivum* L. cv. Turbo) grown in open-top chambers to ozone and water stress. *Environmental Pollution.* 83(3):317–325. doi: 10.1016/0269-7491(94)90153-8.

Fangmeier A, Grüters U, Hertstein U, Sandhage-Hofmann A, Vermehren B, and Jäger HJ. 1996. Effects of elevated CO_2 nitrogen supply and tropospheric ozone on spring wheat. I. Growth and yield. *Environmental Pollution.* 91(3):381–390. doi: 10.1016/0269-7491(95)00042-9.

Fangmeier A, Grüters U, Högy P, Vermehren B, and Jäger HJ. 1997. Effects of elevated CO_2, nitrogen supply and tropospheric ozone on spring wheat—II. Nutrients (N, P, K, S, Ca, Mg, Fe, Mn, Zn). *Environmental Pollution.* 96(1):43–59. doi: 10.1016/S0269-7491(97)00013-4.

Feng Z, and Kobayashi K. 2009. Assessing the impacts of current and future concentrations of surface ozone on crop yield with meta-analysis. *Atmospheric Environment.* 43(8):1510–1519. doi: 10.1016/j.atmosenv.2008.11.033.

Feng Z, Uddling J, Tang H, Zhu J, and Kobayashi K. 2019. Comparison of crop yield sensitivity to ozone between open-top chamber and free-air experiments. *Global Change Biology.* 24(6):2231–2238. doi: 10.1111/gcb.14077.

Fernando N, Panozzo J, Tausz M, et al. 2014. Elevated CO_2 alters grain quality of two bread wheat cultivars grown under different environmental conditions. *Agriculture, Ecosystems and Environment.* 185:24–33 doi: 10.1016/j.agee.2013.11.023.

Fiore AM, Dentener FJ, Wild O, et al. 2009. Multimodel estimates of intercontinental source-receptor relationships for ozone pollution. *Journal of Geophysical Research.* 114:D04301. doi: 10.1029/2008JD010816.

Fitzgerald GJ, Tausz M, O'Leary G, et al. 2016. Elevated atmospheric [CO_2] can dramatically increase wheat yields in semi-arid environments and buffer against heat waves. *Global Change Biology.* 22(6):2269–2284. doi: 10.1111/gcb.13263.

Frei M. 2015. Breeding of ozone resistant rice: Relevance, approaches and challenges. *Environmental Pollution.* 197:144–155 doi: 10.1016/j.envpol.2014.12.011.

Food and Agriculture Organization (FAO). 2018. http://www.fao.org/3/I9553EN/i9553en.pdf (accessed September 27, 2021).

Fuhrer JA, Lehnherr EB, Grandjean A, and Tschannen W. 1989. Effects of ozone on the yield of spring wheat (*Triticum aestivum* L., cv. Albis) grown in open-top field chambers. *Environmental Pollution.* 60(3–4):273–289. doi: 10.1016/0269-7491(89)90109-7.

Goumenaki E, Taybi T, Borland A, and Barnes J. 2010. Mechanisms underlying the impacts of ozone on photosynthetic performance. *Environmental and Experimental Botany.* 69,(3):259–266. doi: 10.1016/j.envexpbot.2010.04.011.

Grange SK, Lee JD, Drysdale WS, et al. 2021. COVID-19 lockdowns highlight a risk of increasing ozone pollution in European urban areas. *Atmospheric Chemistry and Physics.* 21(5):4169–4185. doi: 10.5194/acp-21-4169-2021.

Grossiord C, Buckley TN, Cernusak LA, Novick KA, Poulter B, Siegwolf RT, Sperry JS, and McDowell NG. 2020. Plant responses to rising vapor pressure deficit. New Phytologist. 226(6):1550-1566. doi: 10.1111/nph.16485

Guenther AB, Jiang X, Heald CL, et al. 2012. The Model of Emissions of Gases and Aerosols from Nature version 2.1 (MEGAN2. 1): An extended and updated framework for modeling biogenic emissions. *Geoscientific Model Development.* 5(6):1471–1492. doi: 10.5194/gmd-5-1471-2012.

Hasegawa T, Sakai H, Tokida T, et al. 2013. Rice cultivar responses to elevated CO_2 at two free-air CO_2 enrichment (FACE) sites in Japan. *Functional Plant Biology.* 40(2):148–159. doi: 10.1071/FP12357.

Hill JF. 2012. Early pioneers of photosynthesis research. In *Photosynthesis. Advances in Photosynthesis and Respiration,* Vol 34, eds. J. Eaton-Rye, B. Tripathy, and T. Sharkey, 771–800. Springer, Dordrecht. doi: 10.1007/978-94-007-1579-0_30.

Hofmann DJ, Butler JH, and Tans PP. 2009. A new look at atmospheric carbon dioxide. *Atmospheric Environment.* 43(12):2084–2086. doi: 10.1016/j.atmosenv.2008.12.028.

Högy P, and Fangmeier A. 2008. Effects of elevated atmospheric CO_2 on grain quality of wheat. *Journal of Cereal Science.* 48(3):580–591. doi: 10.1016/j.jcs.2008.01.006.

Högy P, Wieser H, Köhler P, et al. 2009a. Effects of elevated CO_2 on grain yield and quality of wheat: results from a 3-year free-air CO_2 enrichment experiment. *Plant Biology.* 11:60–69 doi: 10.1111/j.1438-8677.2009.00230.x.

Högy, P., C. Zörb, G. Langenkämper, T. Betsche, and A. Fangmeier. 2009b. Atmospheric CO_2 enrichment changes the wheat grain proteome. *Journal of Cereal Science.* 50(2):248–254. doi: 10.1016/j.jcs.2009.06.002.

Hoshika Y, Katata G, Deushi M, Watanabe M, Koike T, and Paoletti E. 2015. Ozone-induced stomatal sluggishness changes carbon and water balance of temperate deciduous forests. *Scientific Reports.* 5(1):1–8. doi: 10.1038/srep09871.

IPCC. 2007. Climate change 2007- the physical science basis: Working group I contribution to the fourth assessment report of the IPCC, Vol. 4. Cambridge University Press. http://www.ipcc.ch/ipccreports/ar4-wg1.htm (accessed November, 20, 2021).

IPCC. 2013. Climate change 2013: The physical science basis: Working Group I contribution to the Fifth assessment report of the Intergovernmental Panel on Climate Change. Cambridge University Press, 2014. https://archive.ipcc.ch/pdf/assessment-report/ar5/wg1/WG1AR5_Frontmatter_FINAL.pdf (accessed November 20, 2021).

IPCC. 2018. Global warming of 1.5°C. An IPCC Special Report on the Impacts of Global Warming of 1.5. https://apps.ipcc.ch/outreach/documents/451/1551801374.pdf (accessed April 10, 2022).

Jablonski LM, Wang X, and Curtis PS. 2002. Plant reproduction under elevated CO_2 conditions: a meta-analysis of reports on 79 crop and wild species. *New Phytologist.* 156(1):9–26. doi: 10.1046/j.1469-8137.2002.00494.x.

Jacob DJ, Logan JA, and Murti PP.1999. Effect of rising Asian emissions on surface ozone in the United States. *Geophysical Research Letters.* 26(14):2175–2178. doi: 10.1029/1999GL900450.

Jain SL, Arya BC, Kumar A, Ghude SD, and Kulkarni PS. 2005. Observational study of surface ozone at New Delhi, India. *International Journal of Remote Sensing.* 26(16):3515–3524. doi: 10.1080/01431160500076616.

Johansson JM, Watne ÅK, Karlsson PE, Karlsson GP, Danielsson H, Andersson C, and Pleijel H. 2020. The European heat wave of 2018 and its promotion of the ozone climate penalty in southwest Sweden. *Boreal Environment Research.* 25:39-50.

Khoder MI. 2009. Diurnal, seasonal and weekdays–weekends variations of ground level ozone concentrations in an urban area in greater Cairo. *Environmental Monitoring and Assessment.* 149(1):349–362. doi: 10.1007/s10661-008-0208-7.

Kim HY, Lieffering M, Kobayashi K, Okada M, Mitchell MW, and Gumpertz M. 2003. Effects of free-air CO_2 enrichment and nitrogen supply on the yield of temperate paddy rice crops. *Field Crops Research.* 83(3):261–270. doi: 10.1016/S0378-4290(03)00076-5.

Kimball BA. 1983. Carbon dioxide and agricultural yield: An assemblage and analysis of 430 prior observations 1. *Agronomy Journal.* 75(5):779–788. doi: 10.2134/agronj1983.00021962007500050014x.

Kimball BA. 2016. Crop responses to elevated CO_2 and interactions with H_2O, N, and temperature. *Current Opinion in Plant Biology.* 31:36–43 doi: 10.1016/j.pbi.2016.03.006.

Kimball BA, and Mauney JR. 1993. Response of cotton to varying CO_2, irrigation, and nitrogen: Yield and growth. *Agronomy Journal.* 85(3):706–712. doi: 10.2134/agronj1993.00021962008500030035x.

Kuehny JS, Peet MM, Nelson PV, and Willits DH. 1991. Nutrient dilution by starch in CO_2-enriched Chrysanthemum. *Journal of Experimental Botany.* 42(6):711–716. doi: 10.1093/jxb/42.6.711.

Kumar A, Nayak AK, Sah RP, Sanghamitra P, and Das BS. 2017. Effects of elevated CO_2 concentration on water productivity and antioxidant enzyme activities of rice (*Oryza sativa* L.) under water deficit stress. *Field Crops Research.* 212:61–72 doi: 10.1016/j.fcr.2017.06.020.

Li Y, Yu Z, Liu X, et al. 2017. Elevated CO_2 increases nitrogen fixation at the reproductive phase contributing to various yield responses of soybean cultivars. *Frontiers in Plant Science.* 8:1546. doi: 10.3389/fpls.2017.01546.

Li K, Jacob DJ., Shen L, Lu X, Smedt ID, and Liao H. 2020. Increases in surface ozone pollution in China from 2013 to 2019: Anthropogenic and meteorological influences. *Atmospheric Chemistry and Physics.* 20(19):11423–11433. doi: 10.5194/acp-20-11423-2020.

Lipper L, Thornton P, Campbell BM, et al. 2014. Climate-smart agriculture for food security. *Nature Climate Change.* 4(12):1068–1072. doi: 10.1038/nclimate2437.

Liu Y, and Wang T. 2020a. Worsening urban ozone pollution in China from 2013 to 2017–Part 1: The complex and varying roles of meteorology. *Atmospheric Chemistry and Physics.* 20(11):6305–6321. doi: 10.5194/acp-20-6305-2020.

Liu Y, and Wang T. 2020b. Worsening urban ozone pollution in China from 2013 to 2017–Part 2: The effects of emission changes and implications for multi-pollutant control. *Atmospheric Chemistry and Physics.* 20(11):6323–6337. doi: 10.5194/acp-20-6323-2020.

Lobell DB, Schlenker W, and Costa-Roberts J. 2011. Climate trends and global crop production since 1980. *Science*. 333(6042):616–620. doi: 10.1126/science.1204531.

Long SP. Ainsworth EA, Leakey ADB, Nösberger J, and Ort DR. 2006. Food for thought: lower-than-expected crop yield stimulation with rising CO_2 concentrations. *Science*. 312(5782):1918–1921. doi: 10.1126/science.1114722.

Lu X, Zhang L, and Shen L, 2019. Meteorology and climate influences on tropospheric ozone: a review of natural sources, chemistry, and transport patterns. *Current Pollution Reports*. 5(4):238-260. doi: 10.1007/s40726-019-00118-3

Manderscheid R, Bender J, Jäger H-J, and Weigel HJ. 1995. Effects of season long CO_2 enrichment on cereals. II. Nutrient concentrations and grain quality. *Agriculture, Ecosystems and Environment*. 54(3):175–185. doi: 10.1016/0167-8809(95)00602-O.

Manderscheid R, Erbs M, and Weigel HJ. 2014. Interactive effects of free-air CO_2 enrichment and drought stress on maize growth. *European Journal of Agronomy*. 52:11–21 doi: 10.1016/j.eja.2011.12.007.

Manning, WJ, Paoletti E, Sandermann JrH, and Ernst D. 2011. Ethylenediurea (EDU): a research tool for assessment and verification of the effects of ground level ozone on plants under natural conditions. *Environmental Pollution*. *159*(12):3283–3293. doi: 10.1016/j.envpol.2011.07.005.

Massman WJ. 1998. A review of the molecular diffusivities of H_2O, CO_2, CH_4, CO, O_3, SO_2, NH_3, N_2O, NO, and NO_2 in air, O_2 and N_2 near STP. *Atmospheric Environment*. 32(6):1111–1127. doi: 10.1016/S1352-2310(97)00391-9.

Miller JD, Arteca RN, and Pell EJ. 1999. Senescence-associated gene expression during ozone-induced leaf senescence in Arabidopsis. *Plant Physiology*. 120(4):1015–1024. doi: 10.1104/pp.120.4.1015.

Mills G, and Harmens H. 2011. *Ozone Pollution: A Hidden Threat to Food Security*. NERC/Centre for Ecology and Hydrology. ISBN: 978-1-906698-27-0.

Mills G, Buse A, Gimeno B, et al. 2007. A synthesis of AOT40-based response functions and critical levels of ozone for agricultural and horticultural crops. *Atmospheric Environment*. 41(12):2630–2643. doi: 10.1016/j.atmosenv.2006.11.016.

Mittal ML, Hess PG, Jain SL, Arya BC, and Sharma C. 2007. Surface ozone in the Indian region. *Atmospheric Environment*. 41(31):6572–6584. doi: 10.1016/j.atmosenv.2007.04.035.

Molotoks A, Smith P, and Dawson TP. 2021. Impacts of land use, population, and climate change on global food security. *Food and Energy Security*. 10(1):e261. doi: 10.1002/fes3.261.

Morgan PB, Mies TA, Bollero GA, Nelson RL, and Long SP. 2006. Season-long elevation of ozone concentration to projected 2050 levels under fully open-air conditions substantially decreases the growth and production of soybean. *New Phytologist*. 170(2):333–343. doi: 10.1111/j.1469-8137.2006.01679.x.

Myers SS, Wessells KR, Kloog I, Zanobetti A, and Schwartz J. 2015. Effect of increased concentrations of atmospheric carbon dioxide on the global threat of zinc deficiency: A modelling study. *The Lancet Global Health*. 3(10):639–645. doi: 10.1016/S2214-109X(15)00093-5.

NAAQS, 2015. 2015 Revision to 2008 Ozone National Ambient Air Quality Standards (NAAQS) Related Documents. https://www.epa.gov/ground-level-ozone-pollution/2015-revision-2008-ozone-national-ambient-air-quality-standards-naaqs (accessed September 30, 2022)

NOAA, ESRL. 2021. Globally averaged marine surface annual mean data. https://gml.noaa.gov/webdata/ccgg/trends/co2/co2_annmean_gl.txt (accessed November 28, 2021).

O'Brien P, Kral-O'Brien K, and Hatfield JL. 2021. Agronomic approach to understanding climate change and food security. *Agronomy Journal*. doi: 10.1002/agj2.20693.

O'Leary GJ, Christy B, Nuttall J, et al. 2015. Response of wheat growth, grain yield and water use to elevated CO_2 under a Free-Air CO_2 Enrichment (FACE) experiment and modelling in a semi-arid environment. *Global Change Biology*. 21(7):2670–2686. doi: 10.1111/gcb.12830.

Osborne SA, Mills G, Hayes F, Ainsworth EA, Büker P, and Emberson L. 2016. Has the sensitivity of soybean cultivars to ozone pollution increased with time? An analysis of published dose–response data. *Global Change Biology*. 22(9): 3097–3111. doi: 10.1111/ gcb.13318.

Pandey AK, Majumder B, Keski-Saari S, et al. 2015. Searching for common responsive parameters for ozone tolerance in 18 rice cultivars in India: Results from ethylenediurea studies. *Science of the Total Environment*. 532:230–238. doi: 10.1016/j.scitotenv.2015.05.040.

Pandey AK, Ghosh A, Agrawal M, and Agrawal SB. 2018. Effect of elevated ozone and varying levels of soil nitrogen in two wheat (*Triticum aestivum* L.) cultivars: Growth, gas-exchange, antioxidant status, grain yield and quality. *Ecotoxicology and Environmental Safety*. 158:59–68 doi: 10.1016/j.ecoenv.2018.04.014.

Paoletti E, and Grulke NE. 2010. Ozone exposure and stomatal sluggishness in different plant physiognomic classes. *Environmental Pollution*. 158(8):2664–2671. doi: 10.1016/j.envpol.2010.04.024.

Permadi DA, and Oanh NTK. 2008. Episodic ozone air quality in Jakarta in relation to meteorological conditions. *Atmospheric Environment*. 42(28):6806–6815. doi: 10.1016/j.atmosenv.2008.05.014.

Pleijel H, Gelang J, Sild E, et al. 2000. Effects of elevated carbon dioxide, ozone and water availability on spring wheat growth and yield. *Physiologia Plantarum*. 108(1):61–70. doi: 10.1034/j.1399-3054.2000.108001061.x.

Prasad PVV, and Djanaguiraman M. 2014. Response of floret fertility and individual grain weight of wheat to high temperature stress: sensitive stages and thresholds for temperature and duration. *Functional Plant Biology*. 41(12):1261–1269. doi: 10.1071/FP14061.

Prasad PVV, Bheemanahalli R, and Jagadish SVK. 2017. Field crops and the fear of heat stress: Opportunities, challenges and future directions. *Field Crops Research*. 200:114–121 doi: 10.1016/j.fcr.2016.09.024.

Prior SA, Runion GB, Rogers HH, and Torbert HA. 2008. Effects of atmospheric CO_2 enrichment on crop nutrient dynamics under no-till conditions. *Journal of Plant Nutrition*. 31(4):758–773. doi: 10.1080/01904160801928364.

Rai R, Rajput M, Agrawal M, and Agrawal SB. 2012. Gaseous air pollutants: a review on current and future trends of emissions and impact on agriculture. *Journal of Scientific Research*. 55:77–102.

Rudorff BFT, Mulchi CL, Daughtry CST, and Lee EH. 1996. Growth, radiation use efficiency, and canopy reflectance of wheat and corn grown under elevated ozone and carbon dioxide atmospheres. *Remote Sensing of Environment.* 55(2):163–173. doi: 10.1016/0034-4257(95)00208-1.

Sarkar A, Rakwal R, Agrawal SB, et al. 2010. Investigating the impact of elevated levels of ozone on tropical wheat using integrated phenotypical, physiological, biochemical, and proteomics approaches. *Journal of Proteome Research* 9(9):4565–4584. doi: 10.1021/pr1002824.

Satapathy SS, Swain DK, Pasupalak S, and Bhadoria PBS. 2015. Effect of elevated [CO_2] and nutrient management on wet and dry season rice production in subtropical India. *The Crop Journal.* 3(6): 468–480. doi: 10.1016/j.cj.2015.08.002.

Saxena PA, Srivastava AN, Tyagi MR, Kaur SI. 2019. Impact of tropospheric ozone on plant metabolism: A review. *Pollution Research Journal.* 38(1):175–180.

Schnell JL, Prather MJ, Josse B, et al. 2016. Effect of climate change on surface ozone over North America, Europe, and East Asia. *Geophysical Research Letters.* 43(7):3509–3518. http://doi.org/10.1002/2016GL068060.

Seinfeld JH, and Pandis SN. 2016. *Atmospheric Chemistry and Physics: From Air Pollution to Climate Change.* John Wiley & Sons, Hoboken, NJ.

Sharma P, Kuniyal JC, Chand K, Guleria RP, Dhyani, PP, and Chauhan C. 2013. Surface ozone concentration and its behaviour with aerosols in the northwestern Himalaya, India. *Atmospheric Environment.* 71:44-53. doi: 10.1016/j.atmosenv.2012.12.042

Sharma A, Ojha N, Pozzer A, Beig G, and Gunthe SS. 2019. Revisiting the crop yield loss in India attributable to ozone. *Atmospheric Environment.* X1:100008. doi: 10.1016/j.aeaoa.2019.100008.

Sicard P, Marco AD, Agathokleous E, et al. 2020. Amplified ozone pollution in cities during the COVID-19 lockdown. *Science of the Total Environment.* 735:139542. doi: 10.1016/j.scitotenv.2020.139542.

Siciliano B, Dantas G, da Silva CM, and Arbilla G. 2020. Increased ozone levels during the COVID-19 lockdown: Analysis for the city of Rio de Janeiro, Brazil. *Science of the Total Environment.* 737:139765. doi: 10.1016/j.scitotenv.2020.139765.

Simpson D, Arneth A, Mills G, Solberg S, and Uddling J. 2014. Ozone—the persistent menace: Interactions with the N cycle and climate change. *Current Opinion in Environmental Sustainability.* 9:9–19 doi: 10.1016/j.cosust.2014.07.008.

Singh A, and Jasrai YT. 2012. Response of crops to elevated atmospheric carbon dioxide. *Proceedings of the Indian National Science Academy.* 78(1):45–49.

Singh AA, and Agrawal SB. 2017. Tropospheric ozone pollution in India: Effects on crop yield and product quality. *Environmental Science and Pollution Research.* 24(5):4367–4382. doi: 10.1007/s11356-016-8178-8.

Slafer GA, and Rawson HM. 1994. Sensitivity of wheat phasic development to major environmental factors: a re-examination of some assumptions made by physiologists and modellers. *Functional Plant Biology.* 21(4):393–426. doi: 10.1071/PP9940393.

Squire OJ, Archibald AT, Griffiths PT, Jenkin ME, Smith D, and Pyle JA. 2015. Influence of isoprene chemical mechanism on modelled changes in tropospheric ozone due to climate and land use over the 21st century. *Atmospheric Chemistry and Physics.* 15(9):5123–5143. doi: 10.5194/acp-15-5123-2015.

Stadtman ER, and Oliver CN. 1991. Metal-catalyzed oxidation of proteins: Physiological consequences. *The Journal of Biological Chemistry.* 266(4):2005–2008. doi: 10.1016/S0021-9258(18)52199-2.

Stevenson DS, Young PJ, Naik V, et al. 2013. Tropospheric ozone changes, radiative forcing and attribution to emissions in the Atmospheric Chemistry and Climate Model Intercomparison Project (ACCMIP). *Atmospheric Chemistry and Physics.* 13(6):3063–3085. doi: 10.5194/acp-13-3063-2013.

Tai APK, and Martin MV. 2017. Impacts of ozone air pollution and temperature extremes on crop yields: Spatial variability, adaptation and implications for future food security. *Atmospheric Environment.* 169:11–21 doi: 10.1016/j.atmosenv.2017.09.002.

Teixeira E, Fischer G, Velthuizen HV, et al. 2011. Limited potential of crop management for mitigating surface ozone impacts on global food supply. *Atmospheric Environment.* 45(15):2569–2576. doi: 10.1016/j.atmosenv.2011.02.002.

Tubiello FN, and Ewert F. 2002. Simulating the effects of elevated CO_2 on crops: Approaches and applications for climate change. *European Journal of Agronomy.* 18(1–2):57–74. doi: 10.1016/S1161-0301(02)00097-7.

UN. 2021. https://www.un.org/en/observances/world-population-day (accessed September 27, 2021).

UNFCCC. 2015. https://unfccc.int/process-and-meetings/the-paris-agreement/the-paris-agreement (accessed December 2, 2021).

Unger N., Zheng Y, Yue X, and Harper KL. 2020. Mitigation of ozone damage to the world's land ecosystems by source sector. *Nature Climate Change.* 10(2):134–137. doi: 10.1038/s41558-019-0678-3.

Uprety DC, Sen S, and Dwivedi N. 2010. Rising atmospheric carbon dioxide on grain quality in crop plants. *Physiology and Molecular Biology of Plants.* 16(3):215–227. doi: 10.1007/s12298-010-0029-3.

Van Dop H, Den Tonkelaar JF, and Briffa FEJ. 1987. A Modeling Study of Atmospheric Transport and Photochemistry in the Mixed Layer during Anticyclonic Episodes in Europe. Part I: Meteorology and Air Trajectories. *Journal of Applied Meteorology and Climatology.* 26(10):1305-1316. doi: 10.1175/1520-0450(1987)026<1305:AMSOAT>2.0.CO;2

Varotsos C, and Cartalis C. 1991. Re-evaluation of surface ozone over Athens, Greece, for the period 1901–1940. *Atmospheric Research.* 26(4):303–310. doi: 10.1016/0169-8095(91)90024-Q.

Wan XY, and Liu JY. 2008. Comparative proteomics analysis reveals an intimate protein network provoked by hydrogen peroxide stress in rice seedling leaves. *Molecular and Cellular Proteomics.* 7(8):1469–1488. doi: 10.1074/mcp.M700488-MCP200.

Wang X, Zheng Q, Yao F, Chen Z, Feng Z, and Manning WJ. 2007. Assessing the impact of ambient ozone on growth and yield of a rice (*Oryza sativa* L.) and a wheat (*Triticum*

balance through redox metabolism (Munns and Tester, 2008; Janská et al., 2010). Molecular and epigenetic regulation is important in the regulation of gene expression in response to environmental stresses (Shinozaki and Yamaguchi-Shinozaki, 1997; Tyagi and Chandra, 2006; Chinnusamy et al., 2007; Hauser et al., 2011; Khraiwesh et al., 2012). As a result of stress, certain genes encode regulatory proteins, including transcription factors (TFs), protein kinases, and phosphatases, as well as effector proteins that code for detoxifying enzymes, antioxidants, and transporters (Sairam and Tyagi, 2004). Tyagi et al. (2001) identified early drought-inducible lysine transport protein/lysine-specific permease encoding gene. A family of *blt 101*-homologous small proteins inducible by chilling/low temperature, salt, and drought was found to be differentially expressed in rice (Tyagi et al., 2005).

TFs are proteins that specifically bind to the promoter region of the genome to facilitate gene transcription and hence aid in gene regulation (Latchman, 1997). TFs control gene expression by acting as *trans*-acting elements that bind to *cis*-acting elements in gene promoters as well as forming a regulatory feedback loop between distinct signaling pathways. Stress-responsive promoter analyses have revealed *cis*- and *trans*-acting elements involved in stress-responsive gene transcriptional responses. Specific TFs control gene expression and are critical for response and adaptation.

Besides TFs, some other proteins, namely, chromatin remodelers, histone deacetylases (HDAC), histone acetyltransferases, methylases, and coactivators, are also essential for gene regulation but they do not have DNA-binding domains. Therefore, these proteins are not considered TFs (Brivanlou and Darnell, 2002). The 3%–10% of all genes in eukaryotic genomes are constituted by TFs (Reece-Hoyes et al., 2005). Through interaction with other components of the transcription apparatus, the TFs may either decrease or increase the transcription of their target genes, thus acting as repressors or activators, respectively (Takagi and Kornberg, 2006). TFs regulate gene expression by various mechanisms (Gill, 2001) that include stabilizing or blocking the binding of RNA polymerase to DNA, catalyzing the acetylation/deacetylation of histone proteins using HAT (histone acetylase) and HDAC activity, respectively (Narlikar et al., 2002), and recruitment of proteins (coactivator or corepressor) to the TF DNA complex (Xu et al., 1999).

16.2 Classification

Based on sequence similarity and the tertiary structure of their DNA-binding domain, TFs are divided into five superclasses (Figure 16.1). The presence of DNA-binding domain is the basic feature of the TF which recognizes a specific sequence in the promoter region of gene and thereby regulates its expression (Kummerfeld and Teichmann, 2006). DNA-binding domain binds to specific sequences of DNA (referred to as response elements) in the promoter region of gene. The major family of DNA-binding domains is as follows: Basic helix loop helix (HLH), Basic leucine zipper (ZIP), C-terminal effector domain of bipartite response regulators, AP2 (Apetela2)/ERF/GCC-box, Helix-turn-helix (HTH), Homeodomain (HD) proteins, Lambda repressor–like SRF (Serum response factor) –like, Paired box, Winged helix, and Zinc fingers. A signal-sensing domain/ligand-binding domain is an optional domain on TFs that senses environmental signals and sends them to the transcription complex, which either upregulates or downregulates gene expression.

In the Arabidopsis genome, more than 1,500 genes encode TFs belonging to various families such as AP2/ERF, MYB/C, NAC, bZIP, WRKY, etc. (Riechmann et al., 2000; Soren et al., 2010). Plant TFs are classified into different families based on their characteristic structural motifs.

16.2.1 bZIP (Basic Region Leucine-Zipper) Proteins

Basic region/ZIP motif (bZIP) TFs control a wide range of developmental processes in plants along with activating genes in response to biotic and abiotic stressors. It is one of the largest families of dimerizing TFs. This family is defined

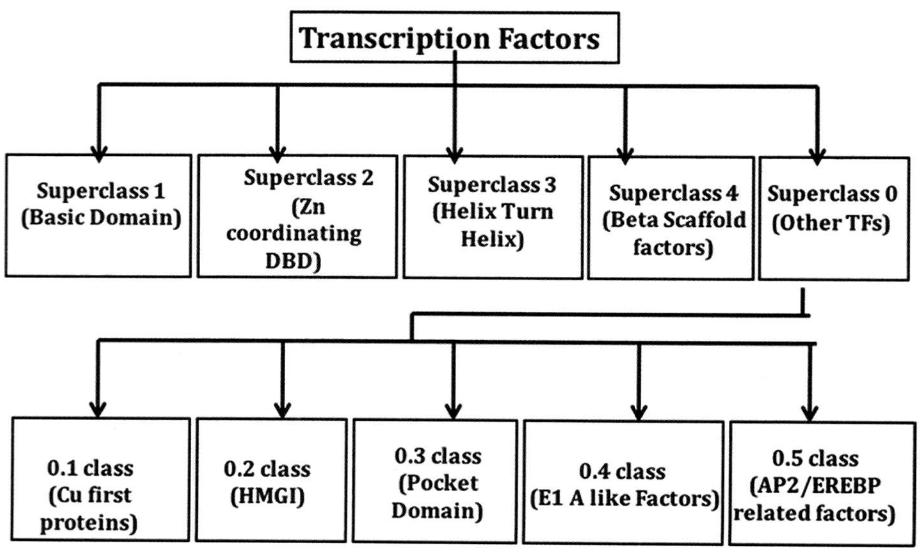

FIGURE 16.1 Classification of transcription factors family.

by a structural domain of 60 amino acids (aa) and consists of 2 subdomains, 1 of which is a basic region, located at the N terminus, consisting of 15 aas (mostly basic), responsible for DNA-binding and another subdomain is a HLH region, located at C-terminus, responsible for dimerization. Structural analyses of MyoD (mammalian bHLH proteins) have confirmed that the basic region interacts with the major groove of DNA via the formation of α-helical structure. HLH region is made up of hydrophobic aas that form two α-helices (amphipathic) separated by a loop region of variable length and sequence and form the interface for hetero and/or homodimerization (Lemmon et al., 1994; Nair and Burley, 2000).

The bZIP proteins are characterized by a DNA-binding and dimerization domain. The bZIP TFs are reported from a wide range of eukaryotes and form one of the largest families of plant TFs (Amoutzias et al., 2008). The motif of this family comprises 25 aas, which are rich in basic aas such as arginine and lysine. These basic residues (positively charged aas) are linked by heptad repeats of leucine residue; therefore, they are named ZIP (Landschulz et al., 1998). The physical and structural studies of this motif confirm that the contact of its basic region with the target DNA sequence forms the α-helical structure (Glover and Harrison, 1995). Two subunits bind DNA by forming a stacked coiled-coil shape for homo- or heterodimerization via contacts between the hydrophobic sides of their helices. Plant bZIPs typically bind to specific sequences such as the 'TACGTA' (A-box), 'GACGTC' (C-box), and 'CACGTG' (G-box); however, nonpalindromic binding sites have also been found (Jakoby et al., 2002).

Lindemose et al. (2013) reported that bZIP TFs are involved in multiple biological processes such as seed germination, seed development, flower development, hormone signaling, sugar signaling, and light signaling. In plants, bZIP TFs are also involved in response to biotic stresses (defense against pathogen) as well as abiotic stresses (extreme temperature, high osmolarity, high salinity, and water deficit) (Jakoby et al., 2002). Approx. 136 bZIPs in *Brassica rapa*, 64 in cucumber and 92 in sorghum, and 141 in barley were identified based on predicted structural features, through genome-wide identification and characterization. Based on several factors such as DNA-binding region, aa sequence, bZIP basic region, and additional conserved motifs 131 in soybean, 75 in Arabidopsis, were identified (Hwang et al., 2016). Overexpression of GhABF2, a bZIP TF, found to improve drought and salt stress tolerance in Arabidopsis and cotton via abscisic acid (ABA) mediated signaling pathways (Liang et al., 2016).

Overexpression of *TaAREB3*, a bZIP family TF of wheat, in Arabidopsis enhanced the ABA sensitivity and also strengthened drought as well as freezing tolerance (Wang et al., 2016). Similarly, overexpression of Arabidopsis bZIP TF *AREB1* in soybean activated cross-signaling responses under water stress and maintained higher photosynthetic rate, water use efficiency, and carboxylation efficiency, resulting in higher grain yield (Fuhrmann-Aoyagi et al., 2021). Another overexpression study of *OsbZIP16ox* in Arabidopsis confirmed its role in drought tolerance at the vegetative stage. During germination, the OsbZIP16ox seedlings displayed reduced sensitivity to mannitol, ABA, and sodium chloride, as well as reduced Reactive Oxygen Species (ROS) generation (Pandey et al., 2018).

The bHLH proteins bind to the core DNA motif by recognizing the consensus hexanucleotide sequence 5′-CANNTG-3′ which is defined as the E-box. E-boxes are also categorized according to the intensity of the two central bases. The palindromic G-box (5′-CACGTG-3′) is one of the most common E-boxes (Toledo-Ortiz et al., 2003). bHLH proteins generate enough diversity to regulate different transcriptional programs through dimerization and recognition of different E-boxes (Fairman et al., 1993). Lc, a protein expressed by the R gene that regulates anthocyanin synthesis in maize, was the first reported plant protein with a bHLH motif (Ludwig et al., 1989). More than 630 bHLH TFs have been identified in different plants such as *Arabidopsis thaliana* L., *Oryza sativa* L., *Nicotiana tabacum* L., and *Vitis vinifera* L. (Carretero-Paulet et al., 2010; Zhang et al., 2015). MYB/bHLH complexes are known to regulate cell wall synthesis, cell death, circadian clock, response to abiotic and biotic stresses, hormone signaling, and the biosynthesis of specialized metabolites (Pireyre and Burow, 2015). There are many examples whereby genetic manipulation of bHLH TFs resulted in improved stress response in plants. One such example is the overexpression of *bHLH122* that resulted in increased drought and salt tolerance in transgenic Arabidopsis plants. On the other side, the *bhlh122* loss-of-function mutants are more sensitive to drought and salinity stress than wild-type (WT) plants (Castilhos et al., 2014). Another example is drought and ABA-responsive protein RD29. It has an ABA-responsive element (ABRE) as well as a dehydration-responsive element (DRE) in its promoter. Many transgenic lines have been created employing the RD29A promoter to boost drought stress tolerance by driving the expression of target genes. RESPONSIVE TO DESICCATION 22 (RD22), another bHLH TF, is involved in the ABA signal transduction pathway in *A. thaliana* under stress circumstances (Castilhos et al., 2014). It was observed that bZIP TF expression is directly related to growth stage and water deficit stress and indirectly to ABA, RWC, and MSI (Ali et al., 2016). In *A. thaliana*, MYC2, a bHLH TF, was reported to regulate salt stress by regulating the proline biosynthesis pathway (Verma et al., 2020).

16.2.2 MYB Proteins

The presence of the MYB domain, which is responsible for DNA binding, distinguishes this family of TFs. Originally, this domain was derived from the proto-oncogene Myb, which has three copies of a 51–53 aa repeat containing three conserved Trp residues spaced at intervals of 18–19 aas to form a hydrophobic core (Ogata et al., 1992). The second and third repeats are required for DNA binding because they cooperate with the Myb-binding sequence of 'AACNG'. The second and third repeats each have three helices, with the recognition helix on the third helix in each (Meshi and Iwabuchi, 1995). The majority of MYB proteins in plants are members of the R2R3-MYB subfamily, which has two repetitions. In diverse species, more than 100 members of this class have been identified. Further, according to the conservation of the DNA-binding domain and aa residues at the C-terminal, it has been divided into 23 subgroups (Dubos et al., 2010). The R2R3-MYB TFs regulate plant primary and secondary metabolism, cellular identity, and abiotic and biotic stress responses.

In comparison to WT, overexpression of rice *OsMYB48-1* increased the expression of ABA biosynthesis (OsNCED4, OsNCED5), early signaling (OsPP2C68, OsRK1), and late response (OsRAB21, OsLEA3, OsRAB16C, and OsRAB16D) genes during drought stress. This confirmed the positive role of *OsMYB48-1* in drought tolerance (Xiong et al., 2014). Overexpression study of *ThMYB8* TF in *Tamarix hispida* maintained K^+/Na^+ homeostasis by regulating Na^+ and K^+ concentrations in plant tissues (Liu et al., 2021).

16.2.3 The HD Proteins

The DNA-binding domain of this family of TFs comprises 60 aas. The HD has three alpha-helices in its 3D structure, the second and third of which form a HTH pattern. The HD consists of loops that interrupt the three alpha-helical segments (designated helix 1–3 from the N terminus). Helices 1 and 2 form an antiparallel HTH unit, while helices 2 and 3 make a HTH unit. The maize Knotted1 (Knl) protein was the first HD protein found in plants. Transposon tagging was used to isolate the knl gene from a maize line with an altered phenotype in leaf development (Vollbrecht et al., 1991). Another class of the HD-ZIP proteins has been identified in Arabidopsis and carrot (Ruberti et al., 1991; Kawahara et al., 1995). Other examples of HD proteins include Plant Homeodomain (PHD) finger-containing proteins, required for normal trichome and root development, and the ATH1 protein, involved in the signal transduction pathway (Quaedvlieg et al., 1995) in Arabidopsis. The overexpression of HD-ZIP TFs has been demonstrated to improve the survival ability of transgenic plants under severe stress conditions. For example, in Arabidopsis and rice, overexpression of *Oshox22* and *Oshox24* genes (HD-Zip I γ-clade) increased ABA content and drought tolerance (Bhattacharjee et al., 2016). Likewise, the overexpression of maize *ZmHDZ4* gene (HD-Zip I γ-clade) in transgenic rice resulted in increased tolerance to drought. *ZmHDZ4*-overexpressing transgenic rice plants showed, relatively, lesser electrolyte leakage, lesser malondialdehyde (MDA) content, and increased proline content under drought as compared to WT plants (Wu et al., 2016). Overexpression of maize *ZmHDZIV14* in Arabidopsis and tobacco enhanced drought and salt tolerance by increasing sensitivity to ABA (Fang et al., 2021). Another study found that ZmZHD9-overexpressing plants were more drought tolerant than WT due to higher RWC, proline content, Superoxide dismutase and Peroxidase activity, and lower Root Electrolyte leakage and MDA content (Zhang et al., 2020).

16.2.4 The MADS-Box Domain-Containing Transcription Factor

They are encoded by the MADS-box gene family. MADS originated from the initials of four identified members, M from MCM1 (involved in cell-specific transcription and pheromone response) of *Saccharomyces cerevisiae*, A from AGAMOUS (involved in flower development) of *A. thaliana*, D from DEFICIENS (involved in flower development) of *Antirrhinum majus*, and S from SRF (involved in transcriptional regulation of immediate early genes) of *Homo sapiens* (Schwarz-Sommer et al., 1990; Dolan and Fields, 1991; Sommer et al., 1990). The MADS-box containing structurally conserved 56 aas was subdivided into two sub-regions: the N terminus region that is rich in hydrophilic and basic residues, involved in DNA binding, and the C-terminus region that is rich in hydrophobic residues. The MADS-box recognized palindromic 10-bp long sequences represented as CArG box, usually with the consensus of 'CC(A/T)$_6$GG'. Although research on MADS-box genes and how they function in developmental processes, particularly floral induction, and flower development, has been conducted, other findings suggest their significance in abiotic stress responses.

OsMADS26, OsMADS22, and *OsMADS55* are some of the MADS-box genes that have a role in stress signaling (Duan et al., 2006). The MADS-box encoding genes mostly involved in the regulation of floral development have been characterized in dicot, monocot, and conifer species such as Arabidopsis, Antirrhinum, Petunia, etc. (Meshi and Iwabuchi, 1995). Overexpression of *OsMADS25* in rice and Arabidopsis enhanced salinity tolerance. Transformants had increased proline content, lowered MDA level, and upregulated expression of genes related to salt stress (Wu et al., 2020).

16.2.5 Proteins with Zinc-Containing Motifs

The various zinc-containing motifs have been reported, in which zinc ion(s) stabilized the tertiary structure of the motif (Klug and Schwabe, 1995). The phrase 'zinc finger' refers to a sequence motif in which histidines and/or cysteines coordinate a zinc atom or atoms and, as a result, produce localized peptide structures that aid in their action. Based on the arrangement of zinc-binding aas, the zinc finger motifs are classified into different classes that are present in various TFs and play a critical role in interaction with other molecules. The zinc finger motifs such as TFIIIA and GATA are part of the DNA-binding domain reported to be directly involved in the recognition of specific DNA regions.

The most famous zinc finger pattern is the TFIIIA or C2H2-type finger, which has two conserved His and two conserved Cys residues that bind tetrahedrally to a zinc ion (Meshi and Iwabuchi, 1995). The finger is made up of ~30 aas with 2 stranded β-sheet (antiparallel) and an α-helical region. Several other hydrophobic aas are also highly conserved in this finger in addition to His and Cys (Meshi and Iwabuchi, 1995). Protein-protein interactions are mediated by zinc finger motifs such as the LIM- and RING-finger types. Plants contain the majority of eukaryotic zinc finger motifs. Additionally, some new motifs such as WRKY and DUF have also been identified in plants. Soren et al. (2012) isolated and characterized a drought-inducible cys3/his-1 zinc finger TF from rice with three distinct features: a nuclear localization signal in Arg 152-Arg 168, a Zn finger domain (185–193 aa), and a conserved domain (71–82 aa), and which belonged to the c-type family of zinc finger proteins.

16.2.6 The HMG (High Mobility Group)-Box Containing Proteins

The HMG-box was originally defined as a ubiquitous, heterogeneous family of small and relatively abundant, chromosomal non-histone proteins found in nuclei that meet certain empirical criteria, such as being extractable in NaCl, soluble in 2%–5% perchloric acid and a high content of charged as well as small-sized residues. The domain of HMG-box reported from eukaryotic organisms is composed of about 75 aa residues. It was first discovered as a distinguishing property of the HMGB type of chromosomal HMG proteins. Three conserved helices constitute the domain's L-shaped fold, according to its structural study. Helixes I and II make up the short arm of the L-shape, whereas helix III and the N-terminal extended strand make up the long arm (Thomas and Travers, 2001; Štros et al., 2007). HMG-box proteins interact with the DNA minor groove, bind to irregular DNA, such as four-way junctions, and bend DNA structures (Meshi and Iwabuchi, 1995). The 10–15 different HMG-box proteins were reported from higher plants (Štros et al., 2007). Differential display of mRNA revealed a drought-inducible cDNA homologous to TF containing HMG domain DNA-binding motif (Tyagi et al., 2002).

16.2.7 Heat Shock Factors (HSFs)

Heat shock genes are regulated by the binding of a special family of TFs called heat shock factors (HSF) to their promoter region. Heat shock genes are triggered in plants by cell differentiation, development, and proliferation, as well as under heat stress. In Arabidopsis, for example, *AtHSFA2* expression was elevated during callus development and root explant growth (Che et al., 2002). HSF stimulated the expression of rice *OsHSFA2a* during panicle and seed development (Chauhan et al., 2011). HSF proteins have a conserved modular structure. A core HTH motif (HTF) is found at the N terminus of the DNA-binding domain, which binds to the heat stress element of the target promoters and thereby stimulates the transcription of stress-inducible genes (Scharf et al., 2012). Leucine-zipper type 2 or 3 hydrophobic heptad repeats are present downstream of the DNA-binding domain, and they are responsible for stress-dependent modulation of activity and DNA binding via changes in the oligomerization state (Morimoto, 1993). A flexible linker connects the oligomerization domain to the DNA-binding domain (Baniwal et al., 2004). Plant HSFs are divided into three classes based on the length of the flexible linker and the number of aa residues inserted into the HR-A/B regions: HSF A, B, and C (Nover et al., 2001; Kotak et al., 2004). According to a comparison of the protein profiles of three chickpea genotypes, Pusa 256, RSG 888, and JG 11, 40 KDa, HSP is expressed differentially in three genotypes (Meena et al., 2014). The heat shock factor HsfA1d from Arabidopsis mitigates the negative effects of heat stress on potato (*Solanum tuberosum* L.) plants by lowering electrolyte leakage and increasing proline content (Shah et al., 2020).

16.2.8 The AP2/ERF Proteins

Over the past decade, the AP2/ERF family has received a great deal of attention. The AP2/ERF family of TFs is unique to plants and plays an important role in plant growth, development, and expression of various stress-responsive genes that are reported to improve abiotic stress tolerance. AP2/EREBP TFs were reported in various crops such as Arabidopsis, rice, maize, barley, soybean, cotton, tobacco, tomato, hot pepper, perennial ryegrass, and several members of Brassicaceae species (Riechmann and Meyrowitz, 1998). Several numbers of AP2/ERF TF were reported such as 200 in Poplar (Zhuang et al., 2008), 291, Chinese cabbage (Song et al., 2013), 116, Moso bamboo (Wu et al., 2015), 171, foxtail millet (Lata et al., 2014), and 178, Sorghum (Srivasta et al., 2010). All the AP2/EREBP TFs reported are identified by about 60 aas–long conserved AP2/ERF DNA-binding domain. AP2/ERF TF regulates the biochemical and physiological responses of plants under drought (Riechmann and Meyerowitz, 1998; Joshi et al., 2016).

16.3 The AP2/ERF Transcription Factor Superfamily

According to their overall structure and the numbers of AP2/ERF DNA-binding domains, the AP2/ERF superfamily is classified into AP2, RAV (related to ABI3/VP1), and DREB (DRE-binding proteins), and ERF major subfamilies (Riechmann et al., 2000; Sakuma et al., 2002; Shigyo and Ito, 2004; Rashid et al., 2012) (Figure 16.2). In Arabidopsis: AP2 subfamily (14 genes), DREB subfamily (56 genes; Group A), ERF subfamily (65 genes; Group B), RAV subfamily (6 genes), and others (4 genes) were reported. Two AP2/ERF domains are repeated in the AP2 family proteins. Rather than roots, this group of genes is expressed in leaves, flowers, and stems (Okamuro et al., 1997). Additionally, to the single AP2/ERF domain, the RAV family members contain the B3 domain, a plant-specific DNA-binding motif conserved in VP1/ABI3. There are two DNA-binding domains in the RAV subfamily: the AP2/ERF domain and the B3 domain. Among the RAV family members, the cDNAs for RAV1 and RAV2 encode proteins with AP2/ERF domains and B3-like domains in Arabidopsis. RAV subfamily proteins are thought to play a variety of roles in higher plants (Kagaya et al., 1999). Plant growth and development are negatively regulated by these proteins in Arabidopsis, and defense-mediated pathways are modulated by them in tomatoes (Hu et al., 2004; Li et al., 2011).

16.4 AP2/ERF Domain Structure

AP2/ERF domain contains about 60–70 aa and participates in DNA binding. The AP2 subfamily is divided into two monophyletic groups i.e., AP2 and AINTEGUMENTA that are involved in several developmental processes in *A. thaliana* (Shigyo and Ito, 2004). As illustrated in Figure 16.3, the AP2 subfamily has two copies of the AP2/ERF domain joined by a conserved 25 aas–long linker. There are significant conservations of aas in this linker segment among all subfamily members and it is essential to AP2/ERF domain functioning. The capacity of the proteins to bind DNA is lost when they are mutated within the linker. It has been shown that multiple linker residues are required for DNA binding in AP2/ERF proteins, which suggests that this region either directly associates with DNA or aligns two copies of AP2/ERF on DNA

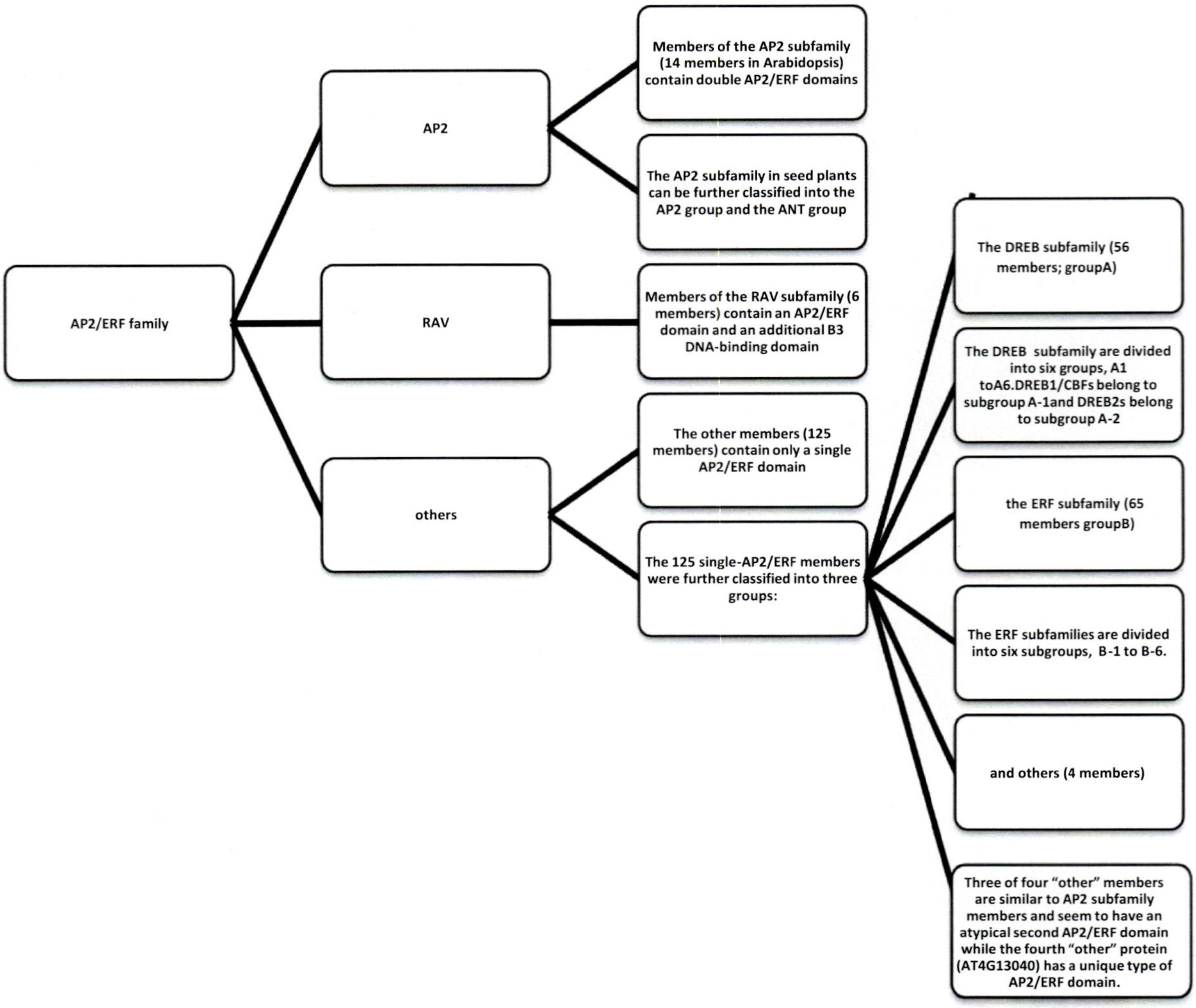

FIGURE 16.2 Summary of the overall classification of AP2/ERF family.

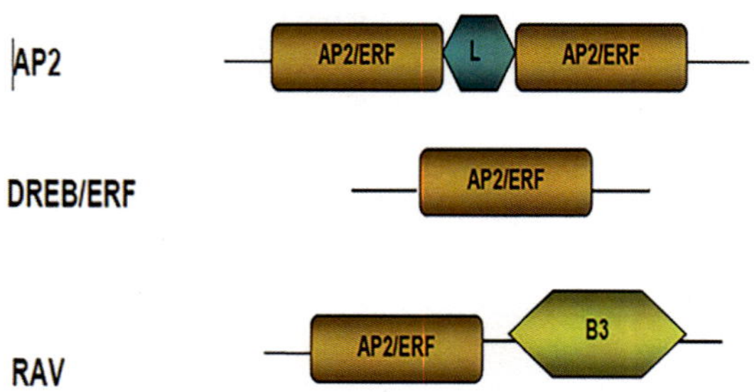

FIGURE 16.3 Domain structure (Giraudat et al., 1992; Kagaya et al., 1999).

(Krizek, 2003). The 60–70 aa AP2/ERF domain is conserved and consists of two regions:

a. The YRG region/element has an N-terminal stretch of roughly 20 aas that is rich in basic and hydrophilic residues. Considering its basic nature, it was believed to function as a DNA-binding protein through direct contact with DNA (Okamuro et al., 1997).
b. The RAYD region/element: A region/element of ~40 aa in the domain's C-terminal sequence consists of 18 aa forming a helix. This helix plays a crucial role in the structure and function of the domain. It is thought that RAYD elements modulate protein-protein interactions via the α-helix or affect DNA binding through interaction between the hydrophobic face of the α-helix and DNA major groove.

ERF subfamily members comprise a single AP2/ERF domain. ERF subfamily is subdivided into two subfamilies based on ERF domain binding to DNA sequences i.e., ERF and CBF/DREB (Sakuma et al., 2002). CBF/DREB subfamily genes with C-repeats (CRTs) encode proteins that identify the *cis*-acting element 'A/GCCGAC' (Yamaguchi-Shinozaki and Shinozaki, 1994), whereas ERF subfamily genes bind to the core motif 'AGCCGCC' (Hao et al., 1998).

Specific residues inside the AP2/ERF domain of DREB proteins possibly decide their capacity to bind the DRE/CRT element (Kizis and Pagès, 2002; Sakuma et al., 2002). Among all AP2/ERF family TFs, the members of the ERF and DREB subfamilies have been extensively studied. By binding to specific *cis*-regulatory sequences in late embryogenesis abundant (LEA), pathogenesis-related (PR), and cold-induced genes, members of the DREB and ERF subfamilies regulate gene expression in response to biotic and abiotic stress.

It is noteworthy that stress-inducible DREBs have been isolated from a wide range of plants, including rape and tomato (Jaglo et al., 2001), maize (Kizis and Pagès, 2002; Qin et al., 2007), rice (Liu et al., 2007), barley (Xue and Loveridge, 2004), and moss (Liu et al., 2007).

16.5 Phylogenetic Tree of the AP2/ERF Family Transcription Factors

The AP2/ERF superfamily of TFs has gained a lot of research in the last decade. AP2/ERF domain was first reported in Arabidopsis (Jofuku et al., 1994). It was hypothesized that the AP2/ERF family derived by HNH-AP2 endonucleases horizontal transfer from bacteria or viruses into plants through homing and transposition (Magnani et al., 2004). It has been demonstrated that TFs with AP2 domains are involved in plant development and stress responses.

In Arabidopsis, ERF family is classified into 12 groups, while in the case of rice the number is 15. ERF gene family analysis has determined around above 100 ERF genes in Arabidopsis and rice (Rashid et al., 2012, Figure 16.4). These 170 putative AP2 genes appear to be induced by one or more stress conditions, such as ABA, salinity, drought, and low temperature. The phylogenetic relationship of rice AP2/ERF family proteins was made using the AP2/ERF domains aa sequences. Utilizing multiple sequence alignments of protein sequence, a phylogenetic tree comprising 170 AP2/ERF genes was constructed using bootstrap tool. The phylogram was split into 11 groups, out of which 4 are major (Group-I to Group IV). Further, group-I is separated into three subgroups: I-a, I-b, and I-c. Group-I proteins are distinguished by the presence of a double domain. In rice, Arabidopsis, *Medicago truncatula*, and *B. napus*, these proteins serve critical roles in vegetative to embryonic growth, seed development, somatic embryogenesis, lateral root creation, floral meristem identity, petal cell identity, flowering time transition, etc. (Klucher et al., 1996; Krizek et al., 2000; Aukerman and Sakai, 2003; Imin et al., 2007).

RAV proteins are proteins that belong to Group II and have an AP2/ERF domain as well as a B3 domain. The remaining genes belong to groups III and IV, which feature a single AP2/ERF domain. Group III genes are crucial in biotic resistance, such as pathogen and disease resistance in crop plants. The Group IV genes are classified as DREB, CBF, and TINY proteins, which are involved in abiotic stress (cold, drought, and salt). DRE/CRT sequences and DREBs play an important role in stress responses as demonstrated by the existence of stress-inducible DREBs in several phylogenetically distinct species. Based on a phylogenetic analysis of 42 rice stress-inducible AP2 genes, six subgroups (I–VI) were identified with diverse signature motifs (Gao et al., 2015; Li et al., 2015). Functional

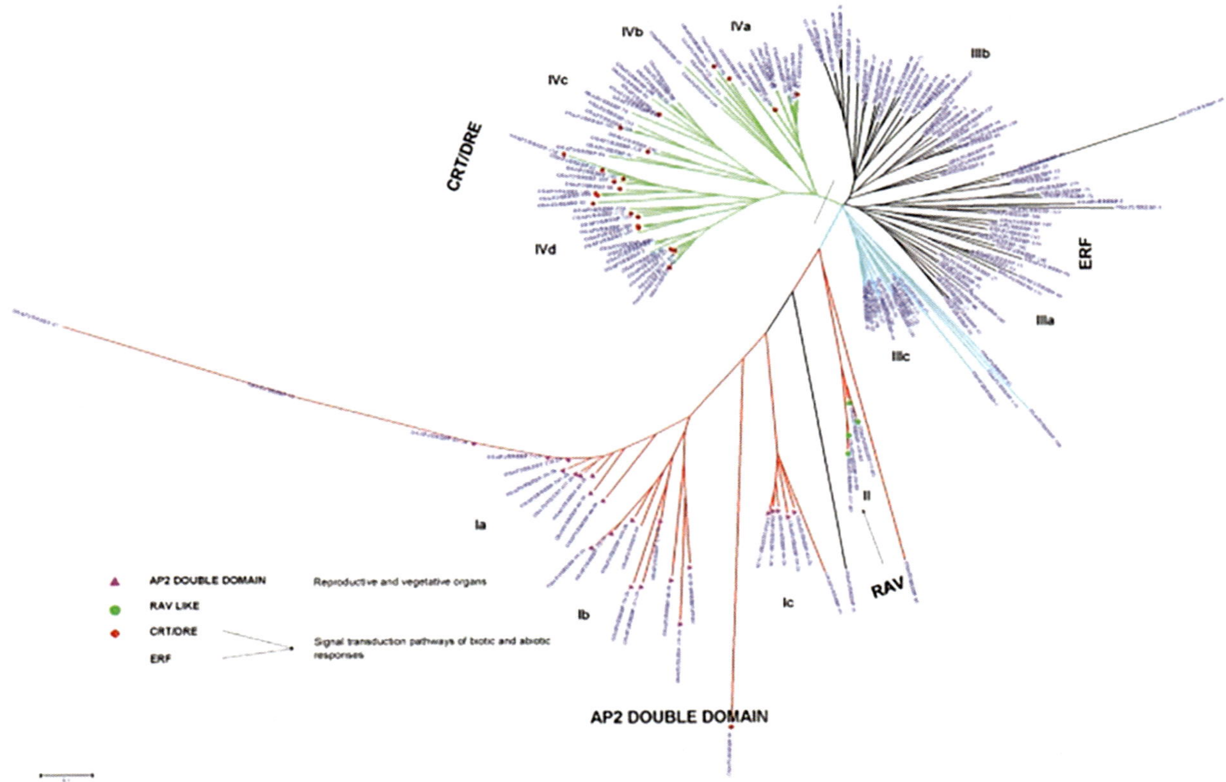

FIGURE 16.4 Phylogenetic tree of rice AP2/ERF family transcription factors (Rashid et al., 2012).

descriptions have been provided for two genes of subgroups I and II, respectively, AP37 and AP59. In rice, constitutive promoter-controlled overexpression of AP37 and AP59 increased its tolerance to drought and high salinity during the vegetative stage. More significantly, the OsCc1:AP37 plants demonstrated improved drought tolerance in the field condition and increased grain output by 16%–57% as compared to control under severe drought while showing no discernible difference under optimal conditions. In the field, however, grain output in OsCc1:AP59 plants was reduced by 23%–43% as compared to controls both under normal and drought circumstances. Microarray analysis demonstrated that AP37 and AP59 upregulate 10 and 38 genes, respectively, as well as 37 genes that are activated by both. According to the findings, the AP37 gene has the potential to enhance rice drought tolerance despite creating undesired growth phenotypes (Licause et al., 2013).

AP2/ERF-N22 is a 964-bp ERF TF from the ERF family that has been identified and characterized by our group. According to the AP2/ERF family's phylogenetic research, AP2/ERF-'N-22' belongs to group Va, along with the SHN clade of AP2/ERF protein that initiates wax formation (Mawlong et al., 2014). Another 1,017 bp gene sequence encoding an AP2/ERF family called AP2/ERF-N22(2) was identified from rice variety N22, translating into a 338-aa-residue protein with no Open Reading Frame (ORF) intron and a molecular weight of 36.58 kDa. Genes contain one AP2 domain of 55 aas, along with a cluster of acidic aas at the C-terminus that may function as a transactivation domain. It belonged to group VI L, which shared characteristics such as the absence of introns in the coding region, an imperfect AP2 domain, a molecular mass of 31–36 kDa, a length of aas between 330 and 338 residues, the presence of conserved motifs CMV, conserved putative MAP kinase, and/or casein kinase I phosphorylation sites, and involvement in cytokinin response under drought stress (Mawlong et al., 2016).

16.6 *Cis*-Acting Elements in Stress-Responsive Gene Expression

Molecular analysis of genes induced under environmental stresses has led to the identification of specific *cis*-regulatory elements that mediate activation. DRE has been implicated in a variety of abiotic stress responses via ABA-dependent and ABA-independent pathways (Busk et al., 1997; Shinozaki and Yamaguchi-Shinozaki, 1997). Moreover, it controls gene expression in response to droughts and other biotic stress. An element (5′-TACCGACAT-3′) was identified in the drought-responsive gene rd29A promoter sequence (also called cor78 and LTI 78). The rd29A gene encodes a protein that functions in stress tolerance and is related to the so-called LEA proteins that are activated throughout embryo development and by various types of stressors in vegetative tissues. The DRE in *rd29A* is required for this gene's expression in response to osmotic stress and low temperature, but not for its activation in response to ABA (Shinozaki and Yamaguchi-Shinozaki, 1997). An ABRE was found in the *RD29A* promoter, which regulated the RD29A gene's ABA-dependent expression (ABRE) (Yamaguchi-Shinozaki and Shinozaki, 1994). Similarly, there is another *cis*-element related to the DRE motif, called the CRT responsive element having core sequence 5′CCGAC3′, and low temperature–responsive element, found in cold-inducible promoters initially identified in three cold-inducible genes *kin1, kin2, rab18* from Arabidopsis and later in *B. napus BN115,* and wheat *WCS120* (Jiang et al., 1996; Ouellet et al., 1998). ABA-dependent responses to osmotic stress were mediated by DRE2 in the maize rab17 promoter. In addition, the *rab17* promoter contained the related DRE1 *cis*-element (5′ACCGAG-3′), which mediates ABA-dependent regulation in the embryo but appears insensitive to stimulation by drought in vegetative tissues (Busk et al., 1997). The AP2/ERF TFs bind to DRE/CRT (*cis*-elements) in the promoter region of stress-related genes and thus regulate their transcription (Mizoi et al., 2012). In abiotic stress responses, the DRE/CRT is a prominent AP2/ERF TF target. Several *cis*-elements that are recognized by the AP2/ERF TF have been identified. ERFs identified such as EREBPs, AtEBP, AtERF1, and Ptis (Pti4/5/6) from tobacco, Arabidopsis, and tomato, respectively, have been demonstrated to bind to the promoter of the pathogen-related gene GCC-box or PR-box (Solano et al., 1998; Brown et al., 2003).

Another study also finds that ERF domain binds the ethylene-responsive gene GCC-box, a DNA region implicated in transcription (Ohme-Takagi and Shinshi, 1995). The binding ability of tomato ERFs (LeERF1–4) to GCC-box varies (Tournier et al., 2003). Using a microarray protein DNA interaction assay, Cheng et al. (2012) demonstrated that the At ERF53 protein interacts with GCC-box and DRE probes as GCC-box and the DRE motif possess 'CCGNC' core sequence. A variety of *cis*-elements that respond to elicitors such as jasmonic acid (JA) have been characterized, JERE (5′-AGACCGCC-3′) which contained the GCC-box motif (Rushton et al., 1996; Menke et al., 1999; Kagaya et al., 1999).

Several *trans*- and *cis*-acting regulatory elements that activate stress-inducible gene expression in ABA-dependent and/or ABA-independent ways have been identified. ABRE (ABA-responsive element; C/GACGTGGC) is a conserved ABA-responsive *cis*-acting element. ABRE interacts with two basic ZIP (bZIP) TF activated by ABA i.e., AREB and ABF, present in the promoter region of numerous ABA-inducible genes (Mundy et al., 1990; Uno et al., 2000). AtMYC2 and AtMYB2 TFs, which are expressed under endogenous ABA accumulation, interact with *cis*-elements of the Arabidopsis RD22 promoter, demonstrating their importance in later stages of the stress response (Abe et al., 1997). CBF/DREB1 and DREB2 are ERF1/AP2 TF that are activated by osmotic stress, regardless of the presence of ABA. Later interact with DRE/CRT motifs in the promoter region of RD29 and also with several stress-inducible promoters (Yamaguchi-Shinozaki and Shinozaki, 1994).

16.7 Stress Responses by AP2/ERF Transcription Factors

Many monocots and dicot plant species have many AP2/ERF TFs, which have a wide range of functions. The AP2/ERF demonstrated to have important roles in the transcriptional regulation of a variety of biological functions, including development and growth, along with diverse responses to environmental stimuli at transcriptional level (Mawlong et al., 2015). AP2 genes have been connected to flower development, spikelet meristem determinacy, leaf epidermal cell identity, and embryo development (Moose and Sisco, 1996; Elliott et al., 1996; Chuck et al., 1998; Boutilier et al., 2002). RAV family members have recently been implicated in ethylene and brassinosteroid response (Alonso et al., 2003; Hu et al., 2004). Many ERF genes, such as *GmERF3, AtERFs,* and *TiERF1* (Berrocal-Lobo et al., 2002), are induced by pathogen infection, whereas *GhERF*4, *JERF*1, *AtERF*4, and *Pti*4 are induced by phytohormones (Samuel Yang et al., 2006). Furthermore, ERF proteins, which act as TFs, can influence the expression of several non-GCC-box defense-regulated genes by modulating the expression of CT-rich, DRE/CRT, CE1, and JERE elements. Crosstalk between plant defense signaling pathways is facilitated by TFs from the ERF subfamily.

Plants' immediate reaction to osmotic stress is transcriptional regulation of gene expression, which is controlled by several TFs and their interactions with *cis*-regulatory sequences, with promoter activation/repression determining the rate of transcription under stressful conditions (Bartels and Sunkar, 2005). The interactions between ABA-dependent TFs and *cis*-acting elements that induce stress-inducible expression of genes have attracted much interest presently. In tobacco and rice, the expression of rab16A from rice was investigated (Yamaguchi-Shinozaki et al., 1990; Ono et al., 1996). Su et al. (1998) found that a combination of one and four copies of an ABRC1 sequence from the barley HVA22 promoter and the minimum rice Act1 promoter strongly induced stress genes by three to eight folds. In rice transformants, the HVA1-like OsLEA3-1 promoter was recently employed, resulting in a ten-fold increase in GUS expression as well as better drought tolerance (Xiao et al., 2007). In rice, RD29A and RAB18 promoters were successfully used to generate transgenic expression under abiotic stress (Kasuga et al., 1999; Karim et al., 2007).

VvERF3b, an ERF gene from *V. vinifera*, is a member of ERF group VIII a. ABA, ethephon, and salicylic acid (SA) all stimulated gene expression, but not NaCl. VvERF3b gene promoter region revealed many crucial *cis*-acting areas that could be identified and bound by TFs implicated in hormone and stress responses. Deletion studies demonstrated that the 5′-flanking sequence from the transcriptional start site 1,047 to 585 was required for high *VvERF3b* expression, but the fragment from 1,324 to 1,047 showed suppression effects. These findings indicate that the promoter contains *cis*-acting elements that could be bound by proteins within a transcription complex to inhibit gene expression (Song et al., 2012). Aside from DREB1s/CBFs and DREB2s, other DREB subgroups have been demonstrated to be stress-responsive and/or confer tolerance against stress. Here's a quick rundown of what we know about these genes right now.

16.7.1 DREB1/CBF (Subgroup A-1) under Cold Stress Tolerance

Sakuma et al. (2002) reported, by regulating various transcriptome and metabolomic alterations, that six Arabidopsis DREB1 members of DREB1s/CBFs subgroup contribute to cold tolerance. DREB1A/CBF3, DREB1B/CBF1, and DREB1C/CBF2 are among those that are rapidly activated in response to cold stress (Stockinger et al., 1997). Transcriptome study of Arabidopsis with DREB1/CBF overexpression indicated that several cold-inducible genes induced (Fowler and Thomashow, 2002; Maruyama et al., 2009a). Cold-inducible proteins, such as LEA proteins and enzymes for sugar metabolism and fatty acid desaturation, are encoded by these putative target genes. Multiple TFs are also upregulated, indicating that transcriptional cascades operate downstream of DREB1s/CBFs. Metabolomics research indicates cold stress causes metabolites such as sugar to accumulate, and these compounds also increase in transgenic Arabidopsis constitutively expressing DREB1s/CBFs (Cook et al., 2004; Maruyama et al., 2009). Overexpression of DREB1/CBF has been reported to provide improved tolerance toward drought in many crops, including soybean (de Paiva Rolla et al., 2014), potato, tomato, chrysanthemum (Iwaki et al., 2013), and rice (Datta et al., 2012). Wang et al. (2008) reported that OsDREB1F was induced by drought and ABA application, while Chen et al. (2008) reported that OsDREB1G overexpression promoted drought tolerance.

Jaglo et al. (2001) found that transgenic Arabidopsis with constitutively overexpressed DREB1/CBFs had dramatically increased tolerance to freezing, drought, and high salt, whereas suppression produced a loss in freezing tolerance. The CRT/DRE was detected by each of these DREB1/CBFs due to their AP2/ERF DNA-binding regions (Baker et al., 1994; Yamaguchi-Shinozaki and Shinozaki, 1994).

16.7.2 Non-Cold-Inducible DREB1s/ CBFs under Abiotic Stress

DREB1D/CBF4, DREB1E/Dwarf, and DREB1F/DDF1 are three DREB1/CBF genes in Arabidopsi*s,* in addition to the three cold-inducible DREB1s/CBFs (Haake et al., 2002). Dehydration and ABA both induce DREB1D/CBF4, while high salinity induces DREB1F/DDF (Haake et al., 2002; Sakuma et al., 2002). DREB1D/CBF4 overexpression resulted in growth retardation, cold- and drought-inducible DREB1/CBF target gene expression, in transgenic Arabidopsis plants (Haake et al., 2002). DREB1F/DDF1 overexpression resulted in dwarf phenotype, RD29A gene expression, and high salinity tolerance. In DREB1F/DDF1 overexpressing plants, active GA amount decreased, but increasing active GA level restored the dwarf phenotype and tolerance to high salinity. Overall, it suggested that non-cold-inducible DREB1s/CBFs influenced downstream gene sets in water stress adaptation that were analogous to those influenced by cold-inducible DREB1/CBFs. Studies by activation tagging of Arabidopsis mutants lines led to the identification of key regulatory genes involved with freezing tolerance. One mutant, freezing tolerant line1 (ftl1-1D), showed increased tolerance as well as dwarfism and blooming delay. The cause for this was DDF1, a gene previously recognized as a regulatory component in salinity signaling. The induced gene encoded a CBF/DREB1 subfamily AP2 TF. In addition to low temperature and salt stress tolerance, ftl1-1D/ddf1 significantly improved drought and heat tolerance (Gilmour et al., 1998).

16.7.3 DREB2 Subgroup under Dehydration and Heat Shock

The DREB2 subgroup comprises eight members in Arabidopsis (Sakuma et al., 2002) and five in rice (Matsukura et al., 2010). The DREB2 subgroup is related to the A-3 subgroup. It comprises ABI4 and is further subdivided into three subclasses. According to large-scale phylogenetic data from 21 angiosperm species, these three subtypes are remained conserved in both Arabidopsis and rice (Matsukura et al., 2010).

DREB proteins from the DREB2 subgroup had been characterized in a variety of other angiosperm species, including eudicot species–like (Díaz-Martn et al., 2005) and grass species–like rice (Dubouzet et al., 2003; Matsukura et al., 2010), wheat (Shen et al., 2003; Egawa et al., 2006; Qin et al., 2007), barley (Xue and Loveridge., 2004), and maize (Qin et al., 2007). Although most of the DREB2 genes were induced under water or heat stress, grass species do induce under cold stress as well (Shen et al., 2003; Egawa et al., 2006; Matsukura et al., 2010).

Subtype 1 contained Arabidopsis 'DREB2A, B, C, E, and H' as well as rice OsDREB2A and OsDREB2B, indicating that subtype 1 comprises the bulk of stress-inducible DREB2s. Subtype 2 included 'DREB2D, DREB2G, and OsDREB2C' from Arabidopsis and rice, respectively, while subtype 3 included 'DREB2F and OsDREB2E' from Arabidopsis and rice, respectively. Even though these two categories had genes in all 21 angiosperm species, their roles are unknown. OsDREB1E, OsDREB1G, and OsDREB2B were found to bind to the CRT/DRE element, particularly in a yeast one-hybrid assay. Drought resistance was improved by overexpressing *Os*DREB*2B* in transgenic Arabidopsis plants (Matsukura et al., 2010).

Rice transgenic plants overexpressing *OsDREB1G* and *OsDREB2B* improved tolerance to water deficit stress tolerance significantly, whereas rice transgenic plants overexpressing *OsDREB1E* improved tolerance to water deficit stress slightly, implying that the *OsDREBs* may play different roles in the stress response pathway (Chen et al., 2008).

16.7.4 Post-Transcriptional Regulation of DREB2s Underwater and High Temperature Stress

Drought and high temperature are two stresses that frequently occur together (Mittler, 2006) in natural ecosystems. Drought and high temperatures together had a greater impact on plant physiology than either condition alone (Rizhsky et al., 2004; Mittler, 2006). According to omics studies in tobacco and Arabidopsis, plant responses to combined drought and high-temperature overlaps are unique from those to a single stress (Rizhsky et al., 2002, 2004).

Drought, salinity, and heat stress–induced DREB2A and DREB2B in Arabidopsis, although heat stress–induced DREB2C later than DREB2A or DREB2B (Sakuma et al., 2002; Rizhsky et al., 2004; Sakuma et al., 2006; Lim et al., 2007). High salinity induces DREB2C, DREB2D, and DREB2F, whereas ABA induces DREB2E, however, their expression is very low (Sakuma et al., 2002). As a result, DREB2A and DREB2B appear to be important DREB2s involved in DRE/CRT-mediated ABA-independent dehydration-inducible gene expression (Nakashima et al., 2000, Sakuma et al., 2002). Contrary to DREB1s/CBFs, ectopic overexpression of DREB2A did not affect plant growth, stress tolerance, or gene expression (Liu et al., 1998), implying that DREB2A is regulated post-translationally. Sakuma et al. (2006) discovered to become constitutively active, DREB2A requires the ablation of the negative regulatory domain located next to its DNA-binding domain. In transgenic plants constitutively expressing DREB2A, growth was retarded, drought tolerance was enhanced, and salinity tolerance was enhanced, but cold tolerance was reduced. In these transgenic plants, many dehydration-responsive genes, such as those producing LEA proteins, were activated. Furthermore, a large number of heat shock–responsive genes were activated (Sakuma et al., 2006). In an unstressed state, post-translational modification of TFs by proteolysis likely played a substantial role in keeping TF low or negligible. Only a few numbers of E3 ubiquitin ligases have been discovered to degrade TFs under dehydration stress. This was established by knockout mutants of certain E3 ligases, which allowed the expression of specific TFs and affected Arabidopsis stress tolerance. When the related E3 ligase genes DRIP1/DRIP2 (Qin et al., 2007) and AIP2 (Zhang et al., 2005) were knocked out, the dehydration tolerance TFs DREB2A, ABI3, and AtERF53 were activated.

Although drought, salt, and osmotic stress all demonstrated a clear elevation of AtERF53 expression, not ABA, JA, or ACC (Cheng et al., 2012). *AtERF53* activated numerous downstream TF genes from the ERF TF family, such as RAP2.4 (Lim et al., 2007), DREB2A (Sakuma et al., 2006), and CBF4 (Haake et al., 2002), as well as many others from other TF families, to establish a network under water-stressed conditions. AtERF53 activated under dehydration had a beneficial role in plant drought signaling and was ubiquitinated under non-stressed conditions for 26S proteasome proteolysis. Once stress signaling occurs, the degradation process may be temporarily slowed, allowing plant cells to obtain sufficient AtERF53 to trigger stress-responsive gene expression.

The PgDREB2A protein, a member of the DREB2 subgroup, has been shown to have an alternative post-translational regulatory mechanism (Agarwal et al., 2010). A pearl millet whole-cell extract phosphorylated an *Escherichia Coli*–recombinant PgDREB2A protein. The phosphorylation, on a Thr residue, lowered the PgDREB2A protein's binding affinity to a DRE/CRT sequence. Although the phosphorylation site was not identified, this study showed that phosphorylation could be used to regulate DREB2s post-translationally. Alternative splicing regulated DREB2 orthologues in grass plants. In grass plants, stress-inducible (subtype 1) DREB2s were separated into two groups: OsDREB2A and OsDREB2B orthologues (Matsukura et al., 2010). The control of OsDREB2B and its orthologues appeared to be aided by an alternative splicing mechanism. In barley (HvDRF1) (Xue and Loveridge, 2004), wheat (TaDREB2) (Egawa et al., 2006), maize (ZmDREB2A) (Qin et al., 2007), and rice (OsDREB2B), genes in the latter clade were revealed to be regulated by alternative splicing (Matsukura et al., 2010). Genes for these proteins had two types of transcripts: one was inactive and had a stop codon before the DNA-binding domain, while the other produced a full-length protein. Active transcripts accumulated at greater induction ratios than inactive transcripts in response to stress signals. A grass species' faster stress response has been attributed to alternative splicing of DREB2 genes that skip the activation phase of the promoter or produce a reservoir of premature DREB2 mRNAs.

Even under extreme temperature stress, no expression of *OsDREB2B1* (Ubi: sGFP-OsDREB2B1) was observed in transgenic rice, leading them to assume that *OsDREB2B1* regulated *OsDREB2B* transcription without impacting plant growth. The constitutively active transcription of *OsDREB2B* pre-mRNA was shown by the accumulation of *OsDREB2B1* under control conditions. *OsDREB2B2* was important for stress response but was detrimental to plant growth under non-stress conditions. As a result, most transcripts were spliced as OsDREB2B1 under non-stress conditions. The stress signal was transduced to a component of the splicing machinery during stress conditions, and the splicing process was modified to actively create OsDREB2B2. Plants could quickly create OsDREB2B2 in response to environmental challenges by using this pathway, which eliminated the time required for transcriptional activation and pre-mRNA buildup (Matsukura et al., 2010).

16.7.5 A-3 Subgroup

ABI4 and its orthologues make up the A-3 subgroup. Even though A-3 group members were related to DREB2s (Nakano et al., 2006; Matsukura et al., 2010), they had unique roles in ABA, sugar, and retrograde signaling. An abi4 mutant, however, was salt insensitive during the germination and seedling stages. As a result, A-3 genes could be implicated in stress response via a distinct mechanism than the DREB2 pathway.

16.7.6 A-4 Subgroup

DREB1s/CBFs and A-4, according to Nakano et al. (2006), shared a conserved motif and belonged to the same group. The activities of A-4 genes were unknown; however, they had a role in stress response configuration. HARDY overexpression (Karaba et al., 2007) altered the shape and architecture of roots, conferring drought tolerance. HRD overexpression results in bigger leaves with more chloroplast-bearing mesophyll cells in Arabidopsis, and an increase in leaf biomass and bundle sheath cells in rice, both of which lead to improved photosynthetic assimilation and efficiency. The findings demonstrated how the Arabidopsis gene could be used to improve rice water use efficiency while also increasing drought tolerance (Karaba et al., 2007). Transgenic rice expressing the Arabidopsis HARDY (HRD) gene improved Water Use Efficiency (WUE), by enhancing photosynthetic assimilation and decreasing transpiration. These drought-tolerant, less-water-using rice plants displayed increased shoot biomass and an increase in root biomass under drought.

16.7.7 A-5 Subgroup

In Arabidopsis (Maruyama et al., 2009), cotton, and soybean, the A-5 subgroup comprised stress-inducible genes (Chen et al., 2008). It includes a subset of proteins with a functional ERF-associated amphiphilic repression (EAR) domain at their C-terminus (Nakano et al., 2006; Chen et al., 2008). Further, transgenic plants overexpressing DREB1A/CBF3 or DREB2A CA showed an increase in gene expression for EAR motif-containing proteins (Maruyama et al., 2009, Sakuma et al., 2006), while DREB1/CBF and DREB2 target genes were reduced in expression under cold and dehydration conditions, respectively (Dong and Liu, 2010). A mutation in any of the genes encoding these EAR motif-containing proteins (related to AP2.1; RAP2.1) increased the expression of these target genes as well as drought and freezing tolerance (Dong and Liu, 2010). Therefore, EAR motif-containing A-5 subgroup proteins may have contributed to negative feedback regulation in the DREB1/CBF and DREB2 pathways. *PpDBF1*, from the moss *Physcomitrella patens*, was another noteworthy gene in the A-5 subgroup (Liu et al., 2007). High salinity, cold, dehydration, hydrogen peroxide, and SA all had a minor effect on PpDBF1, while ABA had a significant effect. PpDBF1 binds to DRE/CRT, and its overexpression in transgenic tobacco increases the synthesis of stress-responsive genes as well as a decreased susceptibility to high salinity. These findings suggest that ancient terrestrial plants had established the control of stress-responsive genes via DREBs via DRE/CRT.

16.7.8 A-6 Subgroup

Stress-inducible genes were also found in the A-6 subgroup. Heat, salinity, and dehydration caused RAP2.4 and RAP2.4B (AT1G22190) to be expressed (Lim et al., 2007). Cold and heat shock were also responsive to RAP2.4 and RAP2.4B, respectively. Multiple aquaporin genes have been identified as targets by a microarray study of RAP2.4 overexpressor and double mutant (RAP2.4 and RAP2.4B) plants, showing that these genes are responsible for water homeostasis (Sharoni et al., 2011). At1G36030 (RAP2.4A) binds to the promoter of the antioxidant enzyme 2-Cys peroxiredoxin. RAP2.4 binds to both the DRE/CRT and the GCC-box in this promoter, which has a CE3-like binding sequence, CACGCGATTC. The redox state of RAP2.4A protein has an impact on its activity and DNA-binding affinity. When compared to the WT, the rap2.4a mutant showed somewhat lower expression of redox-regulated nuclear genes encoding chloroplast proteins as well as enhanced light sensitivity. Overall, it implies that A-6 proteins also play a role in stress responses, but their target genes differ from those of the A3 and A4 subgroups.

16.8 DREB1s/CBFs, DREB2s, and Other Transcription Factors

The differences between DREB1A/CBF3 and DREB2A binding specificities could play a role in the stress-specific expression of DRE/CRT-regulated genes in Arabidopsis. Overexpression of the DREB1/CBF genes enhanced dehydration and cold tolerance (Qin et al., 2007; Dubouzet et al., 2003). Overexpression of DREB2 genes, however, improved tolerance to dehydration but not for cold (Sakuma et al., 2006; Qin et al., 2007). The genes downstream of DREB1A/CBF3 and DREB2A CA partly overlap, according to Maruyama et al. (2009). DREB1A has genes that code for carbohydrate metabolism enzymes, whereas DREB2A CA has genes that code for molecular chaperones (Maruyama et al., 2009). The metabolite profiles of DREB1A-overexpressing plants and plants that were treated with cold were similar, as were the patterns for DREB2A CA-overexpressing plants and dehydration plants (Maruyama et al., 2009), indicating that the differences were mainly due to differences in metabolite accumulation. In vitro and planta, Maruyama et al. (2009) and Sakuma et al. (2006) showed that DREB1A and DREB2A prefer 'A/GCCGACNT' and 'ACCGAC', respectively. Because freezing causes cellular dehydration, it's plausible to assume that some defense mechanisms were common, such as LEA proteins, which were regulated by DRE/CRT (e.g., 'ACCGACNT') and activated by both cold and drought stress signals. Specific DRE/CRT sequences, however, may regulate genes that operate in specific stress circumstances.

Adaptation to a variety of environmental situations may be facilitated by such minor changes in DRE/CRTs and several DREB subfamily genes. It has been shown that the AREB1/ABF2 and AREB2/ABF4 proteins interact with the respective DREB1A, DREB2A, and DREB2C proteins in vivo or in vitro studies (Lee et al., 2006). In developing sunflower embryos, at minor heat shock protein promoter, DRE/CRT and heat shock elements collaborate (Díaz-Martn et al., 2005). HaDREB2 and HaHSFA9, a DREB2-like protein and an HSF, attach to this promoter, resulting in synergistic transactivation. In vitro, and HaHSFA9 interact utilizing the AP2/ERF domain with the C-terminal part of HaDREB2 and HaHSFA9, respectively, containing the activation domain of each protein being identified as contact sites. Drought stress triggered the DREB2A and DREB2B genes, ultimately activating the genes that possess CRT/DRE *cis*-acting regions. Transgenic Arabidopsis overexpressing CBF4 activated the CRT/DRE-responsive element and enhanced the expression of genes involved in cold and drought acclimatization (Haake et al., 2002).

16.9 ERF Subfamily Members in Abiotic Stress Responses

ERF TFs have been demonstrated to be involved in environmental stress responses and have a variety of complicated and flexible expression patterns. Different pathogen and disease-related stimuli (i.e., ET, JA, and SA) signal pathways, such as ET and JA, are integrated by ERF genes. In addition, some ERFs were activated by ABA (Lee et al., 2006; Wang et al., 2012; Zhuang et al., 2013) indicating a crosstalk pathway between abiotic and biotic stress responses. As a result, ERF genes express multifunctional factors that respond to many stresses, integrate various signal transductions pathways, and hence potentially play interaction roles in abiotic and biotic stresses (Zhu et al., 2010; Xiong et al., 2013).

Members of the ERF subfamily TFs were thought to mediate ethylene-related responses, along with abiotic stressors, including drought and excessive salinity (Park et al., 2001). Transgenic studies under overexpression of ERF subfamily TFs revealed enhanced tolerance levels (Park et al., 2001). Several studies reported that ERF subfamily genes were induced by low or high temperature, dehydration, severe salt stress, etc. as per genome-wide expression investigations of AP2/ERF family genes in poplar (*Populus trichocarpa*) (Zhuang et al., 2008), soybean (Zhang et al., 2005), tomato, and rice (Sharoni et al., 2011). Although the significance of ERF subfamily genes in abiotic stress responses is uncertain, they are expected to play a role in ethylene-dependent and independent gene regulation under stress. Yu et al. (2013) reported that ethylene signaling plays a role in cell cycle arrest under osmotic stress. When Arabidopsis is subjected to osmotic stress (mild), ethylene synthesis is induced and the synthesis of ethylene signaling genes (ERF11 and ERF1, ERF2, ERF5, ERF6 from B1 and B3 subgroup, respectively) is increased. Although downstream genes of these ERFs are not known, our findings showed that ERFs had a significant role in plant response to abiotic stress (Li et al., 2015).

Electrophoretic mobility shift assays were utilized to investigate the DNA-binding preferences of ZmDBFs, and it was discovered that both ZmDBF1 and ZmDBF2 bind to the DRE2 region. It was demonstrated that ZmDBF1 acted as an activator of DRE2-dependent transcription of the rab17 promoter by ABA, but ZmDBF2 overexpression acted as a repressor, lowering not only the baseline promoter activity but also the ABA response. The presence of an ABA-dependent mechanism for gene regulation via the CRT/DRE element was suggested by the fact that ABA regulates ZmDBF activity (Kizis and Pagès, 2002).

Under normal conditions, overexpression of the Tsi1 gene triggered many PR genes and resulted in enhanced tolerance to salt and pathogens, implying that Tsi1 may operate as a positive *trans*-acting factor in two distinct signal transduction pathways under abiotic and biotic stresses (Park et al., 2001).

16.10 Validation of a Transcription Factor That Responds to Drought and ABA

A 964-bp genomic DNA sequence, namely, AP2/ERF-N22, was isolated and characterized from drought-tolerant *O. sativa* L. cv. 'N-22' belongs to the AP2/ERF family TF. The coding sequence for Os-AP2/ERF-N22 was 732-bp long and encoded a 243-residue protein with a theoretical isoelectric point of 9.47 and a molecular weight of 25.3 kDa. There were two exons and one intron in the gene. The AP2/ERF-'N-22' protein had a potential nuclear localization signal (KKKFR) from position 5–9, a probable regulatory domain, and a 60-aa AP2 DNA-binding domain. RWC was negatively correlated with epicuticular wax content and AP2/ERF-'N-22' expression (Mawlong et al., 2014). To determine the importance of the gene concerning drought tolerance, AP2/ERF-N-22 sequence homologs were characterized from rice cultivars with varying drought tolerance, such as IR 64, MTU, and Teipei 309. Polymorphism was discovered in positions 249, 424, 619, 735, and 756 of the gene, as well as 168 and 175 of the protein. The AP2/ERF-Jap sequence was identical to the LOC *Os06g40150* locus on chromosome 6 of *O.* L.

Base sequences substitution was reported at two locations 17 and 771 between AP2/ERF-N-22 (Accession # EF 362638) and AP2/ERF-Jap (Mawlong et al., 2018). *Os*-AP2/ERF-N22 TF was cloned in an expression vector. The recombinant protein was expressed in *E. coli* and purified. The 26.74 KDa–fused protein of Os-AP2/ERF–N22 TF was purified and the antibodies against the AP2 TF were custom synthesized and induced expression was further confirmed at protein and transcript level using antibodies against the purified protein (Kumar et al., 2018). *Os-AP2/ERF-N22* was cloned in binary vector pBI121 under a constitutive promoter CaMV35S. A comparison of WT and transgenic revealed no physiological and morphological differences. In comparison to WT, TR demonstrated less wilting and better turgor under WDS. Transformants exhibited physiological abnormalities such as smaller siliques, fewer seeds and branches, and a later flowering time (Mawlong et al., 2015). These results could be attributed to the inserted gene being expressed under a constitutive promoter. To address undesirable phenotypic consequences in Arabidopsis, the gene was transferred into rice by the biolistic method under a drought-inducible promoter RD29A.

Underwater deficit stress, the physio-biochemical analysis revealed higher RWC, MSI, wax content, osmotic potential, radical scavenging activity, root length, yield, and nearly identical chlorophyll stability index, photosynthetic rate, and chlorophyll content, with no phenotypic aberrations and increased drought tolerance (Mawlong et al., 2018).

16.11 Conclusion

To summarize, TFs from the AP2/ERF family have a role in primary and secondary metabolism, growth and development, and responses to environmental stimuli in plants. These can serve as activators or repressors of the expression of stress-responsive genes in the signal transduction pathway, making them interesting targets for genetic engineering and crop breeding.

REFERENCES

Abe, H., Yamaguchi-Shinozaki, K., Urao, T., Iwasaki, T., Hosokawa, D., & Shinozaki, K. (1997). Role of Arabidopsis MYC and MYB homologs in drought-and abscisic acid-regulated gene expression. *The Plant Cell*, 9(10), 1859–1868.

Agarwal, P., Agarwal, P. K., Joshi, A. J., Sopory, S. K., & Reddy, M. K. (2010). Overexpression of PgDREB2A transcription factor enhances abiotic stress tolerance and activates downstream stress-responsive genes. *Molecular Biology Reports*, 37, 1125–1135.

Ali, K., Rai, R. D., & Tyagi, A. (2016). Expression analysis of bZIP transcription factor encoding genes in response to water deficit stress in rice. *Indian Journal of Experimental Biology*, 54, 332–337.

Alonso, J. M., Stepanova, A. N., Leisse, T. J., Kim, C. J., Chen, H., Shinn, P., & Ecker, J. R. (2003). Genome-wide insertional mutagenesis of Arabidopsis thaliana. *Science*, 301(5633), 653–657.

Amoutzias, G. D., Robertson, D. L., Van de Peer, Y., & Oliver, S. G. (2008). Choose your partners: Dimerization in eukaryotic transcription factors. *Trends in Biochemical Sciences*, 33(5), 220–229.

Aukerman, M. J., & Sakai, H. (2003). Regulation of flowering time and floral organ identity by a microRNA and its APETALA2-like target genes. *The Plant Cell*, 15(11), 2730–2741.

Baker, S. S., Wilhelm, K. S., & Thomashow, M. F. (1994). The 5′-region of Arabidopsis thaliana cor15 a has cis-acting elements that confer cold-, drought- and ABA-regulated gene expression. *Plant Molecular Biology*, 24, 701–713.

Baniwal, S. K., Bharti, K., Chan, K. Y., Fauth, M., Ganguli, A., Kotak, S., & von Koskull-DÖring, P. (2004). Heat stress response in plants: A complex game with chaperones and more than twenty heat stress transcription factors. *Journal of Biosciences*, 29(4), 471–487.

Bartels, D., & Sunkar, R. (2005). Drought and salt tolerance in plants. *Critical Reviews in Plant Sciences*, 24(1), 23–58.

Berrocal-Lobo, M., Molina, A., & Solano, R. (2002). Constitutive expression of ETHYLENE-RESPONSE-FACTOR1 in Arabidopsis confers resistance to several necrotrophic fungi. *The Plant Journal*, 29(1), 23–32.

Bhattacharjee, A., Khurana, J. P., & Jain, M. (2016). Characterization of rice homeobox genes, OsHOX22 and OsHOX24, and over-expression of OsHOX24 in transgenic Arabidopsis suggest their role in abiotic stress response. *Frontiers in Plant Science*, 7, 627.

Boutilier, K., Offringa, R., Sharma, V. K., Kieft, H., Ouellet, T., Zhang, L., & van Lookeren Campagne, M. M. (2002). Ectopic expression of BABY BOOM triggers a conversion from vegetative to embryonic growth. *The Plant Cell*, 14(8), 1737–1749.

Brivanlou, A. H., & Darnell, J. E. (2002). Signal transduction and the control of gene expression. *Science*, 295(5556), 813–818.

BrownR. L., Kazan, K., Mcgrath, K. C., Maclean, D. J., & Manners, J. M. (2003). A role for the GCC-Box in Jasmonate-Mediated Activation of the PDF1.2 Gene of Arabidopsis. *Plant Physiology*, 132(2), 1020–1032.

Busk, P. K., Jensen, A. B., & Pagès, M. (1997). Regulatory elements in vivo in the promoter of the abscisic acid responsive gene rab17 from maize. *The Plant Journal*, 11(6), 1285–1295.

Carretero-Paulet, L., Galstyan, A., Roig-Villanova, I., Martínez-García, J. F., Bilbao-Castro, J. R., & Robertson, D. L. (2010). Genome-wide classification and evolutionary analysis of the bHLH family of transcription factors in Arabidopsis, poplar, rice, moss, and algae. *Plant Physiology*, 153(3), 1398–1412.

Castilhos, G., Lazzarotto, F., Spagnolo-Fonini, L., Bodanese-Zanettini, M. H., & Margis-Pinheiro, M. (2014). Possible roles of basic helix-loop-helix transcription factors in adaptation to drought. *Plant Science*, 223, 1–7.

Chauhan, H., Khurana, N., Agarwal, P., & Khurana, P. (2011). Heat shock factors in rice (*Oryza sativa* L.): Genome-wide expression analysis during reproductive development and abiotic stress. *Molecular Genetics and Genomics*, 286(2), 171–187.

Che, P., Gingerich, D. J., Lall, S., & Howell, S. H. (2002). Global and hormone-induced gene expression changes during shoot development in Arabidopsis. *The Plant Cell*, 14(11), 2771–2785.

Chen, J. Q., Meng, X. P., Zhang, Y., Xia, M., & Wang, X. P. (2008). Over-expression of OsDREB genes lead to enhanced drought tolerance in rice. *Biotechnology Letters*, 30(12), 2191–2198.

Cheng, M. C., Hsieh, E. J., Chen, J. H., Chen, H. Y., & Lin, T. P. (2012). Arabidopsis RGLG2, functioning as a RING E3 Ligase, Interacts with AtERF53 and Negatively Regulates the Plant Drought Stress Response. *Plant Physiology*, 158, 363–375.

Chinnusamy, V., Zhu, J., & Zhu, J. K. (2007). Cold stress regulation of gene expression in plants. *Trends in Plant Science*, 12, 444–451.

Chuck, G., Meeley, R. B., & Hake, S. (1998). The control of maize spikelet meristem fate by theAPETALA2-like gene indeterminate spikelet1. *Genes & Development*, 12(8), 1145–1154.

Cook, D., Fowler, S., Fiehn, O., & Thomashow, M. F. (2004). A prominent role for the CBF cold response pathway in configuring the low-temperature metabolome of Arabidopsis. *Proceedings of the National Academy of Sciences of the United States of America*, 101, 15243–15248.

Datta, K., Baisakh, N., Ganguly, M., Krishnan, S., Yamaguchi Shinozaki, K., & Datta, S. K. (2012). Overexpression of Arabidopsis and rice stress genes' inducible transcription factor confers drought and salinity tolerance to rice. *Plant Biotechnology Journal*, 10(5), 579–586.

dePaiva Rolla, A. A., Carvalho, J. D. F. C., Fuganti-Pagliarini, R., Engels, C., Do Rio, A., Marin, S. R. R., & Nepomuceno, A. L. (2014). Phenotyping soybean plants transformed with rd29A: AtDREB1A for drought tolerance in the greenhouse and field. *Transgenic Research*, 23(1), 75–87.

Díaz-Martín, J., Almoguera, C., Prieto-Dapena, P., Espinosa, J. M., & Jordano, J. (2005). Functional interaction between two transcription factors involved in the developmental regulation of a small heat stress protein gene promoter. *Plant Physiology*, 139, 1483–1494.

Dolan, J. W., & Fields, S. (1991). Cell-type-specific transcription in yeast. *Biochimica et Biophysica Acta (BBA)-Gene Structure and Expression*, 1088(2), 155–169.

Dong, C. J., & Liu, J. Y. (2010). The Arabidopsis EAR-motif-containing protein RAP2. 1 function as an active transcriptional repressor to keep stress responses under tight control. *BMC Plant Biology*, 10(1), 1–15.

Dong, J., Teng, W., Buchholz, W. C., & Hall, T. C. (2010). Agrobacterium mediated transformation of javanica rice. *Molecular Breeding*, 2, 267–276.

Duan, K., Li, L., Hu, P., Xu, S. P., Xu, Z. H., & Xue, H. W. (2006). A brassinolide-suppressed rice MADS-box transcription factor, OsMDP1, has a negative regulatory role in BR signaling. *The Plant Journal*, 47(4), 519–531.

Dubos, C., Stracke, R., Grotewold, E., Weisshaar, B., Martin, C., & Lepiniec, L. (2010). MYB transcription factors in Arabidopsis. *Trends in Plant Science*, 15(10), 573–581.

Dubouzet, J. G., Sakuma, Y., Ito, Y., Kasuga, M., Dubouzet, E. G., Miura, S., & Yamaguchi-Shinozaki, K. (2003). OsDREB genes in rice, *Oryza sativa* L., encode transcription activators that function in drought-, high-salt-and cold-responsive gene expression. *The Plant Journal*, 33(4), 751–763.

Egawa, C., Kobayashi, F., Ishibashi, M., Nakamura, T., Nakamura, C., & Takumi, S. (2006). Differential regulation of transcript accumulation and alternative splicing of a DREB2 homolog under abiotic stress conditions in common wheat. *Genes and Genetic Systems*, 81, 77–91.

Elliott, R. C., Betzner, A. S., Huttner, E., Oakes, M. P., Tucker, W. Q., Gerentes, D., & Smyth, D. R. (1996). AINTEGUMENTA, an APETALA2-like gene of Arabidopsis with pleiotropic roles in ovule development and floral organ growth. *The Plant Cell*, 8(2), 155–168.

Fairman, R., Beran-Steed, R. K., Anthony-Cahill, S. J., Lear, J. D., Stafford, W. F., DeGrado, W. F., & Brenner, S. L. (1993). Multiple oligomeric states regulate the DNA binding of helix-loop-helix peptides. *Proceedings of the National Academy of Sciences*, 90(22), 10429–10433.

Fang, P., Yan, H., Chen, F., & Peng, Y. (2021). Overexpression of maize ZmHDZIV14 increases abscisic acid sensitivity and mediates drought and salt stress in Arabidopsis and tobacco. *Plant Molecular Biology Reporter*, 39(2), 275–287.

Fowler, S., & Thomashow, M. F. (2002). Arabidopsis transcriptome profiling indicates that multiple regulatory pathways are activated during cold acclimation in addition to the CBF cold response pathway. *The Plant Cell*, 14(8), 1675–1690.

Fuhrmann-Aoyagi, M. B., de Fátima Ruas, C., Barbosa, E. G. G., Braga, P., Moraes, L. A. C., de Oliveira, A. C. B., & Mertz-Henning, L. M. (2021). Constitutive expression of Arabidopsis bZIP transcription factor AREB1 activates cross-signaling responses in soybean under drought and flooding stresses. *Journal of Plant Physiology*, 257, 153338.

Gao, C., Li, P., Song, A., Wang, H., Wang, Y., Ren, L., Qi, X., Chen, F., Jiang, J., & Chen, S. (2015). Isolation and characterization of six AP2/ERF transcription factor genes in Chrysanthamum nankingense. *International Journal of Molecular Sciences*, 16, 2052–2065.

Gill, G. (2001). Regulation of the initiation of eukaryotic transcription. Essays in biochemistry, 37, 33–43.

Gilmour, S. J., Zarka, D. G., Stockinger, E. J., Salazar, M. P., Houghton, J. M., & Thomashow, M. F. (1998). Low temperature regulation of the Arabidopsis CBF family of AP2 transcriptional activators as an early step in cold-induced COR gene expression. *The Plant Journal*, 16(4), 433–442.

Giraudat, J., Hauge, B. M., Valon, C., Smalle, J., Parcy, F., & Goodman, H. M. (1992). Isolation of the Arabidopsis ABI3 gene by positional cloning. *The Plant Cell*, 4(10), 1251–1261.

Glover, J. M., & Harrison, S. C. (1995). Crystal structure of the heterodimeric bZIP transcription factor c-Fos–c-Jun bound to DNA. *Nature*, 373(6511), 257–261.

Haake, V., Cook, D., Riechmann, J., Pineda, O., Thomashow, M. F., & Zhang, J. Z. (2002). Transcription factor CBF4 is a regulator of drought adaptation in Arabidopsis. *Plant Physiology*, 130(2), 639–648.

Hao, D., Ohme-Takagi, M., & Sarai, A. (1998). Unique mode of GCC box recognition by the DNA-binding domain of ethylene-responsive element-binding factor (ERF domain) in plant. *Journal of Biological Chemistry*, 273(41), 26857–26861.

Hauser, M. T., Aufsatz, W., Jonak, C., & Luschnig, C. (2011). Transgenerational epigenetic inheritance in plants. *Biochimica et Biophysica Acta*, 1809, 459–468.

Hu, Y. X., Wang, Y. H., Liu, X. F., & Li, J. Y. (2004). Arabidopsis RAV1 is down-regulated by brassinosteroid and may act as a negative regulator during plant development. *Cell Research*, 14(1), 8–15.

Hwang, I., Manoharan, R. K., Kang, J. G., Chung, M. Y., Kim, Y. W., & Nou, I. S. (2016). Genome-wide identification and characterization of bZIP transcription factors in Brassica oleracea under cold stress. *BioMed Research International*, 2016, 4376598.

Imin, N., Nizamidin, M., Wu, T., & Rolfe, B. G. (2007). Factors involved in root formation in Medicago truncatula. *Journal of Experimental Botany*, 58(3), 439–451.

Iwaki, T., Guo, L., Ryals, J. A., Yasuda, S., Shimazaki, T., Kikuchi, A., ... & Ohta, D. (2013). Metabolic profiling of transgenic potato tubers expressing Arabidopsis dehydration response element-binding protein 1A (DREB1A). *Journal of Agricultural and Food Chemistry*, 61(4), 893–900.

Jaglo, K. R., Kleff, S., Amundsen, K. L., Zhang, X., Haake, V., Zhang, J. Z., Deits, T., & Thomashow, M. F. (2001). Components of the Arabidopsis C-repeat/dehydration-responsive element binding factor cold-response pathway are conserved in Brassica napus and other plant species. *Plant Physiology*, 127, 910–917.

Jakoby, M., Weisshaar, B., Dröge-Laser, W., Vicente-Carbajosa, J., Tiedemann, J., Kroj, T., & Parcy, F. (2002). bZIP transcription factors in Arabidopsis. *Trends in Plant Science*, 7(3), 106–111.

Janská, A., Maršík, P., Zelenková, S., & Ovesná, J. (2010). Cold stress and acclimation–what is important for metabolic adjustment? *Plant Biology*, 12(3), 395–405.

Jiang, C., Iu, B., & Singh, J. (1996). Requirement of a CCGAC cis: Acting element for cold induction of the BIN115 gene from winter Brassica napus. *Plant Molecular Biology*, 30, 679–684.

Jofuku, K. D., Den Boer, B. G., Van Montagu, M., & Okamuro, J. K. (1994). Control of Arabidopsis flower and seed development by the homeotic gene APETALA2. *The Plant Cell*, 6(9), 1211–1225.

Joshi, R., Wani, S. H., Singh, B., Bohra, A., Dar, Z. A., Lone, A. A., & Singla-Pareek, S. L. (2016). Transcription factors and plants response to drought stress: Current understanding and future directions. *Frontiers in Plant Science*, 7, 1029.

Kagaya, Y., Ohmiya, K., & Hattori, T. (1999). RAV1, a novel DNA-binding protein, binds to bipartite recognition sequence through two distinct DNA-binding domains uniquely found in higher plants. *Nucleic Acids Research*, 27(2), 470–478.

Karaba, A., Dixit, S., Greco, R., Aharoni, A., Kurniawan, R., Trijatmiko., Marsch-Martinez, N., Krishnan, A., Karaba N. N., Udayakumar, M., & Pereira, A. (2007). Improvement of water use efficiency in rice by expression of HARDY, an Arabidopsis drought and salt tolerance gene. *Proceedings of the National Academy of Sciences of the United States of America*, 104(39), 15270–15275.

Karim, S., Aronsson, H., Ericson, H., Pirhonen, M., Leyman, B., Welin, B., ... & Holmström, K. O. (2007). Improved drought tolerance without undesired side effects in transgenic plants producing trehalose. *Plant Molecular Biology*, 64(4), 371–386.

Kasuga, M., Liu, Q., Miura, S., Yamaguchi-Shinozaki, K., & Shinozaki, K. (1999). Improving plant drought, salt, and freezing tolerance by gene transfer of a single stress-inducible transcription factor. *Nature Biotechnology*, 17(3), 287–291.

Kawahara, R., Komamine, A., & Fukuda, H. (1995). Isolation and characterization of homeobox-containing genes of carrot. *Plant Molecular Biology*, 27(1), 155–164.

Khraiwesh, B., Zhu, J. K., & Zhu, J. (2012). Role of miRNAs and siRNAs in biotic and abiotic stress responses of plants. *Biochimica et Biophysica Acta*, 1819(2), 137–48.

Kizis, D., & Pagès, M. (2002). Maize DRE-binding proteins DBF1 and DBF2 are involved in rab17 regulation through the drought-responsive element in an ABA-dependent pathway. *The Plant Journal*, 30(6), 679–689.

Klucher, K. M., Chow, H., Reiser, L., & Fischer, R. L. (1996). The AINTEGUMENTA gene of Arabidopsis required for ovule and female gametophyte development is related to the floral homeotic gene APETALA2. *The Plant Cell*, 8(2), 137–153.

Klug, A., & Schwabe, J. W. (1995). Zinc fingers. *The FASEB Journal*, 9(8), 597–604.

Kotak, S., Port, M., Ganguli, A., Bicker, F., & Von Koskull-Döring, P. (2004). Characterization of C-terminal domains of Arabidopsis heat stress transcription factors (Hsfs) and identification of a new signature combination of plant class A Hsfs with AHA and NES motifs essential for activator function and intracellular localization. *The Plant Journal*, 39(1), 98–112.

Krizek, B. A., Prost, V., & Macias, A. (2000). AINTEGUMENTA promotes petal identity and acts as a negative regulator of AGAMOUS. The Plant Cell, 12(8), 1357–1366.

Krizek, B. A. (2003). AINTEGUMENTA utilizes a mode of DNA recognition distinct from that used by proteins containing a single AP2 domain. *Nucleic Acids Research*, 31(7), 1859–1868.

Kumar, V., Kumar, A., Ali, K., Tewari, K., Garg, N. K., Changan, S. S. & Tyagi, A. (2018). Cloning and heterologous expression of Os-AP2/ERF–N22 drought inducible rice transcription factor in E.Coli. *Indian Journal of Agricultural Sciences*, 88(10), 1515–1520.

Kummerfeld, S. K., & Teichmann, S. A. (2006). DBD: A transcription factor prediction database. *Nucleic Acids Research*, 34(suppl_1), D74–D81.

Landschulz, W. H., Johnson, P. F., & McKnight, S. L. (1988). The leucine zipper: A hypothetical structure common to a new class of DNA binding proteins. *Science*, 240(4860), 1759–1764.

Lata, C., Mishra, A. K., Muthamilarasan, M., Bonthala, V. S., Khan, Y., & Prasad, M. (2014). Genome-wide investigation and expression profiling of AP2/ERF transcription factor superfamily in foxtail millet (*Setaria italica* L.). *PLoS One*, 9(11), e113092.

Latchman, D. S. (1997). Transcription factors: An overview. *The International Journal of Biochemistry & Cell Biology*, 29(12), 1305–1312.

Lee, B. J., Cansizoglu, A. E., Süel, K. E., Louis, T. H., Zhang, Z. & Chook, Y. M. (2006). Rules for nuclear localisation sequence recognition by karyopherin beta 2. *Cell*, 126(3), 543–558.

Lemmon, M. A., Treutlein, H. R., Adams, P. D., Brünger, A. T., & Engelman, D. M. (1994). A dimerization motif for transmembrane α–helices. *Nature Structural Biology*, 1(3), 157–163.

Li, C. W., Su, R. C., Cheng, C. P., You, S. J., Hsieh, T. H., Chao, T. C., Chan, M. T. (2011). Tomato RAV transcription factor is a pivotal modulator involved in the AP2/EREBP-mediated defense pathway. Plant Physiology, 156(1), 213–227.

Li, M.-Y., Xu, Z.-S., Huang, Y., Tian C., Wang, F., & Xiong, A.-S. (2015). Genome wide analysis of AP2/ERF transcription factors in carrot reveals evolution and expression profiles under abiotic stress. *Molecular Genetics and Genomics*. DOI: 10.1007/S00438-015-1061.

Liang, C., Meng, Z., Meng, Z., Malik, W., Yan, R., Lwin, K. M., … & Zhang, R. (2016). GhABF2, a bZIP transcription factor, confers drought and salinity tolerance in cotton (*Gossypium hirsutum* L.). *Scientific Reports*, 6(1), 1–14.

Licause, F, Ohme-Takagi, M, & Perata, P. (2013). AP2/ERF transcription factors: Mediators of stress responses and developmental programs. *New Phytology*, 199, 639–649.

Lim, C. J., Hwang, J. E., Chen, H., Hong, J. K., Yang, K. A., Choi, M. S., Lee, K. O., Chung, W. S., Lee, S. Y., & Lim, C. O. (2007). Over-expression of the Arabidopsis DRE/CRT-binding transcription factor DREB2C enhances thermo tolerance. *Biochemical and Biophysical Research Communication*, 362, 431–436.

Lindemose, S., O'Shea, C., Jensen, M. K., & Skriver, K. (2013). Structure, function and networks of transcription factors involved in abiotic stress responses. *International Journal of Molecular Sciences*, 14(3), 5842–5878.

Liu, Q. K. M., Sakuma, Y., Abe, H., Miura. S., Yamaguchi-Shinozaki, K., & Shinozaki, K. (1998). Two transcription factors, DREB1 and DREB2, with an EREBP/AP2 DNA binding domain separate two cellular signal transduction pathways in drought-and low-temperature-responsive gene expression, respectively, in Arabidopsis. *The Plant Cell Online*, 10(8), 1391–406.

Liu, N., Zhong, N. Q., Wang, G.-L., Li, L. J., Liu, X.-L., He, Y. K., & Xia, G. X. (2007). Cloning and functional characterization of PpDBF1 gene encoding a DRE-binding transcription factor from Physcomitrella patens. *Planta*, 226, 827–838.

Liu, Z. Y., Li, X. P., Zhang, T. Q., Wang, Y. Y., Wang, C., & Gao, C. Q. (2021). Overexpression of ThMYB8 mediates salt stress tolerance by directly activating stress-responsive gene expression. *Plant Science*, 302, 110668.

Ludwig, S. R., Habera, L. F., Dellaporta, S. L., & Wessler, S. R. (1989). Lc, a member of the maize R gene family responsible for tissue-specific anthocyanin production, encodes a protein similar to transcriptional activators and contains the myc-homology region. *Proceedings of the National Academy of Sciences*, 86(18), 7092–7096.

Magnani, E., Sjölander, K., & Hake, S. (2004). From endonucleases to transcription factors: Evolution of the AP2 DNA binding domain in plants. *The Plant Cell*, 16(9), 2265–2277.

Maruyama, K., Sakuma, Y., Kasuga, M., Ito, Y., Seki, M., Goda, H., Shimada, Y., Yoshida, S., Shinozaki, K., & Yamaguchi-Shinozaki, K. (2009a). Identification of cold-inducible downstream genes of the Arabidopsis DREB1A/CBF3 transcriptional factor using two microarray systems. *Plant Journal*, 38, 982–993.

Maruyama, K., Takeda, M., Kidokoro, S., Yamada, K., Sakuma, Y., Urano, K., … & Yamaguchi-Shinozaki, K. (2009b). Metabolic pathways involved in cold acclimation identified by integrated analysis of metabolites and transcripts regulated by DREB1A and DREB2A. *Plant Physiology*, 150(4), 1972–1980.

Matsukura, S., Mizoi, J., Yoshida, T., Todaka, D., Ito, Y., Maruyama, K., Shinozaki, K., & Yamaguchi-Shinozaki, K. (2010). Comprehensive analysis of rice DREB2 type genes that encode a transcription factor involve in the expression of abiotic stress responsive genes. *Molecular Genetics and Genomics*, 283, 185–196.

Mawlong, I., Ali, K., Kurup, D., Yadav, S., & Tyagi, A. (2014). Isolation and characterization of an AP2/ERF-type drought stress inducible transcription factor encoding gene from rice. *Journal of Plant Biochemistry and Biotechnology*, 23(1), 42–51.

Mawlong, I., Ali, K., Srinivasan, R., Rai, R. D., & Tyagi, A. (2015). Functional validation of a drought-responsive AP2/ERF family transcription factor-encoding gene in Arabidopsis. *Molecular Breeding*. DOI: 10.1007/s11032-015-0290-9.

Mawlong, I., Ali, K., & Tyagi, A. (2016). Cloning and characterization of a water deficit stress responsive transcription factor gene from rice. *Indian Journal of Experimental Biology*, 54, 26–36.

Mawlong, I., Ali, K., & Tyagi, A. (2018). Functional validation of a water deficit stress responsive AP2/ERF family transcription factor encoding gene in *Oryza Sativa*. *Indian Journal of Biochemistry and Biophysics*, 55, 17–25.

Meena, N. L., Ali, K., Deshmukh, P. R. & Tyagi, A. (2014). Effect of heat stress on physio-biochemical characteristics of chickpea genotypes. *The Indian Journal of Agricultural Sciences*, 84(3), 401–406.

Menke, F. L., Champion, A., Kijne, J. W., & Memelink, J. (1999). A novel jasmonate and elicitor-responsive element in the periwinkle secondary metabolite biosynthetic gene Str interacts with a jasmonate- and elicitor-inducible AP2-domain transcription factor, ORCA2. *EMBO Journal*, 18(16), 4455–4463.

Meshi, T., & Iwabuchi, M. (1995). Plant transcription factors. *Plant and Cell Physiology*, 36(8), 1405–1420.

Mittler, R. (2006). Abiotic stress, the field environment and stress combination. *Trends in Plant Science*, 11, 15–19.

Mizoi, J., Shinozaki, K., & Yamaguchi-Shinozaki, K. (2012). AP2/ERF family transcription factors in plant abiotic stress responses. *Biochimica et Biophysica Acta (BBA)-Gene Regulatory Mechanisms*, 1819(2), 86–96.

Moose, S. P., & Sisco, P. H. (1996). Glossy15, an APETALA2-like gene from maize that regulates leaf epidermal cell identity. *Genes & Development*, 10(23), 3018–3027.

Morimoto, R. (1993). Cells in stress: Transcriptional activation of heat shock genes. *Science*, 259(5100), 1409–1411.

Mundy, J., Yamaguchi-Shinozaki, K., & Chua, N. H. (1990). Nuclear proteins bind conserved elements in the abscisic acid-responsive promoter of a rice rab gene. *Proceedings of the National Academy of Sciences*, 87(4), 1406–1410.

Munns, R., & Tester, M. (2008). Mechanisms of salinity tolerance. *Annual Review of Plant Biology*, 59, 651–681.

Nair, S. K., & Burley, S. K. (2000). Recognizing DNA in the library. *Nature*, 404(6779), 715–717.

Nakano, T., Suzuki, K., Fujimura, T., & Shinshi, H. (2006). Genome-wide analysis of the ERF gene family in Arabidopsis and rice. *Plant Physiology*, 140(2), 411–32.

Nakashima, K., Shinwari, Z. K., Sakuma, Y., Seki, M., Miura, S., Shinozaki, K., & Yamaguchi-Shinozaki, K. (2000). Organization and expression of two Arabidopsis DREB2 genes encoding DRE-binding proteins involved in dehydration-and high-salinity-responsive gene expression. *Plant Molecular Biology*, 42(4), 657–665.

Narlikar, G. J., Fan, H. Y., & Kingston, R. E. (2002). Cooperation between complexes that regulate chromatin structure and transcription. *Cell*, 108(4), 475–487.

Nover, L., Bharti, K., Döring, P., Mishra, S. K., Ganguli, A., & Scharf, K. D. (2001). Arabidopsis and the heat stress transcription factor world: How many heat stress transcription factors do we need? *Cell Stress & Chaperones*, 6(3), 177.

Ogata, K., Hojo, H., Aimoto, S., Nakai, T., Nakamura, H., Sarai, A., & Nishimura, Y. (1992). Solution structure of a DNA-binding unit of Myb: A helix-turn-helix-related motif with conserved tryptophans forming a hydrophobic core. *Proceedings of the National Academy of Sciences*, 89(14), 6428–6432.

Ohme-Takagi, M., & Shinshi, H. (1995). Ethylene-inducible DNA binding proteins that interact with an ethylene-responsive element. *Plant Cell*, 7, 173–182.

Okamuro, J. K., Caster, B., Villarroel, R., Montagu, M. V., & Jofuku, K. D. (1997). The AP2 domain of APETALA2 defines a large new family of DNA binding proteins in Arabidopsis. *Proceedings of the National Academy of Sciences of the United States of America*, 94, 7076–7081.

Ono, A., Izawa, T., Chua, N. H., & Shimamoto, K. (1996). The rab16B promoter of rice contains two distinct abscisic acid-responsive elements. *Plant Physiology*, 112(2), 483–491.

Ouellet, F., Vazquez, T., & Sarhan, F. (1998). The wheat WCS120 promoter is cold: Inducible in both monocotyledonous and dicotyledonous species. *FEBS Letters*, 423, 324–328.

Pandey, A. S., Sharma, E., Jain, N., Singh, B., Burman, N., & Khurana, J. P. (2018). A rice bZIP transcription factor, OsbZIP16, regulates abiotic stress tolerance when overexpressed in Arabidopsis. *Journal of Plant Biochemistry and Biotechnology*, 27(4), 393–400.

Park, J. M., Park, C. J., Lee, S. B., Ham, B. K., Shin, R., & Paek K. H. (2001). Overexpression of the TobaccoTsi1gene encoding an EREBP/AP2–type transcription factor enhances resistance against pathogen attack and osmotic stress in Tobacco. *The Plant Cell*, 13, 1035–1046.

Pireyre, M., & Burow, M. (2015). Regulation of MYB and bHLH transcription factors: A glance at the protein level. *Molecular Plant*, 8(3), 378–388.

Qin, F., Kakimoto, M., Sakuma, Y., Maruyama, K., Osakabe, Y., Tran, L. S. P., Shinozaki, K., & Yamaguchi-Shinozaki, K. (2007). Regulation and functional analysis of ZmDREB2A in response to drought and heat stresses in Zea mays. *Plant Journal*, 50, 54–69.

Quaedvlieg, N., Dockx, J., Rook, F., Weisbeek, P., & Smeekens, S. (1995). The homeobox gene ATH1 of Arabidopsis is derepressed in the photomorphogenic mutants cop1 and det1. *The Plant Cell*, 7(1), 117–129.

Rashid, M., Guangyuan, H., Guangxiao, Y., Hussain. J., & Xu, Y. (2012). AP2/RF transcription factor in rice: Genome-Wide canvas and syntenic relationships between Monocots and eudicots. *Evolutionary Bioinformatics*, 8, 321–355.

Reece-Hoyes, J. S., Deplancke, B., Shingles, J., Grove, C. A., Hope, I. A., & Walhout, A. J. (2005). A compendium of Caenorhabditis elegans regulatory transcription factors: A resource for mapping transcription regulatory networks. *Genome Biology*, 6(13), 1–12.

Riechmann, J. L., & Meyerowitz, E. M. (1998). The AP2/EREBP family of plant transcription factors. *Journal of Biological Chemistry*, 379(6), 633–646.

Riechmann, J. L., Heard, J., Martin, G., Reuber, L., Jiang, C. Z., Keddie, J., & Yu, G. L. (2000). Arabidopsis transcription factors: Genome-wide comparative analysis among eukaryotes. *Science*, 290(5499), 2105–2110.

Rizhsky, L., Liang, H., & Mittler, R. (2002). The combined effect of drought stress and heat shock on gene expression in tobacco. *Plant Physiology*, 130, 1143–1151.

Rizhsky, L., Liang, H., Shuman, J., Shulaev, V., Davletova, S., & Mittler, R. (2004). When defense pathways collide. The response of Arabidopsis to a combination of drought and heat stress. *Plant Physiology*, 134,1683–1696.

Ruberti, I., Sessa, G., Lucchetti, S., & Morelli, G. (1991). A novel class of plant proteins containing a homeodomain with a closely linked leucine zipper motif. *The EMBO Journal*, 10(7), 1787–1791.

Rushton, P. J., Macdonald, H., Huttly, A. K., Lazarus, C. M., & Hooley, R. (1996). Members of a new family of DNA-binding proteins bind to a conserved cis-element in the promoters of alpha-Amy2 genes. *Plant Molecular Biology*, 29(4), 691–702.

Sairam, R. K. & Tyagi, A. (2004). Physiology and molecular biology of salinity stress tolerance in plants. *Current Science*, 8693), 407–421.

Sakuma, Y. L. Q., Dubouzet, J. G., Abe, H., Shinozaki, K., & Yamaguchi-Shinozaki, K. (2002). DNA-binding specificity of the ERF/AP2 domain of Arabidopsis DREBs, transcription factors involved in dehydration-and cold-inducible gene expression. *Biochemical and Biophysical Research Communications*, 290(3), 998–1009.

Sakuma, Y., Maruyama, K., Osakabe, Y., Qin, F., Seki, M., Shinozaki, K., & Yamaguchi-Shinozakia, K. (2006). functional analysis of an arabidopsis transcription factor, DREB2A, involved in drought-responsive gene expression. *The Plant Cell*, 18, 1292–1309.

SamuelYang, S., Cheung, F., Lee, J. J., Ha, M., Wei, N. E., Sze, S. H., … & Jeffrey Chen, Z. (2006). Accumulation of genome-specific transcripts, transcription factors and phytohormonal regulators during early stages of fiber cell development in allotetraploid cotton. *The Plant Journal*, 47(5), 761–775.

Scharf, K. D., Berberich, T., Ebersberger, I., & Nover, L. (2012). The plant heat stress transcription factor (Hsf) family: Structure, function and evolution. *Biochimica et Biophysica Acta (BBA)-Gene Regulatory Mechanisms*, 1819(2), 104–119.

Schwarz-Sommer, Z., Huijser, P., Nacken, W., Saedler, H., & Sommer, H. (1990). Genetic control of flower development by homeotic genes in Antirrhinum majus. *Science*, 250(4983), 931–936.

Shah, Z., Shah, S. H., Ali, G. S., Munir, I., Khan, R. S., Iqbal, A., & Jan, A. (2020). Introduction of Arabidopsis's heat shock factor HsfA1d mitigates adverse effects of heat stress on potato (*Solanum tuberosum* L.) plant. *Cell Stress and Chaperones*, 25(1), 57–63.

Sharoni, A. M., Satoh, K., & Shimizu, T. (2011). Gene structures, classification and expression models of the AP2/EREBP transcription factor family in rice. *Plant and Cell Physiology*, 52(2), 344–360.

Shen, Y. G., Zhang, W-K., He, S. J., Zhang, J-S., Liu, Q., & Chen, S. Y. (2003). An EREBP/AP2-type protein in Triticum aestivum was a DRE-binding transcription factor induced by cold, dehydration and ABA stress. *Theoritical and Applied Genetics*, 106, 923–930.

Shigyo, M., & Ito, M. (2004). Analysis of gymnosperm two-AP2-domain-containing genes. *Development Genes and Evolution*, 214(3), 105–114.

Shinozaki, K., & Yamaguchi-Shinozaki, K. (1997). Gene expression and signal transduction in water-stress response. *Plant Physiology*, 115, 327–334.

Solano, R., Stepanova, A., Chao, Q., & Ecker, J. R. (1998). Nuclear events in ethylene signaling: A transcriptional cascade mediated by ETHYLENE-INSENSITIVE3 and ETHYLENE-RESPONSIVE-FACTOR1. *Genes Development*, 12, 3703–3714.

Sommer, H., Beltran, J. P., Huijser, P., Pape, H., Lönnig, W. E., Saedler, H., & Schwarz-Sommer, Z. (1990). Deficiens, a homeotic gene involved in the control of flower morphogenesis in Antirrhinum majus: The protein shows homology to transcription factors. *The EMBO Journal*, 9(3), 605–613.

Song, Y., Lin, Y., Tong, S., & Hou, H. (2012). Molecular cloning, promoter analysis, and expression profile of VvERF3b gene in Vitis vinifera. *Biologia Plantarum*, 56(1), 31–36.

Song, X., Li, Y., & Hou, X. (2013). Genome-wide analysis of the AP2/ERF transcription factor superfamily in Chinese cabbage (Brassica rapa ssp. pekinensis). *BMC Genomics*, 14(1), 1–15.

Soren, K. R., Ali, K., & Tyagi, A. (2010). Recent advances in molecular breeding of drought stress in rice. *Indian Journal of Biotechnology*, 9, 233–251.

Soren, K. R., Ali, K., & Tyagi, A. (2012). Cloning and characterization of gene encoding novel zinc finger protein transcription factor from rice. *Indian Journal of Biochemistry and Biophysics*, 49, 36–41.

Srivasta, A., Mehta, S., Lindlof, A., & Bhargava, S. (2010). Over-represented promoter motifs in abiotic stress-induced DREB genes of rice and sorghum and their probable role in regulation of gene expression. *Plant Signaling & Behavior*, 5(7), 775–784.

Stockinger, E. J., Gilmour, S. J., & Thomashow, M. F. (1997). Arabidopsis thaliana CBF1 encodes an AP2 domain-containing transcriptional activator that binds to the C-repeat/DRE, a cis-acting DNA regulatory element that stimulates transcription in response to low temperature and water deficit. *Proceedings of the National Academy of Sciences of the United States of America*, 94, 1035–1040.

ros, M., Launholt, D., & Grasser, K. D. (2007). The HMG-box: A versatile protein domain occurring in a wide variety of DNA-binding proteins. *Cellular and Molecular Life Sciences*, 64(19), 2590–2606.

Su, J., Shen, Q., David Ho, T. H., & Wu, R. (1998). Dehydration-stress-regulated transgene expression in stably transformed rice plants. *Plant Physiology*, 117(3), 913–922.

Takagi, Y., & Kornberg, R. D. (2006). Mediator as a general transcription factor. *Journal of Biological Chemistry*, 281(1), 80–89.

Thomas, J. O., & Travers, A. A. (2001). HMG1 and 2, and related 'architectural' DNA-binding proteins. *Trends in Biochemical Sciences*, 26(3), 167–174.

Toledo-Ortiz, G., Huq, E., & Quail, P. H. (2003). The Arabidopsis basic/helix-loop-helix transcription factor family. *The Plant Cell*, 15(8), 1749–1770.

Tournier, B., Sanchez-ballesta, M. T., Jones, B., Pesquet, E., Regad, F., Latche, A., Pech, J. C., & Bouzayen, M. (2003). New members of the tomato ERF family show specific expression pattern and diverse DNA-binding capacity to the GCC box element. *FEBS Letters*, 550(1–3), 149–154.

Tyagi, A. & Chandra, A. (2006). Isolation of stress responsive Psb A gene from rice (*Oryza Sativa* L.) using differential display. *Indian Journal of Biochemistry & Biophysics*, 43, 244–246.

Tyagi, A., Klueva, N., & Nguyen, H. T. (2001). Isolation of a novel early drought responding partial cDNA sequence from rice by differential display of mRNA. *Journal of Plant Biochemistry and Biotechnology*, 10(1), 39–41.

Tyagi, A., Klueva, N., Zheng, H., & Nguyen, H. T. (2002). Isolation of a novel transcription factor from rice by differential display of mRNA. *Current Science*, 83, 1568–1573.

Tyaqi, A., Klueva, N., Zheng, H., & Nguyen, H. T. (2005). Isolation and characterization of a drought inducible blt101 homologous gene from rice. *Journal of Plant Biochemistry and Biotechnology*, 14(2), 115–119.

Uno, Y., Furihata, T., Abe, H., Yoshida, R., Shinozaki, K., & Yamaguchi-Shinozaki, K. (2000). Arabidopsis basic leucine zipper transcription factors involved in an abscisic acid-dependent signal transduction pathway under drought and high-salinity conditions. *Proceedings of the National Academy of Sciences*, 97(21), 11632–11637.

Verma, D., Jalmi, S. K., Bhagat, P. K., Verma, N., & Sinha, A. K. (2020). A bHLH transcription factor, MYC2, imparts salt intolerance by regulating proline biosynthesis in Arabidopsis. *The FEBS Journal*, 287(12), 2560–2576.

Vollbrecht, E., Veit, B., Sinha, N., & Hake, S. (1991). The developmental gene Knotted-1 is a member of a maize homeobox gene family. *Nature*, 350(6315), 241–243.

Wang, Q., Guan, Y., Wu, Y., Chen, H., Chen, F., & Chu, C. (2008). Overexpression of a rice OsDREB1F gene increases salt, drought, and low temperature tolerance in both Arabidopsis and rice. *Plant Molecular Biology*, 67(6), 589–602.

Wang, Y., Wan, L., Zhang, L., Zhang, Z., Zhang, H., Quan, R., Zhou, S., & Huang, R. (2012). An ethylene response factor OsWR1 responsive to drought stress transcriptionally activates wax synthesis related genes and increases wax production in rice. *Plant Molecular Biology*, 78, 275–288.

Wang, J., Li, Q., Mao, X., Li, A., & Jing, R. (2016). Wheat transcription factor TaAREB3 participates in drought and freezing tolerances in Arabidopsis. *International Journal of Biological Sciences*, 12(2), 257.

Wu, H., Lv, H., Li, L., Liu, J., Mu, S., Li, X., & Gao, J. (2015). Genome-wide analysis of the AP2/ERF transcription factors family and the expression patterns of DREB genes in Moso Bamboo (Phyllostachys edulis). *PLoS One*, 10(5), e0126657.

Wu, J., Zhou, W., Gong, X., & Cheng, B. (2016). Expression of ZmHDZ4, a maize homeodomain-leucine zipper I gene, confers tolerance to drought stress in transgenic rice. *Plant Molecular Biology Reporter*, 34(4), 845–853.

Wu, J., Yu, C., Hunag, L., Wu, M., Liu, B., Liu, Y., & Gan, Y. (2020). Overexpression of MADS-box transcription factor OsMADS25 enhances salt stress tolerance in Rice and Arabidopsis. *Plant Growth Regulation*, 90(1), 163–171.

Xiao, B., Huang, Y., Tang, N., & Xiong, L. (2007). Over-expression of a LEA gene in rice improves drought resistance under the field conditions. Theoretical and Applied Genetics, 115(1), 35–46.

Xiong, A. S., Jiang, H. H., Zhuang, J., Peng, R. H., Jin, X. Fen., Zhu, B., Wang, F., Zhang, J., & Yao, Q. H. (2013). Expression and function of a modified AP2/ERF transcription factor from Brassica napus enhances cold tolerance in transgenic Arabidopsis. *Molecular Biotechnology*, 53(2), 198–206.

Xiong, H., Li, J., Liu, P., Duan, J., Zhao, Y., Guo, X., … & Li, Z. (2014). Overexpression of OsMYB48-1, a novel MYB-related transcription factor, enhances drought and salinity tolerance in rice. *PLoS One*, 9(3), e92913.

Xu, L., Glass, C. K., & Rosenfeld, M. G. (1999). Coactivator and corepressor complexes in nuclear receptor function. *Current Opinion in Genetics & Development*, 9(2), 140–147.

Xue, G. P., & Loveridge, C. W. (2004). HvDRF1 is involved in abscisic acid-mediated gene regulation in barley and produces two forms of AP2 transcriptional activators interacting preferably with a CT-rich element. *The Plant Journal,* 37(3), 326–339.

Yamaguchi-Shinozaki, K., & Shinozaki, K. (1994). A novel cis-acting element in an Arabidopsis gene is involved in responsiveness to drought, low-temperature, or high-salt stress. *The Plant Cell*, 6(2), 251–264.

Yamaguchi-Shinozaki, K., Mino, M., Mundy, J., & Chua, N. H. (1990). Analysis of an ABA-responsive rice gene promoter in transgenic tobacco. *Plant Molecular Biology*, 15(6), 905–912.

Yu, L., Chen, X., Wang, Z., Wang, S., Wang, Y., Zhu, Q., Li, S., & Xiang, C. (2013). Arabidopsis enhanced drought tolerance1/HOMEODOMAIN GLABROUS11 confers drought tolerance in transgenic rice without yield penalty. *Plant Physiology,* 162, 1378–1391.

Zhang, J. Z., Creelman, R. A., & Zhu, J. K. (2005). From laboratory to field. Using information from Arabidopsis to engineer salt, cold, and drought tolerance in crops. *Plant Physiology*, 135, 615–621.

Zhang, J., Liu, B., Li, M., Feng, D., Jin, H., Wang, P., & Wang, H. B. (2015). The bHLH transcription factor bHLH104 interacts with IAA-LEUCINE RESISTANT3 and modulates iron homeostasis in Arabidopsis. *The Plant Cell*, 27(3), 787–805.

Zhang, P., Wei, L., Cao, L., Qiu, X., Fu, J., Wang, G., & Wang, T. (2020). Function analysis of ZmZHD9, a positive regulator in drought stress response in transgenic maize.

Zhu, Q. H., Hoque, M. S., Dennis, E. S., & Upadhyaya, N. M. (2010). Ds tagging of BRANCHED FLORETLESS 1 (BFL1) that mediates the transition from spikelet to floret meristem in rice (*Oryza sativa* L). *BMC Plant Biology*, 3, 6–18.

Zhuang, J. C. B., Peng, R. H., Zhu, B., & Jin, X. F. (2008). Genome-wide analysis of the AP2/ERF gene family in Populus trichocarpa. *Biochemical and Biophysical Research Communications*, 371(3), 468–474.

Zhuang, J., Jiang H. H., Wang, F., Peng, R. H., Yao, Q. H., & Xiong, A. S. (2013). A Rice OsAP23, functioning as an AP2/ERF transcription factor, reduces salt tolerance in transgenic Arabidopsis. *Plant Molecular Biology Reporter*, 31(6), 1336–1345.

17

Impact of Climate Change on Productivity of Field Crops

Shyamashree Roy
Regional Research Station

Rajiv Rakshit
Bihar Agricultural University

CONTENTS

17.1 Climate Change – World Scenario

Growth and development of crops are highly dependent on climatic variables like temperature, rainfall, and relative humidity. For each species, an optimum range of each variables are important. Changing climate is a major challenge in present day's agriculture not only in India but also all over the world. Earth's ecosystem has already been witnessing various changes due to the change in climate. Though gradual change in climate is a normal earth process, the fast changes in climatic variables are a matter of concern. And this concern is no more a hypothetical threat to mankind, we have enough evidences for the global temperature increase and change in the pattern of rainfall during previous 100 years (IPCC, 2001; Balling and Cerveny, 2003). The main culprit behind these changes is the increasing concentration of green house gases (GHGs) in the atmosphere. This may lead to various dangerous changes not only in atmosphere but also are going to affect our future agriculture in manyways (Aggarwal, 2003; Mall et al., 2004). In the fourth assessment of Intergovernmental Panel on Climate Change (IPCC), it was reported that in the past few years, the GHG concentration has increased manyfold affecting seriously soil, atmosphere, water bodies glaciers everything. The IPCC has found that these increases in GHGs caused in a 0.74°C warming of the climate between 1906 and 2005, based on observations and modelling studies (IPCC, 2007a). Droughts and severe precipitation events are becoming more common around the world. This results in an obvious increase in the frequency of warmer days, nights, and heat waves, whereas there is a drop in the frequency of cooler days, nights, and frost. According to IPCC, there will be huge possibilities of global temperature rise in upcoming future, about 0.88°C–3.16°C by 2050, and 1.56°C–5.44°C by 2080 for the Southern Asia (IPCC, 2007b). Rise in temperature also influences some related parameters. Winter seasons will get warmer than the rainy season. Drop in rainfall is expected in the months of December and February, while the rest of the year may show an upward trend.

One very burning example is flooding in Bihar in the last year (August, 2021) which caused losses of human and animal lives and damage of land. Africa witnessed very heavy devastating rainfall which drowned the Republic of Congo, Uganda, and Ethiopia. More than 40,000 wildfire occurred in the USA where more than 4.59 million acres of land was destroyed.

The India Meteorology Department and the Indian Institute of Tropical Meteorology (IITM) in Pune found that temperature, heat waves, glaciers, droughts and floods, and sea level rise are all trending in the same direction as the IPCC, and however the extent of the shift differs. This is reported that at the ending of this century, there is a possibility of 15%–40% increase in India's rainfall pattern and 3°C–6°C increase in average annual temperature (NATCOM, 2004). This total change will affect the lands of Northern part of the country most. This alteration in Climate also influences other parameters like soil erosion rates indirectly by changing rainfall pattern, temperature, and carbon dioxide (CO_2) level (Walthall et al., 2012).

17.2 Milestones of Extreme Weather Worldwide

The year 1998 was declared as the hottest year because of some extreme weather conditions. In this year, Central America witnessed the huge destruction of hurricane, whereas the South Asian countries like China, India, and

DOI: 10.1201/9781003258063-17

Bangladesh faced loss due to flood. Also, heavy ice storm in January knocked Canada and the United Kingdom. In another part of the world, in June 1998, Turkey, Argentina, and Paraguay inundated with flood.

However, crop losses in huge amount were recorded in Maharashtra (India) due to uneven rainfall distribution of monsoon rain. India faced unseasonal rainfall during 1997–1998. In the same year also El Nino occurred in the Pacific Ocean, the strongest one of the last century. Due to the calamity, 110 million people suffered from huge losses and global economy witnessed a financial loss of about US$ 100 billion.

In 2003, extreme heat and cold waves were witnessed by the world. The summer season brought a lot of suffering due to heat wave in European Union. As a place full of diversity, India experienced both heat and cold wave. Northern states like Uttar Pradesh (UP), Bihar, West Bengal, Orissa, and Andhra Pradesh experienced the heat wave. But on the contrary, in winter (December 2002 to January 2003) some other parts of the country India faced extreme cold wave. Sates like Jammu, Punjab, Haryana, Himachal Pradesh, Bihar, UP, and the North Eastern States went through this. Minimum 10%–100% loss of horticultural crops and seasonal crops was recorded in farmers' field. Also in fruit crop, the size and quality were lesser than the normal. But, however, beneficial effect of cool temperature, temperate fruits like apple, peach, plum, and cherry performed better with higher yield. The low-lying areas had more sufferings as the cold air settled and stay for a longer period there (Samra et al., 2004).

In the very next year, in 2004, the Indian state Himachal Pradesh experienced an yield loss of minimum 20% and maximum 60% in different crops like tea, apple, mustard, linseed, potato, wheat vegetables, pea, and rapeseed recorded in March due to high temperature. There was a hike of 2.1°C–7.9°C higher maximum temperature than the normal recorded in Himachal Pradesh. Also, the crops showed different physiological responses in the reproductive phases. For example, the harvesting of potato and wheat was done 15–20 days earlier than the usual. Apple, an important fruit of the state, showed an early blooming by 15 days. May be the reason behind that was the experience of above 26°C temperature for a spell of 17 days, whereas the optimum temperature for blossom and fruit set is 24°C (Prasad and Rana, 2006).

In Kerala, in the same year, a decline was noticed (39%) in yearly cocoa productivity due to a rise in maximum temperature by 1°C–3°C from mid-January to mid-March. Untimely rains and hailstorms destroyed 15,000 ha wheat in UP, Haryana, and Punjab in the Rabi season in 2007 in India. In Kashmir valley, in contrast, heavy precipitation of snow was recorded in 2007 due to western disturbances. In 2007, similarly, it was observed in monsoon, when heavy rains caused floods in many countries (Mexico, August 2007 – Hurricane Dean), including India and Bangladesh. Heavy rains recorded in September led to floods in southern states like Kerala, Andhra Pradesh, and Karnataka. The government declared the year 2007 as the flood year in India. Crop loss was recorded throughout the country due to this. Similar case reported from other countries like Ethiopia, Sudan, Algeria, Uganda, and Kenya. Super cyclone 'Sidr' hit and devastated Bangladesh in November 2007.

In Beijing, China, in February 2007, the highest temperature recorded was 16°C, creating a maximum hike since 1840. After that the coldest and snowiest winter of the city took place in 2008 accompanied by heavy snow prevailing for 3 weeks since 10th of January 2008. Damages to huge farm land were recorded and 22,000 houses were destroyed causing the economic loss of $7.5 billion. The phenomenon called La Nina also may be another reason for extreme snow storms.

17.3 Changes Observed in Climatic Variables That Are Related to Crop Production

- Increase in atmospheric CO_2 level.
- Soil temperature increases in tropical and subtropical belts.
- Evapotranspiration increases very little in the tropical areas, whereas high latitudes show more increases due to a rise in atmospheric temperature increase and longer growing period.
- Tropical areas show higher rainfall with variability.
- Sea level rises slowly and causes floods in river.

Soils, plants, the atmosphere, and animals interact with insects, diseases, and weeds in agricultural systems. Understanding the potential consequences of extreme events on agricultural systems might provide insight into possible adaptive strategies for dealing with extreme occurrences. Climate change has increased the frequency and severity of rainfall episodes; more intense rainfall causes more soil erosion; therefore, fresh efforts to reduce soil erosion will necessitate more aggressive conservation techniques. Drought increases terrain susceptibility to wind erosion of soil due to a lack of precipitation. In India, excessive rainfall has a negative influence on soybean. 'Rain makes grain' – the old agricultural proverb that implies that harvesting of grain is proportionately related to the amount of rainfall received during the growing season. This has long been extended to all crops, including soybeans, which is incorrect. The term 'wetfeet' has long been linked to low soybean yields (United Soybean Board, 2015). The problematic Regur soil of western India is the most detrimental cause for the soybean yields (Simonson, 1953). During the monsoon, in India, when more than 1,000 mm (about 40 inch) of rain takes place on an average of 4-month period, soils can become quickly saturated. The soil is thick and clayey; therefore, aeration is less which leads to lethal root infections and insufficient nitrogen fixation resulting in the yellowing of soybeans, according to Christmas (2000). In Iowa, during 1992 and 1993, an analysis of soybean yields was conducted using National Agricultural Statistics Service (NASS) survey data. There was a drastic variation of yield which was very high in 1992 but went very low in 1993 (year with rainfall much above normal for the whole growing season). The higher yields obtained in 1992, maybe due to a rainfall average of almost 1 inch per week for the whole growing season, compared to over 2 inches per week in 1993 when yields were substantially lower. In 2013, rainfall totalled above 1,200 mm (about 47 inches) in soybean-growing parts of India during the Indian Monsoon season.

17.4 Components of Climate Change

Climate impact assessments, likewise, must make decisions regarding which crops to evaluate. Wheat, maize, rice, and soybean are by far the most prevalent crops considered in published studies. These crops are the primary sources of calories for human and cattle worldwide, as well as in many locations and provide the majority of protein in many locations, either directly or indirectly. Many other foods, however, are essential sources of calories, but less is known about their response to climate change. Rising temperature, an intensified hydrological cycle, growing concentration of CO_2, and enhanced tropospheric O_3 are the four key elements that have affected and will continue to affect agricultural productivity in the coming decades.

To begin with, higher temperature induces faster crop development and, as a result, shorter crop duration, which is usually associated with lower yields. Crops having a C_4 photosynthetic pathway (such as maize and sugarcane) have a greater optimum temperature for photosynthesis than C_3 crops such as rice and wheat (Crafts-Brandner and Salvucci, 2002). Warming during the day can either improve or decrease net photosynthesis (photosynthesis-respiration), but warming at night increases respiration costs with no possible benefit for photosynthesis. The saturation vapour pressure of air increases exponentially as the temperature rises. Warming increases the vapour pressure deficit (VPD) between air and the leaf, which is defined as the simple difference between the saturation vapour pressure and the actual vapour pressure of the air, assuming constant relative humidity. Over wide spatial scales, relative humidity has remained roughly steady in recent decades (Willett et al., 2007) and is expected to alter minimally in the future. Plants lose more water per unit of carbon uptake when VPD rises, resulting in lower water efficiency (Ray et al., 2002). Crop water stress will grow as the frequency of agricultural drought rises. In some areas, expanding irrigation is a potential answer. Plants may be able to escape dryness by obtaining moisture below the surface in instances when groundwater is shallow or medium depth. Crop plants, however, will close their stomata and decrease carbon uptake in response to reduced soil moisture to avoid water stress. During the reproductive stage of cereal crops, water stress can be extremely detrimental (Hatfield et al., 2011). Rising CO_2 concentrations in the atmosphere give some counteracting effects to the otherwise unfavourable effects of rising temperature and decreased soil moisture. For example, increasing CO_2 has a fertilizing effect in C_3 species like wheat, rice, and most fruit and vegetable crops, since higher CO_2 reduces photorespiratory costs in the C_3 photosynthetic pathway. Increased CO_2 also reduces stomatal conductance, resulting in higher water-use efficiency in both C_3 and C_4 crops (Ainsworth and Long, 2005). In the presence of sunshine, air pollutants such as nitrogen oxides, carbon monoxide, and methane react with hydroxyl radicals to generate tropospheric O_3, which damages photosynthetic machinery in all main agricultural plants (Wilkinson et al., 2012). Over the last half-century, researchers have single-mindedly observed the detrimental effects of variation in climatic variables on the production and management of food crops (Parry et al., 2005; Lobell, 2007; Mall et al., 2006; Kalra et al., 2008). Climate models which are on use for research purpose take average monthly or yearly information for estimating temperatures and rainfall. But other aspects like everyday variability or inter-annual variability of climate, frequent heat spells, changes in humidity, solar radiation, and other extreme events have been studied in a limited number of researches (Mearns et al., 1997; Brown and Rosenberg, 1997; Rosenzweig et al., 2002).

Cai et al. (2009) studied and reported that a comparatively drier and warmer summer during the growth season of corn and wetter and warmer before and after seasons may be expected. High degree of mercury and precipitation changes may result in huge variation in soil moisture and yield performances of crop, and bigger deficit in soil moisture and frequent occurrence of crop yield reduction may take place. Water stress would be the most common at anthesis and grain formation. Yield of rainfed corn by 2055 may reduce by 23%–34%, and 32%–70% chances are there that the yield may not touch 50% of the potential yield if sufficient measures are not taken.

Chaudhari et al. (2014) reported that yield of rabi crops like wheat, potato, and rapeseed-mustard showed a sharp negative response with the increasing temperature but in the case of kharif rice, it showed the mixed response. Kumudini et al. (2014), observed that productivity of plant was influenced drastically due to the extremities that occurred during the summer season. The study by Barlow et al. (2015) revealed that in wheat (*Triticum aestivum* L.), higher temperature at the reproductive phase can cause decreased number of matured grains and required number of days for grain filling, whereas frost may be responsible for sterility and abortion of formed grains. In another study, Meehl et al. (2007) revealed that there is faster increase in daily minimum temperatures than daily maximum temperatures which ultimately increases the daily mean temperatures; because of the changes, detrimental effects could be found on grain yield. Erratic monsoon rain in India is the most important reason contributing towards lower crop yields (Isik and Devadoss, 2006). Hence, farming as in India is under extreme threat of climate change. In every 10 years, a temperature hike of 0.3°C–0.8°C was recorded. There may also be a huge jump in rise of temperature by 2°C–4°C by the end of this century with increased rainfall (6%–14%). Various problems like low rainfall, lowering of groundwater levels, and dry spell during monsoon are the direct reasons for variation in yield of food crops thus threatening national food security. In drier parts of the country, draught is the single biggest threat to the efficiency of agriculture (Hasan et al., 2016; Chen et al., 2004).

The growth and development of different plant parts depends directly on photosynthesis, which, in turn, depends on the climatic variables like light, temperature, moisture, and CO_2. The time of production of different plant parts like leaves, stems, and flowers is obviously influenced by temperature as well as day length. Not only this, the next phase of grain development and maturity also depends on these factors. The most important yield attributes grain number and grain weight at harvest, are also dependent on the availability of moisture at anthesis and post-anthesis.

In agriculture, the use of land, crop emergence, and its growth, yield and market are influenced by solar radiation,

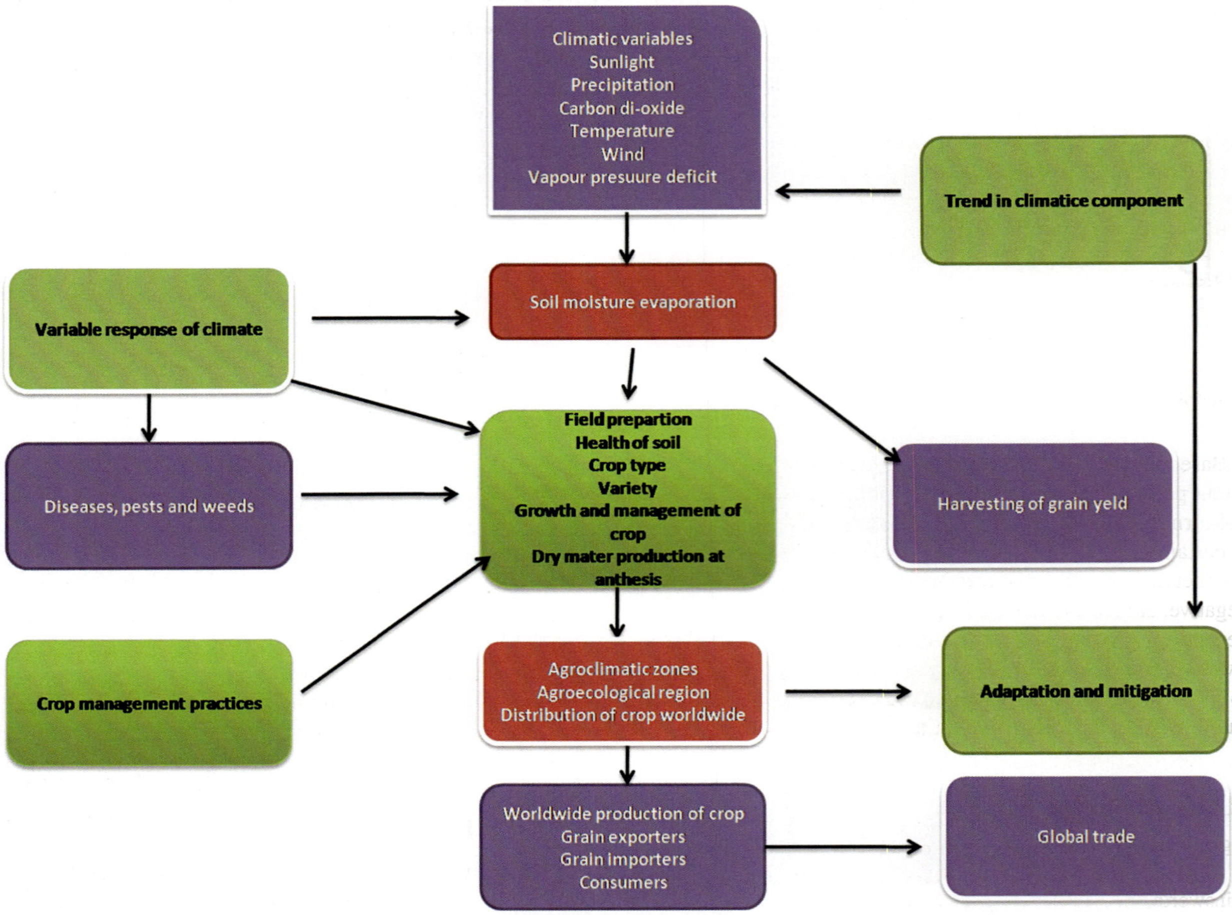

FIGURE 17.1 Impact of climate on total farming society. (Modified from Howden and White (2007), Encyclopedia of Life Support System (EOLSS).)

degrees of temperature, and availability of moisture and other climatic factors. All these interactions are explained by extensively tested computer-based simulation models. These models are very useful in modern-day scientific agriculture for assuming future yield values and also improve our understanding of the complex problems related to crop behaviour (Howden and White, 2007). One important relationship of average daily solar radiation on crop canopy and fractional dry matter of a crop (LI, the light index) described by Professor Nix is shown in Figure 17.2. It depicts when solar radiation is low, its effect on plant growth is more pronounced.

Figure 17.3 shows how low temperature influenced plant growth. With low temperature, growth is also low. As temperature has its own role in every phase from germination to maturity. Report of Mall and Singh (2000) stated over the crop-growing season, very minute changes in the temperature are experienced by the crop. But over the years, these changes appeared as the key contributors of changed weather affecting annual wheat yield.

In Indo Gangetic Plains, Pathak et al. (2003) observed that the potential yields of rice and wheat are decreasing due to increase in the value of minimum temperature. In Delhi, also the minimum and maximum temperature both are showing the upward pattern throughout the year. But solar radiation is decreasing. These changes indicate the trend of warming of the global climate. Reduction in solar radiation is detrimental to crop growth and thus will reduce the agricultural productivity. This is a matter of huge tension in upcoming days that the increasing trend of minimum temperature will increase the energy expense of plant for respiration and further it is going to cause depletion of the net growth and productivity (Aggarwal, 2003).

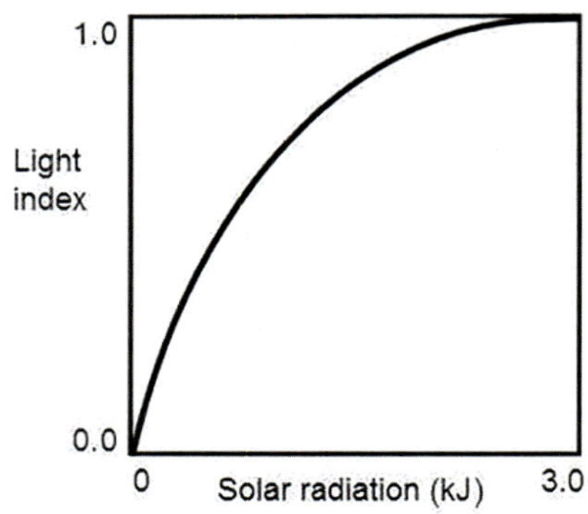

FIGURE 17.2 Light index (LI). (Adopted from Nix (1981).)

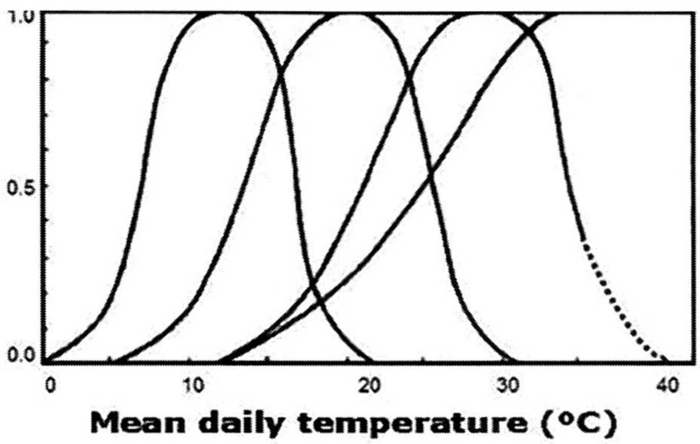

FIGURE 17.3 Thermal index (TI). (Adopted from Nix (1981).)

Based on many research by numerous scientists, from different parts of the world, a conclusion may be drawn that the out-turn of climate change on field crop production is very uncertain and unpredictable. Depending on the movement of the mercury, the overall average impact could be positive or negative. Impacts differ both quantitatively and qualitatively depending on the crop, species, its management, place of cultivation, and season. There may not be an immediate threat to food and agriculture from climate change but coming near future will suffer a lot if it is not taken care of straight away.

REFERENCES

Aggarwal, P. K. 2003. Impact of climate change on Indian agriculture. *Journal of Plant Biology* 30(2): 189–198.

Ainsworth, E. A. and Long, S. P. 2005. What have we learned from 15 years of free-air CO_2 enrichment (FACE)? A meta-analytic review of the responses of photosynthesis, canopy properties and plant production to rising CO_2. *New Phytologist* 165: 351–371.

Balling, Jr, R. C. and Cerveny, R. S. 2003. Compilation and discussion of trends in severe storms in the United States: Popular perception v. climate reality. *Natural Hazards* 29: 103–112.

Barlow, K. M., Christy, B. P., O'Leary, G. J., Riffkin, P. A., Nuttall, J. G. 2015. Simulating the impact of extreme heat and frost events on wheat crop production: A review. *Field Crops Reserch* 171: 109–119.

Brown, R. A. and Rosenberg, N. J., 1997. Sensitivity of crop yield and water use to change in a range of climatic factors and CO_2 concentrations: A simulation study applying EPIC to the central USA. *Agricultural and Forest Meteorology* 83: 171–203.

Cai, X., Wang, D. and Laurent, R. 2009. Impact of climate change on crop yield: A case study of rainfed corn in central Illinois. *Journal of Applied Meteorology and Climatology* 48: 1868–1881.

Chaudhari, K., Oza, M. and Ray, S. 2014. Impact of climate change on yields of major food crops in India. In: *Proceeding of ISPRS Archives XXXVIII-8/W3 Workshop* Ahmedabad, Gujarat, India.

Chen, C. C., Carl, B. A., Schimmelpfennig, D. E. 2004. Yield variability as influenced by climate: A statistical investigation. *Climatic Change* 66: 239–261.

Christmas, E. 2000. Impact excessive rainfall soybeans. https://www.agry.purdue.edu/ext/corn/news/articles.00/Soy_Wet_Feet0705.htm.

Crafts-Brandner, S. J. and Salvucci, M. E. 2002. Sensitivity of photosynthesis in a C_4 plant, maize, to heat stress. *Plant Physiology* 129: 1773–1780.

Hasan, M. M., Sarker, M. A. R., and Gow, J. 2016. Assessment of climate change impacts on Aman and Boro rice yields in Bangladesh. *Climate Change Economics* 7: 1650008.

Hatfield, J. L., Boote, K. J., Kimball, B. A., Ziska, L. H., Izaurralde, R. C., Ort, D., Thomson, A. M. and Wolfe, D. 2011. Climate impacts on agriculture: Implications for crop production. *Agronomy Journal* 103: 351–370.

Howden, S. M. and White, D. H. 2007. Climate and its effects on crop productivity and management. *Soils, Plant Growth and Crop Production*, Encyclopedia of Life Support Systems (EOLSS), pp. i–ix.

IPCC, 2007a. Climate Change 2007: The Physical Science Basis. Summary for Policymakers. Inter-Governmental Panel on Climate Change.

IPCC, 2007b. Climate Change 2007: Climate Change Impacts, Adaptation and Vulnerability. Summary for Policymakers. Inter Governmental Panel on Climate Change.

IPCC (Intergovernmental Panel for Climate Change). 2001, *Climate Change 2001: The Scientific Basis, Contribution of Working Group I to the Third Assessment Report of the Intergovernmental Panel on Climate Change* (Houghton, J. T., Y. Ding, D. J. Griggs, M. Noguer, P. J. van der Linden, X. Dai, K. Maskell and C. A. Johnson (eds.)), Cambridge University Press, Cambridge, UK, 881 pp.

Isik, M. and Devadoss, S. 2006. An analysis of the impact of climate change on crop yields and yield variability. *Applied Economics* 38: 835–44.

Kalra, N., Chakraborty D., Sharma, A., Rai, H. K. Jolly, M., Chander, S., Ramesh Kumar, P., Bhadraray, S., Barman, D., Mittal, R. B., Lal, M., and Sehgal, M., 2008. Effect of increasing temperature on yield of some winter crops in North-West India. *Current Science* 94: 1–10.

Kumudini, S., Andrade, F. H., Boote, K. J., Brown, G. A., Dzotsi, K. A., Edmeades, G. O., Gocken, T., Goodwin, M., Halter, A. L., Hammer, G. L., Hatfield, J. L., Jones, J. W., Kemanian, A. R., Kim, S.-H., Kiniry, J., Lizaso, J. I., Nendel, C., Nielsen, R. L., Parent, B., Stöckle, C. O., Tardieu, F., Thomison, P. R., Timlin, D. J., Vyn, T. J., Wallach, D., Yang, H. S., Tollenaar, M..2014. Predicting maize phenology: Intercomparison of functions for developmental response to temperature. *Agronomy Journal* 106: 2087–2097.

Lobell, D. B. 2007. Changes in diurnal temperature range and national cereal yields. *Agricultural and Forest Meteorology* 145: 229–238.

Mall, R. K. and Singh, K. K. 2000. Climate variability and wheat yield progress in Punjab using the CERES-wheat and WTGROWS models. *Yayu Mandal* 30(3–4): 35–41.

Mall, R. K., Lal, M., Bhatia, V. S., Rathore, L. S. and Singh, R. 2004. Mitigating climate change impact on Soybean productivity in India: A simulation study. *Agricultural and Forest Meteorology* 121(1–2): 113–125.

Mall, R. K. Singh, R., Gupta, A., Srinivasan, G., and Rathore, L. S., 2006. Impact of climate change on Indian agriculture: A Review. *Climatic Change* 78: 445–478.

Mearns, L. O., Rosenzweig, C. and Goldberg, R., 1997. Mean and variance change in climate scenarios: Methods, agricultural applications, and measures of uncertainty. *Climatic Change* 35: 367–396.

Meehl, G. A., Stocker, T. F., Collins, W. D., Gaye, A. J., Gregory, J. M., Kitoh, A., Knutti, R., Murphy, J. M., Noda, A., Raper, S. C. B., Watterson, J. G., Weaver, A. J., Zhao, Z., 2007. Global Climate Projections. In: Solomon, S., Qin, D., Manning, M., Chen, Z., Marquis, M., Averyt, K. B., Tignor, M., Miller, H. L. (Eds.), Cambridge University Press, Cambridge, U.K. and New York, NY.

NATCOM. 2004. India's Initial National Communication to the United Nations Framework Convention on Climate Change. 268 pp. Ministry of Environment and Forests.

Nix, H. A. 1981. Simplified simulation models based on specified minimum data sets: The CROPEVAL concept. In: Berg, A. (Ed.), *Application of Remote Sensing to Agricultural Production Forecasting*. AA Balkema: Rotterdam, pp. 151–169. [Improvement of cropping systems should be based on efficient agricultural information systems based on remote sensing, crop-weather modeling and crop monitoring].

Parry, M., Rosenzweig, C. and Livermore, M., 2005. Climate change, global food supply and risk of hunger. *Philosophical Transactions of the Royal Society B* 360: 2125–2138.

Pathak, H., Ladha, P. K., Peng, S., Das, S., Singh, Y., Singh, B., Kamra, S. K., Mishra, B., Sastri, A. S. R. A. S., Aggarwal, H. P., Das, D. K., and Gupta, R. K. 2003. Trends of climatic potential and on-farm yields of rice and wheat in the Indo-Gangetic Plains. *Field Crops Research* 80: 223–234.

Prasad, R and Rana, R. 2006. A study on maximum temperature during March 2004 and its impact on rabi crops in Himachal Pradesh. *Journal of Agrometeorology* 8(1): 91–99.

Ray, J. D., Gesch, R. W., Sinclair, T. R. and Hartwell, A. L. 2002. The effect of vapor pressure deficit on maize transpiration response to a drying soil. *Plant Soil* 239: 113–121.

Rosenzweig, C., Tubiello, F. N., Rand, G. et al. 2002. Increased crop damage in the US from excess precipitation under climate change. *Global Environmental Change* 12: 197–202.

Samra, J. S., Singh, G. and Ramakrishna, Y. S. 2004. Cold wave during 2002–03 over North India and its effect on crops. *The Hindu*, dated 10th January, p. 6.

Simonson, R. W. 1953. The regur soils of India and their utilization. *Soil Science Society of America Journal* 18(2): 199–203.

United Soybean Board. 2015. Thriving with wet feet. http://unitedsoybean.org/article/thriving-with-wet-feet/.

Walthall, C. L., Hatfield, J., Backlund, P., Lengnick, L., Marshall, E., Walsh, M., Ad-kins, S., Aillery, M., Ainsworth, E. A., Ammann, C., Anderson, C. J., Bartomeus, I., Baumgard, L. H., Booker, F., Bradley, B., Blumenthal, D. M., Bunce, J., Burkey, K., Dabney, S. M., Delgado, J. A., Dukes, J., Funk, A., Garrett, K., Glenn, M., Grantz, D. A., Goodrich, D., Hu, S., Izaurralde, R. C., Jones, R. A. C., Kim, S. -H., Leaky, A. D. B., Lewers, K., Mader, T. L., McClung, A., Morgan, J., Muth, D. J., Nearing, M., Oos- terhuis, D. M., Ort, D., Parmesan, C., Pettigrew, W. T., Polley, W., Rader, R., Rice, C., Rivington, M., Rosskopf, E., Salas, W. A., Sollenberger, L. E., Srygley, R., Stöckle, C., Takle, E. S., Timlin, D., White, J. W., Winfree, R., Wright-Morton, L., and Ziska, L. H., 2012. *Climate Change and Agriculture in the United States: Effects and Adaptation*. USDA Technical Bulletin 1935. Washington, DC, 186 p.

Wilkinson, S., Mills, G., Illidge, R. and Davies, W. J. 2012. How is ozone pollution reducing our food supply? *Journal of Experimental Botany* 63: 527–536.

Willett, K. M., Gillett, N. P., Jones, P. D. and Thorne, P. W. 2007. Attribution of observed surface humidity changes to human influence. *Nature* 449: 710–712.

18

Impact of Climate Change on Growth and Productivity of Major Field Crops

Sagar Maitra and Triptesh Mondal
Centurion University of Technology and Management

Akbar Hossain
Bangladesh Wheat and Maize Research Institute

Rajesh Shriram Kalasare and Subhashisa Praharaj
Centurion University of Technology and Management

Jince Mary Mundonayil Joy
Kerala Agricultural University

Tanmoy Shankar
Centurion University of Technology and Management

Biswajit Pramanick
Dr. Rajendra Prasad Central Agricultural University

Martin Leonardo Battaglia
The Nature Conservancy

Urjashi Bhattacharya
Bidhan Chandra Krishi Viswavidyalaya

Marko Zupanič
University of Maribor

Masina Sairam
Centurion University of Technology and Management

CONTENTS

DOI: 10.1201/9781003258063-18

18.1 Introduction

Climate change is a change in the statistical distribution of weather over a long time. Climate change is expected to affect all aspects of agriculture. Though the impacts of climate change are predicted to vary from place to place and crop to crop, the overall impact is considered negative. Frequent weather anomalies, rise in global temperature, rise in sea level etc. may lead to irrecoverable damage to the agroecosystem altogether. Climate change results in soil health deterioration, poor crop performance, insect-pest insurgence and poor crop environment leading to lower system productivity (Figure 18.1) (Wang et al. 2018; Hossain et al. 2021). The cost of agricultural production is also expected to rise, as the ecosystem services are affected (Scholes 2016). To overcome the threat of climatic aberrations, adaptation and mitigation options are carefully chosen for a specific agroecosystem based on feasibility, compatibility and cost-effectiveness.

There is a need to realize the complexity of climate change to predict its influence on a particular crop or group of crops. Rise in global temperature causes glaciers to melt and increase the sea level resulting in subsequent submergence of landmasses on coasts (Shukla et al. 2017). This in turn will reduce

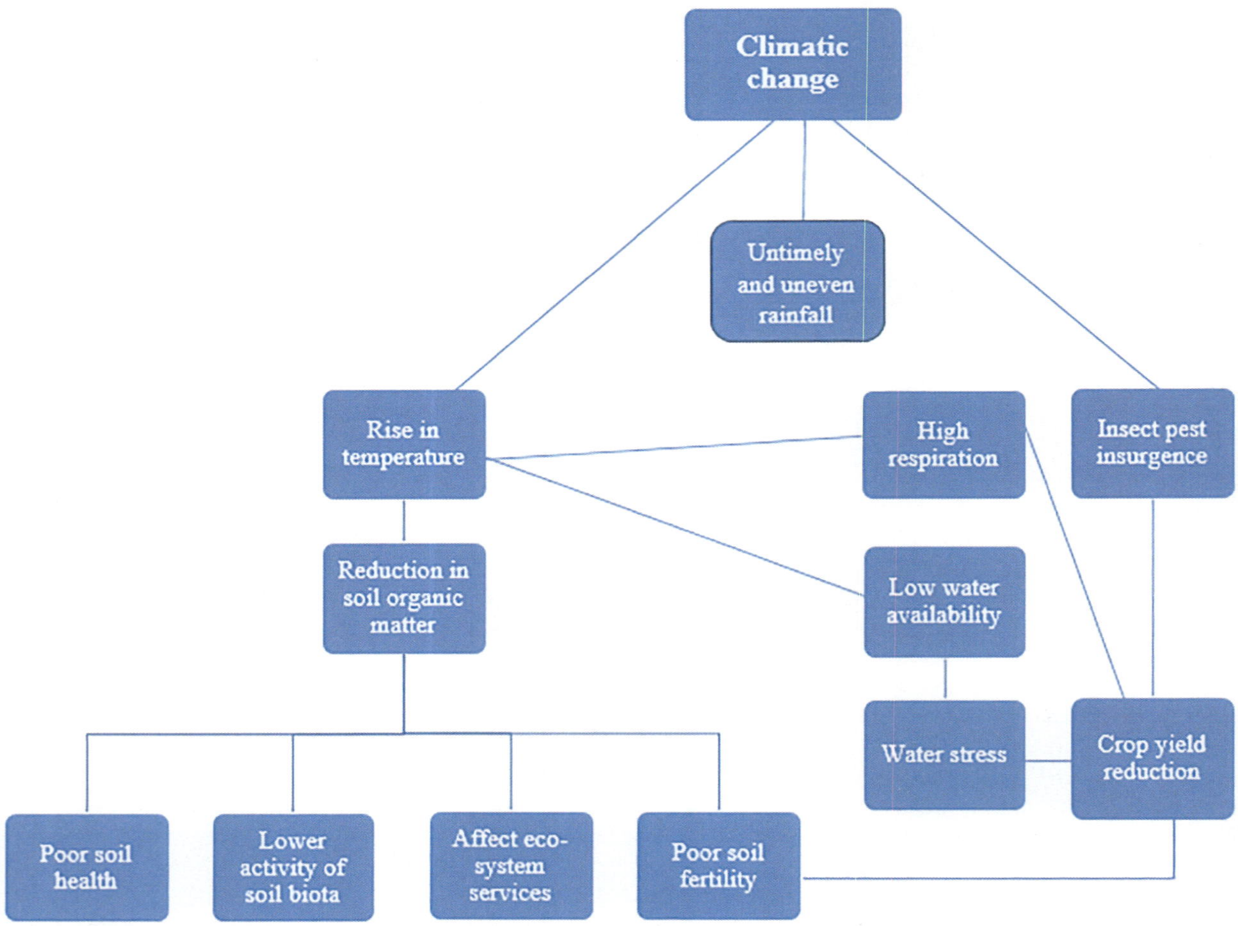

FIGURE 18.1 Effect of climate change on crop productivity.

the productive lands on earth. Similarly, rise in temperature may also lead to higher respiration as compared to photosynthesis resulting in lesser yield. However, understanding the resultant effect is not as simple as the previous two examples, considering the fact that multiple weather parameters and other driving factors give rise to the final effect. For example, rise in carbon dioxide is expected to increase the photosynthetic activities in C_3 crops (Ruan et al. 2012); however, rise in temperature (due to rise in carbon dioxide level) may reverse some of this beneficial effect. Hence, climate change models predicting yield need to consider these highly interactive and extremely complex interrelationships for accurate predictions.

Cereals are one of the most commonly grown crops across the globe. Cereal productivity has a large bearing on the food security of the globe as it is consumed by majority of the people. However, the cereal yield is expected to be influenced by climate alteration. By 2050, around 70%–100% enhancement in cereals supply is required to fulfil the food requirement of the assumed world population of 9.8 billion (Wang et al. 2018). However, meeting this target will be an uphill task especially under the changing climatic scenario. In the present context of climatic aberration, the cereal productivity can be altered by heat or water stress; however, indirectly the abiotic stresses may cause poor nutrient availability, insect pest insurgence etc. (Porter et al. 2014; Bhadra et al. 2021).

18.2 Climate Change Impacts on Growth and Productivity of Major Cereals

As a result of climatic aberration, water deficiency and heat stresses will be too frequent. The flower formation and inflorescence are badly hampered when cereals are subjected to water stress (Winkel et al. 1997). Temperature around 30°C during floret formation can also induce sterility in cereals (Saini and Aspinall 1982). Following are the impacts of climatic aberrations on cereals.

18.2.1 Rice

Climate change has various impacts on rice production systems. The food security of large parts of the world depends on the rice production, as rice is a staple food for many. Negative impacts on the grain yield of rice will directly create major shortages in world's food basket. International Food Policy Research Institute (IFPRI) predicted 15% decrement in irrigated rice yields in developing countries with a 12% increment in the price of rice as a result of climate change by 2050 (Saseendran et al. 2000). In India, rice is cultivated with varying climatic situations. Rice needs warm and moist microclimate as well as bright sunshine for a long period of time. The mean temperature during its growing period differs from 21°C to 37°C (Singh et al. 2010). Rice is mainly cultivated as a rainfed crop in lowland of upland situations in our country, many of which falls under water crisis and heat stress (Coffman 1977). Sudden changes in the environment have both direct and indirect impacts on the yield obtained from paddy plants. This change in climate occurs due to abrupt reduction in forest area and rangelands along with heavy use of fossil fuels. The most commonly available greenhouse gas, i.e., CO_2 concentration has now reached 400 from 280 ppm in the atmosphere (Vaughan et al. 2018). Moreover, submerged and well-managed rice systems release large amounts of methane (CH_4) gas, even though they do carbon sequestration by adding soil organic matter. During the growth of worldwide industrialization during the 1970s, surface temperature also increased at the rate of 0.16°C to 0.18°C/decade (IPCC 2007). A reduction in rice spikelet fertility is observed, when high temperatures prevail at >35°C for a period of about a week during anthesis (Satake and Yoshida 1978). Rice was most susceptible to high temperatures during booting and flowering stages (Farrell et al. 2006). Early dehiscence of anthers occurs and the development of the androecium is also highly affected due to increment in atmospheric temperature at anthesis leading to poor grain setting. Higher respiration losses have a great link to higher temperatures and also make rice less productive (Matthews et al. 1995). A research conducted at IRRI showed that an enhancement in night temperature by 1°C may decrease the grain yield of rice by about 10% (Nakagawa et al. 2003).

18.2.1.1 Adaptation Options

- In this condition, the development of heat-tolerant rice genotypes is of utmost importance to maintain its yield level. The research work for developing heat stress-tolerant high-yielding varieties through cross-breeding technique has already been started in many countries.
- Using advanced genetic tools and markers, heat-tolerant genes can be identified and included in the breeding programme (Elshire et al. 2011).
- In addition to heat tolerance, other stress tolerance traits may be included depending on the major stress the crop is subjected to under particular agroclimatic conditions.
- However, increasing efficient practices such as direct seeding and balanced fertilization could not only reduce methane emission but also improve rice yield (Gartaula et al. 2020).
- Switching to water-efficient cultivation practices such as DSR, aerobic rice and SRI may be suitable as compared to conventionally grown flooded rice.
- Suitable plant growth regulators (PGRs) and hormones may be applied to ameliorate the negative impacts of stress to a certain extent (Maitra et al. 2021).

18.2.2 Wheat

Wheat production in India has started showing a declining trend in proportion to the increasing population. One of the main reasons behind this situation is adverse climatic condition. Presently, many semi-dwarf, heavy-yielding cultivars are used by the farmers but they are still unable to achieve the target production (Pathak et al. 2003). In India, the target for wheat has been fixed to 140 million tonnes of production and 4.75 t/ha grain yield, respectively, by 2050 (ICAR-IIWBR 2015). The problems of small winter, high rainfall before harvest, high wind velocity during the growing period and rainfed condition in different parts of India have been observed. Grain yield of wheat is expected to decrease by 6%–23% by the year 2050 due to climate change (Kumar et al. 2014). Heat stress lowers down the relative water content (RWC) in wheat plants which causes high osmotic potential (Sattar et al. 2020). This crop often comes under terminal heat stress, which puts a severe impact on grain production. Generally, delayed sowing of wheat results in terminal heat stress during the anthesis and grain development (Dwivedi et al. 2017). Lodging is another menace to wheat production. It is induced due to incapability of the plants to stand. Wind gusts and excessive rain at harvest cause root or stem lodging of wheat. Stem lodging and root lodging are the two important factors for decline yield in wheat (Mondal 2020). Lodging decreases yield of wheat more or less 80% (Piñera-Chavez et al. 2016). The reduction of 7%–35% in yield was observed previously by Fischer and Stapper (1987) due to lodging problem at the milking stage.

18.2.2.1 Adaptation Options

- Date of sowing should be in the range of optimum sowing time because atmospheric temperature and soil temperature, both are favourable for better seed germination during this period which is required to maintain optimum stand count initially leading to higher grain yield (Pandey et al. 2010).
- Terminal heat stress has to be reduced by spraying some synthetic compounds such as calcium chloride ($CaCl_2$) (Tan et al. 2011) which enhance the effectiveness of photosynthesis and nitrogen assimilation (Naeem et al. 2018).
- Breeding for stress-tolerant varieties for wheat should be considered with top priority.
- Foliar application of PGRs such as chlormequat chloride, ethephon, and trinexapac-ethyl at certain doses and certain stages reduces the extent of lodging to the minimum. PGRs reduce the shoot length of wheat plants which is mainly achieved by reducing cell elongation and the rate of cell division. These chemicals control wheat lodging by releasing ethylene and by inhibiting gibberellic acid biosynthesis (Mondal 2020).

18.2.3 Maize

Heavy rainfall resulting in waterlogging highly affects maize growth and productivity. Delayed sowing may expose maize plants to excessive rainfall and hence care must be taken to complete the sowing operation within an optimum window to avoid such an unfavourable situation. Rosenzweig et al. (2002) also examined the effects of changing climate on the grain yield of maize with the help of the CERES model. They found that the chances of crop damages due to higher precipitation might be 90% higher in the year 2030 compared to the year 2002. Swaziland has faced some severe droughts and flood situations in the last two decades while withdrawal of significant rainfall at critical crop growth stages has frequently led to a serious decrement in the production of maize (Oseni and Masarirambi 2011). In some areas, increment in productivity may enhance the intensification of maize and associated improvement in rural prosperity. However, decrease in yield is synonymous to disturbance to rural livelihood in developing countries (Jones and Thornton 2003).

18.2.3.1 Adaptation Options

- Waterlogging conditions should not be allowed. Proper drainage facility has to be developed in maize fields.
- Early sowing (preferably 1 month early) should be beneficial in those places where rainfall occurs from December to January.
- Stress-tolerant varieties must be developed.

18.3 Effect of Climate Change on Growth and Yield of Major Fibre Crops

18.3.1 Jute

In the eastern Indo-Gangetic plains of India, the rice-wheat-jute system is commonly adopted because of the availability of suitable cultivars and necessary inputs (Singh et al. 2018). Rainfall and temperature are the two most dominant climatic factors responsible for the growth and development of jute plants. Jute is mainly cultivated as a rainfed crop and requires about 500-mm water during the growing period. Over the last four decades, 40%–50% lower precipitation occurred from the 12th to 15th meteorological week (Singh 2017). The unequal distribution of rainfall causes early season drought in jute, which roots sufficient nutrient uptake by roots, thus reducing jute production (Geethalakshmi et al. 2009). Around 4.9 tonnes of carbon (C) sequestration occurs from 1 tonne of raw jute fibre which is more than many species of forest trees as per the report of Rajagopal and Sanyal (2012). Life cycle assessment (LCA) study on jute and its products by the National Jute Board (NJB) of India revealed that the green jute plants do carbon sequestration most significantly during their growth stage and on an average, 1.8–2.0 Mg/ha of the residue of jute is incorporated yearly to the soils of jute fields. Soil organic C development rate was also satisfactory with a rate of 0.11–0.25 Mg C/ha year when jute is grown in after rice and wheat (Mandal et al. 2007) in eastern India.

18.3.1.1 Adaptation Options

- Integrated crop management in jute includes early anthesis, use of resistant varieties, sowing in lines, physical weed management, fertilizer application as per soil test data, control of new disease and insect infestation and following intercropping instead of monocropping (Singh et al. 2019).

- Climate smart jute farming (CSJF) should be done which is an integrated crop management model for the management of risk of crop failure in farming systems that helps in adaptation, develops resilience to climatic changes and improves the quality of jute fibre production.

18.3.2 Cotton

Cotton is somewhat resilient to high-temperature stress and drought conditions because of its deep root system. This crop shows sensitivity to moisture stress, particularly at flowering and boll formation stages. Although the growth rate of this crop is sensitive to high carbon dioxide levels, the major developmental stages such as the square formation and the boll maturity periods are not reduced by CO_2 concentration (Reddy et al. 1997). Increasing temperature favours plant development unless the day temperature exceeds 32°C (Ton 2011). In the rainfed fields of central India, where atmospheric temperature increment is comparatively low, cotton crop faces low temperatures in the winter season at the end of the reproductive phase (Hebbar et al. 2020). If global warming occurs, fibre yield is expected to decline. Temperature (both high and low) tolerant cultivars will be necessary to sustain cotton production (Reddy et al. 2002). Cotton cultivation also contributes to climate change. It has been reported that cotton cultivation contributes from 0.3% to 1% of world's greenhouse gases (GHGs) emission (Ton 2011).

18.3.2.1 Mitigation Options

In cotton production, some mitigation measures of GHG emission have been described as follows (Ton 2011).

- Reduced tillage of soils in cotton fields to prevent GHG emissions from soil to air;
- Reduced carbon-based fuel machine use and transportation;
- Reduced chemical fertilizer application and especially nitrogenous fertilizers in particular which are responsible for the emission of N_2O;
- Reduced irrigation water application due to its carbon-based fuel footprint;
- Reduced residue burning of cotton.

18.4 Effect of Climate Change on Growth and Yield of Major Pulse Crops

Among pulses, chickpea, pigeon pea and lentil are the majorly cultivated pulses in India and share 39%, 21% and 10% of the total pulse production, respectively (Anonymous 2009). Pulses are the important crops grown under rainfed farming. Climate change events are expected to influence pulses (Basu et al. 2009) and they face various abiotic stresses like drought, shift in monsoon, erratic rainfall, aberrant temperatures, increased solar radiation, soil salinity and inundation of lands. The sensitivity of the crop to various stresses fluctuates greatly, depending on the growth periods in which the crops are exposed to the stress. Delayed sowing due to shifts in monsoon and erratic rainfall makes the crop susceptible to terminal heat and drought stress during its growth period which shortens the vegetative phase and causes forced maturity (Table 18.1).

18.4.1 High-Temperature Stress

Majority of pulse varieties are well adapted to an optimum temperature range falling between 15°C and 35°C but have the ability to tolerate up to a maximum limit of 45°C, though with significant reduction in photosynthesis, chlorophyll content, decreased foliage growth and enhanced pollen sterility with a progressive increase in the temperature beyond 35°C. The heat stress above 35°C often causes irreversible damage to the growth and development of pulses (Wahid et al. 2007). Biological nitrogen fixation in legumes is temperature sensitive (Kurdali 1996). Nodule growth is adversely affected during podding stages if temperature is high, which results in poor or lack of nitrogen accumulation in the leaves and decreasing protein content, and tends to enhance starch content in the seeds. Temperature rises up to 42°C–45°C even for a shorter period may induce forced maturity, leaf senescence, pollen sterility and causes failure of anthesis. Grain filling is adversely affected in young pods that recently set seeds exposed to high temperature above 42°C and as a result of further seed development abruptly terminates and such impaired seeds never restore to their normal conditions even temperature becomes favourable afterwards. Temperature above 45°C is considered to be lethal for a majority of the pulses, as it causes irreversible

TABLE 18.1
Impact of Terminal Drought Stress on some rabi Pulses (Maji et al. 2019)

Name of Crop	Growing Period	Impact	Reduction in Productivity (%)	References
Gram	Flowering and podding	Pod abortion and low pod filling	30	Behboudian et al. (2001)
Gram	Flowering onwards	Reduction in assimilate production, reduced crop duration and poor seed growth	52–72	Davies et al. (1999) and Maji et al. (2019)
Gram	Early podding	Reduced podding period inclusive of enhanced seed abortion	65–93	Shrestha et al. (2006)
Lentil	pod development	Reduction in leaf area and number of flowers, pod and seed with greater seed abortion	70	Shrestha et al. (2006)
Lathyrus	Flowering	Abortion of flowers and pod	87	Gusmao et al. (2012)
Lathyrus	Podding	Reduction in photosynthesis and filled pods with increased seed abortion	66	Kong et al. (2015)

physiological changes at cellular level, macromolecules aggregation, massive destruction of photosynthesis process leading to complete inhibition and no fertilization takes place at this extreme temperature.

- In moong, atmospheric temperature higher than 42°C during *pre-kharif* shows seed hardening due to improper sink development;
- Dehiscence of anthers occurs at high temperature in cowpea;
- Flower initiation, flowering, pollen formation, fertilization and pod setting/development are adversely affected due to heat stress in common bean and pea;
- High night temperatures affect pod filling in common bean, lima bean and cowpea.

18.4.2 Low-Temperature Stress

Base temperature below <5°C for a longer period often becomes lethal for pulses. The flowering process and grain development process, both are very much sensitive to low temperatures.

- Particularly, red gram is very sensitive to temperature aberrations resulting heavy flower drop, stipulated drying and curving of apical leaves under low-temperature stress (<5°C).
- Lentil and gram show cessation of crop growth at lower than 7°C temperature (Basu et al. 2009).

18.4.3 Rainfall

High post-monsoon rainfall during the month of January and February acts as a detrimental factor for *rabi* pulses, particularly, during the reproductive period. Cloudy or rainy weather at flowering and fruiting adversely affected the pod filling and setting of seeds (Sardana et al. 2010). Rainfall during the harvesting stage results in decline in the yield.

18.4.3.1 Adaptation Options

Pulses have very high adaptive capacity due to their very deep root system (e.g. pigeon pea and chickpea), high degree of drought tolerance, phenotypic flexibility and wide range of sensitivity towards variation in photo-thermo periods (Table 18.2).

As nodulation and nitrogen fixation almost terminate at high temperature, exogenous application of biofertilizers could prove beneficial to improve yield under heat stress. Biofertilizers like arbuscular mycorrhizae fungi increase water use efficiency (11%–24%) in field pea under the rainfed condition as previously observed by Kumar et al. (2016).

TABLE 18.2

Photo and Thermo Insensitive Genotypes of Legumes

Pulse Crop	Varieties
Gram	ICCV 92944, ICCV 96030, ICCV 96029
Urd bean	PGRU 95016, BGP 247, IPU 99-89
Red gram	ICPL 90011

Bacterial strain that belongs to the genera of *Azotobactor, Bacillus, Pseudomonas and Azospirillum* produce an enzyme called ACC deaminase which creates longer roots for helping in the uptake of higher amount of water from deeper layers of soil under drought situations, thus improving water use efficiency of the plants (Grover et al. 2011). Pulse varieties tolerant to different stress conditions should be developed and grown in such areas. Some examples are:

- Karnal Chana 1 of chickpea is salinity tolerant.
- Narendra Mung 1 of greengram and Pant G186 of chickpea are tolerant to heat stress.
- Bennur Local, Bahar (pigeon pea) PKG-U03, GP-15, IPU 90-32 IPU 06-12 (black gram), Ganga-8, HUM 1, and ML5 (green gram) varieties have high osmotic adjustment and well adapted to drought conditions.

Diversification of farming through crop diversification through mixed cropping or intercropping or adopting alternative land use systems like agro-forestry, alley cropping or ley farming could reduce the risks associated with farming along with better resource utilization.

Raised bed planting could be adopted in areas with high rainfall. Conservation tillage practices could be followed to conserve soil moisture and increase the availability of soil nutrients to pulses. Stubble mulching can enhance water infiltration and reduce runoff losses. Crop management practices like ridge planting of *kharif* pulses, fertilizers incorporation in furrows behind the seeds and water saving irrigation methods under drought, optimum use of organic manures and chemical fertilizers etc. can help obtain better yield under stressed conditions. Oilseed crops suffer huge yield losses due to abiotic (drought, extreme temperatures, salinity, waterlogging etc.) and biotic stresses (pests and diseases). Drought stress reduces the seed quality, seedling vigour index and germination percentage (Alqudah et al. 2011).

18.5 Effect of Climate Change on Growth and Yield of Major Oilseed Crops

18.5.1 Groundnut

Germination rate and emergence increase with the rise in temperature and soil moisture; however, with the temperatures above optimum, these processes slow down and almost stop at the maximum threshold values. The optimum air temperature required during vegetative growth of peanut is ranged from 25°C to 30°C, while the same for reproductive growth is between 22°C and 24°C (Ong 1984). Long dry spells at the time of peg formation or penetration and grain filling affect the quantity and quality of groundnut produced (Oilseeds Division-MOA 2014). Exposure to the above optimal air and soil temperatures for both short and long duration can show significant yield reduction in peanut (Dreyer et al. 1981). Mohamed et al. (2002) reported a decline in groundnut production in the Sahel region of West Africa in 2025. Reduction in pod moisture content results in lower physiological activity of seeds resulting increased susceptibility to aflatoxin contamination. Drought

TABLE 18.3

Effect of Cardinal Temperature Ranges for some Major Field Crops

Name of the Crop	Base Temperature (°C)	Optimum Temperature (°C)	Highest Tolerable Temperature (°C)
Rice	10.0–12.0	30.0–32.0	36.0–38.0
Wheat	3.0–4.5	23.5–25.0	30.0–32.0
Maize	8.0–10.0	32.0–35.0	42.0–44.0
Jute	10.0–12.0	24.0–37.0	40.0–42.0
Cotton	15.5–16.0	30.0–32.5	42.0–43.5
Major pulses	5.0–10.0	20.0–30.0	40.0–45.0
Groundnut	15.5–16.0	25.0	35.0–40.0
Soybean	4.5–5.0	30.0	40.0
Rapeseed and mustard	0.5-3.0	15.0–25.0	35.0–40.0

could also alter the nutritional quality of groundnut proteins (Kambiranda et al. 2011; Birthal et al. 2012).

18.5.2 Soybean

Climate variation accounts to 26%–34% of the yield variability in soybean (Ray et al. 2015). The reproductive phase of soybean is sensitive to high air temperature (Reddy and Kakani 2007). Variation in rainfall along with high air temperature in the reproductive phase of soybean may decline seed productivity (Thanacharoenchanaphas and Rugchati 2011). Drought stress during seed development reduces the total protein percentage in seeds (Ghanbari et al. 2013). Moisture shortage during the grain formation stage also alters the fatty acids composition influencing oil content and composition.

18.5.3 Rapeseed and Mustard

Climate change could adversely affect the crop growth and yield. Seawater flooding in rapeseed reduces reproductive output (most notably siliqua number) (Hanley et al. 2020). Morrison and Doug (2002) showed that high temperature during flag leaf to anthesis decreases the yield of all the *Brassica spp.* An increase of 3.0°C in atmospheric temperature above normally increased anthesis in groundnut, whereas reduced anthesis in mustard. Warm temperature causes the growth and development faster leading to lesser time for carbon assimilation and biomass accumulation before seed setting ultimately causes reduced grain number and poor yield. Less leaf area together with low radiation use efficiency reduces the net photosynthesis and total dry matter accumulation in mustard. High temperature during crop establishment, cold, fog and intermittent rainfall during the growth period creates adverse effects on the crop growth and show significant yield reductions by physiological abnormalities along with attack of different pests and diseases such as aphids, white fly, stem rot and downy mildew diseases.

18.5.3.1 Adaptation Options

- Breeding of climate-resilient crops helps to adapt to the adverse effects of climatic aberrations.
- Late sowing of mustard results in lower temperature during anthesis compared to normal sown crop, reducing yield losses. Growing long duration variety instead of the existing variety reduces yield losses as maturity occurs late in long duration variety.
- Tolerance to heat stress in crops could be improved through spraying of plant growth retardants, osmoprotectants (OSP) and organic and inorganic fertilizers which are efficient, economically and environmentally viable (Ahanger et al. 2017; Yakhin et al. 2017).
- External application of proline can increase the oil and protein percentages in drought-stressed oilseeds (El Sabagh et al. 2019).

Atmospheric temperature is one of the most important among all the climatic factors which affects the crops mostly if being shifted from optimum. The cardinal temperature ranges for some major field crops have mentioned in Table 18.3.

18.6 Conclusion

Climate change is one of the major constraints in achieving food security. Rise in temperature, uneven and untimely rainfall, salinization, insurgence of insect pests and decline in soil organic matter are some of the negative outcomes of climate change that will result in lower yield and poor nutritional quality of food. In addition to cereals and pulses, oilseed and fibre production is also expected to be adversely influenced by the ill effect of climate change. Unless careful adaptation and mitigation options are planned and timely implemented, it will be nearly impossible to meet the future demand for food, feed and fibre. Understanding climate change effects at phenological, physiological and biochemical helps in developing suitable adaptation options. While recommending adaptation and mitigation strategies, the cost-effectiveness, compatibility and feasibility of the options must be considered. Such plans should be recommended based on crop, agroclimatic condition in which the crop is grown and economic status of the farmers for their better adoption.

REFERENCES

Ahanger, M.A., Akram, N.A., Ashraf, M., Alyemeni, M.N., Wijaya, L. and P. Ahmad. 2017. Plant responses to environmental stresses: From gene to biotechnology. *AoB Plants* 9:plxo25. doi.org/10.1093/aobpla/plx025.

Alqudah, A.M., Samarah, N.H. and R.E. Mullen. 2011. Drought stress effect on crop pollination, seed set, yield and quality. In E. Lichtfouse (ed.), *Alternative Farming Systems, Biotechnology, Drought Stress and Ecological Fertilisation*. Springer, Dordrecht, pp. 193–213.

Anonymous. 2009. Annual Report of National Food Security Mission Pulse Component. Department of Agriculture and Cooperation, Ministry of Agriculture, Govt. of India, New Delhi.

Basu, P.S., Masood, A. and S.K. Chaturvedi. 2009. Terminal heat stress adversely affects chickpea productivity in northern India: Strategies to improve thermo tolerance in the crop under climate change. In *W3 Workshop Proceedings: Impact of Climate Change on Agriculture*, New Delhi: International Society for Photogrammetry and Remote Sensing. 23:189–193.

Behboudian, M.H., Ma, Q., Turner, N.C. and JA Palta. 2001. Reactions of chickpea to water stress: Yield and seed composition. *Journal of Science, Food and Agriculture* 81:1288–1291.

Bhadra, P., Maitra, S., Shankar, T., Hossain, A., Praharaj, S. and T. Aftab. 2021. Climate change impact on plants: Plant responses and adaptations, Chapter 1, In T. Aftab and A. Roychoudhury (Eds), *Plant Perspectives to Global Climate Changes*, Elsevier Inc., Academic Press: Cambridge, MA, pp. 1–24.

Birthal, P.S., Nigam, S. N., Narayanan, A.V. and K.A. Kareem. 2012. Potential economic benefits from adoption of improved drought-tolerant groundnut in India. *Agricultural Economics Research Review* 25:1–14.

Coffman, W.R. 1977. Rice varietal development for cropping systems at IRRI. In *Symposium on Cropping Systems Research and Development for the Asian Rice Farmer*, IRRI, Philippines pp. 359–371.

Davies, S.L., Turner, N.C., Siddique, K.H., Leport, L. and J.A. 1999. Seed growth of desi and kabuli chickpea (*Cicer arietinum* L.) in a short-season Mediterranean-type environment. *Australian Journal of Experimental Agriculture* 39:181–188.

Dreyer, J., Duncan, W.G. and D.E. McCloud. 1981. Fruit temperature, growth rates, and yield of peanuts. *Crop Science* 21:686–688.

Dwivedi, S.K, Basu, S., Kumar, S., Kumar, G., Prakash, V., Kumar, S., Kumar, G., Prakash, V., Kumar, S., Mishra, J.S., Bhatt, B.P., Malviya, N., Singh, G.P. and A. Arora. 2017. Heat stress induced impairment of starch mobilisation regulates pollen viability and grain yield in wheat: Study in Eastern Indo-Gangetic Plains. *Field Crops Research* 206:106–114.

El Sabagh, A., Sorour, S., Morsi, A., Islam, M.S., Ueda, A., Barutçular, C., Arioglu, H. and H. Saneoka. 2019. Effects of drought stress on the quality of major oilseed crops: Implications and possible mitigation strategies: A review. *Applied Ecology and Environmental Research* 17:4019–4043.

Elshire, R.J., Glaubitz, J.C., Sun, Q., Poland, J.A., Kawamoto, K., Buckler, E.S. and S.E. Mitchell. 2011. A robust, simple genotyping-by-sequencing (GBS) approach for high diversity species. *PLoS One* 6:e19379. doi: 10.1371/journal.pone.0019379.

Farrell, T.C., Fox, K.M., Williams, R.L. and S. Fukai. 2006. Genotypic variation for cold tolerance during reproductive development in rice: Screening with cold air and cold water. *Field Crops Research* 98:178–194.

Fischer, R.A., and M. Stapper. 1987. Lodging effects on high-yielding crops of irrigated semidwarf wheat. *Field Crops Research* 17:245–258.

Gartaula, H., Tek, B.S., Arun, K-C., Prasad, G. and L. Badstue. 2020. Gendered impacts of greenhouse gas mitigation options for rice cultivation in India. *Climatic Change* 163:1045–1063. doi: 10.1007/s10584-020-02941-w.

Geethalakshmi, V., Palanismy, K., Aggarwal, P.K. and A. Lakshmanan. 2009. Impact of climate change on rice, maize and sorghum productivity in Tamil Nadu. In P.K. Aggarwal (ed.), *Global Climate Change and Indian Agriculture: Case Studies from the ICAR Network Project*. Indian Council of Agriculture Research (ICAR), New Delhi, pp. 13–18.

Ghanbari, A.A., Mousavi, S.H., Mousapour Gorgi, A. and I.M. Rao. 2013. Effects of water stress on leaves and seeds of bean (*Phaseolus vulgaris* L.). *Turkish Journal of Field Crops* 18:73–77.

Grover, M., Ali, S.Z., Sandhya, V., Rasul, A. and B. Venkateswarlu. 2011. Role of microorganisms in adaptation of agriculture crops to abiotic stresses. *World Journal of Microbiology and Biotechnology* 27:1231–1240.

Gusmao, M., Siddique, K.H., Flower, K., Nesbitt, H. and E.J. Veneklaas. 2012. Water deficit during the reproductive period of grass pea (*Lathyrus sativus* L.) reduced grain yield but maintained seed size. *Journal of Agronomy and Crop Science* 198:430–441.

Hanley, M.E., Francesca C.H., Louise H., and M. Franco. 2020. Simulated seawater flooding reduces oilseed rape growth, yield and progeny performance. *Annals of Botany* 125:247–254.

Hebbar, K.B., M.V. Venugopalan, M.R.K. Rao, G.D. Gadade, S. Chatterji, and C.D. Mayee. 2020. Effect of sowing dates and fertilizer levels on phenology, growth and yield of cotton. *Indian Journal of Plant Physiology* 7:380–383.

Hossain, A., Ali, M.A., Maitra, S., Bhadra, P., Md. Muzahid E. Rahman, M.M.E., Sahin Ali, S., and T. Aftab. 2021. The role of soil microorganisms in plant adaptation to abiotic stresses: Current scenario and future perspectives, Chapter 13, In T. Aftab and A. Roychoudhury (Eds), *Plant Perspectives to Global Climate Changes*. Elsevier Inc., Academic Press, Cambridge, MA, pp. 233–278.

ICAR-IIWBR. 2015. Vision 2050. Indian Council of Agricultural Research-Indian Institute of Wheat and Barley Research, pp. 1–48.

IPCC. 2007. Climate Change-Impacts, Adaptation and Vulnerability. Contribution of Working Group II contribution to the Fourth Assessment Report of the Intergovernmental Panel on Climate Change. Parry, M.L., Canziani, O.F., Palutikof, J.P., Van der Linden, P.J. and C.E. Hanson (Eds.). Cambridge University Press, Cambridge, MA.

Jones, P.G. and P.K. Thornton. 2003. The potential impacts of climate change on maize production in Africa and Latin America in 2055. *Global Environmental Change* 13:51–59.

Kambiranda, D.M., Hemanth, K.N/, Vasanthaiah, R.K., Ananga, A., Basha, S.M., and K. Naik. 2011. Impact of drought stress on peanut (*Arachis hypogaea* L.) productivity and food safety. *Plants and Environment* 1:249–272.

Kong, H., Palta, J.A., Siddique, K.H., Stefanova, K., Xiong, Y.C. and N.C. Turner. 2015. Photosynthesis is reduced, and seeds fail to set and fill at similar soil water contents in grass pea (*Lathyrus sativus* L.) subjected to terminal drought. *Journal of Agronomy and Crop Science* 201:241–252.

Kumar, S.N., Aggarwal, P.K., Rani, D.S., Saxena, R., Chauhan, N. and S. Jain 2014. Vulnerability of wheat production to climate change in India. *Climate Research* 59:173–187.

Kumar, A., Choudhary, A.K. and V.K. Suri. 2016. Influence of AM fungi, inorganic phosphorus and irrigation regimes on plant water relations and soil physical properties in okra (*Abelmoschus esculentus* L.)–pea (*Pisum sativum* L.) cropping system in Himalayan acid alfisol. *Journal of Plant Nutrition* 39:666–682.

Kurdali, F. 1996. Nitrogen and phosphorus assimilation, mobilization and partitioning in rainfed chickpea (*Cicer arietinum L.*). *Field Crops Research* 47:81–92.

Maitra, S., Hossain, A., Banerjee, P. and P. Bhadra. 2021. The role of phytohormones in heat stress tolerance in plants. In S. Fahad, O. Sönmez, S. Saud, D. Wang, C. Wu, M. Adnan and V. Turan (Eds), *Plant Growth Regulators for Climate-Smart Agriculture*. CRC Press, Hoboken, NJ, pp.145–164.

Maji, S., Das, A., Nath, R., Bandopadhyay P., Das, R., and S. Gupta. 2019. Cool season food legumes in rice fallows: An Indian perspective. In *Agronomic Crops*. Springer, Singapore. pp. 561–605. doi: 10.1007/978-981-32-9151-5_25.

Mandal, B., Majumdar, B., Bandyopadhyay, P.K., Hazra, G.C., Gangopadhyay, A., Samantaray, R.N., Mishra, A.K., Chaudhury, J., Saha, M.N. and S. Kundu. 2007. The potential of cropping systems and soil amendments for carbon sequestration in soils under long-term experiments in subtropical India. *Global Change Biology* 13:357–369.

Matthews, R.B., Kropff, M.J. and D. Bachelet (Eds) 1995. *Modeling the Impact of Climate Change on Rice Production in Asia*. International Rice Research Institute, Los Banos.

Mohamed, A.B., Duivenbooden, N.V. and S. Abdoussallam. 2002. Impact of climate change on agricultural production in the Sahel–Part 1: Methodological approach and case study for millet in Niger. *Climatic Change* 54:327–348.

Mondal, T. 2020. Lodging in wheat: Its causes, ill effects and management for higher productivity and profitability. *Agriallis* 2:39–44.

Morrison, M.J., and W.S. Doug. 2002. Heat stress during flowering in summer Brassica. *Crop Science* 42:797–803.

Naeem, M., Naeem, M.S., Ahmad, R., Ihsan, M.Z., Ashraf, M.Y., Hussain, Y. and F. Shah. 2018. Foliar calcium spray confers drought stress tolerance in maize via modulation of plant growth, water relations, proline content and hydrogen peroxide activity. *Archives of Agronomy and Soil Science* 64:116–131.

Nakagawa, H., T. Horie, and T. Matsui. 2003. Effects of climate change on rice production and adaptive technologies. In *Rice Science: Innovations and Impact for Livelihood. Proceedings of the International Rice Research Conference*, Beijing, China, 16–19 September 2002. International Rice Research Institute (IRRI), pp. 635–658.

Oilseeds Division -MOA. 2014. Status paper on oilseeds. Oilseed Division, Department of Agriculture & Cooperation, Ministry of Agriculture, Govt. of India, Krishi Bhawan, New Delhi. pp. 196.

Ong, C.K. 1984. The influence of temperature and water deficits on the partitioning of dry matter in groundnut (*Arachis hypogaea* L.). *Journal of Experimental Botany* 35:746–755.

Oseni, T.O. and M.T. Masarirambi. 2011. Effect of climate change on maize (*Zea mays*) production and food security in Swaziland. *American-Eurasian Journal of Agricultural & Environmental Sciences* 1:385–391.

Pandey, I.B., R.K. Pandey, D.K. Dwivedi, and R.S. Singh. 2010. Phenology, heat unit requirement and yield of wheat (*Triticum aestivum*) varieties under different crop-growing environment. *Indian Journal of Agricultural Sciences* 80:136–140.

Pathak, H., Ladha, J.K., Aggarwal, P.K., Peng, S., Das, S., Singh, Y., Singh, B., Kamra, S.K., Mishra, B., Sastri, A.S.R.A.S., Aggarwal, H.P. Das, D.K. and R.K. Gupta. 2003. Trends of climatic potential and on-farm yields of rice and wheat in the Indo-Gangetic Plains. *Field Crops Research* 80:223–234.

Piñera-Chavez, F.J., Berry, P.M., Foulkes, M.J., Jesson, M.A. and M.P. Reynolds. 2016. Avoiding lodging in irrigated spring wheat. I. Stem and root structural requirements. *Field Crops Research* 196:325–336.

Porter, J.R., Xie, L., Challinor, A.J., Kevern C.S., Howden, M., Iqbal, M.M., Lobell, D.B. and M.I. Travasso. 2014. Food security and food production systems. In *Climate Change 2014: Impacts, Adaptation, and Vulnerability*. Part A: Global and Sectoral Aspects. Contribution of Working Group II to the Fifth Assessment Report of the Intergovernmental Panel on Climate Change [Field, C.B., Barros, V.R., Dokken, D.J., Mach, K.J., Mastrandrea, M.D., Bilir, T.E., Chatterjee, M., Ebi, K.L., Estrada, Y.O., Genova, R.C., Girma, B., E.S. Kissel, E.S., Levy, A.N., MacCracken, S., Mastrandrea, P.R. and White, L.L. (Eds.), Cambridge University Press, Cambridge, United Kingdom and New York, NY, pp. 485–533.

Rajagopal, K., and T. Sanyal. 2012. Sustainable infrastructure development including limited life geosynthetics. *Bangkok: Proceedings of Geosynthetics Asia*, pp. 143–146.

Ray, D.K., Gerber, J.S., MacDonald, G.K. and P.C West. 2015. Climate variation explains a third of global crop yield variability. *Nature Communications* 6:1–9.

Reddy, K.R. and V.G. Kakani. 2007. Screening capsicum species of different origins for high temperature tolerance by *in-vitro* pollen germination and pollen tube length. *Scientia Horticulturae* 112:130–135.

Reddy, K.R., Hodges, H.F., and J.M. McKinion. 1997. A comparison of scenarios for the effect of global climate change on cotton growth and yield. *Functional Plant Biology* 24(6):707–713.

Reddy, K.R., Doma, P.R., Mearns, L.O., Boone, M.Y.L., Hodges, H.F., Richardson, A.G. and V.G. Kakaniand. 2002. Simulating the impacts of climate change on cotton production in the Mississippi Delta. *Climate Research* 22:271–281.

Rosenzweig, C., Tubiello, F.N., Goldberg, R., Mills, E., and J. Bloomfield. 2002. Increased crop damage in the US from excess precipitation under climate change. *Global Environmental Change* 12:197–202.

Ruan, C.J., Shao, H.B. and J.A. da Silva Teixeira. 2012. A critical review on the improvement of photosynthetic carbon assimilation in C3 plants using genetic engineering. *Critical Reviews in Biotechnology* 32:1–21.

Saini H. and D. Aspinall. 1982. Abnormal sporogenesis in wheat (*Triticum aestivum* L.) induced by short periods of high temperature. *Annals of Botany* 49:835–846. doi: 10.1093/oxfordjournals.aob.a086310.

Sardana, V., Sharma, P. and P. Sheoran. 2010. Growth and production of pulses. In W.H. Verheye (Ed.), *Soils, Plant Growth and Crop Production.* Eolss Publishers Co., Ltd., Oxford, UK, Vol. III, pp. 378–416.

Saseendran, S.A., Singh, K.K., Rathore, L.S., Singh, S.V. and S.K. Sinha. 2000. Effects of climate change on rice production in the tropical humid climate of Kerala, India. *Climatic Change* 44:495–514.

Satake, T. and S. Yoshida. 1978. High temperature induced sterility in indica rices at flowering. *Japanese Journal of Crop Sceince* 47:6–17.

Sattar, A., Sher, A., Ijaz, M., Ul-Allah, S., Rizwan, M.S., Hussain, M., Jabran, K. and M.A. Cheema. 2020. Terminal drought and heat stress alter physiological and biochemical attributes in flag leaf of bread wheat. *PLoS One* 15:e0232–974.

Scholes, R.J. 2016. Climate change and ecosystem services. *Wiley Interdisciplinary Reviews. Climate Change* 7:537–550.

Shrestha, R., Turner, N.C., Siddique, K.H., Turner D.W. and J. Speijers. 2006. A water deficit during pod development in lentils reduces flower and pod numbers but not seed size. *Australian Journal of Agricultural Research* 57:427–438.

Shukla, J.B., Verma, M. and A.K. Misra. 2017. Effect of global warming on sea level rise: A modeling study. *Ecological Complexity* 32:99–110.

Singh, A.K. 2017. The potential of jute crop for mitigation of greenhouse gas emission in the changing climatic scenario: A review. *International Journal of Agricultural Sciences,* 13:419–423.

Singh, A.K., Barman, D., Behera, M.S., Mazumdar, S.P., Saha, A.R. and D.K. Kundu. 2018. Impact of climate change on productivity of tropical rice-wheat-jute system under long term fertilizer management in alluvial soils. *International Journal of Current Microbiology and Applied Sciences* 7:1623–1632.

Singh, A.K., Jha, S.K., Majumdar, B., Roy, M.L., Sarkar, S. and A.K. Ghorai. 2019. Impacts of climate smart jute farming on resource use efficiency, productivity and economic benefits in rural Eastern India. *Outlook on Agriculture* 48:75–82.

Singh, C., Singh, P. and R. Singh. 2010. *Modern Techniques of Raising Field Crops* (2nd Edition). Oxford & IBH Publishing Co. Pvt. Ltd., New Delhi.

Tan, W., Meng, Q.W., Brestic, M., Olsovska, K. and X. Yang. 2011. Photosynthesis is improved by exogenous calcium in heat-stressed tobacco plants. *Journal of Plant Physiology* 168:2063–2071.

Thanacharoenchanaphas, K. and O. Rugchati. 2011. Simulation of climate variability for assessing impacts on yield and genetic change of Thai soybean. *World Academy of Science, Engineering and Technology* 59:1484–1488.

Ton, P. 2011. Cotton and climate change: Impacts and options to mitigate and adapt, Technical paper, International Trade Centre (ITC) UNCTAD/WTO, Geneva, p. 32.

Vaughan, M.M., Block, A., Christensen, S.A., Allen, L.H. and E.A. Schmelz. 2018. The effects of climate change associated abiotic stresses on maize phytochemical defenses. *Phytochemistry Reviews* 17:37–49. doi: 10.1007/s11101-017-9508-2.

Wahid, A., Gilani, S., Asraf, M. and M.R. Foolad. 2007. Heat tolerance in plants: An overview. *Environmental and Experimental Botany* 61:199–223.

Wang, J., Vanga, S. K., Saxena, R., Orsat, V. and V. Raghavan. 2018. Effect of climate change on the yield of cereal crops: A review. *Climate* 6:41.

Winkel T., Renno J.F. and W. Payne 1997. Effect of the timing of water deficit on growth, phenology and yield of pearl millet (*Pennisetum glaucum* (L.) R. Br.) grown in Sahelian conditions. *Journal of Experimental Botany* 48:1001–1009. doi: 10.1093/jxb/48.5.1001

Yakhin, O.I., Lubyanov, A.A., Yakhin, I.A. and P.H. Brown. 2017. Biostimulants in plant science: A global perspective. *Frontiers in Plant Science* 7:2049.

19

*Growth, Physiology, Yield, and Yield Attributes of Lentil (*Lens culinaris* Medikus) with Reference to Abiotic Stresses*

Anuj Kumar Singh
S.D. Agricultural University

Brajesh Kumar Mishra
ICAR-DRMR Guwahati

Vijai Pandurangam and J.P. Srivastava
Banaras Hindu University

CONTENTS

19.1 Introduction

Lentil (*Lens culinaris* Medikus) is a cool-season grain legume of family Fabaceae (Papilionaceae), sub-family Faboideae and tribe Vicieae. It is also called as *masoor* or *masser* (India), *masura, renuka, mangalaya* (Sanskrit), *adas* (Arabic), *messer* (Ethopia), *mercimek* (Turkey), *heramame* (Japanese) and *mangu* or *margu* (Parsian) (Sandhu 2007). The name lentil has been derived from the shape of the seeds which is lens (biconcave) shaped.

Lentil (*Lens culinaris* Medikus) is a primitive domesticate among food legumes being cultivated since 7000 BC (Zohary 1972, Erskine 1997, Sehgal et al. 2021) with various uses as food and feed because of its protein-rich grains and straw. Cultivated lentil is the third most important cool-season legume in the world after chickpea and pea. Globally, lentil is cultivated in an area of 3.85 m ha with production of 3.59 mt (Erskine 1997, Kumar et al. 2013). Major geographical regions of lentil production are South Asia and China (44.3%), North America (41%), Central and West Asia and North Africa (CWANA) (6.7%), Sub-Saharan Africa (3.5%) and Australia (2.5%) (Kumar et al. 2013). In India among *rabi* (winter season) pulses, lentil is next to chickpea, being grown in 1.56 m ha with annual production of 1.06 mt. It is generally grown as a rainfed crop. Its cultivation in India is mainly concentrated in UP, Bihar, MP and West Bengal, accounting for more than 80% area and production of the crop (Anonymous 2013). In north eastern parts of the country, lentil is cultivated as a *paira* crop with rice. In most of the lentil-growing areas of the world, drought is a recurrent and detrimental constraint to yield (McWilliam 1986), whereas in the mediterranean environments, lentil crop, which is sown in autumn or winter,

DOI: 10.1201/9781003258063-19

experiences sporadic drought and terminal drought during the vegetative and reproductive growth periods, respectively, coinciding with increasing atmospheric temperatures and decreasing precipitation. In South Asia, drought may also occur during plant establishment, particularly when sowing is delayed and plant roots cannot reach subsoil moisture (Bhattarai et al. 1998, Rahman and Mallick 1998). Generally, lentil is cultivated either on residual soil moisture or on current rainfall, and in both cases, it is frequently subjected to terminal drought that results in low and variable seed yields (Panwar and Srivastava 2012). Nevertheless, crop responds well to irrigation, especially on light textured soils (Singh 1979, Yusuf et al. 1979, Turner 1992, Shrestha 1996, Shrestha et al. 2006a). Being a dry land crop and cultivated by resource-poor farmers, lentil production is limited by moisture stress and unfavourable temperatures. Progress has been made in breeding for tolerance to drought through selection for an appropriate phenology, increased water-use efficiency and winter hardiness through selection for cold tolerance (ICRISAT 2006). As lentils are predominantly grown under rainfed condition, their productivity depends on water conserved in the soil as well as on distribution and frequency of rainfall. Management practices can contribute to increased yield in water-deficit environments; nevertheless, major progress might be realized through genetic improvement (Yadav et al. 2003).

Evolution, diversity and distribution: Lentil is considered as one of the agricultural crops grown from last 8,500 years. South West Asia and northern Syria regions are considered as the centre of origin of this crop. There are a number of wild types of this crop, but *L. orientalis* is considered as the progenitor of the cultivated lentil (Sandhu 2007). Lentil seeds have been found in Egyptian tombs (2,400 BC). During the neolithic period, it spread to Greece and Bulgaria and reached to Near East and Mediterranean regions during Bronze Age. In the 16th century, it reached to new world with Spanish and Portuguese, but in the US, it was introduced during the First World War.

Lentil is a self-pollinating pulse crop. The ploidy level is diploid with seven pairs of chromosomes ($2n = 14$). Genome size of lentil is relatively bigger i.e. about 4.2 Gb and considered as an orphan crop from genomics perspective (Arumuganathan and Earle 1991a). Genus *Lens* includes seven annual species, namely *L. culinaris* ssp. *odemensis* (Ladizinsky); *L. culinaris* ssp. *orientalis* (Boiss) Ponert; *L. ervoides* (Brign) Granade; *L. lamottei* (Czefr); *L. nigricans* (Bieb) Godron and *L. tomentosus* (Ladizinsky). *L. culinaris* ssp *orientalis* is fully cross-compatible with the cultivated species *L. culinaris* ssp *culinaris* and is considered as its putative ancestor (Redden et al. 2007).

The botanical features of cultivated lentil can be described as an annual bushy herb, much branched and softly hairy with slender stems (Duke 1981). The leaves are alternate with four to seven leaflet pairs. Flowers are small, pale blue, purple, white or pink. Lentils exhibit hypogeal mode of germination. The advantage offered by such germination is that it reduces the lethal effects of cold temperatures (freezing) and protects the seedling from desiccating environmental stress conditions. The growth stages between maximum flowering and early pod formation stages witness maximum node number and biomass accumulation (Kurdali et al. 1997, Van Kessel 1994). The plant develops primary, secondary and tertiary branches. Plant population density is the major factor that determines the number of branches produced per plant. The main stem bears 18% of the pods, while the primary and higher order branches bear 52% and 30% of the plant's pods, respectively (Erskine and Goodrich 1991). Based on the seed size and weight, there are two categories of lentils, namely macrosperma (large-seeded) with test weight >50 g per 1,000 seeds and microsperma (small-seeded) with test weight 40 g or lesser per 1,000 seeds. The lentil seeds are marketed as whole seed or de-hulled and split with two major market classes, namely green lentils and red lentils.

Nutritional and neutraceutical value: After soybean and hemp, lentil seeds are the richest source of proteins (22%–35%). Digestibility of protein is very high and does not produce uric acid accumulation in human body (Thavarajah et al. 2007). Seeds are also rich in carbohydrates (60%), dietary fibres (31%), vitamin B_1 (0.86 mg 100 g^{-1}) and minerals such as iron, calcium, phosphorus, magnesium and molybdenum. It does not contain cholesterol. Caloric value varies with cultivars and ranges from 110 to 160 kcal 100 g^{-1} seed (Farshadfar et al. 2008). Nearly one-third energy in seeds is stored in the form of proteins. Seeds are rich in leucine, lysine, threonine and phenylalanine but deficient in sulphur-containing amino acids viz. methionine and cystine. Seeds have a negligible amount of fat and also contain flavanoids which have ability to bind with toxic metals and help in their removal from body (Butu et al. 2014). Seed fibres help in maintaining blood sugar level. Seeds contain trypsin inhibitor compounds which inhibit enzyme trypsin and lower down the digestibility of seed proteins (Liener and Kakade 1980). Nevertheless, the nutritive value of seeds varies with variety, agronomic practices and the environmental conditions prevailing during the growth period (Adsule et al. 1989). In India and the neighbouring countries, lentil is a staple food. It is also used in soups, stews, casseroles and salad dishes, and vegetative parts and poor-quality seeds are good animal feeds (Panwar and Srivastava 2012). Being a leguminous crop, it fixes about 20 kg N/ha and improves soil fertility. Crop is also used for green manuring.

19.2 Physiological Frame Work

19.2.1 Crop Establishment and Development

The lentil plants have many sequences of events that occur when drought stress develops, starting with cellular growth as the most sensitive response, followed by a wide range of biochemical and physiological events (Hsiao 1973, Mishra 2014, Mishra et al. 2016, 2018). The sequences of events which occur at later stages in response to drought at very low water potentials (more negative) are mostly indirect and complex. The first and foremost effect of drought is impaired germination and poor stand establishment (Harris et al. 2002). The observed reduction in plant height is associated with a decline in cell enlargement and increased leaf senescence (Bhat and Srinivasa Rao 2005). In an experiment conducted under controlled conditions in a glasshouse to determine the sensitivity of reproductive development of lentil genotypes of different origins to water deficit, it was demonstrated that water deficit reduced plant height by about 20% compared with well-watered plants (Shrestha et al. 2006b). Drought stress may also cause leaf abscission (Rouhi et al. 2007). Decline in plant height and fresh weight under drought is due to suppression of cell expansion and cell growth due owing to decreased cell turgor pressure (Jaleel et al. 2008).

19.2.2 Crop Phenology: Vegetative and Reproductive Growth

The plant of lentil is slender with soft hairs (pubescence). Plant height varies from 15 to 75 cm. In cultivated species, it varies from 20 to 30 cm. Being indeterminate in its growth, lentil exhibits variation in plant structure and branching pattern. They are either (i) compact, highly branched and bushy with single stem of erect to semi-erect stature; (ii) tall and erect with sparse branching; (iii) semi-tall erect and sparsely branched; (iv) semi-tall sub-compact, moderate to highly branched or (v) short, sub-compact, moderately to highly branched. The ratio of plant height to canopy width is used as a measure of plant structure and is also used to characterize growth habit during germplasm evaluation (Ferguson and Robertson 1999).

The first two leaves of lentil are simple, scale-like fused with two stipules. The next two to three leaves formed at ensuing nodes are bifoliate. The further nodes possess multifoliate leaves that are alternate, compound and pinnate. Two to eight pairs of ovate lanceolate leaflets are present on the rachis of 4–5 cm (Barulina 1930). The rachis sometimes may end in a bristle or simple tendril, especially in the upper leaves or only on leaves that originate just prior to flowering. The tendril length may be almost equal to that of the rachis. Some genotypes are tendril less and the rachis may end with a bristle. The leaves of lentil exhibit variation in greenness from yellowish green, dark green to dark bluish green.

Flowering in lentil is acropetal and the duration of flowering may range from 10 to 40 days depending on genotype and environment. The flowers are present on a single axillary raceme inflorescence with a slender peduncle 2–5.5 cm long. The number of peduncles may vary from 10 to 150 per plant (Malhotra 1974). Each peduncle bears one to four small flowers. The flowers are complete and papilionaceous. The petal colour varies from white, pink, purple, light purplish blue to pale blue according to genotype. They are cleistogamous and exclusively self-pollinated with some cross-pollination (<0.08%) through thrips. Most of the flowers are pollinated by the time the petal expansion is about three-fourth of the sepals. By the second and third day, all open flowers close completely and the petal colour fades away. Pod visually appears after three to four days.

The pods are small, oblong (0.6–2.0 cm in length and 0.3–1.0 cm width) and laterally compressed. During grain filling, in microsperma types, the pod bulges over the seeds due to the convex nature of the seeds, whereas in macrosperma types, the pods remain almost flat. The unripe pods are usually green but the colour varies from purple to violet or spotted according to genotypes. At maturity, the pods become straw-coloured or dull brown. The pods contain one to three seeds.

After harvest, the seeds exhibit dormancy for 3–4 week. The non-dormant seeds imbibe rapidly when placed in moist soil. It imbibes water from 85% (microsperma types) to more than 100% (macrosperma types) of its dry weight. The germination in the field usually occurs in 8–10 days when the temperature is around 20°C or may take 20–25 days if temperature is <10°C.

Lentil plant has a slender taproot system with fibrous lateral roots. Striking genotypic variations are found in the length of primary root and lateral root proliferation. It also varies with the type of soil in which they are grown and differences in soil moisture availability. The root system may be shallow, moderate or deep. In the Indian varieties, HUL 57 and DPL 62, grown under irrigated and rainfed conditions, significant genotypic differences in root lengths have been reported (Figure 19.1). The maximum root length of 96 cm was observed in HUL 57 under irrigated condition at 100 days after sowing. Under the rainfed condition, root length declined. As compared to HUL 57, roots of genotype DPL 62 penetrated deeper layers of soil under rainfed condition.

Being a leguminous plant, it fixes atmospheric nitrogen. The tap root and laterals bear numerous nodules that may be round or elongated. Due to the presence of apical meristem, the nodules become oblong and sometimes bifurcated and multi-lobed. The root nodules are formed as early as from 15 days

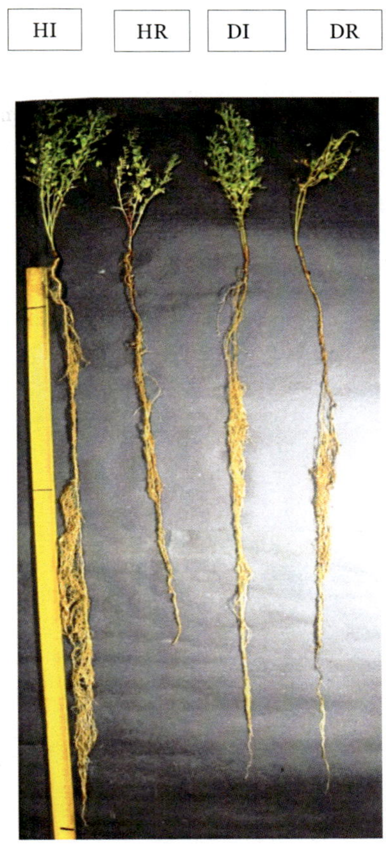

FIGURE 19.1 Root development in lentil genotypes HUL 57 (H) and DPL 62 (D) under irrigated (I) and rainfed (R) conditions after 100 days of sowing. The maximum root length of 95 cm was observed in HUL 57 under irrigated condition. Under rainfed condition, root length declined (Singh 2016).

after seedling emergence. In legumes, the number of nodules is reported to be the maximum during maximum vegetative growth which declines after the onset of flowering. Nodules also decline on account of stress. Under irrigated condition in lentil genotype HUL-57, the maximum nodules (218 plant^{-1}) were observed at 100 DAS; thereafter, it decreased, while in moisture-stressed plants, the maximum nodule number (154 plant^{-1}) was recorded at 80 DAS. In DPL-62, the maximum nodule number (210 plant^{-1}) was at 80 DAS in plants under normal condition, while under stress condition, the maximum nodules plant^{-1}, i.e. 142 was observed at 60 DAS (Singh 2016).

Prolonged maturity duration in lentil may affect yield by increasing the number of high-order branches (secondary and tertiary branches), which carry fewer pods than the main stem and primary branches. However, no significant correlations have been observed between seed yield and the number of branches (Erskine and Goodrich 1991). Prolonged life cycle and delayed maturity is reported to increase plant height and branch length instead of branch number (Erskine and Goodrich 1991). Although longer days to maturity may increase biomass and yield, but it may reduce harvest index (HI), delay maturity and leave the crop susceptible to frost damage (Thomson et al. 1997). In lentil cultivars, maturity varies from 75 to 180 days; nevertheless, it is affected by temperature, rainfall and sowing dates (Saxena 2009). The degree days (DD) required for seedling emergence, flowering and pod formation of lentil are reduced in drier sites and drier years (Thomson et al. 1997). Lentil requires 944 to 1270 heat units (growing DD; GDD) from seeding to maturity. This heat requirement is generally achieved in most parts of the Great Plains (Miller et al. 2002). Cooler temperatures may reduce yield by reducing HI. In drier conditions, early flowering cultivars tend to increase HI by extending pod-filling duration. HI is also improved by late-season drought (Silim et al. 1993). In low-yield potential environments, the flexible growth habit of indeterminate cultivars is considered advantageous for plant survival and yield. In contrast, terminal drought leads to forced maturity and enhanced partitioning of assimilates to seeds by reducing sink-source competition (Silim et al. 1993). Drought stress breeding procedures in lentil often require the identification of traits associated with drought tolerance and their genetic variability. Increased early vigour, early flowering and early maturity have been proved beneficial, especially to escape terminal drought. High-yielding cultivars in Syria were found to accumulate more biomass, flowered early with a shorter grain-filling period even under drought conditions (Silim et al. 1993). Time to flowering accounts for 49% of the variation in seed yield, indicating that avoidance through early flowering is the key to minimizing the effects of drought stress (Silim et al. 1993). In habitats characterized by low average rainfall (Erskine et al. 1994), *L. culinaris* ssp. *orientalis* (wild lentils) are often found but their dry matter accumulation and productivity in terms of seed yield are significantly lower than the other cultivated lentils (Hamdi and Erskine 1996). The lentil species *L. odemensis* (Ladizinsky) was reported to exhibit the least reduction in seed yield when drought and non-drought treatments were compared.

Lentils accumulate higher levels of osmolytes imparting higher degree of osmotic adjustment in comparison to other grain legumes, especially with low moisture availability during flowering (Shrestha et al. 2006b). This indicates that avoidance may not be the only mechanism for drought tolerance in this crop. Deeper rooting has been advocated as a way of increasing legume productivity under moisture-limiting conditions (Buddenhagen and Richards 1998, Turner et al. 2017). Genotypic differences in the length and spread of roots have been reported for lentil but not in association with water usage (Singh 2016, Saxena et al. 1993, Saxena and Hawtin 1981).

Short-duration cultivars of chickpea and field pea have faster early root growth rates, but this is not sustained during seed fill, due to early root senescence leading to loss in effective root length (Saxena et al. 1993). The scenario also holds well for lentils. Earliness in flowering under moisture-deficit environment is considered as an important trait for drought avoidance, but it may have detrimental or negative impact on growth of roots, specifically limiting the length of root growth. Hence, screening for drought tolerance genotypes in terms of avoidance and adaptive mechanisms may be achieved successfully by limiting water supply, but when early flowering is regarded as the main mechanism for drought escape, sufficient care must be taken to ensure that selection is not for major regional and adaptational traits. Rather, screening multiple genotypes under controlled environmental conditions, without interference from other potentially adaptive physiological responses, may prove to be an initial useful step. As with many crops, early sowing has been advocated in many areas to avoid rising temperatures and drought during the reproductive period and to maximize yields. However, early sowing may expose crop to increased weed competitions, diseases and other abiotic stresses.

Nitrogen fixation and assimilation: Under the rainfed/water-deficit environmental conditions, legume crops such as lentil are generally cultivated. Nitrogen fixation, one of the key vital physiological processes in this crop, is reported to be influenced by soil moisture stress and water deficit. Such stresses result in hampered N accumulation, dry matter partitioning and yield (Mishra 2014, Mishra et al. 2016, 2018). Nitrogen is one of the integral components of proteins and is essential for healthy crop growth and optimum physiological development. Nitrogen is also needed to synthesize chlorophyll, an important pigment for photosynthesis. New leaves may contain up to 6% N. It is a very mobile nutrient because it moves from older to newer leaves. It is taken up throughout the growing season. Major fraction of plant N at pre-flowering stage is transported and stored in the leaves. In legumes, the N requirement for pod development is partially met from N stored in the leaves (Hafner et al. 1995). So, proper supply of N is necessary for achieving optimum plant growth and crop productivity under adverse environmental conditions.

Lentil fulfils most of its N requirement through fixation of atmospheric N_2 by *Rhizobia* living in its root nodules symbiotically. Generally, the level of N_2 fixation in legumes depends on host genotypes, *Rhizobial* strains, environment and their interactions. Lentil cultivars have shown genetic variability in their ability to symbiotically fix N_2 (Rennie and Dubetz 1986); therefore, genotypes with high N_2 fixation and high seed yield are desirable for sustainable agriculture. N_2 fixation in legumes is highest between flowering and early pod fill, and all legumes derived about 90% of their N from the atmosphere by 80 days after emergence (Senaratne and Ratnasinghe 1993). Using wheat as a reference crop, it is been reported that lentils accumulate more dry matter and N content during the latter part of the growing season (Van Kessel 1994). The reproductive plant parts are reported to contain higher percentage of N derived from the atmosphere (% Ndfa) than the vegetative components which indicated that N in the reproductive plant parts is derived largely from current N_2 fixation. Lentils continue to fix N until the end of the pod fill stage (Van Kessel 1994). Experiment was carried out in field to assess the source of nitrogen (N_2 fixation, soil and fertilizer), N assimilation, partitioning and mobilization in rainfed lentil at various growth stages using ^{15}N isotopic dilution (Kurdali et al. 1997). The relative importance of these sources are reported to depend upon the factors viz. plant species, genotype, drought stress, plant and soil N status and N_2 fixation ability (Kurdali et al. 1997). Nitrogen for developing pods may be supplied from soil, atmospheric N_2 and from the mobilization of existing N in plant tissues. A study was conducted on five lentil cultivars differing in seed size. The results showed net mobilization of N (43%–94%) from shoots and roots to pods. It was concluded that when soil is poor in nitrogen and atmospheric N-fixation is poor, translocation of N from the pre-existing N pool in vegetative tissues became the major source of N for pod filling (Kurdali et al. 1997, Bhattacharya 2009). However, this phenomenon was affected considerably upon the cultivars.

19.3 Yield Attributes and Yield

Variations in yielding ability in lentils are reported under non-irrigated and irrigated conditions. Under field condition, lentil seed yields are reported to be higher under irrigated than under non-irrigated condition (Bhattacharya 2009). Multiple regressions between seed yield and different traits exhibited that all of them do not have a significant contribution to seed yield. Seed yield determining factors are found to be different under irrigated and non-irrigated conditions. Under non-irrigated condition, lentil seed yield is determined by total dry matter production and its partitioning to seeds day^{-1}, mean temperature and total DD during podding to maturity. Under irrigated condition, nevertheless, lentil seed yield is determined by root dry matter at vegetative stage, total dry matter produced, per day dry matter production and its partitioning during seed filling period and total DD during flowering to podding. It has been reported that various physiological processes differ with growth stages, and their effects on seed yield of any crop plant are different (Sinha 1973). The pod initiation stage is the most critical with respect to moisture stress in different cool-season grain legumes; therefore, life-saving irrigation at this stage results in high yields (Soltani et al. 2001). In the case of less availability of water, the crop may be irrigated two or three times as per the needs, but with lower amounts of water (may be 1/3 or 2/3 of the full irrigation), to realize satisfactorily high biomass and grain yield (Oweis et al. 2004). Besides phenological changes, some genotypes may perform better under moisture stress conditions than others due to differences in physiological processes (Shrestha et al. 2006a, Mishra 2014, Shrestha et al. 2006b, Mishra et al. 2014). Differences in the response of genotypes to water deficits imposed during flowering and podding indicate differences between lentil groups in growth and water use (Shrestha et al. 2006b). Water deficits reduced seed yield up to 60% in crossbreds selected for South Asia and the South Asian cultivar; Simal, but seed yield increased with water deficit at flowering in the West Asian genotype; Cassab. Water deficit during early pod development or seed filling (after embryo cell division) altered both the source and sink strength (reduced seed numbers, shortened the duration of seed filling) to maintain the flux of assimilates to the developing embryo, without affecting seed size or seed growth rate (Shrestha et al. 2006b). There has been considerable work on the effect of water deficits on seed development. Legumes such as lentil with indeterminate growth are able to adjust seed numbers in response to changes in assimilate supply due to water deficit, thus can minimize variation in the assimilate supply to the remaining seeds (Egli 1998, Mishra 2014, Mishra et al. 2014. Therefore, any variation in seed yield in indeterminate species is normally due to a variation in final seed numbers rather than mean seed size or seed growth rate under drought stress condition (Mishra 2014, Mishra et al. 2014). The severity and timing of water stress that occurs during specific growth stages determine the source-sink ratio, specifically the number of seeds and seed size. For example, the number of cells in the cotyledons and the size of the seed are significantly impacted when water deficit occurs at active cell division stages coinciding with seed development phase (Munier-Jolain et al. 1998). As compared to well-watered conditions, plants grown under water-deficit condition exhibited significant variations in the number of flowers, fruiting nodes, pods, seeds, and HI. Higher flower drop and empty pods were observed upon withholding water supply. Variation in physiological trait, such as osmotic adjustment at flowering and podding, was not directly related to seed yield, but seed number was related to seed yield under both well-watered and water-deficit conditions (Mishra 2014, Shrestha et al. 2006b, Mishra et al. 2014). On the basis of single plant study, the HI in lentils is reported to vary from 1% to 60% (Kirthisinghe 1986). However, in field under non-stress and drought conditions, HI varied from 53% to 59%, and 36% to 45%, respectively (Mishra 2014, Shrestha et al. 2006b, Mishra et al. 2014). On evaluating the relation between growth and yield in 20 lentil lines that were grown under eight environments varying in water and photothermal conditions, it was found that under stressful environments, lentil yield and crop growth rate are highly correlated and coupled but the same does not hold true under normal or favourable conditions Lake and Sadras (2021).

19.4 Responses of Lentil to Abiotic Stress

Environmental factors (temperature, water and salinity) affect the production and productivity of lentil. Abiotic stresses negatively impact the crop during vegetative and reproductive stages and ultimately the crop yield declines. Abiotic stresses change the plant morphology, physiology and reproductive developments. In the following section with impact of heat stress, drought, waterlogging and salinity on the growth and development of lentils are described.

19.4.1 High Temperature Stress

Global temperature is increasing due to climate change (IPCC 2014). For cool-season crop temperature, above 25°C is injurious to growth and biochemical functions (Wahid et al. 2007). Lentil being a cool-season crop, it is more susceptible to heat stress. The temperature threshold range for lentil crop is 15°C–30°C (Barghi et al. 2012).

Vegetative growth: Proper seed germination is the need for higher crop productivity. It has been reported that high temperature (35°C–40°C) for 4 hours reduces the germination of lentil and retards growth of seedlings (Chakraborty and Pradhan 2011). The first effect of high temperature is observed on the shoot of plants. High temperature damages the bio-membrane, and homeostasis of cells is impaired. The impaired homeostasis negatively impacts the key physiological functions of plant. Heat stress causes membrane damage, reduces the stomatal conductance, and decreases the plant water status (Sehgal et al. 2017). Under high temperature, membrane damage also leads to chlorophyll loss. High temperature not only leads to chlorophyll reduction but also decreases the photosynthetic efficiency of lentil leaves. Photosynthesis is the physiological function which leads to biomass production of crops. If proper photosynthesis is not occurring during vegetative growth, the biomass of the plant is reduced (Sehgal et al. 2017).

19.4.2 Reproductive Growth

Phenology is the characteristic feature in the life cycle of any crop. Higher temperature during the reproductive stages of any crop reduces the period of reproductive growth, as a result, the number of reproductive units as well as weight of reproductive units decrease causing the reduction in crop yield. In lentil, it is reported that higher temperature fastens the reproductive growth i.e. time of flowering, podding and maturity occurs early (Sehgal et al. 2017, Sita et al. 2017). Fertility and growth of pollens are negatively impacted by higher temperature. Higher temperature reduces the mobilization of dry matter to pod. Metabolism of starch and sucrose in the seeds also decreases under elevated temperature condition.

19.4.3 Quality of Seed

High temperature negatively damages the enzyme activities and cell functions; as a result, the chemical composition of stressed seeds changes. At normal and elevated temperatures, per cent reductions in starch and protein contents of the seed were similar and found in the range of 25%–43% and 26%–41%, respectively. At higher temperature, as compared to normal, significant reduction in fat content (39%–57%) and increase in total sugars content (36%–68%) has been observed. Accumulation of storage proteins belonging to classes including albumins, globulins, prolamins and glutelins decreases in the range of 22%–42% at elevated temperature. Amino acids such as proline glycine, alanine, isoleucine, leucine and lysine contents are reported to increase, whereas levels of other amino acid are reported to decline. Heat stress has been found to decrease the levels of minerals such as Ca (13%–28%), Fe (17%–52%), P (10%–54%), K (12.4%–28.3%) and Zn (36%–59%) in seeds (Sita et al. 2018).

19.4.4 Low-Temperature Stress

Low-temperature stress reduces the potential of germination and seedling emergence, vegetative growth and reproductive growth in lentil. Nodulation in lentil diminishes only below 10°C (Lira et al. 2005). Lentil is relatively more tolerant to chilling stress than pea or chickpea (Murray et al. 1988). Pod formation and seed filling are sensitive to low temperature in lentil (Maqbool et al. 2010).

19.4.5 Waterlogging Stress

Lentil is the most sensitive of all the legumes to waterlogging. It generally does not survive after waterlogging, especially in cool conditions of winter. Even mild waterlogging is an issue, even when foliar disease is kept under control. Flowering is the most sensitive stage to waterlogging in lentil. The response of lentil to waterlogging is similar to its response to low light and low temperatures, all resulting in flower and pod abortion and leaf senescence (GRDC 2018). Under waterlogging, plant root faces oxygen shortage which negatively impacts the physiological functions. It is reported that large-seeded genotypes of lentil have higher root porosity and lesser effect on shoot nitrogen content under waterlogging than the other small-seeded genotypes (Malik et al. 2015). Transient waterlogging reduces the yield of cool-season grain legumes in several parts of the world (Solaiman et al. 2007). Different grain legumes were ranked to tolerance to waterlogging on the basis of reduction in root and shoot growth under the waterlogged condition with respect to controls as follows: faba bean > yellow lupin > grass pea > narrow-leafed lupin > chickpea > lentil > field pea (Solaiman et al. 2007). Potentially waterlogging may cause reductions in N-fixation and nodulation and, if severe enough, leads to loss of nodules. In Nepal, it has been observed that waterlogging under field conditions caused disintegration and loss of nodules in certain lines. Therefore, in such lines, fixation of atmospheric nitrogen nodules is affected. In comparison with cereals, lentil, field pea and chickpea are susceptible to waterlogging at germination and have severely depressed vegetative growth, especially of roots (Crawford 1977). Lake et al. (2021) found that lentil genotypes that maintain higher rates of growth during both during and after the periods of waterlogging exhibit better tolerance to hypoxia conditions when compared to cereal crops like wheat and barley where plant growth rate after and not during waterlogging better correlates to yield responses (Ciancio et al. 2021).

19.4.6 Salt Stress

Salinity is one of the most severe environmental stresses, negatively affecting the productivity of salt-sensitive crop species. Salinity occurs mainly in arid and semi-arid regions, where evaporation considerably exceeds precipitation, such as in West and Central Asia and in Australia, and in coastal areas because of the ingress of seawater (Saxena et al. 1993). Faulty methods of irrigation have also led to salinization of productive lands in several parts of the world (Saxena et al. 1993). Legumes are relatively sensitive to salt, and lentil is comparatively more sensitive than field pea and faba bean and similar to chickpea

(Saxena et al. 1993). Salinity is generally not considered to be a problem for lentil production worldwide (Erskine et al. 1994), although the major lentil-growing areas of the world are regions with a high frequency of saline or sodic soils (Saxena et al. 1993). Seed germination potential is essential to establish the plant population in the field. Salinity decreases the germination percentage, seed water absorption, root and shoot lengths and seedling vigour index. Genotypes responded differently to the varying salinity levels. This approach allows the selection of desirable genotypes at early growth stages, thus enabling more efficient application of various breeding methods to achieve salinity-tolerant lentil genotypes (Foti et al. 2018). Salt stress causes reduced nitrogen fixation in lentil and tolerant genotypes have higher capacity to fix nitrogen than the susceptible lines (Rai and Singh 1999, Maher et al. 2003).

19.4.7 Drought

In India, lentil is cultivated either on residual soil moisture or after irrigating field, and hence due to this in both cases, the crop is frequently subjected to terminal drought during the reproductive stages that result in low and variable seed yields (Panwar and Srivastava 2012). In most countries of the world, cultivation of lentil crop is done under rainfed or water-scarce conditions (Soltani et al. 2001). The osmotic potential at which germination is inhibited varies greatly between genotypes. Seed germination and emergence are reduced even at very less negative osmotic potentials (Singh et al. 2016). Initiation of cell elongation during germination is also highly sensitive to water stress. When crops are grown on soils with decreasing water potential and low water table, lentil genotypes with deeper root system may have little advantage (Singh 2016, Loss and Siddique 1994). In lentil, the biomass of plant decreases under water deficit condition (Mishra 2014). Drought stress decreases plant height, biomass, and number, size and area of leaf, and leaf chlorophyll of lentil (Mishra 2014, Singh 2016). The biomass apportioning between the various parts of the shoot viz. stem, leaves, podwall and seed indicated significant contribution of podwall to seed development; especially during active seed filling stage. During the active seed development stage, podwall stored dry matter is remobilized to contained seeds; as a result, podwall dry weight declines (Figure 19.2). Four Indian genotypes, KLS 218, HUL57, DPL62 and IPL81, grown under irrigated and rainfed conditions, showed genotypic variations for attainment of maximum shoot biomass at different growth stages and it's apportioning.

Source and sink strength are drastically reduced due to the prevalence of moisture deficit, specifically during early pod development and during seed filling stages. These are reflected in terms of reduced photosynthetic rate and leaf area and decline in seed size, mainly by inducing flower and fruit shedding, reducing seed number and shortening the duration of seed filling (Shrestha et al. 2006). Hastening phenological development in plants under water-deficit conditions has also been reported (Mishra et al. 2016), and shortening of time to maturity of lentils grown under rainfed conditions compared to those under irrigated conditions has been well documented (Ghanem et al. 2015). In lentil, early maturity has been correlated with seed yield (Silim et al. 1993). In the dry season of 1988/1989, 49% of the variation in seed yield among lines was accounted for by variation in flowering time. Drought escape has been found to be the key response to drought. For severely drought-prone areas, selection for early flowering is, therefore, required. Variation in flowering time results for 62% of seed yield variation under drought stress with respect to irrigated. Variation in acquisition and partitioning of assimilated nitrogen in different plant parts under irrigated and rainfed conditions in lentil genotypes (Figure 19.3) has been reported (Mishra 2014).

Genotype viz. HUL 57, which efficiently apportioned more N (Figure 19.3) and dry matter (Figure 19.2) to seeds under rainfed condition produced higher and exhibited higher yield stability index under rainfed condition (Mishra 2014).

19.5 Lentil Genomics and Its Application in Crop Improvement

The genome size of lentil is larger i.e. ~4 Gbp (Arumuganathan and Earle 1991b), and it is one of the limiting factors for genomic studies in this crop. The first linkage maps with 10 loci and 18 loci were constructed using isozyme markers by Zamir and Ladizinsky (1984) and Tadmor et al. (1987). Tahir and Muelbauer (1994) used isozyme and morphological markers to construct linkage maps with 20 loci. These low-density linkage maps were due to the limitations in number of morphological and isozyme markers. With the advent of molecular markers, comprehensive linkage maps were constructed using RAPD, AFLP and RFLP markers systems (Eujayl et al. 1998). The first intraspecific lentil map that comprised 114 RAPD, inter-simple sequence repeat and resistance gene analog markers was reported by Rubeena et al. (2003). Duran et al. (2004) published the first lens map using SSR markers. Later with the help of additional SSR makers, more detailed lentil maps were constructed (Hamwieh et al. 2005, Sharpe et al. 2013) and lentil linkage map using SNP markers by Temel et al. (2014). Molecular markers linked to desirable gene(s) as well as QTLs for several quantitative traits have been reported for marker-assisted selection in this crop (Kumar et al. 2015). Morphological markers namely cotyledon (Yc), anthocyanin content in stem (Gs), pod indehiscence (Pi), seed coat pattern (Scp), flower colour (W), radiation frost tolerance locus (Rf), early flowering (Sn) and ground colour of the seed (Gc) were mapped as qualitative markers because they exhibited monogenic dominant mode of inheritance (Eujayl et al. 1998; Duran et al. 2004, Hamwieh et al. 2005, Tullu et al. 2008). With regard to mapping of quantitatively inherited traits, three QTLs for plant height while five QTLs each have been reported for height of the first ramification and flowering time, seven QTLs for pod dehiscence, and one QTL each for shoot number and seed diameter (Duran et al. 2004). Similarly, five QTLs were identified for winter survival and four QTLs reported for winter injury (Kahraman et al. 2004). Molecular markers have also been used for F1 hybrid testing in lentil (Solanki et al. 2010) when the phenotypic diversity is low between the hybridizing parents. The first interspecific map developed from SNP markers using the RIL population

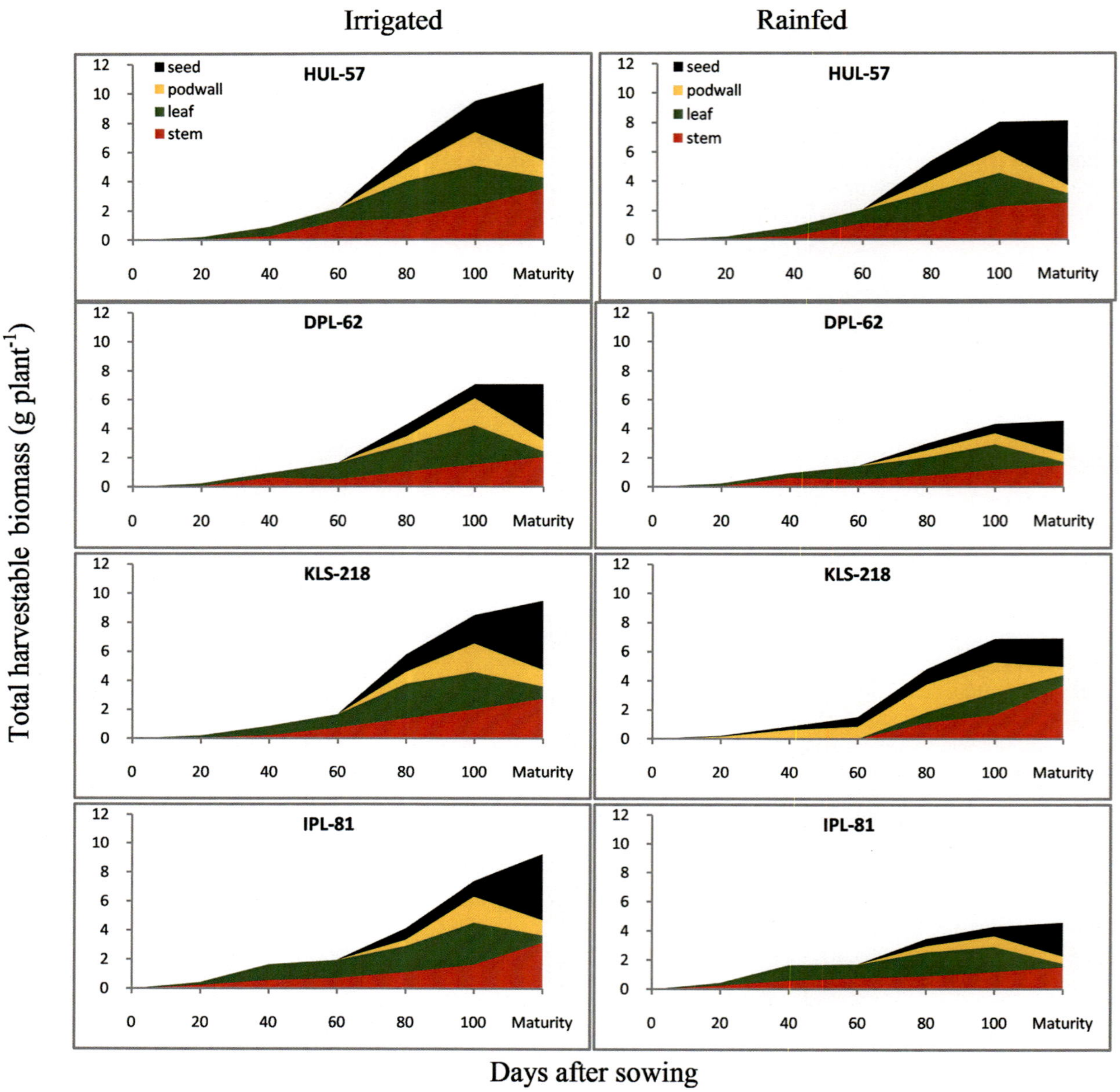

FIGURE 19.2 Total harvestable biomass (g/plant) in four genotypes of lentil under irrigated and rainfed conditions (Mishra 2014). LSD and SEm±values for total biomass of shoot: T (0.11***, ±0.04), G (0.22***, ±0.08), T×G (0.32***, ±0.11), S (0.19***, ±0.07), T×S (0.27***, ±0.10), G×S (0.27***, ±0.10), T×G ×S (0.54***, ±0.20); stem: T (0.05***, ±0.02), G (0.10***, ±0.03), T×G (0.08***, ±0.03), S (NS, ±0.05), T×S (0.12***, ±0.04), G×S (0.12***, ±0.04), T×G×S (0.24***, ±0.09); leaf: T (0.03***, ±0.01), G (0.06***, ±0.02), T×G (NS, ±0.03), S (0.08***, ±0.02), T×S (0.08***, ±0.03), G×S (0.12***, ±0.03), T×G×S (0.15***, ±0.05); podwall: T (0.05***, ±0.02), G (0.10***, ±0.04), T×G (0.15***, ±0.05), S (0.06***, ±0.02), T×S (0.09***, ±0.03), G×S (0.09***, ±0.03), T×G×S (0.18***, ±0.06) and seeds: T (0.15***, ±0.05), G (0.29***, ±0.10), T×G (NS, ±0.15), S (0.18***, ±0.06), T×S (0.25***, ±0.09), G×S (0.25*, ±0.09), T×G×S (NS, ±0.17). LSD, Least Significant Differences; SE±, Standard Error of Means; NS, Not significant; P, Probability level of significance at ***$P \leq 0.001$, **$P \leq 0.001$, *$P \leq 0.005$; T, Treatment (Irrigated and Rainfed); G, Genotype; S, Crop growth stage; T×G; T×S; T×G×S: Interactions of treatment×genotype×crop growth stage, respectively.

gen 7 of the cross between *L. culinaris* cv.Alpo x *L.odemensis* acc ILWL235 helped in identification of candidate genes for seed coat spotting colour, flower colour, stem pigmentation and QTLs for flowering time, seed size and Ascochyta blight resistance in lentil (Polanco et al. 2019). Kumar et al. (2018a) identified functional markers associated with flowering time in lentil. Association mapping analysis identified QTLs for agronomic traits including primary branches per plant, days to maturity, pods per plant, secondary branches per plant, 100 seed weight, yield per plant and reproductive duration (Kumar et al. 2018b).

Next-generation-sequencing–based transcriptome analysis has also accelerated the identification of candidate genes under abiotic and biotic stress conditions in lentil. For heat tolerance, candidate genes related to physiological process and phenotype of pollen, cell wall and secondary metabolism have been

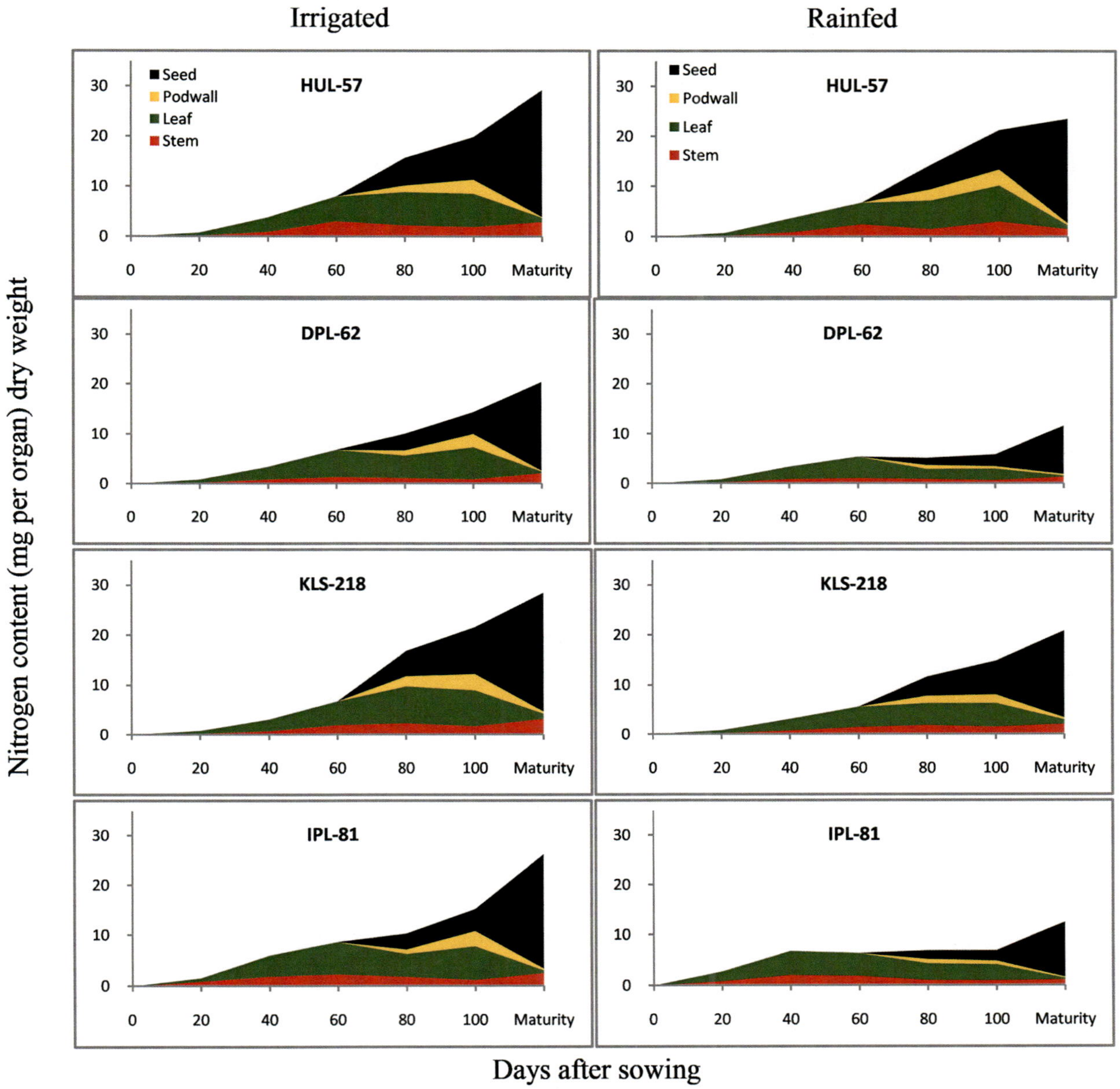

FIGURE 19.3 Nitrogen partitioning in four genotypes of lentil under irrigated and rainfed conditions (Mishra 2014). LSD and SEm± values for total nitrogen in shoot: T (0.40***, ±0.14), G (0.78***, ±0.28), T×G (1.1***, ±0.39), S (0.67***, ±0.25), T×S (0.95***, ±0.34), G×S (0.95***, ±0.35), T×G×S (1.90***, ±0.68; stem: T (0.08***, ±0.03), G (0.15***, ±0.05), T×G (0.21***, ±0.07), S (0.13***, ±0.05), T×S (0.18***, ±0.06), G×S (0.19***, ±0.07), T×G×S (0.36***, ±0.13); leaf: T (0.12***, ±0.04), G (0.24***, ±0.08), T×G (0.34***, ±0.12), S (0.22***, ±0.07), T×S (0.35***, ±0.10), G×S (0.34***, ±0.09), T×G×S (0.58***, ±0.21); podwall: T (0.07***, ±0.03), G (0.15***, ±0.05), T×G (0.21***, ±0.08), S (0.09***, ±0.03), T×S (0.13***, ±0.04), G×S (0.12***, ±0.05), T× G×S (0.26***, ±0.09) and seeds: T (0.70***, ±0.25), G (1.39***, ±0.50), T×G (NS, ±0.71), S (0.85***, ±0.30), T×S (1.20***, ±0.43), G×S (1.19*, ±0.43), T×G×S (NS, ±0.85). LSD, Least Significant Differences; SE±, Standard Error of Means; NS, Not significant; P, Probability level of significance at ***$P \leq 0.001$, **$P \leq 0.001$, *$P \leq 0.005$; T, Treatment (Irrigated and Rainfed); G, Genotype; S, Crop growth stage; T×G; T×S; T×G×S: Interactions of treatment×genotype×crop growth stage, respectively.

identified in lentil through transcriptome analysis of heat-sensitive and heat-tolerant genotypes (Singh et al. 2019). In this study, the genes encoding plasmodesmata callose-binding protein 3 (PDCB), phosphatidylinositol/phosphatidylcholine transfer protein SFH13, CDP-diacylglycerol-glycerol-3-phosphate 3-phosphatidyltransferase 1 chloroplastic, probable glycerol-3-phosphate acyltransferase 2 (GPAT2), O-acyltransferase, and phosphatidylcholine diacylglycerol choline phosphotransferase showed their upregulation under heat stress conditions. A gene-encoding pyruvate phosphate dikinase identified under heat stress produced phosphoenolpyruvate (PEP) metabolite, which is an essential compound of shikimate pathway. This pathway produces secondary metabolites involved in heat tolerance (Singh et al. 2019). The comparative genomic analysis identified orthologous gene loci of ELF3 in lentil. This gene has been found to control differences in photoperiodic responses when compared between wild and domesticated pea (Weller et al. 2012). In lentil, genes for boron tolerance have been identified on the basis of a comparison of SNP 20002998 and SNP 20000246 flanking markers with the genome sequences of *A. thaliana* and *M. truncatula* (Kaur et al. 2014). Association mapping analysis identified EST-SSR markers for flowering time in lentil, and a comparative analysis of associated EST sequence with closely related genome sequences led to identification of candidate genes for flowering time in lentil (Kumar et al. 2018a).

Genetic transformation in lentil is a useful tool to study gene function through gene knockouts or overexpression as well as to develop transgenic plants with desirable traits. This also necessitates an efficient regeneration protocol. Bajaj and Dhanju (1979) reported shoot regeneration from apical meristems. Since then shoot regeneration has been achieved from various explants like stem nodes, cotyledonary nodes, epicotyls, decapitated embryo, embryo axis, immature seeds and cotyledonary petioles by various scientists (Kumar et al. 2015). Rooting of the *in vitro* developed shoots is a major problem in lentil with a few successful reports (Fratini and Ruiz 2003, Sarker et al. 2003). Lentil genetic transformations have been successful by various methods involving Agrobacterium (Lurquin et al. 1998), particle bombardment (Gulati et al. 2002, Mahmoudian et al. 2002) electroporation and biolistic method (Chowrira et al. 1996). The first fertile lentil transgenic plants were developed by Gulati et al. (2002) where the lentil plants were regenerated on MS medium and then micrografted on non-transgenic root stock for transplantation in soil. Khatib et al. (2007) used *Agrobacterium*-mediated transformation of bar gene to develop glufosinate ammonium herbicide-resistant lentil, and later the lentil transgenic transformed with DREB1A gene for drought and salinity tolerance was developed by Khatib et al. (2011). The role and function of SBT1.1 pea subtilase in controlling seed size and tl, tendrilless genes which have been found to control tendril formation were determined by using non-transgenic techniques, like TILLING (D'Erfurth et al. 2012). Efforts are underway to target genes for traits like pod shattering, herbicide tolerance and orobanche tolerance using mutagenic lentil populations. Other non-transgenic methods like RNA interference and virus-induced gene silencing methods are being used to understand the molecular mechanisms of biological nitrogen fixation.

19.6 Crop Simulation Models

Recently crop simulation models were devised to predict the crop production under varying environments. In lentil, LENMOD model, a calibrated model, in which parameters were adjusted by comparing the output of the model with experimental observations from New Zealand (McKenzie and Hill 1990) and the United Kingdom (Andrews et al. 2001), was developed. The drawback of this calibration approach is that it shifts the model to an empirical description and thereby limits the suitability of the model only to a particular environment and geographical area. The LENMOD model also does not include water and nitrogen budgets for the soil, which also poses a major limitation in simulating a high-protein seed legume crop that is grown in stress-prone areas (Soltani and Sinclair 2011).

A non-calibrated, mechanistic and simple model for simulating growth and yield of crops called simple simulation model (SSM) has also been proposed (Soltani and Sinclair 2012). A SSM legumes version of the model that is generic to grain legume species has been proved to be able to predict dry matter accumulation, crop development and water consumption for different species of legumes, namely soybean (Sinclair 2000) and chickpea (Vadez et al. 2013). The quantitative information published with respect to lentil growth and development allows SSM legumes to be developed and used as a simple and transparent lentil model. In general, lentil crop survival is the first and foremost requirement for crop performance, specifically under severe drought-prone areas as well as under low-yielding environments (Sinclair 2000). To upgrade the current model incorporation of a physiologically based sub-model that could deal with very dry conditions such as the ones that are generally encountered by lentil crops growing in South Asia, Africa or the Middle East would be required. Accelerated phonological development in response to soil drying has evidently been shown (Hamdi et al. 1992). SSM legumes also predict similar observation (Soltani and Sinclair 2012).

19.7 Significance of Lentil under Climate Change

Carbon dioxide, the major greenhouse gas, was about 280 ppm in 1880. It is expected to reach 445–640 ppm by 2050 and 720–1,020 ppm by 2100 AD. The global mean temperature is projected to increase by 0.80°C–2.60°C (2050) and 1.40°C–5.80°C (2100) (Raja Reddy and Kakani 2008). All pulses are C_3 plants, therefore, it is expected that global warming could initially impact positively on them.

Desi chickpea varieties (480 DD), dry pea (530 DD) and lentil (540 DD) require fewer DD to reach anthesis, compared with spring wheat (600 DD) (Miller et al. 2002). Dry pea (1010 DD) generally attains maturity sooner than spring wheat (1070 DD), while lentil (1060 DD) matures earlier than spring wheat only in years when near-normal climatic conditions occur. Currently available chickpea and lentil cultivars have an indeterminate growth habit and require physiological stress viz., drought to terminate flowering and induce seed set. However, lentil yield is expected to be affected more by terminal heat stress and drought. Under changing climatic conditions, the frequency of uncertainty in weather conditions will increase; hence, lentil productivity is expected to be also limited by waterlogging, extremes of temperatures and frost. Incidence of diseases and pests is also expected to increase. Therefore, it is necessary to develop climate-resilient genotypes with resistance to multiple biotic and abiotic stresses.

19.8 Future Line of Works

Lentil cultivars for future must be of determinate type. Plants with early phenological development are desirable to protect them from terminal drought. Selection of cultivars that have abiotic stress tolerance is desirable. Identifying cultivars with higher yield potential under increased atmospheric CO_2 and resistance to multiple stresses viz. elevated temperature, terminal heat stress, frost, waterlogging, drought and salinity are required. It would also be desirable to develop cultivars having higher nutritive value and water-use efficiency. To achieve these objectives, it is essential to identify morphological, physiological and biochemical parameters associated with stress tolerance in this crop and to utilize traits associated with stress tolerance to develop location-specific climate-resilient, high-yielding genotypes with high yield stability index using conventional breeding approaches and modern biotechnological tools.

Acknowledgement

The inputs from doctoral theses research works of Dr. Brajesh Kumar Mishra (2014) and Dr. Anuj Kumar Singh (2016), who also co-authored this chapter, are duly acknowledged.

REFERENCES

Adsule, R.N., S.S. Kadam, and D.K. Salunkhe. 1989. Lathyrus bean. In: Salunkhe, D.K. and Kadam, S.S. (Eds.), *CRC Handbook of World Food Legumes: Nutritional Chemistry*, Processing Technoilogy and Utilization. CRC Press, Boca Ralton, FL, Vol. 2, pp. 133–152.

Andrews, M., B.A. McKenzie, A. Joyce, and M.E. Andrews. 2001. The potential of lentil (*Lens culinaris*) as a grain legume crop in the UK: An assessment based on a crop growth model. *Annals of Applied Biology*, 139: 293–300.

Anonymous. 2013. All India Coordinated Research Project on MULLaRP, Project Coordinator's Report. Indian Institute of Pulse Research Kanpur, Uttar Pradesh, India.

Arumuganathan, K., and E.D. Earle. 1991a. Nuclear DNA content of some important plant species. *Plant Molecular. Biology Reporter*, 9: 208–218.

Arumuganathan, K., and E.D. Earle. 1991b. Estimation of nuclear DNA content of plants by flow cytometry. *Plant Molecular Biology Reporter*, 9: 221–231.

Bajaj, Y.P.S., and M.S. Dhanju. 1979. Regeneration of plants from apical meristem tips of some legumes. *Current. Science*, 48: 906–907.

Barghi, S.S., H. Mostafaii, F. Peighami, and R.A Zakaria. 2012. Path analysis of yield and its components in lentil under end season heat condition. *International Journal of Agricultural Researchand Reviews*, 2: 969–974.

Barulina, H. 1930. Lentils of the USSR and other countries. In: *Bulletin of Applied Botany, Genetics and Plant Breeding* (Suppl 40) (Leningrad, English summary), pp. 225–238.

Bhatt, R.M., and N.K. Srinivasa Rao. 2005. Influence of pod load response of okra to water stress. *Indian Journal of Plant Physiology*, 10: 54–59.

Bhattacharya, A. 2009. Yield determination in lentil under non-irrigated and irrigated condition: Phenology, dry matter distribution and yield attributes. *Indian Journal of Agricultural Research*, 43: 1–9.

Bhattarai, A.N., M.P. Bharati, and B.K. Gyawali. 1998. Factors which limit the productivity of cool season food legumes in Bangladesh. In: Summerfield, R.J. (Ed.), *World Crops: Cool Season Food Legumes*. Kluwer Academic Publishers, Dordrecht, Netherlands, pp. 230–234.

Buddenhagen, I.W., and R.A. Richards. 1988. Breeding cool-seasons food legumes for improved performance in stress environments. In: Summerfield, R.J. (Ed.), *World Crops: Cool Seasons Food Legumes*. Kluwer Academic Publishers, Dordrecht, Netherlands, pp. 81–85.

Butu, M., S. Rodino, A. Butu, and M. Butnariu. 2014. Screening of bioflavonoid and antioxidant activity of *Lens culinaris* medikus. *Digest Journal of Nanomaterials and Biostructures*, 9: 519–529.

Chakraborty, U., and D. Pradhan. 2011. High temperature-induced oxidative stress in *Lens culinaris*, role of antioxidants and amelioration of stress by chemical pre-treatments. *Journal of Plant Interaction*, 6: 43–52.

Chowrira, G.M., V. Akella, P.E. Fuerst, and P.F. Lurquin. 1996. Transgenic grain legumes obtained by in planta electroporation-mediated gene transfer. *Molecular Biotechnology*, 5: 85–96.

Ciancio N., D. J. Miralles, G.G. Striker, and L.G. Abeledo. 2021. Plant growth rate after, and not during waterlogging better correlates to yield responses in wheat and barley. *Journal of Agronomy and Crop Science*, 207(2): 304–316.

Crawford, R.M.M. 1977. Tolerance of anoxia and ethanol metabolism in germinating seeds. *New Phytologist*, 79: 511–517.

D'Erfurth, I., C. Le Signor, G. Aubert, M. Sanchez, V. Vernoud, B. Darchy, et al. 2012. A role for an endosperm-localized subtilase in the control of seed size in legumes. *New Phytoogist*, 196: 738–751.

Duke, J.A. 1981. *Handbook of Legumes of World Economic Importance*. Plenum Press, New York, pp. 52–57.

Duran, Y., R. Fratini, P. Garcia, and M. de la Vega Pérez. 2004. An intersubspecific genetic map of Lens. *Theoretical and Applied Genetics*, 108: 1265–1273.

Egli, D.B. 1998. *Seed Biology and the Yield of Grain Crops*, CAB International, Wallingford, UK.

Erskine, W. 1997. Lessons for breeders from landraces of lentil. *Euphytica*, 93: 107–112.

Erskine, W., and W.J. Goodrich. 1991. Variability in lentil growth habit. *Crop Science*, 31: 1040–1044.

Erskine, W., M. Tufail, A. Russell, M.C. Tyagi, M.M. Rahman, and M.C. Saxena. 1994. Current and future strategies in breeding lentil for resistance to abiotic and biotic stresses. *Euphytica*, 73: 127–135.

Eujayl, I., M. Baum, W. Powell, W. Erskine, and E. Pehu. 1998. A genetic linkage map of lentil (*Lens* sp.) based on RAPD and AFLP markers using recombinant inbred lines. *Theoretical and Applied Genetics*, 97: 83–89.

Farshadfar, E., M. Farshadfar, and M. Aghaee. 2008. Genetic variability and path analysis of chickpea landraces. *Journal of Applied Science*, 8: 3951–3956.

Ferguson, M., and L.D. Robertson. 1999. Morphological and phonological variation in the wild relatives of lentil. *Genetic Resources and Crop Evolution*, 46: 3–12.

Foti, C., E.M. Khah, and O.I. Pavli. 2018. Germination profiling of lentil genotypes subjected to salinity stress. *Plant Biology (Stuttgart, Germany)*, 21(3): 480–486.

Fratini, R., and M.L. Ruiz. 2003. A rooting procedure for lentil (*Lens culinaris* Medik.) and other hypogeous legumes (pea, chickpea and lathyrus) based on explant polarity. *Plant Cell Reports*, 21: 726–732.

Ghanem, M.E., H. Marrou, A. Soltani, S. Kumar, and T.R. Sinclair. 2015. Lentil (*Lens culinaris* Medik.) variation in phenology and yield evaluated with model. *Agronomy Journal*, 107: 1967–1977.

GRDC (Grains Research and Development Corporation) Grownotes. 2018. Lentil. Section 14 Environmental Issues. pp 1–10.

Gulati, A., P. Schryer, and A. McHughen. 2002. Production of fertile transgenic lentil (*Lens culinaris* Medik) plants using particle bombardment. *In Vitro Cellular and Developmental Biology-Plant*, 38 (4): 316–324.

Hafner, H., B.J. Ndunguru, A. Bationo, and H. Marschner. 1995. Effect of nitrogen, phosphorus and molybdenum application on growth and symbiotic N_2-fixation of groundnut in an acid sandy soil in Niger. *Nutrient Cycling in Agroecosystems*, 31: 69–77.

Hamdi, A., and W. Erskine. 1996. Reaction of wild species of the genus Lens to drought. *Euphytica*, 91: 173–179.

Hamdi, A., W. Erskine, and P. Gates. 1992. Adaptation of lentil seed yield to varying moisture supply. *Crop Science*, 32: 987–1001.

Hamwieh, A., S. Udupa, W. Choumane, A. Sarker, F. Dreyer, C., Jung, et al. 2005. A genetic linkage map of Lens sp. based on microsatellite and AFLP markers and the localization of fusarium vascular wilt resistance. *Theoretical and Applied Genetics*, 110: 669–677.

Harris, D., R.S. Tripathi, and A. Joshi. 2002. On-farm seed priming to improve crop establishment and yield in dry direct-seeded rice, Pandey, S., Mortimer, M., Wade, L., Tuong, T.P., Lopes, K., Hardy, B., Eds., International Research Institute, Manila, Philippines. pp. 231–240.

Hsiao, T.C. 1973. Plant response to water stress. *Annual Review of Plant Physiology*, 24: 519–570.

ICRISAT 2006. www.ICRISAT.org.

IPCC, Climate Change 2014: Synthesis Report, Contribution of Working Groups I. II and III to the Fifth Assessment Report of the Intergovernmental Panel on Climate Change, Geneva: IPCC, 2014, 151.

Jaleel, C.A., P. Manivannan, G.M.A. Lakshmanan, M. Gomathinayagam, and R. Paneersekvam. 2008. Alterations in morphological parameters and photosynthetic pigment responses of *Catharanthus roseus* under soil water deficits. *Colloids and Surfaces B: Biointerfaces,* 61: 298–303.

Kahraman, A., I. Kusmenoglu, N. Aydin, A. Aydogan, W. Erskine, and F.J. Muehlbauer, 2004. QTL mapping of winter hardiness genes in lentil. *Crop Science,* 44: 13–22.

Kaur S, Cogan NO, Stephens A, Noy D, Butsch M, Forster JW, Materne M. 2014. EST-SNP discovery and dense genetic mapping in lentil (*Lens culinaris* Medik.) enable candidate gene selection for boron tolerance. *Theoretical and Applied Genetics*, 127(3):703–13.

Khatib, F., S. Koudsieh, B. Ghazal, J. Barton, H. Tsujimoto, and M. Baum. 2007. Developing herbicide resistant lentil (*Lens culinaris* Medikus subsp. culinaris) through Agrobacterium-mediated transformation. *Arab Journal of Plant Protection*, 25: 185–192.

Khatib, F., Makris, A., Yamaguchi-Shinozaki, K., Kumar, S., Sarker, A., Erskine, W., et al. 2011. Expression of the DREB1A gene in lentil (*Lens culinaris* Medik. Subsp culinaris) transformed with the Agrobacterium system. *Crop and Pasture Science*, 62: 488–495.

Kirthisinghe, J.P. 1986. Plant to plant variation in harvest index in lentil (*Lens culinaris* L.), DipAgrSc Dissertation, Lincoln College, University of Canterbury, New Zealand.

Kumar S, Rajendran K, Kumar J, Hamwieh A, Baum M. 2015. Current knowledge in lentil genomics and its application for crop improvement. Frontiers in Plant Sciences, 6(78): 1–13.

Kumar S., S. Barpete, J. Kumar, P. Gupta, and A. Sarker. 2013. Global lentil production: Constraints and strategies. SATSA Mukhapatra- Annual Technical Issue. 17, ISSN 0971-975X.

Kumar, J., S. Gupta, R.S. Biradar, P. Gupta, S. Dubey, and N.P. Singh. 2018a. Association of functional markers with flowering time in lentil. *Journal of Applied Genetics,* 59(1): 9–21.

Kumar J, S. Gupta, D.S. Gupta, and N.P. Singh 2018b. Identification of QTLs for agronomic traits using association mapping in lentil. *Euphytica* 214(4):75.

Kurdali, F., K. Kalifa, and M. Al-Shamma. 1997. Cultivar differences in nitrogen assimilation, partitioning and mobilization in rain-fed grown lentil. *Field Crops Research*, 54: 235–243.

Lake, L., and V.O. Sadras. 2021. Lentil yield and crop growth rate are coupled under stress but uncoupled under favourable conditions. *European Journal of Agronomy*, 216: 126266.

Lake, L., N. Izzat, T. Kong, and V.O. Sadras. 2021. High-throughput phenotyping of plant growth rate to screen for waterlogging tolerance in lentil. *Journal of Agronomy and Crop Science*, 207(6): 995–1005. doi: 10.1111/jac.12522.

Liener, I.E. and M.L. Kakade. 1980. Protease inhibitors. In: Liener, I.E., (Ed.), *Toxic Constituents of Plant Food Stuffs*. Academic Press, New York, pp. 7–71.

Lira J.M.A., A.S.T. Lima, J.R.F. Arruda, and D.L. Smith. 2005. Effect of root temperature on nodule development of ben, lentil and pea. *Soil Biology and Biochemistry*, 37: 235–239.

Loss, S.P., and K.H.M. Siddique. 1994. Morphological and physiological traits associated with wheat yield increases in Mediterranean environments. *Advances in Agronomy*, 52: 229–276.

Lurquin, P.F., Z. Cai, C.M. Stiff and E.P. Fuerst. 1998. Half-embryo cocultivation technique for estimating the susceptibility of pea (*Pisumsativum* L.) and lentil (*Lens culinaris* Medik.) cultivars to Agrobacterium tumefaciens. *Molecular Biotechnology*, 9: 175–179.

Maher, L., R. Armstrong, and D. Connor. 2003. Salt tolerant lentils: A possibility for the future? In: *Solution for a better environment: Proceedings of the 11th Australian Agronomy Conference*, Geelong, Victoria, Australia, 2–6 February.

Mahmoudian, M., M. Yucel, and H.A. Oktem. 2002. Transformation of lentil (*Lens culinaris* M.) cotyledonary nodes by vacuum infiltration of Agrobacterium tumefaciens. *Plant Molecular Biology Reporter*, 20: 251–257.

Malhotra, R.S., K.B. Singh, and J.K. Singh 1974. Genetic variability and genotype-environmental interaction studies in lentil. *Journal of Research PAU.* 10: 17–21.

Malik, A.I, T.I. Ailewe, and W. Erskine. 2015. Tolerance of three grain legume species to transient waterlogging. *Annals of Botany*, 88: 579–589.

Maqbool, A., S. Shafiq, and L. Lake 2010. Radiant frost tolerance in pulse crops-a review. *Euphytica*, 172: 1–12.

McKenzie, B.A., and G.D. Hill. 1990. Environmental control of lentils (*Lens culinaris* Medik). *Journal of Agricultural Scienc,* 113: 67–72.

McWilliam, J.R. 1986. The national and international importance of drought and salinity effects on agricultural production. *Australian Journal of Plant Physiology*, 13: 1–13.

Miller, P.R., B.G. McConkey, G.W. Clayton, S.A. Brandt, J.A. Staricka, A.M. Johnston, G.P. Lafond, B.G. Schatz, D.D. Baltensperger, and K.E. Neill. 2002. Pulse crop adaptation in the Northern Great Plains. *Agronomy Journal*, 94: 261–272.

Mishra, B.K. 2014. Growth, phenology, dry matter and nitrogen partitioning in lentil (*Lens culinaris* Medik.) under optimal and moisture stress conditions, Ph.D. Thesis, Banaras Hindu University, Varanasi, India.

Mishra, B.K., J.P. Srivastava, and J.P. Lal. 2014. Drought stress resistance in two diverse genotypes of lentil (*Lens culinaris* Medik.) imposed at different phenophases. *Journal of Food Legumes*, 27(4): 307–314.

Mishra, B.K., J.P. Srivastava, J.P. Lal, and M.S. Sheshshayee. 2016. Physiological and Biochemical Adaptations in lentil genotypes under drought stress. *Russian Journal of Plant Physiology*, 63(5): 695–708.

Mishra, B.K., J.P. Srivastava, and J.P. Lal. 2018. Drought resistance in lentil (*Lens culinaris* Medik.) in relation to morphological, physiological parameters and phenological developments. *International Journal of Current Microbiology and Applied Sciences*, 1: 2288–2304.

Munier-Jolain, N.G., N.M. Munier-Jolain, R. Roche, B. Ney, C. Duthion. 1998. Seed growth rate in grain legumes I.(). Effect of photo assimilate availability on seed growth rate. *Journal of Experimental Botany*, 49: 1963–1969.

Murray, G.A., D. Eser, L.V. Gusta, and G. Eteve. 1988. Winter hardiness in pea, lentil, faba bean and chickpea. In: Summerfield, R.J. (Ed.), *World Crops: Cool Season Food Legumes*. Kluwer Academic Publishers. Dordrecht, Netherlands, pp. 831–843.

Oweis, T., A. Hachum, and M. Pala 2004. Lentil production under supplemental irrigation in a Mediterranean environment. *Agriculture Water Management*, 68: 251–265.

Panwar, J.D.S. and Srivastava, J.P. 2012. *Pulse Productivity: Physiological Constraints*. ICAR, India.

Polanco, C., L.E. de Miera, A.I. González, P. García, R. Fratini, F. Vaquero, F.J. Vences, and M.P. de la Vega. 2019. Construction of a high-density interspecific (*Lens culinaris* x *L. odemensis*) genetic map based on functional markers for mapping morphological and agronomical traits, and QTLs affecting resistance to Ascochyta in lentil. *PLoS One* 14(3): e0214409. doi: 10.1007/s11033-020-05891-9.

Rahman, M.M. and Mallick, R.N. 1998. Factors which limit cool season food legume productivity in Bangladesh. In: Summerfield, R.J. (Ed.), *World Crops: Cool Season Food Legumes*. Kluwer Academic Publishers, Dordrecht, Netherlands, pp. 280–294.

Rai, R., and R.P. Singh. 1999. Effect of salt stress on interaction between lentil (*Lens culinaris*) genotypes and *Rhizobium* spp. Strains: Symbiotic nitrogen fixation in normal and sodic soils. *Biology and Fertility of Soils*, 29: 187–195.

Raja Reddy, K. and Kakani. V.G, 2008. Screening for genetic variability among crops to global environmental changes: Challenges and opportunities. *Proceedings of the Golden Jubilee Conference on Challenges and Emerging Strategies for Improving Plant Productivity*. ISPP and IARI, New Delhi, India, pp. 21–23.

Redden, B., N. Maxted, B. Furman, and C. Coyne. 2007. Lens Biodiversity. In: Yadav, S.S., McNeil, D., and Stevenson, P.C. (Eds.), *Lentil: An Ancient Crop for Modern Times*. Springer, Berlin, Germany, pp. 11–22.

Rennie, R.J., and S. Dubetz. 1986. Nitrogen-15-determined nitrogen fixation in field-grown chickpea, lentil, fababean and field pea. *Agronomy Journal*, 78: 654–660.

Rouhi, V., R. Samson, R. Lemeur, and P. Van Damme. 2007. Photosynthetic gas exchanges characteristics in three different almond species during drought stress and subsequent recovery. *Environment and Experimental Botany*, 59: 117–129.

Rubeena, R. Ford, and P.W.J. Taylor. 2003. Construction of an intraspecific linkage map of lentil (*Lens culinaris* ssp. culinaris). *Theoretical and Applied Genetics*, 107: 910–916.

Sandhu, J.S. and S. Singh. 2007. History and origin. In: Yadav, S., McNeil, D. and Steveson, P.C. (Eds.), *Lentil: An Ancient Crop for Modern Times*. Springer, Berlin, Germany, pp. 1–9.

Sarker, R., B.M. Mustafa, A. Biswas, S. Mahbub, M. Nahar, R. Hashem, et al. 2003. In vitro regeneration in lentil (*Lens culinaris* Medik.). *Plant Tissue Culture*, 13: 155–163.

Saxena, M.C. 2009. Plant morphology, anatomy and growth habit. In: Erskine, W., Maeuhlbauer, F., Sarker, A., and Sharma, B. (Eds), *The Lentil: Botany, Production and Uses*. CABI Publishing, London, UK, pp. 34–46.

Saxena, M.C. and G.C. Hawtin. 1981. Morphology and growth patterns. In: Webb, C. and Hawtin, G. (Eds.), *Lentils*. Royal Commonwealth Agricultural Bureau, Slough, pp. 39–52.

Saxena, N.P., C. Johansen, M.C. Saxena, and S.N. Silim. 1993. Selection for drought and salinity: A case study with chickpea. In: Singh, K.B., and Saxena, M.C., (Eds.), *Breeding for Stress Tolerance in Cool-Season Food Legumes*. John Wiley & Sons, Chichester, UK, pp. 245–270.

Sehgal, A., K. Sita, J. Kumar, S. Kumar, S. Singh, K.H.M. Siddique, and H. Nayyar. 2017. Effect of drought, heat and their interaction on growth, yield and photosynthetic function of lentil (*Lens culinaris* Medikus) genotypes varying in heat and drought sensitivity. *Frontiers in Plant Science*, 17(8): 1776.

Sehgal, A., S. Kumari, A. Rehman, M. Farooq, S. Kumar, R. Yadav, H. Nayyar, S. Singh, and K.H.M. Siddique. 2021. Lentil. In: Sardras, V.O. and Calderini, D.F. (Eds), *Crop Physiology: Case Histories for Major Crops*. Academic Press and Elsevier, Cambridge, MA, pp. 408–428.

Senaratne, R., and D.S. Ratnasinghe. 1993. Ontogenic variation in nitrogen fixation and accumulation of nitrogen in mungbean, blackgram, cowpea, and groundnut. *Biology of Fertile Soil*, 16: 125–130.

Sharpe, A.G., L. Ramsay, L.A. Sanderson, M.J. Fedoruk, W.E. Clarke, L. Rong, et al. 2013. Ancient orphan crop joins modern era: Gene-based SNP discovery and mapping in lentil. *BMC Genomics*, 14: 192.

Shrestha, R. 1996. Report on varietal evaluation of lentil in Khumaltar (1995–96), National Winter Crop Workshop, (Regional Agricultural Research Station: Bhairahawa, Napal).

Shrestha, R., N.C. Turner, K.H.M. Siddique, D.W. Turner, and J. Speijers. 2006a. Water deficit during pod development in lentils reduces flower and pod numbers but not seed size. *Australian Journal Agricultural Research*, 57: 427–438.

Shrestha, R., N.C. Turner, K.H.M. Siddique, and D.W. Turner. 2006b. Physiological and seed yield responses to water deficits among lentil genotypes from diverse origins. *Australian Journal of Agricultural Research,* 57: 903–915.

Silim, S.N., M.C. Saxena, and W. Erskine. 1993. Adaptation of lentil to the Mediterranean environment, I. Factors affecting yield under drought conditions. *Experimental Agriculture,* 29: 9–19.

Sinclair, T.R. 2000. Model analysis of plant traits leading to prolonged crop survival during severe drought. *Field Crop Research,* 68: 211–217.

Singh, P.K. 1979. Symbiotic algal N_2-fixation and crop productivity. *Annual Review of Plant Science*, 1: 37–65.

Singh, A.K. 2016. Root Growth, pod development and nutrient uptake dynamics in lentil [*Lens culinaris* (Medikus)] genotypes under water deficit condition. Ph.D. Thesis, Banaras Hindu University, Varanasi, India.

Singh, A.K., J. P. Srivastava, and J.P. Lal. 2016. Effect of PEG-6000 induced osmotic stress on germination, growth and nutrient uptake of two lentil [*Lens culinaris* (Medikus)] genotypes. *Journal of Food Legumes*, 29(3&4): 188–194.

Singh D, C.K. Singh, J. Taunk, V. Jadon, M. Pal, and K. Gaikwad. 2019. Genome wide transcriptome analysis reveals vital role of heat responsive genes in regulatory mechanisms of lentil (*Lens culinaris* Medikus). *Scientific Reports*, 9(1):1–9.

Sinha, S.K. 1973. Green revolution and break-through in food production in India. *Indian Journal of Agricultural Economics,* XXVIII(2): 26–42.

Sita, K., A. Sehgal, J. Kumar, S. Kumar, S. Singh, K.H.M. Siddique, and H. Nayyar. 2017. Identification of high temperature tolerant lentil (*Lens culinaris* Medikus) genotypes through leaf and pollen traits. *Frontiers in Plant Science*, 19(8): 744.

Sita, K., A. Sehgal, K. Bhandari, J. Kumar, S. Kumar, S. Singh, K.H.M. Siddique, and H. Nayyar. 2018. Impact of heat stress during seed filling on seed quality and seed yield in lentil (*Lens culinaris* Medikus) genotypes. *Journal Food Science & Agriculture*, 98(13): 5134–5141.

Solaiman, Z., T.D. Colmer, S.P. Loss, B.D. Thomson, and K.H.M. Siddique. 2007. Growth responses of cool-season grain legumes to transient waterlogging. *Australian Journal of Agricultural Research,* 58: 406–412.

Solanki, R., S. Singh, and J. Kumar. 2010. Molecular marker assisted testing of hybridity of F1 plants in lentil. *Food Legumes,* 23: 21–24.

Soltani, A., and T.R. Sinclair. 2011. A simple model for chickpea development, growth and yield. *Field Crops Research*, 124: 252–260.

Soltani, A., and T.R. Sinclair. 2012. *Modeling Physiology of Crop Development, Growth and Yield.* CABI Publication, Wallingford, pp. 322.

Soltani, A., F.R. Khooie, K. Ghassemi-Golezani, and M. Moghaddam. 2001. A simulation study of chickpea crop response limited irrigation in a semi-arid environment. *Agriculture Water Management*, 49: 225–237.

Tadmor, Y., D. Zamir, and G. Ladizinsky. 1987. Genetic mapping of an ancient translocation in the genus Lens. *Theoretical and Applied Genetics*, 73: 883–892.

Tahir, M., and F.J. Muehlbauer. 1994. Gene-mapping in lentil with recombinant inbred lines. *Journal of Heredity*, 85: 306–310.

Temel, H.Y., D. Gol, A. Kahriman, and M.B. Tanyolac. 2014. Single nucleotide polymorphism discovery through Illumina- based transcriptome sequencing and mapping in lentil. *Turkish Journal of Agricilture and Forestry*, 38: 1–19.

Thavarajah, D., A. Vandenberg, G. N. George, and I. J. Pickering. 2007. Chemical form of selenium in naturally selenium-rich lentils (*Lens culinaris* L.) from Saskatchewan. *Journal of Agricultural and Food Chemistry*, 55: 7337–7341.

Thomson, D.D., K.H.M. Siddique, M.D. Barr, and J.M. Wilson. 1997. Grain legume species in low rainfall Mediterranean-type environments. I, Phenology and seed yield. *Field Crops Research*, 54: 173–187.

Tullu, A., B. Tar'an, T. Warkentin, and A. Vandenberg. 2008. Construction of an Intraspecific linkage map and QTL analysis for earliness and plant height in lentil. *Crop Science,* 48: 2254–2264.

Turner, N.C. 1992. Crop production on duplex soils: An introduction. *Australian Journal of Experimental Agriculture* 32: 797–800.

Turner, N.C., G.C. Wright, and K.H.M. Siddique 2017. Adaptation of grain legumes (pulses) to water limited environments. *Advances Agronomy*, 1: 193–231.

Vadez, V., A. Soltani, and T.R. Sinclair. 2013. Crop simulation analysis of phenological adaptation of chickpea to different latitudes of India. *Field Crops Research*, 146: 1–9.

Van Kessel, C. 1994. Seasonal accumulation and partitioning of nitrogen by lentil. *Plant and Soil*, 164: 69–76.

Wahid, A., Gelani, S., Ashraf, M., and Foolad, M. R. 2007. Heat tolerance in plants: an overview. *Environmental and Experimental Botany*, 61:199–223.

Weller, J.L., L.C. Liew, V.F.G. Hecht, V. Rajandran, R.E. Laurie, and S. Ridge. 2012. A conserved molecular basis for photoperiod adaptation in two temperate legumes. *Proceedings of the National Academy of Sciences of the United States of America,* 109: 21158–21163.

Yadav, N., J. Lee, J. Kim, J. Shen, M.C. Hu, C. Aldaz, and M.T. Bedford. 2003. Specific protein methylation defects and gene expression perturbations in coactivator-associated arginine methyltransferase 1-deficient mice. *Proceedings of National Academy of Science, USA*, 100: 6464–6468.

Yusuf, M., N.P. Singh, and N.G. Dastane. 1979. Effect of frequency and timings of irrigation on gain yield and water use efficiency of lentil. *Annals of Arid Zone*, 18: 127–134.

Zamir, D., and G. Ladizinsky. 1984. Genetics of allozyme variants and linkage groups in lentil. *Euphytica*, 33: 329–336.

Zohary, D. 1972. The wild progenitor and place of origin of the cultivated lentil (*Lens culinaris* L.). *Economic Botany*, 26:326–332.

20

Micropropagation for Stress Tolerance in Crop Plants: An Overview

Soumen Saha, Khasru Alam, and Tanmay Chowdhury
Raiganj University

Parthadeb Ghosh
University of Kalyani

CONTENTS

20.1 Introduction

Global climate change and a growing population make the development of stress-resistant crops essential. Stress factors are a major complication that reduces the yield capability of modern cultivation. Classical breeding systems and specific management systems remain to perform a significant part in crop development. Since, in numerous cases, these traditional (classical) breeding ways have been unsuccessful to bring the desired consequences (Purohit et al., 1998; Rai et al., 2011). The practices of traditional breeding techniques such as selection, artificial pollination, hybridization, backcross, compound crossing and multiline have been employed, but there were diverse obstacles such as slow, long time, money cost and busyness in building crop resistance (Helaly, 2017). Further, classical breeding faced a little complexity that affected the crossing (hybridization) such as environmental variance notably, low and high relative humidity, likewise low and high temperatures. Therefore, looking at the economy and the urgent need for solutions, now is the time to move on to some simpler approach.

Production and productiveness of innumerable crops are being negatively influenced for the reason that numerous biotic and abiotic pressure. Biotic stresses entering portion of plants is damaged for the reason that several living organisms like viruses, bacteria, fungi, nematodes, insects, parasites and weeds reduce crop yields by 25% globally (Savary et al., 2012). Abiotic stress is a varied component of an individual environment that affects plants (Kayum et al., 2016) and is harmful to plant growth, development and crop yields, as a result decrease in crop productivity by more than 50% (Acquaah, 2007). The damage caused by this pressure is responsible for the huge economic losses worldwide. Abiotic stresses give on a chain of morpho-physiological, biochemical and molecular variation, which negatively affect plant augmentation, development, survival and productiveness (Helaly, 2017). A common mechanism of abiotic factors such as drought and salinity is that they disrupt the water metabolism of plants and therefore inhibit the growth of cell proliferation (Munns and Tester, 2008). It is believed that about 8.6 million hectares of land in India are influenced by salinity (Pathak, 2000).

Plant tolerance to salt stresses is a very complex phenomenon where the primary responsibility is to overcome osmotic and ionic stresses and subsequently to deal with secondary stresses (e.g. oxidative stress). In current years, the use of dexterity ways on in vitro tissue culture has built feasible the progress of biotechnological device to focus on the complex issue of crop development for sustainable agriculture cultivation.

Various agronomic practices can be used to manage various abiotic stresses, including changes in seed sowing, soil improvement and irrigation by quality water. Although, these strategies are never profitable and durable for crop production (Farooq et al., 2009; Hashmi et al., 2015). Thus, it's mandatory to discover a stable explanation toward preserving the various crop plants against stress. The traditional crop breeding programs are being employed to integrate interest inclinable of genes from different genera and species are crossed into the crops instigated to stress

DOI: 10.1201/9781003258063-20

tolerance. In addition, traditional breeding techniques have less benefits and barely produced the desired outcome (Purohit et al., 1998; Rai et al., 2011). Furthermore, biotechnological techniques could be applied as improving the performance of stress tolerable in various crop plants. The genetic metamorphosis requires different transmission of genes (accountable for stress tolerable) coming out of remote gene pools of various plant species for the improvement of stress-tolerable crop plants. Accordingly, genetic engineering technique for the improvement of stress-tolerable plants is possibly an effective route during the development of plant varieties; however, the key limitations are less transformation efficiency, transgene mutation and decreased gene manifestation (Rai et al., 2011). Nowadays, plant micropropagation technique is justified being a beneficial and affordable tool owing to the progress of stress-tolerable plants. This method handles under-regulated laboratory rules with a short time period and limited space retaining huge potentiality for the improvement of stress-tolerable crop varieties and contribute to a greater understanding of the physiology and biochemistry of crop plants cultured under acute climate conditions (Rai et al., 2011; Pérez-Clemente and Gómez-Cadenas, 2012). Applying micropropagation strategies, many stress-tolerable crop plants are found. In this chapter, advances in the evolution of various stress-tolerable plants through in vitro tissue-culture–based methods have been expressed.

20.2 Somaclonal Variation

Somaclonal diversification has been broadly investigated in plants obtained through adventitious regeneration (Jain et al., 1998). The variations produced show chromosome breakdown and rearrangement and point mutations from polyploid and aneuploid to line. Epigenetic forms are also produced. Although it has been suggested that the nature of somaclonal variations varies from that of mutants produced by physiological mutagenesis, it is challenging to explain, as it is based on the sensitivity and characteristics of individual genomes. Although traumatic stress (oxidative bursts) may be challenging to exclude and may even trigger cell distribution and disagreement, it may be desirable to cut down the percent of gross aberrant by taking steps to cut down the implements of oxidative stress, such as by including antioxidants through culture medium. The oxidative stress that may be studied for somaclonal diversification is better supported by the existence of developmental changes, as oxidative stress is still known to cut down DNA methylation (Cassells and Curry, 2001). Revitalization, which takes place in regenerating tissues in general and in cultures, may come from in hypomethylation of DNA due to oxidative stress. Some modifications of the advanced stem in vitro may be due to genetic variations in the embedding tissue. Aberrant mitosis in vitro under the guide of oxidative stress may again describe genetic vulnerability. In the conditions of in vitro development, collection may occur in difficulty due to selection pressure, loss of capacity, varied character of secondary metabolism etc. Somaclonal areas may suffer hyperhydration and exhibit linguistic variations and exaggerations from genetically stable buds and nodal implants as found lines in culture.

In addition to strategies to decrease somaclonal diversity below economically adequate thresholds, micropropagation needs fresh proposals to adequately identify vulnerability in their cultures, either due to crop genotyping or due to a radical transformation in axillary bud propagation pathways. The latter is accompanied by the callus formation on bud explants. If the capital investment of a ploidy analyzer (unrefined flow cytometer) is not prohibited, it may be useful for the principle of universal somaclonal variant (description of somaclonal variation) polyploid description in some genotypes. But this technique may not be comprehensively predictable. Genetic fingerprinting, specifically those based on Amplified Fragment Length Polymorphism (AFLPs) or microsatellite primers, can be accepted on the condition that even this system cannot identify volatile locations that may differ from 157 mutate to a large number such as Pelargonium×in the domestic 'Grand Slam'. Arguably, the most valuable way to evaluate protocols for somaclonal diversification is to store and raise production samples and hope for identifications. This can be aided by computerized image evaluation (Cassells et al., 1999). Somaclonal variations can be detected through a variety of techniques, such as morpho-physiological, biochemical and cytological or molecular detection techniques. The two main ways used to identify somaclonal variants are screening and cell selection. Screening strategies require the monitoring of much number of cells or rejuvenate plants as identifying individual's form. Mutants can easily be distinguished from cell culture much more than the whole plant population for different characteristics. This is a consequence of immense number of cells that could be quickly and efficiently evaluated for mutant traits. Many times screening different types of plants is very difficult and usually impossible (Skirvin, 1978). Mutants are effectively selected for various disease resistances, develop nutritional value, with adaptive stress situations, e.g. saline, soils, less temperature, harmful metals and prevention to herbicides and to enhance the biogenesis of plant commodity utilized in the medicinal or manufactured purpose. Profitable and extensive screening is employed to isolate cell clones that produce large amounts of biochemicals (Vasil, 1994).

In the cell selection process, a suitable is used to allow only the survival of the selected and /or to grow the variant cells. A successful selection approach has been developed to recover the prevention genotype against different toxins, pesticides, fungicides, excess salt concentrations etc. (Ammirato and Styer, 1985). While the selection force permission the mutant cells to outlive/division, it's referred to as positive selection. Conversely, in the case of negative selection, wide type cell division in normally impaired by a selection agent, viz., arsenate. Mutant cells are protected from counter-selection agents because they are unable to divide the cells. These cells were later recovered by removing the counter-choice agent (Predieri, 2001).

20.3 Abiotic Stress Tolerant of Plants in in vitro Selection

Abiotic stress is a large comprehensive issue that defines crop production. Stress factors are a major complication that defines the yield capacity of modern cultivation. Abiotic stresses like salinity and heavy metals are the main sources of crop damage. This pressure is growing due to pollution, decreasing availability of good quality water and soil erosion. Stress creates a nutritional imbalance in the plant, which occurs in reduced water intake and enhanced toxicity, which in turn reduces the product.

The presence of the proper selection agent is the determining reason for the optimum result. Various research have shown that in vitro culture, single or coupled with mutagenesis, catalyze with basic or biological agents, can be used as a potential source for commercial cultivation of genetic variability and growth of new cultivars (Helaly, 2017). It has been observed at various times that the environments of in vitro cultures may be mutagenic and that plant organisms may regenerate through organ cultures, calli, protoplasts and somatic embryos again occasionally exhibited phenotypic/genotypic variations. Abiotic-stress-tolerant genotype is selected through in vitro selection stress technique which is the most widely used method. It is based on in vitro culture of plant cells, tissues/organs, and it helps out with selective agents, which allow the selection and reproduction of plants with the desired character. A list of different species of biotic stress prevention strategies through plant screening and in vitro selection is shown in Table 20.1.

20.3.1 Drought Stress

All humans are directly dependent on plants for food (Farooq et al., 2011). The potential for higher yields is undeniably important in solving the growing global starvation give. Since the crop has a complicated characteristic and is highly affected across the environment, it can cause strong damage due to drought, which is a common stress in most arid and semi-arid areas. Drought stress affects vegetation and varies from morphological to molecular density at all phenological developmental stages. Therefore, drought tolerability is one of the major elements of yield durability, and its development is a major challenging task for geneticists and breeders. The goal of any crop breeding program is to improve the yield, adapt to changing climatic conditions and restructure the genotypes that have a balanced food quality. Drought stress limits crop productivity by reducing tanning, leaf growth, seed weight and seed count. To increase agricultural productivity, crop plants need to be created to use the available moisture efficiently. Water potentials between 1.0 and 2.0 MPa cause drought stress, which hinders plant growth and development unless precautions are in place to reduce or prevent water loss (Hale and Orcutt, 1987). Tissue culture provides a unique opportunity to study the cellular response to drought stress by enabling plant cells to grow in defined water potential.

Polyethylene glycol (PEG) is used as the common ingredient for drought tolerance in in vitro selection. This type of technique has been applied to numerous plants (Bouslama and Scapaugh, 1984). Petkova et al. (1995) and Lestari (2005) established a positive relationship between seed germination capacity and plant growth under stress in media containing PEG. It is related to Dragisga et al. (1996) who found the same results that PEG could be applicable after in vitro selection to treat osmotic stress.

Drought is an environmentally complex stress factor, which can occur at different intensities and over time, impairing the increase and uplift of the crop cycle. Water scarcity is a multifarious pressure that affects plants at different layers of their physiological organization (Yordanov et al., 2000). Growing up tolerable to abiotic stress, such as future droughts, is necessary for worldwide food safety. The implements of the scarcity are required to develop with atmosphere diminishing and enhancing H_2O shortages. Drought concerned the potential reduction of environmental water, driven by the potential gradient of water flowing from plant cells, and cellular dehydration occurs in consequence of osmotic pressure. Destruction of intracellular H_2O loses the bi-layer constitution of the membrane, disrupts cellular metabolism (Mahajan and Tuteja, 2005), embarrasses photosynthesis and induces an increase in ROS. It also reduce stomatal closure by reducing Co_2 uptake and reducing photosynthetic activity and again preventing H_2O transport.

20.3.2 Cold

Cold stress limited the agricultural production of crops in hilly areas, which is a major environmental factor. Plant responds and adapts to stressful situations at the molecular and cellular layers in addition to the physiological and biochemical layers for its survival. Although, the expression of different genes in different plants is induced by different stresses. Different research showed appraised various plant species with various stresses like tolerant drought, salinity and cold. Nevertheless, very few details are given about the way employed to appraise the stress responsive, which can lead to a little confusing epilog about studies from an agronomic or physiological point of view (Sanghera and Wani, 2008).

Low temperatures often affect plant outgrowth and crop productivity, leading to serious crop injury (Xin and Browse, 2001). Plant tolerance varies in cold (0°C–15°C) and frozen (<0°C) temperatures. Plants in temperate climates generally discriminat to be cold tolerant with variable degress, and exposure to non-freezing temperate was increases their freezing tolerance, this method is known as cold affection (Levitt, 1980), which is related to biochemical and physiological exchange (Shinozaki et al., 1996; Thomasho, 1998; Gilmour et al., 2000), and finally reveal in gene expression, biomembrane lipid composition and accumulation of little molecules (Yamaguchi-Shinozaki and Shinozaki, 2006).

There has been limited progress in improving cold tolerance in crop plants due to inadequate genetic variableness of cold tolerance or deficiency of cold-tolerant–cultivated germplasm (Duncan and Carrow, 1999). Somaclonal selection on plant stress tolerance is a powerful tool to increase genetic variability (Karp, 1995). Earlier research has indicated noble cold tolerance in agricultural crops through somaclonal selection (Bertin and Bouharmont, 1997; Kendall et al., 1990; Lazar et al., 1988). For example, the experiment by Kendall et al. (1990) shows that winter wheat (*Triticum aestivum*) regenerate from callus and thereafter shows significant improvement in cold tolerance (below freezing) at open temperatures. Some plants survive this cryoselection process due to their own genetic

TABLE 20.1

Screening and in vitro Selection of some Economically Important Crop Plants for Various Abiotic Stress (Salt, Drought, Low Temperature, Metal and Ultraviolet) Tolerance During Last 20 Years

Plant Species	Stress	Selecting Agents	References
Glycine max (soya bean)	Salt	NaCl and $CaCl_2$	Liu and Staden (2000)
Oryza sativa (rice)	Salt	NaCl	Shankhdhar et al. (2000)
Tagetes minuta (Mexican marigold)	Drought	Mannitol	Mohamed et al. (2000)
Echinochloa colona (jungle rice)	Metal	Cr and Ni	Samantaray et al. (2001)
Lycopersicon esculentum (tomato)	Salt	NaCl	Kripkyy et al. (2001)
O. sativa (rice)	Drought/ chilling/Al	PEG	Biswas et al. (2002)
G. max L. Merrill (Soybean)	Salt	NaCl	Essa, (2002)
Brassica oleracea (cauliflower)	Salt	NaCl	Elavumoottil et al. (2003)
Dendrocalamus strictus (bamboo)	Salt	NaCl	Singh et al. (2003)
Diplachne fusca (kallar grass)	Salt	NaCl	Nanakorn et al. (2003)
Fragaria×ananassa (strawberry)	Salt	NaCl	Dziadczyk et al. (2003)
Helianthus annuus (sunflower)	Salt	NaCl	Davenport et al. (2003)
O. sativa (rice)	Salt	NaCl	Lee et al. (2003)
T. aestivum (wheat)	Salt	NaCl	Zair et al. (2003)
Triticum turgidum (wheat)	Salt	NaCl	Munns and James, (2003)
O. sativa (rice)	Drought/chilling/Al	PEG/low temperature/ $AlC_{l3}{\cdot}6H_2O/Al(SO_4)_3{\cdot}18H_2O$	Roy and Mandal, (2005)
Saccharum sp. (sugarcane)	Salt	NaCl	Gandonou et al. (2006)
Saccharum sp. (sugarcane)	Drought	Mannitol	Errabii et al. (2006)
G. max L. Merrill (soybean)	Salt	NaCl	Datta et al. (2006)
Wheat and other cereals	Salt	NaCl	Munns et al. (2006)
Chrysanthemum morifolium (chrysanthemum)	Salt	NaCl	Hossain et al. (2007)
Cynodon transvaalensis×Cynodon dactylon (Bermuda grass)	Salt	NaCl	Lu et al. (2007)
Nicotiana tabacum (tobacco)	Metal	Cu	Rout and Sahoo, (2007)
Solanum tuberosum (potato)	Salt	NaCl	Queiros et al. (2007)
Prunus cerasus×Prunus canescens (cherry rootstock)	Salt	NaCl	Erturk et al. (2007)
N. tabacum (tobacco)	Salt	NaCl	Rout et al. (2008)
Vigna radiata (mungbean)	Salt	NaCl	Hassan et al. (2008)
Ipomoea batatas L.(sweet potato)	Salt	NaCl	Dasgupta et al. (2008)
Peanut cultivars	Salt	NaCl	Mungala et al. (2008)
I. batatas (sweet potato)	Salt	NaCl	He et al. (2009)
Citrus aurantium (sour orange)	Salt	NaCl	Koc et al. (2009)
O. sativa (rice)	Salt	NaCl	Shanthi et al. (2010)
Medicago sativa L. (Alfalfa)	Salt	NaCl	Campanelli et al. (2013)
Tomato	Salt	NaCl	Seth and Kendurkar (2015)
G. max L. Merrill (Soybean)	Salt	NaCl	Hasan et al. (2016)
Solanum lycopersicum L. (Tomato)	Salt	NaCl	Rashed et al. (2016)
Tomato	Salt	NaCl	Zaki and Yokoi (2016)
T. aestivum L. (Wheat)	Salt	NaCl and Na_2SO_4	Oyiga et al. (2016)
S. tuberosum L. Cv. Challisha (Potato Challisha)	Salt	NaCl	Biswas et al. (2017)
Zea mays L. (Maize)	Salt	NaCl	Saputro et al. (2017)
O. sativa L. cv. MR263 (Rice)	Salt	NaCl	Atabaki et al. (2018)
O. sativa L. (Rice)	Salt	NaCl and $CaCl_2$	Kakar et al. (2019)
T. aestivum L. (Wheat)	Salt	NaCl or Na_2SO_4	Dadshani et al. (2019)
Citrus macrophylla (Alemow)	Salt	NaCl	Pérez-Jiménez and Pérez-Tornero (2020)
Phoenix dactylifera L. (date palm)	Salt	NaCl	Al-Khateeb et al. (2020)
S. tuberosum L. (potato)	Salt	NaCl	Ahmed et al. (2020)
Chenopodium quinoa Willd. (Quinoa)	Salt	NaCl	Derbali et al. (2020)

variants, which suggests that selection at the callus level may make a significant contribution to the hereditary progress of stress tolerance, although few epigenetic diversity can occur (Kendall et al., 1990). In addition, Bertin and Bouharmont (1997) indicated increased cold toleration in rice (*Oriza sativa*), which uses somaclonal changes and in vitro selection. Cold toleration of warm-season turfgrass can be elevated by in vitro selection under cold treatment for somaclonal changes. However, in vitro choosing is not well documented for the somaclonal variations of cold resistance in turfgrass species.

20.3.3 Salt

Salt stress is considered to be the maximum significant as abiotic stress, which can have injurious effects on plant growth and productivity by changing the morpho-physiological, biochemical and molecular levels of plants (Khan et al., 2009; Syeed et al., 2011). Salinity is deliberated to be the maximal useful limitation for the production of various crops in the soil, especially in arid and semi-arid areas of the globe (Zaman et al., 2005).

The usual consequence of the concentration of large amounts of salts (i.e. ions, in most cases Na^+) in soil fluidity is impaired H_2O, which reduce nutrient receiving and decreased photosynthetic activity and growth. Additionally, high salinity can lead to infavorable calcium cation (Ca^{2+}) or potassium ion to sodium ion cations, membrane lipid peroxidation and severe intracellular sodium ion (Na^+) concentrations (Levitt, 1980). Salt stress tolerance is obviously most significant for secondary oxidative stress tolerance; therefore, a number of salt toleration species not only increase the activity of antioxidant enzymes but also assimilate them in response to salt stress as opposed to salt-sensory species. A pressure of a salt acquires an unknown value for the ionic stress component at the cellular level of a certain signal path. One study found that in *Arabidopsis*, the salt-hypersensitive pathway involves Ca^{2+}-dependent protein kinase (Zhu, 2000). In addition, salt tolerance such as vacuolar sodium ion (Na^+)/hydrogen ion (H^+) defense Arabidopsis thaliana sodium/hydrogen antiporter gene (AtNHX) or more relation K^+ transporter (HKT) determinants have been significantly recognized in *Arabidopsis* (Rus et al., 2001; Yokoi et al., 2002).

Several researchers have tried to create an abiotic tolerant plant using tissue culture. For this, a large number of tissue and cell scanning techniques (such as shoot cultures, suspension cultures and callus culture) are required, which produce the ability to tolerate high levels of salt (NaCl) in a production mechanism. To this solution, several researchers have developed resolutions on agricultural species, including some basic plant layouts (Winicov, 1991; Johnson and Smith, 1992).

20.3.4 Heat

Plants perform their important functions at optimally characteristic temperatures, regulated by the limited thermosensitivity of the plant species. Temperatures over and down optimal array strongly affect cell structure and functions, such as membrane unity, enzyme activity and cell separation. As a mild stress, the thermo-tolerance of pretreatment plants can increase significantly with temperatures outside the optimum range. Thus, cold and heat adaptation is possible (Thomashow, 1999). Temperature stresses could be classified as frozen stress, heat stress and cold stress.

The effects of increased transpiration and oxidative stress are severely harmful to plant species, such as tissue collapse and plant death, and even plant survival and well-being are affected by high temperatures. In these situations, different types of events occur such as the synthesis of small heat shock proteins that act as a molecular chaperone in high abundance (Vierling, 1997), increase in Ca^{2+} influx and ROS accumulation in different parts of the cell (Doke et al., 1996). The suboptimal temperature, where the plant reduces the activity of its enzymes and possibly suffers from a lack of water availability, is called cold stress; however, the temperature is more than the freezing point of water. Metabolic processes, membrane spreading value, weaken enzyme response, membrane formation and connected transportation belongings are mainly contacted by cold stress (Tippmann et al., 2006; John et al., 2016).

The plant undergoes freezing stress, when the temperature is below zero degrees. The structures of the protoplast membrane are damaged, mechanically stricken and eventually killed by expanded ice crystals due to intracellular water accumulation; and severe dehydration of the protoplasm occurs due to extracellular deposits. Cold-induced 'cold-regulated gene' (COR), encoded by a category called hydrophilic polypeptides and the added protein stabilizes the membrane in opposition to freeze-induced blow, thereby aiding in frozen adaptation (Artus et al., 1996). Presumably by ethylene induction, defensive antifreeze proteins accumulate afterwards cold and drought treatment (Yu et al., 2001; Yu and Griffith, 2001; Griffith et al., 2005). Until now, two signal pathways, the ABA-independent pathway (Lang et al., 1994) and the ABA-dependent pathway (Capel et al., 1997), have been indicated to be activated by the release of cold-regulated genes in plants.

20.3.5 Toxin

The toxic effects of salt also depend on the species and duration of the plant. Plants cultivated in saline soils have different ionic structures and concentrations of dissolved salts. The pressure of salt also reduces the amount of total sugar. At the molecular level, there are many adaptive mechanisms for salt tolerance such as (i) ion homeostasis and (ii) detoxification or repair. These processes control stress damage (Zhu, 2002). Sodium homeostasis plays an important role in salt stress tolerance.

The growing dependability of agriculture on chemical fertilizers and wastewater watering, but fast industrialization has resulted in the entry of poison metals into agricultural soils, adversely affecting the soil environment.

20.4 Crop Plants and Biotic Stresses

Plants fight different types of biotic stresses, such as fungi, viruses, bacteria, nematodes and insects. The components of this biotic stress cause of variety of disorders, pollution and crop damage and which finally affect crop production. However, many strategies have been developed through various research methods to reduce biotic stress. Biotic stress could be reduced by examining the genetic mechanisms of these stress-causing elements. Genetically changes plants have shown significant success against biotic stress, through the development of impermeable resistant varieties of crop plants.

20.5 Conclusion

Based on the current situation, it is expected that the Earth's temperature will increase by 3°C–5°C in the next 50–100 years. Due to the constant rise in temperature and erratic rainfall, the difference between flood and drought is constantly in consideration. The anthropological functions, viz. too much fertilizer, improper watering and exploiting of metallic properties, may contribute to excessive salt stress. Under these conditions, plants face more frequent, simultaneous both stress. It is the responsibility of plant breeders to develop stress-tolerable varieties to ensure food security and farmer safety. The molecular function should be undertaken at the genetic level to generate structures to inhibit particular types of stress in plants. Unless a responsive mechanism in opposition to both stresses is created, plants will be subjected to such stress and will eventually show to be a serious risk to global agriculture.

REFERENCES

Acquaah, G. 2007. *Principles of Plant Genetics and Breeding*. Blackwell, Oxford.

Ahmed, H.A.A., Şahin, N.K., Akdoğan, G., Yaman, C., Köm, D., Uranbe, S. 2020. Variability in salinity stress tolerance of potato (*Solanum tuberosum* L.) varieties using in vitro screening. *Ciência e Agrotecnologia* 44:e004220.

Al-Khateeb, S.A., Al-Khateeb, A.A., Sattar, M.N., Mohmand, A.S. 2020. Induced in vitro adaptation for salt tolerance in date palm (*Phoenix dactylifera* L.) cultivar Khalas. *Biological Research* 53: 37.

Ammirato, P.V., Styer, D.J. 1985. Strategies for large scale manipulation of somatic embryos in suspension culture. In: *Biotechnology in Plant Science: Relevance to Agriculture in the Eighties*, M. Zaitlin, P. Day, A. Hollaender (Eds.), pp. 161–178. Academic Press, New York.

Artus, N.N., Uemura, M., Steponkus, P.L., Gilmour, S.J., Lin, C., Thomashow, M.F. 1996. Constitutive expression of the cold-regulated Arabidopsis thaliana COR15a gene affects both chloroplast and protoplast freezing tolerance. *Proceedings of the National Academy of Sciences* 93: 13404–13409.

Atabaki, N., Nulit, R., Kalhori, N., Lasumin, N., Sahebi, M., Abiri, R. 2018. In vitro Selection and Development of Malaysian Salt-Tolerant Rice (*Oryza sativa* L. cv. MR263) Under Salinity. *Acta Scientific Agriculture* 2: 08–17.

Bertin, P., Bouharmont, J. 1997. Use of somaclonal variation and in vitro selection for chilling tolerance improvement in rice. *Euphytica* 96: 135–142.

Biswas, J., Chowdhury, B., Bhattacharya, A., Mandal, A.B. 2002. In vitro screening for increased drought tolerance in rice. *In Vitro Cellular & Developmental Biology* 38: 525–530.

Biswas, M.S., Islam, M.R., Zakaria, M. 2017. Evaluation of Indigenous Potato Challisha (*Solanum tuberosum* L. Cv. Challisha) Somaclonals Tolerance to Salinity In Vitro. *The Journal of Tropical Life Science* 7: 77–82.

Bouslama, M., Schapaugh, W.T. 1984. Stress tolerance in soybean. I. Evaluation on three screening techniques for heat and drought tolerance. *Crop Science* 24: 993–937.

Campanelli, A., Ruta, C., Morone-Fortunato, I., Mastro G.D. 2013. Alfalfa (*Medicago sativa* L.) clones tolerant to salt stress: In vitro selection. *Central European Journal of Biology* 8: 765–776.

Capel, J., Jarillo, J.A., Salinas, J., Martinez-Zapater, J.M. 1997. Two homologous low- temperature-inducible genes from Arabidopsis encode highly hydrophobic proteins. *Plant Physiology* 115: 569–576.

Cassells, A.C., Curry, R.F. 2001. Oxidative stress and physiological, epigenetic and genetic variability in plant tissue culture: Implications for micropropagators and genetic engineers. *Plant Cell Tissue and Organ Culture* 64: 145–157.

Cassells, A.C., Joyce, S.M., Curry, R.F., McCarthy, T.F. 1999. Detection of economically important variability in micropropagation. In: *Plant Biotechnology and In Vitro Biology in the 21st Century: Current Plant Science and Biotechnology in Agriculture*, A. Altman, M. Ziv, S. Izhar (Eds.), vol. 36. Springer, Dordrecht. doi: 10.1007/978-94-011-4661-6_56.

Dadshani, S., Sharma, R.C., Baum, M., Ogbonnaya, F.C., Léon, J., and Ballvora, A. 2019. Multi-dimensional evaluation of response to salt stress in wheat. *PLoS One* 14: e0222659.

Dasgupta, M., Sahoo, M.R., Kole, P.C., Mukherjee, A. 2008. Evaluation of orange-fleshed sweet potato (*Ipomoea batatas* L.) genotypes for salt tolerance through shoot apex culture under in vitro NaCl mediated salinity stress conditions. *Plant Cell Tissue and Organ Culture* 94: 161–170.

Datta, A.K., Hossain, M.A., Robin, A.B.M.A.H.K., Raffi, S.A., Hossain, M.A. 2006. Screening of soybean genotypes for salinity tolerance at seedling stage. *Bangladesh Journal of Crop Science* 17: 155–161.

Davenport, S.B., Gallego, S.M., Benavides, M.P., Tomaro, M.L. 2003. Behaviour of antioxidant defense system in the adaptive response to salt stress in *Helianthus annuus* L. cells. *Plant Growth Regulation* 40: 81–88.

Derbali, W., Goussi, R., Koyro H.W., Abdelly, C., Manaa A. 2020. Physiological and biochemical markers for screening salt tolerant quinoa genotypes at early seedling stage. *Journal of Plant Interactions* 15: 27–38.

Doke, N., Miura, Y., Sanchez, L.M., Park, H.J., Noritake, T., Yoshioka, H., Kawakita, K. 1996. The oxidative burst protects plants against pathogen attack: Mechanism and role as an emergency signal for plant bio-defence-a review. *Gene* 179: 45–51.

Dragisga, R., Djilianov, D., Dencher, P., Atanassov, A. 1996. In vitro selection for somatic tolerance in alfalfa (*Medicago sativa* L.). *Bulgarian Journal of Plant Physiology* 22: 30–39.

Duncan, R.R., Carrow, R.N. 1999. *Seashore Paspalum: The Environmental Grass*. Wiley, Hoboken, NJ.

Dziadczyk, P., Bolibok, H., Tyrka, M., Hortynski, J.A. 2003. In vitro selection of strawberry (*Fragaria×ananassa* Duch.) clones tolerant to salt stress. *Euphytica* 132: 49–55.

Elavumoottil, O.C., Martin, J.P., Moreno, M.L. 2003. Changes in sugars, sucrose synthase activity and proteins in salinity tolerant callus and cell suspension cultures of *Brassica oleracea* L. *Biologia Plantarum* 46: 7–12.

Errabii, T., Gandonou, C.B., Essalmani, H., Abrini, J., Idomar, M., Senhaji, N.S. 2006. Growth, proline and ion accumulation in sugarcane callus cultures under drought- induced osmotic stress and its subsequent relief. *African Journal of Biotechnology* 5: 1488–1493.

Erturk, U., Sivritepe, N., Yerlikaya, C., Bor, M., Ozdemir, F., Turkan, I. 2007. Responses of the cherry rootstock to salinity in vitro. *Biologia Plantarum* 51: 597–600.

Essa, T.A. 2002. Effect of Salinity Stress on Growth and Nutrient Composition of Three Soybean (*Glycine max* L. Merrill) Cultivars. *Journal of Agronomy and Crop Science* 188: 86–93.

Farooq, M., Wahid, A., Kobayashi, N., Fujita, D., Basra, S.M.A. 2009. Plant drought stress: Effects, mechanisms and management. *Agronomy for Sustainable Development* 29: 185–212.

Farooq, J., Khaliq, I., Kashif, M., Ali, Q., Mahpara, S. 2011. Genetic analysis for relative cell injury percentage and some yield contributing traits in wheat under normal and heat stress conditions. *Chilean Journal of Agricultural Research* 71: 511–520.

Flexas, J., Bota, J., Loreto, F., Cornic, G., Sharkey, T.D. 2004. Diffusive and metabolic limitations to photosynthesis under drought and salinity in C3 plants. *Plant Biology* 6: 269–279.

Gandonou, C.B., Errabii, T., Abrini, J., Idaomar, M., Senhaji, N.S. 2006. Selection of callus cultures of sugarcane (*Saccharum* sp.) tolerant to NaCl and their response to salt stress. *Plant Cell Tissue and Organ Culture* 87: 9–16.

Gilmour, S.J., Sebolt, A.M., Salazar, M.P., Everard, J.D., Thomashow, M.F. 2000. Overexpression of the Arabidopsis CBF3 transcriptional activator mimics multiple biochemical changes associated with cold acclimation. *Plant Physiology* 124: 1854–1865.

Griffith, M., Lumb, C., Wiseman, S.B., Wisniewski, M., Johnson, R.W., Marangoni, A.G. 2005. Antifreeze proteins modify the freezing process in plant. *Plant Physiology* 138: 330–340.

Hale, M.G., Orcutt, D.M. 1987. *The Physiology of Plant under Stress*. John Wiley & Sons Inc., New York.

Hasan, M.K., Nasiruddin, K.M., Al-Amin M., Hossain, A.K.M.S. 2016. In vitro screening of Soybean genotypes under salinity stress. *International Journal of Applied Sciences and Biotechnology* 4: 207–212.

Hashmi, U., Shafqat, S., Khan, F., Majid, M., Hussain, H., Kazi, A.G., John, R., Ahmad, P. 2015. Plant exomics: Concepts, applications and methodologies in crop improvement. *Plant Signaling & Behavior* 10: e976152.

Hassan, N.M., Serag, M.S., El-Feky, F.M., Nemat Alla, M.M. 2008. In vitro selection of mung bean and tomato for improving tolerance to NaCl. *Annals of Applied Biology* 152: 319–330.

He, S., Han, Y., Wang, Y., Zhai, H., Liu, Q. 2009. In vitro selection and identification of sweet potato (*Ipomoea batatas* (L.) Lam.) plants tolerant to NaCl. *Plant Cell Tissue and Organ Culture* 96: 69–74.

Helaly, A.A. 2017. strategies for improvement of horticultural crops against abiotic stresses. *Journal of Horticulture* 4:e107. doi: 10.4172/2376-0354.1000e107.

Hossain, Z., Mandal, A.K.A., Datta, S.K., Biswas, A.K. 2007. Development of NaCl tolerant line in *Chrysanthemum morifolium* Ramat. through shoot organogenesis of selected callus line. *Journal of Biotechnology* 129: 658–667.

Jain, S.M., Brar, D.S., Ahloowalia, B.S. 1998. *Somaclonal Variations and Induced Mutations in Crop Improvement*. Kluwer Academic Publishers, Dordrecht.

John, R., Anjum, N.A., Sopory, S.K., Akram, N.A., Ashraf, M. 2016. Some key physiological and molecular processes of cold acclimation. *Biologia Plantarum* 60: 603–618.

Johnson, D.W., Smith, S.E. 1992. Response to NaCl of alfalfa plants regenerated from non- saline callus cultures. *Plant and Soil* 143: 311–315.

Kakar, N., Jumaa, S.H., Redoña, E.D., Warburton, M.L., Reddy, K.R. 2019. Evaluating rice for salinity using pot-culture provides a systematic tolerance assessment at the seedling stage. *Rice* 12: 57.

Karp, A. 1995. Somaclonal variation as a tool for crop improvement. *Euphytica* 85: 295–302.

Kayum, M.A., Kim, H.T., Nath, U.K., Park, J.I., Kho, K.H., Cho, Y.G., I-Sup, N. 2016. Research on biotic and abiotic stress related genes exploration and prediction in *Brassica rapa* and *B. oleracea*: A review. *Plant Breeding and Biotechnology* 4: 135–144.

Kendall, E.J., Qureshi, J.A., Kartha, K.K., Leung, N., Chevrier, N., Caswell, K., Chen, T.H.H. 1990. Regeneration of freezing tolerant spring wheat (*Triticum aestivum* L.) plants from cryoselected callus. *Plant Physiology* 94: 1756–1762.

Khan, M.A., Shirazi, M.U., Khan, M.A., Mujtaba, S.M., Islam, E., Mumtaz, S., Shereen, A., Ashraf, M.Y. 2009. Role of proline, K/Na ratio and chlorophyll content in salt tolerance of wheat (*Triticum aestivum* L.). *Pakistan Journal of Botany* 41: 633–638.

Koc, N.K., Bas, B., Koc, M., Kusek, M. 2009. Investigations of in vitro selection for salt tolerant Lines in Sour Orange (*Citrus aurantium* L.). *Biotechnology* 8: 155–159.

Kripkyy, O., Kerkeb, L., Molina, A., Belver, A., Rodrigues, R.P., Donaire, P.J. 2001. Effects of salt-adaptation and salt-stress on extracellular acidification and microsome phosphohydrolase activities in tomato cell suspensions. *Plant Cell Tissue and Organ Culture* 66: 41–47.

Kusumi, K., Hirotsuka, S., Kumamaru, T., Iba, K. 2012. Increased leaf photosynthesis caused by elevated stomatal conductance in a rice mutant deficient in SLAC1, a guard cell anion channel protein. *Journal of Experimental Botany* 63: 5635–5644.

Lang, V., Mantyla, E., Welin, B., Sundberg, B., Palva, E.T. 1994. Alterations in water status, endogenous abscisic acid content, and expression of rab18 gene during the development of freezing tolerance in *Arabidopsis thaliana*. *Plant Physiology* 104: 1341–134.

Lazar, M.D., Chen, T.H.H., Gusta, L.V., Kartha, K.K. 1988. Somaclonal variation for freezing tolerance in a population derived from norstar winter wheat. *Theoretical and Applied Genetics* 75: 480–484.

Lee, S.Y., Lee, J.H., Kwon, T.O. 2003. Selection of salt-tolerant doubled haploids in rice anther culture. *Plant Cell Tissue and Organ Culture* 74: 143–149.

Lestari, E.G. 2005. Seleksi in vitro untuk ketahanan terhadap kekeringan pada tanaman padi. [Disertasi]. Bogor: Sekolah Pascasarjana, Institut Pertanian Bogor.

Levitt, J. 1980. Responses of plants to environmental stress. In: *Chilling, Freezing, and High Temperature Stress*. Academic Press, New York 365.

Liu, T., Staden, J.V. 2000. Selection and characterization of sodium chloride-tolerant callus of *Glycine max* (L.) Merr cv. Acme. *Plant Growth Regulation* 31: 195–207.

Lu, S., Peng, X., Guo, Z., Zhang, G., Wang, Z., Wang, C., Pang, C., Fan, Z., Wang, J. 2007. In vitro selection of salinity tolerant variants from triploid bermuda grass (*Cynodon transvaalensis*×*C. dactylon*) and their physiological responses to salt and drought stress. *Plant Cell Reports* 26: 1413–1420.

Mahajan, S., Tuteja, N. 2005. Cold, salinity, and drought stress: An overview. *Archives of Biochemistry and Biophysics* 444: 139–159.

Mohamed, M.H., Harris, P.J.C., Henderson, J. 2000. In vitro selection and characterisation of a drought tolerant clone of Tagetes minuta. *Plant Science* 159: 213–222.

Mungala, A.J., Radhakrishnan, T. Dobaria, J.R. 2008. In vitro Screening of 123 Indian peanut cultivars for sodium chloride induced salinity tolerance. *World Journal of Agricultural Sciences* 4: 574–582.

Munns, R., James, R.A. 2003. Screening methods for salinity tolerance: A case study with tetraploid wheat. *Plant and Soil* 253: 201–218.

Munns, R., Tester, M. 2008. Mechanisms of salinity tolerance. *Annual Review of Plant Biology* 59: 651–681.

Munns, R., James, R.A., Lauchli, A. 2006. Approaches to increasing the salt tolerance of wheat and other cereals. *Journal of Experimental Botany* 57: 1025–1043.

Nanakorn, M., Jiamjetjaroon, W., Suwanawong, S., Wongwattana, C., Shim, I.S. 2003. In vitro selection of salt-tolerant cell lines in kallar grass (*Diplachnefusca* L.). *Weed Biology and Management* 3: 49–52.

Oyiga, B.C., Sharma, R.C., Shen, J., Baum, M., Ogbonnaya, F.C., Leon, J., Ballvora, A. 2016. Identification and characterization of salt tolerance of wheat Germplasm using a multivariable screening approach. *Journal of Agronomy and Crop Science* 202: 472–485.

Pathak, P.S. 2000. Agro forestry: A tool for arresting land degradation. *Indian Farming* 49: 15–19.

Pérez-Clemente, R.M., Gómez-Cadenas, A. 2012. In vitro tissue culture, a tool for the study and breeding of plants subjected to abiotic stress conditions. In: *Recent Advances in Plant in vitro Culture*, A. Leva, L. Rinaldi (Eds.), pp. 30–91. BoD – Books on Demand, Norderstedt, Germany.

Pérez-Jiménez, M., Pérez-Tornero, O. 2020. In vitro plant evaluation trial: Reliability test of salinity assays in citrus plants. *Plants* 9: 1352.

Petkova, D., Nedjialkov, D., Djilianov, D. 1995. Early screening for drought tolerance in cultivatead alfalfa. *Bulgarian Journal of Agricultural Science* 1: 429–432.

Predieri, S. 2001. Mutation induction and tissue culture in improving fruits. *Plant Cell Tissue and Organ Culture* 64: 185–210.

Purohit, M., Srivastava, S., Srivastava, P.S. 1998. Stress tolerant plants through tissue culture. In: *Plant Tissue Culture and Molecular Biology: Application and Prospects*, P.S. Srivastava (Ed.), pp. 554–578. Narosa Publishing House, New Delhi.

Queiros, F., Fidalgo, F., Santos, I., Salema, R. 2007. In vitro selection of salt tolerant cell lines in *Solanum tuberosum* L. *Biologia Plantarum* 51: 728–734.

Rai, M.K., Kalia, R.K., Singh, R., Gangola, M.P., Dhawan, A.K. 2011. Developing stress tolerant plants through in vitro selection: An overview of the recent progress. *Environmental and Experimental Botany* 71: 89–98.

Rashed, M.R.U., Roy, M.R., Paul, S.K., Haque, M.M. 2016. In vitro screening of salt tolerent genotypes in tomato (*Solanum lycopersicum* L.). *Journal of Horticulture* 3: 1–8.

Rout, G.R., Sahoo, S. 2007. In vitro selection and plant regeneration of copper-tolerant plants from leaf explants of *Nicotiana tabacum* L. cv. Xanthi. *Plant Breeding* 126: 403–409.

Rout, G.R., Senapati, S.K., Panda, J.J. 2008. Selection of salt tolerant plants of *Nicotiana tabacum* L. through in vitro and its biochemical characterization. *Acta Biologica Hungarica* 59: 77–92.

Roy, B., Mandal, A.B. 2005. Anther culture response in indica rice and variations in major agronomic characters among the androclones of a scented cultivar, Karnal local. *African Journal of Biotechnology* 4: 235–240.

Rus, A., Yokoi, S., Sharkhuu, A., Reddy, M., Lee, B.H., Matsumoto, T.K., Koiwa, H., Zhu, J.K., Bressan, R.A. Hasegawa, P.M. 2001. AtHKT1 is a salt tolerance determinant that controls Na+ entry into plant roots. *Proceedings of the National Academy of Sciences, USA* 98: 14150–14155.

Samantaray, S., Rout, G.R., Das, P. 2001. Induction, selection and characterization of Cr and Ni-tolerant cell lines of *Echinochloa colona* (L.) Link in vitro. *Journal of Plant Physiology* 158: 1281–1290.

Sanghera, G.S., Wani, S.H. 2008. Innovative approaches to enhance genetic potential of rice for higher productivity under temperate conditions of Kashmir. *Journal Plant Science and Research* 24: 99–113.

Saputro, T.B., Dianawati, S., Sholihah, N.F., Ermavitalini, D. 2017. Genetic diversity of improved salt tolerant Calli of Maize (*Zea mays* L.) using RAPD. *AIP Conference Proceedings* 1854: 020033.

Savary, S., Ficke, A., Aubertot, J.N., Hollier, C. 2012. Crop losses due to diseases and their implications for global food production losses and food security. *Food Security* 4: 519–537.

Seth, R., Kendurkar, S.V. 2015. In vitro screening: An effective method for evaluation of commercial cultivars of tomato towards salinity stress. *International Journal of Current Microbiology and Applied Science* 4: 725–730.

Shankhdhar, D., Shankhdhar, S.C., Mani, S.C., Pant, R.C. 2000. In vitro selection for salt tolerance in rice. *Biologia Plantarum* 43: 477–480.

Shanthi, P., Jebaraj, S., Geetha, S. 2010. In vitro screening for salt tolerance in Rice (*Oryza sativa*). *Electronic Journal of Plant Breeding* 1: 1208–1212.

Shinozaki, K., Yamaguchi-Shinozaki, K. 1996. Molecular responses to drought and cold stress. *Current Opinion in Biotechnology* 7: 161–167.

Singh, M., Jaiswal, U., Jaiswal, V.S. 2003. In vitro selection of NaCl-tolerant callus lines and regeneration of plantlets in a bamboo (Dendrocalamus strictus Nees.). *In Vitro Cellular & Developmental Biology - Plant* 39: 229–233.

Skirvin, R.M. 1978. Natural and induced variation in tissue culture. *Euphytica* 27: 241–266.

Syeed S., Anjum, N.A., Nazar R., Iqbal N., Masood A., Khan N.A. 2011. Salicylic acid- mediated changes in photosynthesis, nutrients content and antioxidant metabolism in two mustard (*Brassica juncea* L.) cultivars differing in salt tolerance. *Acta Physiologiae Plantarum* 33: 877–886.

Thomashow, M.F. 1998. Role of cold-responsive genes in plant freezing tolerance. *Plant Physiology* 118: 1–8.

Thomashow, M.F. 1999. Plant cold acclimation: Freezing tolerance genes and regulatory mechanisms. *Annual Review of Plant Biology* 50: 571–599.

Tippmann, H.F., Schlüter, U., Collinge, D.B. 2006. Common themes in biotic and abiotic stress sig nalling in plants. In: *Floriculture, Ornamental and Plant Biotechnology*, J.A. Teixeira da Silva (Ed.), pp. 52–67. Global Science Books, Ikenobe.

Vasil, I.K. 1994. Automation of plant propagation. *Plant Cell Tissue and Organ Culture* 9: 105–108.

Vierling, E. 1997. The small heat shock proteins in plants are members of an ancient family of heat induced proteins. *Acta Physiologiae Plantarum* 19: 539–547.

Winicov, I. 1991. Characterization of salt-tolerant alfalfa (*Medicago sativa* L.) plants regenerated from salt tolerant cell lines. *Plant Cell Reports* 10: 561–564.

Xin, Z., Browse, J. 2001. Cold comfort farm: The acclimation of plants to freezing temperatures. *Plant Cell and Environment* 23: 893–902.

Yamaguchi-Shinozaki, K., Shinozaki, K. 2006. Transcriptional regulatory networks in cellular responses and tolerance to dehydration and cold stresses. *Annual Review of Plant Biology* 57: 781–803.

Yokoi, S., Quintero, F.J., Cubero, B., Ruiz, M.T., Bressan, R.A., Hasegawa, P.M., Pardo, J.M. 2002. Differential expression and function of Arabidopsis thaliana NHX Na^+/H^+ antiporters in the salt stress response. *The Plant Journal* 30: 529–539.

Yordanov, I., Velikova, V., Tsonev, T. 2000. Plant response to drought, acclimation, and stress tolerance. *Photosynthetica* 38: 171–186.

Yu, X.M., Griffith, M. 2001. Winter rye antifreeze activity increases in response to cold and drought, but not abscisic acid. *Physiologia Plantarum* 112: 78–86.

Yu, F., White, S.B., Zhao, Q., Lee, F.S. 2001. HIF-1α binding to VHL is regulated by stimulus-sensitive proline hydroxylation. *Proceedings of the National Academy of Sciences* 98: 9630–9635.

Zair, I., Chlyah, A., Sabounji, K., Tittahsen, M., Chlyah, H. 2003. Salt tolerance improvement in some wheat cultivars after application of in vitro selection pressure. *Plant Cell Tissue and Organ Culture* 73: 237–244.

Zaki, H.E.M., Yokoi, S. 2016. A comparative in vitro study of salt tolerance in cultivated tomato and related wild species. *Plant Biotechnology* 33: 361–372.

Zaman, B., Niazi, B.H., Athar, M., and Ahmad, M. 2005. Response of wheat plants to sodium and calcium ion interaction under saline environment. *International Journal of Environmental Science & Technology* 2: 7–12.

Zhu, J.K. 2000. Genetic analysis of plant salt tolerance using Arabidopsis. *Plant Physiology* 124: 941–948.

Zhu, J.K. 2002. Salt and drought stress signal transduction in plants. *Annual Review of Plant Biology* 53: 247–273.

21

Specialty Traditional Rice Landraces: Its Nutraceutical and Therapeutic Potentiality for Human Health

Priyabrata Roy
University of Kalyani
International Rice Research Institute

Tanima Saha
University of Kalyani

CONTENTS

21.1 Introduction

Plant development, agricultural crop productivity, and the quality of plant products harvested for human or animal sustenance are all influenced by a variety of biotic and abiotic stressors. Nutrient deficiency, an imbalanced nutrient supply, and pollution are examples of abiotic stresses, as are climatic variables like as drought, flooding, heat waves, or cold temperatures. Climate change makes climatic conditions (especially extreme events) increasingly important, and this might alter plant nutrition. Aside from individual stress reactions, combinations of stresses (for example, drought and heat, drought and pests, drought, and an imbalanced nutrition supply) must be considered. Field crops like rice are considered as a significant staple food for Asians. The broad term 'nutraceutical' is used to describe foods having therapeutic properties with health benefits for human beings. Its therapeutic properties have been clearly described in various folk medicinal literatures (Umadevi et al., 2012). Rice has been used since ancient times for many reasons and was one of the essential components of oral rehydration solution (ORS). Rice is an important part of festivals and rituals in almost all Asian countries (Deb, 2005). Recent health benefits of rice have brought attention to the rise of lifestyle-associated diseases such as cancer, cardiovascular disease, and diabetes (de Oliveira et al., 2020). Several laboratory tests have already shown that rice and its by-products exhibit anticancerous effects and cure certain diseases such as diabetes, kidney problems, and high cholesterol (Tan and Norhaizan, 2017; de Oliveira et al., 2020). While rice is known as a starchy food, it contains proteins that are important for normal metabolic processes with significant amino acids, fatty acids, fiber, many minerals, and vitamins (Gul et al., 2015). Scientific evidence backs up some of the therapeutic properties of rice (Deb et al., 2015; Ashraf and Lokanandan, 2017).

DOI: 10.1201/9781003258063-21

As an example, the antioxidant properties of rice have cancer prevention abilities, and some traditional rice varieties having low glycemic index (GI) can significantly decrease blood glucose levels and diabetics (Ranilla et al., 2010). Because of its nourishing and moisturizing effects on skin and hair, rice starch extracts and bran oil are also used in a wide range of cosmetics and other products (Charoen et al., 2011). Many of these nutraceutical and medicinal properties of rice differ greatly depending on the rice types, and studies have shown that colored rice such as red and black rice of folk varieties are much better than high-yielding rice varieties (Deepa et al., 2008). For example, the Rakktashali red rice has been identified as a food and medicine in the old Ayurvedic treatises (Ahuja et al., 2008). In folk medicinal literature, in the treatment of various diseases, a therapeutic explanation of various different rice varieties, for example, Sashtika, Sali, and dried rice has been included. A number of rice varieties are known to treat dermal problems, high pressure, fatigue, and leukorrhea, as a health tonic and for lactation with other medicinal properties, even today in various parts of India (Deb, 2005). Kerala's well-known Nayavara rice is widely used as a body-enhancing thing in Ayurvedic treaties to avoid toxins and postpone aging (Joseph et al., 2007). Pigmented rice varieties (dark and red) are enriched in minerals (Sen Gupta et al., 2017) and polyphenols, having cell reinforcement properties. There is a low GI for traditional rice varieties such as basmati, which is used in weight-decreasing diets (Srinivasa et al., 2013). The rice ORS is stronger than the glucose ORS and thus included in recommendations from the World Health Organization (WHO). Indian nutritionists have normally opposed junk foods and have been continually suggesting the utilization of fresh, balanced, and functional foods. There are sincere efforts to discover novel food-based ways to reduce the frequency of life-related diseases (Chaudhary et al., 2018) by improving mineral substances, cancer prevention agents, and food glycemic records. Despite the fact that the amount of protein in rice is low, its dietary advantage is high (Chaudhary and Tran, 2001). The old literature of rice-developing Asian nations has attributed some restorative properties to rice. They utilize red rice yeast for the treatment of different diseases. In folk literature of Malaysia, cooked rice has been used for eye treatment and for use in internal body tissue infection. Rice hulls in Cambodia are considered helpful for the treatment of dysentery and rice hulls (3 months of age) are diuretic. Rice bran (tikitiki) is a very good source of vitamin B, which is processed in the Philippines and used to prevent and treat beriberi (Ahuja et al., 2008). In India, rice has served as an eccentric since old times for its exceptional characteristics and properties which are referenced in antiquated Indian writings and old stories (Deb, 2005). Rice is an excellent source for enriching our body with micronutrients, excluding toxic metabolites, fortifying, revitalizing, and energizing the body, regulating blood pressure, preventing skin diseases and premature aging due to its antioxidant property (Umadevi et al., 2012). Thousands of Indian tradition rice landraces which are confined with marginal farmers and conservators should be clinically validated for their restorative properties. Most significantly, efforts need to be made to preserve these landraces by public awareness campaigns, which help to compete with the current market of high-yield varieties. Through this review, the authors intend to critically discuss the effects of stress on rice while focusing mostly on the nutraceutical attributes of traditional rice landraces which are genetically stress tolerant by nature.

21.2 Rice Grain Structure and Characterization

Rice grain is made of hull consisting of lemna and palea, and the caryopsis (Juliano and Bechtel, 1985). There are three layers of endosperm: aleurone, subaleurons, and the starchy or inner endosperm. The pericarp is associated mainly with pigments (Juliano and Bechtel, 1985). Hull contains ~20% of the rice dry weight, however, depending on the various rice varieties, values vary from 16% to 28%. In a decorticated rice grain, on a dry weight basis, 1%–2% is pericarp, aleurone plus nucellus and seed coat 4% to 6%, germ 1%, scutella 2%, and endosperm 90% to 91% (Juliano, 1972). Aleuron layer is abundant in protein, phytate bodies, and lipid bodies, varying from 1 and 5 cell layers (Del Rosario et al., 1968; Tanaka et al., 1973; Tanaka et al., 1977). Nutrients like vitamins, minerals, and phenolic compounds are mainly deposited in rice bran (Sen Gupta et al., 2017).

Characterization of different rice varieties is important for studying the agronomical, morphological, physical, and chemical properties. A glimpse of diverse Indian traditional rice landraces is shown in Figure 21.1, conserved by Basudha in its *Vrihi* seed bank (http://cintdis.org/vrihi/) which is India's largest non-governmental indigenous rice germplasm preservation bank. It is also important to study the results of diverse agricultural processes like handling, harvesting, threshing, and drying on rice varieties (Abano and Amoah, 2011). The length, width, and weight of rice grains are essential choosing physical attributes. For fragrant rice grains, the shape and the size of rice are essential parameters directly linked to product quality (Zheng et al., 2007). The thickness of the different rice grains determine the cooking quality. The most critical depiction for quality confirmation of rice grain is compound depiction of different cultivars reliant on sogginess, starch, protein, fat, and so forth. Protein and sugar (essentially starch) are the two critical components of rice, with practically 8% and 80% use, respectively (Marshall and Wordsworth, 1994). Additionally, rice contains a small proportion of fat. Cooking time and grain quality depend on the characteristics of the starch in the rice grain. Of 416 folk rice varieties stored in Vrihi's seed bank, morphological and agronomic characterizations have been reported (Deb, 2005). In West Bengal, Hunter (1876–1881), who recorded the rice varieties grown in different seasons, in different landscapes and different climatic conditions, gave an account of individual rice varieties. After the introduction of modern "high yielding varieties" in the 1960s, various rice varieties were pushed to obliteration. A survey conducted by Dr. Debal Deb in 1995 recorded that in Southern Bengal 137 folk rice varieties were still surviving with the marginal farmers. *Vrihi*, the largest non-governmental rice seed bank in India, established in 1997 for the farmers, coordinated by

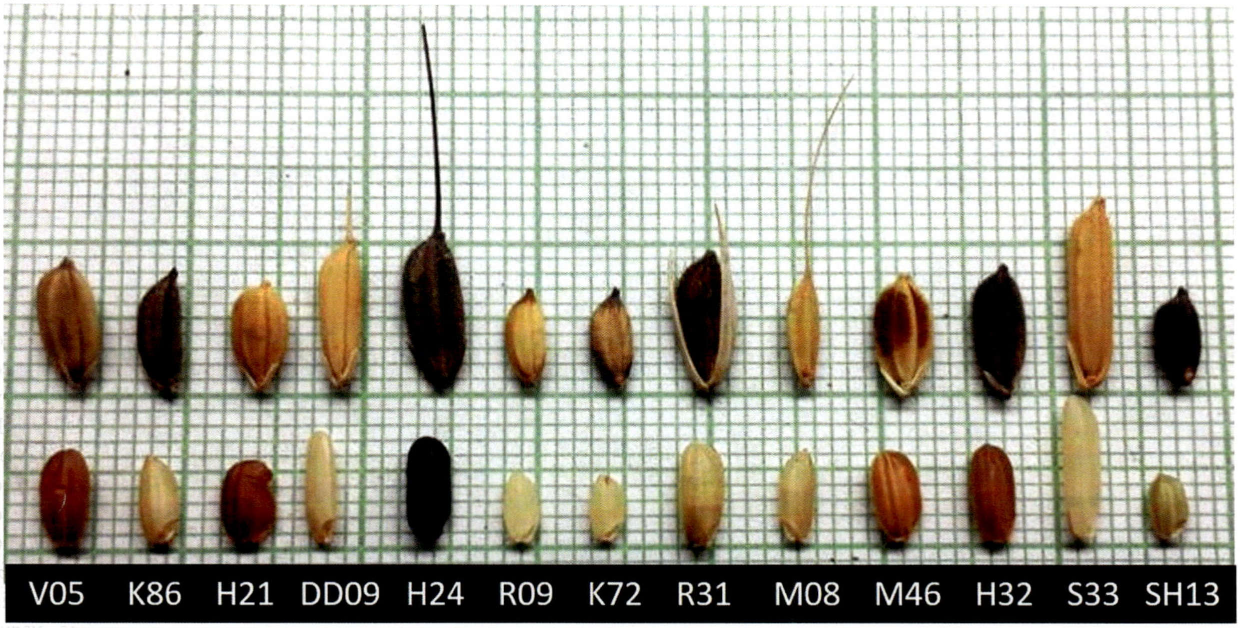

FIGURE 21.1 Glimpse of Indian traditional rice landraces conserved by Basudha (*Vrihi* seed bank).

TABLE 21.1

List of the Use of Various Molecular Markers in Conventional Rice-Genetic Divergence

Marker Name	Remarks	Varieties Tested	References
Amplified fragment length polymorphism	Detection of genomic restriction fragment by PCR amplification	49	Prashanth et al. (2002)
		33	Aggarwal et al. (2002)
Inter simple sequence repeat	Single primer based on SSR motif	42	Joshi et al. (2000)
		84	Cao et al. (2006)
	Phylogenetic relationship of rice germplasms	17	Filiz et al. (2018)
Random amplified polymorphic DNA	Random primers for PCR amplification	18	Patra and Chawla (2010)
		75	Pervaiz et al. (2010)
		8	Skaria et al. (2011)
Restriction fragment length polymorphism	Based on restriction digestion and hybridization with probe	71	Olufowote et al. (1997)
		67	Kojima et al. (2005)
Sequence characterized amplified regions	RAPD marker termini sequenced for designing longer primer	26	Singh et al. (2011)
Simple sequence repeats	Based on tandem repeat flanking sequence	11	Herrera et al. (2008)
		21	Raihan et al. (2016)
	Molecular fingerprinting	53	Chen et al. (2017)
	Genetic diversity and association mapping technique	172	Park et al. (2019)
Single nucleotide polymorphism	DNA sequence differs by single base	6	Nasu et al. (2002)
	Genome-wide SNPs by GBS of the parental varieties	12	Tang et al. (2016)

its founder Dr. Debal Deb, is now conserving, demonstrating agro-ecological practices, and documenting more than 1,400 traditional rice landraces to promote cultivation and characterization of folk rice varieties along with the obsolete practices of non-commercial seed exchange. Molecular markers have been used to recognize the different group and the plant genetic resources. Genome examinations using different markers have shown that the inherited association between different varieties of rice can change at a very basic level (Virk et al., 2000). A list of the use of various molecular markers in conventional rice-genetic divergence is provided in Table 21.1.

21.3 Rice Grain Quality and Effects of Stress

For over three billion people in Asia, rice is the major source of calories and protein. The majority of people favor white rice, which is a milling form of raw or brown rice (Juliano and Villareal, 1993). The ultimate protein composition of brown and white rice is influenced by both nitrogen supply and plant nitrogen status. Grain protein is linked to the majority of the quality features such as translucency, grain whiteness, stickiness, and chewiness (Blumenthal et al., 2008).

It has been reported that grain protein is positively correlated with milled rice translucent properties, but negatively associated with grain whiteness (Amagliani et al., 2017). That's why, with proper nitrogen fertilizer management, we could achieve the ideal levels of grain quality. Physical features such as size and shape, homogeneity, and elongation properties are all important quality criteria. The grain phenotypes, nutritional value, and cooking quality all play a role in its overall quality. Rice grain also contains phytic acid, oxalic acid, etc. some secondary metabolite compounds, considered as anti-nutritional components. Phytate is an antinutrient component in rice grains, and its high concentration hinders trace element absorption in humans and non-ruminant animals (Wang et al., 2009). Abiotic stressors have an impact on rice yield and quality (Moulick et al., 2016). As rice yield and milling quality are linked, the effects are resulted in a declination of rice production. Increased temperature has a negative impact on rice yield as well as a significant reduction in the market value of rice. According to the studies, rice yield and quality both influence the economic value of rice from field to mill and mill to market. As a result, any negative effects on rice at any stage of growth will result in a lower market value (Fahad et al., 2019). Climate change is reducing the amount of paddy rice available for milling and has a negative impact on the national and global markets. Climate change has had an impact not just on paddy rice production but also on milling quality, which influences the edible rice output. High temperatures reduce the market value of broken rice by increasing the quantity of broken kernels. In the United States and Asian markets, broken rice fetches about 60% of the price of head rice (USDA NRCS, 2012). Rice's actual response to expected warming due to climate change was not totally explained by preliminary studies, which demonstrated a warming influence on all-day temperature regime. Later, researchers discovered that warming during the day or at night has a substantial impact on crop development and output (Oh-e et al., 2007). Rice yield, rice grain quality, etc. all suffered as a result of night-time warming after anthesis (Dong et al., 2014). Warming also reduced the milling quality of brown rice, milled rice, and head rice yield by 4.3%, 7.2%, and 45.1% for Japonica rice, respectively (Dong et al., 2014). The effects of night-time post-anthesis warming on leaf photosynthetic efficiency and increased night metabolism rates are negative (Dong et al., 2014).

21.4 Use of Different Metabolomic Approaches for Rice Grain Quality Assessment

Phytochemicals related to rice grain quality include aroma and aroma volatiles; amino acids; vitamins that contribute to nutritional properties, and active biomolecules such as phenolic compounds rich in pigmented rice imparting functionality, or health characteristics, all together can be analyzed by metabolomics (Chaves and Broeckling, 2020). The analytical technique for the study of the minor composition of molecules in biological materials is metabolomics. The methods used for the processing of knowledge and the range of metabolites (Ortmayr et al., 2016) may be primarily divided into targeted and untapped. Untargeted metabolomics provides broad coverage of metabolites, and the tests can be either used for research or for generating hypotheses. There is no need for extensive knowledge on the therapeutic properties of reactant compounds, and the metabolomic approaches can be used for more intensive testing of the compound or pathway (Lee et al., 2017). Multiple quantitative characteristics and biotic and abiotic environmental factors influence aromatic and nutraceutical features of rice grain quality. The genetic study of the rice metabolome would combine the quantitative trait loci (QTL) with the number of compounds to decide how metabolism is regulated. This data is used to classify markers in breeding programs in a quality-focused marker-assisting selection strategy. GC-TOF MS (Mass spectrometry) metabolomic analyzes of Chinese rice germplasm set identified nutritional compounds including different phytochemicals, specific to the molecular markers analyzed by association maps. GC-TOF-MS hybrid one- and two-dimensional methods were used to test the metabolic phenotyping of rice variants in the global range (Kusano et al., 2007). The two major Japanese and Indica subspecies of rice have distinct differences and qualitative features in genome sequences. Untargeted metabolomic testing using mass spectrophotometric platforms in grains from 100 *japonica* and *indica* varieties have revealed various metabolomes unique to species (Hu et al., 2014). In reproductive strategy (Calingacion et al., 2017), QTLs which have been calculated using a genomics method linked to both yield and organic compound profile could be used. A genome-wide association analysis was performed using a wide variety of HPLC-MS/MS metabolomics, including accessions of both *indica* and *oryza* subspecies. The metabolic trait candidate genes have been identified (Chen et al., 2014). Targeted, untargeted HPLC-MS methods have been used in related interaction studies (Gong et al., 2013; Matsuda et al., 2015). Due to the small coverage possible of the metabolite using common platforms, complementarity and greater coverage was achieved with multiple platforms. In studying brown rice grains, it was possible to classify the relationships between quality traits and metabolites with complementary platform metabolomics (GC-TOF-MS, HPLC-QTOF-MS, and other TOF-MS platforms). Compared to single systems, integrated data sets provided better overall prediction (Matsuda et al., 2012). Studies in metabolic diversity compared GC-MS metabolomics of rice found in diverse phytochemicals and sterol ingredients that affect rice taste and nutritional quality and affect discrimination between genotypes (De and Nag, 2014; Ranjitha et al., 2019). Among pigmented and nonpigmented rice, multivariate data analysis through HPLC-MS and non-target GC-MS test of targeted anthocyanine identified varieties of *indica* rice with a different color profiles (Frank et al., 2012).The unregulated GC-MS study classifies organic acids, carbohydrates, sugar spirits as well as unsaturated and saturated fats that have helped to differentiate between typical Thai rice germplasm, such as colored and non-colored genotypes (Wongsa et al., 2018). The ready-to-eat item, like cooked rice, can be characterized using an unspecified metabolomics of UPLC-Q-ToF-MS from multiple varieties. Multivariate metabolite profile analysis contributes to sample classification by subspecies (Heuberger et al., 2010). 2-acetyl-1-pyrroline was considered to be the main compound responsible for the fragrance of the aromatic rice varieties. In order to obtain a better interpretation of biosynthetic pathway, a static head space array of volatiles accompanied by 2D-GC-TOF-MS has been carried out (Daygon et al., 2017). After cooking, scented rice releases aroma volatiles, which were collected and analyzed by integrating GC–O and GC-MS techniques in various fragrant rice varieties. Aroma-active substances, including sulfur chemical compounds, are identified in jasmine rice (Mahattanatawee and Rouseff, 2014). There are a number of active substances whose quality alters with seasons, and various maturation spans of cultivars were tested with Nuclear Magnetic Resonance (NMR)-based metabolomics. The specific build-in metabolite profiles (Song et al., 2018) have discovered an increase in saccharose, amino acid, and free fatty acid in early maturing grains. Flavonoids are found to increase the health of many food items with restricted presence in unpigmented rice grains, except tricin, a flavonoid known to develop on rice bran with a potential to inhibit colon cancer cell development. Newly developed genetically modified (GM) rice lines which accumulated various bioactive phytochemicals in seeds, their contents are tracked by the targeted HPLC-MS study (Ogo et al., 2013). HPLC-QQQ-MS analysis revealed that overexpressing *Arabidopsis thaliana* folate biosynthesis genes has achieved a 100 times higher folate content in milled rice in comparison to raw rice. Rice endosperm is rich in folate, containing four times higher amount than that of one adult's daily demand for folate (Storozhenko et al., 2007). The prevalence of several chronic diseases, including cancer and cardiovascular disorders, is further decreased by the intake of rich-carotenoidal foods, and many carotenoids are an important precursor of vitamin A. Kim et al. (2012) described that a GC-TOF-MS was used to quantify lipophilic compounds, beta-carotene, and non-trans-genes in a GM rice. Similarly, the GC-TOF-MS was untargeted and could not distinguish between pigmented and white rice from the beta-carotene transgenic line and its predecessor line (Kim et al., 2013). GC-MS metabolomics examined for the sck gene, the cowpea trypsin inhibitor (mainly coleopteran pest protection), showed a different profile of metabolites than non-GM (Zhou et al., 2009). The key components of rice seed lipids are phospholipids that have had an effect on consistency. The environmental factors that define each rice-growing country seem to strongly influence these molecules. Therefore, for the target study of phospholipids in white rice of various origins, high-performance lipidomic methods by direct infusion tandem mass spectrometry (MS/MS) were used (Lim et al., 2017; Long et al., 2017). Targeted HPLC-Q-TOF-MS study of two genotypes of black rice in two locations showed that phenolic compounds of black rice had an effect on the genotype and the regional compounds (Dittgen et al., 2019). Heat and drought can lead to lower yield and quality of rice grains. GC-MS metabolomics have studied rice varieties with various tolerance levels due to combined drought and heat stress. Certain metabolites were produced differentially, despite cultivar tolerance, cultivar common responses (Lawas et al., 2019). Degenkolbe et al. (2013) studied the drought tolerance level for indica and japonica cultivars for their genetic make-up through GC-MS and found major variations among the cultivars. LC-IT-TOF-MS lipidomycin showed that the rice lipid metabolism was regulated under different inorganic environmental conditions and was correlated with the accumulation of glucuronosyl diacylglycerol (Okazaki et al., 2013). The discovery of several primary and secondary metabolites has emerged from the use of metabolomics as a tool for understanding plant stress adaptability and tolerance. Any variation in the metabolic pool closely reveals the stress-responsive features (Choudhury et al., 2021). As a result of such a remarkable and significant result, metabolomics' knowledge and practical applicability require further intensive study inputs. One such strategy is next-generation metabolomics, which provides practical insight into plant responses to abiotic stressors. The identification and characterization of novel features appear to be critical in countering abiotic stress in plants, given the growing influence of climate change, which has resulted in a significant increase in the occurrence of abiotic stresses (Choudhury et al., 2021).

Despite the sensitive and insightful methods available for detection, compound identification is still a novel and complex area. Metabolomic methods continue to evolve, and each further advancement will allow to develop more complex systems which will help to understand the relation between the composition of the small molecules and the quality of rice grain.

21.5 Nutritional Composition and Health Benefit of Rice

The nutritional contents of rice change comprehensively on the soil characters and the environmental conditions on which they are grown. A large portion of the essential nutritional components are provided by rice grain in the human body as an important nutritional staple food. Some components of the rice bran fraction have been known to demonstrate biological effects, apart from rice being the source of energy. Rice is the world's leading source of carbohydrates, proteins, lipids, minerals, and vitamins for the population. Humans need to consume a certain amount of macronutrients, trace elements (Fe, Zn, Cu, Co, etc.), and vitamins, for normal growth and health during their life (Welch and Graham, 2004). Rice grain contains most of these important components for humans; some occurs in large quantity, while others only in small amounts (Table 21.2). Nutritional components in rice can broadly be divided into the following sections discussed below.

21.5.1 Carbohydrate

Carbohydrate is the major portion of the rice grain. The starch content of rice varies from 72% to 82% (Frei et al., 2003), and around 90% in the milled rice. There are two structural components of starch, viz., amylose and amylopectin. Dietary fiber consists of cellulose, hemicellulose, galactooligosaccharides, pectins, resistance starch, lignin, and substances related with non-starch polysaccharides and tannins (AACC, 2001). Dietary fiber, mostly insoluble, makes up 2.9%–4.0% of milled rice, 0.7%–2.3% of cooked rice, and 17%–29% of rice bran (Juliano and Bechtel, 1985; Abdul-Hamid and Luan, 2000). One of the Indian medicinal rice varieties, Njavara has the total dietary fiber content of 8.08%, higher than that of other two modern rice varieties (Deepa et al., 2008).

TABLE 21.2
Nutrient Composition of Brown Rice Milled Rice and Rice Bran

Nutrients	Brown Rice	White Rice (Milled)	Rice Bran
Carbohydrate (g%)	73–76	77–78	34–62
Protein (g%)	7.1–8.3	6.3–7.1	11.3–14.9
Fat (g%)	1.6–2.8	0.3–0.5	15.0–19.7
Ash (g%)	1.0–1.5	0.3–0.8	6.6–9.9
Energy (kcal)	363–385	349–373	399–476
Thiamine (mg/100g)	0.29–0.61	0.02–0.11	1.20–2.40
Riboflavin (mg/100g)	0.04–0.14	0.02–0.06	0.18–0.43
Niacin (mg/100g)	3.5–5.3	1.3–2.4	26.7–49.9
Vitamin E (mg/100g)	0.90–2.50	0.075–0.30	2.60–13.3
Nicotinic acid (mg/100g)	4.4–6.2	0.8–2.6	-
Ca (mg/100g)	10–50	10–30	30–120
K (mg/100g)	120–340	14–120	-
Na (mg/100g)	3.1–17.6	2.2–8.5	-
Fe (mg/100g)	0.7–5.4	0.2–2.7	8.6–43.0
Zn (mg/100g)	1.5–2.2	0.3–2.1	4.3–25.8
Mn (mg/100g)	1.3–4.2	1.0–3.3	-

Source: Saleh et al. (2019) and Roy et al., (2021).

21.5.2 Lipid

The amount of crude lipid of milled rice ranged between 2.76% and 3.84% on dry mass basis, depending upon various rice varieties, and was influenced by the cooking conditions (Kitta et al., 2005). Oleic (18:1) and linoleic (18:2) acids were the most abundant unsaturated fatty acids, followed by palmitic acid (16:0). Over 90% of unsaturated fats were made up of these three unsaturated fats (Kitta et al., 2005). The crude lipid content of darkly colored rice types was found to be higher than that of red and white kinds (Frei and Becker, 2005). The lipid bodies of the embryos and aleurone layers, as well as the sub-aleurone layer of the endosperms, contain the majority of the lipids present in rice grains. Rice bran oil (20–25 wt%), which is high in phytochemicals, is safe to consume. During the milling process, the majority of these phytochemicals are lost (Patel and Naik, 2004). Usage of rice grain oil may help to protect the liver from oxidative damage by lipid peroxidation (Nam et al., 2008).

21.5.3 Protein

The protein content in rice ranged between 4.3% and 18.4%, with a mean of 9.5% (Champagne et al., 2004). However, most of the protein may not be utilized for assimilation. The types/classes of protein used for nutrition, and can be ingested into the system, may be called as soluble protein. Most of the proteins in rice grain are found in the bran and embryo portion (Schramm et al., 2007). These are mostly the storage proteins, like albumin and globulin, (Hamada, 1997) and they have functional properties with high nutritional values. Protein bodies are sequestered inside the cell. In rice, there are four types of proteins in grain as demonstrated by their solubility and extraction processes (Shewry and Casey, 1999). The proteins, albumins, globulins, prolamins, and glutelins are dissolved in water, salt, and alcohol. Rice grains are rich in amino acids, the main components of the proteins, lysine, arginine, glutamic acid, aspartic acid, tyrosine, etc. Rice grain proteins have high potential as nutritional supplements (Tang et al., 2003). Some of the nutritional components are suitable for infant food formulations, some are known to be a hypoallergenic (Helm and Burks, 1996) and some supplements have potentiality.

21.5.4 Vitamins

Rice contains water- and fat-soluble vitamins in the grain (USDA National Nutrient Database for Standard Reference, 2016). Rice is a good source of vitamin B complex (Deepa et al., 2008). Previously, it was under consideration that rice does not contain vitamin A or its precursor, known as Beta-Carotene (Ye et al., 2000). Nevertheless, recent findings indicate that traditional rice varieties contain Pro-vitamin A (Valarmathi et al., 2015). As vitamins are mainly present in bran, they are commonly found in the bran of unpolished rice (brown, red, and black rice) than in the milled rice. According to Tan et al. (2005), majority of the unpolished rice varieties showed beta-carotene and lutein peaks. Frei and Becker (2005) identified that black upland rice varieties from the Philippines had more beta-carotene values that the red/brown or colorless varieties. Beta-carotene is a precursor of vitamin A. Its deficiency can lead to severe visual impairment, respiratory diseases, and measles. A GM rice variety "Golden Rice" has been introduced to tackle the vitamin A-deficiency problem (Paine et al., 2005). However, the ecological as well as food security-related aspects of the GM rice variety are under a big question. Moderate contents of such B vitamins in black rice varieties might be helpful for human health and nutrition.

21.5.5 Metal Micronutrients

In the context of human health and nutrition, rice grains were found to accumulate metals (Prasad, 1998; Sen Gupta et al., 2017). Low levels of metals such as Fe, Zn, Cu, and Mn in rice and other cereals are thought to be good and sufficient in preventive nutrition. While cereals contain nutrients such as Fe, Cu, Zn, and Mn, many other toxic metals in cereals (such as As and Hg) are highly toxic and pose health risks such as cancer and central nervous system disruption. According to Moulick et al. (2018), seed priming with Se performs a dual role of growth promotion and antagonism in a more practical and farmer-friendly manner to mitigate As-induced toxicity and enhance growth in rice seedlings. Previously, seed priming with Se was found to improve plant growth and yield by bringing down As absorption and controlling oxidative stress in rice seedlings by enhancing antioxidant build-up (Moulick et al., 2016). When compared to 1.0 mg Se/L, seed primed with 0.8 mg Se/L was more effective in boosting rice germination and growth (Moulick et al., 2016). Several traditional rice varieties have high nutritional and therapeutic qualities in folk remedies. Many of them are believed to treat anemia during and after pregnancy (Deb, 2005). Many are also found to contain high amounts of iron (Deb, 2015; Sen Gupta et al., 2017). Some rice varieties (Kalo bora, Kelas, Garibsal, Parmaisal, and Kabirah-sal) have recently been documented for being able to absorb micronutrients like, Fe, Cu, Zn, and Mn (Sen Gupta et al., 2017). Among them, some varieties with therapeutic properties are classified in folk (Deb, 2005). Most of these varieties have Fe and Zn concentrations up to 13 times higher than transgenic metal-fortified rice (Deb, 2014). For example, the IR68144-2B-2-2-3 transgenic high-iron rice produced at the International Rice Research Institute in the Philippines contains 9.8 mg/kg of iron during feeding experiment (Hass et al., 2005). In the presence of metal ions in cereal crops, oxidative stress is crucial in their mobilization from soil to grains. Computational (molecular docking) analysis using Cu–Zn–SOD, Mn–SOD, and CAT as target enzymes revealed that SA may efficaciously interfere with the complexation and interaction of Cr^{6+} and Zn^{2+} ions with the active site of Cu–Zn–SOD (Mazumder et al., 2022).

21.5.6 Volatile and Aroma Compounds

Aroma of rice – a critical quality parameter can decide the market value of rice before it is tasted. The aroma, consisting of taste and smell, is a key sensory characteristic of consumers. The fragrance is considered by modern rice consumers to be the important criteria. Gas chromatography (GC) has gained remarkable developments and improvements and, in particular, has made it much easier to classify and quantify volatile aroma compounds (VACs) in mixtures of the sample matrix when coupled with MS known as GC-MS and has significantly enhanced our knowledge of rice aroma science. Based on the assessment of available literature, more than 500 VACs have been reported to date in various cultivars of fragrant and non-fragrant rice (Verma and Srivastav, 2016, 2018). The first VAC, 2-acetyl-1-pyrroline, was reported by Buttery et al. (1982), and the same authors identified it in aromatic rice in 1983 because of its volatile influence. Five major classes of rice aromas are listed in Table 21.3. It can also be attributed to a mixture of compounds in fixed proportions if no specific aroma of cooked rice is caused by a single chemical compound (Verma and Srivastav, 2020a). Anti-infective, antibiotic, anti-tumor, anti-allergic, anti-ulcer, anti-inflammatory, anti-arthritic, and nutraceutical properties are biological properties of VACs. They are also known for their cardiovascular antioxidant function and can also cure many degenerative diseases (Devi and Arumughan, 2007). As they hold body cholesterol levels at low lipoprotein densities (Patel and Naik, 2004), VACs are also known to lower cholesterol levels over several years. The dietary use of cereal-dependent foods leads to lower blood pressure and increases the sensitivity of insulin, which contributes to a lower risk of heart disease (Truswell, 2002). Cyanidin-3-glycoside has been found to be metabolized into cyanidine and protocatechuic acid based on studies. These metabolites are used to reduce lipopolysaccharide (LPS)-induced inflammation (Hassimotto et al., 2008). An investigation was conducted in which a group of individuals were exposed to volatile organic compounds for a certain period of time. Over-exposure to VACs has been shown to contribute to human neurobehavioral disorders, as some of them bind to hemoglobin and impair its ability to pass oxygen to various parts of the body, leading to heart and brain disorders (Koren et al., 1992). VACs isolated from rice bran have antioxidant and chemopreventive properties (Jasti et al., 2020).

21.5.7 Bioactive Phytochemicals

Rice is a source of a number of bioactive compounds, including flavonoids, carotenoids, phytosterols, vitamins, μ-oryzanol, phenolic acids, tricin, etc., which are nutrients and phytochemicals. These compounds include a wide variety of molecules which are biologically active. Higher levels of most compounds are present inside the bran fraction (Verma and Shrivastava, 2020b). Rice's health advantages attributed to its varied phenolic components have not been thoroughly investigated. However, various processing, treatment, and bioprocesses technologies have been discussed to boost the bioactive constituents in rice (Kasote et al., 2021). Bioactive compounds found in many cellular constituents induce oxidative stress. In the end, this leads to other diseases such as heart disease, cancer, aging, and inflammatory diseases (Miller et al., 2000; Stratil et al., 2007). They are identified in many natural ingredients, such as cereal crops and other food products, which are becoming a lucrative alternative and have harmful effects on the health and prevention of many diseases, in contrast to antioxidant compounds. Black rice was found to have high antioxidant activity mainly caused by anthocyanins (Goffman and Bergman, 2004) and phenolics and flavonoids (Ghasemzadeh et al., 2018). A study conducted by Hosoda et al. (2018) showed the higher antioxidant

TABLE 21.3
Classification of Rice Aroma Volatiles

Types of Aroma Compounds	Chemical Compounds Responsible for the Aroma
Green	Aldehydes, (E)-2,(E) 4-hexadienal, (E)-2-hexenal, Decanal, Geranyl acetone, (E)-2-octenal, (E)-3 octen-2-onea, 2-heptanone, 2-methyl-2 pentanol, Benzaldehyde, Methyl heptanoate, some alcohol and Ketones, and n-hexanal
Fruity/floral	(E)-2,(E)-4-hexadienal, (E)-3-octen-2-onea, 2-hexanone, 2-nonanone, 2-undecanone, 6-methyl-5-hepten-2-one, Heptanone, Ketones, Methyl heptanoate, n-heptanal, n-nonanal, and n-octanol
Roasty	2, 3-octanedione
Nutty	(E,E)-2,4-nonadienal, 2-pentyfuran, 4-vinyl guaiacol, and Benzaldehyde
Bitter	Benzaldehyde and Pyridine

TABLE 21.4

Fatty Acid Composition in Aromatic and Non-Aromatic Rice Varieties

		Saturated fatty acids			Unsaturated fatty acids		
Rice Varieties		**Myristic Acid (CT%)**	**Palmitic Acid (CT%)**	**Stearic Acid (CT%)**	**Oleic acid (CT%)**	**Linoleic acid (CT%)**	**Linolenic acid (CT%)**
Gopal Bhog	Aromatic	1.55	18.55	2.28	41.75	34.88	0.99
Govind Bhog		4.60	31.91	2.87	25.06	35.57	-
Badshah Bhog		0.27	19.15	2.84	44.00	32.01	0.89
Kalanamak		1.37	16.13	3.97	45.71	32.81	-
Swetganga		-	19.60	-	44.21	36.19	-
Khushboo		-	26.19	-	36.23	37.58	-
PB 1509		1.22	24.63	2.60	39.50	31.26	0.7
Basmati		0.7	21.4	2.8	46.4	26.6	1.1
Sarbati	Non-aromatic	-	2.60	-	49.14	46.99	1.27
Chak-hao		0.41	28.28	1.99	38.80	29.65	0.61
Jaya		1.1	19.6	3.5	47.3	2.8	2.4

Source: Khatoon and Gopalkrishna, (2004); Verma and Shrivastava (2017); Krishnan et al., (2021).
CT, Content.

potentiality in ten cultivars of pigmented rice grains. Red rice also demonstrated greater antioxidant potentiality than black rice. Phenolics (in particular, red rice proanthocyanidins and black rice anthocyanins) and flavonoids are source of antioxidant function (Huang and Lai, 2016). Some research articles have shown that black rice bran has shown substantial antioxidant activity compared to red and brown rice bran. Research by Ghasemzadeh et al. (2018) revealed in a subsequent analysis that the free fraction of black rice bran had a larger IC_{50} value than those of red and white varieties. Phytochemicals observed in many rice varieties have many potential pathways to prevent a specific form of cancer (Chatthongpisut et al., 2015; Forster et al., 2013; Henderson et al., 2012). Red rice could positively impede the proliferation of cancer cells (Upanan et al., 2019). The rich fraction of proanthocyanidine was extracted from the germ and bran of the red rice and applied to HepG2 cells (Upanan et al., 2019). Fatty acid compositions of some aromatic and non-aromatic rice varieties are provided in Table 21.4. As an antidiabetic, the rice phytochemical compound oryzanol has been suggested as having a direct correlation between insulin sensitivity and adiponectin, playing an important role in type-2 diabetes (Ohara et al., 2009). Son et al. (2011) showed that oryzanol could regulate the production of insulin and glucose, normalize the function of liver enzymes, and thus reduce the risks of high-fat diet-induced hyperglycemia. Bioactive compounds of many pigmented and nonpigmented rice cultivars have gained new opportunities and increased their use in nutraceuticals and food industry.

21.6 Unique Rice Landraces with Nutraceutical Potentiality

In India, rice has achieved a special status since ancient times for its unique properties. Several folk medicinal literature have described the exceptional properties of rice. It is viewed as the best food among all cereals. Different authors, for example, Charaka, Susruta, Kautilya, Varahamihira, and Panini in their separate treatises have mentioned this quality. They discussed rice under the cereal food items with therapeutic properties. Charaka mentioned various rice varieties that were developed in various seasons and their impact on the human physiology. Susruta described the properties of different varieties of rice developed under various conditions. No other cereals, including wheat, have received such a great amount of consideration in these treatises (Nene, 2005). A traditional rice landrace from Bengal, Garib-sal is reported to be a unique landrace, is capable of accumulating up to 15 mg/kg of silver from the soil (Deb et al., 2015). In folk medicine of Bengal, Garib-sal was once recommended by indigenous people as a diet for patients with gastro-intestinal infections (unpublished). The presence of silver in this rice variety may have a therapeutic potentiality by killing pathogenic microbes in the human gut. Several traditional rice varieties containing iron, riboflavin, and other nutritional components have been recognized in South Asia (Deb, 2017). Some special rice varieties of Tamil Nadu like Pichha vari and Karthigai samba and Dudhsar of West Bengal are believed to enhance milk production in lactating mothers (Deb et al., 2015). A few landraces of West Bengal, as Kelas and Bhutmoori, are believed to cure anemia in women (Deb, 2005). Ongoing research demonstrates that the content of iron and zinc in numerous traditional rice landraces are essentially higher than that of improved cultivars (Anandan et al., 2011). Different micronutrients like vitamin E (α-tocopherol), the B vitamins (riboflavin, thiamin, and niacin), iron, zinc, and certain alkaloids have been identified in different varieties and developed by indigenous farmers over hundreds of years (Deb, 2005). Good amount of β-carotene (provitamin-A) is found in the bran of some pigmented rice varieties from the Philippines, Malaysia, Vietnam, and Thailand (Frei and Becker, 2005). In Tamil Nadu and Kerala, people recommend utilization of Nyavara rice for treating patients suffering from

FIGURE 21.2 Pictorial representation of the importance of the traditional rice landraces.

neurotic issue (Deepa et al., 2008). Biochemical examination reveals that this rice contains oridine (Juliano, 1993), which has neuro-restorative property. In Bengal, Kabiraj-sal rice is suggested for convalescing patients (Deb, 2005). In comparison to white rice, Zhang et al. (2005) reported that colored rice contains more anthocyanin molecules and other phenolic compounds. Anthocyanin is found in the bran layer of rice as a group of flavonoids and is known to have some beneficial effects on human health. A commonly known anthocyanin found in rice bran is acetylated procyanidin, which is reported to show free radical scavenging activity (Oki et al., 2005). Because of its phenolic compounds, which are highly selective to inhibit the production of reactive cell-damaging free radicals, rice is also a health-promoting food that reduces the risk of cancer, oxidative lipid damage, and heart disease (Dutta et al., 2012). Butsat and Siriamornpun (2010) reported that a Thai rice variety (Khao Dawk Mali 105) has higher total phenolic acids in the rice-milling fraction and demonstrates both radical scavenging and ferric-reducing antioxidant potential, that could be considered as valuable food additive. An investigation carried by Hansakul et al. (2010) on Thai rice cultivars showed that flavonoid content was higher in colored rice samples than the white varieties.

Indigenous rice-genetic diversity in India is reported to harbor vast amount of genetic wealth, including nutritional and medicinal properties and aroma (Das and Oudhia, 2001). These indigenous landraces are potent source of many bioactive nutrient components. There are very low amounts of essential micronutrients in commonly consumed polished rice, leading to the occurrence of micronutrient deficiency upon continuous intake, which justifies the scenario of genetic engineering development of micronutrient and vitamin-fortified rice (Ye et al., 2000). Rice contains a lot of bioactive phytochemicals that have been identified as strong antioxidants (Lai et al., 2009), apart from the characteristic features mentioned in indigenous agricultural knowledge. The importance of such indigenous rice landraces has been represented in Figure 21.2. Geographical conditions, environmental and genetic factors are known to influence the vitamins and other nutritional factors in rice varieties (Miller and Engel, 2006). There are many such rice landraces with astonishing medicinal properties, described in folk medicine, awaiting detailed scientific investigation and validation, which will provide benefit to modern medicine while ensuring community health and nutritional security.

21.6.1 Black Rice as a Potent Nutraceutical

The white rice grain is affected by random mutation and chemically induced mutation according to Jeng et al. (2012). Three specific genes (Ra, Rc, and Rd) regulate grain coloration. Ra gene regulates the pigments of pericarp color over the white one. The Rc gene produces brown pericarp color where the Rd gene is not present. The Rd gene does not develop any color alone, but it is formed by the fusion of both Rc and Rd genes. Black rice has become increasingly common among pigmented rice varieties due to its high nutraceutical properties. Black rice is especially classified as a potent functional food because it contains high levels of bioactive molecules (Tananuwong and Tewaruth, 2010). Increased consumption of white rice, with higher risk of glucose homeostasis, has shown to create a substantial rise in risk of type-2 diabetes (Soriguer et al., 2013). A variety of enhanced functionality, such as antioxidant activity (Roy et al., 2021), endothelial cell safety, prevention and anti-cardiovascular disease, and anticancer functions, have also been documented for black rice (Tanaka et al., 2012; Shao et al., 2014). The US Health Association and the American Cancer

Society recommend its use (USA Rice Federation, 2018). Trinin and tricin derivatives from black rice isolates have bioactive compounds with antioxidant function, lipid peroxidation inhibition, antiviral activity, anti-inflammatory, and strong anticancerous activity, sparing activity on vitamin E in an erythrocyte membrane (Duyi et al., 2017). Shao et al. (2018) indicated that free black rice phenolic acids are transported throughout the body into the stomach and intestine after they are consumed, with health benefits such as low-density oxidation and reduction of lipoprotein cholesterol. By the action of enzymes in the intestine after digestion, bound phenolic acids can be processed (Saura-Calixto et al., 2007). There are some beneficial health impacts of ferulic acid extracted from black rice (Srinivasan et al., 2007). The increase in hypercholesterolemics and hyperlipidemia associated with the reduction of lipogenic hepatic enzymes in hyperlipidemic rats was shown by cyanidine-3 glucoside. These pigments have been shown to be effective in lowering cholesterol in the human body (Lee et al., 2008) and reducing allergy symptoms (Choi et al., 2007). Black rice extract is a natural antioxidant that protects biological molecules such as fatty acids, protein, and DNA and can reduce oxidative stress. These compounds can have a single effect or influence on biological functions synergistically (Shao et al., 2018), which indicates that one or more antioxidant foods may have an increased health benefit than any one of the antioxidants alone. Black rice has health-promoting and disease prevention properties. Several studies are going on to improve the quality and health-enhancing properties of black rice for their future implementations.

21.6.2 Some Selected Research Studies on the Nutraceutical Potentialities of Specialty Rice Landraces

Abiotic stress, on the other hand, causes significant damage to rice plants and includes key stressors such as drought, cold, salinity, and high temperatures. Excess soil salt is a major abiotic stressor influencing the formation and production of a variety of crops, including rice. Rice can typically withstand a small quantity of saltwater without harming its growth or output. However, the different kinds and lines of rice used, as well as the stage at which they are grown, have a substantial impact (Hasanuzzaman et al., 2013). According to Lee et al., indica rice has higher seedling tolerance than japonica rice (2003). Rice is classified as a salinity-sensitive cereal at an early stage of development, which decrease its production efficiency at maturity. As a potent source of essential nutrients, rice grain also contains several components with significant bioactivity and associated health advantages. Therefore, some traditional varieties of rice can now be portrayed as functional foods. Rice-ORS has been shown to be appropriate to enhance intestinal absorption in acute diarrhea by inhibiting chloride channels due to intestinal loss minimization (Goldberg and Saltzman, 1996). Pigmented rice landraces have been extensively studied and antioxidant properties are found in their anthocyanin content and flavonoids (Zhang et al., 2005). Table 21.5 provides a brief summary of the studies on the potential health benefits of traditional rice. These rice varieties have shown to decrease the atherosclerotic plaque significantly than the white rice in rabbits (Ling et al., 2001). The parboiled red rice varieties of Sri Lanka have lower GI than white rice and have been suggested for diabetics (Hettiarachchi et al., 2001). Rhoades (2003) reported that there is a lower GI for a certain number of traditional rice varieties. One of them is Basmati rice. In Basmati, which is rich in cystine that leads to iron absorption, metallothione-like protein has been discovered; this biotechnological approach is used in the improvement of Fe-rich rice (Chaudhary and Tran, 2001). Those landraces that have some additional health benefits than the standard ones can be identified by the term "medicinal rice". Varieties, landraces, cultivars, and wild ancestors known by indigenous farmers for their therapeutic potentiality are under investigations for their special features and medicinal properties. Njavara is a medicinal rice of Kerala, extensively used in Ayurveda for the treatment of neurodegenerative diseases and in rejuvenation therapy. Phytochemical assessments and spectroscopic examinations of the diethyl ether extract of methanolic concentrate of Njavara rice bran gave some important compounds specifically, tricin, and two rare flavonolignanstricin, namely 4′-O-(erythro-β guaiacylglyceryl) ether and tricin 4-O-(threo-β-guaiacylglyceryl) ether. This was the first report on the presence of tricin at significant levels in Njavara and occurrence of the two flavonolignans in rice (Mahanta and Sikia, 2016). Whole rice grain and rice bran are rich in different nutritionally important chemical components and contain a significant amount of mineral elements (Gurpreet and Sogi, 2007; Sen Gupta et al., 2017). Rice by-products are used periodically in developing countries (Sharif et al., 2014). A study on the phytochemistry of rice grain has shown that tocopherols, tocotrienols, and oryzanols contain many distinctive bioactive compounds. These compounds have a significant potential for cancer prevention compared to other rice bran compounds, and the benefits as bioactive compounds are used to increase the stability of different foods in storage. It has been found that cholesterol-lowering capacity of rice bran oil is better than its envision of fatty acid profile. This recommends the hypocholesterolemic impact of rice bran oil (Moldenhauer et al., 2003). The risk of cardiovascular disease at the onset of oxidative stress is high (Marini et al., 2003). Over the past two decades, researchers have extensively studied the bran fraction of rice for its medicinal and therapeutic potential, such as anti-atherogenic, cholesterol-lowering, and antioxidant properties (Jariwalla, 2001; Parrado et al., 2003; Ha et al., 2005). Research has shown that the fundamental component in rice bran with anti-atherogenic potential is γ-oryzanol. Naturally, γ-oryzanol with vitamin E has shown free radical action, which is useful to protect cells from oxidative pressure (Kennedy and Burlingame, 2003). Wilson et al. (2007) observed the cholesterol-lowering capability of trans-ferulic acid, γ-oryzanol, and rice bran oil. A few studies on oncogenic events have utilized active components from rice bran. For example, a successful investigation was carried out by Kong et al. (2009) with rice bran polyphenols to study the growth inhibitory effects on the human colorectal adenocarcinoma SW480. Similarly, Revilla et al. (2013) showed that enzymatic extraction of rice bran has anticancer potentiality against leukemic cell lines. Black rice is composed of different types of nutrients and phytochemicals, including

TABLE 21.5

Studies on Potential Health Benefit of Rice

Health Benefits	Objectives and Findings	References
Antioxidant effect	Under *in vitro* condition, different solvent extraction methods were employed to compare antioxidant activity of brown rice and bran, among which ethyl acetate extract showed higher DPPH and ABTS scavenging activity than other extracts	Kim et al. (2013)
	Germinated brown rice and brown rice proved to enhance the antioxidant effect and immune functions in aged mice compared to white rice when fed with diet contained 50% of rice under In vivo condition	Wu Chen et al. (2013)
	Germinated rice and brown rice reduced the plasma glucose level, preserved liver enzymes, and increased the total antioxidant activity in type-2 diabetic rat under In vivo condition	Imam et al. (2012)
Antidiabetic effects	Roasted brown rice showed antibiotic effect, higher insulin sensitivity, and lowering the concentration of fasting plasma triacylglycerol compared to white rice when type-2 diabetic mice were fed with the diet (30% rice) under *In vivo* condition	Matsumoto et al. (2016)
	An investigation was carried out to check the blood glucose lowering effect and bioactive components of colored rice. It was found in both the cases that when type-2 diabetic rats were fed diet contained colored rice and under In vitro condition when HEPG 2 cells were treated with GABA, oryzanol, and phenolic extract of rice, it reduced the plasma glucose level and weight potentially than metformin	Imam and Ismail (2013)
Anticancer effect	*In vivo*, 1,2DMH-induced colon carcinogenesis rats were fed diets containing varying concentrations of rice for 17 weeks and significant reduction in the number of aberrant crypt foci was observed along with their distribution	Li et al. (2011)
	In an investigation on the cytotoxic effect of rice bran, *In vitro*; human acute lymphoblastic leukemia Jurkat cells were showed apoptotic dead when treated with varying concentrations of fermented rice bran extract	Horie et al. (2016)
	Germinated brown rice in combination with probiotics inhibited colorectal carcinogenesis in rats (in vivo) by increasing the antioxidant capacity	Lin et al. (2019)
Anti-inflammatory effect	When *In vitro* human colon cancer cells were treated with pulse-electric-field–assisted brown rice extract, it enhanced the production of bioactive compounds which increased the antioxidant activity and reduced the cell toxicity compared to the untreated extracts	Quagliariello et al. (2016)
	In vivo; CCl_4 hepatotoxicity induced rats showed higher hepatoprotective effect by decreasing the liver inflammation, when fed with germinated brown rice for 8 weeks	Wunjuntuk et al. (2016)
Neuroprotective effects	Varying concentration of ethyl acetate extracts of germinated brown rice, showed higher phenolic content and antioxidant activity, and did not show any damage on *In vitro* human neuronal cell up to the concentration of 20 ppm. Also, the extract (10 ppm) prevented H_2O_2-induced apoptotic changes in the cell	Azmi et al. (2013)
Against Alzheimer's disease	The therapeutic effect of water-pressurized pigmented rice showed to reduce the amyloid-β in the brain which is a major protein responsible for the disease	Okuda et al. (2018)

Source: Saleh et al. (2019).

anthocyanin and flavonoids (Xia et al., 2006). The effects of an anthocyanin-rich extract on atherosclerotic plaque vulnerability and serum lipid level of Chinese black rice, as well as total serum antioxidant level, were investigated by Xia et al. (2006). In more than 500 traditional indica rice landraces, Sen Gupta et al. (2017) investigated the heavy metal micronutrient compositions. It was found that several rice landraces contain greater amounts of heavy metal micronutrients in the grains. Black rice extracts exhibit an antioxidant property by inhibiting cellular ROS and malondialdehyde (Lee et al., 2014). Cyanidin-3-glucoside, the main anthocyanin compound of black rice, suppresses the occurrence of high-fat-diet-induced obesity (Tsuda et al., 2003) and increases hyperglycemia and insulin sensitivity via Adenosine Monophosphate (AMP) activated protein kinase in type-2 diabetic mice (Takikawa et al., 2010). In another study in coronary heart disease patients, the black rice pigment fraction (BRF) reduced the plasma levels of soluble vascular cell adhesion molecule-1 (sVCAM-1), soluble CD40 ligand (sCD40L), and high-sensitive C-reactive protein (hs-CRP) (Wang et al., 2007). At present, rice is being analyzed from different aspects – beyond its conventional use as a staple food. Its nutritional quality and medicinal values have made rice unique among cereals.

21.7 Conservation of Rice-Genetic Diversity

Traditional rice landraces must be saved from extinction because they are rich in nutraceuticals, and many of them have medicinal properties (Deb, 2005). Cultivating and consuming traditional rice landraces, as well as researching and disseminating folk knowledge about the importance of local seeds to human well-being, will aid in the preservation of traditional rice landraces. We must assist marginal farmer groups and seed conservators in preserving traditional rice varieties by emphasizing the importance of including diverse rice genotypes in our diets. Intellectual property rights (IPRs), which provide monopolistic private rights over seeds, should be replaced by a system that protects farmers' exclusive rights to save, use, maintain, and freely exchange seeds. Farmers created all of the seeds and distributed them for free to the rest of the world. We must focus on ensuring that seeds are not subject to IPRs. In an era of open markets and modernization, corporate control over traditional rice-genetic diversity is one of the most serious threats to biodiversity loss, which ensures the planet's and humans' safety, health, nutrition, and sustainability. We can enjoy a variety of foods because of traditional seed exchange systems for farmers and local innovation processes; we must therefore ensure the continuity of these systems and pathways.

21.8 Conclusion

India has a wide range of diversity of folk rice landraces from thousands of years which are still surviving with several types of stress-induced environments. Genetic and nutritional diversity as contained in traditional rice landraces and their wild relatives with their unique morphological characters and stress-tolerant properties provide the bedrock of evolution for cultivated rice (*Oryza sativa*) and their differentiation into various cultivars for adapting to different environments. Nevertheless, this rice-genetic diversity began to decline after the 1970s when modern varieties were introduced with the main objective of grain yield improvement in the Indian agricultural system. The urgent need is for these conventional species to be actively commercialized and promoted, especially among the younger generation, by increasing public awareness. Clinical validation of their medicinal values is important in order to develop the global market. In order to prevent diseases linked to lifestyle, the promotion and protection of this national patrimony as a safe food is essential. In folk medicine, several rice landraces with extraordinary medical properties await comprehensive scientific study and confirmation, which will support contemporary medicine while maintaining community health and nutritional security. Furthermore, more research is needed to extract data from integrated omics databases in order to better understand abiotic stress tolerance causes and build large-spectrum abiotic stress-tolerant agricultural production.

Acknowledgments

The authors are grateful to Basudha's *Chair* Dr. Debal Deb (http://cintdis.org/basudha/) for sharing his research experiences as well as the published literatures, which helped a lot to understand rice biology and review write-up. PR acknowledges Dr. Deb's continued generous support and encouragement during his Ph.D. tenure. PR is thankful to Indian Council of Medical Research, Govt. of India for funding support (file no. 2019-6997). The authors are thankful to the University of Kalyani for research support.

Conflict of Interest

The authors declare no conflict of interest.

REFERENCES

AACC (American Association for Clinical Chemistry). 2001. The definition of dietary fiber. *Cereal Foods World*, 46, 112.

Abano, E. E. and Amoah, K. K. 2011. Effect of moisture content on the physical properties of tiger nut (*Cyperus esculentus*). *Asian Journal of Agricultural Research*, 5, 56–66.

Abdul-Hamid, A. and Luan, Y. S. 2000. Functional properties of dietary fibre prepared from defatted rice bran. *Food Chemistry*, 68, 15–19.

Aggarwal, R. K., Shenoy, V. V., Ramadevi, J., Rajkumar, R. and Singh, L. 2002. Molecular characterization of some Indian Basmati and other elite rice genotypes using fluorescent AFLP. *Theoretical and Applied Genetics*, 105, 680–690.

Ahuja, U, Ahuja S. C., Thakrar, R., and Singh, R. K. (2008) Rice: A nutraceutical. *Asian Agri-History*, 12, 93–108.

Amagliani, L., O'Regan, J., Kelly, A. L., and O'Mahony, J. A. 2017. Composition and protein profile analysis of rice protein ingredients. *Journal of Food Composition and Analysis*, 59, 18–26.

Anandan, A., Rajiv, G., Eswaran, R. and Prakash, M. 2011. Genotypic variation and relationships between quality traits and trace elements in traditional and improved rice (*Oryza sativa* L.) genotypes. *Journal of Food Science*, 76(4), H122–H130.

Ashraf, A. M., and Lokanadan, S. 2017. A review of rice landraces in India and its inherent medicinal values-the nutritive food values for future. *International Journal of Current Microbiology and Applied Sciences*, 6(12), 348–35.

Azmi, N. H., Ismail, N., Imam, M. U., and Ismail, M. 2013. Ethyl acetate extract of germinated brown rice attenuates hydrogen peroxide-induced oxidative stress in human SH-SY5Y neuroblastoma cells: Role of anti-apoptotic, pro-survival and antioxidant genes. *BMC Complementary & Alternative Medicine*, 13, 177.

Blumenthal, J. M., Baltensperger, D. D., Cassman, K. G., Mason, S. C., and Pavlista, A. D. 2008. Importance and effect of nitrogen on crop quality and health. In: *Nitrogen in the Environment*. Academic Press: Cambridge, MA, pp. 51–70.

Butsat, S., and Siriamornpun, S. 2010. Antioxidant capacities and phenolic compounds of the husk, bran and endosperm of Thai rice. *Food Chemistry*, 119(2), 606–613.

Buttery, R. G., Lilng, L. C., and Juliano, B. O. 1982. 2-Acety-1-pyrroline: An important aroma component of cooked rice. *Chemical Industry*, 23, 958–959.

Calingacion, M., Mumm, R., Tan, K., Quiatchon-Baeza, L., Concepcion, J. C., Hageman, J. A., … and Hall, R. D. 2017. A multidisciplinary phenotyping and genotyping analysis of a mapping population enables quality to be combined with yield in rice. *Frontiers in Molecular Biosciences*, 4, 32.

Cao, Q., Lu, B. R., Xia, H. U. I., Rong, J., Sala, F., Spada, A., and Grassi, F. 2006. Genetic diversity and origin of weedy rice (*Oryza sativa* f. spontanea) populations found in northeastern China revealed by simple sequence repeat (SSR) markers. *Annals of Botany*, 98(6), 1241–1252.

Champagne, E. T., Bett-Garber, K. L., McClung, A. M., and Bergman, C. 2004. Sensory characteristicsof diverse rice cultivars as influenced by genetic and environmental factors. *Cereal Chemistry*, 81, 237–243.

Charoen, R., Jangchud, A., Jangchud, K., Harnsilawat, T., Naivikul, O., and McClements, D. J. 2011. Influence of biopolymer emulsifier type on formation and stability of rice bran oil-in-water emulsions: Whey protein, gum arabic, and modified starch. *Journal of Food Science*, 76(1), E165–E172.

Chatthongpisut, R., Schwartz, S. J., and Yongsawatdigul, J. 2015. Antioxidant activities and antiproliferative activity of Thai purple rice cooked by various methods on human color cancer cells. *Food Chemistry*, 188, 99–105.

Chaudhary, R. C., and Tran, D. V. 2001. Specialty rices of the world: A prologue. In: *Specialty Rices of the World: Breeding, Production, and Marketing*. FAO, Rome, Italy; and Oxford & IBH Publishing Co. Pvt. Ltd., New Delhi, India, pp. 3–14.

Chaudhary, R. C., Tran, D. V., and Duffy, R. 2001. Speciality rices of the world: Breeding, production and marketing. Food and Agriculture Organization of the United Nations (FAO).

Chaudhary, A., Gustafson, D., and Mathys, A. 2018. Multi-indicator sustainability assessment of global food systems. *Nature Communications*, 9(1), 1–13.

Chaves, F. C., and Broeckling, C. D. 2020. Metabolomics for rice grain quality. In: *The Future of Rice Demand: Quality beyond Productivity* (Eds. de Oliveira, A. C., Pegoraro, C., and Viana, V. E.). Springer, Cham, pp. 495–531.

Chen, W., Gao, Y., Xie, W., Gong, L., Lu, K., Wang, W., … and Zhang, W. 2014. Genome-wide association analyses provide genetic and biochemical insights into natural variation in rice metabolism. *Nature Genetics*, 46(7), 714–721.

Chen, C., He, W., Nassirou, T. Y., Nsabiyumva, A., Dong, X., Adedze, Y. M. N., and Jin, D. 2017. Molecular characterization and genetic diversity of different genotypes of *Oryza sativa* and *Oryza glaberrima*. *Electronic Journal of Biotechnology*, 30, 48–57.

Choi, S. P., Kang, M. Y., Koh, H. J., Nam, S. H., & Friedman, M. 2007. Antiallergic activities of pigmented rice bran extracts in cell assays. *Journal of Food Science*, 72(9), 719–726.

Choudhury, S., Sharma, P., Moulick, D., and Mazumder, M. K. 2021. Unrevealing metabolomics for abiotic stress adaptation and tolerance in plants. *Journal of Crop Science and Biotechnology*, 1–15.

Das, G. K., and Qudhia, P. 2001. Rice as medicinal plant in ghhatiisgarh (India): A survey. *Agricultural Science Digest*, 21(3), 204–205.

Daygon, V. D., Calingacion, M., Forster, L. C., De Voss, J. J., Schwartz, B. D., Ovenden, B.,,… and Fitzgerald, M. A. 2017. Metabolomics and genomics combine to unravel the pathway for the presence of fragrance in rice. *Scientific Reports*, 7(1), 1–12.

de Oliveira, A. C., Pegoraro, C., and Viana, V. E. (Eds.). (2020). *The Future of Rice Demand: Quality Beyond Productivity*. Springer International Publishing, New York.

De, B., and Nag, G. 2014. Metabolic diversity in the grains of Indian varieties of rice. *Comptes Rendus Biologies*, 337(4), 283–293.

Deb, D. 2005. *Seeds of Tradition, Seeds of Future: Folk Rice Varieties from East India*. Research Foundation for Science Technology & Ecology, New Delhi.

Deb, D. 2014. In: *Multicultural Knowledge and the University* (Ed. Alvares, C.). Multiversity/Citizens International, Mapusa, Goa, India.

Deb, D., Sen Gupta, S., and Pradeep, T. 2015. A profile of heavy metals in rice (*Oryza sativa* ssp. indica) landraces. *Current Science*, 109(3), 407.

Deepa, G., Singh, V., and Naidu, K. A. 2008. Nutrient composition and physicochemical properties of Indian medicinal rice: Njavara. *Food Chemistry*, 106, 165–171.

Degenkolbe, T., Do, P. T., Kopka, J., Zuther, E., Hincha, D. K., and Köhl, K. I. 2013. Identification of drought tolerance markers in a diverse population of rice cultivars by expression and metabolite profiling. *PLoS One*, 8(5), e63637.

Del Rosario, A. R., Briones, V. P., Vidal, A. J. and Juliano, B. O. 1968. Composition and endosperm structure of developing and mature rice kernel. *Cereal Chemistry*, 45, 225–235.

Devi, R. R., and Arumughan, C. (2007). Antiradical efficacy of phytochemical extracts from defatted rice bran. *Food and Chemical Toxicology*, 45(10), 2014–2021.

Dittgen, C. L., Hoffmann, J. F., Chaves, F. C., Rombaldi, C. V., Colombari Filho, J. M., & Vanier, N. L. 2019. Discrimination of genotype and geographical origin of black rice grown in Brazil by LC-MS analysis of phenolics. *Food Chemistry*, 288, 297–305.

Dong, W., Chen, J., Wang, L., Tian, Y., Zhang, B., Lai, Y., … and Guo, J. 2014. Impacts of nighttime post-anthesis warming on rice productivity and grain quality in East China. *The Crop Journal*, 2(1), 63–69.

Dutta, A. K., Gope, P. S., Banik, S., Makhnoon, S., Siddiquee, M. A., and Kabir, Y. 2012. Antioxidant properties of ten high yielding rice varieties of Bangladesh. *Asian Pacific Journal of Tropical Biomedicine*, 2(1), S99–S103.

Duyi, S., Baran, A., and Chandra, D. S. 2017. Pigmented rice a potential source of bioactive compounds: A review. *International Journal of Food Science & Technology*, 52(5), 1073–1081.

Fahad, S., Noor, M., Adnan, M., Khan, M. A., Rahman, I. U., Alam, M., … and Nasim, W. (2019). Abiotic stress and rice grain quality. In: *Advances in Rice Research for Abiotic Stress Tolerance* (Eds. Hasanuzzaman, M., Fujita, M., Nahar, K., and Biswas, J. K.). Woodhead Publishing, Sawston, United Kingdom, pp. 571–583.

Filiz, E., Uras, M. E., Ozyigit, I. I., Sen, U., and Gungor, H. 2018. Genetic diversity and phylogenetic analyses of Turkish rice varieties revealed by ISSR markers and chloroplast trnl-F region. *Feb-Fresenius Environmental Bulletin*, 27(12), 8351.

Forster, G. M., Raina, K., Kumar, A., Kumar, S., Agarwal, R., Chen, M.-H., et al. 2013. Rice varietal differences in bioactive bran components for inhibition of colorectal cancer cell growth. *Food Chemistry*, 141, 1545–1552.

Frank, T., Reichardt, B., Shu, Q., and Engel, K. H. 2012. Metabolite profiling of colored rice (*Oryza sativa* L.) grains. *Journal of Cereal Science*, 55(2), 112–119.

Frei, M., and Becker, K. 2005. Fatty acids and all-trans- β -carotene are correlated in differently colored rice landraces. *Journal of the Science of Food and Agriculture*, 85, 2380–2384.

Frei, M., Siddhuraju, P., and Becker, K. 2003. Studies on the in vitro starch digestibility and the glycemic index of six different indigenous rice cultivars from the Philippines. *Food Chemistry*, 83, 395–402.

Ghasemzadeh, A., Karbalaii, M. T., Jaafar, H. Z. E., and Rahmat, A. 2018. Phytochemical constituents, antioxidant activity, and antiproliferative properties of black, red, and brown rice bran. *Chemistry Central Journal*, 12(1), 17.

Goffman, F. D., and Bergman, C. J. 2004. Rice kernel phenolic content and its relationship with antiradical efficiency. *Journal of the Science of Food and Agriculture*, 84, 1235–1240.

Goldberg, E. D., and Saltzman, J. R. C. 1996. Rice inhibits intestinal secretions. *Nutrition Reviews*, 54, 36–37.

Gong, L., Chen, W., Gao, Y., Liu, X., Zhang, H., Xu, C., … and Luo, J. 2013. Genetic analysis of the metabolome exemplified using a rice population. *Proceedings of the National Academy of Sciences*, 110(50), 20320–20325.

Gul, K., Yousuf, B., Singh, A. K., Singh, P., and Wani, A. A. 2015. Rice bran: Nutritional values and its emerging potential for development of functional food: A review. *Bioactive Carbohydrates and Dietary Fibre*, 6(1), 24–30.

Gurpreet, K. C., and Sogi, D. S. 2007. Functional properties of rice bran protein concentrates. *Journal of Food Engineering*, 79, 592–597.

Ha, T., Han, S., Kim, S., Kim, I., Lee, H., and Kim, H. 2005. Bioactive components in rice bran oil improve lipid profiles in rats fed a highcholesterol diet. *Nutrition Research*, 25(6), 597–606.

Hamada, J. S. 1997. Characterization of protein fractions of rice bran to devise effective methods of protein solubilization. *Cereal Chemistry*, 74(5), 662–668.

Hansakul, P., Srisawat, U., Kaendee, N., and Panunto, W. 2010. Determination of phenolic compounds, flavonoids, and antioxidant activities in crude water extracts of Thai red and white rice cultivars. *Planta Medica*, 76(12), P133.

Hasanuzzaman, M., Nahar, K., Fujita, M., Ahmad, P., Chandna, R., Prasad, M. N. V., and Ozturk, M. 2013. Salt stress in plants. In: *Enhancing Plant Productivity under Salt Stress: Relevance of Poly-Omics*. Springer, New York, pp. 113–156.

Hass, J. D., Beard, J. L., Murray-Kolb, L. E., del Mundo, A. M., Felix, A., and Gregorio, G. 2005. Iron-biofortified rice improves the iron stores of non-anemic Filipino women. *Journal of Nutrition*, 35, 2823–2830.

Hassimotto, N. M. A., Genovese, M. I., and Lajolo, F. M. 2008. Absorption and metabolism of cyanidin-3-glucoside and cyanidin-3-rutinoside extracted from wild mulberry (*Morus nigra* L.) in rats. *Nutrition Research*, 28(3), 198–207.

Helm, R. M. and Burks, A. W. 1996. Hypoallergenicity of Rice Protein. *Cereal Foods World*, 41, 839–843.

Henderson, A. J., Ollila, C. A., Kumar, A., Borresen, E. C., Raina, K., Agarwal, R., et al. 2012. Chemopreventive properties of dietary rice bran: Current status and future Prospects1,2. *Advances in Nutrition*, 3, 643–653.

Ghneim Herrera, T., Posso Duque, D., Pérez Almeida, I., Torrealba Núñez, G., Pieters, A. J., Martinez, C. P., & Tohme, J. M. (2008). Assessment of genetic diversity in Venezuelan rice cultivars using simple sequence repeats markers. *Electronic Journal of Biotechnology*, 11(5), 3–4.

Herrera, T., Duque, D., Almeida, I., Núñez, G., Pieters, A., Martinez, C., and Tohme, J. 2008. Assessment of genetic diversity in Venezuelan rice cultivars using simple sequence repeats markers. *Electronic Journal of Biotechnology*, 11(5), 3–4.

Hettiarachchi, P., Jiffery, M. T. M., Jansz, E. R., Wickramasinghe, A. R., and Fernando, D. J. S. 2001. Glycaemic index of different varieties of rice grown in Sri Lanka. *Ceylon Medical Journal*, 46, 11–14.

Heuberger, A. L., Lewis, M. R., Chen, M. H., Brick, M. A., Leach, J. E., and Ryan, E. P. 2010. Metabolomic and functional genomic analyses reveal varietal differences in bioactive compounds of cooked rice. *PLoS One*, 5(9), e12915.

Horie, Y., Nemoto, H., Itoh, M., Kosaka, H., and Morita, K. 2016. Fermented brown rice extract causes apoptotic death of human acute lymphoblastic leukemia cells via death receptor pathway. *Applied Biochemistry & Biotechnology*, 178, 1599–1611.

Hosoda, K., Sasahara, H., Matsushita, K., Tamura, Y., Miyaji, M., and Matsuyama, H. 2018. Anthocyanin and proanthocyanidin contents, antioxidant activity, and in situ degradability of black and red rice grains. *Asian-Australasian Journal of Animal Sciences*, 31(8), 1213–1220.

Hu, C., Shi, J., Quan, S., Cui, B., Kleessen, S., Nikoloski, Z., … and Wang, J. 2014. Metabolic variation between japonica and indica rice cultivars as revealed by non-targeted metabolomics. *Scientific Reports*, 4, 5067.

Huang, Y. P., and Lai, H. M. 2016. Bioactive compounds and antioxidative activity of colored rice bran. *Journal of Food and Drug Analysis*, 24, 564–574.

Imam, M. U., and Ismail, M. 2013. Nutrigenomic effects of germinated brown rice and its bioactives on hepatic gluconeogenic genes in type 2 diabetic rats and HEPG2 cells. *Molecular Nutrition & Food Research*, 57, 401–411.

Imam, M. U., Azmi, N. H., Bhanger, M. I., Ismail, N., and Ismail, M. 2012. Antidiabetic properties of germinated brown rice: A systematic review. *Evidence-Based Complementary and Alternative Medicine*, 2012, 1–12.

Jariwalla, R. J. 2001. Rice-bran products: Phytonutrients with potential applications in preventive and clinical medicine. *Drugs under Experimental and Clinical Research*, 27, 17–26.

Jasti, T., Chintala, R. K., and Kolluru, V. C. 2020. Biological functions of volatile compounds extracted from the rice bran: A review. *Biomedical and Pharmacology Journal*, 13(1), 213–231.

Jeng, T. L., Lai, C. C., Ho, P. T., Shih, Y. J., and Sung, J. M. 2012. Agronomic, molecular and antioxidative characterization of red- and purple-pericarp rice (*Oryza sativa* L.) mutants in Taiwan. *Journal of Cereal Science*, 56(2), 425–431.

Joseph, J., Francies, R. M., Zachariah, G., and Av, S. K. 2007. Characterisation of navara (*Oryza sativa* L.) a traditional medicinal rice of Kerala for qualitative traits. *Indian Journal of Agricultural Research*, 41(4), 267–271.

Joshi, S. P., Gupta, V. S., Aggarwal, R. K., Ranjekar, P. K., and Brar, D. S. 2000. Genetic diversity and phylogenetic relationship as revealed by inter-simple sequence repeat (ISSR) polymorphism in the genus *Oryza*. *Theoretical and Applied Genetics*, 100(8), 1311–1320.

Juliano, B. O. 1993. Rice in human nutrition (No. 26). International Rice Research Institute.

Juliano, B. O., 1972. *Physicochemical Properties of Starch and Protein in Relation to Grain Quality and Nutritional Value of Rice: IRRI Rice Breeding*. IRRI, Los Banos, Philippines, pp. 389–405.

Juliano, B. O. and Bechtel, D. B. 1985. The rice grain and its gross composition. In: *Rice, Chemistry and Technology* (Ed. Juliano, B. O.), 2nd edition. AACC, St Paul, MN, pp. 17–57.

Juliano, B. O., and Villareal, C. P. 1993. Grain quality evaluation of world rices. International Rice Research Institute.

Kasote, D., Tiozon Jr, R. N., Sartagoda, K. J. D., Itagi, H., Roy, P., Kohli, A., … and Sreenivasulu, N. (2021). Food processing technologies to develop functional foods with enriched bioactive phenolic compounds in Cereals. *Frontiers in Plant Science*, 12, 771276.

Kennedy, G. and Burlingame, B. 2003. Analytical, nutritional and clinical methods analysis of food composition data on rice from a plant genetic resources perspective. *Food Chemistry*, 80, 589–596.

Khatoon, S., & Gopalakrishna, A. G. (2004). Fat-soluble nutraceuticals and fatty acid composition of selected Indian rice varieties. Journal of the American Oil Chemists' Society, 81(10), 939–943.

Kim, J. K., Ha, S. H., Park, S. Y., Lee, S. M., Lim, S. H., ... and Cho, H. S. 2012. Determination of lipophilic compounds in genetically modified rice using gas chromatography–time-of-flight mass spectrometry. *Journal of Food Composition and Analysis*, 25(1), 31–38.

Kim, J. K., Park, S. Y., Lee, S. M., Lim, S. H., Kim, H. J., Oh, S. D., ... and Ha, S. H. (2013). Unintended polar metabolite profiling of carotenoid-biofortified transgenic rice reveals substantial equivalence to its non-transgenic counterpart. *Plant Biotechnology Reports*, 7(1), 121–128.

Kim, J. Y., Seo, W. D., Park, D. S., Jang, K. C., Choi, K. J., Kim, S. Y., ... and Han, S. I. 2013. Comparative studies on major nutritional components of black waxy rice with giant embryos and its rice bran. *Food Science & Biotechnology*, 22, 121–128.

Kitta, K., Ebihara, M., Iizuka, T., Yoshikawa, R., Isshiki, K., and Kawamoto, S. 2005. Variations in lipid content and fatty acid composition of major non-glutinous rice cultivars in Japan. *Journal of Food Composition and Analysis*, 18, 269–278.

Kojima, Y., Ebana, K., Fukuoka, S., Nagamine, T. and Kawase, M. 2005. Development of an RFLP-based rice diversity research set of germplasm. *Breeding Science*, 55, 431–440.

Kong, C. K., Lam, W. S., Chiu, L. C., Ooi, V. E., Sun, S. S., and Wong, Y. S. 2009. A rice bran polyphenol, cycloartenyl ferulate, elicits apoptosis in human colorectal adenocarcinoma SW480 and sensitizes metastatic SW620 cells to TRAIL-induced apoptosis. *Biochemical Pharmacology*, 77(9), 1487–1496.

Koren, H. S., Graham, D. E., and Devlin, R. B. 1992. Exposure of humans to a volatile organic mixture. III. Inflammatory response. *Archives of Environmental Health: An International Journal*, 47(1), 39–44.

Krishnan, V., Awana, M., Raja Rani, A. P., Bansal, N., Bollinedi, H., Srivastava, S., ... & Praveen, S. (2021). Quality matrix reveals the potential of Chak-hao as a nutritional supplement: a comparative study of matrix components, antioxidants and physicochemical attributes. Journal of Food Measurement and Characterization, 15(1), 826–840.

Kusano, M., Fukushima, A., Kobayashi, M., Hayashi, N., Jonsson, P., Moritz, T., ... and Saito, K. 2007. Application of a metabolomic method combining one-dimensional and two-dimensional gas chromatography-time-of-flight/mass spectrometry to metabolic phenotyping of natural variants in rice. *Journal of Chromatography B*, 855(1), 71–79.

Lai, P., Li, K. Y., Lu, S., and Chen, H. H. 2009. Phytochemicals and antioxidant properties of solvent extracts from Japonica rice bran. *Food Chemistry*, 117(3), 538–544.

Lawas, L. M. F., Li, X., Erban, A., Kopka, J., Jagadish, S. K., Zuther, E., and Hincha, D. K. 2019. Metabolic responses of rice cultivars with different tolerance to combined drought and heat stress under field conditions. *GigaScience*, 8(5), giz050.

Lee, K. S., Choi, W. Y., Ko, J. C., Kim, T. S., and Gregorio, G. B. 2003. Salinity tolerance of japonica and indica rice (*Oryza sativa* L.) at the seedling stage. *Planta*, 216(6), 1043–1046.

Lee, J.-C., Kim, J.-D., Hsieh, F.-H., and Eun, J.-B. 2008. Production of black rice cake using ground black rice and medium-grain brown rice. *International Journal of Food Science and Technology*, 43(6), 1078–1082.

Lee, S. M., Youngmin, C., Sung, J., Kim, Y., Jeong, H. S., and Lee, J. 2014. Protective effects of black rice extracts on oxidative stress induced by tert-butyl hydroperoxide in HepG2 cells. *Preventive Nutrition and Food Science*, 19(4), 348–352.

Lee, S., Singh, D., Jang, E. S., Shin, H. W., Moon, B. S., and Lee, C. H. 2017. Time-resolved comparative metabolomes for Koji fermentation with brown-, white-, and giant embryo-rice. *Food Chemistry*, 231, 258–266.

Li, S. C., Chou, T. C., and Shih, C. K. 2011. Effects of brown rice, rice bran, and polished rice on colon carcinogenesis in rats. *Food Research International*, 44, 209–216.

Lim, D. K., Long, N. P., Mo, C., Dong, Z., Cui, L., Kim, G., and Kwon, S. W. 2017. Combination of mass spectrometry-based targeted lipidomics and supervised machine learning algorithms in detecting adulterated admixtures of white rice. *Food Research International*, 100, 814–821.

Lin, P. Y., Li, S. C., Lin, H. P., and Shih, C. K. 2019. Germinated brown rice combined with *Lactobacillus acidophilus* and *Bifidobacterium animalis* subsp. *Lactis* inhibits colorectal carcinogenesis in rats. *Food Science and Nutrition*, 7, 216–224.

Ling, W. H., Cheng, Q. X., Ma, J., and Wang, T. 2001. Red and black rice decrease atherosclerotic plaque formation and increase antioxidant status in rabbits. *Journal of Nutrition*, 131, 1421–1426.

Long, N. P., Lim, D. K., Mo, C., Kim, G., and Kwon, S. W. 2017. Development and assessment of a lysophospholipid-based deep learning model to discriminate geographical origins of white rice. *Scientific Reports*, 7(1), 1–10.

Mahanta, C. L., and Sikia, S. 2016. Bioactive compounds of rice as health promoters. In: *Natural Bioactive Compounds from Fruits and Vegetables as Health Promoters Part II* (Eds. da Silva, L. R., and Silva, B. M.). Bentham Science Publishers, Netherlands, pp. 221–234.

Mahattanatawee, K., and Rouseff, R. L. 2014. Comparison of aroma active and sulfur volatiles in three fragrant rice cultivars using GC–Olfactometry and GC–PFPD. *Food Chemistry*, 154, 1–6.

Marini, F., Balestrieri, F., Bucci, R., Magri, A. L. and Marini, D. 2003. Supervised pattern recognition to discriminate the geographical origin of RBOs: A first study. *Microchemical Journal*, 74, 239–248.

Marshall, W. G., and Wordsworth, J. I. 1994. *Rice Science and Technology*. Marcel Dekker, Inc., New York, pp. 237–259.

Matsuda, F., Okazaki, Y., Oikawa, A., Kusano, M., Nakabayashi, R., Kikuchi, J., ... and Saito, K. 2012. Dissection of genotype–phenotype associations in rice grains using metabolome quantitative trait loci analysis. *The Plant Journal*, 70(4), 624–636.

Matsuda, F., Nakabayashi, R., Yang, Z., Okazaki, Y., Yonemaru, J. I., Ebana, K., ... and Saito, K. 2015. Metabolome-genome-wide association study dissects genetic architecture for generating natural variation in rice secondary metabolism. *The Plant Journal*, 81(1), 13–23.

Matsumoto, K., Yasuyoshi, E., Nishi, K., Honda, Y., Nakaya, M., and Kitamura, S. 2016. Resistant starch-rich *wx/ae*, brown rice prevents insulin resistance and hypertriglyceridaemia in type 2 diabetic NSY mice. *Journal of Functional Foods*, 23, 556–564.

Mazumder, M.K., Sharma, P., Moulick, D. et al. Salicylic acid ameliorates zinc and chromium-induced stress responses in wheat seedlings: a biochemical and computational analysis. CEREAL RESEARCH COMMUNICATIONS 50, 407–418 (2022).

Mazumder, M. K., Sharma, P., Moulick, D. et al. 2022. Salicylic acid ameliorates zinc and chromium-induced stress responses in wheat seedlings: A biochemical and computational analysis. *Cereal Research Communications*, 50, 407–418.

Miller, A., and Engel, K. H. 2006. Content of γ-oryzanol and composition of steryl ferulates in brown rice (*Oryza sativa* L.) of European origin. *Journal of Agricultural and Food Chemistry,* 54(21), 8127–8133.

Miller, H. E., Rigelhof, F., Marquart, L., Prakash, A., and Kanter, M. 2000. Antioxidant content of whole grain breakfast cereals, fruits and vegetables. *Journal of the American College of Nutrition*, 19(sup3), 312S–319S.

Moldenhauer, K. A., Champagne, E. T., McCaskill, D. R. and Guraya, H. 2003. Functional products from rice. In: *Functional Foods* (Ed. Mazza G.). Technomic Publishing, Lancaster, UK, pp. 71–89.

Moulick, D., Ghosh, D., and Santra, S. C. 2016. Evaluation of effectiveness of seed priming with selenium in rice during germination under arsenic stress. *Plant Physiology and Biochemistry,* 109, 571–578.

Moulick, D., Santra, S. C., and Ghosh, D. 2018. Seed priming with Se mitigates As-induced phytotoxicity in rice seedlings by enhancing essential micronutrient uptake and translocation and reducing As translocation. *Environmental Science and Pollution Research,* 25(27), 26978–26991.

Nam, Y. J., Nam, S. H. and Kang, M. Y. 2008. Cholesterol-lowering efficacy of unrefined bran oil from the pigmented black rice (*Oryza sativa* L cv. Suwon 415) in hypercholesterolemic rats. *Food Science and Biotechnology*, 17(3), 457–463.

Nasu, S., Suzuki, J., Ohta, R., Hasegawa, K., Yui, R., Kitazawa, N., ... and Minobe, Y. 2002. Search for and analysis of single nucleotide polymorphisms (SNPs) in rice (*Oryza sativa, Oryza rufipogon*) and establishment of SNP markers. *DNA Research*, 9(5), 163–171.

Nene Y. L. 2005. Rice research in South Asia through ages. *Asian Agri-History*, 9, 85–106.

Ogo, Y., Ozawa, K., Ishimaru, T., Murayama, T., and Takaiwa, F. 2013. Transgenic rice seed synthesizing diverse flavonoids at high levels: A new platform for flavonoid production with associated health benefits. *Plant Biotechnology Journal*, 11(6), 734–746.

Ohara, K., Uchida, A., Nagasaka, R., Ushio, H., and Ohshima, T. 2009. The effects of hydroxycinnamic acid derivatives on adiponectin secretion. *Phytomedicine*, 16, 130–137.

Oh-e, I., Saitoh, K., and Kuroda, T. 2007. Effects of high temperature on growth, yield and dry-matter production of rice grown in the paddy field. *Plant Production Science,* 10(4), 412–422.

Okazaki, Y., Otsuki, H., Narisawa, T., Kobayashi, M., Sawai, S., Kamide, Y., ... and Saito, K. 2013. A new class of plant lipid is essential for protection against phosphorus depletion. *Nature Communications*, 4(1), 1–10.

Oki, T., Masuda, M., Nagai, S., Take'ichi, M., Kobayashi, M., Nishiba, Y., ... and Sato, T. 2005. Radical-scavenging activity of red and black rice. In: *Rice Is Life: Scientific Perspectives for the 21st Century. Proceedings of the World Rice Research Conference*, Tsukuba, Japan, 4–7 November 2004, pp. 256–259. International Rice Research Institute (IRRI).

Okuda, M., Fujita, Y., Katsube, T., Tabata, H., Yoshino, K., Hashimoto, M., and Sugimoto, H. 2018. Highly water pressurized brown rice improves cognitive dysfunction in senescence-accelerated mouse prone 8 and reduces amyloid beta in the brain. *BMC Complementary & Alternative Medicine,* 18, 110.

Olufowote, J. O., Xu, Y., Chen, X., Park, W. D., Beachel, H. M., Dilday, R. H., Goto, M., and McCouch, S. R. 1997. Comparative evaluation of within-cultivar variation of rice (*Oryza sativa* L.) using microsatellite and RFLP markers. *Genome*, 40, 370–378.

Ortmayr, K., Causon, T. J., Hann, S., and Koellensperger, G. 2016. Increasing selectivity and coverage in LC-MS based metabolome analysis. *TrAC Trends in Analytical Chemistry*, 82, 358–366.

Paine, J. A., Shipton, C. A., Chaggar, S., Howells, R. M., Kennedy, M. J., Vernon, G., ... and Drake, R. 2005. Improving the nutritional value of Golden Rice through increased provitamin A content. *Nature Biotechnology*, 23(4), 482–487.

Park, J. R., Yang, W. T., Kwon, Y. S., Kim, H. N., Kim, K. M., and Kim, D. H. 2019. Assessment of the genetic diversity of rice germplasms characterized by black-purple and red pericarp color using simple sequence repeat markers. *Plants*, 8(11), 471.

Parrado, J., Miramontes, E., Jover, M., Marquez, J. C., Mejias, M. and Collantes de Teran, L. 2003. Prevention of brain protein and lipid oxidation elicited by a water soluble oryzanol enzymatic extract derived from rice bran. *European Journal of Nutrition*, 42, 307–314.

Patel, M. and Naik, S. N. 2004 Gamma-oryzanol from rice bran oil: A review. *Journal of Scientific and Industrial Research*, 63, 569–578.

Patra, N. and Chawla, H. S. 2010. Biochemical and RAPD molecular markers for establishing distinctiveness of basmati rice (*Oryza sativa*) varieties as additional descriptors for plant variety protection. *Indian Journal of Biotechnology*, 9, 371–377.

Pervaiz, Z. H., Rabbani, M. A., Shinwari, Z. K., Masood, M. S., and Malik, S. A. 2010. Assessment of genetic variability in rice (*Oryza sativa* L.) germplasm from Pakistan using RAPD markers. *Pakistan Journal of Botany*, 42(5), 3369–3376.

Prasad, A. S. 1998, Zinc in human health: An update. *Journal of Trace Elements in Experimental Medicine*, 11, 63–87. doi: 10.1002/(SICI)1520-670X(1998)11:2/3<63::AID-JTRA2>3.0.CO;2-5.

Prashanth, S. R., Parani, M., Mohanty, B. P., Talame, V., Tuberosa, R., and Parida, A. 2002. Genetic diversity in cultivars and landraces of Oryza sativa subsp. indica as revealed by AFLP Markers. *Genome*, 45, 451–459.

Quagliariello, V., Iaffaioli, R. V., Falcone, M., Ferrari, G., Pataro, G., and Donsi, F. 2016. Effect of pulsed electric fields - assisted extraction on anti-inflammatory and cytotoxic activity of brown rice bioactive compounds. *Food Research International,* 87, 115–124.

Raihan, A., Nahiyan, A. S. M., Rahman, A., and Rahman, L. 2016. Molecular characterization of selected M5 lines of rice after TILLING for salinity tolerance using 20 SSR primers. *Biotechnology Journal International*, 1–11.

Ranilla, L. G., Kwon, Y. I., Apostolidis, E., and Shetty, K. 2010. Phenolic compounds, antioxidant activity and in vitro inhibitory potential against key enzymes relevant for hyperglycemia and hypertension of commonly used medicinal plants, herbs and spices in Latin America. *Bioresource Technology*, 101(12), 4676–4689.

Ranjitha, H. P., Gowda, R., Nethra, N., Amruta, N., and Kandikattu, H. K. 2019. Biochemical and metabolomics on rice cultivars. *Rice Science*, 26(3), 189–194.

Revilla, E., Santa-Marıa, C., Miramontes, E., Candiracci, M., RodrıguezMorgado, B., Carballo, M., Bautista, J., Castano, A. and Parrado, J. 2013. Anti-proliferative and immunoactivatory ability of an enzymatic extractfrom rice bran. *Food Chemistry*, 136, 526–531.

Rhoades, A. 2003. Basmati rice: The quality grain. www.allcreatures.org.

Roy, P., Deb, D., Pradeep, T., Talai-Mukhopadhyay, S., Sinha, A. K., and Saha, T. 2021. Comparative analyses of the nutraceutical potentialities of selected Indian traditional black rice (*Oryza sativa* L.) landraces. *ORYZA-An International Journal of Rice*, 58(2), 295–309.

Saleh, A. S., Wang, P., Wang, N., Yang, L., and Xiao, Z. 2019. Brown rice versus white rice: Nutritional quality, potential health benefits, development of food products, and preservation technologies. *Comprehensive Reviews in Food Science and Food Safety,* 18(4), 1070–1096.

Saura-Calixto, F., Serrano, J., and Goñi, I. 2007. Intake and bioaccessibility of total polyphenols in a whole diet. *Food Chemistry*, 101(2), 492–501.

Schramm, R., Abadie, A., Hua, N., Xu, Z., and Lima, M. 2007. Fractionation of the rice bran layer and quantification of vitamin E, oryzanol, protein, and rice bran saccharide. *Journal of Biological Engineering*, 1, 1–9.

Sen Gupta, S., Baksi, A., Roy, P., Deb, D., and Pradeep, T. 2017. Unusual accumulation of silver in the aleurone layer of an Indian rice (*Oryza sativa*) landrace and sustainable extraction of the metal. *ACS Sustainable Chemistry & Engineering*, 5(9), 8310–8315.

Shao, Y., Xu, F., Sun, X., Bao, J., and Beta, T. 2014. Identification and quantification of phenolic acids and anthocyanins as antioxidants in bran, embryo and endosperm of white, red and black rice kernels (*Oryza sativa* L.). *Journal of Cereal Science*, 59(2), 211–218.

Shao, Y., Hu, Z., Yu, Y., Mou, R., Zhu, Z., and Beta, T. 2018. Phenolic acids, anthocyanins, proanthocyanidins, antioxidant activity, minerals and their correlations in non-pigmented, red, and black rice. *Food Chemistry*, 239, 733–741.

Sharif, M. K., Butt, M. S., Anjum, F. M., and Khan, S. H. 2014. Rice bran: A novel functional ingredient. *Critical Reviews in Food Science and Nutrition*, 54, 807–816.

Shewry, P. R., and Casey, R. 1999. *Seed Proteins.* Kluwer Academic Publishers, Dordrecht, The Netherlands, pp. 1–10.

Singh, B. U., Rao, K. V., and Sharma, H. C. 2011. Comparison of selection indices to identify sorghum genotypes resistant to the spotted stemborer Chilopartellus (Lepidoptera: Noctuidae). *International Journal of Tropical Insect Science*, 31, 38–51.

Skaria, R., Sen, S., and Munner, P. 2011. Analysis of genetic variability in rice varieties (*Oryza sativa* L.) of Kerala using RAPD markers. *Genetic Engineering and Biotechnology Journal*, 2011, 1–24.

Son, M. J., Rico, C. W., Nam, S. H., and Kang, M. Y. 2011. Effect of oryzanolandferulic acid on the glucose metabolism of mice fed with a highfat diet. *Journal of Food Science*, 76, 7–10.

Song, E. H., Jeong, J., Park, C. Y., Kim, H. Y., Kim, E. H., Bang, E., and Hong, Y. S. 2018. Metabotyping of rice (*Oryza sativa* L.) for understanding its intrinsic physiology and potential eating quality. *Food Research International*, 111, 20–30.

Soriguer, F., Colomo, N., Olveira, G., García-Fuentes, E., Esteva, I., Ruiz de Adana, M. S., ... Rojo-Martínez, G. 2013. White rice consumption and risk of type 2 diabetes. *Clinical Nutrition*, 32(3), 481–484.

Srinivasa, D., Raman, A., Meena, P., Chitale, G., Marwaha, A., and Jainani, K. J. 2013. Glycaemic index (GI) of an Indian branded thermally treated Basmati rice variety: A multi centric study. *Journal of the Association of Physicians*, 61, 716–720.

Srinivasan, M., Sudheer, A. R., and Menon, V. P. 2007. Ferulic acid: Therapeutic potential through its antioxidant property. *Journal of Clinical Biochemistry and Nutrition*, 40(387), 92–100.

Storozhenko, S., De Brouwer, V., Volckaert, M., Navarrete, O., Blancquaert, D., Zhang, G. F., ... and Van Der Straeten, D. 2007. Folate fortification of rice by metabolic engineering. *Nature Biotechnology*, 25(11), 1277–1279.

Stratil, P., Klejdus, B., and Kubáň, V. 2007. Determination of phenolic compounds and their antioxidant activity in fruits and cereals. *Talanta*, 71(4), 1741–1751.

Takikawa, M, Inoue, S., Horio, F., and Tsuda, T. 2010. Dietary anthocyanin-rich bilberry extract ameliorates hyperglycemia and insulin sensitivity via activation of AMP activated protein kinase in diabetic mice. *Journal of Nutrition*, 140, 527–533.

Tan, B. L., and Norhaizan, M. E. 2017. Scientific evidence of rice by-products for cancer prevention: Chemopreventive properties of waste products from rice milling on carcinogenesis in vitro and in vivo. *BioMed Research International*, 2017, 1–18.

Tan, J., Baisakh, N., Oliva, N., Parkhi, V., Rai, M., Torrizo, L., Datta, K., and Datta, S. K. 2005. The screening of rice germplasm, including those transgenic rice lines which accumulate β-carotene in their polished seeds, for their carotenoid profile. *International Journal of Food Science and Technology*, 40, 563–569.

Tanaka, K., Yoshida, T., Asada, K., and Kasai, Z. 1973. Subcellular particles isolated from aleurone layer of rice seeds. *Archives of Biochemistry and Biophysics*, 155, 136–143.

Tanaka, K., Ogawa, M., and Kasai, Z. 1977. The rice scutellum. II. A comparison of scutellar and aleurone electron-dense particles by transmission electron microscopy including energy-dispersive X-ray analysis. *Cereal Chemistry*, 54, 684–689.

Tanaka, J., Nakamura, S., Tsuruma, K., Shimazawa, M., Shimoda, H., and Hara, H. 2012. Purple rice (*Oryza sativa* L.) Eetract and its constituents Inhibit VEGF-induced angiogenesis. *Phytotherapy Research*, 26(2), 214–222.

Tananuwong, K., and Tewaruth, W. 2010. Extraction and application of antioxidants from black glutinous rice. *LWT - Food Science and Technology*, 43(3), 476–481.

Tang, S., Hettiarachchy, N., Horax, R., and Eswaranandam, S. 2003. Physicochemical properties and functionality of rice bran protein hydrolyzate prepared from heat stabilized defatted rice bran with the aid of enzymes. *Journal of Food Science*, 68, 152–157.

Tang, W., Wu, T., Ye, J., Sun, J., Jiang, Y., Yu, J., ... and Wan, J. 2016. SNP-based analysis of genetic diversity reveals important alleles associated with seed size in rice. *BMC Plant Biology*, 16(1), 1–11.

Truswell, A. S. 2002. Cereal grains and coronary heart disease. *European Journal of Clinical Nutrition*, 56(1), 1–14.

Tsuda, T., Horio, F., Uchida, K., Aoki, H., and Osawa, T. 2003. Dietary cyanidin 3-O-beta Dglucosiderich purple corn color prevents obesity and ameliorates hyperglycemia in mice. *Journal of Nutrition*, 133, 2125–2130.

Umadevi, M., Pushpa, R., Sampathkumar, K. P., and Bhowmik, D. 2012. Rice-traditional medicinal plant in India. *Journal of Pharmacognosy and Phytochemistry*, 1(1), 6–12.

Upanan, S., Yodkeeree, S., Thippraphan, P., Punfa, W., Wongpoomchai, R., and Limtrakul Dejkriengkraikul, P. 2019. The proanthocyanidin-rich fraction obtained from red rice germ and bran extract induces HepG2 hepatocellular carcinoma cell apoptosis. *Molecules*, 24(4), 813.

USA Rice Federation. (2018). Available online: http://www.usarice.com. (Accessed on 20.08.2020).

USDA, NRCS. (2012). The PLANTS database. http://plants.usda.gov.

USDA. (2016) National Supplement Database for Standard Reference.

Valarmathi, R., Raveendran, M., Robin, S., and Senthil, N. 2015. Unraveling the nutritional and therapeutic properties of 'Kavuni'a traditional rice variety of Tamil Nadu. *Journal of Plant Biochemistry and Biotechnology*, 24(3), 305–315.

Verma, D. K., and Srivastav, P. P. 2016. Extraction technology for rice volatile aroma compounds. In: *Food Engineering* (Eds. Meghwal, M., and Goyal, M. R.). Apple Academic Press, New Jersey, pp. 281–328.

Verma, D. K., and Srivastav, P. P. 2017. Proximate composition, mineral content and fatty acids analyses of aromatic and non-aromatic Indian rice. *Rice Science,* 24(1), 21–31.

Verma, D. K., and Srivastav, P. P. 2018. Introduction to rice aroma, flavor, and fragrance. In: *Science and Technology of Aroma, Flavor, and Fragrance in Rice*. Apple Academic Press, , New Jersey, pp. 21–52.

Verma, D. K., and Srivastav, P. P. 2020a. A paradigm of volatile aroma compounds in rice and their product with extraction and identification methods: A comprehensive review. *Food Research International*, 130, 108924.

Verma, D. K., and Srivastav, P. P. 2020b. Bioactive compounds of rice (*Oryza sativa* L.): Review on paradigm and its potential benefit in human health. *Trends in Food Science & Technology*, 97, 355–365.

Virk, P. S., Zhu, J., Newbury, H. J., Bryan, G. J., Jackson, M. T., and Ford-Lloyd, B. V. (2000). Effectiveness of different classes of molecular marker for classifying and revealing variation in rice (*Oryza sativa*) germplasm. *Euphytica*, 112(3), 275–284.

Wang, Q., Han, P., Zhang, M., Xia, M., Zhu, H., Ma, J., Hou, M., Tang, Z., and Ling, W. 2007. Supplementation of black rice pigment fraction improves antioxidant and antiinflammatory status in patients with coronary heart disease. *Asia Pacific Journal of Clinical Nutrition*, 16(Suppl 1), 295–301.

Wang, H., Li, M., Zhang, J., Shi, Y., Lee, J., and Piao, Z. 2009. Phytic acid content in different parts of grain and its correlation with rice quality traits. *Chinese Journal of Rice Science,* 23(2), 215–218.

Welch, R. M., and Graham, R. D. 2004. Breeding for micronutrients in staple food crops from a human nutrition perspective. *Journal of Experimental Botany*, 55(396), 353–364.

Wilson, T., Nicolosi, R., Woolfrey, B. and Kritchevsky, D. 2007. Rice bran oil and oryzanol reduce plasma lipid and lipoprotein cholesterol concentrations and aortic cholesterol ester accumulation to a greater extent than ferulic acid in hypercholesterolemic hamsters. *Journal of Nutritional Biochemistry*, 18(2), 105–112.

Wongsa, P., Landberg, R., and Rattanapanone, N. 2018. Chemical compositions and metabolite profiling of rice varieties from Chiang Rai province Thailand. *Chiang Mai Journal of Science*, 45, 2703–2714.

Wu, F., Chen, H., Yang, N., Duan, X., Jin, Z., and Xu, X. 2013. Germinated brown rice enhances antioxidant activities and immune functions in aged mice. *Cereal Chemistry,* 90(6), 601–607.

Wunjuntuk, K., Kettawan, A., Charoenkiatkul, S., and Rungruang, T. 2016. Anti-fibrotic and anti-inflammatory effects of parboiled germinated brown rice (*Oryza sativa* 'KDML' 105) in rats with induced liver fibrosis. *Journal of Functional Foods,* 26, 363–372.

Xia, X., Ling, W., Ma, J., Xia, M., Hou, M., Wang, Q., Zhu, H., and Tang, Z. 2006. An anthocyanin-rich extract from black rice enhances atherosclerotic plaque stabilization in apolipoprotein E–deficient mice. *The Journal of Nutrition,* 136(8), 2220–2225.

Ye, X., Al-Babili, S., Klöti, A., Zhang, J., Lucca, P., Beyer, P., and Potrykus, I. 2000. Engineering the provitamin A (β-carotene) biosynthetic pathway into (carotenoid-free) rice endosperm. *Science*, 287(5451), 303–305.

Zhang, M.-W., Guo, B.-J., Chi, J.-W., Wei, Z.-C., Xu, Z.-H., Zhang, Y., and Zhang, R.-F. 2005. Antioxidations and their correlation with total flavonoid and anthocyanin contents in different black rice varieties. *Scientia Agricultura Sinica*, 38(7), 1324–1331.

Zheng, Y. F., Sun, G. P., and Chen, X. G. 2007. Characteristics of the short rachillae of rice from archaeological sites dating to 7000 years ago. *Chinese Science Bulletin*, 52(12), 1654–1660.

Zhou, J., Ma, C., Xu, H., Yuan, K., Lu, X., Zhu, Z., ... and Xu, G. 2009. Metabolic profiling of transgenic rice with cryIAc and sck genes: An evaluation of unintended effects at metabolic level by using GC-FID and GC–MS. *Journal of Chromatography B*, 877(8–9), 725–732.

22

Computational Modeling and in silico *Approaches in Understanding Abiotic Stress Responses in Field Crops*

Ananya Ali
Serampore Girls' College

CONTENTS

22.1 Introduction

Productivity is of immense importance for agricultural crop production as well as the quality. Stress in plants affects their productivity hugely. Researchers worldwide are interested to identify the causes of plant stress and their amelioration.[1–3] Plant stresses have been defined as the condition(s) that can hamper normal plant growth and metabolic homeostasis.[4] Among abiotic stresses, extensive studies have been found on temperature stresses, water and salt stresses, sulfur and phosphorus stresses, oxidative stresses, heavy metal ion stresses and combinations of them. Several experimental studies have been continuing for long to model the stress responses of plants, due to several external stimuli, either natural or anthropogenic causes. Modern approaches incorporated biochemical analyses, molecular biology techniques and also the applications of bioinformatics.[2–10] Biochemical and molecular studies revealed that several stress response proteins were generated to cope with the stressful condition(s). Gene expression patterns have also been studied, keeping plants in stressful environments. During such studies, several authors have applied extensive statistical analyses to find out the significance level of their results. Tools like Statistical Analysis System (SAS) and R have been found to be useful during those works.[9–14] MATLAB and other software for data analyses have also been utilized for image data analyses.[15]

Incorporating computational modeling and in silico approaches will not only ease the laborious efforts of wet-lab experiments but also have been proven to be a lot cost-effective by narrowing down the expectable outcomes. Ultimately dry-lab analyses will reduce the temporal cost that will help to find out stress responses of plants, much earlier than it used to be.[16–18] Data revealed from biochemical and molecular techniques need to be analyzed deeply, to get the idea of underlying mechanisms, well. The sequences of genetic materials as well as proteins need to be studied thoroughly to find out meaningful correlations. The structural aspects of proteins are also helpful as being indicative toward their functions. Studies from different laboratories, worldwide, are not only relying upon several bioinformatics tools for analyzing their data but also inventing newer computational tools to do that.[15,19–25]

This review will concentrate to understand what tools, from bioinformatics and computational biology, are widely being used and what their utilities are.

22.2 Databases

Databases are the online depositories of information and two types[18,23,26–28]:

1. **Sequence databases**: Sequence databases are important to find out the sequences of DNA, RNA and proteins from several organisms. Sequences can be searched, using appropriate keywords like name of the gene(s) or protein(s), using appropriate filters like name of the organism(s) or year or the publications etc. Some of the very useful sequence databases are
   i. **NCBI GenBank**[29,30]: National Center for Biotechnology Information (NCBI) is responsible for the GenBank database since October 1992. They maintain all known sequences of proteins and nucleotides from National Institute of Health (NIH). Sequences have unique entries that include summarized descriptions of the sequences, source organism scientific names,

DOI: 10.1201/9781003258063-22

taxonomy, the date of publication etc. This database has shared collaboration with EMBL DDBJ, etc. Other database from other parts of the world.

ii. **Uniprot**[31]: UniProt Archive (UniParc) is one among the most comprehensive, non-redundant protein sequence databases. Each unique sequence has been stored only once with an appropriate identifier, to get rid of redundancy. As well, a cross-referencing with other databases for each entry has been established that helps researchers by mimicking a search against UniProt as the search against all the cross-linked databases.

iii. **Protein Information Resources (PIR)**[32]: PIR provides the largest and annotated protein sequence database. This database also allows the user to search sequences of their interest by sequence similarity and other methods. Context cross-referencing with its own entries is the feature available at only PIR sequence database, allowing researchers to find out relationships among molecules such as enzyme-substrate interaction, activation and regulation cascades.

iv. **Swiss-Prot**[33]: Since 1988, the department of Medical Biochemistry of the University of Geneva and the EMBL Data Library is maintaining collaboratively Swiss-Prot, an annotated protein sequence database. This database sourced their sequences from PIR, EMBL Nucleotide Sequence Database and also from literature. Along with minimal redundancy, integration with other databases is also maintained.
   - Many studies have utilized these databases to retrieve their sequence of interest. Also, the NCBI BioSystems database centralizes and cross-links all the data from other databases to increase further the usefulness of these resources.

2. **Structural databases**: As indicated by the name, these databases provide us with the structure of the molecule of our interests. People often used databases like:

   i. **PDB**[34]: Protein Data Bank (PDB) database has a huge storage of protein structure, resolved by either X-ray crystallography or Nuclear Magnetic Resonance (NMR) or other methods like cryoEM or modeling, etc. This database allows us to not only have the sequence of our molecule of interest but to visualize how it looks three dimensionally (3D). This PDB structure file is often required for further analyses, associated with structure, like docking with ligands and analyzing and understanding the ligand-receptor or enzyme-substrate-like interactions.

   ii. **CATH**[35]: This database not only stores protein structures but also classify them based on class (C), architecture (A), topology (T) and homology superfamily (H). This structural family analysis by CATH helps us to find out the distinct features of protein structure space. As a result, this will help researcher to find out the structure-function/evolution relationships between known and novel protein structures.

   iii. **SCOP**[36]: Structural Classification of Proteins (SCOP) database that delivers detailed and comprehensive information about protein structure and evolutionary relationships of proteins with known 3D structures. It not only have all the proteins from PDB database but also the proteins with known structure but the 3D co-ordinate file is still lacking at PDB.
      - Search can be done by the name of the protein(s) of our interest and can be filtered by methods of studying (X-ray crystallography or NMR), resolution etc. It is possible that some protein sequences, available in the sequence database, will not have a resolved structure yet. At that time, utilizing the sequences we can model the structure for further analyses, involved with structures of our protein of interest. Several online tools and servers like SWISS-MODEL will come in aid that time. The mostly used algorithm for modeling 3D structure of a protein is homology based; i.e. by predicting more similar sequences will have more common patterns in structures. Many researchers have utilized such tools to model their protein of interest and the structure of which is yet to be studied experimentally.[37,38]

3. **Inorganic chemical databases**: Searching metal ions and their chemical structure is also very useful during the studies of metal stresses in plants. For those types of works,

   i. **NCBI PubChem**[39]: This is an open repository for chemical structures with their biological tests' results. It was launched in 1994, under the NIH Molecular Libraries Roadmap Initiative that was with an interest to discover chemical probes, potential to modulate gene products' activities, through high-throughput screening of small molecules. Their target was to make this information readily available to bio-medical researchers.

   ii. **ChEMBL**[40]: Along with small molecule structures, ChEMBL harbors information regarding bioactivities like binding, structure-activity relationships and ADMET (absorption, distribution, metabolism, excretion, toxicity) properties, too. Access to this information helps researchers understand the working principles and properties of the active molecules of their interest in a system.

   iii. **ChemDB**[41]: ChemDB is a public database with small molecules' 3D structures, predicted solubility and annotated entries. This is one of the great tools for not only bioinformatics work but also for chemoinformatics.

- Etc. quite popular databases are here to aid the work of not only bioinformatics people but also wet-lab researchers.

All these types of databases were actually helpful during sampling of the required materials, upon which our further computational and bioinformatics work will be done and output will be analyzed.

22.3 Working with Sequences

Sequences, either retrieved from databases or being generated as sequencing experiments, need to be analyzed for several purposes and by different methods. A most common method for sequence analyses is sequence alignment that is employed to find out sequence homologies (how likely two different sequences have originated from a common ancestor sequence), between different sequences. Several well-known algorithms like the Needleman-Wunsch algorithm for semi-global alignments and Smith-Waterman algorithms for local alignments have been used for this purpose.[26,27] Online tools like:

i. **NCBI BLAST**[42,43]: are a well-utilized tool that helps to find out homologous sequences from the above-mentioned databases against a query sequence. The Basic Local Alignment Search Tool (BLAST) program can be used through a web interface or as a standalone tool to compare the query sequence of our interest. BLAST, a heuristic, finds short similarities between two sequences and tries to align those 'hot spots'. Appropriate alignment and homologies of protein sequences help a lot to understand the evolutionary details.

   Also multiple sequence alignment (MSA) tools are there to compare between two sequences. MSA, being a high computationally complex problem, employs lots of random methods for improving the accuracy of the methods. Along with genetic algorithm, progressive alignment, guide tree, iterative approaches and hidden Markov model-based alignments etc. are quite common algorithms to solve MSA problems. Tools like:

ii. **CLUSTALW**[44,45]: Clustal programs are helpful for automatic multiple sequence alignment of sets of nucleotides or amino acid sequences. ClustalW is mostly used as a command line program by web servers and automatic batch systems. As an outcome, it provides not only NSA scores but also can highlight conserved domain(s) along different sequences and also can generate phylogenetic tree based on sequence similarity scores.

22.4 Working with Structures and Structural Quality Assessment

We have seen earlier that modeling of a sequence to 3D structure is possible by different servers and software; for a sequence, the structure of which is yet to be solved experimentally. Also, there are several other online servers to help us docking molecules with each other like: attaching a heavy metal with a stress protein and trying to find out how stress protein interacts with a heavy metal ion. These servers and tools have a spectrum of algorithms to dock two (or more) chemical molecules together to study their interaction(s). People have utilized tools like[18,24,26,27]:

i. **Molegro Virtual Docker (MVD)**[46]: Along with AutoDock programs, MVD allows us to perform docking of molecules as a protein-ligand docking simulation. MVD also helps researchers to analyze the binding site and interactions by providing MolDoc scores that help to understand the chemical interactions between a ligand and its receptor.

ii. **Z dock server**[47,48]: This has been used quite frequently to do such studies, by many researchers. Based on the rigid body docking program, ZDOCK and M-ZDOCK predict the structure of protein-protein interactions and provide us with the interacting or binding surface for docking analyses of the interacting molecules. ZDOCK provides a set of outcomes or interacting partners of the same pair of molecules and we need to go for different binding parameters or scoring to find out the most suitable structure among them.

While working with structures, i.e. either modeling of a single structure from the sequences or docked structures of two different molecules, we need to know the 3D physico-chemical validity of structures by means of steric-hindrance. To check the stereo-chemical properties of docked molecules people have frequently utilized:

i. **PROCHECK**[49]: PROCHECK program helps compute a set of stereo-chemical parameters for the protein molecule of our interest. Then it compares those parameters with 'ideal' values that have been found from well-resolved protein structures. If any significant deviation is found, that will be marked as 'Unusual'.

ii. **Verify 3D**[50]: This program relies upon the compatibility of a protein model to the 3D profile of its own amino acid sequences. The 3D profile is generated by computing the atomic co-ordinates of the structure. If the structure is more accurate the matching of 3D profile is high with its sequence. With lower structural accuracy matching will decrease.

22.5 Gene Expression Data Analyses

Gene expressions are required to study to understand the molecular mechanisms of how plants respond to stresses. Different studies have shown various bioinformatics tools, which have been employed to study the genetic expression of different plants under stressed condition(s) and the interactions among genes.[51–53]

i. **Gene expression omnibus (GEO)**[54]: GEO is quite like high-throughput gene expression data in a public repository. With its flexible and open design, it helps in submission, storage and retrieval of data from a huge data set of gene expression and hybridization experiments.
ii. **Entrez Gene**[55]: For gene-specific information, NCBI maintains this database. Their unique integer identifier for each entry is species specific. The GeneID is also tied to RefSeq (Reference Sequence database), and other consortia like Gene ontology and other databases of NCBI.
    - Etc. are some well-utilized tools for that.

22.6 Other Tools

Stress response phenotype analyses are currently quite a well-practiced method in this arena of research. People have not only used different computational tools but also devised newer ones to make this approach more lucid and easy.[13,15,19,51,56,57] We are going to describe below some of such tools:

i. **Virtual plants**[58,59]: This tool can help biologists who are not trained in computer science, too well like a professional bioinformatist. This helps people to fish out lists of genes, microarray experiments and gene networks to address several questions related to molecular mechanisms of plant stress physiology. Virtual plant gives the freedom to visualize, integrate and analyze genomic data as a system biology perspective. Eventually, this tool enables researcher to build a testable hypothesis from genomic data.
ii. **LiDAR**[60]: Though this technology is still new and developing, people have shown this method is quite accurate during crop phenotyping, under drought stress. Light detection and ranging (LiDAR), active remote sensing technology, works by measuring the time of flight of an emitted laser pulse within the sensor and the target. The results of this technique have shown till now a greater potential for non-destructive crop phenotyping. Developers are building further improved algorithms to analyze the LiDAR data by incorporating deep learning methods and other information to extract crop phenotypes more efficiently.
iii. **HTPheno**[61]: This is an open-source pipeline to analyze images from plants for phenotypic analyses. It has a connection to various other phenotyping systems and thus can be used by broader communities of researchers.
iv. **PhenoImage**[62]: This is open-source high-throughput phenotyping (HTP) cross-platform, based on the graphical user interface (GUI). This is helpful for people who have very little understanding or knowledge of programming. This also analyses plant images to fish out phenotypic characteristics.
    - There are other tools for image analyses like Plant Computer Vision (PlantCV) and Canopy reconstructions[63,64] which are quite well utilized. All of these have successfully helped people to analyze the stress response phenotype of plants while comparing with the control set. Several image analysis tools and algorithms have made it easier for researchers to study the stress responses in plants. I have already mentioned earlier that tools like MATLAB and RStudio have been well utilized to run different image analyses software and/or algorithms.

Along with these, with the advent of next-generation sequencing, a flock of newer algorithms, servers, software etc. tools are at our service. To align more efficiently numerous short read sequences (short DNA/RNA sequences, output of sequencing work) of nucleotides to a reference, a bunch of new tools like:

i. **BLAT**[65]: The BLAST-Like Alignment Tool (BLAT) is a more accurate and 500 times faster tool for mRNA/DNA alignments and 50 times faster for protein alignments. Though it has similarities with BLAST, it is quite different in many aspects like (i) indexing of databases is done apriori to the linear scanning through the query sequences, (ii) it triggers extensions on any number of perfect or near perfect hits, (iii) uses special code for introns in RNA/DNA alignment, (iv) 'Unsplices' mRNA onto the genome to provide a single alignment, using the mRNA only once. Along with these and other properties, BLAT is a very useful tool to align a huge set ('Big Data'?) of sequencing data against other databases with known sequences, within an effectively shorter span of time.
ii. **Bowtie**[66]: As people have recognized short read alignment to the reference genome as the first step of comparative genomic workflows, including variant detection and digital gene expression; packages like Bowtie which are able to align a large set of sequence reads to a reference sequence, ultrafast and in a memory-efficient way, are surely quite helpful tool. This also aligns reads after indexing them and that is how it achieves speed and memory efficiency, using Burrows-Wheeler Transform.
iii. **Short Read Mapping Package (SHRiMP)**[67]: This is an algorithm set that can map short reads despite the presence of numerous polymorphisms in the sequences. This is based upon a fast mapping technique and a statistical model for false positive hits.
iv. **TopHAT**[68]: TopHat is open-source software, with an efficient read mapping algorithm, for aligning reads

from RNA sequencing experiments to a reference genome despite of known splice site.

v. **RMAP**[69]: Smith et al. introduced RMAP algorithm, and it becomes one among the earliest program for mapping reads from sequencing work. This was with an important incorporation of using quality scores directly into the mapping process, to find out the best alignment.

So, we can understand from the discussion, above, that from sequence analyses tools to visualization of biological macromolecules and their assemblies to image analyses to plant phenomics, in every field computational modeling and in silico analyses are making its way to make research work easier and data meaningful.

22.7 Discussions

Comparative genomics and metabolomics studies have been quite common to understand stress responses in plant physiology. High-throughput analyses of large-scale parallel analyses of gene expression, mutation analyses and discovery of novel genes are the need of hour. Empirical functional genomics thus needs a complement analytical tool set that can help to analyze the huge date, generated by them. This is where computational modeling and bioinformatics tools will come in aid. As mentioned earlier these techniques are rapid, cost-effective and quite reliable to monitor plants' health and growth parameters. Though being indirect, these methods have facilitated the research work well accurate, till now.[17,28,70,71]

Current days are providing a hint for some more powerful newer techniques like deep learning. These powerful techniques can generate models with more accurate prediction if provided with enough data and also learn 'smartly' from direct input data like DNA sequences or pixel intensities of microscopic images. So in future, a huge domain of biology, including plant stress physiology, is being expected to not only rely on computational and bioinformatic tool but also will enrich them. It's the time when biologists and computer scientists need to join their hands for a prosperous and enlightening future.

REFERENCES

1. Moulick, D., Sarkar, S. & Awasthi, J. P. Rice grain quality traits: Neglected or less adressed? *Rice Res. Qual. Improv. Genomics Genet. Eng.* (2020). doi:10.1007/978-981-15-4120-9.
2. Hussain, T. M. et al. Recent advances in salt stress biology: A review. *Biotechnol. Mol. Biol. Rev.* **3**, 8–13 (2008).
3. Potters, G., Horemans, N. & Jansen, M. A. K. The cellular redox state in plant stress biology: A charging concept. *Plant Physiol. Biochem.* **48**, 292–300 (2010).
4. Shulaev, V., Cortes, D., Miller, G. & Mittler, R. Metabolomics for plant stress response. *Physiol. Plant.* **132**, 199–208 (2008).
5. Yamaguchi-Shinozaki, K. & Shinozaki, K. Transcriptional regulatory networks in cellular responses and tolerance to dehydration and cold stress. *Annu. Rev. Plant Biol.* **57**, 781–803 (2006).
6. Sharma, S. S. & Dietz, K.-J. The significance of amino acid-derived molecules in plant responses and adaptation to heavy metal stress. *J. Exp. Bot.* **57**, 711–726 (2006).
7. Demidchik, V. Mechanism of oxidative stress in plants: from classical chemistry to cell biology. *Environ. Exp. Bot.* 212–228 (2015).
8. Agehan, M., Greenham, K., CMockler, T. & McClung, C. R. Transcriptional networks: Crops, clocks and abiotic stress. *Curr. Opin. Plant Biol.* 39–46 (2015).
9. Moulick, D., Ghosh, D. & Chandra Santra, S. Evaluation of effectiveness of seed priming with selenium in rice during germination under arsenic stress. *Plant Physiol. Biochem.* **109**, 571–578 (2016).
10. Moulick, D., Santra, S. C. & Ghosh, D. Seed priming with Se alleviate As induced phytotoxicity during germination and seedling growth by restricting As translocation in rice (Oryza sativa L c.v. IET-4094). *Ecotoxicol. Environ. Saf.* **145**, 449–456 (2017).
11. Moulick, D., Santra, S. C. & Ghosh, D. Rice seed priming with Se: A novel approach to mitigate As induced adverse consequences on growth, yield and As load in brown rice. *J. Hazard. Mater.* **355**, 187–196 (2018).
12. Moulick, D., Santra, S. C. & Ghosh, D. Effect of selenium induced seed priming on arsenic accumulation in rice plant and subsequent transmission in human food chain. *Ecotoxicol. Environ. Saf.* **152**, 67–77 (2018).
13. Fahlgren, N. et al. A versatile phenotyping system and analytics platform reveals diverse temporal responses to water availability in Setaria. *Mol. Plant* **8**, 1520–1535 (2015).
14. Mazis, A. et al. Application of high-throughput plant phenotyping for assesing biophysical traits and drought response in two oak species under controlled environment. *For. Ecol. Manage.* **465**, 118101 (2020).
15. Boughton, B. A., Thinagaran, D., Sarabia, D., Bacic, A. & Roessner, U. Mass spectrometry imaging for plant biology: A review. *Phytochem. Rev.* **15**, 445–488 (2016).
16. Markowetz, F. All biology is computational biology. *PLoS Biol.* **15**(3), 4–7 (2017). doi:10.1371/journal.pbio.2002050.
17. Jones, W., Alasoo, K., Fishman, D. & Parts, L. Computational biology: Deep learning. *Emerg. Top. Life Sci.* **1**, 257–274 (2017).
18. Choudhury, S., Moulick, D. & Mazumder, M. K. Secondary metabolites protect against metal and metalloid stress in rice: An in silico investigation using dehydroascorbate reductase. *Acta Physiol. Plant.* **43**, 1–10 (2021).
19. Lin, Y. LiDAR: An important tool for next generation phenotyping technology of high potential for plant ohenomics. *Comput. Electron. Agric.* **119**, 61–73 (2015).
20. Hartler, J., Tharakan, R., Köfeler, H. C., Graham, D. R. & Thallinger, G. G. Bioinformatics tools and challenges in structural analysis of lipidomics MS/MS data. *Brief. Bioinform.* **14**, 375–390 (2013).
21. Angermueller, C., Pärnamaa, T., Parts, L. & Stegle, O. Deep learning for computational biology. *Mol. Syst. Biol.* **12**, 878 (2016).
22. Jansson, S. & Douglas, C. J. Populus: A model system for plant biology. *Annu. Rev. Plant Biol.* **58**, 435–458 (2007).

23. Mazumder, M. K., Moulick, D. & Choudhury, S. Iron (Fe^{3+})-mediated redox responses and amelioration of oxidative stress in cadmium (Cd^{2+}) stressed mung bean seedlings: A biochemical and computational analysis. *J. Plant Biochem. Biotechnol.* **4**, 49–60 (2021).
24. Choudhury, S., Moulick, D., Borah, A., Saikia, P. & Mazumder, M. K. In search of drugs to alleviate suppression of the host's innate immune responses against SARS-CoV-2 using a molecular modeling approach. *Silico Pharmacol.* **9**, 1–9 (2021).
25. Li, H. et al. Modern deep learning in bioinformatics. *J. Mol. Cell Biol.* **12**, 823–827 (2020).
26. Krane, D. E. & Raymer, M. L. *Fundamental Concepts of Bioinformatics.* Benjamin Cummings: San Francisco (2003).
27. Gromiha, M. M. *Protein Bioinformatics.* Academic Press: Cambridge, MA (2010).
28. Lee, H. C. et al. Bioinformatics tools and databases for analysis of next-generation sequence data. *Brief. Funct. Genomics* **11**, 12–24 (2012).
29. Benson, D., Lipman, D. J. & Ostell, J. GenBank. *Nucleic Acids Res.* **21**, 2963–2965 (1993).
30. Geer, L. Y. et al. The NCBI BioSystems database. *Nucleic Acids Res.* **38**, 492–496 (2009).
31. Leinonen, R. et al. UniProt archive. *Bioinformatics* **20**, 3236–3237 (2004).
32. Barker, W. C. et al. The Protein Information Resource (PIR). *Nucleic Acids Res.* **28**, 41–44 (2000).
33. Boeckmann, B. The swiss-prot protein sequence data bank. *Nucleic Acids Res.* **19**, 2247–2249 (1991).
34. Berman, H. M. The Protein Data Bank: A historical perspective. *Acta Crystallogr. A.* 88–95 (2008). doi: 10.1107/S0108767307035623.
35. Orengo, C. A. et al. CATH: A hierarchic classification of protein domain structures. *Structure* **5**, 1093–1109 (1997).
36. Murzin, A. G., Brenner, S. E., Hubbard, T. & Chothia, C. SCOP: A structural classification of Protein databse for the investigation of sequences and structures. *J. Mol. Biol.* **247**, 536–540 (1995).
37. Arnold, K., Bordoli, L., Kopp, J. & Schwede, T. The SWISS-MODEL workspace: A web-based environment for protein structure homology modelling. *Bioinformatics* **22**, 195–201 (2006).
38. Eswar, N. et al. Tools for comparative protein structure modeling and analysis. *Nucleic Acids Res.* **31**, 3375–3380 (2003).
39. Wang, Y. et al. PubChem: A public information system for analyzing bioactivities of small molecules. *Nucleic Acids Res.* **37**, 623–633 (2009).
40. Gaulton, A. et al. ChEMBL: A large-scale bioactivity database for drug discovery. *Nucleic Acids Res.* **40**, 1100–1107 (2012).
41. Chen, J., Swamidass, S. J., Dou, Y., Bruand, J. & Baldi, P. ChemDB: A public database of small molecules and related chemoinformatics resources. *Bioinformatics* **21**, 4133–4139 (2005).
42. Hurlbert, S. H. The nonconcept of species diversity: A critique and alternative parameters. *Ecology* (1971). doi: 10.2307/1934145.
43. 20200330_161747.
44. Chowdhury, B. & Garai, G. A review on multiple sequence alignment from the perspective of genetic algorithm. *Genomics* **109**, 419–431 (2017).
45. Thompson, J. D., Gibson, T. J. & Higgins, D. G. Multiple sequence alignment using ClustalW and ClustalX. *Curr. Protoc. Bioinforma.* **2**, 1–22 (2003).
46. Bitencourt-Ferreira, G. & de Azevedo Jr, W. F. Molegro virtual docker for docking. In: de Azevedo Jr, W. F. (Ed.) *Docking Screens for Drug Discovery.* Springer: New York, pp. 149–167 (2019).
47. Pierce, B. G. et al. ZDOCK server: Interactive docking prediction of protein-protein complexes and symmetric multimers. *Bioinformatics* **30**, 1771–1773 (2014).
48. Chen, R., Li, L. & Weng, Z. ZDOCK: An initial-stage protein-docking algorithm. *Proteins Struct. Funct. Genet.* **52**, 80–87 (2003).
49. Laskowoski, R., MacArthur, M. & Thornton, J. PROCHECK: Validation of protein-structure coordinates. *J. Appl. Crystallogr.* **16**, 284–284 (1983).
50. Luthy, R., Bowei, J. & Einsenberg, D. Verify3D: Assessment of protein models with three-dimensional profiles. *Methods Enzymol.* **277**, 396–404 (1997).
51. Tardieu, F. Virtual plants: Modelling asa tool for the genomics of tolerance to water deficit. *Trends Plant Sci.* **8**, 9–14 (2003).
52. Zhao, W., Shafiq, S., Berr, A. & Shen, W.-H. Genome wide gene expression profilling to investigate molecular phenotypes of Arabidopsis mutants deprived in distinct histone methyltransferase and demythylases. *Genomics Data* **4**, 143–145 (2015).
53. Koryachko, A. et al. Computational approaches to identify regulators of plant stress response using high-throughput gene expression data. *Curr. plant Biol.* **3**, 20À29 (2015).
54. Edgar, R., Domrachev, M. & Lash, A. E. Gene expression omnibus: NCBI gene expression and hybridization array data repository. *Nucleic Acids Res.* **30**, 207–210 (2002).
55. Maglott, D., Ostell, J., Pruitt, K. D. & Tatusova, T. Entrez gene: Gene-centered information at NCBI. *Nucleic Acids Res.* **39**, 52–57 (2011).
56. Fahlgren, N., Gehen, M. A. & Baxter, I. Lights, camera, action: Hight-throughput plant phenotyping is ready for a close up. *Curr. Opin. Plant Biol.* **24**, 93–99 (2015).
57. Li, L., Zhang, Q. & Huang, D. A review of imaging techniques for plant phenotyping. *Sensors* **14**(11), 20078–20111 (2014).
58. Katari, M. S. et al. Virtual plant: A software platform to support systems biology research. *Plant Physiol.* **152**, 500–515 (2010).
59. Wilson, P. A. & Chakrabrty, S. The virtual plant: A new tool for the study and management of plant diseases. *Crop Prot.* **17**, 231–239 (1998).
60. Su, Y. et al. Evaluating maize phenotype dynamics under drought stress using terrestrial lidar. *Plant Methods* **15**, 1–16 (2019).
61. Hartmann, A., Czauderna, T., Hoffmann, R., Stein, N. & Schreiber, F. HTPheno: An image analysis pipeline for high-throughput plant phenotyping. *BMC Bioinformatics* 12 (2011).
62. Zhu, F. et al. PhenoImage : An open-source graphical user interface for plant image analysis. *Plant Phenome J.* **4**, 1–13 (2021).

63. Burgess, A. J., Retkute, R., Pound, M. P., Mayes, S. & Murchie, E. H. Image-based 3D canopy reconstruction to determine potential productivity in complex multi-species crop systems. *Ann. Bot.* **119**, 517–532 (2017).
64. Burgess, A. J., Retkute, R., Herman, T. & Murchie, E. H. Exploring relationships between canopy architecture, light distribution, and photosynthesis in contrasting rice genotypes using 3D canopy reconstruction. *Front. Plant Sci.* **8**, 1–15 (2017).
65. Kent, W. J. BLAT: The BLAST-like alignment tool. *Genome Res.* **12**, 656–664 (2002).
66. Langmead, B. Aligning short sequencing reads with Bowtie. *Curr. Protoc. Bioinforma.* 1–14 (2010). doi:10.1002/0471250953.bi1107s32.
67. Rumble, S. M. et al. SHRiMP: Accurate mapping of short color-space reads. *PLoS Comput. Biol.* **5**, 1–12 (2009).
68. Trapnell, C., Pachter, L. & Salzberg, S. L. TopHat: Discovering splice junctions with RNA-Seq. *Bioinformatics* **25**, 1105–1111 (2009).
69. Smith, A. D. et al. Updates to the RMAP short-read mapping software. *Bioinformatics* **25**, 2841–2842 (2009).
70. Cushman, J. C. & Bohnert, H. J. Genomic approaches to plant stress tolerance. *Curr. Opin. Plant Biol.* **3**, 117–124 (2000).
71. Sankaran, S., Mishra, A., Ehsani, R. & Davis, C. A review of advanced techniques for detecting plant diseases. *Comput. Electron. Agric.* **72**, 1–13 (2010).

23

Flood Stress Prevalence and Assessment of Farmers' Preparedness with Resilient Rice Varieties: A Review in Context of Eastern India

Swati Nayak, Kuntal Das, Malay K. Bhowmick, Sk Mosharaf Hossain, Sudhanshu Singh, Uma Maheshwar Singh, and Vikas Kumar Singh
International Rice Research Institute

CONTENTS

DOI: 10.1201/9781003258063-23

23.1 Introduction

The global rice production stands at 700 million tons (mt). Still, more than 800 million people go hungry every year (IRRI, 2019). Rice is one of the major staple foods in Asia, which contributes to around 90% of the world's rice production (Haefele et al., 2012; Rao et al., 2017). Irrigated rice environments contribute majorly to rice production, while the rice-based systems under rainfed environments are less favorable. Such unfavorable environments cater to the livelihood for 100 million farm families. Nearly 40% of total rice-cultivated land is categorized under rainfed (lowland or upland) ecosystem. Approximately 3.5 million hectares (m ha) of lands under rice cultivation are classified as flood-prone or under the deep-water situation, whereas it covers about 46 m ha in Asia (Maclean et al., 2002). In rainfed rice, the most important yield-limiting factor is variation in rainfall pattern in terms of quantity and distribution. Uncertainties and changes in rainfall distribution result in severe floods and droughts in several places across the world (Depledge, 2002).

Rainfed lowlands are characterized by low and fluctuating productivity, limited access and use of quality inputs, large number of resource-poor farmers, fragmented land holdings, more reliance on traditional varieties, poor seed replacement, use of lowquality seeds, weak information, and delivery network, with associated slow adoption and diffusion of improved technologies. In India, more than 40 million people depend on rainfed lowland rice areas which face various biophysical stresses like submergence, drought, and/or salinity/sodicity (Srivastava et al., 2016). Compared to the input-intensive irrigated system (5.0 t/ha), average rice productivity in the rainfed lowland system is quite low (<2.5 t/ha), and it is considerably lower and unstable in flood-prone areas (1.5 t/h). Despite this, due to the abundance of rich soil and freshwater resources, rainfed ecosystems have significant potential to contribute and increase overall food production to fulfill the ever-increasing demands for rice supply (Bailey-Serres et al., 2012; Ismail et al., 2013; Singh et al., 2013).

In India, rice is grown in a total area of about 45 m ha, including about 13 m ha under rainfed lowland (17%), 3 m ha under deep water (7%), and 1 m ha under coastal saline zone (2%). Nearly 40% of the area under rice cultivation in the country, especially in eastern India, is highly vulnerable to recurrent flooding and induced crop losses. This affects 28 mt of annual rice production, vulnerable to risks, posing a perilous concern for food security (Singh, 2006; Singh et al, 2012). Nearly 30% of India's rice-growing areas are at risk with catastrophic crop destruction due to the ominous result of persistent flooding (Dar et al., 2013). With the trend in the recent era, due to the climate change scenario, the frequency and intensity of flooding are predicted to rise over time. As a result, rainfed crop production along with the livelihoods of millions of farmers in flood-affected areas would be severely impacted. Managing the floodplain ecosystem through diverse technological interventions, thus, becomes crucial to food security and improve the livelihood of the rice farmers. This calls for an understanding and evaluation of the development, dissemination, deployment, and adoption of resilient technologies through research and extension efforts for sustainable rice production in a fragile ecosystem.

In eastern India, rice is the major food crop, and rice farming is an important source of livelihood. Poor productivity is often experienced and majorly associated with climatic stresses, leading to poor income and poverty. Even with the abundance of surface water and groundwater susceptibility to floods, waterlogging and droughts affect people's lives and livelihoods in many ways in eastern India. Unpredictable flood risks also become a constraint for getting return on investment for farmers, discouraging them to further invest in cost-intensive practices be the use of optimum farm inputs, mechanization, or better crop management practices. This creates a vicious economic trap within the farming community through low productivity and unstable income. The only use of stress-tolerant rice varieties (STRVs) cannot significantly contribute to the complete productivity gains or avert the overall crop losses unless equipped with best management practices or optimal use of needful inputs. It has been studied that farmers while growing submergence tolerant rice variety like Swarna-Sub1 in eastern India have reported more investment by 9–10% in their cultivation practice, 11% in fertilizer use, and 6.3% in using better crop establishment methods (Dar et al., 2013). This is in addition to the aversion of yield loss in flooding situation or having yield advantage to farmers check like Swarna. As evidenced from field-level demonstrations, ensuring the availability of flood-tolerant rice varieties along with conforming management interventions would be a game-changer for smallholder farmers in the floodplains. Rice genotypes with Sub1 gene offer yield advantages under flood stress conditions (Singh et al., 2009; Sarkar and Panda, 2009). In flood-prone areas, rice-based production systems mostly include wet season rice and dry season *rabi* crops, *boro* rice, or rice-fallows. Improved production technologies including high-yielding STRVs can substantially contribute and usher a significant transformation of these unfavorable flood-prone ecosystems into favorable rice-based cropping systems.

The development of improved flood-tolerant varieties has been profusely beneficial for rice farmers. But the technological needs, choices and adoption scenarios vary widely among the farmers. Only a single improved variety cannot essentially cater to the varying needs of farmers across diverse geographies with different socio-cultural contexts and preferences (IRRI, 2008; Bairagi et al., 2018). In view of faster dissemination, availability, and adoption of the improved Sub1 varieties to the farmers, there is a strong need for efficient delivery and extension mechanisms, targeted positioning of these new varieties, and enabling policies.

23.2 Floodplains and Their Distribution

Floodplains represent the flat lands which are situated nearer or connected to rivers and streams and are mostly subjected to periodic or recurring water inundation due to floods. Majority of the floodplains are *'flood-prone'* and thus become risky and uncertain for developmental activities. The floodplains have highly heterogeneous ecosystems, having a complex coexistence of small water channels, backwaters, hillocks, ridges, and depressions (mostly wetlands). These are connected to the rivers or canals which bring flood water along with associated fine sediments and nutrients to the floodplains. These areas are generally characterized by poverty and reduced food security at family and community levels. Many farmers with their inherent traditional knowledge follow their own thoughts in the selection of crop varieties, crop establishment, and management practices in these areas.

Flood-prone environments in South and Southeast Asia are characterized by diverse growing conditions, including variation in rainfall amount and duration; flood depth, duration, frequency, and time; drought time, frequency, and duration; and soil type and terrain. Over 20 m ha of rice fields become submerged each year in South and Southeast Asia, and more than 100 million people live and work in this flood-prone ecosystem. Flooding may persist for 1–10, and it may be day(s) up to 6 months of water stagnation at various depths. Another type of stress that affects crop yields is the daily submergence due to changes in tidal schedule. Rainfed shallow submergence and medium-deep waterlogged lowlands, deep-water situations, and tidal wetlands are the affianced features of flood-prone rice-growing areas (IRRI, 1984).

Though extreme precipitation events are common in many parts of the country, resulting in torrential rains and flash floods, it is mostly exacerbated in the eastern region. The southwest monsoon in India brings torrential rains and causes all rivers and water bodies to overflow their banks, causing low to severe floods in the surrounding areas based on the frequency and intensity. Nearly 40% of India's total rice-cultivated land is rainfed, with eastern India accounting for more than 70% of the country's rainfed rice. Rainfed upland makes up roughly 23% of the overall rainfed area, while rainfed lowland makes up the remaining 77%. Approximately 17% of rainfed lowlands are prone to flooding (Pathak et al., 2018). In India, around 49.82 m ha of land have been identified as flood-prone (NITI Aayog, 2021). Out of these, about 10–12 m ha get affected by flood each year (NAAS, 2013). Lowland rice is grown on around 14 m ha (32% of the total area) in India, primarily in eastern India (Pathak et al., 2018). Due to irregular precipitation, these areas are frequently exposed to drought and flooding. The flood-prone rice environment in eastern India covers roughly a total area of 2.5 m ha (Pathak et al., 2018). Eastern India, which includes Assam, Bihar, Chhattisgarh, Eastern Uttar Pradesh, Jharkhand, Odisha, and West Bengal, covers ~21.85% of India's total geographical area and supports 34% of the country's human population and 31% of its animal population (NAAS, 2013). Bihar is more prone to floods (nearly 70% area is flood-affected, especially Northern Bihar). West Bengal, as a part of the Bengal delta, has a long history of flooding and flooding affects 42.3% of the state's total area (NAAS, 2013). Odisha has a history of natural disasters, with floods and cyclones severely affecting the coastal districts. Flooding is also a serious concern in the Brahmaputra and Barak basins, as well as other minor river sub-basins in Assam's floodplains, which account for 39.58% of the state's total land area (Government of Assam, 2022).

23.3 Flooding Types and Extent of Losses

Flooding is the most frequent natural disaster worldwide causing immense loss to life and livelihood. One of the greatest abiotic challenges to crop growth and agricultural output is stress caused by soil flooding and deeper water inundation. Rice farming and flooding are interconnected. Flooding causes composite stresses which result in a responsive change within the plant cells (Pierdomenico et al., 2011). As a result, flood stress also acts as a powerful driver for adaptive evolution of plants. Several fundamental chemicals build up and cause changes within plants, such as ethylene, CO_2, O_2, reactive oxygen species (ROS), and phytotoxins which produce visual effects of flooding stress. The types of flooding and the affected plant parts have a big influence on the changes in these chemicals.

23.3.1 Types of Flooding

Flooding may be broadly divided into two types: (i) waterlogging, which affects just the root system within the soil, and (ii) submergence, which affects sections or the whole shoot (Sasidharan et al., 2017). However, flooding can be simply classified into the following types in accordance with the duration and depth of its occurrence.

23.3.1.1 Flash Flooding

For a short period (approximately 1–2 weeks), flash flooding occurs because of strong rainfall or sudden overflowing of surrounding rivers and streams, resulting in complete waterlogging of plants (Vergara et al., 2014). Low-lying lands, as well as areas near streams and rivers, are vulnerable and frequently experience flash floods. Flash flooding can happen at any time during crop growth stages, and despite the shallow water depth, flash flooding can cause significant damage.

23.3.1.2 Stagnant Flooding

Stagnant flooding can happen either an aftermath of a flash flood, or on its own due to accumulated water in an area (Collard et al., 2013). Water does not retreat quickly in case of stagnant flooding, and it stays in the field at a level of

50–60 cm for several weeks to months (Mallik et al., 1995), where certain sections of the plant shoot are visible above the flooded water. During the monsoon season, when excess rainwater overflows into the areas near water channels, stagnant flooding is a regular phenomenon which is aggravated with improper drainage systems connected to waterways. Stagnant flooding is very unpredictable like flash floods and usually happens in coastal areas during extreme cyclonic events such as hurricanes and tropical storms. Rice fields in coastal zones, which are divided as medium-low and lowlands, are prone to severe stagnant flooding conditions.

23.3.1.3 Deep-Water Flooding

Deep-water flooding may last over a month, depending on geography and meteorological circumstances. In this condition, water levels may soar up to 4 m from ground (Singh et al., 2017). To escape complete waterlogging, rice plants adapt to this environment and speed up their development continuously with the rise in flood water levels. During deep-water flooding, a section of the plant always tends to stay above the accumulated water level. As a result, deep-water rice avoids full water inundation by extensive growth through internodal elongation that uses a lot of carbohydrates. With the increase in water level during this type of flooding, rice plants can grow in terms of height at the rate of 25 cm per day and can reach up to 5 m (Jackson and Ram, 2003; Singh et al., 2017). In India, most of the deep-water rice growing areas are majorly located in eastern parts and in states like Assam, Bihar, Odisha, West Bengal, and Uttar Pradesh (Bin Rahman and Zhang, 2016).

23.3.2 Extent of Crop Damage Due to Flood Stress

The amount of damage caused by flooding depends on the depth and length of plant inundation, as well as land topography (Mishra et al., 2010), intensity of floodwater (Das et al., 2009), and time and frequency of flood occurrence. When the crops are typically in an active growth stage and the farmers incur considerable expenditures on farm inputs, the level of damage by flood incidents is most severe which is generally irreversible. Farmers often miss out a growing season because of floods since flooded land cannot be used right away. As a result, farmers usually suffer significant losses during the rainy season, especially when floods hit late in the season in low and medium land under rice cultivation. Rice production in rainfed lowlands is frequently damaged by flooding or complete waterlogging. Moreover, during the inundated phase, the rice crop is stressed at various stages of development due to the limited gaseous exchange, effusion of soil nutrients, susceptibility to pests and diseases, mechanical damage, and low light (Greenway and Gibbs, 2003; Singh et al., 2009). Frequent flooding in rainfed lowland areas during monsoon months causes full drowning of the rice crop within roughly 10–15 days, to which most lowland rice cultivars are susceptible (Nishiuchi et al., 2012).

About 86% of production loss in Asia's agriculture sector is due to floods (FAO, 2017). During the monsoon, the enormous rice-growing areas of South and Southeast Asia, including Bangladesh, India, Indonesia, Myanmar, Thailand, and Vietnam, are vulnerable to unprecedented flash floods (Sarkar et al., 2006). Flooding affects around 22 m ha of rainfed lowland areas in South and Southeast Asia, including nearly 6.2 m ha of rice-cultivated areas in India (Azarin et al., 2017; Dar et al., 2017). A total of 15 m ha of rainfed lowland rice are impacted by frequent short-term flash floods (Singh et al., 2017), resulting in a $1 billion economic loss in South and Southeast Asia (Mackill et al., 2012). Altogether, the flood-prone ecosystem comprises around 7% of worldwide rice-cultivated land and contributes 4% of global rice production (Yang et al., 2017).

23.4 Mechanism of Flood Tolerance in Rice

Excess water during flooding has a substantial impact on rice crop, mainly because it hinders the diffusion of gasses between the plant and its surroundings (Armstrong, 1980; Voesenek and Bailey-Serres, 2015). Because O_2 and CO_2 cannot efficiently diffuse via stomata and cell walls under water, flooding generates an energy crisis within plant cells. This results in lack of oxygen inside flooded plant tissues, which limits heterotrophic energy generation. Furthermore, photosynthesis gets limited in waterlogged plant parts due to reduced CO_2 availability, and restricted radiation, more so with enhanced water turbidity. Generally, the O_2 content quickly drops in flooded plant portions without continued photosynthesis, resulting in hypoxic circumstances (Lee et al., 2011).

23.4.1 Quiescence and Escape Strategies

Rice plants exhibit two prominent mechanisms in response to excess water presence: low oxygen quiescence and low oxygen escape (Luo et al., 2011). The low oxygen quiescence approach conserves energy within plants, reduces growth levels, and delays energy crises (Sauter, 2000; Bailey-Serres and Voesenek, 2008), all of which are helpful in deep-water situation for survival (Sauter, 2000; Bailey-Serres and Voesenek, 2008). Low oxygen escape, on the other hand, is characterized by increased shoot elongation in plants to reach above the water-air interface to restore air contact under shallow water depth (Bailey-Serres and Voesenek, 2008).

23.4.2 Flooding Tolerance Mechanism

Being a semi-aquatic plant, rice crop adapts to various forms of flooding through an array of physiological and biochemical processes. Many investigations on flood-tolerant rice genotypes have led to a better understanding of flood stress adaptation mechanisms.

23.4.2.1 Tolerance Mechanism at Seed Germination Level

Waterlogging above the soil normally takes place during initial rainfall after sowing of seeds, particularly in improperly leveled lands. 'Anaerobic germination' (AG) is the term considered for germination that occurs under inundated conditions, where seeds usually sprout in environments with no air or limited availability of oxygen. In flood-prone locations, sluggish seed germination, unequal and delayed seedling establishment, and heavy weed infestation are some of the aftereffects that limit rice development (Ismail et al., 2009; Ismail et al., 2012). Almost all rice cultivars are extremely vulnerable to floods and are badly harmed during germination (Angaji et al., 2010; Miro and Ismail, 2013; Singh et al., 2017). AG is a complicated characteristic influenced by several gene families that are involved in a variety of critical activities such as starch breakdown, glycolysis, fermentation, and other biochemical and metabolic processes (Ismail et al., 2012). As a result, AG is detrimental to uniform germination and improved crop establishment in flood-prone areas (Magneschi and Perata, 2009; Septiningsih et al., 2013).

Some varietal seeds may sprout in water without or with very little oxygen, resulting in anoxic or hypoxic conditions, and subsequent transition to anaerobic metabolism (Taylor, 1942). Because of special metabolic changes, AG-tolerant rice genotypes germinate and grow faster, with improved seedling vigor under low oxygen level. When compared to their intolerant counterparts, AG-tolerant cultivars demonstrate rapid elongation of coleoptile, expression of 'expansin' genes, and a reduction in peroxidase activity and cell division (Magneschi and Perrata, 2009). Under anoxic circumstances, these genes are substantially up-regulated during germination under water (Huang et al., 2000; Lasanthi-Kudahettige et al., 2007). Gene expression studies show that AG-tolerant rice genotypes have all the major enzymes needed in normal germination. Enzymatic studies demonstrate that anaerobic respiration enzymes are highly active during coleoptile elongation, implying that they play a critical part in the mechanism of germination to adapt during the flooded situation (Ismail et al., 2009).

23.4.2.2 *Tolerance Mechanism at Vegetative Stage*

The underwater plant part exhibits limited or no elongation during the vegetative stage, which helps to prevent energy dissipation and stores valuable carbohydrate reserves that are usually used later for recovery and regenerative growth when the flood water recedes (Almeida et al., 2003). Stem elongation interferes with physiological energy maintenance systems, and thus survival of plants under total inundation is reduced. Das et al. (2005) investigated the reaction of four rice types to total flooding for up to 10 days and found that IR42, Sabita, and Hatipanjari elongated more under water. Studies on rice plants after total flooded conditions reveal that plants bearing the Sub1 allele limits the gibberellic acid signaling (Fukao and Bailey-Serres, 2008). Because of ethylene released during submergence, intolerant genotypes display stem elongation and chlorosis, leading to apparent damage following de-submergence (Kawano et al., 2002; Ella et al., 2003). Non-elongating cultivars withstand submersion by preventing photo-damage and retaining a high chlorophyll content in their leaves (Sone et al., 2011).

23.4.2.3 *Tolerance Mechanism at Stagnant Flooding*

Plants elongate when in contact with air, and the process of tolerance to stagnant water in rice differs significantly from that of flash floods. Recent research on rice genotypes that differ in their sensitivity to stagnant water at a depth of 50 cm reveals that the tolerant genotypes are fundamentally tall or considerably elongated with the rise in water level and are characterized by higher tillering, reduced glucose depletion, and higher fertility of spikelet (Kato et al., 2014; Vergara et al., 2014). Under static floods, the IRRI (International Rice Research Institute) has found important quantitative trait loci (QTLs) for yield, yield-dependent components, and other critical agronomic parameters.

23.4.2.4 *Tolerance Mechanism at Deep-Water Flooding*

One of the most essential properties of deep-water rice is stem elongation in response to rising flood water level. Deep-water rice cultivars can grow up to 25 cm each day to keep at least a portion of their growing tip above the water level and sustain photosynthetic contact with air and light (Bailey-Serres and Voesenek, 2008). When a rice plant is drowned completely in water, the ethylene concentration in the stems rises, triggering gibberellic acid synthesis and internode elongation, but ethylene has no effect on internode elongation in non-deep-water rice varieties (Metraux and Kende, 1983; Hattori et al., 2009). In flooded deep-water conditions, ethylene is also responsible for the activation of adventitious roots of rice plants. Gibberellic acid and ethylene work together to facilitate the formation of adventitious roots. Gibberellic acid activity is particularly essential for root emergence (Steffens et al., 2006).

23.5 Traditional Flood-Tolerant Rice Varieties

Grain yield, grain quality, biotic and abiotic stress tolerance, and input use efficiency vary greatly among traditional rice cultivars or landraces. Farmers prefer growing landraces for their local adaptations in a variety of lowland conditions. These landraces or cultivars are often tall, prone to lodging, photoperiod-sensitive, and low yielders. Even though they are inherently low-yielding, nevertheless a few of these cultivars predominate in flood-prone locations due to their moderate level of flood tolerance (Singh et al., 2014). Several landraces have been identified as tolerant to complete submergence (Patra et al., 2018). Some important traditional rice landraces of eastern states for rainfed shallow lowland, submergence, waterlogging, semi-deep, and/or deep-water situations have been listed in Table 23.1. However, only a few rice cultivars have been identified as possible landraces, with the ability to withstand submergence in water depths of up to 80 cm for 10–12 days during the early vegetative stage of crop growth (Patra et al., 2018). 'FR 13A' is a noteworthy selection from the Dhalaputia landrace in Odisha (Patra et al., 2018). Through the introgression of the Sub1 gene into the background of mega rice varieties that are otherwise sensitive to flooding stress, several improved rice varieties have been developed using 'FR 13A' as an effective donor. Almost all deep-water and floating rice types have varying degrees of elongability – a genetic feature that is stable (Thakur, 2004).

TABLE 23.1

Rice Landraces for Rainfed Shallow Lowland, Submergence, Waterlogging, Semi-Deep, and/or Deep-Water Situations in Eastern India

State	Landraces	References
Assam	Adolia Bao, Ahiniabao Amana Bao, Borjahinga, Dhepa Bao, Herepi Bao, Ikarasali, Jul Bao, Kalangi Bao, Kekoa Bao, Lati Sali, Mahsuri, Maguri Bao, Manohar Sali, Moimorsingia Bao, Neghari Bao, Ogri Sali, Ranga Bao, Rangoonsali, Sialasali, Tulsi Sali	Sharma (2013) and Singh et al. (2004)
Bihar	Bakol, Barogar, Dasmi, Desaria, Jessoria group, Jogar, Kalma, Maghnath, Salmot, Sohar, Sugara	Singh et al. (2004)
Chhattisgarh	Sausari Dhan, Tulsighati, Matko Dhan, Dandras, Painbudi, Saldhenti, Bhursi Dhan, Gadakhuta Dhan	Dwivedi et al. (2017)
Jharkhand	Agin Sar, Bhorang Sar, Jhaliar Geanda, Kalamdani, Khanika Sar, Sambalpuria	Sinha (2016)
Odisha	Ande Karma, Atiranga, Bhundi, Biesik, Boga Bordhan, Chakia 59, Champakali, Chaula pakhia, CN 540, Dhola Badal, Dhusara, FR 13A, FR 43B, Ganga Siuli, Janki, Kalaketki, Kalaputia, Kanawar, Khadara, Khajara, Khoda, Kolasali, Kusuma, Madhukar, Manasarovar, Nahng Tip, Nali Baunsagaja, Ravana, Rongasali, S 22, Sail Badal, Sarumuli, SL276, Solpona, Telgri	Patra et al. (2018) and Reddy et al. (2013)
Uttar Pradesh	Aamgaud, Agahari, Balgani, Balmlot, Dhaneshwar, Gauria, Ghoghari, Goanth, Kabara, Sainger, Sugaparkhi, Venaga	Singh et al. (2004)
West Bengal	Achra, Alakkachari, Asgo, Auskali, Barh Avarodhi, Badrasail 2, Bakui, Bhasamanik, Bhashakalmi, Bhelki, Dubraj, Gabura, Harmanona, Hetomari, Jabra, Jaldhapa, Jalkamini, Jaldubi, Jalaplaban, Jhingasail, Kaimsail, Kalakhersail, Kalawriji, Kalma, Kariwai 1, Katarangi, Katiputur, Kerala Sundari, Kumargore, Langalmuthi, Matla, Meghi, Motajata, Narayankamini, NC1281, Nizersail, Panibhasha, Panikalash, Salkele, Sholey, Sadamota, Talmugur, Tilakkachari	Dana et al. (2013), Dana et al. (2017), and Goswami et al. (2017)

Traditional rice cultivars grown in flood-prone coastal locations are tolerant to both salinity and submergence, although they are poor yielders. Some of the prominently used varieties are Bhaluki, Bhurarata, Chettivirippu, Getu, Kalarata, Kalundai samba, Kamini, Karekagga, Korgut, Kuthiru, Patnai, Nona Bokra, Pichaneelu, Pokkali, Rupsal, Sathi, Talmugur, Vikas, etc. (Chattopadhyay et al., 2018; Bhowmick et al., 2020). During the wet season, cultivars such as Nonasail (Sel.), SR 26B, BKN, Sadamota, and Kalamota Selection are appropriate for low (15–25 cm water depth) to medium (25–50 cm or above water depth) land situations (Bhowmick et al., 2020). Because uncontrolled waterlogging and poor drainage are the main concerns, taller cultivars or that may elongate with rising water level are suited for these conditions (Singh et al., 2011; Kato et al., 2014). Farmers prefer long-duration varieties (160–170 days) for lowlands to avoid harvesting in standing water, but they prefer certain genotypes with medium height (100–115 cm) and medium duration (140–150 days) for medium lands with water depths of 20–30 cm for most of the wet season when there are no flooding problems (Burman et al., 2018).

Traditional landraces are valuable genetic resources that contribute to ecological balance; therefore preserving them is critical for future food security. It would be possible to identify more genes/QTLs from varied genetic resources, characterize, and use them to generate improved genotypes with greater flood, salinity, or even multiple stress tolerance, as well as other desired traits. Climate-smart rice varieties with many benefits can be developed using trait combinations in superior backgrounds.

23.6 Rice Varietal Improvement for Flood Tolerance

Significant progress has been made in creating flood-tolerant rice varieties utilizing traditional and molecular breeding processes to minimize the damaging effects of various flood occurrences. Conventional breeding has only been partially successful in combining high levels of flooding resistance with high grain output. Traditional rice cultivars that can withstand flooding are available and are being grown by rice growers. There is still a lot to be done in terms of breeding and delivering more resilient and productive varieties.

23.6.1 Conventional Breeding Approaches

23.6.1.1 Identification of Flood-Tolerant Genetic Resources at Germination

Rice genotypes with different AG tolerance potentials have been identified and it was found that the tolerant cultivars perform better under flooding during germination (Ray et al., 2016). Twelve rice genotypes with enhanced AG ability were identified after evaluating 662 rice accessions from the IRRI and International Network for the Genetic Evaluation of Rice gene banks (Yamauchi et al., 1993). In another study, a total of 359 rice accessions were investigated and found growth variations in coleoptile with a water depth of 20 cm (Jiang et al., 2004). At IRRI, over 8,000 rice genotypes were evaluated, and only a small percentage (0.23%) were found to be resistant to germination stress, which was reduced to 0.06% after further testing in repeated experiments (El-Hendawy et al., 2011). More study into rice genetic diversity for anaerobic conditions during germination from germplasm sources is critical and necessary. Many genetic studies have recently been done with more intensity to uncover the molecular mechanism behind AG (Septiningsih et al., 2012). Some AG tolerant varieties have been found and are being used in breeding and genetic investigations (Septiningsih et al., 2012; Toledo et al., 2015). Under anaerobic conditions, morpho-physiological variations of 88 lowland rice landraces were reported, and three rice landraces were found to have stronger AG tolerance capacity than other rice genotypes (Barik et al., 2019).

23.6.1.2 Identification of Genetic Resources for Flooding Tolerance at Post Germination Stage

Traditional rice varieties and landraces have essential adaptation features that allow them to thrive in a variety of harsh conditions (Rabara et al., 2014). Farmers use a variety of traditional rice types that are popular in flood-prone habitats, even though they are low-yielding. Rice landraces are subjected to severe water fluctuation, and they provide a source of genetic variety for breeding efforts aimed at improving rice flooding resistance.

23.6.1.3 Identification of Genetic Resources for Flash Flooding Tolerance

Long back in the 1970s, the IRRI in the Philippines began testing rice genetic resources for tolerance against flash flooding (Vergara and Mazaredo, 1975). Simultaneously, many rice genotypes were investigated at IRRI through systematic screening in controlled submerged habitats, leading to the identification of two resilient extremely resistant varieties, FR13A and FR43B (Oladosu et al., 2020), which were selected as pure lines from traditional rice landraces collected from Odisha, India. The farmer's varieties 'Dhalaputia' and 'Bhetnasia' were used to create FR13A and FR43B. From the 1970s to the present, FR13A has become the most well-known and acknowledged flood-tolerant rice variety, with rice breeders all over the world adapting it (Azarin et al., 2017; Singh et al., 2017). After a rigorous screening of 15,000 rice germplasm at the ICAR-National Rice Research Institute (NRRI), Cuttack, barely 0.1% of rice genotypes survived after 12 days of total water inundation (Sarkar and Bhattacharjee, 2011). Goswami et al. (2017) conducted a comprehensive physiological investigation of 27 indigenously known rice landraces and wild rice from West Bengal and Odisha under 14 days of total waterlogging. From 1978–79, the ICAR-NRRI in Cuttack had an outstanding submergence tolerance screening facility (Paul and Bhattacharya, 1980).

23.6.1.4 Identification of Genetic Resources for Stagnant Flooding Tolerance

A few studies have been conducted to assess rice's ability to withstand stagnant flood situations. Vergara et al. (2014) investigated 626 rice accessions over the course of 3 years under static flooding conditions. In a field experiment research, 577 genotypes were found to have different resistance during stationary floods. Under stationary flooding, Mallik (1995) discovered that genotypes with restricted elongation had superior survival and yield performance. Under the 'rainfed lowland breeding program' at IRRI, some prospective breeding lines were found tolerant to both stagnant flooding and submergence (Mackill et al., 2010). At IRRI, rice germplasm was also screened for tolerance under water stagnation to identify acceptable genotypes (Kato et al., 2014). Eighty elite rice genotypes were examined under stagnant flooding, and several of the Sub1 introgressed high-yielding varieties were found to be more resistant to stagnant flooding than the checks (Kato et al., 2020). Since 1960, the ICAR-NRRI in Cuttack has focused on screening and characterization of rice germplasm, as well as the creation of rice varieties and agricultural strategies for stationary flooding tolerance in rice. Through numerous experiments, several scientists contributed significantly to the discovery of rice genetic resources under stationary floods (medium-deep to semi-deep water situations) (Sarkar, 2016).

23.6.2 Modern Breeding Approaches

23.6.2.1 Molecular Markers and QTL for Flooding Tolerance

Discovery of molecular markers, QTL analysis, precise genetic mapping, and subsequent gene cloning responsible for flood tolerant genes have been made possible with the recent advances in molecular biology and marker technology. For the features connected to flood tolerance in rice, the molecular basis has been widely investigated. Deep-water elongation was the first of these features to catch the attention of plant breeders and geneticists because of its distinctive quality. While studying the inheritance pattern of stem elongation in deep-water rice, Ramiah (1940) discovered that internode elongation is governed by duplicated genes and is classified as ef1 and ef2 (elongation factor) using the usual breeding approach. Following that, by assessing the progeny for flood tolerance, partial dominance and five to six relevant genes for the floating of deep-water rice were discovered using allele cross (Hamamura and Kupkanchankul, 1979). The discovery of a QTL controlling flooding, the Sub-1 gene, was a watershed moment in the history of flood tolerance breeding (Xu and Mackill, 1996). A short chromosomal area had been narrowed down to identify the genomic fragment encoding the Sub1 QTL. The gene was eventually identified as an ethylene response factor (ERF) when it was cloned (Xu et al., 2006). This genomic locus was further mapped on chromosome 9 (Xu et al., 2000), and the locus was narrowed by employing a fine-scale physical mapping (Xu et al., 2006). On chromosomes 1, 2, 9, and 12, Septiningsih et al. (2012) discovered four new QTLs. Hattori et al. (2008) found that QTL on chromosome 12 is the most efficient for deep-water internode elongation. In QTL analysis of deep-water rice, the most relevant phenotyping characteristic is elongation under water. During the seedling stage, Nagai et al. (2014) discovered two novel QTLs, qTIL2 and qTIL4, that affect total internode elongation. The IRRI has carried out and discovered the mapping of significant QTLs linked with AG tolerance (Angaji et al., 2010; Nagai et al., 2012).

23.6.2.2 Marker-Assisted Breeding

Molecular marker-assisted selection (MAS) is a technique for enhancing crop development programs for various abiotic and biotic stress tolerances (Ahmad et al., 2017; Mantri et al., 2014; Wei et al., 2009). MAS is a method of selecting a genetic determinant or determinants of a characteristic of interest, such as abiotic stress tolerance, tolerance/resistance for disease and pest, production, and/or quality, by using a marker (Prabhu et al., 2009). Using molecular markers, this technique selects plants with genomic areas implicated in the production of characteristics of interest. For the discovery and availability of a wide range of molecular markers and rich molecular genetic maps in agricultural plants, MAS can now be used to study characteristics influenced by key genes and QTLs (Choudhary et al., 2008). Targeted genes to be strategically transferred and the distance between the flanked markers and the target gene are two criteria that influence the effectiveness of MAS (Perumalsamy et al., 2010). MAS is gaining popularity because it can increase plant breeding efficiency by allowing for accurate transfer of genomic areas of interest and faster recovery of the recurrent parent genome (Wijerathna, 2015). Individual plants can be selected depending on their genotype during the selection stage using MAS. Conventional phenotypic screening cannot discriminate between homozygous and heterozygous plants for most characteristics. MAS can be utilized to help in parent selection, increasing the success of backcross breeding, and improving sex-limited characteristics (Zhou et al., 2007). In the rice crop, a marker-assisted backcrossing (MABC) breeding technique is primarily used to incorporate the desired QTL from agronomically non-standard donors into mega varieties and to create stress-tolerant genotypes. This strategy is more precise and efficient than the previous conventional breeding method (Collard and Mackill, 2008). The discovery of the Sub1 gene had established the basis for MAS flooding tolerance (Xu et al., 2006; Ismail et al., 2013; Azarin et al., 2017; Tiwari, 2018). Table 23.2 lists popular Sub1-introgressed mega rice varieties produced by MABC for flood tolerance.

There are several advantages of using MAS in rice breeding. It may be easier than phenotypic screening, for example, and so saves time, effort, and resources. Furthermore, MAS selection may be performed at the seedling stage, allowing unwanted plant genotypes to be promptly removed (Khan et al., 2015). The benefits of using markers include increased speed, consistency, efficiency, biosafety, and the potential to tilt the odds in our favor even when dealing with complicated features.

23.6.2.3 Genome-Wide Association Mapping

From a large germplasm aggregate, an alternate QTL mapping method called association mapping is employed to conduct the association analysis. Association mapping in a population is based on linkage disequilibrium or the non-independence of alleles. The use of sparse simple sequence repeat (SSR) and single-nucleotide polymorphism (SNP) markers in association mapping studies has been proven to be successful in detecting marker-trait relationships in rice (Qiu et al., 2015; McCouch et al., 2016). With the introduction of high-throughput sequencing and SNP chip technology, genome-wide association studies (GWASs) using high-density molecular markers are gaining popularity in rice genetics (Yu et al., 2014). GWAS is a strong technique for fully understanding the genetic basis of complex variables that has proved fruitful, particularly for rice (Qiu et al., 2015).

TABLE 23.2

Introgressed Mega Rice Varieties Developed through MABC

Name of Sub-1 Variety	Donor Parent	Recurrent Parent	Origin	Year of Release	References
BR11-Sub1	IR49830	BR11	Bangladesh	2010	Iftekharuddaula et al. (2016)
Swarna-Sub1	IR49830	Swarna	India	2009–2010	Iftekharuddaula et al. (2016), Tiwari (2018), and Neeraja et.al. (2007)
Samba Mahsuri-Sub1	IR49830	Samba Mahsuri	India	2011, 2013	Sarkar and Panda (2009)
CR1009-Sub1	IR40931	CR1009/Savitri	India	2017	Azarin et al. (2017)
Gayatri-Sub1	IR49830	Gayatri	India	2016	Sarkar et al. (2017)
Sarala-Sub1	IR49830	Sarala	India	2016	Sarkar et al. (2017)
Varshadhan-Sub1	IR49830	Varshadhan	India	2016	Sarkar et al. (2017)
Pooja-Sub1	IR49830	Pooja	India	2016	Sarkar et al. (2017)
Pratikshya-Sub1	IR49830	Pratikshya	India	2017	Sarkar et al. (2017)
Ciherang-Sub1	IR49830	Ciherang	Indonesia	2012–2013	Singh et al. (2017) and Septiningsih et al. (2015)
Thadokkam1-Sub1/TDK1-Sub1	IR40931	Thadokkam1/TDK1	Laos	2017	Dar et al. (2013)
IR 64-Sub1	IR40931	IR 64	Philippines	2009	Sarkar and Panda (2009)
PSBRc 18-Sub1	IR 64-Sub1	PSB Rc 18	Philippines	2017	Singh et al. (2017) and Septiningsih et al. (2015)
KDML 105-Sub1	IR40931	KDML 105	Thailand	2017	Toojinda et al. (2005) and Siangliw et al. (2003)
AS996-Sub1	IR40931	AS996	Vietnam	2016	Azarin et al. (2017) and Luu et al. (2012)
Bacthom 7-Sub1	IR40931	Bacthom 7	Vietnam	2016	Linh et al. (2013)

23.7 Improved Flood-Tolerant Rice Varieties and Their Dissemination

23.7.1 Breeder Seed Indent Reflection

India has 14 m ha of rice-cultivated area under rainfed lowland which is considerably prone to the flood risk. In eastern India, 2.5 m ha from Odisha, Assam, Bihar, and eastern Uttar Pradesh experience flood almost every year with substantial crop damage (Pathak et al., 2018). With the IRRI-pioneered Sub1 technology and subsequent varietal development, a handful of varieties have been bred by ICAR institutes in close research collaboration with IRRI. Sub1 varieties that have been successfully integrated into the national seed chain are CR 1009-Sub1, Swarna-Sub1, Samba-Sub1, and BINA dhan 11 (Odisha Annual Report, IRRI, 2022). Besides, the research efforts for introgression of Sub1 varieties into other popular varieties to make them flood-resilient are rapidly progressing. Recently, IR 64, one of the mega varieties developed by IRRI, has been rebred with Sub1 gene introgression. Likewise, Pooja, a widely cultivated variety under the rainfed lowland conditions in Odisha, has very recently been added with Sub1 gene to make it more efficient in flood-prone coastal districts of Odisha. There are about 14 Sub1 varieties released by the national system and in the seed chain in India (Table 23.3). Maximum indent for breeder seed (BS) of the Sub1 varieties comes from eastern India; it is the largest flood-prone rice-growing region in the country. In total, about ten states periodically put indent for the Sub1 BS and six of them are in eastern India (SeedNet Portal, 2022).

On-field better performance of these varieties against their non-Sub1 variants has triggered farmers' demand, and this can be amply observed from BS indent trend over the last 7 years. Large share of the BS indent was by flood-prone states like Assam, Odisha, Bihar, Uttar Pradesh, Tamil Nadu, Chhattisgarh, and West Bengal (SeedNet Portal, 2022). An increased BS indent is, therefore, suggestive of progressive and wide-scale adoption of Sub1 varieties. The spatial spread of these varieties, as indicated by the BS indent by at least ten states (Table 23.4) in India, bears testimony to the acceptance and adoption of these varieties. Since, both in India and Bangladesh, 4 mt of rice are forgone every year due to flood (IRRI, 2014), the positive impact of Sub1 varieties on productivity is a strong adoption trigger (Emerick et al., 2015).

TABLE 23.3

Spread of Notified Sub1 and Submergence/Flood-Tolerant rice Varieties in India

Notification Year	Product Name (Released/ Notified)	Stress Type	Duration (Days)	Yield (q/ha)	UP	WB	OD	TRP	ASS	GUJ	TN	KNT	AP
2010	Amara (MTU-1060)	Flood tolerant	115–120	60–65									1
2011	Chinsurah Rice (CNI 383-5-11)	Partial submergence tolerance	120–128	55–57		1	1			1	1		1
2011	Reeta (CR Dhan 401)	Submergence tolerant	145–150	40–41		1	1				1		1
2011	CR Dhan 501	Submergence tolerant	152	40	1				1				
2015	Samba Mahsuri-Sub1	Submergence tolerant	135–140	45–50	1								
2015	BINA dhan 11	Submergence tolerant	120	50–54		1		1	1				
2015	BINA dhan 12	Submergence tolerant	125–130	42–45		1		1	1				
2016	CR1009-Sub1	Submergence tolerant	150–155	55–60							1		
2017	Bheema/Dheera (MTU 1140)	Flood tolerant	145–150	60									1
2018	CO 43-Sub1	Submergence tolerant	135–140	40.4			1				1	1	1
2018	Ranjit-Sub1	Submergence tolerant	145–150	50–55					1				
2018	Bahadur-Sub1	Submergence tolerant	145–150	55					1				
2009	Swarna-Sub1 (CR 2539-1)	Submergence tolerant	140–145	60–65	1		1						
2008	AAUDR-1	Submergence tolerant	85–90	40–45						1			
				TOTAL	**3**	**4**	**4**	**2**	**5**	**2**	**4**	**1**	**5**

UP, Uttar Pradesh; WB, West Bengal; OD, Odisha; TRP, Tripura; ASS, Assam; GUJ, Gujarat; TN, Tamil Nadu; KNT, Karnataka; AP, Andhra Pradesh.

TABLE 23.4
BS Indent (*q*) of Sub1 Varieties Since 2016

Name of Sub1 Variety	Year of Release	2016	2017	2018	2019	2020	2021	2022	Total
CR 1009-Sub1	2013	14.80	45.25	6.05	6.10	9.20	23.45	49.75	154.60
Swarna-Sub1	2010	232.90	553.20	251.80	314.70	210.85	106.65	79.50	1749.60
BINA dhan 11	2015	0	0.40	14.85	28.25	9.30	15.88	24.50	93.18
Samba-Sub1	2015	24.30	134.86	4.05	23.45	14.30	14.30	23.30	238.56

Source: SeedNet Portal, GoI.

23.7.2 Innovation in Varietal Positioning

In order to facilitate a rapid uptake of newly released rice varieties, a strong technology transfer model should be in place. Such technology transfer model requires innovative extension approaches that put farmers at the center but target all seed chain stakeholders strategically. IRRI in South Asia (especially India and Bangladesh) had pioneered the design of a range of demonstration methods such as head-to-head trial, integrated cluster demonstration, varietal cafeteria, dealer-led demonstration, etc. IRRI uses these advanced seed extension tools in Bangladesh and East Indian states like Odisha, Assam, Uttar Pradesh, and Bihar. IRRI had demonstrated and promoted Sub1 varieties in India and Bangladesh through these advanced seed extension approaches. These methods have been instrumental in field validation, participatory varietal selection, and mass awareness generation about Sub1 varieties. These approaches play a catalyzing role in influencing seed system stakeholders to observe and advance best-rated Sub1 varieties in their regions. A brief account of these extension methods is explained below.

23.7.2.1 On-Farm Trials

On-farm trial (OFT) method, which is executed in an individual farmer's field, has the most optimal condition to give reliable results for assessing varietal performance. In this comparative method, test varieties are grown adjacent to each other in the same plot with a benchmark and farmers' check variety under the same crop management practices. Out of these varieties, the benchmark is the most competitive relatively newer product available and farmers' check is the oldest product which needs to be replaced. The newest varieties developed are the test varieties which need to be evaluated to be able to see genetic gain in the farmers' fields. The varietal performance result is contextual and realistic enough to take a decision on its positioning in the seed chain. Apart from generating evidence around varietal performance, these trials can also act as a unique way of demonstrating varietal performance in the farmers' fields and provide a comparative learning experience for farmers, neighbors, and more stakeholders visiting the plot. It can serve as one of the critical extension management tools to create awareness which is driven by data and designed statistically. Head-to-head trial, triadic experiment, and strip trial are effective OFTs to validate a variety in farmers' field and generate evidence-based actual performance.

23.7.2.2 Cluster Demonstration

Cluster demonstration is typically a collective large-scale demonstration by a homogenous group of farmers (10–20 or more) who put together their contiguous yet small plots to make a large patch of 5–10 ha and demonstrate an improved variety. Cluster demonstration aims at creating mass awareness about the worth of an environment-specific modern variety, sensitizing the visiting stakeholders, and creating visibility for the new varieties. At times, these cluster demonstration sites are used as seed production plots, where farmers are trained on quality seed production, and they demonstrate the varieties as seeds. Such arrangement is focused when a variety (e.g., Swarna-Sub1) is already found to be suitable in an ecology but seeds are not adequately available through formal channels. In such a situation, this semi-formal system of cluster demonstration-cum-seed production plots helps quick production and spread of the target varieties.

23.7.2.3 Varietal Cafeteria

Varietal cafeteria is a display of 15–20 promising varieties in one large plot replicated along with their checks. Farmers' representatives of the Agriculture Department of Government, seed producers, public and private seed suppliers, distributors, and researchers are invited to observe and assess comparative performance of these varieties. Both quantitative data on yield and yield-dependent parameters and qualitative data on stakeholder rating of few basic traits are recorded in the varietal cafeteria.

23.8 Adoption Dynamics and Farmers' Preference for Improved Rice Varieties

Adoption of a new variety is determined by the interplay of a host of factors (Figure 23.1), which can be classified as market and social factors. While market factors are strongly linked and influence production, marketability, and profitability, social factors like seed security and socio-economic conditions indicate farmers' capacity to adopt a variety (Kumar et al., 2021). While popularizing the Sub1 varieties in eastern India and Bangladesh, the strong adoption trigger is their flood tolerance ability. It acts as a strong yield protector in the face of field inundation that causes a significant yield penalty. Flood tolerance of Sub1 varieties along with a marginal yield gain under normal conditions are appealing for farmers. It has been amply substantiated in several studies that yield potential of a variety is the strongest adoption enabler, besides added benefits like submergence tolerance, especially in flood-affected areas. If a variety is packaged as a good yield and tolerant to submergence, then it is highly likely to be diffused among target farmers at a faster rate. The performance of these varieties

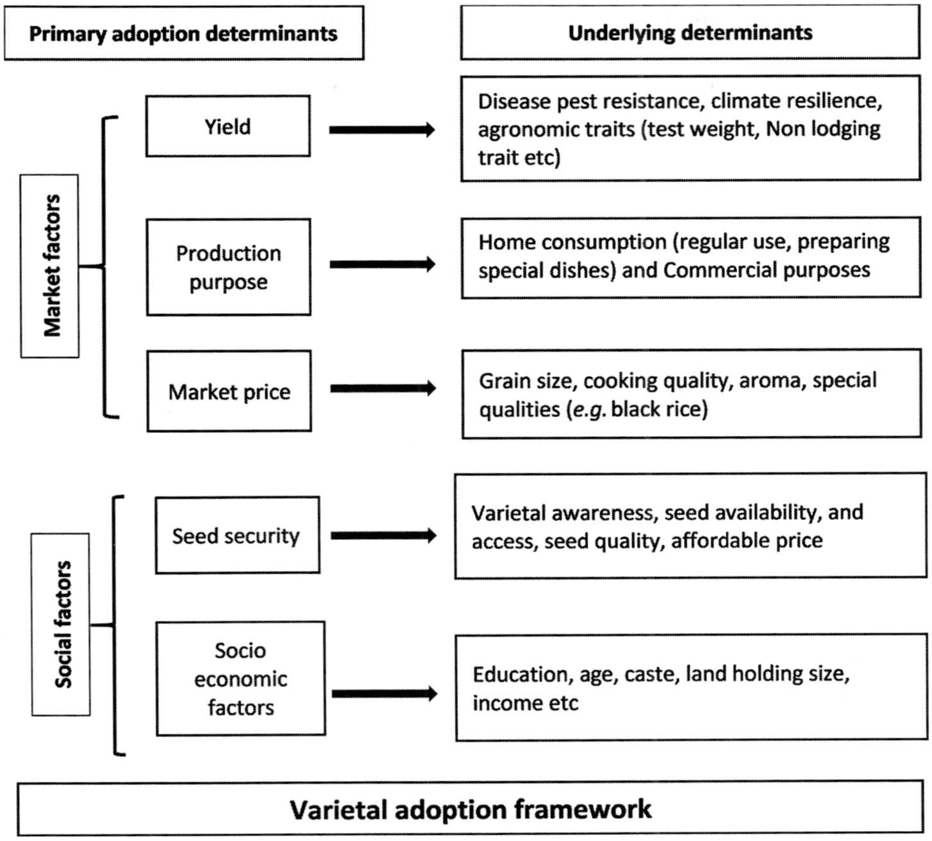

FIGURE 23.1 Conceptualized varietal adoption framework.

can be further improved by adopting appropriate management techniques in both the nursery and the main field (Bhowmick et al., 2014).

In states like Odisha, Bihar, Assam, etc., the state governments have taken significant initiatives to indent BS for multiplication and diffusion of these varieties. Following multiplication and distribution of the seeds, seed security parameters have been improved considerably, and it has boosted the adoption. The massive awareness program undertaken by the state governments under different schemes has been instrumental in catalyzing the scale of varietal information and insights at farmers' level, thus translating into an enhanced adoption.

From 2017 to 2020, IRRI in collaboration with eastern Indian states (Odisha, Assam, Uttar Pradesh) demonstrated three major Sub1 varieties *viz.* Swarna-Sub1, CR 1009-Sub1, and BINA dhan 11. Innovative methods like integrated cluster demonstrations along with on-farm and participatory research trials were employed to validate, evaluate, and advance these varieties through formal and informal seed systems. In view of capacity limitation and the informal seed system, IRRI has been extending huge efforts to handle individual farmers, and through their collectives such as self-help group (SHG), farmer producer organization (FPO), farmer producer company (FPC) to support them with quality seed production technologies and linkages of early generation seeds (EGS). Such massive efforts have translated into spatial spread of these varieties. An adoption study has reaffirmed that, following farmers' exposure to the improved varieties in flood-prone ecosystems, 70% of the farmers adopted BINA dhan 11 in three consecutive years. The adoption rate is 84%, both for CR 1009-Sub1 and Swarna-Sub1 (Figure 23.2). Notably, 75% of farmers who participated in the demonstration program in 2017 now continue to grow at least one Sub1 variety fitting into their ecology (Odisha Annual Reports, IRRI, 2022).

Dar et al. (2013) reported that STRVs protect farmers from potential yield penalty, and it triggers varietal adoption significantly in South Asia. The overall experiment and adoption with Sub1 varieties in India and Bangladesh have shown the progressive adoption trend in target ecology (Dar et al., 2013). A variety suitable for advancement in flood-affected rice ecosystems should therefore possess submergence tolerance traits along with all other commonly preferred agronomic and market-demanded qualities. In addition, dissemination of information and insights about the Sub1 varieties is important. A study by Bairagi et al. (2018) in Bangladesh has revealed that only one-third of farmers in 2015 have adopted the Swarna-Sub1 even though it is the most suitable variety in the flood-prone ecosystem. Nearly half of the non-adopters did not have information and insights about the benefits of this Sub1 variety. Thus, innovative extension tools should ensure improved communication and build farmers' confidence about the variety.

23.8.1 Seed Scaling: A Case Study on Sub1 Varieties in Odisha, India

Odisha, one of the leading rice-producing states in India, has 30% (1.15 m ha) of its rice area prone to frequent floods,

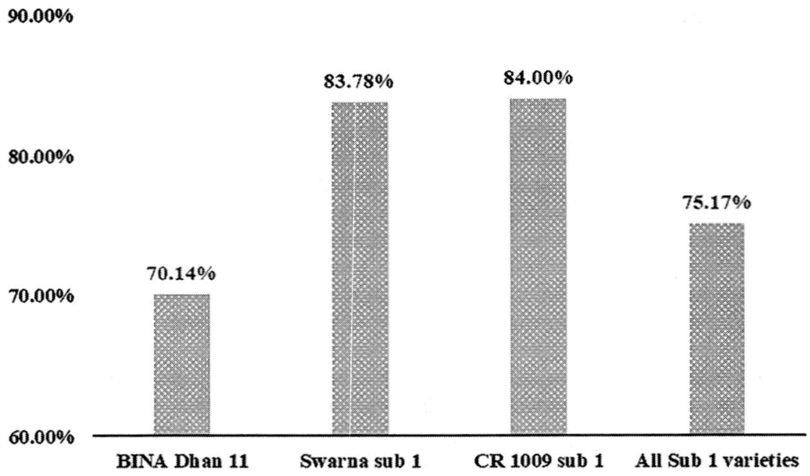

FIGURE 23.2 BINA dhan 11, Swarna-Sub1, CR 1009-Sub1, All Sub1 varieties in eastern India.

particularly in the coastal districts (Odisha Agriculture Statistics, 2019). IRRI, in collaboration with the state government and other international donors, runs projects in the state on testing and scaling of Sub1 varieties in the target ecosystem. The State Department of Agriculture with the state-owned seed corporation is the biggest producer and distributor of rice seeds within Odisha. During this period, IRRI with all its cutting-edge extension mechanism evaluated and demonstrated several Sub1 varieties like Swarna-Sub1, BINA dhan 11, CR 1009-Sub1, and Samba Mashuri-Sub1. Of these four major Sub1 varieties experimented by IRRI, the first three were evaluated as suitable and beneficial for advancement. Subsequently, state-owned seed multiplication agencies were supported with BS linkages. Resultantly, the seed sale of these varieties spiked sharply in the following years. The progressive increase in the adoption can be observed from the seed sale data of the State Department of Agriculture.

23.8.1.1 *Swarna-Sub1*

This variety has been derived from the mega variety 'Swarna' with incorporation of the Sub1 gene. Owing to its flood tolerance ability, this variety displayed a huge rise in seed sale in the last 5 years after large-scale on-farm experiment and demonstration, though there are some inter-year fluctuations (Figure 23.3). In 2019, Swarna-Sub1 contributed to an extent of 10.25% of total seed sale by the Agriculture Department.

23.8.1.2 *BINA dhan 11*

BINA dhan 11 was released in India as part of the 'seeds without borders' initiative fostered by IRRI. The Sub1 trait, excellent yield (5.5 t/ha), medium duration, and its suitability in two major seasons (wet and dry) are the important factors which make the variety popular among farmers in a quick time. It was found to be extremely suitable in rainfed shallow lowland areas with the frequent occurrence of flash floods. The sale backed by extensive demonstration, awareness creation, and BS access, picked up quickly (Figure 23.4). In a span of just 4 years, BINA dhan 11 constituted 5.3% of total seed sale in the state (Seed Sale Data, Government of Odisha, 2020). The rapid uptake and high adoption of BINA dhan 11 in Odisha is a classic case of a variety, perfectly matching with the farmers' requirements and preferences.

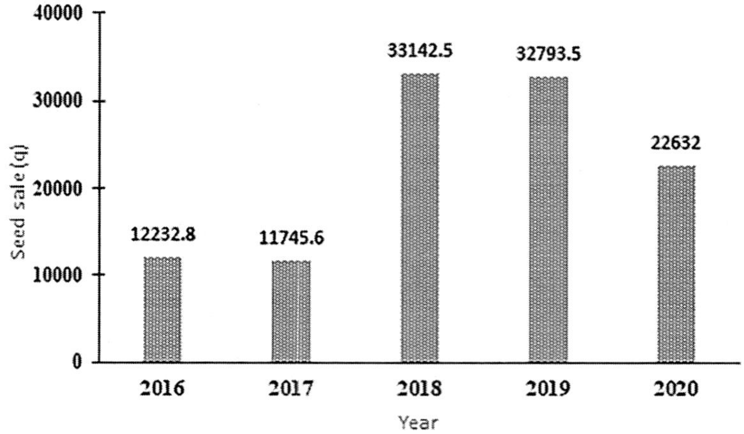

FIGURE 23.3 Seed sale (q) trend of Swarna-Sub1 in Odisha, India. (Compiled Seed Sale Data, DoA, Odisha.)

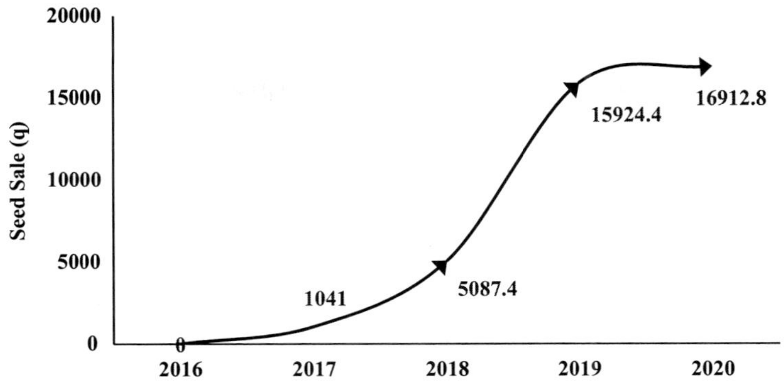

FIGURE 23.4 Seed sale trend of BINA dhan 11 in Odisha, India. (Compiled Seed Sale Data, DoA, Odisha.)

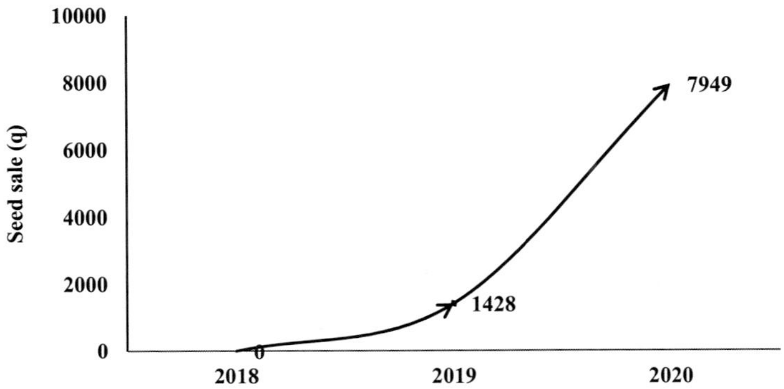

FIGURE 23.5 Seed sale trend of CR 1009-Subl in Odisha, India. (Compiled Seed Sale Data, DoA, Odisha.)

23.8.1.3 CR 1009-Sub1

This is a long-duration Subl variety, ideally recommended for low-lying areas characterized by submergence. In the last 5 years, following testing and popularization of this variety, the seed sale has picked up considerably affirming high adoption of this variety (Figure 23.5). The high yield of this variety is another strong adoption trigger.

23.8.2 Dissemination Learning from Odisha Initiative

Rice Subl varieties need appropriate positioning that targets a fitting ecosystem for demand creation. First, a Subl variety may not be a preferred choice where flood is not recurrent and destructive. Therefore, correct characterization of rice ecosystems should be an important exercise before a Subl variety is positioned and popularized. GIS applications and field validation may prove to be assistive in characterization assignment. Second, a rapid assessment of non-Subl traits of the variety needs to be completed to understand farmers' acceptance of the variety. Third, when the variety is found well accepted by the farmers from on-farm experimentations, the access to EGS should be ensured for a fast-paced seed production and distribution. Finally, there are some inherent capacity deficits of formal seed chains, and on account of those constraints, there is shortage of seeds in the system. For a rapid and aggressive promotion of a variety, seed production should be decentralized. The community networks (SHG, FPC, *etc.*) when encouraged and supported adequately, the seed production and distribution processes take place at a faster pace.

23.9 Future Perspectives

23.9.1 Product Management, Positioning, and Sustained Adoption

In rainfed lowland ecosystems across South Asia, Subl varieties prove to be a boon for millions of rice farmers. The research breakthrough of Subl gene identification has tremendously helped in protecting the farmers' interest even in the face of frequent submergence. However, the full potential of Subl varieties in India has not yet been realized, and the reasons are complex and multifaceted. In respect of product development, introgression of Subl gene in mega rice varieties should encompass careful consideration of no yield penalty. When the flood-tolerant gene is integrated, other preferred traits in the variety should remain unaltered so as not to disturb the trait-specific adoption dynamics. Quite often, as witnessed in eastern India, the product positioning plans and efforts are fragile and fragmented and hardly follow any market segment principles. Therefore, to develop and advance the Subl varieties, the market should be appropriately segmented based on ecosystem, varietal duration, crop establishment method, and grain qualities. A product with such a planned market segment is likely to reach target farmers at a faster pace, and thus, it improves the adoption and benefits the farmers and consumers. Since Sub1 varieties are supposedly identical in their yield or perform better in non-stress condition against the original variety (non-sub1), they can be positioned even in non-stress condition. In eastern India, especially in Odisha, the significant yield advantage of sub1 varieties in non-stress condition over their non-sub1 counterparts has been observed and documented.

Access to EGS is the key to fast dissemination of a variety. The multiplying agencies (formal, semi-formal) and individuals should have provision of adequate BSs for further multiplication and distribution among farmers. Presently, limitation of timely BS access remains a constraint for augmented product diffusion. For faster cross-country spread of a Sub1 germplasm, the 'seeds without borders' agreement and arrangement can be followed for the supply of nucleus seeds. IRRI, on this front, has taken phenomenal strides and facilitated cross-border seed exchange (India-Bangladesh, including Sub1 variety) under the aegis of 'seeds without borders'. National system needs to ensure adequate supply of BSs (EGS) to the government and non-government agencies who endeavoring to position for advancing Sub1 varieties.

The importance of formal and informal systems in diffusing varieties among farmers is enormous. There are several reported cases where farmers adopted but later discontinued Sub1 varieties on account of non-availability of seeds. In such a scenario, strategic promotion of both formal and farm-managed seed systems is vital. In countries like India, where emergence of farmers collectives (like Farmer producer Organisation (FPO), Farmer producer Company (FPC), Self Help Group (SHG), etc.) is fast and effective, production and distribution can be routed through such entities through forging an apt entrepreneurship model. Such decentralized seed production models will greatly complement the formal seed system which are often constrained by under-capacity and system shortcomings in South Asian countries.

23.9.2 Multicomponent Stress-Tolerant Variety Development

Abiotic stresses are the result of adverse climate change, fueling the ever-increasing need for food for the fast-growing populations (Lesk et al., 2016). Drought and flood are the most common abiotic stresses affecting rice output in rainfed areas, particularly in India's eastern region. High rainfall in a short period of time during vegetative crop growth, as well as low rainfall or early monsoon departure, may cause flooding or protracted dry spells, resulting in significant agricultural output decrease (Lobell et al., 2011). Flooding and drought may occur at various phases of rice plant growth within the same season. According to current trends, rainfed shallow lowland areas experience excessive precipitation during the early crop growth stage, resulting in flooding, followed by a dry spell, resulting in drought at the middle (flowering) or terminal (maturity) phases of crop growth. Thus, unpredictability in rainfall pattern, intensity, and frequency are the key variables contributing to the unexpected occurrence of drought and flood situations in rice-based agricultural systems. In this position, rice varieties with both drought and flood tolerance may give crop assurance to farmers and in places where dry phase flooding occurs in repeated patterns under rainfed environments. MABC for abiotic stress has evolved over the previous decade from just boosting yield under abiotic stress to combining high yield potential with good yield under various stressors. Drought- and submergence-tolerant rice varieties derived from marker-assisted breeding will help to reduce yield losses associated with cultivation in drought-flood-prone rainfed lowland areas, provide farmers with yield assurance, and encourage MABC programs to develop better varieties tolerant to multiple abiotic and biotic stresses.

23.9.3 Use of Remote Sensing and Eco-Spatial Modeling for Targeting

Remote Sensing (RS) and Geographical Information System (GIS) technologies (generally named geospatial) have been extensively employed in recent years for obtaining geomorphological information (Soni, 2016; Rai et al., 2017) for flood monitoring (Smith, 1997), flood mapping, and flood hazard zonation (Bisht et al., 2016). This technology can deal with multidimensional qualities as well as intrinsic spatial components (Sharma et al., 2018). It is utilized not just for visual realization but also for additional assessment of potential flood damage (Clarck, 1998). The RS/GIS technology is used to identify the areas, which are most vulnerable to flooding by determining the interrelationships between morphometric characteristics. Advances in RS technology and GIS aid in real-time monitoring, early warning, and rapid damage assessment during flood catastrophes. GIS is a technique that floodplain managers may use to detect flood-prone locations. Identified flood-prone locations can be targeted for mitigation with tougher floodplain management techniques by overlaying or intersecting different geographical layers. RS may be particularly helpful for flood control by generating a wider scale assessment of the general flood status within a river basin and identifying locations at the highest danger and in need of quick assistance.

Satellite data provides a synoptic perspective and highly relevant environmental information at a variety of sizes, from entire continents to details of a few meters. As anomalies in a time series, the satellites can identify the early phases of these occurrences. Images are accessible at frequent short time intervals and can be used to forecast both fast and slow disasters. One critical piece of information is the kind and degree of the damage caused by floods in flood-prone areas. RS and co-spatial models provide a near-perfect assessment of adaptive flood risk management strategy at the district level in India during the major rice planting season. RS data can be used in conjunction with land use data to obtain reliable estimates of seed requirement needs, which can improve government departments' readiness to procure necessary stress-tolerant rice seeds in high-demand areas. In a study, the flood recurrence map of India over the 18-year period (2001–2018) revealed two major hotspots as the Ganges and the Brahmaputra (Koppa and Amarnath, 2021). The study also indicated that during the monsoon season, the primary tributaries of these rivers in Bihar, Uttar Pradesh, Assam, and West Bengal cause significant flooding every year.

23.9.4 Stress Molecular Sensing and Signaling

Signaling is critical for plant life and is specific enough to detect floods reliably. A plant may use a multitude of indicators to correctly predict a flooding occurrence. Among various flood-adaptive features, ethylene is a prominent signal trigger (Sasidharan and Voesenek, 2015). In contrast to continuous changeable oxygen levels, rapid ethylene buildup occurs regardless of flooding circumstances which makes it a trustworthy indicator. For a long time, it was unclear whether oxygen sensing happened via direct processes or by an indirect influence of oxygen on metabolism. Plants sense oxygen via the oxygen-dependent degradation of Group VII Ethylene Response Factors (ERFVIIs) (Gibbs et al., 2011; Licausi et al., 2011). Changes in CO_2 levels may likewise also function as a flooding signal. However, even though numerous potential functions have been proposed (Greenway et al., 2006), there is presently no evidence for a substantial involvement in plant flooding responses (Colmer et al., 2006). Plants have minimal control over these signals, which are caused mostly by the limitation of gas diffusion underwater. However, an adaptive decrease in oxygen consumption rate has been proposed (Zabalza et al., 2009). Little is known about the interactions between oxygen and ethylene. In this regard, ROS generation during flooding is analogous, since it happens in an uncontrolled manner, primarily at the electron transport chain and peroxisomes. A particular degree of uncontrolled ROS, together with ethylene-regulated scavenging, is necessary in aerenchyma and adventitious root development to trigger cell death in certain cells (Steffens and Rasmussen, 2016). When taken together, ROS is a signaling intermediary in hypoxia sensing and ethylene-regulated root growth, rather than a trigger to gauge water flooding.

Plants clearly employ a plethora of cues to assess a flooding event and initiate suitable responses that extend its survival. The diverse signal exchanges during and after inundation reveal that the role of these signals' changes throughout time. Little is known about the dynamics of signaling during recovery and how this relates to response initiation during floods. Furthermore, electrophysiological alterations occur within minutes of hypoxia/anoxia, although their physiological relevance is unknown (Wang et al., 2017). Too little is known about the time dynamics of flooding signals and identifying this through research will be a future task in order to assess the key positive and negative regulatory feedback loops under actual flooded situations and how they impact plant response and survival.

23.10 Conclusion

Flooding is a significant and recurrent abiotic stress that occurs because of current climate changes induced by global warming. Flood-prone parts of India in general, and eastern states undergo frequent submergence, inflicting considerable agricultural destruction. The discovery of the Sub1 locus in the flood-tolerant Indian rice landrace sparked interest in marker-assisted breeding of this locus into giant rice variants. For example, the improved rice variety 'Swarna-Sub1' can offer roughly a 45% increase in productivity following flood exposure when compared to existing kinds. A variety of flood-tolerant cultivars have been developed and released, and they are gaining appeal among rice producers in flood-prone areas. However, technology requirements, varietal selections, adaptation scenarios, and complying management approaches differ greatly among farmers as well as between locales and ecologies. New products developed by research programs need enough validation through scientific evidence along with feedback through the end users. In the evolving domain of robust data-driven product validation, testing in farmers' fields and designing farmers' feedback systems are steadily contributing to the varietal development process. Various critical data and feedback collected through farmers' fields and research trials very well support the development of desired products by breeding programs. Trials and demonstrations in farmers' field conditions have generated comparative results for breeders and very reliable product targeting insights for specific regions before large-scale adoption of these varieties. This has also given impetus to the product improvement process keeping the choices open for farmers. Managing the floodplain environment for rainfed rice cultivation requires a full understanding and evaluation to create, disseminate, and implement relevant technologies depending on need through more research and extension activities for improved rice varieties under such difficult ecology. Flood-tolerant genes, proteins, and metabolites may be found using the omics technique and then genetically engineered into the host to create flood-tolerant crop types suited for rice-based systems.

REFERENCES

Ahmad, F., Akram, A., Farman, K. 2017. Molecular markers and marker assisted plant breeding. Current status and their applications in agricultural development. *J. Environ. Agric. Sci.*, 11:35–50.

Almeida, A.M., Vriezen, W.H., Van der Straeten, D. 2003. Molecular and physiological mechanisms of flooding avoidance and tolerance in rice. *Russian J. Plant Physiol.* 50:743–751.

Angaji, S.A., Septiningsih, E.M., Mackill, D.J., Ismail, A.M. 2010. QTLs associated with tolerance of flooding during germination in rice (*Oryza sativa* L.). *Euphytica* 172:159–168.

Armstrong, W. 1980. Aeration in higher plants. In: *Advances in Botanical Research* (Kader, J.-C., Delseny, M., Ed.). Elsevier, London, UK, Vol. 7, pp. 225–332. ISBN 978-0-12-005907-2.

Azarin, K.V., Usatov, A.V., Kostylev, P.I. 2017. Molecular breeding of submergence-tolerant rice. *Annu. Res. Rev. Biol.* 1–10. doi: 10.9734/ARRB/2017/35616.

Bailey-Serres, J., Voesenek, L.A.C.J. 2008. Flooding stress: Acclimations and genetic diversity. *Annu. Revi. Plant Biol.* 59:313–339.

Bailey-Serres, J., Lee, S.C., Brinton, E. 2012. Waterproofing crops: Effective flooding survival strategies. *Plant Physiol.* 160:1698–1709.

Bairagi, S., Bhandari, H., Das, S.K., Mohanty, S. 2018. Impact of submergence-tolerant rice varieties on smallholders' income and expenditure: Farm-level evidence from Bangladesh. *Selected Paper Prepared for Presentation at the Agricultural & Applied Economics Association Annual Meeting*, Washington, D.C., August 5–7.

Barik, J., Kumar, V., Lenka, S.K., Panda, D. 2019. Genetic potentiality of lowland indigenous indica rice (Oryza sativa L.) landraces to anaerobic germination potential. *Plant Physiol. Rep.* 24(2):249–261. doi: 10.1007/s40502-019-00441-3.

Bhowmick, M.K., Dhara, M.C., Singh, S., Dar, M.H., Singh, U.S. 2014. Improved management options for submergence-tolerant (Subl) rice genotype in flood-prone rainfed lowlands of West Bengal. *Am. J. Plant Sci.* 5:14–23. http://dx.doi.org/10.4236/ajps.2014.51003.

Bhowmick, M.K., Srivastava, A.K., Singh, S., Dhara, M.C., Aich, S.S., Patra, S.R., Ismail, A.M. 2020. Realizing the potential of coastal flood-prone areas for rice production in West Bengal: Prospects and challenges. In: *New Frontiers in Stress Management for Durable Agriculture* (Rakshit, A., Singh, H.B., Singh, A.K., Singh, U.S., Fraceto, L., Eds.). Springer Nature, Singapore, pp. 543–576.

Bin Rahman, A.N.M.R., Zhang, J.H. 2016. Flood and drought tolerance in rice: Opposite but may coexist. *Food Energy Secur.* 5:76–88.

Bisht, S., Sharma, S., Chaudhry, S. 2016. Flash flood risk susceptibility in Gagas River Watershed - Kumaun Lesser Himalaya. *Int. J. Adv. Remote Sens GIS* 5(5):1709–1725.

Burman, D., Maji, B., Singh, S., Mandal, S., Sarangi, S.K., Bandyopadhyay, B.K., Bal, A.R., Sharma, D.K., Krishnamurthy, S.L., Singh, H.N., delosReyes, A.S., Villanueva, D., Paris, T., Singh, U.S., Haefele, S.M., Ismail, A.M. 2018. Participatory evaluation guides the development and selection of farmers' preferred rice varieties for salt- and flood-affected coastal deltas of South and Southeast Asia. *Field Crop Res.* 220:67–77.

Chattopadhyay, K., Reddy, J.N., Pradhan, S.K., Patnaik, S.S.C., Marndi, B.C., Swain, P., Nayak, A.K., Anandan, A., Chakraborty, K., Sarkar, R.K., Bose, L.K., Katara, J.L., Parameswaram, C., Mukherjee, A.K., Mohapatra, S.D., Poonam, A., Mishra, S.K. and Korada, R.R. 2018. Genetic improvement of rice for multiple stress tolerance in unfavorable rainfed ecology. In: *Rice Research for Enhancing Productivity, Profitability and Climate Resilience* (Pathak, H., Nayak, A.K., Jena, M., Singh, O.N., Samal, P., Sharma, S.G., Eds.). ICAR-National Rice Research Institute, Cuttack, Odisha, India, pp. 122–137.

Choudhary, K., Choudhary, O.P., Shekhawat, N.S. 2008. Marker assisted selection: A novel approach for crop improvement. *Am. Eurasian J. Agric.* 1:26–30.

Clarck, M.J. 1998. Putting water in its place: A perspective on GIS in hydrology and water management. *Hydrol. Processes* 12:823–834.

Collard, B.C., Mackill, D.J. 2008. Marker-assisted selection: an approach for precision plant breeding in the twenty-first century. *Philos. Trans. R. Soc. Lond. B Biol. Sci.* 363(1491):557–572. doi: 10.1098/rstb.2007.2170.

Collard, B.C.Y., Septiningsih, E.M., Das, S.R., Carandang, J.J., Pamplona, A.M., Sanchez, D.L., Kato, Y., Ye, G., Reddy, J.N., Singh, U.S., Iftekharuddaula, K.M., Venuprasad, R., Vera-Cruz, C.N., Mackill, D.J., Ismail, A.M. 2103. Developing new flood-tolerant varieties at the International Rice Research Institute (IRRI) towards 2025. *SABRAO J. Breed. Genet.* 45:42–56.

Colmer, T.D., Cox, M.C., Voesenek, L.A. 2006. Root aeration in rice (Oryza sativa): Evaluation of oxygen, carbon dioxide, and ethylene as possible regulators of root acclimatizations. *New Phytol.* 170:767–777.

Dana, I., Chatterjee, S., Kundu, C. 2013. Twenty years of achievements of the EIRLSBN at the Rice Research Station, Chinsurah. In: *EIRLSBN: Twenty Years of Achievements in Rice Breeding* (Collard, B.C.Y., Ismail, A.M., Hardy, B., Eds.). International Rice Research Institute, Los Baños, Philippines, pp. 53–64.

Dana, I., Bhowmick, M.K., Kundu, C. 2017. Thirty years of research accomplishments on lowland rice breeding in West Bengal. In: *Contemporary Issues in Agriculture, Environment and Rural Development in India* (Moktan, A., Ed.). New Delhi Publishers, Mohan Garden, New Delhi, India, pp. 97–104.

Dar, M.H., de Janvry, A., Emerick, K., Raitzer, D., Sadoulet, E. 2013. Flood-tolerant rice reduces yield variability and raises expected yield, differentially benefitting socially disadvantaged groups. *Sci. Rep.* 3:3315. doi:10.1038/srep03315.

Dar, M.H., Chakravorty, R., Waza, S.A., Sharma, M., Zaidi, N.W., Singh, A.N., Singh, U.S., Ismail, A.M. 2017. Transforming rice cultivation in flood prone coastal Odisha to ensure food and economic security. *Food Sec.* 9(4):711–722.

Das, K.K., Sarkar, R.K., Ismail, A.M. 2005. Elongation ability and nonstructural carbohydrate levels in relation to submergence tolerance in rice. *Plant Sci.* 168:131–136.

Das, K.K., Panda, D., Sarkar, R.K., Reddy, J.N., Ismail, A.M. 2009. Submergence tolerance in relation to variable floodwater conditions in rice. *Environ. Exp. Bot.* 66(3):425–434.

Depledge, J. 2002. Climate Change in Focus: The IPCC Third Assessment Report. Royal Institute of International Affairs, Briefing Paper, New Series. 29, February 2002.

Dwivedi, A.P., Chand, P., Mishra, A., Singh, S.R.K., Athare, T. 2017. Identification of traditional rice varieties in Chhattisgarh: An institutional arrangement. *Eco. Env. Cons.* 23: S313–S320.

El-Hendawy, S.E., Sone, C., Ito, O., Sakagami, J.I. 2011. Evaluation of germination ability in rice seeds under anaerobic conditions by cluster analysis. *Res. J. Seed Sci.* 4(2):82–93.

Ella, E.S., Kawano, N., Yamauchi, Y., Tanaka, K., Ismail, A.M. 2003. Blocking ethylene perception during submergence reduced chlorophyll degradation and improved seedling survival in rice. *Funct. Plant Biol.* 30:813–819.

Emerick, K., de Janvry, A., Sadoulet, E., Dar, M.H. 2015. Technological innovations, downside risk, and the modernization of agriculture. *Am. Econ. Rev.* 106(6):1537–1561.

FAO. 2017. The future of food and agriculture-Trends and challenges. Rome. Retrieved from http://www.fao.org/3/a--i6583e.pdf.

Fukao, T., Bailey-Serres, J. 2008. Submergence tolerance conferred by Sub1A is mediated by SLR1 and SLRL1 restriction of gibberellin responses in rice. *Proc. Natl. Acad. Sci. U S A* 105:16814–16819.

Gibbs, D.J., Lee, S.C., Isa, N.M., Gramulia, S., Fukao, T., Bassel, G.W., Correia, C.S., Corbineau, F., Theodoulou, F.L., Bailey-Serres, J. 2011. Homeostatic response to hypoxia is regulated by the N-end rule pathway in plants. *Nature* 479:415–418.

Goswami, S., Kar, R.K., Paul, A., Dey, N. 2017. Genetic potentiality of indigenous rice genotypes from Eastern India with reference to submergence tolerance and deep-water traits. *Curr. Plant Biol.* 11(12):23–32.

Government of Assam. 2022. Flood and erosion problems. Department of Water Resources, Government of Assam. https://waterresources.assam.gov.in/portlets/flood-erosion-problems.

Government of India. 2022. Ministry of Agriculture and Farmers' Welfare. GOI, *Seednet Portal*, access on March, 2022. https://seednet.gov.in/.

Government of Odisha. 2020. Odisha Agriculture Statistics. https://agriodisha.nic.in/content/pdf/ACTIVITY%20REPORT%202019-20.pdf.

Greenway, H., Gibbs, J. 2003. Mechanisms of anoxia tolerance in plants. II. Energy requirements for maintenance and energy distribution to essential processes. *Funct. Plant Biol.* 30:999–1036.

Greenway, H., Armstrong, W., Colmer, T.D. 2006. Conditions leading to high CO_2 (>5 kPa) in waterlogged-flooded soils and possible effects on root growth and metabolism. *Ann. Bot.* 98:9–32.

Haefele, S.M., Ismail, A.M., Johnson, D.E., Vera Cruz, C.M., Samson, B. 2012. Crop and natural resource management for climate-ready rice in unfavorable environments: Coping with the adverse and creating opportunities. In: *Responding to Changing Climate in Unfavourable Rice Environments* (Johnson, D.E., Haefele, S.M., Hardy, B., Eds.). *Proceedings of Mini-Symposium of the Consortium for Unfavorable Rice Environments*, May 04, 2010, Siem Reap, Cambodia. International Rice Research Institute, Los Baños, Philippines, pp. 19–31.

Hamamura, K., Kupkanchanakul, T. 1979. Inheritance of floating ability in rice. *Jpn. J. Breed.* 29(3):211–216. doi: 10.1270/jsbbs1951.29.211.

Hattori, Y., Nagai, K., Mori, H., Kitano, H., Matsuoka, M., Ashikari, M. 2008. Mapping of three QTLs that regulate internode elongation in deep-water rice. *Breed. Sci.* 58:39–46.

Hattori, Y., Nagai, K., Furukawa, S., Song, X.J., Kawano, R., Sakakibara, H. 2009. The ethylene response factors SNORKEL1 and SNORKEL2 allow rice to adapt to deep-water. *Nature* 460:1026–1030.

Huang, J., Takano, T., Akita, S. 2000. Expression of α-expansin genes in young seedlings of rice. *Planta*. 211:467–473. doi: 10.1007/s004250000311.

Iftekharuddaula, K.M., Ghosal, S., Gonzaga, Z.J. 2016. Allelic diversity of newly characterized submergence-tolerant rice (Oryza sativa L.) germplasm from Bangladesh. Genet. Resour. Crop Evol. 63(5):859–867.

IRRI. 1984. Terminology for Rice Growing Environments. International Rice Research Institute (IRRI), Philippines.

IRRI. 2008. From genes to farmers' fields: Waterproof rice set to make waves in South Asia, published 21 November 2008. http://beta.irri.org/news/index.php/press-releases/from-genes-to-farmers-fieldswaterproof-rice-set-to-make-waves-in-south-asia.html.

IRRI. 2019. World Food Day. A healthier rice-based diet for people and the planet. https://www.irri.org/world-food-day-2019-rice-zero-hunger.

IRRI. 2022. Climate Change-Ready Rice. Accessed in March. https://www.irri.org/climate-change-ready-rice.

Ismail, A.M., Ella, E.S., Vergara, G.V., Mackill, D.J. 2009. Mechanisms associated with tolerance to flooding during germination and early seedling growth in rice (*Oryza sativa*). *Ann. Bot.* 103:197–209 doi: 10.1093/aob/mcn211.

Ismail, A.M., Johnson, D.E., Ella, E.S., Vergara, G.V., Baltazar, A.M. 2012. Adaptation to flooding during emergence and seedling growth in rice and weeds, and implications for crop establishment. *AoB Plants* pls019.

Ismail, A.M., Singh, U.S., Singh, S., Dar, M.H., Mackill, D.J. 2013. The contribution of submergence-tolerant (Sub1) rice varieties to food security in flood-prone rainfed lowland areas in Asia. *Field Crops Res.* 152:83–93 doi: 10.1016/j.fcr.2013.01.007.

Jackson, M.B., Ram, P.C. 2003. Physiological and molecular basis of susceptibility and tolerance of rice plants to complete submergence. *Ann Bot.* 91(2):227–241.

Jiang, L., Hou, M.Y., Wang, C.M., Wan, J.M. 2004. Quantitative trait loci and epistatic analysis of seed anoxia germinability in rice (*Oryza sativa*). *Rice Sci.* 11(5–6):238–244.

Kato, Y., Collard, B.C., Septiningsih, E.M., Ismail, A.M. 2014. Physiological analyses of traits associated with tolerance of long-term partial submergence in rice. *AoB Plants.* 6:58.

Kato, Y., Collard, B.C.Y., Septiningsih, E.M., Ismail, A.M. 2020. Increasing flooding tolerance in rice: combining tolerance of submergence and of stagnant flooding. *Ann. Bot.* 124(7):1199–1210. doi: 10.1093/aob/mcz118.

Kawano, N., Ella, E., Ito, O., Yamauchi, Y., Tanaka, K. 2002. Metabolic changes in rice seedlings with different submergence tolerance after desubmergence. *Environ. Exp. Bot.* 47:195–203.

Khan, M.H., Dar, Z.A., Dar, S.A. 2015. Breeding strategies for improving rice yield-a review *Agric. Sci.* 6:467–478 doi: 10.4236/as.2015.65046.

Koppa, N., Amarnath, G. 2021. geospatial assessment of flood-tolerant rice varieties to guide climate adaptation strategies in India. *Climate* 9:151. doi: 10.3390/cli9100151.

Kumar, A., Tripathi, G., Joshi, P.K. 2021. Adoption and impact of modern varieties of paddy in India: Evidence from a nationally representative field survey. *J. Agribus. Dev. Emerging Econ.* 11:255–279. doi: 10.1108/JADEE-11-2019-0198.

Lasanthi-Kudahettige, R., Magneschi, L., Loreti, E., Gonzali, S., Licausi, F., Novi, G. 2007. Transcript profiling of the anoxic rice coleoptile. *Plant Physiol.* 144:218–231 doi: 10.1104/pp.106.093997.

Lee, S.C., Mustroph, A., Sasidharan, R., Vashisht, D., Pedersen, O., Oosumi, T., Voesenek, L.A.C.J., Bailey-Serres, J. 2011. Molecular characterization of the submergence response of the Arabidopsis thaliana ecotype Columbia. *New Phytol.* 190:457–471.

Lesk, C., Rowhani, P., Ramankutty, N. 2016. Influence of extreme weather disasters on global crop production. *Nature* 529:84. doi: 10.1038/nature16467.

Licausi, F., Kosmacz, M., Weits, D.A., Giuntoli, B., Giorgi, F.M., Voesenek, L.A., Perata, P., van Dongen, JT. 2011. Oxygen sensing in plants is mediated by an N-end rule pathway for protein destabilization. *Nature* 479:419–422.

Linh, L.H., Linh, T.H., Xuan, T.D., Ham, L.H., Ismail, A.M., Khanh, T.D. 2012. Molecular Breeding to Improve Salt Tolerance of Rice (Oryza sativa L.) in the Red River Delta of Vietnam. Int. J. Plant Gen. 66:949038

Lobell, D.B., Bänziger, M., Magorokosho, C., Vivek, B. 2011. Nonlinear heat effects on African maize as evidenced by historical yield trials. *Nat. Clim. Chang.* 1:42. doi: 10.1038/nclimate1043.

Luo, F.L., Nagel, K.A., Scharr, H., Zeng, B., Schurr, U., Matsubara, S. 2011. Recovery dynamics of growth, photosynthesis and carbohydrate accumulation after desubmergence: a comparison between two wetland plants showing escape and quiescence strategies. *Ann Bot.* 107:49–63.

Luu, T.N.H., Luu, M.C., Ismail, A.M., Ham, L.H. 2012. Introgression the Salinity Tolerance QTLs Saltol into AS996, the Elite Rice Variety of Vietnam. Am. J. Plant Sci. 3:981–987.

Mackill, D.J., Ismail, A.M., Pamplona, A.M. 2010. Stress tolerant rice varieties for adaptation to a changing climate. *Zuowu, Huanjing Yu Shengwu Zixun.* 7:250–259.

Mackill, D.J., Ismail, A.M., Singh, U.S., Labios, R.V., Paris, T.R. 2012. Development and rapid adoption of submergence tolerant (Subl) rice varieties. *Adv. Agron.* 115:299–352.

Maclean, J.L., Dawe, D., Hardy, B., Hettel, G.P. 2002. Rice Almanac. Los Baños (Philippines), Bouaké (Côte d'Ivoire), Cali (Colombia), Rome (Italy); IRRI, WARDA, CIAT, FAO. 253.

Magneschi, L., Perata, P. 2009. Rice germination and seedling growth in the absence of oxygen. *Ann. Bot.* 103:181–196. doi: 10.1093/aob/mcn12.

Mallik, S. 1995. Rainfed Lowland Rice Agricultural Research for High-Risk Environments. Rice germplasm evaluation and improvement for stagnant flooding. International Rice Research Institute, Manila, Philippines, pp. 97–109.

Mantri, N., Patade, V., Pang, E. 2014. Recent advances in rapid and sensitive screening for abiotic stress tolerance. In: *Improvement of Crops in the Era of Climatic Changes* (Ahmad, P., Wani, M.R., Azooz, M.M., Phan Tran, L.-S., Eds). Springer, New York, pp. 37–47.

McCouch, S.R., Wright, M.H., Tung, C.W., Maron, L.G., McNally, K.L., Fitzgerald, M. 2016. Open access resources for genome-wide association mapping in rice. *Nat. Commun.* 7:10532. doi: 10.1038/ncomms10532.

Metraux, J.P., Kende, H. 1983. The role of ethylene in the growth response of submerged deep-water rice. *Plant Physiol.* 72:441–446.

Miro, B., Ismail, A.M. 2013. Tolerance of anaerobic conditions caused by flooding during germination and early growth in rice (*Oryza sativa* L.). *Front Plant Sci.* 4:269.

Mishra, A.K., Erickson, K., Harris, M., Hallahan, C., Uematsu, H. 2010. Determinants of farm household income diversification in the United States. Evidence from Farm-Level Data. *Paper Presented at Agricultural and Applied Economics Association Annual Meeting*, July 25–27, Denver, Colorado, USA.

NAAS. 2013. Water Use Potential of Flood-affected and Drought-prone Areas of Eastern India. Policy Paper No. 60, National Academy of Agricultural Sciences, New Delhi, India.

Nagai, K., Kuroha, T., Ayano, M., Kurokawa, Y., Angeles-Shim, R.B., Shim, J.H. 2012. Two novel QTLs regulate internode elongation in deep-water rice during the early vegetative stage. *Breed. Sci.* 62:178–185.

Nagai, K., Kondo, Y., Kitaoka, T. 2014. QTL analysis of internode elongation in response to gibberellin in deep-water rice. *AoB Plants* 6:28. doi:10.1093/aobpla/plu028.

Nishiuchi, S., Yamauchi, T., Takahashi, H., Kotula, L., Nakazono, M. 2012. Mechanisms for coping with submergence and waterlogging in rice. *Rice* 5:2.

NITI Aayog. 2021. Report of the Committee Constituted for Formulation of Strategy for Flood Management Works in Entire Country and River Management Activities and Works Related to Border Areas (2021–2026). National Institution for Transforming India, New Delhi, NITI Aayog, January 2021. 99.

Odisha Annual Report, IRRI. 2022. Philippines accessed on 11.03. http://books.irri.org/Odisha-Annual-Report-2019-2020.pdf.

Oladosu, Y., Rafii, M.Y., Arolu, F. 2020. Submergence tolerance in rice: Review of mechanism, breeding and, future prospects. *Sustainability* 12(4):1632. doi: 10.3390/su12041632.

Pathak, H., Samal, P., Shahid, M. 2018. Revitalizing rice-systems for enhancing productivity, profitability and climate resilience. In: *Rice Research for Enhancing Productivity, Profitability and Climate Resilience* (Pathak, H., Nayak, A.K., Jena, M., Singh, O.N., Samal, P., Sharma, S.G., Eds.). ICAR-National Rice Research Institute, Cuttack, Odisha, India, pp. 1–17.

Patra, B.C., Marndi, B.C., Sanghamitra, P., Samantaray, S., Umakanta, N., Katara, J.L. 2018. Rice genetic resources: Collection, conservation, maintenance and utilization. In: *Rice Research for Enhancing Productivity, Profitability and Climate Resilience* (Pathak, H., Nayak, A.K., Jena, M., Singh, O.N., Samal, P., Sharma, S.G., Eds.). ICAR-National Rice Research Institute, Cuttack, Odisha, India, pp. 18–36.

Paul, D., Bhattacharya, D. 1980. Simulation of waterlogged situation in normal field for screening of varieties by engineering skill [rice, India]. *Note. Oryza.* 17:251–253.

Perumalsamy, S., Bharani, M., Sudha, M., Nagarajan, P., Arul, L., Sarawathi, R. 2010. Functional marker-assisted selection for bacterial leaf blight resistance genes in rice (*Oryza sativa* L.). *Plant Breed.* 129:400–406. doi: 10.1111/j.1439-0523.2009.01705.x.

Pierdomenico, P., William, A., Laurentius, A., & Voesenek, C.J. 2011. Plants and flooding stress. *New Phytol.*, 190:269–273.

Prabhu, A.S., Filippi, M.C., Silva, G.B., Silva-Lobo, V.L., Morais, O.P. 2009. An unprecedented outbreak of rice blast on a newly released cultivar BRS Colosso in Brazil. In: *Advances in Genetics, Genomics and Control of Rice Blast* (Wang, G.L., Valent, B., Eds). Springer Science, Dordrecht, pp. 257–267.

Qiu, X., Pang, Y., Yuan, Z., Xing, D., Xu, J., Dingkuhn, M. 2015. Genome-wide association study of grain appearance and milling quality in a worldwide collection of indica rice germplasm. *PLoS One* 10:145577. doi: 10.1371/journal.pone.0145577.

Rabara, R.C., Ferrer, M.C., Diaz, C.L. 2014. Phenotypic diversity of farmers' traditional rice varieties in the Philippines. *Agron* 4(2):217–241. doi: 10.3390/agronomy4020217.

Rai, P.K., Chaubey, P.K., Mohan, K., Singh, P. 2017. Geoinformatics for assessing the inferences of quantitative drainage morphometry of the Narmada Basin in India. *Appl. Geomatics* 9(3):1–23. doi: 10.1007/s12518-017-0191-1.

Ramiah, K. 1940. Floating habit in rice. *Indian J. Agric. Sci.* 11:1–8.

Rao, A.N., Wani, S.P., Ramesha, M.S., Ladha, J.K. 2017. Rice production systems. In: *Rice Production Worldwide* (Chauhan, B.S., Jabran, K., Mahajan, G., Eds.). Springer International Publishing AG, New York, pp. 185–205.

Ray, S., Vijayan, J., Sarkar, R.K. 2016. Germination Stage Oxygen Deficiency (GSOD): An emerging stress in the era of changing trends in climate and rice cultivation practice. *Front Plant Sci.* 7:671.

Reddy, J.N., Sarkar, R.K., Patnaik, S.S.C. 2013. Rice breeding and an overview of the EIRLSBN at the coordinating center, CRRI, Cuttack. In: *EIRLSBN: Twenty Years of Achievements in Rice Breeding* (Collard, B.C.Y., Ismail, A.M., Hardy, B., Eds.). International Rice Research Institute, Los Baños, Philippines, pp. 7–29.

Sarkar, R.K. 2016. Stagnant flooding tolerance in rice: endeavors and achievement. NRRI Research Bulletin No. 11, ICAR-National Rice Research Institute, Cuttack. 48.

Sarkar, R.K., Bhattacharjee, B. 2011. Rice genotypes with SUB1 QTL differ in submergence tolerance, elongation ability during submergence and re-generation growth at re-emergence. *Rice (N. Y.).* 5(1):7. doi: 10.1007/s12284-011-9065-z.

Sarkar, R.K., Panda, D. 2009. Distinction and characterization of submergence tolerant and sensitive rice cultivars, probed by the fluorescence OJIP rise kinetics. *Func. Plant Biol.* 36:222–233. doi: 10.1071/FP08218.

Sarkar, R.K., Reddy, J.N., Sharma, S., Ismail, A.M. 2006. Physiological basis of submergence tolerance in rice and implications for crop improvement. *Curr. Sci.* 91(7):899–906.

Sasidharan, R., Voesenek, L.A.C.J. 2015. Ethylene-mediated acclimations to flooding stress. *Plant Physiol.* 169:3–12.

Sasidharan, R., Bailey-Serres, J., Ashikari, M., Atwell, B.J., Colmer, T.D., Fagerstedt, K., Fukao, T., Geigenberger, P., Hebelstrup, K.H., Hill, R.D. 2017. Community recommendations on terminology and procedures used in flooding and low oxygen stress research. *New Phytol.* 214:1403–1407.

Sauter, M. 2000. Rice in deep water: 'how to take heed against a sea of troubles'. *Naturwissenschaften* 87:289–303.

Septiningsih, E.M., Sanchez, D.L., Singh, N., Sendon, P.M., Pamplona, A.M., Heuer, S. 2012. Identifying novel QTLs for submergence tolerance in rice cultivars IR72 and Madabaru. *Theor. Appl. Genet.* 124:867–874.

Septiningsih, E.M., Collard, B.C.Y., Heuer, S., Bailey-Serres J., Ismail, A.M., Mackill, D.J. 2013. Applying genomics tools for breeding submergence tolerance in rice. In: *Translational Genomics for Crop Breeding* (Varshney, R.K., Tuberosa, R., Eds.). Wiley-Blackwell, New York, pp. 9–30.

Septiningsih, E.M., Hidayatun, N., Sanchez, D.L., Nugraha, Y., Carandang, J., Pamplona, A.M., Collard, B.Y.C., Ismail, A.M., Mackill, D.J. 2015. Accelerating the development of new submergence tolerant rice varieties: the case of Ciherang-Sub1 and PSB Rc18-Sub1. Euphytica 202:259-268.

Sharma, K.K. 2013. Rice for stress environments in Lakhimpur District of Assam. In: *EIRLSBN: Twenty Years of Achievements in Rice Breeding* (Collard, B.C.Y., Ismail, A.M., Hardy, B., Eds.). International Rice Research Institute, Los Baños, Philippines, pp. 87–98.

Sharma, A., Singh, P., Rai, P.K. 2018. Morphotectonic analysis of Sheer Khadd River basin using geospatial tools. *Spatial Inf. Res.* doi: 10.1007/s41324-018-0185-z.

Siangliw, M., Toojinda, T., Tragoonrung, S., Vanavichit, A. 2003. Thai jasmine rice carrying QTLch9 (SubQTL) is submergence tolerance. Ann. Bot. 91:255–261.

Singh, B.N. 2006. Rainfed lowland rice improvement: current status and future stratégies. *J. Rice Res.* 1(1):104–112.

Singh, R.K., Dwivedi, J.L., Thakur, R., Mallik, S., Ahmed, T. 2004. Rice biodiversity and genetic wealth of the flood-prone environment in eastern India. In: *Rice Research and Development in the Flood-prone Ecosystem* (Bhuiyan, S.I., Abedin, M.Z., Singh, V.P., Hardy, B., Eds.). *Proceedings of the International Workshop on Flood-Prone Rice Systems*, Gazipur, Bangladesh, 9–11 January 2001. International Rice Research Institute, Los Baños, Philippines, pp. 115–128.

Singh, S., Mackill, D.J., Ismail, A.M. 2009. Responses of Sub1 rice introgression lines to submergence in the field: yield and grain quality. *Field Crops Res.* 113:12–23 doi: 10.1016/j.fcr.2009.04.003.

Singh, S., Mackill, D.J., Ismail, A.M. 2011. Tolerance of longer-term partial stagnant flooding is independent of the SUB1 locus in rice. *Field Crop Res.* 121:311–323.

Singh, K.M., Jha, A., Meena, M., Singh, R. 2012. Constraints of rainfed rice production in India: An overview. In: *Innovations in Rice Production* (Shetty, P.K., Hegde, M.R., Mahadevappa, M., Eds.). National Institute of Advance Studies, Indian Institute of Science Campus, Bangalore, pp. 71–84. Available at SSRN: https://ssrn.com/abstract=2210401 or http://dx.doi.org/10.2139/ssrn.2210401.

Singh, U.S., Dar, M.H., Singh, S., Zaidi, N.W., Bari, M.A., Mackill, D.J., Collard, B.C.Y., Singh, V.N., Singh, J.P., Reddy, J.N., Singh, R.K., Ismail, A.M. 2013. Field performance, dissemination, impact and tracking of submergence tolerant (Sub1) rice varieties in South Asia. *SABRAO J. Breed. Genet.* 45:112–131.

Singh, S., Mackill, D.J., Ismail, A.M. 2014. Physiological basis of tolerance to complete submergence in rice involves genetic factors in addition to the SUB1 gene. *AOB Plants.* 6:60. doi: 10.1093/aobpla/plu060.

Singh, A., Septiningsih, E.M., Balyan, H.S., Singh, N.K., Rai, V. 2017. Genetics, physiological mechanisms and breeding of flood-tolerant rice (*Oryza sativa* L.). *Plant Cell Physiol.* 58(2):185–197.

Sinha, H. 2016. Rediscovering the traditional paddy varieties in Jharkhand: Conservation priority in hybrid rice era. *J. Rural Dev.* 35(2):285–307.

Smith, L.C. 1997. Satellite remote sensing of river inundation area, stage, and discharge: A review. *Hydrol. Process* 11:1427–1439.

Sone, C., Ito, O., Sakagami, J.I. 2011. Characterizing submergence survival strategy in rice via chlorophyll fluorescence. *J. Agron. Crop Sci.* 198:152–160.

Soni, S. 2016. Assessment of morphometric characteristics of Chakrar watershed in Madhya Pradesh India using geospatial technique. *Appl. Water Sci.* doi: 10.1007/s13201-016-0395-2.

Srivastava, A.K., Singh, S., Dar, M.H., Singh, U.S. 2016. Stress-tolerant rice varieties and conforming management practices for intensification of rainfed ecosystem in India. *SATSA Mukhapatra Annu. Tech. Issue* 20:1–14.

Steffens, B., Rasmussen, A. 2016. The physiology of adventitious roots. Plant Physiol. 170: 603–617.

Steffens, B., Wang, J., Sauter, M. 2006. Interactions between ethylene, gibberellin and abscisic acid regulate emergence and growth rate of adventitious roots in deep-water rice. *Planta* 223:604–612.

Taylor, D.L. 1942. Influence of oxygen tension on respiration, fermentation and growth of wheat and rice. *Am. J. Bot.* 29:721–738. doi: 10.2307/2437726.

Thakur, R. 2004. Genetic improvement of flood-prone rice: Where are we today and what are the future prospects? In: *Rice Research and Development in the Flood-Prone Ecosystem* (Bhuiyan, S.I., Abedin, M.Z., Singh, V.P., Hardy, B., Eds.). *Proceedings of the International Workshop on Flood-Prone Rice Systems*, Gazipur, Bangladesh, 9–11 January 2001. International Rice Research Institute, Los Baños, Philippines, pp. 151–162.

Tiwari, D.N. 2018. A critical review of submergence tolerance breeding beyond Sub1 gene to mega varieties in the context of climate change. *IJASRE* 4(3):140–148.

Toledo, A.M.U., Ignacio, J.C.I., Casal, C. 2015. Development of improved Ciherang-Sub1 having tolerance to anaerobic germination conditions. *Plant Breed. Biotechnol.* 3(2):77–87. doi: 10.9787/PBB.2015.3.2.077.

Toojinda, T., Tragoonrung, S., Vanavichit, A., Siangliw, J.L., Pa-In, N., Jantaboon, J., Siangliw, M., Fukai, S. 2005. Molecular breeding for rainfed lowland rice in the Mekong Region. Plant Prod. Sci. 8:330–333.

Vergara, B., Mazaredo, A. 1975. Screening for resistance to submergence under greenhouse conditions. *Scientific Meeting of the Crop Science Society of the Philippines Bacolod City (Philippines)*, 8 May 1975.

Vergara, G.V., Nugraha, Y., Esguerra, M.Q., Mackill, D.J., Ismail, A.M. 2014. Variation in tolerance of rice to long-term stagnant flooding that submerges most of the shoot will aid in breeding tolerant cultivars. *AoB Plants* 6:55. doi: 10.1093/aobpla/plu055.

Voesenek, L.A.C.J., Bailey-Serres, J. 2015. Flood adaptive traits and processes: An overview. *New Phytol.* 206:57–73.

Voesenek, L., Vriezen, W.H., Smekens, M., Huitink, F., Bogemann, G.M., Blom, C. 1997. Ethylene sensitivity and response sensor expression in petioles of Rumex species at low O_2 and high CO_2 concentrations. *Plant Physiol.* 114:1501–1509.

Wang, F., Chen, Z.H., Shabala, S. 2017. Hypoxia sensing in plants: on a quest for ion channels as putative oxygen sensors. *Plant Cell Physiol.* 58:1126–1142.

Wei, B., Jing, R., Wang, C. 2009. Dreb1 genes in wheat (*Triticum aestivum* L.): Development of functional markers and gene mapping based on SNPs. *Mol. Breed.* 23(1):13–22. doi: 10.1007/s11032-008-9209-z.

Wijerathna, Y.M.A.M. 2015. Marker assisted selection: Biotechnology tool for rice molecular breeding. *Adv. Crop Sci. Technol.* 3:187. doi: 10.4172/2329-8863.1000187.

Xu, K., Mackill, D.J. 1996. A major locus for submergence tolerance mapped on rice chromosome. *Mol. Breed.* 2:219–224.

Xu, K., Xu, X., Ronald, P.C., Mackill, D.J. 2000. A high–resolution linkage map of the vicinity of the rice submergence tolerance locus Sub1. *Mol. Gen. Genet.* 263:681–689.

Xu, K., Xu, X., Fukao, T., Canlas, P., Maghirang-Rodriguez, R., Heuer, S., Ismail, A.M., Bailey-Serres, J., Ronald, P.C., Mackill, D.J. 2006. Sub1A is an ethylene-response-factor-like gene that confers submergence tolerance to rice. *Nature* 442(7103):705–708.

Yamauchi, M., Aguilar, A., Vaughan, D., Seshu, D. 1993. Rice (*Oryza sativa* L.) germplasm suitable for direct sowing under flooded soil surface. *Euphytica* 67(3):177–184. doi: 10.1007/BF00040619.

Yang, S.Y., Wu, Y.S., Chen, C.T., Lai, M.H., Yen, H.M., Yang, C.Y. 2017. Physiological and molecular responses of seedlings of an upland rice ('Tung Lu 3') to total submergence compared to those of a submergence-tolerant lowland rice ('FR13A'). *Rice* 10:42.

Yu, H., Xie, W., Li, J., Zhou, F., Zhang, Q. 2014. A whole-genome SNP array (RICE6K) for genomic breeding in rice. *Plant Biotechnol. J.* 12:28–37 doi: 10.1111/pbi.12113.

Zabalza, A., van Donge, J.T., Froehlich, A., Oliver, S.N., Faix, B., Gupta, K.J., Schmälzlin, E., Igal, M., Orcaray, L., Royuela, M. 2009. Regulation of respiration and fermentation to control the plant internal oxygen concentration. *Plant Physiol.* 149:1087–1098.

Zhou, L., Wang, J.K., Yi, Q., Wang, Y.Z., Zhu, Y.G., Zhang, Z.H. 2007. Quantitative trait loci for seedling vigor in rice under field conditions. *Field Crop. Res.* 100:294–301 doi: 10.1016/j.fcr.2006.08.003.

24

Determination and Quantification of Chemical Pollutants by Spectroscopic Techniques: A Possible Use in Chemical Analysis of Food Crops under Stress

Subrata Mahanta, Sapana Sinha, and Jaya Shree
National Institute of Technology (NIT) Jamshedpur

CONTENTS

24.1 Introduction

Any matter qualitatively deteriorating the composition of air with respect to the average composition of it in such a way that an adverse effect on humans and the ecosystem evolves is a pollutant, which builds up the basis of environmental pollution [1–3]. This matter can be transformed into any form like solid particles, liquid droplets or gases. These can be originated naturally or due to man-made activities. And these pollutants arise primarily due to physical processes in the environment, while some were released on further interaction or reaction of primary pollutants with each other or together. Some examples of air pollutants are carbon monoxide gas and sulphur dioxide which are released via vehicles or factories. Nowadays, a topical burning issue of pollution due to the indiscriminate regular use of fossil fuels is deleteriously influencing humanity as well the other mammals or living beings globally [4].

Environmental pollution is defined as, "the contamination of the physical and biological components of the earth/atmospheric system to such an extent that normal environmental processes are adversely affected." Organic pollutants are the energies or naturally occurring substances that can be considered to be contaminants when they exceed the natural limit. Overuse of natural resources more than its recovering rate results in the pollution of air, water and land. Environmental pollutions are of several kinds such as water, air, soil, noise and light. Certain phenomena cause damage in the living organisms. A dramatic and drastic change was found in the interaction of the public health, environmental medicine and the environment [1].

Over the years, experimental techniques in the laboratory found a significant modification for accounting toxic hazardous impact of pollutants in the environmental medium. The ubiquitous nature of the pollutants has fascinated very much with the help of different spectroscopic techniques. Spectroscopic technique plays a pivotal for estimating the pollutants [5–10]. A wide range of technologies and instruments were used for both sampling and determination of the concentration levels of different components of air pollutants. These

DOI: 10.1201/9781003258063-24

include chromatography, infrared spectroscopy, fluorometry, spectrophotometry and atomic absorption spectroscopy (AAS). Pollutants, which result in and cause air pollution, are generally released from industries and cause a lot of pollution.

Pollution extent in any particular area can be highlighted by the determination of the pollutants, and it is helpful in recovering the environment by implementing several control measures. The concentration of air pollutants is high in urban area as compared to rural areas because the urban areas are situated nearer to industries. Qualitative estimation of the air pollutants components can be easily done via various novels adequate as well as analytical techniques. It helps to deal with the atmospheric processes and to understand the interactive range of specific pollutants. It is also helpful in estimating the exposure rate and the capturing of pollutants by human being.

Many different sources contributing to air pollution have an adverse impact on human beings and the ecosystem. The emission of pollutants through automobiles, industries, power plants and factories include toxic gaseous pollutants, which majorly contribute to air pollution. The analytical methods are useful in measuring air pollutants.

Among the common gaseous pollutants, the most general are sulphur dioxide, hydrogen sulphide, hydrogen fluoride, nitrogen oxides, chlorine, hydrogen chloride, hydrogen cyanide, ammonia, ozone and carbon monoxide. With infrared activity, as well as the paramagnetic property, spectroscopic techniques are valid in probing the different pollutants such as CO, CO_2, NO_2 and SO_2. For low concentration of NOX and SO_2, a chemi-luminescent method can be applied. When an air sample contains a component which is easily soluble in water, then it requires a fully extractive system.

In this book chapter, we will discuss the different spectroscopic instruments, like Fourier transform infrared spectroscopy (FTIR), fluorescence spectroscopy, AAS, inductively coupled plasma mass spectrometry (ICP-MS) and Raman spectroscopy. These spectroscopic techniques can be used in finding out the pollutants and their amounts in different natural sources.

24.2 Ultraviolet-Visible (UV-VIS) and Fluorescence Spectroscopy

24.2.1 UV-VIS

UV-VIS spectroscopy is one of the important spectroscopic methods which quantify the analytical sample, especially different organic and inorganic samples in the multi-disciplinary area. This technique helps in examining and identifying chemicals [11–13]. The *Principle* behind UV-VIS spectroscopy states that on absorbing UV light and visible light by certain chemical compounds, there is the production of specific spectra as the result [11]. There will be bonding electrons and non-bonding electrons, which show absorption of energy in specific range of UV or visible light region.

On absorption of light by a particular matter, there is excitation and de-excitation resulting in the production of a particular spectrum. The concept revolving around this term explains the absorption of energy by UV radiation which is in the range of particular energy between ground state energy and high-energy state. The *principle* of UV-VIS spectroscopy is based on the "Beer-Lambert Law", according to which specific monochromatic beam of light passes into the sample solution of an analyte, and then the diminished radiative power in accordance with thickness of solution is analysed with respect to incident radiation and concentration (Figure 24.1).

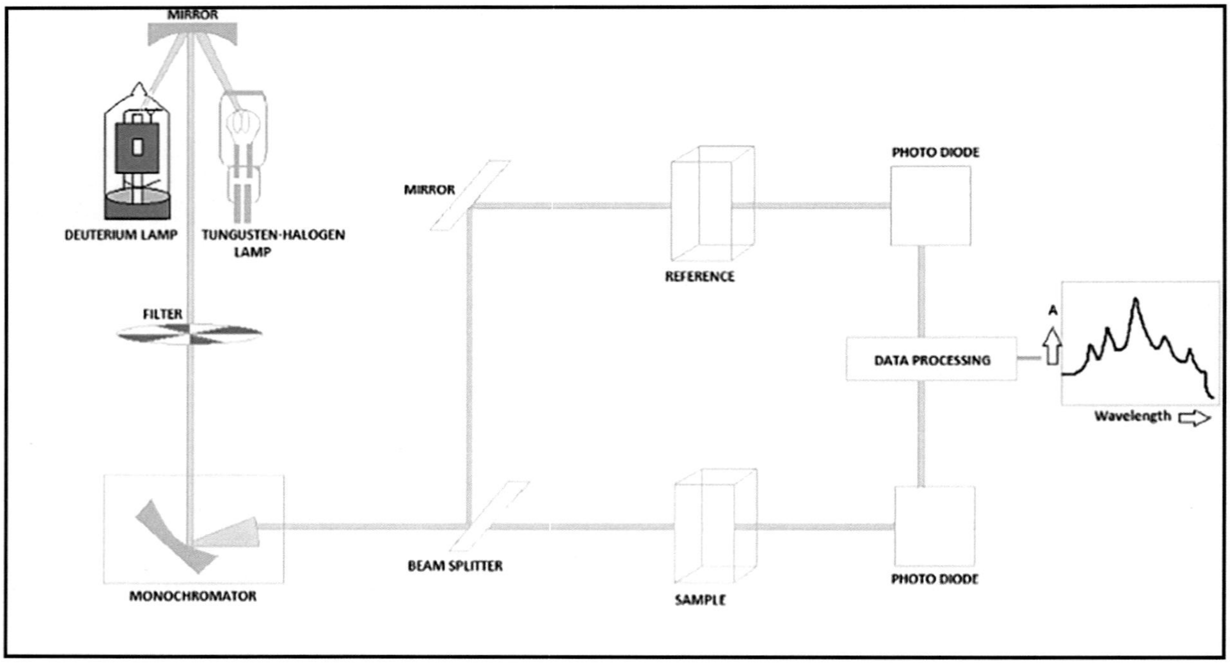

FIGURE 24.1 Schematic representation of UV-visible spectrophotometer.

Generally, for UV-VIS absorption spectroscopy, the sample concentration has to be prepared within the range of 10^{-4} to 10^{-6} mol/dm^3, and optical density was kept <0.3 in order to avoid the aggregate formation and reabsorption parameters. Sample is placed in a quartz cuvette having a path length of 1 cm for the measurements after baseline correction to avoid the absorption effect of contamination in solvents. The schematic representation of the arrangement of the optical components in the UV-VIS absorption spectrophotometer is given in Figure 24.1. In this absorption spectrophotometer, lights from a deuterium discharge lamp (240–340 nm) and iodine tungsten lamp source are steered and selected with suitable optics and grating to the reference and sample cell and finally detected by a photo-detector.

Beer-Lambert Law can be stated through this equation:

$$OD = \log I_0/I = \varepsilon cl$$

Here

OD is the optical density.

I_0 is the initial intensity of radiation.

I is the intensity of transmittance radiation.

ε is the molar extinction coefficient at a certain wavelength in mol^{-1}dm^{3}cm^{-1}.

c is the molar concentration of the sample in mol/dm^3.

l is the path length in cm.

Total absorbance of a sample (A_T) is the sum of absorbances of each of n compounds that give the direct possibility analysis of multicomponent samples [11]:

$$A_T = \varepsilon_1 c_1 l + \varepsilon_2 c_2 l + \varepsilon_3 c_3 l + \cdots + \varepsilon_n c_n l$$

It can be understood from the Beer-Lambert law, which states that the number of absorbing molecules which have a particular wavelength range has a very high light absorption limit, and this signifies *the fundamental principle of UV-VIS spectroscopy*. UV-VIS spectrophotometer is being used to find out the hazardous pollutant elements in the environment.

24.2.2 Fluorescence Spectroscopy

After absorbing radiation or a photon, a molecule reaches its higher excited state [13]. The excited state lifetime of a molecule is very short and varies in the range of nanoseconds (10^{-9} s) to picoseconds (10^{-12} s). Within this time period, there can occur a variety of photophysical and photochemical processes, which can be described using Jablonski diagram (Figure 24.2) [11]. Generally, photoexcitation of a molecule shows certain variations in its geometrical and electronic structures, resulting in changing of different interactions with molecules in accordance with the environment. There is enhancement of molecular interactions in the excited electronic states, and it is found that excitation energy is used partly as a driving force for further reaction leading to various different photochemical and photophysical processes.

The light from an excitation source (a xenon lamp for this case) guided by several optical components passes through a monochromator to discriminate the exciting light at each frequency. Then this light is used to excite the samples. After excitation, the fluorescent light from the sample emits in all directions. The fluorescence light is collected at 90° in accordance with incident beam of light and thus decreases the transmission or reflection of incident light that reaches detector. When fluorescent light reaches into a second monochromator, it passes through the detector photomultiplier tube.

The instrumental setup is described in Figure 24.3.

Recent progress can be observed through UV-VIS and fluorescence spectroscopy in the detection of environmental

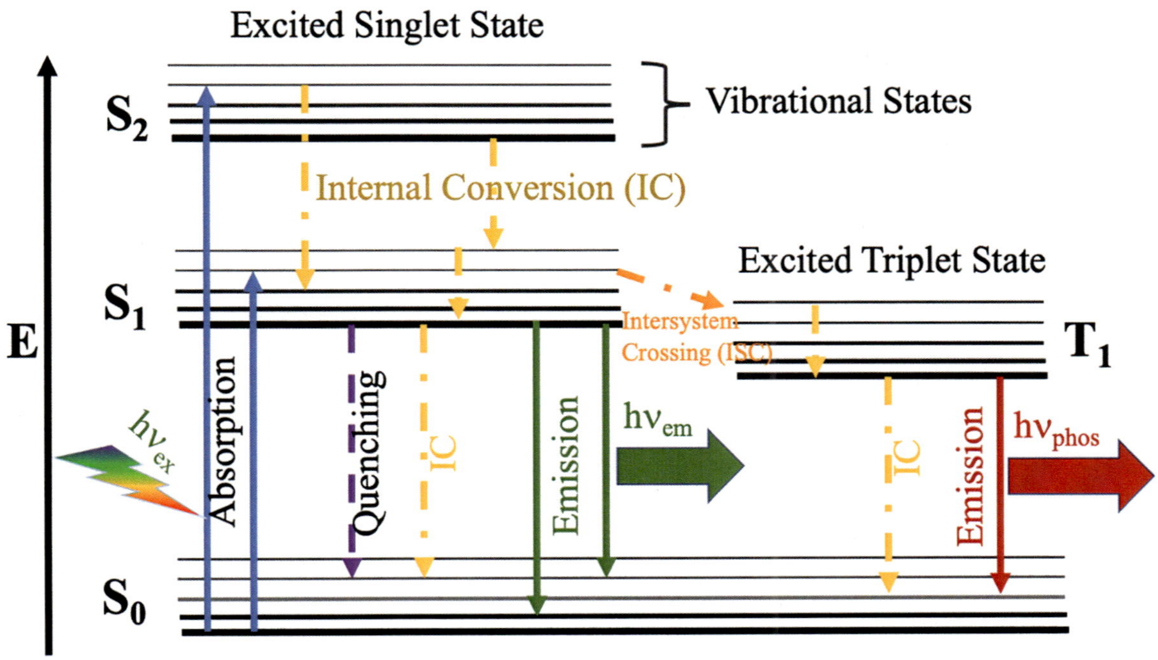

FIGURE 24.2 Schematic representation of Jablonski diagram.

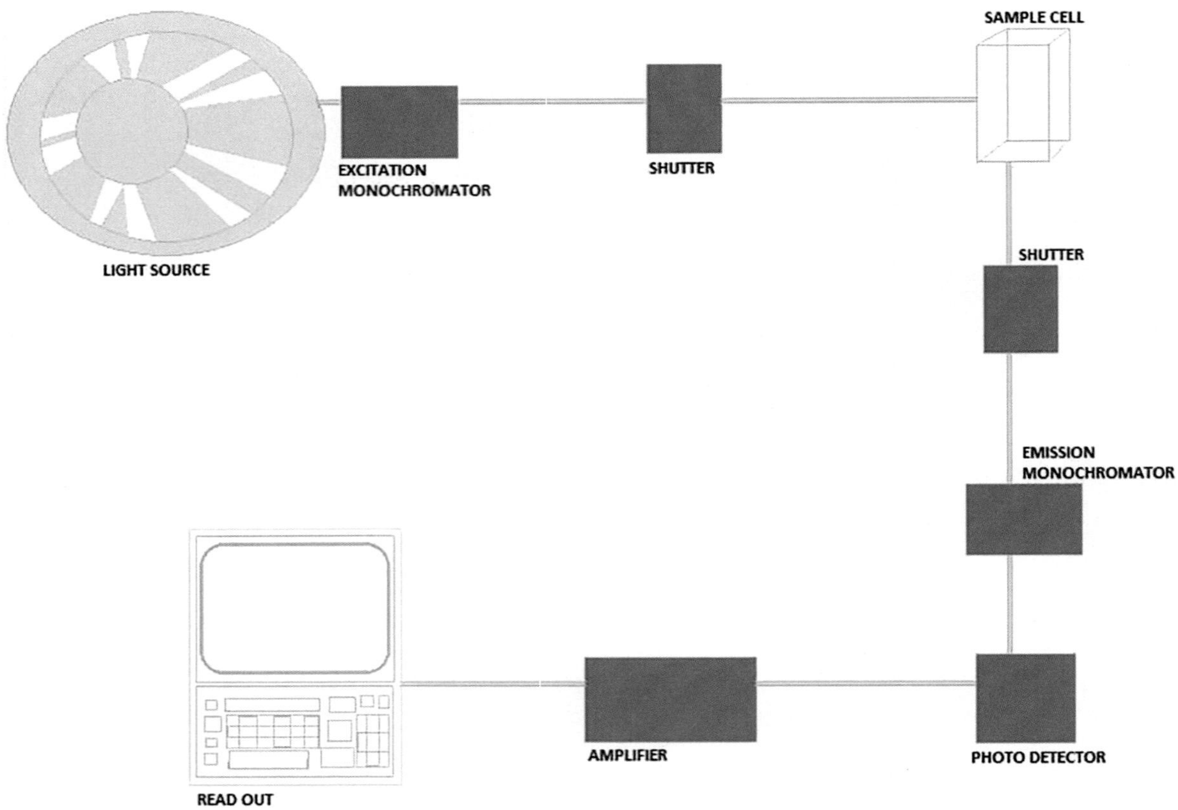

FIGURE 24.3 Schematic representation of spectrofluorometer, excitation polariser (Ex. Pol.) and emission polariser (Em. Pol.).

pollution which is mainly observed due to wheat straws, which are found in the most reactive dissolved organic matter (DOM) of soil organic matter content enhances the agricultural soil qualities implying greater production, reuse, renewability and good economic health globally [14]. Dissolved Organic Matters (DOM) helps to measure the elemental (C, P, N etc.) content that signifies the balance and the long-term accumulation causing a certain impact on the natural sources of the environment. DOM has an easy complexation tendency with heavy metals, which further helps in transformation and migration of it. DOM coming from pig manure gets amended with maize straws and complexes with Cu and Zn heavy metals. Thus, DOM collected from different sources will form complexes with Cd heavy metals via *Microcystis aeruginosa*. Higher presence of Cd^{+2} in contaminated soil enhances the sorption mechanism through DOM amended with wheat straw [15]. Chinese secondary environment signifies quality standards of acidic (red soil, Jiangxi province) and basic (loess soil, Shaanxi province) soil as 0.3 and 0.6 mg/kg Cd, respectively.

Through precise observation of greenhouse experiment, synthesised four different soil samples labelled Ro (red soil + wheat straw), Lo (loess soil + wheat straw), L (loess soil + wheat straw + 0.6 mg/kg Cd) and R (red soil + wheat straw + 0.3 mg/kg Cd) were tested on 0, 30, 90, 180, 300th day, respectively [14,16].

From Table 24.1, dissolved organic carbon (DOC) in L samples is highest as well the [DOC] of L is greater than Lo; similarly, [DOC] of R is greater than Ro signifying Cd^{2+} activation with DOM release. So, from Table 24.1, we can see that the L with 0.6 mg/kg Cd^{2+} has higher [DOC] than R with 0.3 mg/kg Cd^{2+}, which concludes that there is a positive relationship between Cd^{2+} and [DOC] [14]. From Table 24.2 [14], we can see that in an initial stage, [DOC] increases due to the large DOM release, but in the second stage, it is consumed by microorganisms, indicating decrement in [DOC]. And in Tables 24.1 and 24.2, we can see that as the DOM amount in soil decreases, [DOC] increases and shows a positive relationship with Cd^{2+} presence, which means that DOM novel complexation tendency binds the available Cd^{2+} heavy metals, and in general, the free DOM in soil decreases. DOC-based specific UV-VIS absorbance (SUVA) is at 254 nm. The average slope ratio (SR) value for R soil is found to be 2.97, which is much higher than the L soil with 2.75 due to the harsh influence on the molecular weight of DOM by natural properties such as organic carbon content, Cationic Exchange Capacity (CEC), pH etc. of soil. In Table 24.3 [14], we can see that Lo has strong aromaticity results during wheat decomposition than Ro and L or R soil; SUVA at 254 nm <3 Lmg/C m indicates the hydrophilic fraction of DOM.

The larger abundance of C1 in L than Lo and in R than Ro indicates that the DOM-originating organic acid materials associate with additional Cd^{2+} heavy metals, while C2 and C3 have a larger abundance for Lo or Ro than the L or R. In Table 24.4, we can see that fluorescence intensity (FI) value is in between 1.2 and 1.5, which indicates the DOM source as terrestrial and allochthonous. Biological index (BIX) value shows the immediate change in the biological components such as the albuminoid one during this wheat straw decomposition period.

TABLE 24.1
DOC contents in different soil samples are listed here.

S. No.	Type of Soil Sample	[DOC] in g/kg Soil
01	L (0.6 mg/kg Cd^{2+})	0.22–0.34
02	Lo	0.14–0.20
03	R (0.3 mg/kg Cd^{2+})	0.13–0.17
04	Ro	0.10–0.20

In Tables 24.2 and 24.4, we can see that with the change in the amount of DOM activation, accordingly the Cd^{2+} in soils changes the value of FI during the wheat straw decomposition period. In the later stages, the aromatic matter dominant recalcitrant microorganisms create variation in the fluorescence group.

SR
- Shorter wavelength: 175–195 nm
- Longer wavelength: 350–400 nm

FI:
- Rate at emission wavelength: 470 and 520 nm
- Rate at excitation wavelength: 310 nm

BIX:
- FI at emission wavelength of 380 nm with respect to 430 nm
- FI excitation wavelength of 310 nm

Humification index (HIX):
- FI at excitation wavelength of 300 nm With respect to
- FI at emission wavelength of 435 nm wrt 480 nmz

TABLE 24.2
Variation in the [DOC] at different decomposition stages are listed here.

S. No.	Different Stages	Decomposition Period	[DOC]
01	1st stage	0–30	Linear increase
02	2nd stage	30–90	Decrease
03	3rd stage	90–180	Increase in L and Ro then, rapid decrease
04	4th stage	180–300	Little change in Lo and R then remains constant (0.13–0.15) g/kg

TABLE 24.3
SUVA 254 contents in different soil samples are listed here.

S. No.	Type of Soil Sample	SUVA 254 in L mg/C m
01	L (0.6 mg/kg Cd^{2+})	2.44–2.75
02	Lo	2.83–4.04
03	R (0.3 mg/kg Cd^{2+})	2.16–2.71
04	Ro	1.04–2.94

TABLE 24.4
HIX,FI and BIX contents at different decomposition stages are listed here.

S. No.	Decomposition Period	HIX	FI	BIX
01	0–30	1.14–1.04	1.30–1.12	0.78–0.91
02	30–90	1.14–1.04	1.30–1.12	0.78–0.91
03	90–180	1.14–1.04	1.30–1.12	0.91–0.81
04	180–300	1.13	1.35	0.91–0.81

24.2.3 FTIR

In near infrared (NIR), there is no need for pre-treatment of samples, which characterises the techniques as very fast and non-invasive [17–23]. It helps, on the one hand, to measure the surface area, organic matter content, cation-exchange capacity, concentrations of carbon and nitrogen fractions in organic and inorganic matter, moisture content etc. On the other hand, fluorescence spectroscopy has the speciality to distinguish between the organic matter components and crop residue components derived from the soil [24]. And it can be measured from the water extract from the soil rather than the solid samples. The quantitative analysis of fertilised soils is easy because there will be an increase in the amount of microbial biomass, which makes a huge change in vegetation composition.

FTIR refers to a technique. It helps in quantifying some specific compounds of a given unknown mixture. It also helps in the identification of solids, liquids and gases. It is a powerful method as it explains the types of chemical bonds existing in a molecule and thus produces a spectrum of infrared absorption similar to a particular molecular "fingerprint" [25].

24.2.3.1 Basic Principle

Molecular bonds show vibrations at various ranges of frequencies depending on the different types of elements considered and their participation in bond formation. The frequencies corresponding to the ground state (lowest frequency) and several excited states (higher frequencies) can be explained by quantum mechanics [25]. To increase the frequency of the vibration of a molecule, we need to excite that bond of the molecule by absorbing the energy of light radiation. For the transition to take place, we need light energy between the ground state and the first excited state (Figure 24.4).

The FTIR is a useful technique to study and decipher the surface and chemical chemistry in specific types of organic molecules [25]. In FTIR, spectroscopy is used to identify the different functional groups of organic compounds. Using this information, one can easily decipher the presence of different functional groups present in organic molecules. FTIR has been

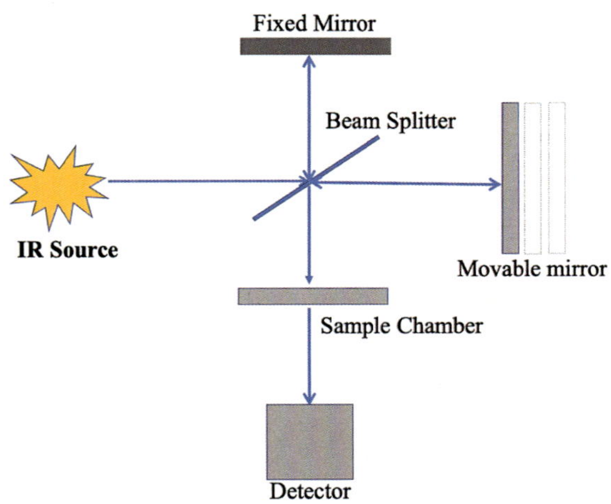

FIGURE 24.4 Schematic representation of FTIR.

used in multi-disciplinary areas. We can use FTIR in agricultural research like soil analysis and food analysis, and agricultural products.

The researcher cannot give a proper explanation of physiochemical-associated properties without understanding any functional group, which has changed the surface behaviours and membrane performances of different types of organic compounds. Every organic compound has different types of functional groups. For example, paddy is made of starch. In this chapter, we can see FTIR techniques as an important tool for monitoring the change of the membrane in surface chemistry. FTIR techniques help to check the performance of a specific membrane by monitoring durability and stability.

FTIR helps in justifying changes in the properties of membranes and their performances in various other applications.

24.2.4 Raman Spectroscopy

Raman spectroscopy is one of the most powerful techniques to probe molecular vibrations, and it is a scattering technique. This effect is named after Sir C.V. Raman, who discovered it in 1928. The schematic representations of the Raman effect are shown in Figure 24.5. There are several applications of Raman spectroscopy in different fields. This can be used in determining the hazardous elements in the environment [26–28].

Raman spectroscopy technique of iron oxides and (oxy) hydroxides with low laser power and its applications can be seen in environment-related magnetic studies [29]. Various iron oxides are generally found in the soil, but distinguishing them from each other via magnetic properties is a huge challenge, implying the necessity for introducing the application of Raman spectroscopy induced with low laser power as to avoid easy transformation of minerals as well due to their feature of poor light scattering. Magnetite and maghemite can be identified easily while the lepidocrocite and ferrihydrite distinctions are difficult via Raman spectroscopy. The dry powdered sample is placed on the glass slide under the microscope. The goethite, haematite, heated goethite, lepidocrocite, siderite, ferrihydrite, magnetite, maghemite and wuestite are assessed via Raman spectroscopy. The goethite has the combination of the strong peaks of 385 cm^{-1} with 244, 299, 480, 548 and 681 cm^{-1} at a laser power of 0.1 mW. But the Rustrel jaune gets destroyed if it is treated under the power of 0.01 mW. The largest or strong peak of 385 cm^{-1} combines with the broad peak of 471 cm^{-1}, and the peaks at 300, 550 cm^{-1} are weak and broad. The yellow ochre shows pure goethite spectra. Rustrel havanne is more stable due to no change on increasing of 0.1 mW than Rustrel jaune, which can be destroyed under 1 mW. Haematite bands, which are visible at 225, 245, 291, 411, 500, 611, 1,321 cm^{-1}, and additional bands can be seen due to the impurities as silica and alumina constitute about 4% in the haematite. The band at 820 cm^{-1} is due to the magnesium iron silicate (olivine). Natural haematite is shown by red ochre due to low crystallinity that broadens up the bands. At 671 cm^{-1}, new peak is visible, and 0.30–35 nm of grain size shows the magnetite here via thermomagnetic measurements that are not shown by the X-ray Diffraction (XRD) methods. Red ochre is not identified at 0.1 mW, which indicates the mineral transformation and at high laser power; it transforms the haematite into maghemite.

Heated goethite on treatment at 700°C transforms the synthetic goethite in order to give the spectra the same as found for the haematite. The haematite bands between the range 670 and 690 cm^{-1} indicate the maghemite broadband. At low laser power, good spectra can be obtained when the sample is illuminated for a time range of less than 1 min. Lepidocrocite shows the band at 250, 348, 528, 379, 650 cm^{-1} stable at 0.1 mW laser power. Lepidocrocite is heated under 400°C in order to give the maghemite. Siderite is stable at moderate laser power of about 1mW showing bands at 184, 287, 731 and 1,090 cm^{-1}.

XRD shows that siderite on heating above 700°C shows the spectra that have both the magnetite and wuestite. Ferrihydrite shows two Raman spectra: one is natural which shows two lines and the other one is synthetic one which shows six lines. The natural ferrihydrite shows a strong band at 710 cm^{-1}, which has the organic matter that exhibits fluorescence. The synthetic ferrihydrite shows three bands at 370, 510, 710 cm^{-1} with the implication of 0.1-mW laser power. It was found via spectra that at 0.01 mW, those bands shown at 0.1 mW are weak, and at 1 mW, spectra change. At a 1 mW of laser power as well on further increment of the laser power, the transformation of haematite gives both of the ferrihydrites. If the

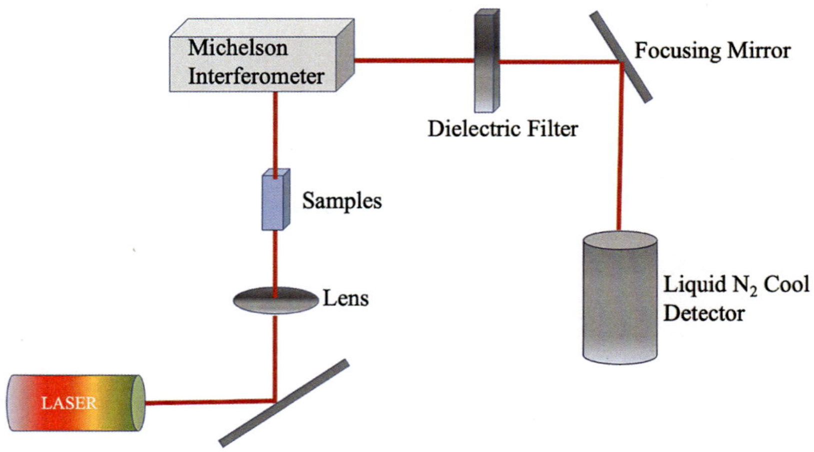

FIGURE 24.5 Schematic diagram of Raman spectrophotometer.

correct timing and laser power are not controlled then the probability of incorrect data will get increased rapidly. If magnetite is exposed with the high laser power then the position of the bands gets varied. The three bands appear at 300, 532, 661 cm^{-1}, while at 0.1 Mw, only 661 cm^{-1} band is found, which also shows variation in the position if the wavelength and the power are changed. The first band appears at above 670 cm^{-1}, and the additional bands are found at below 500 cm^{-1}, which is due to the transformation in the haematite; with this, the shift towards the higher wavenumber is concluded due to this transformation, in general. The second band appears at 532 and 550 cm^{-1}. Then, the two samples with different surface qualities as smooth and rough are assessed under the laser power of 0.1 mW, showing bands at 310, 540 and 670 cm^{-1}, which coincides with the bands of standard material, while at 0.01 mW, only strong band at 670 cm^{-1} is visible. One with smooth surface is stable, and the other with a rough surface initiates for transformation at 1 mW.

Laser-induced oxidation can be differentiated from the transformation effect due to the pyrolytic and photolytic effects. And whatever the grain size is, the band position is found to be same. Maghemite is associated with the silica for composing the sample. Natural ferrihydrite showing two lines is heated at 700°C in order to produce maghemite. At 0.01 mW, the whole experiment is done in order to avoid the maghemite conversion into haematite at 1 mW. Maghemite shows the bands at 350, 512, 664, 726, and 1,330 cm^{-1}, and the small peaks are present at 224, 294, and 411 cm^{-1} in haematite, but in absence of haematite, still the peak at 1,330 cm^{-1} is visible, which is termed as grey spectrum. And the peak at 512 cm^{-1} is shifted to 520 cm^{-1} due to the presence of silica. Grey spectrum can be seen as broadband at 1,360 cm^{-1} with a band in addition at 1,660 cm^{-1}, but these broad and weak bands can be observed even in the absence of silica, maghemite or haematite, which indicates that the band appearance is independent of silica, maghemite or haematite. These broad and weak bands are found for the coal and the graphite, which indicates that the band that appeared in the maghemite case is due to the burning of the organic matter producing the carbon products.

The transformation of ferrihydrite into haematite with the presence of maghemite shows bands at 1,320 and 1,560 cm^{-1}, while in different literatures, the bands are found at 350, 512, 605 and 730 cm^{-1} which shows unreliability for the presence of maghemite, which is further confirmed by treating the sample at high temperature of 400°C, where the magnetisation of the sample is high and transforms the lepidocrocite from maghemite to haematite. At 1 mW, maghemite spectra are found to be same as for haematite. Wuestite associated with magnetite is found when siderite is heated at 700°C. At laser power below 1 mW, wuestite has similar spectra as the magnetite has, exhibiting that the mineral wuestite transforms mostly to magnetite than haematite. XRD does not show the bands other than the bands at 670 and 595 cm^{-1}.

Sometimes the magnetite and maghemite as well as lepidocrocite and ferrihydrite cannot be distinguished via thermomagnetic measurement; hence, Raman spectroscopy is required. Due to similar transformation temperature of ferrihydrite as well as lepidocrocite, the sample of these iron oxides is prepared with hydrogen peroxide. The thermomagnetic measurement of goethite found a confiscation of TiO_2 phase. Haematite and magnetite can be distinguished via magnetic methods. At low temperature, the susceptibility behaviour of large particles of magnetite and maghemite shows Verwey transition at 120°C. The particles that are of superparamagnetic size don't show Verwey transition, and the small-sized particles don't show the bands below 650 cm^{-1}, while magnetite and maghemite can be distinguished easily due to the double peak appearance at 700 and 1,330 cm^{-1} for magnetite, which is not seen in case of maghemite. Sample of magnetite and maghemite is prepared by adding anaerobic iron-reducing bacteria. A broadband appeared at 673 and 722 cm^{-1}. Above 1,300 cm^{-1}, the weak maghemite bands appear due to the presence of magnetite too. Magnetite and maghemite show bands at 540, 309, 344 and 490 cm^{-1}, respectively. One band at 401 cm^{-1} is similar to that of goethite due to the additional iron oxyhydroxide mixtures. On a year-long storage of bacteria-enriched sample, it is concluded that the magnetite and maghemite are not formed due to bacteria, although they can be oxidised via bacteria.

Recent progress can be seen in surface-enhanced Raman spectroscopy, which helps in detecting various kinds of environmental pollution, which is a very powerful analytical tool for the detection of environmental pollutants (heavy metals, organic pollutants, pathogens) due to its advantageous features like sensitivity, selectivity, operation and rapidity. Surface-enhanced Raman scattering (SERS) represents its result in detection via vibrational spectroscopic fingerprints [30–33]. The additional advantage of SERS is its applicability for the sample detection with low concentration and invasive nature. The SERS technique is a miniaturised version of Raman spectrometer, which is rapid in operation as well as cost-effective.

Due to human activities and some natural disasters, nowadays, environmental pollution is a very significant issue globally. Sometimes some unknown chemicals are released into the environment accidentally, and this leads to the importance for the requirement in the SERS reproducibility and portability. Sensitivity can be enhanced by pre-concentration, which is more reliable in comparison to the nanoparticle fabrication techniques. The SERS has a large surface area, which consists of hot spots, capable of expressing the single molecule due to its strongly enhanced locality. SERS signals obtained by the environmental pollutants are nearly the same, so it is very difficult to distinguish the multiple chemicals and molecular composition of the sample, rendering the researchers with a challenging situation (Figure 24.6). This can be enhanced by the modification through specific receptors like polymer and aptamers that will specifically capture the target in order to produce the specific signals. Another way to improve the selectivity is by using the magnetic probe coated with silver or gold nanoparticle capped with antibodies. So, there are various methods of target capturing with the help of various specific substrates; that's why the SERS instrument is associated with separation techniques. Reproducibility is very difficult to achieve due to the aggregation of nanoparticles in colloidal solution by providing very less time for producing reliable and accurate SERS signals.

Nanofabrication improves the reproducibility of the SERS signals. Sometimes the chemicals released in the environment

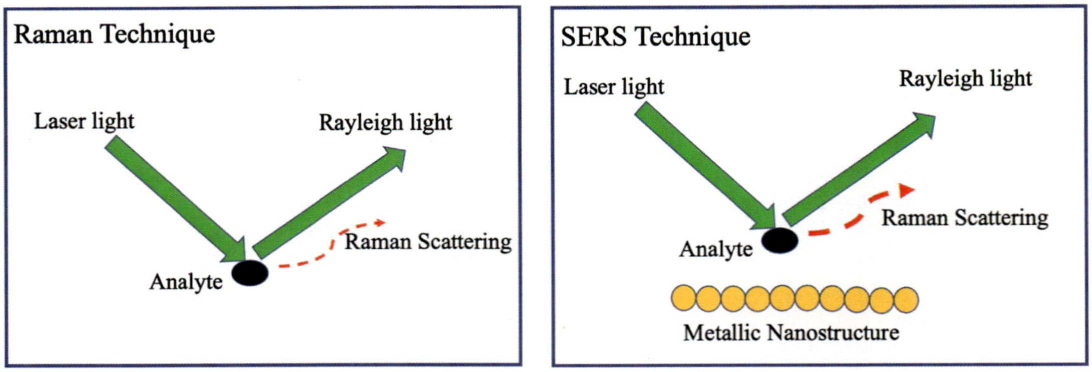

FIGURE 24.6 The principles behind Raman and surface-enhanced Raman spectroscopy or surface-enhanced Raman scattering (SERS).

accidentally render the need for the portability features. The colloidal substrates are less portable than the solid substrates. Electron beam lithography coupling with the bidirectional SERS sensors controls the density of the hot spots. The SERS sensors based on the paper fabrication done via different means are very feasible and portable for industrial as well as field environmental pollutant detection. SERS activity gets reduced due to the oxidation of the metal nanoparticle explored through the paper-based sensors. Due to the resulting features of SERS techniques like specificity, selectivity, reproducibility and portability, it is very important to generate the spectra for the qualitative and quantitative organic pollutants detection, pathogens and heavy metal ions.

The poly aromatic hydrocarbons (PAHs) and polychlorinated biphenyls (PCBs) are obtained via coal or crude oil combustion and the plasticisers or fire retarders into the environment, respectively, increasing the quantity of pollutants in the environment driving the technique for the removal of it that can be advantageously detected by the SERS method or coupling of it for the efficient result. The metal nanostructures coated by the functional groups can be used to fabricate active substrates that are useful for the detection of pesticides. This host and guest interaction leads towards the target capturing based on the cavity size of the host. This method is based on the size that gives rise to the prevention from the interruption of the other pollutants leading towards the specificity of the SERS instruments. Silica and alumina shell coating prevent the agglomeration of the gold or silver nanoparticles, which extends the stability of the SERS instrument detection time range. Precise detection of the pesticides can be achieved by homogeneous mixing of the colloidal SERS substrate with the analytes. The polydimethylsiloxane coupled with microfluidic SERS and SERS microdroplet sensor is used for the detection of the methyl parathion and paraquat pesticides, respectively.

The optofluidic coupling with SERS sensors does multiplexed detection of the fungicides as well as the detection with the little amount of water sample indicating good portability. PAHs have poor affinity for the metallic surface, while the alkyl-coated metal substrates generate the hydrophobicity leading towards the adsorption of PAHs as well the separation of it from the solution. Lucigenin enables the intensified PAHs detection at the interparticle junction due to the ability to form the intermolecular cavity extending the limit of detection (LOD) about 1 nM. Another persistent organic pollutant PCBs can be detected by SERS somewhat similar to the techniques used for PAHs detection. First, the nanospheres are coated with the silver nanoparticles with one more layer over it of decanethiol useful for the 50-pM PCBs detection without any optimisation of the instrument. Another way is by using silver nanoplate filmed with the dithiocarbamate layer for detecting PCBs. Generally, it was found that the PCBs and PAHs are present in environmental samples in very trace amount with the other active SERS pollutants too, showing the need for different or improved SERS techniques. SERS is also useful in the detection of explosives which is an issue of national security. It was found that gold nanoparticles agglomerate with the TNT explosives due to the Meisenheimer complex formation, and 2 pico molar (pM) is enough sample amount to obtain the information without undergoing any interruption by other nitro aromatic compounds. Low-cost facile paper-based SERS sensor when fabricated by gold nanoparticles can be used for the analysis of explosives. The dendritic silver nanosubstrates can be used to find out the other portions too, like oestrogens, antibodies and pigments at the ppb level of detection of limit. Recyclable active substrates of gold nanoparticles not only act as SERS detectors but also as a photocatalytic degrader of organic pollutant.

To overcome the disadvantage of pollutant mixture detection, the SERS technique is coupled with separation techniques as well as some pre-concentration ways that are very useful for the assessment of polar organic pollutants due to adsorption of them on the surface, selectively. By coupling the SERS with the facile thin layer chromatography (TLC), the substituted aromatic hydrocarbon can be detected. The inorganic class of pollutants generally expressed as heavy metal ions were found in the environment in the form of a monatomic metal ions and oxy metal ions. The oxy metal ions can be easily assessed by SERS, but monatomic metal ions were not due to their small scattering cross sections, which are unable to show the vibrational spectra. Arsenate (As^{5+}) and arsenate (As^{3+}) are the species that are mainly found under the criteria of oxy metal ions. As^{3+} can be assessed sensitively at ppb level by the glutathione-modified silver nanoparticles due to the glutathione bond on the surface of silver, and furthermore, there is also formation of oxygen-arsenic linkage. As^{5+} can be detected by the fabrication of a close-packed array of polymer coated with silver nanoparticles. Uranium (UO_2^{2+}), another heavy metal ion found in the environment, is SERS active, which gives the peaks at 700 and 800 cm^{-1}.

It is seen that UO_2^{2+} SERS signals can be obtained by the base-catalysed or acid-catalysed modification of the silver-doped sol-gel film up to the LOD level of 10^{-8} M, which can be further improved by the addition of 2-(5-bromo-2-pyridylazo)-5-diethylaminophenol and N-(2-mercaptopropionyl) glycine up to 10^{-8} to 10^{-9} M. For good reproducibility with a low concentration of UO_2^{2+} sample, an addition of (aminomethyl)-phosphonic acid modified by the gold nanoparticles can be done. Without any surface modification, the UO_2^{2+} with 20-ng/mL concentration can be estimated only by the silver-nanoparticle-modified polypropylene filter substrates. Another toxic monoatomic metal ion Hg^{2+} for the environment is very difficult to detect directly by SERS technique, so some indirect and direct processes lead for the estimation of it. Generally, Hg^{2+} in water sample detected via droplet-based microfluidics in which the water droplet contaminated by Hg^{2+} is reported by the rhodamine B of SERS substrate that is adsorbed by gold nanoparticles. However, due to greater affinity of Hg^{2+} for gold nanoparticles, the rhodamine B is displaced such that the SERS signals are decreased. This "turn-off" mechanism expresses the quantitative result of the Hg^{2+} presence.

In spite of the "turn on" mechanism displayed by the Hg^{2+} due to its strong interaction of it with thymine-thymine base pairs deduced via a single DNA-modified gold microshell, these types of metal- and ligand-specific interactions not only enhance the SERS signals but also reduce the background signals found due to other species. Sometimes metal ions coexist in the samples, which convey the idea for the development of the multiplex detection corresponding to different metal ions present in the sample which is generally possible if the ligand act as both reporters and the binder. Al^{3+}, Mn^{2+}, Fe^{3+}, Cu^{2+}, Zn^{2+} and Pb^{2+} metals multiplex detection can be done via the coupling of the SERS technique with the quantum calculation with the assistance of 1-(2-pyridylazo)-2-naphthol and Eriochrome Black T due to chelation exhibited through the vibrational spectra. Another way of multiplex detection is by using gold nanowire on-chip SERS sensor specifically for Hg^{2+}, Ag^{2+} and Pb^{2+} with the LOD <50 nm. 4-Mercaptobenzoic-acid-modified gold nanoparticle is coated inside the glass surface of the capillary in order to deduce the metal ions on the basis of their different equilibrium constants. In detection of heavy metal ions, wrong results are generally obtained due to a variation of parameters like nanoparticle aggregation, biomolecule denaturation and nonspecific binding.

Due to SERS sensitive fingerprint spectra, SERS technique is an important tool to deduce the pathogens which plays a key role for the several diseases. The pathogen samples are coated with colloidal gold and silver nanoparticles which are studied on the glass slide or cuvette, but this method is uncontrollable for the reproducibility due to the random distribution of nanoparticle on the surface of the pathogens. It is found that periodic nanoparticles in cluster arrays and quasi-3D gold plasmonic nanostructure arrays are capable to improve the problem of reproducibility and the low amount of pathogen in sample quantification by pre-concentration quickly by Fe_3O_4-Au core-shell nanoparticles (Figure 24.7).

24.2.5 AAS

AAS is considered in measuring different concentrations prepared by metallic elements in different samples [9,34–37]. The Burning of fossil fuels and automobile exhaust because of man's increasing activities and more demands have caused a huge amount of environmental pollution, mainly due to economic activities and growing industrialisation. Burning of carbon-based fossil fuels and various petroleum industry activities are responsible for causing large extent of environment-related pollutions. Several studies have focussed to show that there are heavy metals such as lead (Pb), chromium (Cr), cadmium (Cd), nickel (Ni) and manganese (Mn) amongst others, which cause various kinds of diseases. It is seen that heavy metals contribute by acting as systematic toxins with specific nephrotoxic, neurotoxic, phytotoxic and teratogenic effects. Heavy metals have the capability to directly influence neurotransmitter production and utilisation. It also alters numerous metabolic body processes. Systems which are designed in which heavy metals can induce dysfunction and impairment include various eliminative pathways (kidneys, colon, skin), endocrine gastrointestinal, energy production pathways, immunity, enzymatic pathways, nervous (central) and peripheral, urinary and reproductive, which provide various harmful effects on human beings. Diseases caused are kidney disease, chronic bronchitis, abdominal pain, pulmonary oedema, convulsions, liver damage, cancer of the lung and nasal sinus ulcer problems and

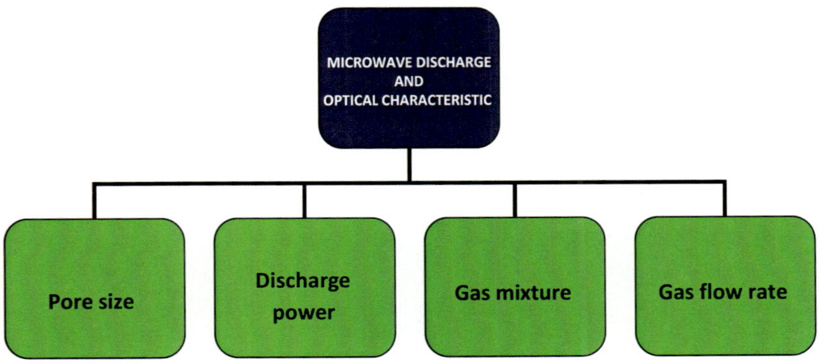

FIGURE 24.7 Dependence of electrical and optical characteristics in microwave discharge by different factors.

at last death. The Burning of fossil fuels is the main source of environmental pollution because of its contribution to global warming and climate change.

Electromagnetic radiation coming from a light source is used in this spectroscopic method, and wavelengths will be absorbed differently with distinct elements. It provides the knowledge of concentrations of a specific element of the material, or liquid, which is being taken for the test. The schematic diagram of the AAS is given below (Figure 24.8):

This method is being used nowadays as a useful tool for the determination of metallic elements like arsenic, lead and mercury present in the pollutants. It is found that around 75% of earth's chemical elements comprise metals, which occur naturally in the world around us. Some metals can also act as contaminants when they are used in a material and when present as content. Therefore, metallic content measurement plays an important role in various processes. AAS is a benchmark technique, which is a useful technique that helps to know various kinds of metallic elements for many years. This is because of its greater sensitivity than other methods with very less limitations. We can directly do analysis for liquid samples. This technique helps by working very rapidly, efficiently and economically as a testing method for very small sample sizes.

In the clinical analysis field, AAS plays a crucial role in testing and examining metals composition in whole blood, saliva, plasma, muscle and brain tissue, the liver and hair. One of the examples where this method can be used is in measuring mercury levels in fish. As we know that mercury is very toxic to humans and animals, it is present as a metal in the environment.

Mercury can be considered as a pollutant as it arises from various sources such as cement production, power plants and metal processing. Mercury when emitted into the air settles into water or onto land where it can be washed into water bodies. There are certain microorganisms which convert mercury into methylmercury $[CH_3Hg]^+$ when deposited. It is of a highly toxic nature that can be seen in fish, shell fish and animals that eat fish. Therefore, mercury causes huge air pollution and also contaminates rivers, lakes and oceans. Biomagnification, also known as biological magnification, is a process which is caused because of the accumulation of a toxic chemical by an organism from water and food exposure at successively higher levels in a food chain. This process increases the amount of methylmercury in tuna, swordfish and sharks. We can test fish samples rapidly using various advanced methodologies of AAS.

24.2.6 ICP-MS

This is one of the fastest multi-element trace element analysis procedures. The primary application of inductively coupled plasma mass spectrometry (ICP-Q-MS) is the accurate, rapid and precise determination of trace elements in solid and liquid samples. Isotopic determinations and speciation studies are some of the other applications of this technique. This powerful modern technique can rapidly help in measuring trace elements in short detection limits (sub-parts per trillion levels), and additionally, detect major and minor elements for the same analytical run.

24.2.7 ICP-MS Applications

The following is the list of applications that have been analysed by ICP-MS:

- **Archaeological applications**: Raw materials.
- **Biological applications**: Hair, shells, teeth.
- **Food applications**: Milk, juices, beverages, food packaging.
- **Health science applications**: Pathology, metabolism studies, medicines.
- **Earth science applications**: Minerals, soil, waters.
- **Environmental applications**: Gases (car exhaust), dust, gases from landfills, organic wastes, paint.
- **Industrial applications**: Steel, superconductors, sulphides.

ICP-MS combines ICP torch and MS interface for performing elemental analysis by ion generation. The ICP generates a very

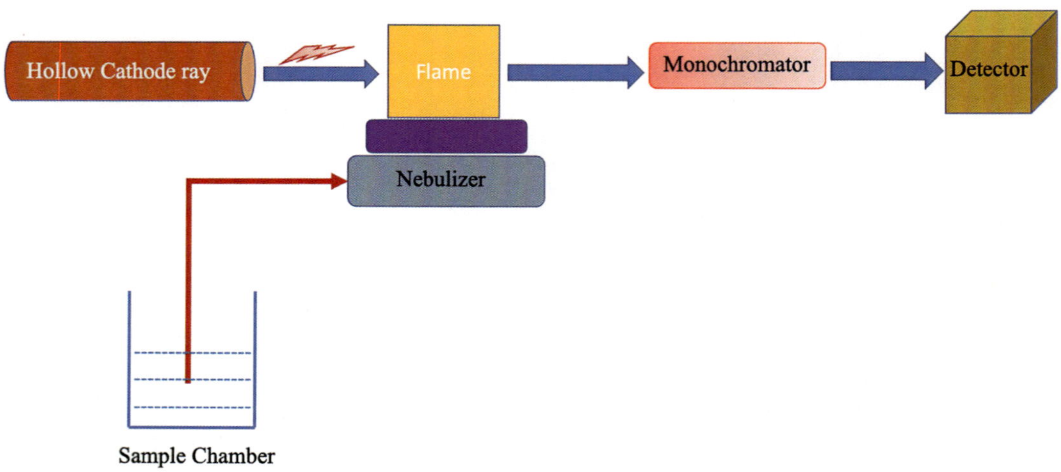

FIGURE 24.8 Schematic diagram of atomic absorption spectroscopy.

high-temperature plasma (at 10,000 K), through which the pre-treated sample passes. Plasma is a highly ionised stream of gas containing equal numbers of electrons and cations, which is conductive and affected by magnetic fields. The high temperature that the plasma attains in the system helps to ionise the elements and thus detect them. The mass spectrometer is used as a detector.

We can see ICP-MS schematic diagram in Figure 24.9, which comprises an ion source (ICP), a sampling interface, an ion lens, a mass spectrophotometer and a detector. The ICP, working as the ion source, can ionise over 90% of elements, and is found to be an ideal ioniser for mass spectrometry. The generated ions go through the sampling interface into the mass analysis unit. These two metallic cones are present in the sampling interface unit – the sampling cone (orifice radius ~ 0.5–1 mm) and the skimmer cone (radius ~ 0.5–1 mm), between which a rotary gear pump ventilates at a condition of several hundred pascals (Pa). Ions drawn by the sampling and skimmer cones converge to the mass spectrometer and pass through the ion lens. The ion lens and mass spectrometer unit are aired to 10^{-3} and 10^{-4} Pa, respectively, by a turbo molecular pump. Lastly, ions sorted by the mass spectrometer are detected by an ion detector.

Thus, problems arise during the measurement of the matrix components in ICP-MS, one of which is spectral interference (cool plasma). As a result of cool plasma, molecular ions of elements in the sample increase, and Argon ions decrease. There can also be a strong desensitisation effect due to matrix formation, so it is impractical to use for environmental samples. Therefore, other methods need to be applied to reduce spectral interference. Several molecular and molecular oxide ions have a pronounced effect in this regard. The water (H_2O) in the sample produces large amounts of oxide ions, so limiting water content decreases oxide ion production significantly. Additionally, optimising plasma conditions with sampling interface shape limits oxide ion production.

At the end of each section, a critical assessment on merits, demerits and possibilities will enhance the quality of the write-up. A note on the LOD of every method will be an added advantage. As an example of the applications of AAS, ICP-MS is in recent literature for the quantification of pollutants. Inorganic and organic pollutants determination from agro environment is required.

The entire focus should be directed/oriented around agro environment and also on how nuclear magnetic resonance (NMR) can be effectively utilised in plant metabolite quantification, limitation, prospect etc.

24.3 Spectroscopic Applications in Pollutant Determination

Spectroscopic techniques have long been the most powerful tool for investigating various kinds of pollutants which are harmful to human beings and animals. We will discuss different kinds of applications where specific spectroscopic techniques can be used for the determination of certain pollutants. They fall under the following categories:

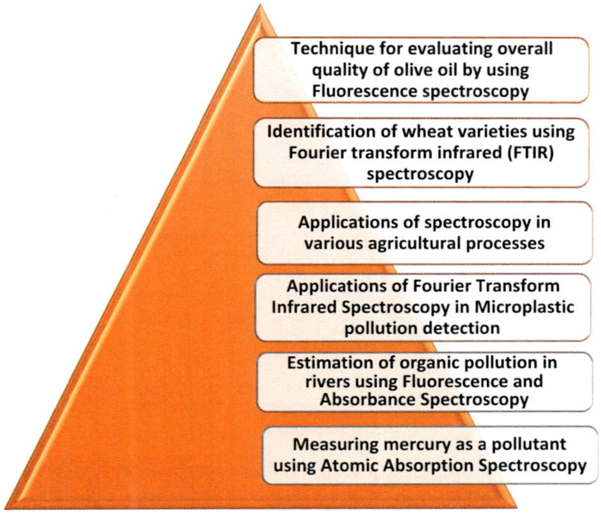

Now, we will discuss the above applications in detail:

24.3.1 Technique for Evaluating of the Overall Quality of Olive Oil by Using Fluorescence Spectroscopy

The spectroscopy can be used to examine oil quality such as oils in natural and oxidised state, with wavelength range emissions of 300–800 nm as an excitation radiation. Three bands can be seen for extra virgin olive oil: one low intensity doublet at 440 and 455 nm; one strong band at 525 nm; and of medium intensity at 681 nm. For characterisation of olive oils on the basis of quality parameters, fluorescence technique can be used by combining with multivariate analysis.

For detecting virgin olive oils, which are obtained from the defective ones(high acidity), we can use fluorescence spectroscopy which provides us the way to differentiate, and quantify, the levels of oil quality with the help of spectrofluorimetric method combined with multivariate analysis. Fluorescence spectroscopy helps us by providing analytical information that can be useful for various applications.

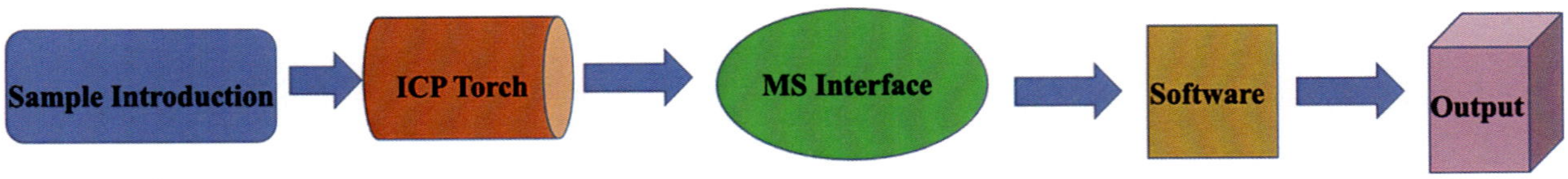

FIGURE 24.9 ICP-MS schematic diagram.

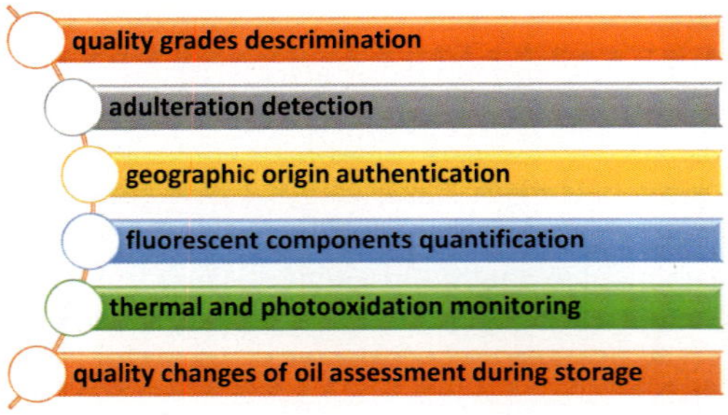

24.3.2 Identification of Wheat Varieties Using FTIR Spectroscopy

It is known that wheat is a component of human diet and is considered as a rich source of carbohydrates as it also contains vitamins, proteins, vitamins and phytochemicals. Different quality parameters of wheat can be analysed using FTIR, which provides the quick result with a variety of sampling techniques. FTIR provides information in the form of peaks on the basis of different functional groups. The value of moisture, protein, fat, ash, carbohydrates and hardness of grain can be determined using FTIR, on the basis of different peaks.

FTIR spectroscopic analysis helps to examine differences in grains formed from different wheat varieties. Chemical mapping in collaboration with FTIR can be applied to different agricultural products, e.g., wheat, flax stems, grains and soya bean seeds. FTIR can also be used to examine the cell wall of cereal grains and the composition of hard and soft milling wheat with the help of FTIR Imaging in conjugation with multivariate statistical analysis.

FTIR provides an excellent means in the detection of a wide variety of functional groups. FTIR is a powerful tool and is a time-saving method that provides information of chemical composition and physical state of the whole sample.

FTIR spectrometric analysis can be done using the following parameters and steps:

a. Preparation of sample:

   At first, we have to collect flour of wheat variety, and these samples taken are converted into pallets. Then, these pallets are cut into thin sections. These thin sections are oven dried and then spectra can be recorded on FTIR.

b. FTIR measurements:

   FTIR spectra can be recorded for the following contents:

   - Moisture is determined by FTIR in mid-infrared range, and peaks of water are observed in the range of 1,640–3,300 cm^{-1}.
   - Samples containing crude fat can be observed having peaks at 1,600–1,700 cm^{-1} and 1,550–1,570 cm^{-1}.
   - Crude protein peaks of amide I band can be observed at 1,600–1,700 cm^{-1} and amide II at 1,550–1,570 cm^{-1}.
   - Peaks of ash content can also be determined with the help of FTIR.
   - Spectra of starch sample can be recorded using FTIR in regions below 800, 800–1,500 cm^{-1} (the fingerprint region), the region between 2,800 and 3,000 cm^{-1} (which shows C-H stretch region) and finally the region between 3,000 and 3,600 cm^{-1} (which shows O-H stretch region).

FTIR is a non-destructive, rapid analytical technique that gives correct measurement, which requires no external calibration.

24.4 Applications of Spectroscopy in Various Agricultural Processes

High-tech instrumentation is required for supplying sufficient and qualified agricultural products [40]. Spectrometry methods can be used for field monitoring and quality control. Visible/NIR spectroscopy (NIRS) was introduced by Abdel-Nour et al. [41]. It is a powerful and fast technique used to measure a large number of chemical and physical properties of agricultural products. The sample's diffused reflectance can be obtained using a spectrophotometer, and a reflectance spectrum is formed on exposing a given sample with frequencies which provide the information of the intensity of light as a function of wavelength.

Mass spectrometry is a powerful tool and can be used for the identification of unknown materials. This technique is based on certain properties, such as ion's motion, a charged particle, in an electric or magnetic field.

- **Field monitoring:** Spectroscopy is a powerful tool for quantification of phytopigment as different groups of pigments absorb different light wavelengths. Quick measurements of green biomass can be obtained, and index is produced for examining the vegetation, such as the normalised difference vegetation index, using a portable spectroradiometer. It is a light measurement tool for measuring both wavelength and amplitude of the emitted light from a light source.
- **Quality control:** Mass spectrometry can be used for bacterial detection and differentiation that are caused by its capability when joined with chromatographic separation for identification of particular compounds in complex mixtures. Fatty acids, proteins and complex lipids can be used in distinguishing bacterial samples. A comprehensive analysis of the coffee headspace brew can be carried using proton-transfer-reaction mass spectrometry.

Determination of moisture concentrations in semolina pasta can be done by NIR reflectance spectroscopy.

24.4.1 Applications of FTIR Spectroscopy in Microplastic Pollution Detection

FTIR can be used to evaluate microplastic (MP) pollution and can be used to explain and characterise various polymer types of MPs. The MPs which are present in different forms such as sediments, water (marine and freshwater), biota, dust, wastewater treatment plants and salts are further classified according as follows:

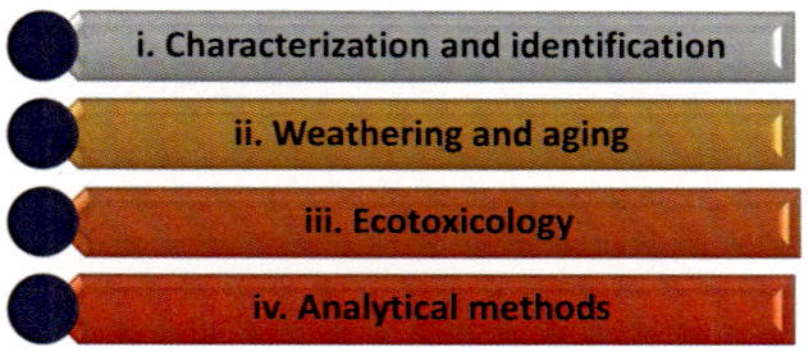

FTIR spectroscopy measures IR radiation that is absorbed by MP sample, which allows for studying molecular composition. An infrared spectrum provides a unique fingerprint of an MP.

Sample where absorption peaks corresponding to the frequencies of vibration between the bonds of different atoms for making up the materials. Therefore, FTIR helps to determine the structure of a given polymer.

24.4.2 Estimation of Organic Pollution in Rivers Using Fluorescence and Absorption Spectroscopy

Spatial variability of the pollution can be estimated using fluorescence and absorption spectroscopy using certain parameters such as absorbance, FI, dissolved organic carbon, biochemical oxygen demand (BOD), chemical oxygen demand (COD) and other water quality parameters.

BOD and COD can be used to examine organic contents present in water samples. BOD requires 5 days for incubating and evaluating the sample. On the other hand, COD uses chemical oxidation for quantification over several hours.

24.4.3 Measuring Mercury as a Pollutant Using AAS

Mercury pollutant arises from different kinds of processes such as power plants, metal processing and cement production. Mercury when comes in contact with air settles into water bodies or on land and gets converted into methylmercury $[CH_3Hg]^+$ using certain microorganisms, which is harmful to fish, shell fish and animals. Therefore, AAS can be used to test mercury contents in fish samples.

24.5 Conclusions

We can detect different pollutant elements as well as quantities in natural sources using different instrumental techniques, but the most reliable techniques are the spectroscopic techniques for quantitative and qualitative determination of chemical pollutants in different natural sources of the environment for the promotion of sustainable impact and management of natural sources. We have reported different spectroscopic instruments and techniques like ICP-MS, AAS, UV-VIS absorption spectroscopy, fluorescence spectroscopy and Raman spectroscopy.

UV-VIS works according to the principle of UV light absorption and visible light absorption by various chemical compounds, and certain spectra can be seen, which helps in the identification of chemicals. According to the fundamental principle of UV-VIS spectroscopy, molecules which show absorption of light under a specified range of wavelength must have very high absorption limit in accordance with Beer-Lambert Law. Various recent progresses and greenhouse experiments are done using UV-VIS and fluorescence spectroscopy for detection of environmental pollution due to the presence of wheat straws, easily available in dissolved organic matter of soil content, which enhances agricultural soil qualities and helps to measure the elemental (C, P, N etc.) content. NIR technique is used for measuring the specific surface area, clay content, moisture content, cation-exchange capacity etc. On the other hand, fluorescence spectroscopy can distinguish between organic matter components and crop residue components of soil.

Recent progresses in "surface-enhanced Raman spectroscopy" can be seen in examining environmental pollutants such as heavy metals and pathogens because of its unique features like sensitivity, selectivity, operationality and rapidity. SERS technique enhances Raman scattering as there is adsorption of molecules on rough metal surface or by nanostructures.

For the measurement of different concentrations and levels of metallic elements existing in samples, AAS can be used. AAS can be used in the clinical analysis for testing and examining metals in whole blood, urine, plasma etc. Mercury levels in fish can be measured. ICP-MS is found in archaeological applications, health sciences, biological applications etc. for measuring trace elements within short-duration time limits. Many efforts and researches were done in recent decades to search for alternative methods for determination of rice quality parameters and other agricultural products rapidly. NIRS can be used for the determination of various quality parameters rapidly.

Acknowledgement

The authors gratefully acknowledge Dr. Debojyoti Moulick for constant support in writing this book chapter.

REFERENCES

1. Brunekreef, B, and Holgate, S.T. 2002. Air pollution and health, *The Lancet* 360:1233–1242.
2. Kim, K.-H., Jahan, S.R., and Kabir, E. 2013. A review on human health perspective of air pollution with respect to allergies and asthma, *Environment International* 59:41–52.
3. Ali, M.U., Lin, S., Yousaf, B., Abbas, Q., Munir, M.A.M., Rashid, A., Zheng, C., Kuang, X., and Wong, M.H. 2022. Pollution characteristics, mechanism of toxicity and health effects of the ultrafine particles in the indoor environment: Current status and future perspectives, *Environmental Science and Technology* 52:436–473.
4. Guttikunda, S.K., Goel, R., and Pant, P. 2014. Nature of air pollution, emission sources, and management in the Indian cities, *Atmospheric Environment* 95:501–510.
5. Begum, R., Farooqi, Z.H., Naseem, K., Ali, F., Batool, M., Xiao, J., and Irfan, A. 2018 Applications of UV/Vis spectroscopy in characterization and catalytic activity of

noble metal nanoparticles fabricated in responsive polymer microgels: A review, *Critical Reviews in Analytical Chemistry*, 48:503–516.

6. Guzmán, E., Baeten, V., Pierna, J.A.F., and García-Mesa, A. 2015. Evaluation of the overall quality of olive oil using fluorescence spectroscopy, *Food Chemistry*, 173:927–934.
7. Andrade, M.C., Winemiller, K.O., Barbosa, P.S., Fortunati, A., Chelazzi, D., Cincinelli, A., and Giarrizzo, T. 2019. First account of plastic pollution impacting freshwater fishes in the Amazon: Ingestion of plastic debris by piranhas and other serrasalmids with diverse feeding habits, *Environmental Pollution* 244:766–773.
8. Hao, J., Han, M.-J., Han, S., Meng, X., Su, T.-I., and Wang, Q.K. 2015. SERS detection of arsenic in water, *Journal of Environmental Sciences* 36:152–162.
9. Palkendo, J.A., Kovach, J., and Betts, T.A. 2014. Determination of wear metals in used motor oil by flame atomic absorption spectroscopy, *Journal of Chemical Education* 91:579–582.
10. Alomar, T.S., Habila, M.A., AlMasoud, N., Alothman, Z.A., Sheikh, M., and Soylak, M. 2021. Biomass-derived adsorbent for dispersive solid-phase extraction of Cr(III), Fe(III), Co(II) and Ni(II) from food samples prior to ICP-MS detection, *Applied Sciences* 11(17):7792.
11. Lakowicz, J.R. 2010. *Principles of Fluorescence Spectroscopy*, 3rd edition. Springer: Berlin, Germany, 2006. Correct 5th printing 2010.
12. Mahanta, S., Singh, R.B., Kar, S., and Guchhait, N. 2008. Evidence of coupled photoinduced proton transfer and intramolecular charge transfer reaction in para-N, N-dimethylamino orthohydroxy benzaldehyde: Spectroscopic and theoretical studies, *Chemical Physics* 354:118–129.
13. Mahanta, S., Singh, R.B., and Guchhait, N. 2009. Study of protein–probe interaction and protective action of surfactant sodium dodecyl sulphate in urea-denatured HSA using charge transfer fluorescence probe methyl ester of N, N-dimethylamino naphthyl acrylic acid, *Journal of Fluorescence* 19:291–302.
14. Zhang, A., Zhou, X., Li, M., and Wu, H. 2017. Impacts of biochar addition on soil dissolved organic matter characteristics in a wheat-maize rotation system in Loess Plateau of China. *Chemosphere* 186:986–993.
15. Zsolnay, A. 2003. Dissolved organic matter: Artefacts, definitions, and functions. *Geoderma,* 113:187–209.
16. Han, N., and Thompson, M.L. 1999. Copper-binding ability of dissolved organic matter derived from anaerobically digested biosolids, *American Society of Agronomy, Crop Science Society of America, and Soil Science Society of America* 28:939–944.
17. Veerasingam, S., Ranjani, M., Venkatachalapathy, R., Bagaev, A., Mukhanov, V., Litvinyuk, D., Mugilarasan, M., Gurumoorthi, K., Guganathan, L., Aboobacker, V.M., and Vethamony, P. 2021. Contributions of Fourier transform infrared spectroscopy in microplastic pollution research, *Environmental Science and Technology* 51:2681–2743.
18. Rinnan, R., and Rinnan, Å. 2007. Application of near infrared reflectance (NIR) and fluorescence spectroscopy to analysis of microbiological and chemical properties of arctic soil, *Soil Biology and Biochemistry* 39:1664–1673.
19. Visser, T. 2006. Infrared spectroscopy in environmental analysis. In: *Encyclopedia of Analytical Chemistry* (Eds. R.A. Meyers and M.P. Miller). John Wiley & Sons, Inc.: Hoboken, NJ. doi: 10.1002/9780470027318.a0832.
20. Mecozzi, M., and Nisini, L. 2019. The differentiation of biodegradable and non-biodegradable polyethylene terephthalate (PET) samples by FTIR spectroscopy: A potential support for the structural differentiation of PET in environmental analysis, *Infrared Physics & Technology* 101:119–126.
21. Mauro, M., Pietroletti, M., and Monakhova, Y. B. 2016. FTIR spectroscopy supported by statistical techniques for the structural characterization of plastic debris in the marine environment: Application to monitoring studies. *Marine Pollution Bulletin* 106:155–161.
22. Andrade, M.C., Winemiller, K.O., Barbosa, P.S., Fortunati, A., Chelazzi, D., Cincinelli, A., and Giarrizzo, T. 2019. First account of plastic pollution impacting freshwater fishes in the Amazon: Ingestion of plastic debris by piranhas and other serrasalmids with diverse feeding habits, *Environmental Pollution* 244:766–773.
23. Chen, Y., Wen, D., Pei, J., Fei, Y., Ouyang, D., Zhang, H., and Luo, Y. 2020. Identification and quantification of microplastics using Fourier-transform infrared spectroscopy: Current status and future prospects, *Current Opinion in Environmental Science & Health* 18:14–19.
24. Samokhvalov, A. 2020. Analysis of various solid samples by synchronous fluorescence spectroscopy and related methods, *Talanta* 216:120944.
25. Pavia, D.L., Lampman, G.M., and Kriz, G.S. 1979. *Introduction to Spectroscopy: A Guide for Students of Organic Chemistry*. W.B. Saunders Co: Philadelphia.
26. Li, D.-W., Zhai, W.-L., Li, Y.-T., and Long, Y.-T. 2014. Recent progress in surface enhanced Raman spectroscopy for the detection of environmental pollutants. *Microchim Acta* 181:23–43.
27. Das, R.S., and Agrawal, Y.K. 2011. Raman spectroscopy: Recent advancements, techniques and applications. *Vibrational Spectroscopy* 57:163–176.
28. Hao, J., Han, M.-J., Han, S., Meng, X., Su, T.-I., and Wang, Q.K. 2015. SERS detection of arsenic in water. *Journal of Environmental Sciences* 36:152–162.
29. Hanesch, M. 2009. Raman spectroscopy of iron oxides and (oxy)hydroxides at low laser power and possible applications in environmental magnetic studies, *Geophysical Journal International* 177:941–948.
30. Halvorson, R.A. and Vikesland P.J. 2010. Surface-enhanced Raman spectroscopy (SERS) for environmental analyses, *Environmental Science & Technology* 44:7749–7755.
31. Hao, J., Han, M.-J., Han, S., Meng, X., Su, T.-I., and Wang, Q.K. 2015. SERS detection of arsenic in water. *Journal of Environmental Sciences* 36:152–162.
32. Muehlethaler, C., Leona, M., and Lombardi, J.R. 2016. Review of surface enhanced Raman scattering applications in forensic science. *Analytical Chemistry* 88:152–169.
33. Song, D., Yang, R., Long, F., and Zhu, A. 2019. Applications of magnetic nanoparticles in surface-enhanced Raman scattering (SERS) detection of environmental pollutants. *Journal of Environmental Sciences* 80:14–34.
34. Sneddon, J. 2002. *Advances in Atomic Spectroscopy*. Elsevier Science: Amsterdam, p. 244.